The Wiley Blackwell Companion
to Political Geography

Wiley Blackwell Companions to Geography

Wiley Blackwell Companions to Geography is a blue-chip, comprehensive series covering each major subdiscipline of human geography in detail. Edited and contributed by the disciplines' leading authorities each book provides the most up to date and authoritative syntheses available in its field. The overviews provided in each *Companion* will be an indispensable introduction to the field for students of all levels, while the cutting-edge, critical direction will engage students, teachers, and practitioners alike.

Published

A Companion to Feminist Geography
Edited by Lise Nelson and Joni Seager

A Companion to Environmental Geography
Edited by Noel Castree, David Demeritt, Diana Liverman, and Bruce Rhoads

A Companion to Health and Medical Geography
Edited by Tim Brown, Sara McLafferty, and Graham Moon

A Companion to Social Geography
Edited by Vincent J. Del Casino Jr., Mary Thomas, Ruth Panelli, and Paul Cloke

The Wiley-Blackwell Companion to Human Geography
Edited by John A. Agnew and James S. Duncan

The Wiley-Blackwell Companion to Economic Geography
Edited by Eric Sheppard, Trevor J. Barnes, and Jamie Peck

The Wiley-Blackwell Companion to Cultural Geography
Edited by Nuala C. Johnson, Richard H. Schein, and Jamie Winders

The Wiley Blackwell Companion to Tourism
Edited by Alan A. Lew, C. Michael Hall, and Allan M. Williams

The Wiley Blackwell Companion to Political Geography
Edited by John Agnew, Virginie Mamadouh, Anna J. Secor, and Joanne Sharp

Also available:

The New Blackwell Companion to the City
Edited by Gary Bridge and Sophie Watson

The Blackwell Companion to Globalization
Edited by George Ritzer

The Handbook of Geographic Information Science
Edited by John Wilson and Stewart Fotheringham

The Wiley Blackwell Companion to Political Geography

Edited by

John Agnew, Virginie Mamadouh,
Anna J. Secor, and Joanne Sharp

WILEY Blackwell

This edition first published 2017
© 2015 John Wiley & Sons Ltd

Edition History: John Agnew, Virginie Mamadouh, Anna J. Secor, and Joanne Sharp.
(Hardback: 9781118725887 - 2015)

Registered Office(s)
John Wiley & Sons, Inc., 111 River Street, Hoboken, NJ 07030, USA
John Wiley & Sons Ltd, The Atrium, Southern Gate, Chichester, West Sussex, PO19 8SQ, UK

Editorial Office
9600 Garsington Road, Oxford, OX4 2DQ, UK

For details of our global editorial offices, customer services, and more information about Wiley products
visit us at www.wiley.com.

Wiley also publishes its books in a variety of electronic formats and by print-on-demand. Some content
that appears in standard print versions of this book may not be available in other formats.

Library of Congress Cataloging-in-Publication Data

The Wiley Blackwell companion to political geography / edited by John Agnew,
Virginie Mamadouh, Anna J. Secor, and Joanne Sharp.
     pages   cm – (Wiley Blackwell companions to geography)
  Includes bibliographical references and index.
  ISBN 978-1-118-72588-7 (cloth) ISBN 978-1-119-10765-1 (paper)
1. Political geography. I. Agnew, John A., editor. II. Mamadouh, Virginie, 1963– editor.
III. Secor, Anna Jean, editor. IV. Sharp, Joanne P., editor. V. Title.
  JC319.C646 2015
  320.1′2–dc23
                                                        2015011286

A catalogue record for this book is available from the British Library.

Cover Image: © REUTERS/Yannis Behrakis
Cover Design: Wiley

Set in 9/12.5pt Sabon by SPi Global, Pondicherry, India
10 9 8 7 6 5 4 3 2 1

Contents

Notes on Contributors

Paul C. Adams is Associate Professor of Geography at the University of Texas at Austin, USA. His research addresses place images in the media, the historical geography of communication technologies, geopolitical discourses, and the integration of communication technologies into particular places. He has published articles in the *Annals of the AAG, Progress in Human Geography*, and *Political Geography*, among other journals. His books include *The Ashgate Research Companion to Media Geography* (co-edited with Jim Craine and Jason Dittmer, Ashgate, 2014), *Geographies of Media and Communication* (Wiley-Blackwell, 2009), *Atlantic Reverberations* (Ashgate, 2007), *The Boundless Self* (Syracuse University Press, 2005), and *Textures of Place* (co-edited with Steven Hoelscher and Karen Till, University of Minnesota Press, 2001). He is the founder of the Communication Geography Specialty Group of the AAG.

John Agnew is Distinguished Professor of Geography at the University of California, Los Angeles. He has taught at a number of universities including Syracuse University, the University of Chicago, and the University of Siena. He has authored or co-authored numerous books including Berlusconi's Italy: Mapping Contemporary Italian Politics (2008) and Globalization and Sovereignty (2009). He is co-editor of the *Wiley Blackwell Companion to Human Geography* (2011).

Anne-Laure Amilhat Szary, PhD, is a full Professor at Grenoble-Alpes University, France, and a member of the Institut Universitaire de France. A political geographer dedicated to border studies, her latest research concerns the interrelations between art and culture, in and about contested places. She is a founding member of the antiAtlas of borders collective (http://www. antiatlas.net/en/), an art-science project. Her most recent book, *Borderities: The Politics of Contemporary Mobile* (Palgrave Macmillan 2015) was co-edited with F. Giraut.

Marco Antonsich is a senior lecturer in Human Geography at the Loughborough University, UK. His work lies at the intersection between territory, power, and identity, exploring the production of Western geopolitical discourses; the relationship between territory and identity in the age of globalization at multiple scales; and how togetherness in diversity is theorized and lived within contemporary multicultural societies. Funded by various institutions (US National Science Foundation; NATO and Italian National Research Council; CIMO-Finland; and the

European Commission), his work has appeared in leading academic journals: *Progress in Human Geography*, *Political Geography*, *European Urban and Regional Studies*, *European Journal of Social Theory*, and *Annals of the Association of American Geographers*, among others. He holds a PhD in Political Geography from the University of Trieste, Italy and a PhD in Geography from the University of Colorado at Boulder, USA.

Joshua E. Barkan is an Associate Professor of Geography at the University of Georgia, USA, where he studies the intersection of law, political economy, and social and political thought. His writing focuses on the relationships between corporations and sovereign, disciplinary, and biopolitical power. He recently published his first book, *Corporate Sovereignty: Law and Government under Capitalism* (University of Minnesota Press, 2013).

Kath Browne is a Reader in Human Geography at the University of Brighton, UK. Her work exists on the interstices of gender, sexualities, and geographies. Her scholarship includes resistances to LGBT equalities, lesbian, gay, bisexual and trans lives, womyn's separatist spaces, pride events, and queer, feminist, and participatory methodologies. Her most recent book, *Ordinary in Brighton: LGBT, Activisms and the City* (Ashgate, 2013), was co-authored with activist researcher Leela Bakshi. Research with Catherine Nash has exposed transnational resistances to LGBT equalities. She is currently working on an Economic and Social Research Council grant that explores what makes life liveable for LGBTQ people.

Brett Christophers is associate professor at the Institute for Housing and Urban Research and the Department of Social and Economic Geography at Uppsala University, Sweden. His research ranges widely across the political and cultural economies of Western capitalism, in both historical and contemporary perspectives. Particular interests include money, finance, and banking; housing and housing policy; urban political economy; markets and pricing; accounting, modeling, and other calculative practices; competition and intellectual property law; and the cultural industries and the discourse of "creativity."

Simon Dalby is CIGI Chair in the Political Economy of Climate Change at the Balsillie School of International Affairs and Professor of Geography and Environmental Studies at Wilfrid Laurier University, Canada. He is author of *Creating the Second Cold War* (Pinter/Guilford, 1990), *Environmental Security* (University of Minnesota Press, 2002) and *Security and Environmental Change* (Polity, 2009).

Cristina Del Biaggio teaches at the University of Geneva and has been invited researcher at the Institute of European Studies, University of Amsterdam, The Netherlands (2013–2014). She obtained her PhD at the University of Geneva in 2013. In her thesis she studied processes of regional institutionalization through a qualitative analysis of networks of local actors that took form from the 1990s in the Alps. Her current postdoctoral research relates to borders and migrations in Europe.

Patricia Ehrkamp is an Associate Professor in the Department of Geography at the University of Kentucky, USA. She researches contemporary processes of immigration, citizenship, and democracy in the United States and Europe. Her research examines expectations for immigrant assimilation in the context of exclusionary discourses about Islam in Western Europe, attending to the relationship between religion, gender, secularism, and democracy. Most recently, she completed a US National Science Foundation–funded research project, "Places of Worship and the Politics of Citizenship: Immigrants and Communities of Faith in the U.S. South."

Jennifer L. Fluri is an Associate Professor in the Department of Geography at the University of Colorado-Boulder, USA. Her research examines geopolitics, gender politics, and the geo-economics of international military aid and development interventions in South Asia. Her research appears in several peer-reviewed academic journals. Her forthcoming book, co-authored with Rachel Lehr, explores intimate geopolitics through several different entanglements between Americans and Afghans, and the various currencies from gender to grief that have manifested from discursive framing of 9/11 and the US-led war in Afghanistan. She is currently working on a new project that explores the role of Afghan women's organizations in Afghanistan's political and economic transitions.

Sara Fregonese is Birmingham Fellow in the School of Geography, Earth and Environmental Studies and the Institute for Conflict, Cooperation and Security at the University of Birmingham, UK. Her research concerns the mutual influence between geopolitics and the urban environment. She researches and publishes on urban warfare, radicalization and social cohesion, uprising and protest, and has over ten years' research experience in Beirut, Lebanon. She co-authored *The Radicals' City: Urban Environment, Polarization, Cohesion* (Ashgate, 2013).

Kathryn Furlong is an Assistant Professor in Geography at the Université de Montréal, Canada. She holds the Canada Research Chair in Urban, Water and Utility Governance. Her most recent work looks at the shifting nature of public utility corporations in Colombia and The Netherlands.

Jouni Häkli is Professor of Regional Studies at the University of Tampere, Finland. He is Vice Director of the Research Center of Relational and Territorial Politics of Bordering, Identities and Transnationalization (RELATE), and leads the Space and Political Agency Research Group (SPARG). He specializes in spatial and social theory, border studies, transnationalization, and political agency and subjectivity.

Malene H. Jacobsen is a PhD student in the Department of Geography at the University of Kentucky, USA. Her research focuses on the experiences of forced migrants with transnational migration management institutions that straddle the Middle East and Europe, with particular emphasis on Syrian asylum seekers migrating to Denmark.

Alex Jeffrey is a University Lecturer in Human Geography at the University of Cambridge, UK. His research has examined the geographies of state-building after conflict, in particular in the former Yugoslavia. In 2013 he published *The Improvised State: Sovereignty, Performance and Agency in Dayton Bosnia* (Wiley-Blackwell). He is currently undertaking research exploring the geographical implications of war crimes trials.

Andrew E.G. Jonas is Professor of Human Geography at Hull University, UK. His PhD is from The Ohio State University under the supervision of Kevin R. Cox. His latest book, co-authored with Eugene McCann and Mary Thomas, is *Urban Geography: A Critical Introduction* (Wiley-Blackwell). His co-edited books include *The Urban Growth Machine: Critical Perspectives Two Decades Later* (SUNY Press, 1999), *Interrogating Alterity* (Ashgate, 2010), and *Territory, State and Urban Politics* (Ashgate, 2012).

Kirsi Pauliina Kallio is Academy Fellow at the University of Tampere, Finland, Research Center of Relational and Territorial Politics of Bordering, Identities and Transnationalization (RELATE), Space and Political Agency Research Group (SPARG). Her areas of interest in political geographies include youthful agency, subject formation and socialization in the early years, and children's rights.

Sara Koopman is an Assistant Professor of Geography at York University in Toronto, Canada. She does research with solidarity movements to support their efforts to change the relationships between global North and South. Her recent work is on international protective accompaniment in Colombia. Her current research looks at the travels of stories from conflict zones, shared to build solidarity and peace. She aims to speak to dynamics in humanitarianism, development, and peacebuilding more generally. She blogs at decolonizingsolidarity.blogspot.com

Merje Kuus is Professor of Geography at the University of British Columbia. Her current research focuses on geopolitics, diplomacy, and transnational policy processes. She is the author of *Geopolitics and Expertise: Knowledge and Authority in European Diplomacy* (Wiley, 2014) and *Geopolitics Reframed: Security and Identity in Europe's Eastern Enlargement* (Palgrave, 2007); she also co-edited (with Klaus Dodds and Joanne Sharp) *The Ashgate Research Companion to Critical Geopolitics* (Ashgate, 2013).

Andrew M. Linke is Research Assistant Professor in the Department of Geography at the University of Utah, USA. He earned his PhD in 2013 from the University of Colorado-Boulder Department of Geography. His research interests are political geography, political violence, climate change and conflict, African politics, spatial statistics, and GIS. Among other journals, his research has been published in *Political Geography, Global Environmental Change*, and *International Interactions*.

Virginie Mamadouh is Associate Professor of Political and Cultural Geography at the University of Amsterdam and an editor of the international academic journal *Geopolitics*. Her research interests are in European geopolitics, new media and multilingualism. She is co-editor of *The Theory and Practice of Institutional Transplantation* (with Martin de Jong and Kostas Lalenis, 2002), *Critical Essays in Human Geography* (with J. Agnew, 2008), and *Urban Europe: Fifty tales of the city* (with A. van Wageningen, 2016).

Lauren Martin is an Academy of Finland Postdoctoral Researcher in the Department of Geography at the University of Oulu, Finland. She has published research on US immigrant family detention, border enforcement, and homeland security. Her current research explores the commercialization of border security and the commoditization of mobile bodies in the United States and European Union.

Katie Meehan is an Assistant Professor of Geography at the University of Oregon, USA. Her research and teaching interests focus on water, technology, and climate adaptation in cities, the role of infrastructure in development, and the social studies of transdisciplinary science in Latin America.

Claudio Minca is Professor of Cultural Geography and Head of the Cultural Geography Department at Wageningen University, The Netherlands. His current research centers on three major themes: tourism and travel theories of modernity; the spatialization of (bio)politics; and the relationship between modern knowledge, space, and landscape in postcolonial geography. His most recent books are *On Schmitt and Space* (with R. Rowan, Routledge, 2015), *Moroccan Dreams* (with L. Wagner, I.B. Tauris, 2015), *Real Tourism* (with T. Oakes, Routledge, 2011), *Social Capital and Urban Networks of Trust* (with J. Häkli, Ashgate, 2009).

Sami Moisio is Professor of Spatial Planning and Policy in the Department of Geosciences and Geography at the University of Helsinki, Finland. His research interests include critical geopolitics, political geographies of Europeanization, state spatial transformation, and the geopolitics of the knowledge-based society.

Olivia C. Molden is a master's student in the Department of Geography at the University of Oregon. Her current research investigates the contemporary role of ancient and traditional water infrastructure in urban development and modernization efforts in Kathmandu, Nepal. She grew up in South Asia.

Martin Müller is Swiss National Science Foundation Professor in the Department of Geography at the University of Zurich, Switzerland. His theoretical interests comprise discourse theories, actor-network theory (ANT), socio-materiality, and the psychoanalytic approach to fantasy and desire. He does research on the planning and organization of mega-events as well as on natural disturbances.

Luca Muscarà is Associate Professor at the Università del Molise, Italy. His research interests include the history of geographical thought, political geography, geopolitics, and urban geography. Honorary Fellow of the Société de Géographie, he is on the scientific board of *Limes* and Nomisma's OSSS. He has published an intellectual biography of Gottmann, a collection of essays on borders, and co-authored with Agnew the second edition of *Making Political Geography* (Rowman & Littlefield, 2012).

Catherine J. Nash is a Geography Professor at Brock University. Her research interests include geographies of sexualities, and feminist/queer and trans geographies, mobilities, and digital technologies. Her work examines the historical geographies of Toronto's gay village, queer women's bathhouse spaces, new LGBTQ neighborhoods, and methodologies and pedagogical issues. Recent work considers intergenerational changes in sexual and gendered identities, new LGBT mobilities, new social media and LGBT urban places, and transnational oppositions to LGBTQ rights in Canada and the UK. She is also working with Andrew Gorman-Murray on changing LGBT neighborhoods in Sydney, Australia and Toronto, Canada.

Emma S. Norman is Chair of the Science Department/Native Environmental Science Program at Northwest Indian College, Bellingham, Washington State, USA. She is also a Research Associate with the Smithsonian Institution, National Museum of the American Indian, USA, and a long-term collaborator with the Program on Water Governance at the University of British Columbia, Canada.

John O'Loughlin is Professor of Geography at University of Colorado-Boulder, USA. He obtained his PhD in 1973 from Pennsylvania State University. He has been editor of *Political Geography* since 1981 and *Eurasian Geography and Economics* since 2001, and his research interests are in the geographies of conflict, including the relationship between climate change and conflict in Sub-Saharan Africa and post-Soviet conflicts. He has also published on the electoral geography of Nazi Germany, regional political geographies of Ukraine, geopolitical views of Russians, and social capital in post-Communist societies.

Anssi Paasi is Professor of Geography at the University of Oulu, Finland. He has published widely on political geographical concepts and processes (e.g., borders, territory, spatial identities) and power–knowledge relations in academia. His books include *Territories, Boundaries and Consciousness* (Wiley, 1996). He has recently co-edited *The Sage Handbook of Progress in Human Geography* (Sage, 2014) and *The New European Frontiers* (Cambridge Scholars Publishing, 2014).

Joe Painter is Professor of Geography at Durham University in the UK, where he teaches urban and political geography. His research interests include the geographies of citizenship and the state, and urban and regional governance and politics. His books include *Political Geography* (co-authored with Alex Jeffrey, Sage, 2009) and *Spatial Politics: Essays for Doreen Massey* (co-edited with David Featherstone, Wiley-Blackwell, 2013).

Marcus Power is a Professor of Human Geography at the University of Durham, UK. His research interests include postsocialist transformations in Southern Africa; critical geographies and genealogies of (post)development; postcolonial geographies of Lusophone Africa; vision, visuality, and geopolitics; and the terms of China–Africa engagement. He is co-author of *China's Resource Diplomacy in Africa: Powering Development?* (Palgrave, 2012).

Clionadh Raleigh is a Professor of Political Geography at the University of Sussex, UK. Her work is focused on subnational patterns of political violence across Africa, governance in developing states, the relationship between environmental change and insecurity, and complex emergencies. She is the creator and director of the Armed Conflict Location and Event Data Project (ACLED).

Michael Samers (BA Clark, MS Wisconsin, DPhil Oxford University) is an Associate Professor of Economic and Urban Geography in the Department of Geography at the University of Kentucky, USA, having previously held positions at the University of Liverpool and the University of Nottingham. His research interests include the political-economic, urban, and labor market dimensions of migration/immigration, as well as alternative forms of economic activity. In 2013–14, he held a Fulbright Fellowship at CERAPS, Université de Lille II. He has also served as Editor of *Geoforum*, consulted for the Belgian Federal Science Policy Office and the UK Home Office, and has appeared a number of times on BBC radio and BBC4 television. He is the author of numerous articles, book chapters, and several books, including *Migration* (Routledge, 2010), which has been translated into Italian and Korean.

Anna J. Secor is Professor of Geography at the University of Kentucky and the Hajja Razia Sharif Sheikh Professor of Islamic Studies. Her research focuses on theories of space, politics, and subjectivity. Recently she has developed ideas of topology in geography by engaging the work of Lacan, Deleuze, and Agamben. Her research on Islam, state, and society in Turkey has been funded by the National Science Foundation.

Joanne Sharp is Professor of Geography at the University of Glasgow. Her research interests are in feminist, postcolonial, cultural and political geographies. She is the author of *Geographies of Postcolonialism: Spaces of Power and Representation* (2009) and editor of *The Ashgate Research Companion to Critical Geopolitics* (with Klaus Dodds and Merje Kuus, 2013).

Michael Shin is an Associate Professor in the Department of Geography at the University of California, Los Angeles, USA. His research interests include electoral geography, applied spatial analysis, and Italian politics and society.

Simon Springer is an Associate Professor in the Department of Geography at the University of Victoria, Canada. His research explores the political, social, and geographical exclusions that neo-liberalism engenders in the global South, emphasizing its geographies of violence. He cultivates a cutting-edge approach to human geography by foregrounding a radical revival of anarchist philosophy. His books include *To Make the Colossus Tremble! Anarchist Geography and Spatial Emancipation* (University of Minnesota Press, 2016), *Violent Neoliberalism: Development Discourse and Dispossession in Cambodia* (Palgrave Macmillan, 2015), and *Cambodia's Neoliberal Order: Violence, Authoritarianism, and the Contestation of Public Space* (Routledge, 2011).

Tristan Sturm is a Lecturer at Queen's University Belfast. He wrote his dissertation at UCLA on American Christian Zionists in Israel and Palestine. His current research explores competing apocalyptic environmental discourses.

Farhana Sultana is Associate Professor of Geography at the Maxwell School of Citizenship and Public Affairs of Syracuse University, USA. Previously, she taught at King's College London and worked at the United Nations Development Programme (UNDP). Farhana has a BA (*Cum Laude*) in Geosciences and Environmental Studies from Princeton University, and an MA and a PhD in Geography from University of Minnesota, where she was a MacArthur Fellow. She has broad and interdisciplinary research interests in critical development geographies, water governance, feminist geography, political ecology, and climate justice. She has published on an array of topics in journals such as Annals of the Association of American Geographers, Transactions of the Institute of British Geographers, The Professional Geographer, Society and Space, Gender Place and Culture, among many others. Her latest book is *The Right to Water: Politics, Governance and Social Struggles* (Routledge, 2012), where she is also a Faculty Affiliate in a number of other programs and departments.

James Tyner is a Professor of Geography at Kent State University, USA. He is the author of 14 books and numerous articles and book chapters. His 2009 book, *War, Violence, and Population*, received the Association of American Geographer's Meridian book award for its outstanding contribution to geography. His most recent book (2016) is *Violence in Capitalism: Devaluing Life in an Age of Responsibility* (University of Nebraska Press).

Herman van der Wusten is professor emeritus of the University of Amsterdam, The Netherlands, where he taught political geography. His dissertation was on Irish resistance against the political unity of the British Isles, 1800–1921. He has studied Dutch electoral politics in a comparative perspective and has been active in the field of international relations. Current main interests are political center formation in Europe, EU governance, and diplomacy.

Chih Yuan Woon is an Assistant Professor at the Department of Geography, National University of Singapore. His research interests lie in the areas of critical geopolitics, geographies of peace and non-violence, and social movements. He has worked with non-governmental organizations and "rebel" groups in the Philippines on issues at the interface of terrorism, violence, and sustainable peace. His most recent publications have appeared in journals such as *Progress in Human Geography, Geopolitics,* and *Transactions of the Institute of British Geographers.*

Chapter 1

Introduction

John Agnew[1], Virginie Mamadouh[2], Anna J. Secor[3], and Joanne Sharp[4]

[1]University of California at Los Angeles, California, USA
[2]University of Amsterdam, The Netherlands
[3]University of Kentucky, Lexington, Kentucky, USA
[4]University of Glasgow, Glasgow, Scotland, UK

In the summer of 2014, Kobani became the epicenter of struggles to redraw the political map of the world region known as the Middle East. Located in Syria at the Turkish border, Kobani is routinely described as a Kurdish city (of roughly 50,000 inhabitants before the war) because the majority of its population is ethnic Kurds. It is also known by its Arabic name – Ayn al-Arab (the Arabs' spring in Arabic, spring as in "well") – a toponym that the Syrian regime imposed in its Arabization campaign of the 1980s. Kobani has been part of the Syrian state since it was created as a French protectorate under the authority of the League of Nations after the First World War and the dismantling of the Ottoman Empire. In fact, the Treaty of Sèvres (1920) had also proposed an Ottoman successor state that would have encompassed most of the region's ethnic Kurds, but Kurdistan did not survive the Turkish revolution and the final Peace Treaty signed in Lausanne in 1923. With the lands that would have been Kurdistan ultimately divided between Turkey, Syria, Iraq, and Iran, Kurdish nationalism has been variously incited and repressed in the region ever since.

Before becoming the epicenter of conflict, for two years Kobani had been part of Rojava, the territories controlled by Kurdish forces in the north and northeast of Syria during the Syrian civil war. In September 2014, Kobani came under siege by Islamic State (IS), a religious movement that emerged during the Iraq war against Western forces. By the time of the siege, IS had gained control of significant territories in northern Iraq and northern Syria. These were regions where autonomous Kurdish authorities had become more or less well established during the Iraq war and the Syrian civil war respectively. In claiming these territories, IS forces attempted to cleanse them of populations hostile to its rule, such as Christian minorities and Yezidis, either by forcing them to convert to IS's particular interpretation and practice of Sunni Islam, or through practices of displacement, massacre, and the sexual enslavement of women. With IS on the move, hundreds of thousands of Syrians fled over the nearby Turkish border. Yet because of its own "Kurdish question" and its opposition to claims for Kurdish autonomy in eastern Turkey, the Turkish government was slow to heed its Western allies' calls

The Wiley Blackwell Companion to Political Geography, First Edition.
Edited by John Agnew, Virginie Mamadouh, Anna J. Secor, and Joanne Sharp.
© 2015 John Wiley & Sons Ltd. Published 2017 by John Wiley & Sons Ltd.

for intervention in Kobani. By early October 2014, Kobani had become the place where the struggle for the future of Kurdistan (in Syria and beyond) was being fought, while the geopolitical stakes of the war against Islamic State became clear: The outcome of this struggle had the potential to redraw the political map of the region.

Kurdish rule in Kobani and the IS siege on the city both need to be understood in relation to broader historical and material developments. The Syrian civil war, which had led to Kobani becoming part of the Kurdish territory of Rojava, evolved from protests challenging the regime of President Bashar al-Assad that began in March 2011. These protests were directly connected to the uprisings that rocked the region and collectively became known as the Arab Spring, a ferment that led to regime change or civil war in several Arab states, including Tunisia, Egypt, Libya, and Syria. The international community intervened selectively in these events (sometimes militarily, as in Libya), but largely seemed unable (and sometimes unwilling) to act to protect human lives and human rights in the region or to promote democracy and security. By 2014, as Syria was plunged deep into civil war and the power of IS grew, some young people from Western countries (many of whom had no previous links with the region) traveled to Syria to support the movement. This flow of fighters revealed the interconnections between the Arab world and Europe and, at the same time, generated widespread sentiments of insecurity that echoed the anxieties of the early 2000s following the 9/11 attacks on New York and Washington and, later on, those on Madrid and London – anxieties that were stoked by the swelling flows of refugees into Europe.

The tragic and highly complex situation of Kobani and its connections to historical, regional, and global dynamics illustrate the importance of geopolitical literacy and the potential for political geographical analysis to illuminate the dynamics and contingencies of current affairs. The complex play of alliances between states and non-state actors, of international organizations and transnational networks, is unintelligible without an understanding of the history and geography of intergroup relations and the geopolitical battles that have shaped the political map of the region. Moreover, we can only begin to make sense of these historical and geographical features with the analytical tools that political geography provides for us. The key concepts presented in the first section of this volume are mighty instruments for analyzing competing claims to power, statehood and sovereignty, the role of territory and borders, the loss of security, and the invocation of justice, as well as more general issues of citizenship, scale, and governance.

The chapters collected in the second section, "Theorizing Political Geography," provide a series of lenses offering different perspectives; in reading these, one might ask oneself what a feminist geography, a children's geography, or a postcolonial geography of a situation might look like. How, for the case of Kobani, might these various approaches bring different aspects of the conflict under the spotlight? How might they open up opportunities for different resolutions, through different (territorial) arrangements? Finally, the themes presented in the other sections of the *Companion* – written before the Summer 2014 confrontation between Islamic State and Kurdish forces in Northern Syria – also provide tools for obtaining a better grasp of specific dimensions of ongoing events. These chapters present state-of-the-art political geographical approaches to nationalist movements (think of Kurdish parties), religious movements (IS), social movements (grassroots protests against Assad's regime), social media (the mobilization of the Kurdish diaspora in Western Europe), electoral geography (the difficult organization of elections in newly established states and postautocratic regimes), sexual politics (the sexual enslavement of girls and women in conquered territory), migration (the large flows of Syrian refugees into Lebanon and Turkey), imperialism and world views (the resistance to Western influences in the Arab world), the role of regional institutions (such as the Arab League and the European Union), urban materialities (the fate of Kobani does indeed echo that of many cities under siege in recent history, such as Beirut), and more. Some of the themes might seem less directly relevant to the particular case of Kobani, but even then they do shape the context in which the international community operates when it weighs up the possibility of intervening in the region, how and at what cost. Indeed, the rise of the BRICS economies – Brazil, Russia, India, China, and South Africa – and the demand for resources make the stability of the region a global security issue, on top of the moral issues regarding human rights; however, the financial and climate crises limit the (material) capabilities of many states to act.

The scholarly work covered by the *Companion* does not provide ready-made solutions to the tragic events in and around Kobani. Political geography is no magic. The production of academic knowledge is a slow and painstaking process and its circulation is characterized by serious oddities (addressed in the last chapter of this volume) that should not prevent academics from making sure that their expertise informs their politics and that their voices are heard in public debate. More importantly, engagement with political geography can greatly enhance anyone's ability to make sense of ongoing events in order to develop their own opinions and boost their agency in the issue – an engagement that necessarily takes place under the constraints of existing conditions, although understanding these constraints better is a necessary step toward empowerment and change.

Introducing political geography

This book is a new edition of a *Companion to Political Geography* published more than a decade ago (Agnew, Mitchell, & Toal 2003). It focuses on recent developments in the field. For much of its history, the subdiscipline of political geography has been centered on the study of the state and its territory. At the time when geography was being established as an academic discipline in Western universities at the end of the nineteenth and into the twentieth century, political geography – a term coined by French statesman Turgot in the eighteenth century and established by German geographer Friedrich Ratzel with his landmark volume *Politische Geographie* (1897) – was at the heart of the production of geographical knowledge in service to imperialist and nation-building projects (Godlewska & Smith 1994). Geography at the time was largely understood in physical terms and the political was generally restricted to questions of the state (Agnew et al. 2003). Political geographical analysis thus involved explaining the success and actions of states and their elites based on their physical locations and resources. At the turn of the century, British geographer and politician Halford Mackinder offered a global model linking world history to geography in an influential lecture at the Royal Geographical Society. He warned that the arrival of the railway had made the British Empire, as a sea power, increasingly vulnerable to threats from control over the "heartland" of Eurasia and that Germany as a land power (after the Russian Revolution and German defeat in the First World War he substituted Russia) and potential allies in Eastern Europe could replace the British as the dominant world power (Mackinder 1904). Geopolitics, a term coined by Swedish political scientist Rudolf Kjellén (1916) to capture what he saw as the geographical basis of world politics, emerged as a metonym for political geography and an expression of an organicist conception of the state and interimperialist rivalry (Parker 1985). Because geopolitics as statecraft later became associated with justifications for German territorial expansionism and ethnic cleansing in Eastern Europe during the Second World War, the latter part of the twentieth century saw not only "geopolitics" but also political geography being pushed aside in favor of more supposedly objective fields of study.

That the trajectory of political geography in the second part of the twentieth century and into the twenty-first is one of revitalization and diversification is routinely signaled in readers, handbooks, and textbooks (such as Agnew & Mamadouh 2008; Cox, Low, & Robinson 2008; Flint & Taylor 2011; Agnew & Muscará 2012) and plainly demonstrated in the pages of the core journals of the discipline (*Political Geography*; *Geopolitics*; *Territory, Politics, Governance*; *Space and Polity*; *Environment and Planning D*; *Antipode*) and in the reports in *Progress in Human Geography*.

However, the renewal of political geography began in the 1960s and was initially premised on development of the spatial science of electoral geography (Cox 1969; Taylor & Johnston 1979) – a field that continues to gain in depth today (Warf & Leib 2011). By the 1980s, critical approaches, including Marxism and "world systems theory" (Wallerstein 1979), began to reorient the subdiscipline toward questions of inequality, dependency, and social justice (Taylor 1982; Smith 1984). As political geographical research gained momentum, questions of borders, territory, political identity, power, and resistance emerged as central to its constitution (Wastl-Walter 2011). And while geopolitics re-entered the US foreign policy lexicon with Henry Kissinger's use of the adjectival form in the 1970s (Hepple 1986), its study was reborn in geography in the late 1990s, but this time from a critical perspective that problematized powerful geographical framings of world politics and statecraft (O Tuathail 1996; Dodds & Atkinson 2000). This strand of critique proliferated, producing new fields of study (Dodds, Kuus, & Sharp 2013) such as popular geopolitics (Sharp 1993; Dittmer 2010), and at the same time converging with other critical approaches such as feminist political geography, which had begun calling into question the premises of masculinist political geography since the beginning of the 1990s (Kofman and Peake 1990; Staeheli 1996; Staeheli, Kofman, & Peake 2004). In the decades that have followed, feminist political geography has been a significant force in the expansion of the subdiscipline, prying open the question of how "the political" is spatially constituted and pressing against scalar hierarchies to include questions of embodied political practice (Smith 1992; Blunt 2000; Marston 2000). Geopolitics, once defined as "the geographical basis of world power," has thus, for some political geographers, become a way of thinking about (world) politics as constituted through everyday spatial practice and experiences (Thrift 2000; Fregonese, this volume).

To the extent that the traditional concepts of states and territory remain important foci for political geography, they have also been significantly rethought. The state is no longer the empty container or billiard ball of international relations (Agnew 1994; Jeffrey this volume). Political geographers have unpeeled the onion of sovereignty, examining its contingency and improvisation (Jeffrey 2013), its divergence from state power and territoriality (Agnew 2005), and the paradoxes of sovereign exceptionalism (Mitchell 2006; Minca 2007; Secor 2007; Mountz 2013; Barkan, this volume). Territory, that other traditional term of political geography, has undergone new genealogical critique that calls into question its logics and constitution (Elden 2009, 2013; Painter 2010; Del Biaggio, this volume). Other concerns that have likewise been central to political geography since its inception have also taken on a new life. As the border becomes unmoored from its traditional mappings (Paasi 1996; Price 2000) and reconceptualized as a "technology of spatial or socio-spatial division" (Amilhat Szary, this volume), border studies have not faded from political geography but instead come to form an increasingly vibrant field. Scale, a concept that has not been the exclusive provenance of political geography but has nonetheless played a prominent role in shaping the subfield (Flint & Taylor 2011), has likewise been subject to rigorous debate, the upshot of which has been the emergence of new understandings of its utility and limitations for political geographical work (Marston, Jones, & Woodward 2005; Jonas, this volume). In short, political geography has not so much departed from its central themes as continued to work through them.

At the same time, political geographers today are deeply engaged at the heart of the subdiscipline with questions that Halford Mackinder could never have foreseen. Some of these concerns, such as environmental geopolitics, were also prominent in the previous volume of the *Companion to Political Geography* (Agnew et al. 2003), but have only become more acute under current conditions (Barnett & Adger 2007; Raleigh & Urdal 2007; Dalby, this volume). Likewise, the previous volume also featured political and social movements; with the Arab Spring, Occupy Wall Street, and anti-austerity protests in Europe punctuating the second decade of the twenty-first century, these engagements remain highly relevant today, at the same time as the explosion of information and communication technologies calls forth new perspectives (Adams, this volume; Koopman, this volume). Our new volume also demonstrates how political geography has continued to grapple with currents that circulate more widely than the subdiscipline. For example, as religion becomes subject to greater attention across the social sciences and in geography, political geographers too have begun to move beyond categories such as "fundamentalism" and "ethno-religious nationalism" (Appleby 2003) to rethink religion in relation to politics, place, and identity (Hopkins, Kong, & Olson 2013; Sturm, this volume). Similarly, growing attention to children's geographies across the discipline has pried open political geography to encompass the agency and subjectivity of children in ways that productively challenge scalar logics and the constitution of the political (Kallio & Häkli, this volume). And intersecting with work in cultural geography, political geographers have found that they have an important role to play in crafting an understanding of the political as not only a product of representations and representational practices, but also imbued with materiality, the non-human, and the affective (Müller, this volume).

The first edition of the *Companion to Political Geography* was a landmark statement about the nature of the subdiscipline that also sought to set the research agenda for political geography. A decade after the publication of that edition and seven years later than that of the *Sage Handbook of Political Geography* (Cox et al. 2008), the second edition of the *Companion* aims to account for the intellectual and worldly changes that have taken place in and around political geography. The impacts of Osama bin Laden's attacks on the United States in 2001 were new to the authors writing for the first edition and there was no way of anticipating their full implications for geopolitical praxis (Martin, this volume; Muscarà, this volume). Additionally, the rise of China as a geoeconomic superpower has begun to influence the field (Power, this volume), as have new concerns about financial crises (Christophers, this volume), the (geo)politics of nature and resources (Furlong & Norman, this volume), and the neoliberal, security, and biopolitical processes associated with migration (Samers, this volume). Intellectually, the practice-based turn in geography has had a significant influence, as have concepts ranging from postcolonialism (Woon, this volume) and the notion of the Anthropocene (Dalby, this volume) to new understandings of the role of non-human actors in networks of power (Müller, this volume; Fregonese, this volume; Painter, this volume). This new edition thus represents a fresh look at the subdiscipline of political geography overall and our changing world.

Outline of the book

There are always multiple ways of dividing up a subdiscipline and so, by definition, there are limitations to every chapter outline. Here we have attempted to combine coverage of the major concepts in political geography while ensuring sufficient flexibility to acknowledge change and dynamism in the field. Our first section, "Key Concepts in Political Geography," is the longest. Here we have asked contributors to address those concepts that we consider to be foundational. Historically, political geographers have focused on the most overtly spatialized concepts, such as **borders and boundaries, scale, territory, sovereignty**, the **state**, and other scales of governance (**federalism/multilevel governance**), and the geographies of **conflict** at different scales and across varied landscapes. Each of these is the focus of a chapter in this section, all of which draw out the traditions of each concept and the ways in which political geographers have engaged with the ideas more recently. Less overtly geographical concepts are also covered in this section, with authors outlining the spatial processes driving the concepts and their geographical implications. Although it has a long history in the understanding of the politics of the state, the concept of **security** has become ever more important to the practices of statecraft, as technological and bureaucratic practices are increasingly used to maintain protection from a range of extra-state dangers (from people to pathogens). **Violence, justice**, and **power** are at the heart of understanding the political process. Political geographers' interest in power was stimulated by engagement with the ideas of Michel Foucault in the 1990s, which generated work engaging with much more complex and even "positive" or "creative" forms of power. This interest has continued and the multiplicity of the operations of power through different political and apparently apolitical systems has also led to re-examinations of our understanding of violence and justice as concepts, which have implications and effects seemingly contradictory to what more formal practices of politics seem to suggest. More attention has been paid to the population of the state and people's relation to the state and to each other, centering around the issue of **citizenship**. Consideration of the ways in which various populations are drawn into politics and managed by states has also been stimulated by Foucault's work and, more recently, the influence of Agamben has led to a new interest in **biopolitics**.

The second section, "Theorizing Political Geography," shifts focus from the concepts themselves to the ways in which geographers have understood them. Clearly, there is considerable overlap with the chapters in the first section, in that those explored the different ways in which their concepts had been dealt with in political geography. However, the authors of the chapters in the second section have foregrounded the ways in which their terms have been reconceptualized and the new paths that have been explored. This applies to **spatial analysis** and the rich application of new technical tools to expand and visualize this mode of analysis. Although conventionally regarded as an approach at the service of state and empire, political geography has also had a vital and dynamic tradition of **radical political geographies**, which has consistently attempted to challenge the goals and political heart of the discipline. In many ways paralleling this, one of the most significant retheorizations in the last 30 years or so has been the shift – in academic geography, if not in the wider world – from a focus on classic **geopolitics** to an interest in **critical geopolitics**. This recognizes the impact on political geography of poststructural theories and what has come to be known as the discipline's "cultural turn." In more recent years, and notably since the previous edition of this collection, both **feminist** political geography and **postcolonial** political geography have increased in prominence significantly, offering further critique and broadening our understandings of where "the political" is located, and drawing our attention more closely to the everyday and the cultural as inherently political realms. While both feminist and postcolonial approaches point to agents of political change who have been marginalized from previous articulations of political geography, others have pointed to **children** as another group whose role in the remaking of political life needs to be taken seriously.

At the heart of political geography is a recognition that where political processes unfold is central to the nature and outcomes of these processes. Thus, our third section, "Doing Politics," considers the how and where of political geographies. **Electoral geographies** perhaps most clearly draw out the geographies of political processes. As they recognize, formal spheres of politics are very powerful in the reproduction of political processes and identities, so we have an enduring interest in **nations and nationalism** and in **regional institutions**. However, although these are widely regarded as formal political institutions, the chapters highlight the sophistication of analysis that recognizes the mundane and apparently informal as equally important to the formal practices of politics. Politics are also made in opposition to the formal institutions of state and region, of course, and so political geographers have offered critique of the **imperialism** of dominant power, and have closely examined the role of **social movements**, including those involved in **religious movements** or engaged in **sexual politics**. New geographies have emerged recently that affect both the remaking of place and identity and the geopolitical scale. Perhaps most obvious is the **rise of the BRICS** powers, and especially the growing political and economic influence of China, in global and regional political geographies. However, perhaps even more disruptive has been the rise of **social media** as a technology that has facilitated both new political identities and communities, while also providing new networks for organizing direct action and the ability to disseminate images of oppression and resistance across the world without the mediation of big news organizations.

While the "cultural turn" undoubtedly had a significant impact on political geography, especially in attuning research to the power of particular representations of the world, it is important not to lose sight of the importance of matter and physical things that are caught up in political networks (or assemblages). The fourth section, "Material Political Geographies," seeks to put things (and not just processes and representations) center stage. The **more-than-representational** turn has put material at the heart of political geography, seeking to theorize the role and even agency of the non-human actors within political networks and driving the political process. Of course, to a certain extent political geography has long been interested in material, as is made clear in the chapter on **resources**. Similarly, chapters on **political ecology** and the **environment** highlight the changing ways in which nature and environmental issues have been understood by political geographers, and the effects that this has had on the wider environment, most especially in the Anthropocene. Clearly, political geographies of all sorts, however drawn through discourse and representation, are embedded in significant material consequences: Both **financial crises** and geographies of **mobility and migration** are, at heart, understandings of the movement of things and bodies, and the very real impacts of flows and imaginations of states and boundaries on the lives of those people caught up in them, whether economic migrants seeking a better life, or those affected by financial crisis and losing theirs. Thus, a focus on **everyday** political geographies is vital to understanding the impacts of the sometimes abstracted or large-scale processes that are the focus of much political geography.

Our final section, "Doing Political Geography," contains just one chapter, **Academic capitalism and the geopolitics of knowledge**, which seeks to turn the critical approach of the preceding chapters on the discipline within which we work, to reflect on the kinds of political geographies that we are remaking through our professional practices and our publishing performances.

References

Agnew, J. (1994). The territorial trap: The geographical assumptions of international relations theory. *Review of International Political Economy* 1(1): 53–80.

Agnew, J. (2005). Sovereignty regimes: Territoriality and state authority in contemporary world politics. *Annals of the Association of American Geographers* 95(2): 437–461. doi:10.1111/j.1467-8306.2005.00468.x

Agnew, J., & Mamadouh, V. (2008). Introduction. In Agnew, J., & Mamadouh, V. (eds), *Politics. Critical Essays in Human Geography*. Aldershot: Ashgate, xiii–xxiv.

Agnew, J., & Muscarà, L. (2012). *Making Political Geography*, 2nd edn. New York: Roman and Littlefield.

Agnew, J., Mitchell, K., & Toal, G. (2003). *A Companion to Political Geography*. Oxford: Blackwell Publishing.

Appleby, R.S. (2003). Fundamentalist and nationalist religious movements. In Agnew, J., Mitchell, K., & Toal, G. (eds), *A Companion to Political Geography*. Oxford: Blackwell Publishing, 378–392.

Barnett, J., & Adger, W.N. (2007). Climate change, human security and violent conflict. *Political Geography* 26(6): 639–655. doi: 10.1016/j.polgeo.2007.03.003

Blunt, A. (2000) Embodying war: British women and domestic defilement in the Indian "mutiny", 1857–8. *Journal of Historical Geography* 26: 403–428.

Cox, K.R. (1969). The voting decision in a spatial context. *Progress in Geography* 1: 83–117.

Cox, K., Low, M., & Robinson, J. (eds) (2008) *The Sage Handbook of Political Geography*. London: Sage.

Dittmer, J. (2010). *Popular Culture, Geopolitics, and Identity*. New York: Roman and Littlefield.

Dodds, K., & Atkinson, D. (eds) (2000). *Geopolitical Traditions: A Century of Geopolitical Thought*. London: Routledge.

Dodds, K., Kuus, M., & Sharp, J.P. (eds) (2013). *The Ashgate Research Companion to Critical Geopolitics*. Farham: Ashgate.

Elden, S. (2009). *Terror and Territory: The Spatial Extent of Sovereignty*. Minneapolis, MN: Minnesota University Press.

Elden, S. (2013). *The Birth of Territory*. Chicago, IL: University of Chicago Press.

Flint, C., & Taylor, P.J. (2011). *Political Geography, World-economy, Nation-state and Locality*, 6th edn. Harlow: Prentice Hall.

Godlewska, A., & Smith, N. (eds) (1994). *Geography and Empire*. Oxford: Blackwell.

Hepple, L.W. (1986). The revival of geopolitics. *Political Geography Quarterly* 5(S): s21–s36.

Hopkins, P., Kong, L., & Olson, E. (2012). *Religion and Place: Landscape, Politics and Piety*. New York: Springer.

Jeffrey, A. (2013). *The Improvised State: Sovereignty, Performance and Agency in Dayton Bosnia*. West Sussex: Wiley-Blackwell.

Kjellén, R. (1916). *Staten som lifsform*. Stockholm: Politiska handböcker.

Kofman, E., & Peake, L.J. (1990). Into the 1990s: A gendered agenda for political geography. *Political Geography Quarterly* 9(4): 313–336.

Mackinder, H.J. (1904). The geographical pivot of history. *Geographical Journal* 23(4): 421–437.

Marston, S. (2000). The social construction of scale. *Progress in Human Geography* 24(2): 219–242.

Marston, S.A., Jones, J.P., III, & Woodward, K. (2005) Human geography without scale. *Transactions* 30(4): 416–432. doi:10.1111/j.1475-5661.2005.00180.x

Minca, C. (2007). Agamben's geographies of modernity. *Political Geography* 26(1): 78–97.

Mitchell, K. (2006). Geographies of identity: The new exceptionalism. *Progress in Human Geography* 30(1): 95–106. doi:10.1191/0309132506ph594pr

Mountz, A. (2013). Political geography I: Reconfiguring geographies of sovereignty. *Progress in Human Geography* 37(6): 829–841. doi:10.1177/0309132513479076

O Tuathail, G. (1996). *Critical Geopolitics*. Minneapolis, MN: University of Minnesota Press.

Paasi, A. (1996). *Territories, Boundaries and Consciousness: The Changing Geographies of the Finnish-Russian Border*. Chichester: John Wiley & Sons Ltd.

Painter, J. (2010). Rethinking territory. *Antipode* 42(5): 1090–1118. doi:10.1111/j.1467-8330.2010.00795.x

Parker, G. (1985). *Western Geopolitical Thought in the Twentieth Century*. London: Croom Helm.

Price, P. (2000). Inscribing the border: Schizophrenia and the aesthetics of Aztlan. *Social and Cultural Geography* 1(1): 101–116.

Raleigh, C., & Urdal, H. (2007). Climate change, environmental degradation and armed conflict. *Political Geography* 26(6): 674–694.

Ratzel, F. (1897). *Politische Geographie*. Munich: R. Oldenbourg.

Secor, A. (2007). An unrecognizable condition has arrived: Law, violence and the state of exception. In Gregory, D., & Pred, A. (eds), *Violent Geographies: Fear, Terror and Political Violence*. London: Routledge, 37–53.

Sharp, J.P. (1993). Publishing American identity: Popular geopolitics, myth and the Reader's Digest. *Political Geography* 12(6): 491–503.

Smith, N. (1984). *Uneven Development*. London: Blackwell.

Smith, N. (1992). Contours of a spatialized politics: Homeless vehicles and the production of geographic scale. *Social Text* 33: 54–81.

Staeheli, L. (1996). Publicity, privacy, and women's political action. *Environment and Planning D: Society and Space* 14(5): 601–619.

Staeheli, L.A., Kofman, E., & Peake, L.J. (eds) (2004). *Mapping Women, Making Politics: Feminist Perspectives on Political Geography*. New York, Routledge.

Taylor, P.J. (1982). A materialist framework for political geography. *Transactions of the Institute of British Geographers* NS7: 15–34.

Taylor, P.J., & Johnston, R. (1979). *Geography of Elections*. Harmondsworth: Penguin.

Thrift, N. (2000). "It's the little things." In Atkinson, D., & Dodds, K. (eds), *Geopolitical Traditions: A Century of Geopolitical Thought*. London: Routledge, 380–387.

Wallerstein, I. (1979). *The Capitalist World-Economy*. Cambridge: Cambridge University Press.

Warf, B., & Leib, J. (eds) (2011). *Revitalizing Electoral Geography*. Farnham: Ashgate.

Wastl-Walter, D. (ed.) (2011). *The Ashgate Research Companion to Border Studies*. Aldershot: Ashgate.

Key Concepts in Political Geography

Chapter 2

Boundaries and Borders

Anne-Laure Amilhat Szary

Grenoble-Alpes University & CNRS-PACTE, Grenoble, France

Boundaries have been a key component of the modern territorial vocabulary. They have origins in Euclidean thinking, which sees space as two-dimensional. They also fit in well with Kantian rationality, which considers space and time as two symmetrical conditions of human experience. They can be considered "a generic term for the linear spatial discontinuity" (Fall 2005). Borders constitute both a restriction and an expansion of the semantic field of boundaries, and in this sense it is hard to discuss one term without examining the other. Borders have indeed been endowed with a stronger political use than boundaries, having been "invented" in modern times to express a specific balance of territorialized powers. They also invite a more than linear approach, enabling the possibility of borderlands. This chapter will reflect on all of these terms.

Recent research in the context of globalization suggests the processual nature of geographical objects and their ongoing adaptation to new technologies that change the nature of well-established spatial notions such as distance and proximity. Thus, any definition of boundary-making has to strive to account for complex elements interacting in a technology-dependent context. Relativity and chaos theories on the one hand, and phenomenology and agency thinking on the other, invite us to reformulate our notion of space and the relations that it fosters. For those reasons, postmodernity requires us to reconceptualize boundaries and borders whose initial definition is definitely modern.

Being delimitations of space, boundaries and borders present a first paradox, that of being symbolic components of our environment that have a high impact on our lives. The second paradox that they share is the fact that our main experience of them is by confronting or crossing them. This may be what makes the recent challenge that two climbers set themselves in 2011–12 – to follow as extensively as possible the border of an Alpine country – so exceptional (Amilhat Szary 2013a). In a renewed way, by using their own bodies to attest to and question the existence of the conventional line, the climbers pose a complex question: Can we experience a border or a boundary as traced on a map? Boundaries may appear to be more

The Wiley Blackwell Companion to Political Geography, First Edition.
Edited by John Agnew, Virginie Mamadouh, Anna J. Secor, and Joanne Sharp.
© 2015 John Wiley & Sons Ltd. Published 2017 by John Wiley & Sons Ltd.

mundane than borders, as more people have probably had the chance to experience them personally – be it physically, through their presence on urban fringes in "park boundary" signs, or mentally, since history books are replete with references to "boundary treaties" that make boundaries achieve a material consistency in the mind of every citizen. It is, however, much more difficult to touch a border, even when it is a material barrier or a wall, either because surveillance devices often restrict access to it or because, as in the case of the mountain fringes that our climbers decided to explore, they were not meant to be reached.

Language appears as an important filter to understand the dynamics of bordering. The Roman etymology of the word boundary or border differs from the Saxon one. The direct translation of boundary in languages derived from Latin, such as "frontière" in French or "frontera" in Spanish, has its roots in "front" and attests to the rivalries and battles that took place before linear devices known as boundaries were in use. The mythological history of the Western world stained with blood the first tracing of a border. It is worth noting that it was because of a sacred furrow, plowed to distinguish the soil of the city to be, Rome, from that of its unknown exterior, that Remus was killed by his twin brother Romulus. This deed was in reaction to Remus's mocking of the boundary that Romulus had just traced. While Romulus was lifting the plow to mark a threshold, Remus deliberately walked over the line to denounce its weakness. Romulus responded to Remus's symbolic violence with a physical and irreversible act: One brother suppressed the other by declaring the possible territorialization of his identity through the establishment of a spatial limit.

Its Saxon etymology suggests that the "boundary" or "border" could be "that which binds," a place of friction or meeting where alterity is negotiated. Probably more than boundaries, borders are a kind of space where the relationship with otherness can be developed in such a way as to allow for identity-building and place-making. This socio-anthropological definition of borders emphasizes the complex relations of the spatial divides with distance. In line with the Simmelian approach, by which social links are considered to originate from the connections between that which is close and that which is far, simultaneously fixed and mobile (Simmel 1972 [1908], especially Chapter 9), it is possible to define a border as a way of "putting distance in proximity" (Groupe Frontière et al. 2004).

The German word "Grenze" seems to open up yet another semantic field of reference. It comes from a Polish term meaning "milestone," used to demark private property before it bore a more political power. The word was coined in the context of large borderlands where the line only existed as a mental construction tied to punctual signs in the landscape. However, curiously, in the European context, those different words have evolved "in a similar direction, that of a dematerialization through the metaphorical meaning of the linear limit" (Lask 1995: 68).

The past two decades have borne witness to two opposing movements in territorial thinking. On the one hand, the first expected the "end of history" to entail the symmetrical "end of territory" (Badie 1995), as the flows of globalization indeed relied on a decline in barriers. A former director of the World Trade Organization was thus able to advocate for a "world without walls" (Moore 2003) at the very moment when rebordering processes were about to gain new legitimacy, as security became a key issue in international relations following the events of 9/11. On the other hand, the same period was marked by an "obsession with borders" (Foucher 2012 [2007]) shared by policy-makers, politicians, and scholars. The renewal of border studies in the 1990s was built on questioning of the linear component of spatial divides, and the shifting of boundaries from political science and international relations toward the field of critical studies. Historical knowledge of border demarcation took an important step (Foucher 1986; Sahlins 1996; Nordman 1998) after the Second World War, renewing the consideration of borders through the research lines set by the Annales school of history (Ancel 1938). A few authors led the way (Rumley & Minghi 1991; Donnan & Wilson 1994: Paasi 1996) to reactivate the theoretical insights of the previous generation (Guichonnet & Raffestin 1974; Raffestin 1974; Prescott 1978) and analyze the complex relations between power and space that borders "reified" (Raffestin 1974). The input from this first generation of what was becoming structured as "border studies" was extremely well summed up in the widely cited paper by Newman and Paasi (1998). Structurally, this was pushed forward by the birth of connecting structures such as the Association of Borderland Studies (founded in 1976), organizer of annual conferences and of a specialized publication, the *Journal of Borderland Studies* (launched in 1986), and the BRIT (Borders in Transition) network initiating its conferences in 1994, lately followed by other regular border forums such as the Borderscapes conferences (2006–14).

Since then, border research has expanded in two divergent directions, one focusing on the "cross-boundary" cooperation processes and socio-spatial issues in borderlands, and the other on the securitization of border crossings and the renewal of demarcation processes. Meanwhile, theoretical thinking about boundedness was much called for, without enough success, since this call was often answered through the prism of modeling (Brunet-Jailly 2005; Hamez et al. 2013) rather than an overall theory of borders. The recent and concomitant release of two companion books (Wastl-Walter 2012; Wilson & Donnan 2012) marked yet another stage of the maturation of this scientific field and opened the way for renewed scholarship on territorial restructuring, now in possession of the epistemological tools enabling them to tackle border and boundary together. Interestingly, for the first time an important research grant has been awarded to a European consortium with the objective of reshaping a conceptual framework for border studies (the EUBORDERSCAPES project, led by J. Scott; see in particular the extensive state of the art produced by the team: http://www.euborderscapes. eu/fileadmin/user_upload/EUBORDERSCAPES_State_of_Debate_Report_1.pdf, accessed 13 March 2015). This expansion has also witnessed a shift from the two original geographical cradles of border analysis – that is, the United States–Mexico divide and the European borders – toward a wider horizon that includes non-Western bordering practices and opens the way for postcolonial border studies.

However, recent spatial practices have shaken the foundation of bordering. The projection of bordering practices, both inside and outside national territories, as well as the multiplication of agency in the process may lead to a questioning of the very notion of linear limits: Is the everywhere border still a border? This questioning goes along with the multiplication of exceptional statuses, whether in the economic sense (with the development of extra-territorial zones, as in Nyíri 2009) or in the political one (with the emergence of all kinds of camps: Minca 2005; Le Cour Grandmaison, Lhuilier, & Valluy 2007). These do indeed bring about shifts in the scale and complexity of time/space relations, which are at the heart of future research.

This chapter will begin with the historical development of bounded thinking. This retrospective view will emphasize the correlation between limit-building, world-compartmentalizing, and the linear shape of boundaries (notably on maps). Subsequently, the evolution of the spatiality of limits will be considered, from linear to topological metrics, before opening a discussion of the ontology of the mobile border and its consequences for our spatial alphabet.

Bounded thinking

A historical review reveals that the invention of physical limits is linked to the process of sedentary settling. The action of materializing compartmentalization in space may date back to the days when the first Sumerian cities were fortified with walls. Such a broad perspective indicates that national borders, even if they have strongly contributed to changing our more general representations of boundaries, are a very recent phenomenon that goes back only a few centuries. The great novelty of the modern (post-Renaissance) period has been the transition away from a medieval multibounded world, since the feudal system was based on interpersonal allegiances that did not require spatial contiguity, and territorial limits were not the most meaningful. The political order that emerged in modern Europe was built on the necessity of finding a balance between political organization and territory. It is thus generally acknowledged that borders (defined as state limits) were "invented" in 1648, with the signing of the Treaties of Westphalia that marked the end of the Thirty Years' War in Europe and the search for a balance of power guaranteed by territorial stability (See the title of Foucher's 1986 book, which is indicative of the constructive paradigm that historians were discovering at that time, whereby they were engaged in the analysis of all kinds of "inventions."). Although these treaties were far from ushering in an era without war, they did open up a bounded way of thinking about modernity. The extent of a state's power was thus to be limited by the extent of a rival political domain, on agreement over the partition of space symbolized by a common separation line that would be called a border.

This led to the possibility of an "international" order, negotiated between nation-states that were in the process of consolidation, while the idea of empire was supposedly left behind, at least concerning the European space (regional empires survived, notably Austria-Hungary, but the idea of a European empire was left behind). The modern nation-state was thus created on the notion of territorial sovereignty (Elden 2013), which in turn gave rise to a long-lasting tautology whereby territory is defined by state, state by sovereignty, and sovereignty by territory.

Just as they were drawing their external borders, the recently established states undertook a concomitant reorganization of their territories in order to confirm their rule over them, which was done through a process of internal division. The emerging administrative grids can be understood as a form of political rationality aimed at ensuring population control. For those who have no personal relation to the people in power and for whom order is symbolic, administrative boundaries have a mundane materiality. Foucault (2004) has stressed that governing techniques, among which administrative divisions play an important rule, have taken over previous sovereignty structures, and his work will help us to draw political links between boundaries and borders.

The generalization of what we could call "territorial thinking" took time. The eighteenth century marked a turning point in the success with which various governments established the administrative grid of their territories through the conjunction of effective networks such as tax collection or postal distribution, achievements based on better physical infrastructures and regional restructuring. The quest for the "territorial optimum," or the best possible division of space, also dates to this period. The French Revolution indeed included the proposal of a new map that offered to replace all previous territorial (ecclesiastical, judicial, etc.) subdivisions by a series of geometric figures. Ultimately, however, France was not divided into squares: New administrative boundaries were attained through the invention of a new level of administration, that of the department ("département"), but its boundaries were negotiated according to the spatial rationality of the time – that is, the quest for natural limits – and many French departments bear the names of the rivers that define their outline (Ozouf-Marignier 1989).

This example highlights the importance of the tool that allows any theory to be enacted. In the case of any kind of boundary or border, the function of a map is much more than illustrative. One could even say that the map has constituted the condition of existence of the political convention that it was meant to represent. If placing a boundary on the map makes it real, it is the existence of the map that makes it possible for the limit to exist as such, notably as a linear device. The first known European boundary treaties are indeed contemporaneous with the first precise regional maps, which enabled decision-makers to provide their governments with tangible evidence of their decisions on the localization of international limits. In this context, the act of drawing a boundary relied on the elements that *did* figure on those maps, and notably on topographical features. For example, rivers appeared as curves, and mountains were represented by a series of small triangles linked together in order to suggest the then fashionable idea of "chains." These two kinds of representational abstractions prepared the ground for the tracing of lines.

The origin of the "natural border" concept harks back to these first border maps and happens to constitute an eighteenth-century coincidence: It was not only technically easier to draw the boundary on a line on the map, but politically also very convenient. As they considered themselves to be reigning over a piece of land in the name of God, the sovereigns of the time were very interested in having their kingdoms rely on naturalized territorial limits. If their borders were supported by topographical elements such as rivers or mountains, which were themselves outcomes of divine creation, then they could claim that their authority also resulted from this divine order. And this would reveal a useful way to deflect questions concerning the justification of their kingdoms' borders (Debarbieux 1997). An analysis of the modern context in which boundaries emerged also allows us to define them as both technical tools and performative creations, according to the double status of maps laid out by Dennis Cosgrove (1999). From the moment they were laid on the vellum of the map, boundaries were given a consistency that retroactively justified the power that had ordered the map. Boundaries were born of this visual condition. This ingrained mapping condition of boundaries and borders may help us to better comprehend the difficulty of understanding territorial constructions in regions that were devoid of mapping culture (or at least of Cartesian drawing culture). African borders are certainly not more artificial than others, but they overlap a political order that was not traditionally established on a "mappable" sovereignty. This may explain why, in some regions, collective representations have clashed so much with territorial borders.

The mobile dimension of space and the correlated necessity to examine borders in a more dynamic way have always existed, but this necessity has in general been obscured by methods of analysis that privilege stable elements over labile configurations. Borders have often been defined as palimpsests: those old manuscripts that kept the traces of vanished words when new scribes had written on them. Foucher (1991: 43) describes them as "time inscribed within space," testifying in the present to the complex restructurings of the past. The genealogical method that Foucault deploys can be very useful in apprehending the bypass that has for so long compelled us to perceive boundaries as stable. When Foucault retrospectively looks to ancient Greece for the origins of governmentality, he distinguishes this period from the rest of Western history as being the only true territorial era, "the Greek god [being] a territorial god, an intra-muros god" (Foucault 2004: 129). He opposes the Greek polity to the rest of the Mediterranean, where shepherds managed mobile flocks, and he points out that the Greeks only managed to produce this political scheme at the very small scale of scattered cities. On the basis of these historical hints, we may question the linear model that we have taken for granted over the past few centuries and ask ourselves whether, perhaps, borders may only have been a Greek illusion.

From linear to topological limits

Boundaries and borders were reinvented during the modern era to define a political order that was conceived as stable, or at least as one that was undergoing a process of constant, linear stabilization. It reflected an ontological perspective that divided objects into exclusive categories, separated from each other, that very few political geographers challenged, except for Gottmann, who was one of the first to emphasize the importance of flows in understanding the organization of the world (Gottmann 1952). Indeed, it has recently been stressed that the inclusion of the word "order" in "b/order" was not a coincidence (Van Houtum, Kramsch, & Ziefhofer 2004). Nevertheless, the contemporary global era has proved the opposite: Borders of all kinds are very unstable. A whole new set of borders has emerged during the disintegration process of Cold War geography; more generally, territorial restructuring is also highly visible on all scales and in very diverse cultural settings (Antheaume & Giraut 2005).

However, the coining of the "b/ordering" word play grounds the term within one linguistic context, exclusive of others, since it does not work in Roman or Germanic languages. The globalization of scientific production and the domination of English writing do not stand alone to explain the success of this expression. The way in which English relies more readily on verbs than on nouns makes it apt for describing the labile condition of contemporary socio-spatial objects. Over the past 25 years, border studies have established that boundaries are never given spatial settings. They are always undergoing "debordering" and "rebordering" processes that can affect them in any point of space and at any time, simultaneously. This dialectical process means that there exist simultaneous conditions for the opening up of any boundary, notably linked to the globalization of trade and the increase of flows of human beings, goods, capital, and ideas, as well as for the closing down of any boundary, in order to ensure the security of those flows through control and filtering. It follows that any given border or boundary is undergoing both opening and closing processes at the same time in the same place.

This can happen because the forms and functions of borders do not spatially coincide any more. Given the dominance of functional spatial analysis in border studies, this kind of understanding of borders has taken some time to emerge. Examining the controlling and filtering roles of any territorial limit too closely can indeed prevent us from questioning the complexity of its spatial forms. The security paradigm that imposed itself in the media and academic agendas after 9/11 has led many authors to try to trace the evolution of controlling functions and to document their apparent dematerialization. In various situations, border crossings are no longer controlled when a line is crossed, but through ever more sophisticated means of technological surveillance (Pickering & Weber 2006) in various network hubs, such as at bus and train stations, airports, ports, and logistical centers.

In this process, the border is "becoming a set of discontinuous dots" (Bonditti 2005: 10), which lead us to shift our thinking from the place of border encounters to the spatialization of the complex interactions that comprise the bordering act. This, in turn, leads us to redirect the focus "from boarding places to data sets"; in other words, from place to digital information, from topography to topology. This computer and communications technology gap not only compels us to consider the territorial dissemination of border functions, projecting borders inward to and outward from national spaces, but unveils the possible "pixelization" of any kind of boundary line (Bigo & Guild 2005). The possible decomposition of any border into sets of data and binary codes that are infinitely malleable could, however, mean the end of the notion of a boundary, because of its obvious dislocation. Could the border be that pervasive? The nomadic horizon traced by Deleuze and Guattari (2008 [1972]) to get out of the previous analytical dead ends of structuralism should probably not be interpreted too literally. Pixelization is derived from the sophistication of both technological border control devices and the networked organizations that manage them. However, the mobile lenses that we now need in order to understand borders may reveal a necessary step toward the understanding of territorial restructuring: data flows intersecting in ways that produce new grids (Vukov & Sheller 2013).

Whether we are looking at debordering or rebordering processes, is it possible to localize the seemingly ubiquitous border? This quest is not merely the whim of a social scientist. On the very open borders within Europe's Schengen Area, we can witness numerous initiatives to physically demarcate and symbolically erase the lines. From the transformation of customs checkposts into tourist shops or museums to the erection of public monuments and the organization of cultural events, borderland inhabitants work on the preservation of the boundary landscape (Amilhat Szary 2012). The materiality of the border enacted by infrastructure, such as walls or barriers, may seem easier to assess on closing borders, even if most of the border control is made through a network of "smart" devices, from video cameras to digital sensors and robots. This technology apparatus is not devoid of its own territoriality (Popescu 2011), since it is linked to various infrastructures that are mobilized to ensure the continuity of the network. The latter, submarine cables and satellite relaying, even impose a three-dimensional vision of border dimensions, if not a more topological approach to boundary analysis.

Emphasizing this orientation, when the US Department of Homeland Security tries to discern what "the border of the future" could be, it describes a "continuum framed by air, sea and land dimensions" (Salter 2004: 76). This notion of an absolute extremity of the national territory disclaims the linear nature of the border without canceling its spatiality. It also paves the way for the activation of bordering processes in parts of our ecosphere that had for a long time escaped territorial management and its effective implementation, in other words, sea and air. Maritime borders have indeed become an active frontier of international law and geopolitics, since GPS (Global Positioning System) and satellite technologies allow surveillance of the legal framework set up by the United Nations Convention on the Law of the Sea (UNCLOS, signed in Montego Bay in 1982 but ratified only in 1994). These borders may become the scene for active claims that could be linked to land conflicts (for example, the conflict over the waters of the disputed area of Western Sahara) or multidimensional interests, notably when issues of illegal migration interact with security agendas. In the Mediterranean, the border map is complicated by the sinuous delineation of the often rocky coastline, the presence of multiple islands, the intense global flows that intersect with military surveillance, and illegal crossings for human and drug trafficking. This is where the nongovernmental organization Watch the Med has set up a framework to monitor cases of migrant deaths at the borders of the European Union that constitute a violation of the human rights supposedly guaranteed by the neighboring states, subsequently engaging in a complex legal discussion of the consequences of territorial sovereignty in a multilayered Europe.

These recent evolutions of "boundedness" appear, once again, to be firmly associated with the tools and instruments that contemporary technology makes available in a determined historical context. Changing our lenses, this context makes the contributions of cultural studies (Price 2000) crucial to all border scholars. As soon as one tries to shift the focus from the institutions to the people who constitute the border by living at and around it, it becomes apparent that the separation line is not essential. It splits into a multitude of fragments, both functional and ideal. When the Chicana poet and activist Gloria Anzaldúa claimed that she "was" the border (Anzaldúa 1987), she was one of the first to stress the performative dimension of border territorialities and to point out that the embodiment of an institutional convention means more than the mere designation of a political status. In the Foucauldian sense, the border proves to function as a complex set of apparatuses, a "dispositif" where the political and the physical intersect. More than any other kind of boundary, the border becomes a place where identities are continuously at play, both at the area of human interaction and in the multiple locations where the data sets are processed.

Beyond network analysis, it can be useful to mobilize non-Euclidean perspectives to offer innovative interpretations of border complexity (Lévy 2008; Martin & Secor 2014). However, if control technologies have made it very clear that the border is no longer "a continuous line demarcating the territory and sovereign authority of the state, enclosing its domain" (Walters 2006: 193), but rather a more complex, reticular, and mobile set of filtering devices, they have not entirely denied its materiality. This paradox hints at the fact that the de/rebordering phenomena are not equivalent to processes of de/reterritorialization. If debordering chronologically coincided with a moderation of state power, it does not necessarily imply a disaffection with territorial stakes. To understand the relations between spatial processes and the possible existence of bounded categories within which they are deployed, reterritorialization is an essential hypothesis. The materiality of bordering may never have been all that important, since the debordering/rebordering processes have been made clear. For that reason, it could be said that contemporary boundaries evidence a hyper-territorialization process that is not at odds with the existence of multiple networks that endow space with topological metrics.

Ontology of the mobile border

The disjunction between the functions and shapes of borders illustrates the difficulties that relational geography faces in making sense of boundaries. What kinds of limits can exist between spatial objects if the very notion of division is tough to unpack? Answers to these questions are of critical political importance. As Balibar points out, "when borders cease to be purely exterior realities … [and] if they are no longer on the *verge* of politics but … *objects*, or better said *things* within the space of politics itself" (Balibar 1996: 374), what do they say about our capacity to organize ourselves collectively? Since territorial sovereignty provides the grounds for state-building and also for the kind of social contract that democracy requires, can democracy survive border disorder?

Early postcolonial approaches to border issues stressed the positive aspects of in-between-ness that borders tend to foster. For example, Price states that "by its very fluidity, the border constitutes a potential-laden space where old power relations can be reworked and perhaps made less oppressive" (Price 2000: 105). However, the flexibility of in-between-ness may only be an illusion, as borderlands are regional constructs that surround a linear device. Since the linear dimension of political limits has not totally disappeared, there is still space for liminal conditions of sociality. These can be seen positively as places of multiple exchanges that benefit from the border as a resource without colliding with its separating function. As Raffestin puts it, "territoriality constitutes the stumbling block of any division" (Raffestin 1980: 153); indeed, border life generally illustrates that linear divides are conventions that allow for multiple interpretations of the field. The theoretical backgrounds for migration and mobilities analysis have had to be reframed to take into account this multiplication of border experiences, and the fact that crossing can no longer simply be defined as a move between two or more points (Richardson 2013).

However, the personalization of borders through the individualization of boundary experiences and definitions may make it too difficult to design a collective social being. If the task of defining identity and alterity is left to individuals alone, they may not be able to endure the violence of the process: Borders, having been invented to materialize state protection over individuals, are becoming the place where the same people now find themselves exposed, without mediation, to the violence of globalization. In *The Archeology of Knowledge*, Foucault (1969) writes that there is anxiety in the thinking of the other, because thinking about discontinuity can be fearful. Border culture, if it exists (Amilhat Szary 2013b), endorses some very violent components, since the constant exposure to alterity, the experience of "différance" (Derrida 1972), leaves every individual facing a constant need to redefine their identity (or identities), with no certainty about the political outcome of this quest.

In a sense, the spatial diffusion of the border and its propensity for ubiquity make it impossible to differentiate an inside from an outside, and therefore to separate insiders from outsiders. Such confusion goes very much against the idea of territorial exceptionality and the possibility of policing these intermediary spaces, even in the perspective that Agamben (2005) adopts. In this latter view, resistance to a system built on such bordering frames can no longer reside in crossing lines, even by illegal means. The only way to pervert such a spatial organization, Razac (2013) suggests, would be to inhabit the line. Finding ways to spatialize the mobile border allows the condensing of an argument in order to denounce its conditions of existence. This idea is also at the core a collective of artists' project: Decolonizing Architecture (DAAR), based in Palestine, has produced an installation based on what they found within the space traced on the map by the Oslo agreements. Less than 1 millimeter on the official paper, the space covered by the line traced with a red pen measures more than 5 meters in reality but has no legal status (http://www.decolonizing.ps/site/battir-3/, accessed 13 March 2015). Basing their project on the official line, they have found exceptional places, including a baroque migrant house, a personal enclave, and the unfinished parliament of Palestine, over which they place the line in a very powerful statement that denounces the partition of the land they claim.

Whether it originates in the proliferation of flows resulting from globalization (Sheller & Urry 2006) or in the wearing of mobile lenses to analyze them (Retaillé 2005; Büscher, Urry, & Witchger 2011; Söderström et al. 2013), the waning of a fixed and integrated border line calls for a renewal of border studies. The conceptualization of these changes should be able to account for the emerging dissociation and crystallization processes at work and shed light on their meaning. One could say that the border is becoming a complex assemblage (Allen & Cochrane 2010; Anderson & McFarlane 2011), or a "socio-technical network" (Latour 2005), and consider it as a "collective entity associating objects, actors (migrant or border police), places and regulations" (Fourny 2013). However, we tend to think that this kind of interpretation erases the very politics of border places and bordering processes. Shifting the focus from institutions to practices by emphasizing the "vernacularisation" of borders (Perkins & Rumford 2013) does bring the individual back into borders, but not necessarily their subjectivity (Mezzadra & Neilson 2013).

The apparently oxymoronic expression "mobile border" was able to impose itself as a growing number of scholars were confirming that the contemporary border could only be grasped through its portativity, considering a change of focus toward the individual and their personalization of a mobile device (Cuttitta 2006; Weber 2006; Steinberg 2009; Agier 2013; Richardson 2013; Jones & Johnson 2014). Ongoing research is suggesting a renewed approach to border studies through the coining of a new term, "borderities," built on a close reading of Foucault's writings and deriving from governmentality (Amilhat Szary & Giraut 2015). By "borderity" we mean any technology of spatial or socio-spatial division; it could be defined as the governmentality of territorial limits. Although initially defined as a technology of power, borderity may also appear as a differentiated social and political quality. Examining how political subjects can be disabled and enabled, the proposed "borderities" approach illuminates the question of how borders can be the site of both power and counter-power.

An ontological perspective can help us formulate proposals to inaugurate possible forms of theorizing a "mobile border." If ontology resides in Aristotle's method of hierarchical classification of the world, it may seem paradoxical to mobilize it to propose a dynamic vision of spatial relations. It is, however, because ontology allows us to question physics and metaphysics in combination that it answers some of the philosophical questions raised by the recourse to topologies. Yet any perspective that allows both the subject and the object to be taken into account had, up until recently, focused on a very stable state of being. A mobile ontology would therefore allow time and instability to be instilled into the analysis of borders. This could be the only way to underscore the fact that any kind of place where an object and a subject are to be thought of together is a theater of unequal relations and domination processes.

Conclusion

Linear borders have not disappeared, although it has been made clear that their linearity represents only one aspect of their complex definition. The diversification of the forms and functions of borders helps us to grasp the need for elaborating strong yet nuanced proposals to think about spatial boundaries and social divides, none of which has been erased by the de/rebordering processes. Understanding contemporary borders and the implications of their penetration into many aspects of our social, spatial, and political relations indeed requires a strong epistemological effort. Only such an effort can allow us to escape the traditional and tautological relationship between territory, state, and borders that always defines one term with another and thereby forecloses the possibility of questioning the politics of mobile limits.

The complexity of contemporary borders imposes the necessity that the first boundaries to be crossed to understand them be those of our own academic fields of activity. Border studies is developing as a multidisciplinary field of knowledge, bringing political geography together not only with other social sciences, but also with many other cognitive approaches. Because this is never easy, some mediation may be called for, such as the development of experimental work with artists. This was at the core of the project undertaken by the French collective that gave itself the name antiAtlas of borders (http://www.antiatlas.net/en/, accessed 13 March 2015). The potential of art–science explorations certainly resides in their call for "undiscipline," since "the inscription of research within specialized disciplines may make invalid or unthinkable questioning what could precisely come out of the refusal of objects and methods that disciplines acknowledge" (Loty 2005: 252, quoted by Mekdijan and al. 2014). Putting border thinking at this crossroads makes borders, more than ever, important laboratories to evidence the evolutions of both spatial thinking and contemporary political transformations (Mezzedra & Neilson 2013).

References

Agamben, G. (2005). *State of Exception: Homo Sacer II, 1*. Chicago, IL: University of Chicago Press.

Agier, M. (2013). *La condition cosmopolite: L'anthropologie à l'épreuve du piège identitaire*. Paris: La Découverte.

Allen, J., & Cochrane, A. (2010). Assemblages of state power: Topological shifts in the organization of government and politics. *Antipode* 42: 1071–1089. doi:10.1111/j.1467-8330.2010.00794.x

Amilhat Szary, A.-L. (2012). Border art and the politics of art display. *Journal of Borderlands Studies* 27: 213–228. doi:10.1080/08865655.2012.687216

Amilhat Szary, A.-L. (2013a). Towards experimental mountaineering? Perspective on two tours of Alpine borders (Lionel Daudet/John Harlin, 2011–12)/Vers un alpinisme expérimental? Deux tours des frontières alpines en perspective, Lionel Daudet/John Harlin, 2011–12. *Journal of Alpine Research/ Revue de géographie alpine*, 101. doi:10.4000/rga.2131/doi:10.4000/rga.2125

Amilhat Szary, A.-L. (2013b). Cultura de Fronteras. In Nates Cruz, B. (ed.), *Frontera, Fronteras*, ed. B. Nates Cruz, Manizales, Colombia: Editorial Universitaria de Caldas, 43–60.

Amilhat Szary, A.-L., & Giraut, F. (2015) *Borderities: The Politics of Contemporary Mobile Borders*. Basingstoke: Palgrave Macmillan.

Ancel, J. (1938). *Géographie des frontières*. Paris: Gallimard.

Anderson, B. & McFarlane, C. (2011). Assemblage and geography. *Area* 43: 124–127. doi:10.1111/j.1475-4762.2011.01004.x

Antheaume, B., & Giraut, F. (2005). *Le territoire est mort, vive les territoires*. Paris: IRD.

Anzaldúa, G. (1987). *Borderlands/La Frontera: The New Mestiza*. San Francisco, CA: Aunt Lute Books.

Badie, B. (1995). *La fin des territoires: Essai sur le désordre international et sur l'utilité sociale du respect*. Paris: Fayard.

Balibar, E. (1996). Qu'est-ce qu'une frontière? In Balibar, E. (ed.), *La crainte des masses: Politique et philosophie avant et après Marx*, Paris: Galilée, 371–380.

Balibar, E. (2002). What is a border? *In Politics and the Other Scene*, London: Verso, 75–86.

Bigo, D., & Guild, E. (2005). *Controlling Frontiers: Free Movement into and within Europe*. London: Ashgate.

Bonditti, P. (2005). Biométrie et maîtrise des flux: Vers une "géo-technopolis du vivant en mobilité." *Cultures & Conflits* 58: 131–154. http://conflits.revues.org/1825, accessed 13 March 2015.

Brunet-Jailly, E. (2005). Theorizing borders: An interdisciplinary perspective. *Geopolitics* 10: 633–645. doi:10.1080/14650040500318449

Büscher, M., Urry, J., & Witchger, K. (2011). Introduction. In Büscher, M., Urry, J., & Witchger, K. (eds), *Mobile Methods*, London: Routledge, 1–19.

Cosgrove, D. (1999). *Mappings*. London: Reaktion Books.

Cuttitta, P. (2006). Points and lines: A topography of borders in the global space. *Ephemera: Theory & Politics in Organization*, 1. http://www.ephemerajournal.org/sites/default/files/6-1cuttitta.pdf, accessed 13 March 2015.

Debarbieux, B. (1997). La montagne comme figure de la frontière. *Le Globe-Genève* 137: 145–166.

Deleuze, G., & Guattari, F. (2008 [1972]). *L'anti-Oedipe: Capitalisme et schizophrénie 1*. Paris: Editions de Minuit.

Derrida, J. (1972). La différance. *In Marges de la philosophie*, Paris: Editions de Minuit, 8.

Donnan, H., & Wilson, T.M. (1994). An anthropology of frontiers. In Donnan, H., & Wilson, T.M. (eds), *Border Approaches: Anthropological Perspectives on Frontiers*, Lanham, MD: University Press of America.

Elden, S. (2013). *The Birth of Territory*. Chicago, IL: University of Chicago Press.

Fall, J. (2005). *Drawing the Line: Boundaries, Identity and Hybridity in Transboundary Spaces*. Aldershot: Ashgate.

Foucault, M. (1969). *L'Archéologie du savoir*. Paris: Gallimard.

Foucault, M. (2004). *Sécurité, territoire, population: Cours au Collège de France 1977–78.* Paris: Gallimard-Seuil.

Foucher, M. (1986). *L'invention des frontières.* Paris: Fondation pour les Etudes de Défense Nationale.

Foucher, M. (1991). *Fronts et frontières: Un tour du monde géopolitique.* Paris: Fayard.

Foucher, M. (2012 [2007]). *L'obsession des frontières.* Paris: Perrin.

Fourny, M.-C. (2013). The border as liminal space: A proposal for analyzing the emergence of a concept of the mobile border in the context of the Alps/La frontière comme espace liminal: Proposition pour analyser l'émergence d'une figure de la frontière mobile dans le contexte alpin. *Journal of Alpine Research/Revue de géographie alpine* 101. doi:10.4000/rga.2120/doi:10.4000/rga.2115

Gottmann, J. (1952). *La politique des Etats et leur géographie.* Paris: Armand Colin.

Groupe Frontière, Arbaret-Schulz, C., Permay, J.-C., Reitel, B., Selimanovski, C., Sohn, C., & Zander, P. (2004). La frontière, un objet spatial en mutation. *EspacesTemps.net,* 29 October. http://www.espacestemps.net/en/articles/la-frontiere-un-objet-spatial-en-mutation-en/, accessed 13 March 2015.

Guichonnet, P., & Raffestin, C. (1974). *Géographie des frontières.* Paris: PUF.

Hamez, G., Amilhat Szary, A.-L., Paris, D., Reitel, C., & Walther, O. (2013). Modelling and benchmarking of borders/Modèles de frontières: Des frontières modèles? *Belgéo,* 1. http://belgeo.revues.org/10659, accessed 13 March 2015.

Jones, R., & Johnson, C. (2014). *Making the Border in Everyday Life.* London: Ashgate.

Lask, T. (1995). Grenze/frontière: Le sens de la frontière. *Quaderni,* 27: 65–78.

Latour, B. (2005). *Reassembling the Social: An Introduction to Actor-Network-Theory.* Oxford: Oxford University Press.

Le Cour Grandmaison, O., Lhuilier, G., & Valluy, J. (2007). *Le retour des camps? Sangatte, Lampedusa, Guantanamo…* Paris: Autrement.

Lévy, J. (2008). Topologie furtive. *EspacesTemps.net,* 28 February. www.espacestemps.net/articles/topologie-furtive/, accessed 14 March 2005.

Loty, L. (2005). Pour l'indisciplinarité. In Douthwaite, J., & Vidal, M. (eds), *The Interdisciplinary Century: Tensions and Convergences in 18th-century Art, History and Literature,* Oxford: Voltaire Foundation, 245–259.

Martin, L., & Secor, A. (2014). Towards a post-mathematical topology. *Progress in Human Geography* 38(3): 420–438. doi:10.1177/0309132513508209

Mekdjian, S., Amilhat Szary, A.-L., Moreau, M., Nasruddin, G., Deme, M., Houbey, L., & Guillemin, C. (2014). Figurer les entre-deux migratoires: Une expérience scientifique et artistique d'ateliers de cartographie participative. *Carnets de géographes,* 6: n.p. http://www.carnetsdegeographes.org/carnets_terrain/terrain_07_01_Mekdjian.php, accessed march 27th, 2015

Mezzadra, S. and B. Neilson, Eds. (2013). Border As Method, or, the Multiplication of Labor. Durham, NC, Duke University Press.

Minca, C. (2005). The return of the camp. *Progress in Human Geography* 29: 405–412. doi:10.1191/0309132505ph557xx

Moore, M. (2003). *A World without Walls: Freedom, Development, Free Trade and Global Governance.* Cambridge: Cambridge University Press.

Newman, D., & Paasi, A. (1998). Fences and neighbours in the postmodern world: Boundary narratives in political geography. *Progress in Human Geography* 22: 186–207. doi:10.1191/030913298666039113

Nordman, D. (1998). *Frontières de France.* Paris: Gallimard.

Nyíri, P. (2009). Extraterritoriality. Foreign concessions: The past and future of a form of shared sovereignty. *EspacesTemps.net,* 23 November. http://www.espacestemps.net/articles/extraterritoriality-pal-nyiri/, accessed 13 March 2015.

Ozouf-Marignier, M.-V. (1989). *La formation des départements: La représentation du territoire français à la fin du XVIIIe siècle.* Paris: Éditions de l'École des hautes études en sciences sociales.

Paasi, A. (1996). *Territories, Boundaries and Consciousness: The Changing Geographies of the Finnish-Russian Border.* Chichester: John Wiley & Sons Ltd.

Perkins, C., & Rumford, C. (2013). The politics of (un)fixity and the vernacularization of Borders. *Global Society* 27: 267–282. doi:10.1080/13600826.2013.790784

Pickering, S., & Weber, L. (2006). *Borders, Mobility and Technologies of Control*. New York: Springer.

Popescu, G. (2011). *Bordering and Ordering the Twenty-First Century: Understanding Borders*. Lanham, MD: Rowman & Littlefield.

Prescott, J.R.V. (1978). *Boundaries and Frontiers*. London: Croom Helm.

Price, P. (2000). Inscribing the border: Schizophrenia and the aesthetics of Aztlán. *Social and Cultural Geography* 1: 101–116. doi:10.1080/14649369950133511

Raffestin, C. (1974). Eléments pour une problématique des régions frontalières. *L'espace géographique*, 1: 12–18.

Raffestin, C. (1980). *Pour une géographie du pouvoir*. Paris: Litec.

Razac, O. (2013). La gestion de la perméabilité. *L'espace politique* 20: n.p. doi:10.4000/espacepolitique.2711

Retaillé, D. (2005). L'espace mobile. In Anttheaume, B., & Giraut, F. (eds), *Le territoire est mort. Vive les territoires!* Paris: IRD, 175–201.

Richardson, T. (2013). Borders and mobilities: Introduction to the special issue. *Mobilities* 8(1): 1–6. doi: 10.1080/17450101.2012.747747

Rumley, D., & Minghi, J. (1991). The border landscape concept. In Rumley, D., & Minghi, J. (eds), *The Geography of Border Landscapes*, London: Routledge, 1–14.

Sahlins, P. (1996). *Frontières et identités nationales: La France et l'Espagne dans les Pyrénées depuis le XVIIe siècle*. Paris: Belin.

Salter, M.B. (2004). Passports, mobility, and security: How smart can the border be? *International Studies Perspectives*, 5(1): 71–91. doi:10.1111/j.1528-3577.2004.00158.x

Sheller, M., & Urry, J. (2006). The new mobilities paradigm. *Environment and Planning A* 28: 207–226. doi:10.1068/a37268

Simmel, G. (1972 [1908]). *Individuality and Social Forms*, trans. D.N. Levine. Chicago, IL: University of Chicago Press.

Söderström, O., Ruedin, D., Randeria, S., D'Amato, G., & Panese, F. (2013). *Critical Mobilities*. Lausanne: EPFL Press.

Steinberg, P.E. (2009). Sovereignty, territory, and the mapping of mobility: A view from the outside. *Annals of the Association of American Geographers* 99: 467–495. doi:10.1080/00045600902931702

Van Houtum, H., Kramsch, O., & Ziefhofer, W. (2004). *B/ordering Space*. London: Aldershot.

Vukov, T., & Sheller, M. (2013). Border work: Surveillant assemblages, virtual fences, and tactical counter-media. *Social Semiotics* 23: 225–241. doi:10.1080/10350330.2013.777592

Walters, W. (2006). Border/control. *European Journal of Social Theory* 9: 187–203. doi:10.1177/1368431006063332

Wastl-Walter, D. (2012). *Companion to Border Studies*. Farnham: Ashgate.

Weber, L. (2006). The shifting frontiers of migration control. In Pickering, S., & Weber, L. (eds), *Borders, Mobility and Technologies of Control*, Dordrecht: Springer, 21–44.

Wilson, T.M., & Donnan, H. (2012). *A Companion to Border Studies*. New York: Wiley-Blackwell.

Chapter 3

Scale

Andrew E.G. Jonas
University of Hull, England, UK

At a basic level, scale is a way of describing the size of an area on a map relative to that of other areas. Some maps strive to represent information at the global scale, while others seek to represent smaller scales like a city or a district within it. Additionally, and especially in political geography, scale is a concept that is increasingly deployed to represent the variety of ways in which political power is spatially constituted. Many of the political organizations that geographers study, such as supranational institutions, nation-states, city regions, regional special-purpose districts, and local governments, occupy areas covering different scales (e.g., international, national, regional, urban, local, and so on) and have scalar-specific powers. For example, the state typically comprises a territorial hierarchy of powers organized around national, regional, and/or local branches. Each branch or level of the state, in turn, exercises its powers across a territorial jurisdiction, such as the entire country, a province or region, or an urban district. Scale is a way of describing the geographical extent of territory covered by each level of government as well as the allocation of powers across, within, and between the state's territorial structures (Cox 1990; Cox & Jonas 1993).

Scale is further used as a shorthand reference to social, environmental, economic, and political processes having some sort of spatial expression. For example, a reference to *national* politics means those political processes taking place throughout a nation-state, and thereby delimiting the national as a meaningful territory or political space. Most, if not all, democratic states operate national elections, which are conducted under specific rules depending on how the nation-state is constituted as a territorial-political ideal (Jonas 2002). In the United States, the presidential candidate who gains a majority of states in the national Electoral College is usually elected President. By way of contrast, in the United Kingdom, which includes the provincial territories of England, Scotland, Northern Ireland, and Wales, the Prime Minister is elected indirectly on the basis of being the leader of the political party that wins a majority of seats in the UK parliament (or forms a winning coalition with another party). In a referendum held in Scotland in September 2014, Scottish voters were asked the question: "Should Scotland be an independent country?" The final

The Wiley Blackwell Companion to Political Geography, First Edition.
Edited by John Agnew, Virginie Mamadouh, Anna J. Secor, and Joanne Sharp.
© 2015 John Wiley & Sons Ltd. Published 2017 by John Wiley & Sons Ltd.

result was 44.7% "Yes" and 55.5% "No," with a turnout of 84.6% of the eligible voting population. If the outcome had been different (that is, a majority voting "Yes"), Scottish voters would in future have voted for representatives in the national parliament of Scotland, but would no longer have participated in elections held in the remainder of the United Kingdom. In this context, scale helps to convey the complicated ways in which political processes, such as national elections, map onto space and how, in turn, processes operating at other scales (e.g., the rise of nationalist sentiments in a province) can eventually lead to the establishment of new territorial structures (i.e., scales) of the state (e.g., an independent Scotland).

The concept of scale should not be conflated with that of level (even though these two concepts are sometimes used interchangeably). When political scientists refer to levels of government, it is generally assumed that each level represents a discrete sphere of political action and is governed by level-specific rules, activities, and modes of coordination, as with the example of national elections (Pierre & Peters 2000). In well-developed political systems, such as those found within the European Union, each level often interacts with other levels within what are known as multilevel governance systems (Marks 1993). However, the concept of multilevel governance does not capture satisfactorily the variety of ways in which political processes are constituted around different territorial structures of the state and governance. When political geographers refer to scale – or the related idea of the "politics of scale" (Jonas 1994; Swyngedouw 1997; Cox 2002) – they are acknowledging not only that political processes reflect how power is distributed hierarchically (e.g., through the different levels of the state), but also that struggles within or across state territory can produce new scales of political organization and governance, which do not map directly onto the corresponding levels of the state.

The purpose of this chapter is to examine the variety of ways in which scale has been used both as a descriptive device and as a conceptual category in political geography. In the first section, I examine how political geographers initially sought to map political space at different scales and identify scalar-dependent political processes. I then show how scale is increasingly deployed as a conceptual device for representing the political dynamics shaping territory. While some political geographers are interested in how territorial structures of the state are organized at different scales, others seek to examine non-state scales and associated forms of contentious politics. Occasionally there have been disagreements about the value of scale as a concept, and these will be explored in the penultimate section. In conclusion, I suggest that despite such disagreements, political geographers find scale to be a useful way of representing and explaining complex geographies of political, social, environmental, and economic change.

Cartographic and quantitative approaches to scale in political geography

In geography, scale has traditionally conveyed two quite distinct ideas about how space is represented (for details, see Herod 2011). Firstly, scale is a tool deployed in cartography for measuring distances on a map. Secondly, scale has become a shorthand descriptor of processes that have a sort of geographical scope. In political geography, scale has been used in both contexts: in cartographic representations of territory; and in the identification and analysis of political processes operating at certain spatial scales.

Most maps have a standardized scale that allows the reader to convert map distance into a precise measurement of the distance between actual objects. In this respect, maps try to represent real scales, which could conceivably range in size from an urban district to the entire globe. A cursory inspection of some classic political geography textbooks and monographs would reveal that maps have been valuable tools for representing political space at a variety of spatial scales, ranging from the international and national to the regional and urban. For instance, geopolitical maps have sought to represent how the international scale can be divided into "geostrategic regions" (Cohen 2003). Political geographers in addition have mapped political spaces at progressively smaller scales, such as civil divisions inside countries or incorporated and unincorporated parts of a metropolitan area (Glassner & de Blij 1989). Finally, attention has been paid to the ways in which map scales can be manipulated in order to distort and misrepresent political power (Monmonier 1991). Analysis of these "propaganda maps" or "persuasive maps" can reveal much about the state of geopolitical rivalries at different time periods and in diverse spatial contexts (Boria 2008).

Alongside mapping political territories, geographers have sought to describe how political processes function at different scales. This has usually meant treating the national scale as the basic geographical entity from which political power emanates. For example, Hartshorne (1950) divided political space into two sets of functional relations, each covering a specific scale of state territorial organization. On the one hand, he identified the external relations of the state as those international and interregional state functions that shape large-scale geopolitical and geoeconomic organizations. On the other hand, internal state relations are functions primarily involved in shaping the territorial organization of regions and political divisions within the state. Evidently, Hartshorne was not merely interested in mapping national state boundaries and borders; he was also concerned with identifying how different political processes operate around a hierarchy of territorial scales. Over the years, a number of key political geography texts have similarly used scale as a way of organizing discussions of different political processes and problems, such as welfare and locational conflict (Cox 1979) or the role of nation-states and localities in the world system (Flint & Taylor 2011).

The growth of quantitative methods in geography in the 1960s and 1970s inspired further developments in how scale could be used to analyze and explain the spatial dimensions of political processes. In the field of electoral geography, there has been considerable interest in why the outcomes of regional and local elections often differ from spatial patterns of voting observed at the national level. Pioneering studies conducted by Kevin Cox (1968, 1970) sought to account for such differences by identifying spatial associations between different social and political variables. Cox discovered that regional and local voting outcomes seemed to be influenced by the spatial proximity of voters rather more than by non-spatial factors such as social class or party affiliation, prompting him to propose scalar-specific spatial concepts such as the "neighborhood effect" (Cox 1972) and "neighborhood activism" (Cox & McCarthy 1982) in order to explain different forms of local political behavior. Cox (1973) further showed that other political problems, such as the metropolitan fiscal disparity issue in the United States, seem to exhibit scalar-dependent attributes, which in this instance could be systematically explained in terms of local exclusionary politics. The findings of quantitative political geographical studies such as these suggest that scale can be more than a useful mapping tool, but that it also helps to show that space has certain causal attributes, albeit it has taken time for this idea to become properly embedded within the subfield (see Johnston & Pattie 2012).

From scale effects to scalar politics

Claims made about the respective contributions of local social context and national political party affiliation in shaping political behavior prompted occasional exchanges between geographers and political scientists, with the latter seemingly unmoved by the former's assertions about the power of place and region in shaping politics (see, for example, discussions in Agnew 1987, 1996; McAllister 1987). Yet within political geography itself there was little corresponding consideration of the ways in which the analysis of political processes and geographical territories could be melded into a unified and systematic approach to scale; not, that is, until an important intervention by Peter Taylor, who drew on world systems theory in order to outline what he called a "materialist framework for political geography" (Taylor 1982). Taylor argued that taken-for-granted political geographical scales of analysis, namely the world economy, the nation-state, and the city or locality, represent different political constructions of space, which he described as, respectively, the "scale of reality," the "scale of ideology," and the "scale of experience" (Taylor 1982). In his view, scale was absolutely central to the development of political geographical theory.

Taylor's materialist political economy of scale soon inspired other political geographers to be much more explicit in showing how concepts of scalar politics could be used to animate studies of the social and political construction of territory. For example, Smith (1984) showed how the global space of capitalism is internally differentiated into territories that correspond more or less with the international, national, and urban scales. Each scale, in turn, expresses various spatial contradictions and tensions around uneven development and also provides socio-regulatory solutions over the short to medium term.

Growing recognition of socio-regulatory issues surrounding globalization and the geography of uneven development has further prompted interest in the rescaling of the state (Brenner 2004; Swyngedouw 2000). State rescaling broadly describes how powers shift up, down, and across the state territorial hierarchy and, in turn, how powers once located at one scale coalesce around a different or entirely new scale. There is some evidence that the rescaling of the state has empowered its local and regional branches, although whether or not this is a genuine response to local and regional political demands rather than a function of national or international geopolitical priorities is an ongoing topic of discussion in the literature (Jones 1997; Moisio 2011). Referring to the regionalism question in England, Jones and MacLeod (2004) draw a distinction between regional spaces, where institutions have emerged to address territorial competition within the global economy, and spaces of regionalism, which might include insurgent regionalist movements. In a study of California's New Regionalism, Jonas and Pincetl (2006) find that the rescaling of the state around regions sometimes can be a bottom-up process led by organized local business interests, rather than a top-down solution imposed by the state. Finally, and returning to the UK context, Harrison (2008) contrasts top-down or "centrally orchestrated" regionalism with bottom-up or "regionally orchestrated" centralism. This contrast is a useful way of capturing the relative balance of political power between, respectively, national and regional branches of the state in different contexts.

The study of state rescaling processes has prompted new ways of conceptualizing how political interests map onto state territorial structures at different scales. In one especially influential contribution, Kevin Cox (1998) draws a distinction between "spaces of dependence" and "scales of engagement." On the one hand, the concept of "spaces of dependence" is a reference to the various forms of social and material-*cum*-economic attachment to place. "Scales of engagement," in comparison, are the territorial structures and institutional arrangements on which political actors might draw in order to reproduce their "spaces of dependence." In positing this distinction, Cox's intention is to show that political struggles often occur around scales that are not necessarily predetermined by the presence of state territorial structures. Where suitable scales of the state do not exist, then actors who are dependent at those (or other) scales will organize politically to ensure that appropriate state scalar capacities are constructed. However, in a commentary on Cox's thesis, Judd offers a note of caution in suggesting that "[w]hen scales are absent, important political consequences ensue – and political agents will typically find it difficult and sometimes impossible to replace the scales that have not been constructed by the state" (Judd 1998: 30).

Cox's particular contribution to the state rescaling literature is insightful in two respects. First, it demonstrates that scalar politics are often fought around the creation of new territorial divisions of responsibility and power inside the state, which can differ from existing territorial structures of the state. Second, it shows how contentious political processes often involve place-bound groups marshalling powers and resources residing outside the local state territorial context. Here, Cox anticipates relational thinking about scale, to which we now turn.

Unbounding scale and politics: The rise of the relational perspective

Most scales with which political geographers deal seem to occupy fixed hierarchies, implying that higher scales (of the state) dominate or dictate what happens at lower scales; that power, in effect, tends to cascade downward through the state hierarchy (for a critique of this idea, however, see Bulkeley 2005). This creates a fundamental challenge of how to represent political processes that can be simultaneously hierarchical and stretched out across or within state territory. A short example might help to illustrate some of the issues at stake.

Consider European Union (EU) laws or directives, such as those pertaining to the regulation of territories having ecologically sensitive habitats and ecosystems (see Gibbs, While, & Jonas 2007). Such laws are issued at the international level, interpreted at the national level, and implemented and enforced locally, reflecting how powers and responsibilities are allocated across the different hierarchical levels within the EU. Nevertheless, local enforcement of EU legislation usually requires some degree of coordination between governmental and non-governmental agencies across a territory that is subject to a directive (for example, an estuarine region within a member EU state). In this case a new scale of governance might be required, such as a regional body having appropriate powers both to bring together local stakeholders and to implement the directive within the country and territory in question. While scale helps to describe the territory covered by the new governance entity (i.e., regional governance), other related concepts such as "scales of engagement" might in addition be useful for representing formative political processes operating at this and other scales, including the nation state and the EU (Gibbs et al. 2007).

In seeking to confront the challenge represented by hierarchical thinking about scale, some scholars have sought to represent political processes as networked, topological, or "flat" (Latham 2002; Marston, Jones, & Woodward 2005). These are different ways of saying that political struggles do not occupy pregiven scales in a state territorial hierarchy, but instead take place and unfold around specific sites distributed across the landscape. Such sites are, in turn, connected to other sites through extended social networks. However, relying exclusively on a concept like network poses the additional challenge of how best to represent political struggles around socio-spatial relations, which can assume many different forms at the same time.

Jessop, Brenner, and Jones (2008) argue that researchers often privilege one spatial form (e.g., scale) at the expense of others (e.g., place, network, or territory). However,

[e]ach falls into the trap of conflating a part (territory, place, scale, or networks) with the whole (the totality of sociospatial organization), whether due to conceptual imprecision, an overly narrow analytical focus, or the embrace of an untenable ontological (quasi-)reductionism. This trap

is notoriously present in methodological territorialism, which subsumes all aspects of socio-spatial relations under the rubric of territoriality. This is manifested, for instance, in state-centric approaches to globalization studies and in narrowly territorialist understandings of cities, states, and the world economy. (Jessop et al. 2008: 392)

For Jessop, Brenner, and Jones, the answer to this challenge lies not in the abandonment of scale, but rather in how it can be used alongside other concepts like territory, network, and place to represent political processes in space more effectively.

Further pointing out some shortcomings of "territorialist" approaches, Painter (2008) argues that geographers are now engaged in a very different kind of mapping in which power and politics are examined relationally. Relational thinking about scale recognizes that powers exercised within a territory also reside in social structures that extend outwith the territory. Take, for example, the regional scale. While many regions have readily identifiable jurisdictional boundaries, the political and economic capacities of such regions are not necessarily confined to within those boundaries. Instead, regional capacities increasingly draw on social relations and political structures extending well beyond the jurisdictional limits of the region, connecting it to wider state spaces and the global economy at large (Allen & Cochrane 2007; Prytherch 2010). It follows that scale can refer to the spatial extent of both the territory in question (e.g., the regional scale) and its wider social relations (e.g., the international or national divisions of labor of which the region is a part).

Besides informing recent debates about ontology and epistemology, scale has increasingly been used to represent contentious political processes occurring outside traditional territorial scales like the city, the nation-state, and the global economy (Jonas 1994). Indeed, it is refreshing to see plenty of work emerging in political geography on non-traditional political scales and social movements such as those pertaining to the body (Herod 2011), labor relations (Gough 2010), indigenous rights (Silvern 1999), peace movements (Miller 1994), community gardening (Smith & Kurtz, 2003), climate change (Bulkeley 2005), and so forth. By linking the analysis of social movements to the politics of scale, such studies have helped to move political geography away from the idea that the nation-state's controlling influence is unmediated by politics occurring around other scales.

Indeed, space has opened up for political geographers to develop a far richer language of scale and political empowerment. For example, Smith (1993) deploys the metaphor of "jumping scales" to demonstrate how political actors, such as homeless people, can project a political message well beyond the physical scale of everyday existence. Likewise, concepts such as "scalecraft" (Fraser 2010) and "scale maneuvering" (Kythreotis & Jonas 2012) have helped to animate different dimensions of scalar politics. A related idea is how socially marginalized individuals and groups deploy "scale frames" (Kurtz 2002) for promoting social and environmental justice. Framing political action and developing organizational structures around different scales can be important for securing resources and empowering social movements as these try to negotiate with or around the corresponding scales of the state. When one spatial scale is closed off for action, another may quickly emerge as a new strategic terrain. In these respects, scale is not a fixed arena for political action; instead, it is constituted in and through the dynamics of political struggles (Swyngedouw 1997). It appears, then, that today's political geographers are in a much stronger position than their predecessors when it comes to demonstrating how spatial formations at one geographical scale are reproduced, come into conflict with, and/or are transformed by formations at other scales.

Some conclusions

This chapter has explored some of the contrasting ways in which political geographers deploy scale in order to animate and analyze political processes. Scale is a way of showing how territory is organized into spatial hierarchies, ranging from the international and national to the regional and local. It has also become a useful concept for revealing the spatial dynamics of political struggles. In the future, we might well see political geographers using concepts like territory and scale interchangeably and perhaps even together. Part of this is due to the complex nature of representing different spatial forms of political organization in an increasingly interconnected yet profoundly divided world. It also reflects some challenges and ambiguities associated with applications of the concept of territory itself. Notably, the gradual erosion of state power and sovereignty by economic globalization and citizen protests has demanded new concepts with which to examine new and emerging spaces of politics. The concept of scale helps to cement the idea that such spaces can be differentiated not only hierarchically, but also horizontally in space.

Despite some important progress on the development of a concept of scalar politics, political geographers are still quite a long way from achieving a consensus about the difference that scale makes to political processes. For some, scale has arguably replaced territory as a concept for explaining how non-state space is politically organized. For others, scale seems to create its own problems in terms of how it represents political struggles. Nevertheless, scale remains useful for shedding light on the production and contestation of territorial structures both within the state and outside it. With regard to the former, ongoing research examines the rescaling of the state and the emergence of neoliberal territorial forms. In the latter case, attention is turning to the analysis of the rescaling of citizenship and contentious politics. Here, scale refers less to pregiven levels of the state and more to the manner in which social relations and political processes extend across, coagulate around, or compress within space. While this demands of its proponents a high degree of critical reflexivity in how scale is deployed, a convincing and sustained case that scale is not a useful concept in political geography has yet to be made.

References

Agnew, J.A. (1987). *Place and Politics: The Geographical Mediation of State and Society*. London: Allen and Unwin.

Agnew, J.A. (1996). Mapping politics: How context counts in electoral geography. *Political Geography* 15: 129–146.

Allen, J., & Cochrane, A. (2007). Beyond the territorial fix: Regional assemblages, politics and power. *Regional Studies* 41: 1161–1175. doi:10.1080/00343400701543348

Boria, E. (2008). Geopolitical maps: A sketch history of a neglected trend in cartography. *Geopolitics* 13: 278–308.

Brenner, Neil. (2004). *New State Spaces: Urban Governance and the Rescaling of Statehood*. New York: Oxford University Press.

Bulkeley, H. (2005). Reconfiguring environmental governance: Towards a politics of scales and networks. *Political Geography* 24: 875–902.

Cohen, S.B. (2003). *Geopolitics of the World System*. Lanham, MD: Rowman and Littlefield.

Cox, K.R. (1968). Suburbia and voting behavior in the London metropolitan area. *Annals of the Association of American Geographers* 58: 111–127.

Cox, K.R. (1970). Geography, social contexts and voting behavior in Wales, 1861–1951. In Allardt, E., & Rokkan, S. (eds), *Mass Politics: Studies in Political Sociology*, New York: Free Press, 117–159.

Cox, K.R. (1972). The neighborhood effect in urban voting response surfaces. In Sweet, D.C. (ed.), *Models of Urban Structure*, Lexington MA: Lexington Books, 159–176.

Cox, K.R. (1973). *Conflict, Power and Politics in the City: A Geographic View*. New York: McGraw-Hill.

Cox, K.R. (1979). *Location and Public Problems: A Political Geography of the Contemporary World*. Chicago, IL: Maaroufa Press.

Cox, K.R. (1990). Territorial structures of the state: Some conceptual issues. *Tijdschrift voor Economische en Sociale Geografie* 81: 251–266.

Cox, K.R. (1998). Spaces of dependence, scales of engagement and the politics of scale, or: Looking for local politics. *Political Geography* 17: 1–23.

Cox, K.R. (2002). "Globalization", the "regulation approach" and the politics of scale. In Harod, A., & Wright, M. (eds), *Geographies of Power: Placing Scale*, Oxford: Blackwell, 85–114.

Cox, K.R., & Jonas, A.E.G. (1993). Urban development, collective consumption and the politics of metropolitan fragmentation. *Political Geography* 12: 8–37.

Cox, K.R., & McCarthy, J.J. (1982). Neighbourhood activism as a politics of turf: A critical analysis. In Cox, K.R., & Johnston, R.J. (eds), *Conflict, Politics and the Urban Scene*, London: Longmans, 196–210.

Flint, C., & Taylor, P.J. (2011). *Political Geography: World-Economy, Nation-state and Locality*, 6th edn. Harlow: Prentice Hall.

Fraser, A. (2010). The craft of scalar practices. *Environment and Planning A* 42: 332–346.

Gibbs, D.C., While, A., & Jonas, A.E.G. (2007). Governing nature conservation: The European Union Habitats Directive and conflict around estuary management. *Environment and Planning A* 39: 339–358.

Glassner, M.I., & de Blij, H.J. (1989). *Systematic Political Geography*, 4th edn. New York: John Wiley & Sons Ltd.

Gough, J. (2010). Workers' strategies to secure jobs, their uses of scale, and competing economic moralities: Rethinking the "geography of justice." *Political Geography* 29: 130–139.

Harrison, J. (2008). Stating the production of scales: Centrally orchestrated regionalism and regionally orchestrated centralism. *International Journal of Urban and Regional Research* 32: 922–941.

Hartshorne, R. (1950). The functional approach in political geography. *Annals of the Association of American Geographers* 40: 95–130.

Herod, A. (2011). *Scale*. London: Routledge.

Jessop, B., Brenner, N., & Jones, M. (2008). Theorizing sociospatial relations. *Environment and Planning D: Society and Space* 26: 389–401. doi:10.1068/d9107

Johnston, R.J., & Pattie, C.J. (2012). Kevin Cox and electoral geography. In Jonas, A.E.G., & Wood, A. (eds), *Territory, the State and Urban Politics: A Critical Appreciation of the Selected Writings of Kevin R Cox*, Farnham: Ashgate, 23–44.

Jonas, A.E.G. (1994). The scale politics of spatiality. *Environment and Planning D: Society and Space* 12: 257–264.

Jonas, A.E.G. (2002). Local territories of U.S. government: From ideals to politics of place and scale. In Agnew, J.A., & Smith, J.M. (eds), *American Space/American Place: Geographies of the Contemporary United States*, Edinburgh: Edinburgh University Press, 108–149.

Jonas, A.E.G., & Pincetl, S. (2006). Rescaling regions in the state: The new regionalism in California. *Political Geography* 25: 482–505.

Jones, M. (1997). Spatial selectivity of the state? The regulationist enigma and local struggles over economic governance. *Environment and Planning A* 29: 831–864.

Jones, M., & MacLeod, G. (2004). Regional spaces, spaces of regionalism: Territory, insurgent politics and the English question. *Transactions of the Institute of British Geographers* 29: 433–452. doi:10.1111/j.0020-2754.2004.00140.x

Judd, D.R. (1998). The case of the missing scales: A commentary on Cox. *Political Geography* 17: 29–34.

Kurtz, H.E. (2002). The politics of environmental justice as a politics of scale. In Herod, A., & Wright, M. (eds), *Geographies of Power: Placing Scale*, Oxford: Blackwell, 249–273.

Kythreotis, A.P., & Jonas, A.E.G. (2012). Scaling sustainable development? How voluntary groups negotiate spaces of sustainability governance in the United Kingdom. *Environment and Planning D: Society and Space* 30: 381–399.

Latham, A. (2002). Retheorizing the scale of globalization: Topologies, actor-networks, and cosmopolitanism. In Herod, A., & Wright, M. (eds), *Geographies of Power: Placing Scale*, Oxford: Blackwell, 115–144.

Marks, G. (1993). Structural policy and multi-level governance in the EC. In Cafruny, A.W., & Rosenthal, G.G. (eds), *The State of the European Community*. Vol. 2, *The Maastricht Debates and Beyond*, Boulder, CO: Lynne Rienner, 391–410.

Marston, S.A., Jones, J.P., III, & Woodward, K. (2005). Human geography without scale. *Transactions of the Institute of British Geographers* 30: 416–432. doi:10.1111/j.1475-5661.2005.00180.x

McAllister, I. (1987). Social context, turnout, and the vote: Australian and British comparisons. *Political Geography Quarterly* 6: 17–30.

Miller, B. (1994). Political empowerment, local–central state relations, and geographically shifting political opportunity structures: The Cambridge, Massachusetts, peace movement. *Political Geography* 13: 393–406.

Moisio, S. (2011). Political geographies of the state and scale. *Political Geography* 30: 173–174.

Monmonier, M. (1991). *How to Lie with Maps*. Chicago, IL: University of Chicago Press.

Painter, J. (2008). Cartographic anxiety and the search for regionality. *Environment and Planning A* 40: 342–361.

Pierre, J., & Peters, B.G. (2000). *Governance, Politics and the State*. Houndsmills: Macmillan.

Prytherch, D.L. (2010). "Vertebrating" the region as a networked space of flows: Learning from the spatial grammar of Catalanist territoriality. *Environment and Planning A* 42: 1537–1544.

Silvern, S.E. (1999). Scales of justice: Law, American Indian treaty rights and the political construction of scale. *Political Geography* 18: 639–668.

Smith, C.M., & Kurtz, H. (2003). Community gardens and politics of scale in New York City. *Geographical Review* 93: 193–212.

Smith, N. (1984). *Uneven Development*. New York: Blackwell.

Smith, N. (1993). Homeless/global: Scaling places. In Bird, J., Curtis, B., Putnam, T., Robertson, G., & Tickner, L. (eds), *Mapping the Futures: Local Cultures, Gobal Change*, London: Routlege, 87–110.

Swyngedouw, E. (1997). Neither global nor local: "Glocalisation" and the politics of scale. In Cox, K.R. (ed.), *Spaces of Globalisation: Reasserting the Power of the Local*, New York: Guilford, 137–166.

Swyngedouw, E. (2000). Authoritarian governance, power, and the politics of rescaling. *Environment and Planning D: Society and Space* 18: 63–76.

Taylor, P.J. (1982). A materialist framework for political geography. *Transactions of the Institute of British Geographers, NS* 7: 15–34.

Chapter 4

Territory beyond the Anglophone Tradition

Cristina Del Biaggio

University of Amsterdam, The Netherlands; University of Geneva, Switzerland *

"Territories still matter," even in a globalized world, affirms Antonsich (2009: 789). Paradoxically, territory has never been discussed as much as in the last decades, exactly when international relations seem less and less regulated by it (Badie 1996: 114). This is especially paradoxical if the term territory is considered to coincide with state territory, as is mostly the case in Anglo-Saxon geography. It is less paradoxical for Francophone geographers, because their conception of *territoire* has never been so strictly linked with national boundaries as has that of Anglophone geography, in which, as suggested by Antonsich, "territory, rather than being explored under the new conditions of globalization, has simply been discarded with the nation-state itself" (2009: 795). In fact, only a few scholars have written on territory in Anglophone geography, notably Soja (1971), Gottmann (1973), and Sack (1983; see Paasi 2003; Elden 2010a, 2010b; Dell'Agnese 2013).

Territory, this "thing" that Cox considers the "central concept in political geography" (Cox 1991), is not quite the same for scholars writing in English and those in French. And this statement goes beyond a simple lexical difference. It is not only that the Francophone *territoire* is often translated by the English "place" and not by "territory," as common sense would suggest. It is also because the difference is epistemological and, as such, this difference between *territoire* and territory opens the door to new research perspectives that will be partly uncovered by this article.

Francophone perspectives on the concept of territory have been and are broad. In fact, Klauser recalls in his introduction to the special issue of *Environment and Planning D*, paying tribute to Raffestin's conception of territoriality, how Raffestin's work aimed to construct a theory of territoriality, which, he writes, is "ultimately a 'theory of the real'" (Klauser 2012: 109–110). Raffestin's and other Francophone geographers' attempts to grasp the relation between human beings and the world through territory thus reflect a

* address for correspondance

The Wiley Blackwell Companion to Political Geography, First Edition.
Edited by John Agnew, Virginie Mamadouh, Anna J. Secor, and Joanne Sharp.
© 2015 John Wiley & Sons Ltd. Published 2017 by John Wiley & Sons Ltd.

broader conception of territory compared to that of Anglophone scholars. The latter concentrate on two main ideas:

- The first is a "[reading] of territoriality ... concerned, predominantly, with the study of geopolitical strategies of control/defence of space and with the resulting political-territorial arrangements" (Klauser 2012: 110). This leads Taylor to declare that "across the whole of our modern world, territory is directly linked to sovereignty to mould politics into a fundamentally state-centric social process" (Taylor 1994: 151; see also Dell'Agnese 2013). A territorial state is a "simple" container of power, wealth, culture, and society (Taylor 1994: 152; see also Elden 2010b: 757; Taylor 1995: 1).
- The second Anglophone notion is an interpretation of territory as a transposition of ethological theories into the social sciences, following Ardrey's book *The Territorial Imperative* (Ardrey 1966; see also Murphy 2012: 159). This idea was notably brought into geography by scholars such as Soja (1971) and Gottmann (1973; Murphy 2012: 159), but also by Sack, who, however, "treated human territoriality as fundamentally distinct from animal territoriality in that the former is not the product of instinct but is instead a culturally situated process intended to achieve particular political and social ends" (Murphy 2012: 160–161).

Yet, as Debarbieux (1999: 34) points out, in French as in English territory has the same Latin etymology and the term had a similar evolution in the two contexts: It first took a juridical-political meaning (the territory of the state) and then an ethological one (the area appropriated by an animal or a group of animals). It was only in the 1970s and 1980s that the meaning split. While Anglophone geographers did not detach territory from the state, Francophone geographers considered *territoire* to have multifaceted connotations. The Francophone tradition has thus been richer concerning *territoire* from the 1970s onward.

Anglo-Saxon scholars rediscovered only recently the relevance of territory, while the French *territoire* has been enriching research for more than 30 years. In order to uncover the "Francophone way" of understanding territory, I start by analyzing the Alpine regional construction process. In this section, the relevance of territories is examined, introducing the (not necessarily) opposite notion of networks. Building from this concrete example, I discuss in which ways territories have to be considered as bounded, though not necessarily state-bounded, spaces. The third section discusses the suitability of considering the Francophone *territoire* as equivalent to the Anglophone place. Both *territoires* and places are conceived as entities capable, especially due to the propensity of actors to build networks, of escaping the "territorial trap" (Agnew 1994). However, if epistemologically place in English had the same effect as *territoire* in French – that is, it helped in thinking of spatial units beyond their jurisdictional meaning – place and *territoire* cannot be considered as equivalent. *Territoire* has a longer history in Francophone geography, though it has yet to come to fruition in the Anglophone study of territory. Thus, in the fifth section, the differences between the Francophone *territoire* and the English territory will be discussed, emphasizing the Francophone meaning. The sixth section exposes a sort of typology, proposed by Giraut (2008), in which territory is tied between two extremities: a specific notion, linked to the state, and a buzzword, linked to the social and cultural uses of space. This last section of the chapter will show how *territoire* allowed Francophone geography to become a social science and to leave behind its state-centric conception.

The Alpine case study

The social and political processes taking place in the Alps can be interpreted within the more general framework of the shrinking of the national level, which made possible new scalar configurations (Swyngedouw 2004: 132) and alternative perimeters of cooperation (Häkli 2008: 475). In the Alps, one of the most interesting phenomena to take place since the 1990s is the establishment of networks of local political actors. The story starts in 1991, when the eight Alpine states signed an international treaty called the Alpine Convention. The European Union (EU) subsequently funded a six-year program (2007–13) to promote transnational cooperation in the Alps in order to encourage the main principle of the convention, sustainable development.

These initiatives, identified by some scientists as "top-down" (Bätzing 1994), set a framework that enabled local, "bottom-up" projects to gain importance. They often took the form of pan-Alpine networks, involving, among other elements, municipalities, cities, ski resorts, protected areas, and enterprises. Members of these networks are local political actors acting at a given common scale, in this case inside the Alpine Convention's limits, and transcending existing administrative national borders. While proposing a new mode of interaction, the network stemming from those activities and institutions also fixed new territorial borders, those of the Alpine region as defined by the Alpine Convention. These processes, "from above" and "below," illustrate Paasi's (2003: 112) theoretical point when he suggests that it is exactly the combination of top-down and bottom-up processes that creates territories.

These two different forms of modus operandi, vertical and horizontal, taking place in the Alps are useful for understanding the links between networks and territories. Networks, indeed, have an effect on geography. The issue is analyzed by the French geographer Fourny, in her research on the Alpine Town of the Year network. Fourny observed, by looking at the rhetoric of the network, that a shift in status of the *territoire* of reference had taken place. The Alpine Town of the Year network refers to the Alps in order to justify its common action and its role in the management of that space. In doing so, Fourny (1999: 179–180) says, the Alpine towns, connected via a network, contribute to building the Alpine *territoire* politically, to creating a public space, an object that will be collectively debated. As a result of the network's activity, a process of redefinition of the *territoires* of action, and so a redefinition of borders, is also taking place, in parallel to the renegotiation of collective identities. This can be seen in the activities of the EU's cooperation projects, such as INTERREG. In fact, as Bray (in Keating 2004: 12) admits, they "have helped to redefine borders as complex zones in which multiple identities can be expressed and negotiated" and where identities are performed in actions and projects.

While performing networked projects and actions, new types of horizontal cooperation that are no longer linked to old territorial units draw new geographies, as Leitner (in Marston, Jones, & Woodward 2005: 417) argues: "transnational networks represent new modes of coordination and governance, a new politics of horizontal relations that also have a distinct spatiality." Or, as Bulkeley (2005: 888) underlines, networks' activities are not outside their boundaries, in "the ways in which they operate and the ways in which they are framed, configured and crystallized." If this is true, Allen and Cochrane (2007) would suggest that actors, in order to be able to govern these "transgressing entities," also need somehow to fix those new spaces of action. This creates a tension between the necessity to spread activities beyond given boundaries and the need to fix these same activities in order to govern them. This conception of territories and networks recalls Bulkeley's (2005: 888) dual observation that on the one hand scalar boundaries are fluid and contested, and, on the other, that networks are at the same time, and contrary to what is commonly thought, bounded. This recognition, so Bulkeley hopes, "may provide the basis for further constructive dialogue" between the two concepts (2005: 888).

The two notions of networks and territories seem to coexist in, as Bulkeley calls them, "new networked arenas" (2005: 897). These confirm that "geographical scales and networks of spatial connectivity can be seen as mutually constitutive rather than mutually exclusive aspects of social spatiality" (Bulkeley 2005: 888). Indeed, Bulkeley stresses the fact that "networks, scales and territories are not alternatives, but are intimately connected in both a politics of scale, and in creating new arenas of political authority and legitimacy" (2005: 896). She employs the example of how climate change is governed, but other examples could be used to illustrate this link between networks and territories. This brings us back to the idea of Elissalde (2002: 195), who argues that in some ways territories are networks, without, however, denying that fluid and unbounded spatial arrangements do not require "greater fixity and boundedness" (Murphy 2012: 170).

The concept of "scaled networks" proposed by Leitner, Sheppard, and Sziarto (2008b: 287; Leitner, Pavlik, & Sheppard 2008a: 162) seems suitable to address Bulkeley's criticism of the dual vision of scales and networks. In the case of the Alps, the scale of the Alpine Convention is the reference used by local political actors, but their activities are anchored in networks connecting different points of the Alpine scale. Thus, it is useful to view the Alps as not a covering but rather a spanning geographical space (Leitner et al. 2008a: 162; Leitner et al. 2008b: 287), since it combines processes of scaling with processes of networking. It demonstrates that it is useful to think in terms of a co-construction of networks and territories.

Transgressing bounded spaces?

The idea of considering networks and territories as co-constructed entities is possible only if territory is not necessarily conceived as state bounded, an idea that seemed unimaginable for Anglophone geographers, at least until the arrival of the idea of a relational territory and space, defended, among others, by Massey (2004; see Dell'Agnese 2013). Yet, territory can be considered in terms of bounded space, although not necessarily state bounded. In that case a question raised by Elden (2010a: 12–13) remains unanswered: What is this (bounded) space and how are these boundaries possible? One can find two answers in the literature.

First, this space could be the unit of reference in a world imagined to be a patchwork formed by bi-dimensional, non-overlapping geometric forms, where every unit presents an internal integrity (or homogeneity) and a distinct identity (Painter 2009: 57, 2010: 1091). This is not without analogy to observations made in animal societies, where territory is exclusive to members of the same species and is limited by a boundary (Bonnemaison 1981: 253). This is normally the vision of the world held by scholars who consider territory as a state prerogative, where territories are demarcated by clear boundaries (Painter 2010: 1094). So, the integrity of this space would be provided by sharing the same national space. This would be the vision favored by Anglophone geographers.

The second possibility better matches the Francophone understanding of *territoire* and could be seen as the area of daily practices and relations. In that case, geographical limits are defined by the surface where those take place (Raffestin in Bonnemaison 1981: 260). Francophone scholars share this view with other geographers from the Anglo-Saxon tradition. Cox, for instance, does not consider territory to be limited only by jurisdictional boundaries. For him, its delimitation could be understood in a broader sense as "bounded zones" capable of containing any social relations (Cox 1991: 5–6). Territory, in this sense, is the container of localized social and/or (non-invariably state) power relations (Cox 1991: 6; Agnew 1999: 503).

If both options are valid, especially taking into account the epistemological tradition into which they are inserted, nonetheless problems arise when they are considered simultaneously, as Jaillet (2009: 115) does by saying that "*territoire* designates at the same time a political circumscription and the group's living space." Yet, these two areas are not always spatially equivalent. It is for that reason that Anglophone geographers turned their back on territory and preferred "place" instead: exactly because territory was considered as a bounded space; that is, bounded by national borders (Antonsich 2009: 790). This is in fact one of the plausible reasons that Painter cites to explain the rationale for Anglophone geographers, feeling some "embarrassment" with regard to territory, deciding to opt for other concepts (Painter 2010: 1091). Building from this point, Anglophone geographers distinguish the "sense of territory" from the "sense of place" by giving more importance to the second than the first option:

> The sense of place literature places little emphasis on the specific boundedness of place. The sense of territory, however – at least as tied to regimes of territorial legitimation – is inextricably tied to the modern state system, and as such bears the imprint of the system's territorial logic. (Murphy 2002: 197–198)

Territoire = place?

It was in the 1970s and 1980s that *territoire* gained importance in Francophone geography, which corresponded to the moment when the symbolic dimension of *territoire* started being essential in geography, when researchers began to think in terms of appropriation and *espace vécu* (lived space). It is from that moment onward that Anglophone researchers converged on the concept of "place." Place, in the Anglo-Saxon world, gave geographers the possibility of introducing the social, cultural, and political dimensions of space and provided a critique of political territory, of its rigid delimitation and its state control (Debarbieux 1999: 42). This enabled Debarbieux (1999: 42) to say that the meanings given to the term "place" in Anglophone geography recall the innovations occurring in Francophone geography with *territoire*.

"Place" and "sense of place" thus helped Anglophone geographers to go beyond the "territorial trap," a term that Agnew coined to acknowledge geographical assumptions about states: particularly that these are fixed units of sovereignty, that there is a polarity between "domestic" and "foreign" policies, and that states are simply "containers of societies" (Agnew 1994). However, as Elden (2010b: 757) reminds his readers, the "trap" is not the territory itself, it is rather "certain ways of thinking about territory." And, as Elden (2010b: 760) regrets, the "territorial trap" has been avoided by simply not being mentioned in scientific texts instead of being critically interrogated. So, it is important to "highlight the mistaken assumption that the spatialities of state power and state territory are homomorphic" (Painter 2010: 1095). Analysis of territorial networks is one of these "escape routes from the 'territorial trap'" that Bulkeley (2005: 881) identifies.

If, as discussed above, epistemologically place in English has the same effect as *territoire* in French, place is often translated into French by *lieu*, a concept that requires further explanation. The link between *lieu* and *territoire* in Francophone geography is well taken up in a paper by Debarbieux (1995: 14) devoted to this issue: "Metaphorically, the *lieu* symbolizes the *territoire*, but the *lieu* is as well a metonym, or, more exactly, a synecdoche, the whole, *territoire*, can be said by its parts, the *lieu*." Debarbieux's theory meets Di Méo's (1998: 110), who argues that the difference between *lieu* and *territoire* is given by their scale and by their "geographical readability": *Territoire* is abstract, ideal, lived, and felt more than visually detected and limited; *territoire* includes the *lieux*, which are defined, as opposed to *territoires*, by their striking reality due to their "*valeur d'usage*" (use value). Yet, Di Méo (1998: 108) continues, if *lieux* differ from *territoire* on those points, they converge in the fact that both are spaces qualified by society (or "semiotized," as Raffestin would say). Debarbieux (1995: 14–15) uncovers the link between *lieu* and *territoire* in a similar way, stating that

> A territory [*un territoire*] is a social construct that connects a material base made of a geographical space [*un espace géographique*] to a system of values that gives multiple and combined meanings to each component of this space (the places [*lieux*], but also the spacing [*espacements*] and the discontinuities it encompasses.

Lieux as Di Méo means them – that is, as areas of daily practice – are considered to be relatively small: They are defined by the contiguity of the points and webs comprising them, by the co-presence of human beings and things that convey a spatial meaning (Di Méo 1998: 108). *Lieux* can be so dense with meaning that they connect at the same time two geographical scales: that of the *emplacement* (location) and that of the *territoire* to which they refer (Debarbieux 1995: 14). *Lieux*, thus, are simultaneously not only fragments of *territoires*, but also figures able to reveal their quintessence (Debarbieux 1995: 14). As conceived by Debarbieux and Di Méo, *lieu* thus entails an essential difference from "place": "Place [*lieu*], unlike territory [*territoire*] abolishes distance; while geographical territory abhors bordering [*bornage*], place draws its substance from it" (Di Méo 1998: 108).

Networked territories and territorial networks

Thus, territory can be conceived as escaping boundaries in general and national boundaries in particular:

> Spatial practices, the ways in which space is produced and used, have changed profoundly. In particular, both territorial states and non-state actors now operate in a world in which state boundaries have become culturally and economically permeable to decisions and flows emanating from networks of power not captured by singularly territorial representations of space. (Agnew 1994: 72)

The most emblematic example underscoring this tendency is the overused expression of "global" or "world economy," terms employed to indicate that monetary flows are circulating worldwide without being stopped by any state border. Indubitably, these socio-political trends influence the ways in which social scientists in general and geographers in particular acknowledge the links between territories and networks, although these links are conceived in a different way in Anglophone and Francophone geography.

One of the main and most interesting Anglophone contributors to this debate is certainly Painter, who discussed the issue in two papers (2009, 2010), in which he defends the thesis that territory and network are not "as is often assumed, incommensurable and rival principles of spatial organization, but are intimately connected" (Painter 2010: 1093–1094). In Francophone geography, the possibility that networks and territories are connected, or highly integrated, has a longer heritage. Already in 1981, Bonnemaison (1981: 254) wrote that "territoriality covers both what is fixed and what is mobile, in other words, itineraries as much as *lieux*." Three years later, Raffestin and Turco (1984: 45) affirmed almost the same idea with their definition of *territoire* as produced from space through the networks, circuits, and flows projected by social groups. Elissalde (2002: 195) builds on articles published in the 1980s to argue, 20 years later, that "a geography of *territoires* cannot be limited on the study of delimited or nested surfaces; *territoires* are networks," adding "… and not only for nomads" (that is, in Western societies as well). Elissalde (2002: 197) finds it as futile to oppose *territoire* and *réseaux* (networks) as it is to avoid imagining blurred boundaries and overlapping territories. This conception of territories, however, is only possible if they are considered beyond their jurisdictional meaning.

The Alpine case study shows the fruitfulness of this approach, since it pushes scholars to conceive of territories not only as mere "given spatial entities" fixed by administrative units, but as constructed and flexible portions of space. This conception, however, has been more deeply analyzed by Francophone than by Anglophone geographers. An exception is Painter's article, in which he proposes to "rethink territory" by using a large number of Francophone sources long ignored by Anglophone scholars (Klauser 2012: 107). By doing so, he fills a gap that Fall (2007) attributes to the fact that, mostly for institutional reasons, the prosperous theorists of *territoire*, among them Swiss geographer Raffestin, never dared to go beyond Francophone boundaries. But what is this "Francophone conception" of territories? The aim of what follows is to answer this question.

The Francophone *territoire* versus the Anglophone territory

Concerning territory, two different paths have been followed, two separate ways not profiting from possible mutual fertilization. As Chamussy (2003: 168) declared, "There still does not exist, for the moment, an English equivalent for the word *territoire* as it is understood by Francophone geographers." So, Painter's (2010: 1090) motto "territory is back" only makes sense in Anglophone geography, because in the Francophone tradition, *territoire* never disappeared. Painter, however, seems to be aware of his Anglo-centrism when he writes that "until recently the concept of territory has not received the same level of attention, at least in the Anglophone literature" (Painter 2010: 1091); although territory was a key concept in an article he wrote with Bialasiewicz and Elden (Bialasiewicz et al. 2005), in which they analyze it through the lens of the Treaty Establishing a Constitution for Europe. Thanks to this example, the three scholars found it useful to understand territory in a more "francophone way"; that is, going beyond his state-centric conception. They underline how territory is central to the process of European integration and how exactly European integration makes it possible to transcend "existing notions of territory, particularly those associated with the nation-state" (Bialasiewicz et al. 2005: 335).

So, what do Francophone geographers mean by *territoire*? Historically, *territoire* broke into Francophone geography when cultural and symbolic dimensions were introduced into the concept following the growing interest of social and political geographers in concepts such as "power," "spatial control," "differentiation," "domination," and "social appropriation" (Alphandéry & Bergues 2004; Claval 1996: 96; Debarbieux 2003: 38). The emphasis on *territoire* corresponds to geography's claim to belong to the social sciences and to distance itself from a naturalist or mathematical conception of geography (Douillet 2003: 215). *Territoire* replaced the concept of *région* (region) first and *espace* (space) later (Chamussy 2003: 167; Debarbieux 2003: 36–37) and does not cleave to the idea that *territoire* necessarily finishes where states do, as is the case for the English "territory" (Debarbieux 2003: 35, 42). The Anglophone perspective is well summarized in this statement: "Territory represents the extent of the sovereign power of the state" (Forsberg 2003: 13). The invitations of Sack (1983), Cox (1991), and Agnew (1999) to break with the necessity of analyzing power and control via the prism of the state have yet to come to fruition in the Anglophone study of territory.

Tied between two poles: A specific notion or a buzzword?

Giraut summarizes quite well this "richer" Francophone tradition, identifying two poles toward which the concept is tied: a specific notion, corresponding to the area of the (national) state, as for Anglophone geographers, and a buzzword, corresponding to a non-specified area (Giraut 2008: 59). These two tendencies are also distinguished by Alphandéry and Bergues (2004): a territory stemming from the *maillage historique* (traditional grid) on the one hand and, on the other, a territory taking different forms in space, which is produced and transformed by people and groups of people. The latter approach is a more diffused and less institutionalized way of conceiving of *territoire* and it covers the idea that a *territoire*, insofar as it is such, has to be appropriated by individuals or groups of individuals. The appropriation can be concrete and/or symbolic (Bourdeau 1991: 30).

Giraut, in his analysis, put the idea of appropriation not as a possible shape that territory can take (*maillage historique* or space transformed by people, as for Alphandéry and Bergues), but as the concept used by cultural geographers to analyze identities or by political geographers to signify power and control. Francophone geographers do not always conceive of power as necessarily linked to state power, since power is conceived as inherent to every social relation (Tizon 1996: 27; Giraut 2008: 60; Ozouf-Marignier 2009: 35).

"Appropriation" also played an important role in the debates between the partisans of *espace* (space) and the supporters of *territoire*. The first, *espace*, is primarily used by planners and technocrats who consider it as a *donné* (a "given thing"), something "flat," "uniform," and "without mystery" (Bonnemaison 1981: 260). The second, *territoire*, is privileged by geographers and is considered to be appropriated, invested with affect and subjectivity; it is *vécu* (lived) (Bonnemaison 1981: 260). In this sense, *territoire* represents the socialization of *espace* (Bourdeau 1991: 29; Klauser 2012: 111). The role of humans and groups of humans is thus crucial in this movement from "space" to "territory," since territories are built by humans through technical actions and discursive practices (Claval 1996: 97). Raffestin usefully suggests that territorial arrangements constitute a "semiotization of space"; that is, a space, the material world, progressively transformed into *territoire* (Raffestin 1986a: 181, 1986b: 94).

These conceptions are akin to the three orders of *territoire* suggested by Di Méo (1998: 108): materiality; individual psyche (an emotional and presocial relation of human beings with the Earth); and collective, social, and cultural representations. Geographers rarely use the second order, but the other two can be frequently found in the literature, since geographers insist on the dual dimension, material and ideal, ecological and symbolic, of *territoire* (Claval 1996: 97; Tizon 1996: 21; Debarbieux 1999: 36; Elissalde 2002: 195). This is also the case for the three dimensions of territory distinguished by Hassner: material, symbolic, and functional (Hassner in Paasi 2003: 109). And the inclusion of immateriality and representation is exactly what distinguishes a *territoire* from a Euclidian space.

"Territory is what people make it to be"

In the social sciences territory is a useful tool for introducing the logics of agents (Ozouf-Marignier 2009: 34), since, as Knight suggests, it is "actions that give territory meaning" (Knight 1982: 517). This is also Paasi's main point (2003: 110) in his contribution to the first edition of this *Companion*, since he considers territories to be "social processes in which social space and social action are inseparable." This idea of "social action" allowed Francophone geographers, as Giraut (2008: 57) argues, to shift the focus from state territory to a territory in the hands of individuals and multiple collectivities. Territory becomes the place where action and social thoughts are possible, while entering into contact with, transforming, and "deforming" (Di Méo 1991: 145) materiality (Barel in Marié 2004: 90). In that sense, what is interesting about *territoire* is that it opens up for geographers the possibility of inverting their emphasis from materiality to immateriality (or "semiosphere," using Raffestin's vocabulary); that is, from space to the instruments and codes of actors leaving traces in the territory (Raffestin 1986b: 94). It was through the awareness of the capacity of individuals and collectivities to model territories that a shift took place in Francophone geography: from a territory linked to its national referent to a territory of belongings, projects, and individual and collective practices (Giraut 2008: 57).

It was indeed when the idea of *projet* started circulating among geographers that *territoire* became a "fetish object," as Giraut describes it (2008: 61), not only for cultural and political geographers but also for economic geographers in France, who from the 1950s to the "territorial turn" in the 1980s (Benko 2008: 38) preferred conceptualizations of space to those of territory, finding space more useful for their abstract and quantitative analysis of economic phenomena (Benko 2008). Yet, the economic crisis in the 1970s and the subsequent idea that "development" cannot be stirred from above provided an opportunity to argue that the solution to the crisis would be to advocate for territorialized production and local development, often qualified as a "territorial development" (Giraut 2008: 61) and supported by local claims (Debarbieux 1999: 38). This is emblematically summarized in a sentence uttered by the French minister of planning in 1997 and reported by Benko (2008: 41): "There are no territories in crisis, there are only territories without projects." Thus, economic geographers have been interested from that moment on in how specialized districts could boost the economy and in what manner territorial resources could generate added value.

Cultural geographers instead stress the first aspect that Giraut distinguishes when he defines appropriation; that is, the symbolic dimension of *territoire*. Bonnemaison, for instance, argues that the symbolic relation between culture, which other scholars call "representations" (Claval 1996: 102; Tizon 1996: 21) or "imaginary" (Tizon 1996: 21; Corboz 2001: 214), and space, or materiality, is realized through the *territoire* (Bonnemaison 1981: 254). In that sense, *territoire* should be considered a material and symbolic mediator between a group and its culture (Bourdeau 1991: 41); it is a "*savant mélange*," a clever mix, of materiality and ideal (Tizon 1996: 21), in which identity plays an important role. Claval suggests that identities are built from the representations that transform some portions of the humanized space into territories (Claval 1996: 102). From a similar perspective, Bourdeau argues that *territoires* and shared cultures comprise the main components of collective identities: If *territoire* represents the spatial and temporal dimensions of identity, culture reflects the historical, mnemonic, and symbolic ones (Bourdeau 1991: 42). This leads Bourdeau to posit that *territoire* is at the same time the cultural mirror of an identity and the identity mirror of a culture (Bourdeau 1991: 42); a mirror keeping outside the Other, alterity (Piveteau 1995: 114).

We thus come back to Giraut (2008: 59), who situates "identity" as the pivotal concept for cultural geographers to describe territorial "appropriation." However, for Bonnemaison and Cambrezy (1996: 13) for instance, territory is not stirred by a material but by a cultural principle of appropriation; that is, by "belonging." The idea of appropriation and belonging refers to the original use of territory – the translation of what ethologists observed in the animal realm into social realities. Taking advantage of considerations made in animal groups, social scientists were called to analyze the means that social subjects implement to control space (Claval 1996: 95). Yet, natural scientists consider territory to be an environment from which animals cannot escape, while human beings are able to thanks to culture and through a process of semiologization (Raffestin 1986c: 76). Following Tizon (1996: 34), human societies, compared to animal societies that define territories as spaces of exclusion, can modify their territories and transform them, following their aspirations, into places of social differentiation (or even segregation) or, on the contrary, into places of gathering and belonging. It is by gathering together that people can use the inherent potentialities of space and transform it into *territoire*, by implementing those potentialities in projects (Bourdeau 1991: 29). So, as Aase sums it up: "Territory is what people make it to be" (in Forsberg 2003: 10; see also Dell'Agnese 2013: 118). This is, obviously, radically different from the Taylorian conception of territories and territoriality, considered as the "geographical link between states and nations" (Taylor 1995: 3).

Conclusion

The Alpine case suggests that, even today (or perhaps *especially* now), in an era in which states seem to be undermined (or at least reshaped) by transnational networks, there is interest in continuing to foster the concept of territory. Territories did not disappear; they are still "inescapable principles of social life" (Antonsich 2009: 801) and they remain a central dimension of understanding the ways in which "'living together' is produced, organized, contested and negotiated" (Antonsich 2009: 801). Our understanding of them should change and take into account the fact that "if territories are portions of relational space, and not portions of abstract homogenised spaces, the quality of their interactions is not an (inescapable) outcome of the essentialised characteristics of homogenised populations, but a consequence of the sum of interactions within, and among, individuals" (Dell'Agnese 2013: 122–123); or, I would add, among collectivities of individuals.

Bonnemaison is quite clear on that point when he writes:

> The increasing of mobility and the diminishing of the "Westphalian" function of *territoire* did not dispossess it from every meaning or necessity. In the contemporary world the need still subsists, although *territoire* takes different forms and responds to multiple functions. (Bonnemaison & Cambrezy 1996: 10)

In that sense, scholars are now called to analyze *how* ideologies are changing as territorial logics are challenged (Murphy 2002: 198) and to reinterrogate territory (Elden 2010a: 20).

These are times of "territorial complexity" (Giraut & Antheaume 2005: 29), of the blooming of "new regions." In Europe at least, the number of what Deas and Lord (2006) call "unusual arrangements" – that is, territorial arrangements not linked to the territorial state – considerably increased. Following these two scholars, in Europe there are already 146 regions transcending territorially bounded entities. These need a framework that allows for their conceptualization, as the entire EU integration processes needs it (Bialasiewicz et al. 2005; Clark & Jones 2008). The Francophone *territoire* could turn out to be the right tool to enrich this framework and to give substance to the "more work on territory" that Elden (2010b) advocates. Thus, Anglophone (and other) geographers could build on what French geographers call *territoire* and take advantage of new perspectives to frame together a new and fascinating theory of "territory."

References

Agnew, J. (1994). The territorial trap: The geographical assumptions of international relations theory. *Review of International Political Economy* 1(1): 53–80. doi:10.1080/09692299408434268

Agnew, J. (1999). Mapping political power beyond state boundaries: Territory, identity, and movement in world politics. *Millennium: Journal of International Studies* 28(3): 499–521. doi:10.1177/03058298990280030701

Allen, J., & Cochrane, A. (2007). Beyond the territorial fix: Regional assemblages, politics and power. *Regional Studies* 41(9): 1161–1175.

Alphandéry, P., & Bergues, M. (2004). Territoires en questions: Pratiques des lieux, usages d'un mot. *Ethnologie française* 1: 5–12.

Antonsich, M. (2009). On territory, the nation-state and the crisis of the hyphen. *Progress in Human Geography* 33(6): 789–806. doi:10.1177/0309132508104996

Ardrey, R. (1966). *The Territorial Imperative: A Personal Inquiry into the Animal Origins of Property and Nations*. New York: Atheneum.

Badie, B. (1996). La fin des territoires westphaliens. *Géographie et cultures* 20: 113–118.

Bätzing, W. (1994). Die Alpenkonvention – Ein internationales Vertragswerk für eine nachhaltige Alpenentwicklung auf dem mühevollen Weg der politischen Realisierung. In Franz, H. (ed.), *Gefährdung und Schutz der Alpen*, Vienna: Österreichische Akademie der Wissenschaften, 185–204.

Benko, G. (2008). La géographie économique: Un siècle d'histoire. *Annales de géographie* 664(6): 23–49. doi:10.3917/ag.664.0023

Bialasiewicz, L., Elden, S., & Painter, J. (2005). The constitution of EU territory. *Comparative European Politics* 3(3): 333–363. doi:10.1057/palgrave.cep.6110059

Bonnemaison, J. (1981). Voyage autour du territoire. *Espace géographique* 10(4): 249–262. doi:10.3406/spgeo.1981.3673

Bonnemaison, J., & Cambrezy, L. (1996). Le lien territorial entre frontières et identités. *Géographie et cultures* 20: 7–18.

Bourdeau, P. (1991). *Guides de haute montagne: Territoire et identité: Recherches sur la territorialité d'un groupe professionnel*. Grenoble: Revue de géographie alpine.

Bulkeley, H. (2005). Reconfiguring environmental governance: Towards a politics of scales and networks. *Political Geography* 24(8): 875–902.

Chamussy, H. (2003). Le territoire, notion heuristique ou concept opératoire? In de Bernardy, M., & Debarbieux, B. (eds), *Le territoire en sciences sociales: Approches disciplinaires et pratiques de laboratoire*, Grenoble: Publications de la MSH-Alpes, 167–182.

Clark, J., and Jones, A. (2008). The spatialities of Europeanisation: Territory, government and power in "EUrope." *Transactions of the Institute of British Geographers* 33(3): 300–318. doi:10.1111/j.1475-5661.2008.00309.x

Claval, P. (1996). Le territoire dans la transition à la postmodernité. *Géographie et cultures* 20: 93–112.

Corboz, A. (2001). *Le territoire comme palimpseste et autres essais*. Besançon: Editions de L'Imprimeur.

Cox, K.R. (1991). Redefining "territory." *Political Geography Quarterly* 10(1): 5–7. doi:10.1016/0260-9827(91)90023-N

Deas, I., & Lord, A. (2006). From a new regionalism to an unusual regionalism? The emergence of non-standard regional spaces and lessons for the territorial reorganization of the state. *Urban Studies* 43(10): 1847–1877.

Debarbieux, B. (1995). Le lieu, fragment et symbole du territoire. *Espaces et sociétés* 80A(1): 13–36.

Debarbieux, B. (1999). Le territoire: Histoires en deux langues. A bilingual (his-)story of territory. In Chivallon, C., Ragouet, P., & Samers, M. (eds), *Discours scientifiques et contextes culturels: géographies françaises et britanniques à l'épreuve postmoderne*, Bordeaux: Maison des Sciences de l'Homme d'Aquitaine, 33–46.

Debarbieux, B. (2003). Le territoire en géographie et en géographie grenobloise. In de Bernardy, M., & Debarbieux, B. (eds), *Le territoire en sciences sociales: Approches disciplinaires et pratiques de laboratoire*, Grenoble: Publications de la MSH-Alpes, 35–50.

Dell'Agnese, E. (2013). The political challenge of relational territory. In Featherstone, D., & Painter, J. (eds), *Spatial Politics. Essays for Doreen Massey*, Chichester: Wiley-Blackwell, 115–132.

Di Méo, G. (1991). *L'homme, la société, l'espace*. Paris: Editions Economica.

Di Méo, G. (1998). De l'espace aux territoires: Eléments pour une archéologie des concepts fondamentaux de la géographie. *L'information géographique* 62(3): 99–110. doi:10.3406/ingeo.1998.2586

Douillet, A.-C. (2003). Le "territoire" en science politique au regard des autres sciences sociales. In de Bernardy, M., & Debarbieux, B. (eds), *Le territoire en sciences sociales: Approches disciplinaires et pratiques de laboratoire*, Grenoble: Publications de la MSH-Alpes, 207–224.

Elden, S. (2010a). Land, terrain, territory. *Progress in Human Geography* 34: 799–817. doi:10.1177/0309132510362603

Elden, S. (2010b). Thinking territory historically. *Geopolitics* 15(4): 757–761.

Elissalde, B. (2002). Une géographie des territoires. *L'information géographique* 66(3): 193–205.

Fall, J.J. (2007). Lost geographers: Power games and the circulation of ideas within francophone political geographies. *Progress in Human Geography* 31(2): 195–216. doi:10.1177/0309132507075369

Forsberg, T. (2003). The ground without foundation: Territory as a social construct. *Geopolitics* 8(2): 7–24. doi:10.1080/714001038

Fourny, M.-C. (1999). Affirmation identitaire et politiques territoriales des villes alpines. *Revue de géographie alpine* 87(1): 171–180.

Giraut, F. (2008). Conceptualiser le territoire. *Historiens et géographes* 403: 57–67.

Giraut, F., & Antheaume, B. (2005). Au nom du développement, une (re)fabrication des territoires. In Antheaume, B., & Giraut, F. (eds), *Le territoire est mort. Vive les territoires!* Paris: IRD Editions, 9–36.

Gottmann, J. (1973). *Significance of Territory*. Charlottesville, VA: University Press of Virginia.

Häkli, J. (2008). Re-bordering spaces. In Cox, K., Low, M., & Robinson, J. (eds), *Handbook of Political Geography*, London: Sage, 471–482.

Jaillet, M.-C. (2009). Contre le territoire, la "bonne distance." In Renard, J. (ed.), *Territoires, territorialité, territorialisation: Controverses et perspectives*, Rennes: Presses Universitaires de Rennes, 115–121.

Keating, M. (2004). *Regions and Regionalism in Europe*. Cheltenham: Edward Elgar.

Klauser, F.R. (2012). Thinking through territoriality: Introducing Claude Raffestin to Anglophone socio-spatial theory. *Environment and Planning D: Society and Space* 30(1): 106–120. doi:10.1068/d20711

Knight, D.B. (1982). Identity and territory: Geographical perspectives on nationalism and regionalism. *Annals of the Association of American Geographers* 72(4): 514–531.

Leitner, H., Pavlik, C., & Sheppard, E. (2008a). Networks, governance, and the politics of scale: Inter-urban networks and the European Union. In Herod, A., & Wright, M.W. (eds), *Geographies of Power: Placing Scale*, Oxford: Blackwell, 274–303.

Leitner, H., Sheppard, E., & Sziarto, K.M. (2008). The spatialities of contentious politics. *Transactions of the Institute of British Geographers* 33(2): 157–172. doi:10.1111/j.1475-5661.2008.00293.x

Marié, M. (2004). L'anthropologue et ses territoires. *Ethnologie française* 34(1): 89–96.

Marston, S.A., Jones, J.P., & Woodward, K. (2005). Human geography without scale. *Transactions of the Institute of British Geographers* 30(4): 416–432. doi: 10.1111/j.1475-5661.2005.00180.x

Massey, D. (2004). Geographies of responsibility. *Geografiska Annaler* 86(1): 5–18.

Murphy, A.B. (2002). National claims to territory in the modern state system: Geographical considerations. *Geopolitics* 7(2): 193–214. doi:10.1080/714000938

Murphy, A.B. (2012). Entente territorial: Sack and Raffestin on territoriality. *Environment and Planning D: Society and Space* 30(1): 159–172.

Ozouf-Marignier, M.-V. (2009). Le territoire, la géographie et les sciences sociales: Aperçus historiques et épistémologiques. In Vanier, M. (ed.), *Territoires, territorialité, territorialisation: Controverses et perspectives*, Rennes: Presses Universitaires de Rennes, 31–35.

Paasi, A. (2003). Territory. In Agnew, J., Mitchell, K., & Toal, G. (eds), *A Companion to Political Geography*, Oxford: Blackwell, 109–122.

Painter, J. (2009). Territoire et réseau: Une fausse dichotomie? In Vanier, M. (ed.), *Territoires, territorialité, territorialisation: Controverses et perspectives*, Rennes: Presses Universitaires de Rennes, 57–66.

Painter, J. (2010). Rethinking territory. *Antipode* 42(5): 1090–1118. doi:10.1111/j.1467-8330. 2010.00795.x

Piveteau, J.-L. (1995). Le territoire est-il un lieu de mémoire? *Espace géographique* 24(2): 113–123. doi:10.3406/spgeo.1995.3364

Raffestin, C. (1986a). Ecogenèse territoriale et territorialité. In Auriac, F., & Brunet, R. (eds), *Espaces, jeux et enjeux*, Paris: Fayard, 173–175.

Raffestin, C. (1986b). Territorialité: Concept ou paradigme de la géographie sociale? *Geographica Helvetica* 41(2): 91–96.

Raffestin, C. (1986c). Punti di riferimento per una teoria della territorialità umana. In Copeta, C. (ed.), *Esistere e abitare: Prospettive umanistiche nella geografia francofona*, Milano: Franco Angeli, 75–89.

Raffestin, C., & Turco, A. (1984). Espace et pouvoir. In Bailly, A. (ed.), *Les concepts de la géographie humaine*, Paris: Masson, 45–50.

Sack, R.D. (1983). Human territoriality: A theory. *Annals of the Association of American Geographers* 73(1): 55–74. doi:10.1111/j.1467-8306.1983.tb01396.x

Soja, E. (1971). *The Political Organization of Space*. Washington, DC: Commission on College Geography, Association of American Geographers.

Swyngedouw, E. (2004). Scaled geographies: Nature, place, and the politics of scale. In McMaster, R.B., & Sheppard, E.S. (eds), *Scale and Geographic Inquiry: Nature, Society, and Method*, London: Blackwell, 129–151.

Taylor, P.J. (1994). The state as container: Territoriality in the modern world-system. *Progress in Human Geography* 18(2), 151–162. doi:10.1177/030913259401800202

Taylor, P.J. (1995). Beyond containers: Internationality, interstateness, interterritoriality. *Progress in Human Geography* 19(2), 1–15. doi:10.1177/030913259501900101

Tizon, P. (1996). Qu'est-ce que le territoire? In Di Méo, G. (ed.), *Les territoires du quotidien*, Paris: L'Harmattan, 17–34.

Chapter 5

Sovereignty

Joshua E. Barkan

University of Georgia, Athens, Georgia, USA

The last decade has witnessed renewed academic interest in questions of sovereignty. Much of it, however, reads like a horror story, with scholars charting sovereignty as it goes through its death throes and comes out in monstrous new forms. As many of our most pressing political issues – climate change, economic integration, global security – concern transnational processes, state sovereignty lurches on as a figure from another order. Yet, what makes the study of sovereignty so captivating is that even as global forces and processes continue to buffet states, sovereign power remains a structuring influence on politics across and between local, national, and global scales. Although many *sovereign states* at times appear ineffectual at containing the global dynamics of capital, warfare, and population movements, *sovereign power* continues to express itself across the political landscape. Rather than diminishing, sovereign power (and not simply state sovereignty) is in many ways intensifying in its control over populations and territory with ever more frequent episodes of paroxysmal violence.

As this distinction between state sovereignty and sovereign power indicates, sovereignty is not simply a characteristic of national territories and was never as coherent a political force as the image of a world divided into neatly partitioned nation-states suggested. Sovereignty has always been a contested term in the political lexicon of the West, used by those advocating certain spatial and institutional orders. In spite of the declarative statements of political theorists – from Jean Bodin's (1992: 1) claim that "sovereignty is the absolute and perpetual power of a commonwealth" to Carl Schmitt's (1985 [1922]: 5) contention that the "sovereign is he who decides on the exception" – the meaning of the term has never been settled. Although sovereignty suggests the highest political authority and is usually assumed to be a property of nation-states, it has carried contradictory and even incommensurate meanings. For instance, sovereignty has referred to forms of political power over others while also being used to describe the autonomy of political entities from external control. In liberal regimes sovereignty is embodied in the legal system and the rule

The Wiley Blackwell Companion to Political Geography, First Edition.
Edited by John Agnew, Virginie Mamadouh, Anna J. Secor, and Joanne Sharp.
© 2015 John Wiley & Sons Ltd. Published 2017 by John Wiley & Sons Ltd.

of law, yet others characterize sovereign power as a political power beyond the reach of law. That distinction refers to a more fundamental problem concerning the ways in which sovereignty emerged as a political concept. Whereas early modern political uses of sovereignty referenced the absolute monarchs of Europe and harked back to theological concepts of divine power, the bourgeois revolutions of the eighteenth century transposed the term into a language of popular sovereignty that uneasily linked sovereignty to democratic forms of rule. Moreover, these transformations occurred alongside alternative anti-colonial articulations of sovereignty, including indigenous struggles for tribal sovereignty in settler colonial countries like the United States and movements for national sovereignty in the context of decolonization.

This complexity only increases when we consider sovereignty from an explicitly geographical perspective. This is for at least two reasons. First, theories of sovereignty have always contained an implicit spatiality (Sitze 2010) in which sovereign power is imagined in relation to places, territories, and jurisdictions. Yet, because theories of sovereignty are themselves polemical and political, attempting to define and support various types of regimes, the implicit spatiality of sovereign power is not simply descriptive. Instead, theories of sovereignty have always been about constituting the spatiality of politics. Second, because this implicit spatiality orients our acting in the world but is also unevenly realized, the geographies of sovereign power are highly variegated, *but in ways that are often difficult to see or grasp*. This is never more so than when we study sovereignty through the lens of law and the state. Forms of sovereignty that are presented as governing national territories often rely on a host of micropolitical spaces and processes that make up and compose the daily practices of authority and rule. For this reason, much of the most interesting work on the geography of sovereign power focuses not on the nation-state but on what Alison Mountz (2013: 830) has referred to as the "'gray' zones" – such as island detention centers or border security stations – "through which sovereign power operates and is produced."

In what follows, I map out four ways in which scholars have approached the complex histories and geographies of sovereign power. Because of the pervasive sense that sovereignty is ceasing to function as it once did, almost all contemporary critical approaches identify a gap between the hypostatized image of sovereign nation-states and the daily functioning of political power. In this sense, all of these approaches differ from the conceptualization of sovereignty, used in the formal arenas of law, diplomacy, and politics, in which sovereign power is coterminous with the state and its territory. Questions remain, however, concerning the meaning of this gap. For some scholars the distance between our theories of sovereignty and the actual functioning of political power compels more nuanced empirical and analytical accounts that jettison sovereignty as an unhelpful term. For others, the gap signifies a theological remainder or a repressed element in our politics that still exercises a profound influence on contemporary power relations and thus has to be dealt with both analytically and politically. In this regard, I argue that we should not treat accounts broadly grounded in either critical social science and ethnography, on the one hand, or political theology and political ontology, on the other, as progressively more accurate representations of sovereignty. Rather, we should recognize the very different ways in which these approaches conceptualize sovereignty and sovereign power, and what those diverse conceptualizations tell us about the spatial and social organization of political power. In the final section of this chapter, I gauge these distinctions while also indicating some future avenues of research.

Regimes of sovereignty

For at least a generation, scholars have been noting two problems with the assumption that sovereignty is a universal property of nation-states. The first concerns the concept of sovereignty itself. Although sovereignty has been central in the development of Western political thought, recent approaches in the social sciences highlight the concept's ambiguity. International relations theorist Stephen Krasner (1999) notes four types of sovereign power that states exercise. The philosophical tradition of Bodin and Hobbes considered the organization of government within a state, and was thus concerned with *domestic sovereignty* as the ultimate authority within a polity. Krasner contrasts this notion with others. *International legal sovereignty* refers to the ability of entities to be recognized by other states in the international state system, and has included both internationally recognized non-state entities as well as governments that are out of power and cannot exercise effective control over state territory or populations. *Westphalian sovereignty* is grounded in the principle that a state is the highest power within its political territory and thus has the ability and right to control the structures of government and authority without external interference. Finally, *interdependent sovereignty* emphasizes the forms of authority associated primarily with globalization and the ability of states to control the cross-border movements of everything from people and capital to pollution and disease. Krasner argues that the leaders of states strategically engage with the norms characterizing the multiple forms of sovereign authority, averring the centrality of some while periodically shirking others, as they pursue their interests within the asymmetric relation of the international state system. Given these multiple definitions and strategic uses, as well as the inability of most states to exercise basic elements of Westphalian sovereignty, Krasner contends that sovereignty functions less as a governing norm or a real fact of world affairs than as a kind of myth to which elites appeal in a form of "organized hypocrisy" that structures international relations.

For Krasner, then, sovereignty fails to capture the real ways in which states and their leaders operate in the international system. Yet, a second problem concerns the numerous political actors and processes whose modes of power and authority are not contained within state territory. Whereas Krasner conceptualizes states and state elites as the fundamental agents shaping international politics, political geographers and critical geopolitics scholars have argued that the exclusive focus on states – whether through the lens of sovereignty or not – gives us a partial view of politics. Geographical thinkers have noted that one aspect of the problem is that academic fields such as political science, international relations, political sociology, and law take the territorially organized nation-state as an unproblematic assumption grounding both political analysis and policy recommendations (Agnew 1994; Brenner 1999). Geographers have also demonstrated that state functions – from the provision of welfare to the regulation of capital, labor, national security, and the environment – are increasingly located in new spatial and scalar configurations (Glassman 1999; Brenner 2004; Sparke 2006). However, it is only recently that geographers have begun explicitly to consider the significance of sovereignty to these transformations.

John Agnew's (2005, 2009) work is notable in this respect. Engaging in debates about the growing disjunction between effective sovereignty and state territory, Agnew suggests shifting our focus to the multiform ways in which that sovereign power is exercised geographically. In terms of authority, this implies recognizing the diverse institutions that exercise legitimate forms of control in society besides territorially bounded nation-states. In terms of geography, it means examining the other spatial forms that authority can take besides state territory, including networked forms of political power that flow through specific nodes or locales rather than across bounded terrestrial space.

To draw out the analytical importance of these claims, Agnew suggests that sovereignty regimes (and not simply state sovereigns) can be situated within a typology that considers both the territorial dynamics of sovereign power and its reliance on central authority. Adopting Michael Mann's (1984) discussion of despotic and infrastructural power, Agnew arrives at a matrix of sovereignty regimes characterized by multiple combinations of strong or weak forms of central state authority, which he links to despotic power, against consolidated or open forms of state territoriality, linked to infrastructural power. The four resulting ideal types of sovereignty regimes – classic (strong state, consolidated territory), globalist (strong state, open territory), integrative (weak state, consolidated territory), and imperialist (weak state, open territory) – can then be used to characterize the diverse ways in which authority and territory have been historically and geographically organized. Agnew draws on examples ranging from the management of currencies to the regulation of international migration to show how wide variances in policy correspond to different types of regimes.

What is critical about these analytical approaches to sovereignty is their circumspection relative to the concept. Both accounts go beyond claims of a contemporary decline in state sovereignty to question whether sovereignty was ever a particularly good way of characterizing political power in world politics. Although Krasner and Agnew differ markedly in their understandings of the fundamental forces shaping international politics – with Agnew critiquing Krasner's state-centric account of politics as a "territorial trap" (Agnew 1994) – they both characterize sovereignty as a narrative or fiction to which diverse actors appeal as they pursue their interests in the uneven landscape of international affairs. The actual workings of states and non-state actors who exercise authority bear little resemblance to state sovereignty as conventionally understood.

These conclusions are supported by Lauren Benton's (2010) work on sovereignty and European empires, which deserves a wider reading in geography. Benton demonstrates the importance of law and geography to the extension of European colonialism, but also notes its patchwork and negotiated character. Instead of a smoothly spreading imperium that increasingly engulfs areas of the map over time, Benton presents the extension of European empires as a halting process characterized by imperial agents' lack of knowledge of the spaces and people that they were attempting to control. In this context, law served as an "epistemological framework" (2010: 21) in which colonial powers attempted to organize and make sense of the various spaces they encountered, from rivers, oceans, islands, and mountains to the numerous enclaves and imperial zones that were reoccurring features of colonial legal geography. And, of course, this entailed a variety of negotiated relations with extra-European legal structures and people. These relations could entail outright domination, certainly, but also reliance on local systems of law, significant instances of resistance, and extensive forms of legal creativity and experimentation that deployed imperial law in novel ways.

The central importance of Benton's work is that it disrupts a retroactive linkage between sovereignty and territory, which she suggests became tightly intertwined only in the nineteenth century and was then projected back onto the entire history of European empires. Instead of sovereignty emerging out of whole cloth in the territorial nation-states of Europe and then exported abroad, Benton helps us understand sovereignty as composed of highly variegated expressions of authority in a dizzying array of spatial forms. In some cases, sovereignty was a "tubulous" power that adhered to European subjects as they explored and attempted to control imperial trade routes, often over watery spaces like rivers and oceans (Benton 2010: 110).

In other cases, sovereignty was connected with zones or clusters of imperial authority, but its delegation across national territory could be highly differentiated. For instance, Benton discusses the various classifications of sovereignty and quasi-sovereignty that colonial powers asserted over enclaves such as Native American reservations in the United States, Basutoland in southern Africa, or the princely states in the hill regions of India. In these places, colonists developed a wide array of legal categories to characterize their own authority while also accommodating indigenous forms of rule that they were unable to subdue or eliminate.

In her displacement of state territory and her emphasis on the uneven geographies of sovereign power, Benton's arguments mirror those of Agnew and Krasner, while also marking a decisive shift. Whereas Agnew and Krasner treat sovereignty as a myth used to secure state authority, Benton emphasizes the *active* aspects of sovereignty as an *assertion* of legal power and a form of legal politics. Sovereignty, in this account, is neither a property of a regime nor simply a narrative that states deploy. Rather, it is an ideological and institutional form that is itself always under construction. In other words, there is no ideal-typical form of state sovereignty that stands behind or outside of the fragmented, incomplete, or "anomalous" legal spaces of imperial sovereignty. Sovereignty, for Benton, does not precede its articulation. Colonialism entailed a continuous process of legal experimentation in which the concept and institutions of sovereignty were created and transformed. As Benton puts it, "the legal and political practices (and, yes, institutions) constituting imperial sovereignty might be seen as shaping discourses about the rule of law, generating a varied set of available narratives about the special virtues and qualities of law in different periods" (Benton 2009: 120). The result is much greater attention to the construction of legal concepts and institutions through the process of legal politics.

New spaces of sovereignty

Although not directly engaged with Benton, much of the contemporary work on sovereignty in geography supports her claims, focusing on the constitution of sovereign power through the production of uneven political landscapes. As Mountz (2013: 830) has suggested, part of the timing of this work results from geopolitical developments over the last decade, particularly events associated with the aftermath of the terrorist attacks of September 11, 2001 and the ensuing US Global War on Terror. Scholars examined the extensive use of extraterritorial detention centers along with the practices of extraordinary rendition, indefinite detention, and torture in relation to US sovereignty (Gregory 2006, 2007; Hannah 2006; Paglen & Thompson 2006; Reid-Henry 2007; Elden 2009). They have also considered the ways in which sovereign power was rearticulated around and through the bodies and places targeted by international aid and development agencies in the wake of these conflicts (Fluri 2009, 2012). The US prison in Guantánamo Bay has become paradigmatic, as the decision to house prisoners captured in the US War on Terror in that location was explicitly connected with legal argument that the camp was outside the territorial-based jurisdiction of US federal district courts (Reid-Henry 2007: 629). Scholars also noted that the spatial location of the prison – in Amy Kaplan's (2005: 832) useful description "a liminal national space, in, yet not within, Cuba, but at the same time a 'bit of American territory'" – emerged from a long history of imperialism (see also Gregory 2006; Reid Henry 2007).

Taken as a whole, this research reiterates the role of law and politics in the creation of sovereignty, but extends the discussion to incorporate the centrality of state violence in the process. In addition to their pressing political urgency, all of these cases are interesting because they involve states legally designating people and places as targets of state violence. Moreover, this legal designation happens through both the extension and the withdrawal of normal legal protocols, as well as through the overlapping and uneven jurisdictional claims of competing legal systems. These developments, however, have not been confined to the War on Terror. Geographers have noted a growing continuity between military tactics using legal means to classify groups and spaces as exceptional and the domestic policing of people and places in relation to issues such as migration, trade, and border security (for overviews, see Martin & Mitchelson 2009; Hyndman 2012; Lloyd, Mitchelson, & Burridge 2012). As such, these processes are part of the disarticulation of state sovereignty into what Aihwa Ong termed "graduated" forms of sovereign power (Ong 2000). For Ong, the term captured the range of techniques – from military strategies to governmental projects of surveillance and regulation – that states use to classify populations differentially, often along ethno-racial lines, in relation to the development projects of industrializing Asian states. When read alongside geographical accounts (see Mountz 2011), graduated sovereignty highlights that state practices not only surveil and classify individuals, but they do so in novel geographical ways. New geographies of sovereign power bear little resemblance to the comprehensive controls over national territory ascribed to state sovereignty and include the extensive use of islands, oceans, walled spaces, and other barriers to manage the movement of things and individuals across space and territory.

Two examples are particularly illustrative: the policing of migration, on the one hand, and that of commodities and capital, on the other. The large literatures on these topics consider the movement of people and goods in a variety of places, but they converge in their focus on the state production of spatial barriers that do not attempt to limit as much as regulate and differentiate the movement of people and things in ways that reinforce state power. In terms of migration, research on "immigration geopolitics" (Coleman 2007) demonstrates a profound reworking of sovereignty in the post–9/11 context. In the United States, this shift has been characterized by the devolution of federal oversight over immigration to local police forces (through mechanisms like the 287(g) program) as well as the shifting location of immigration policing from national borders to internal spaces within the nation-state (Coleman 2007, 2009; Coleman & Kocher 2011; Varasnyi 2008). These shifts in sovereignty occur alongside and through surveillance and disciplinary techniques targeted at the bodies of migrants (L. Martin 2012). Without reducing the historical and geographical specificity of these studies, scholars have noted related policies of population differentiation, discipline, and detention in other regional and national contexts, as well as in zones such as island detention centers (see Mountz 2010, 2011).

The diverse attempts to monitor, classify, and contain migrants in relation to national security contrasts deeply with the ways in which spaces and spatial boundaries are opened relative to the movement of capital and commodities. Matthew Sparke (2006), for instance, has examined the way in which the very same technologies of surveillance developed after 9/11 to police undocumented migrants have fostered the transnational movement of business through "smart border" programs marketed to a transnational business class. Craig Martin (2012) has similarly explored the production of networks of interconnectivity through logistics management as an expression of geopower. As with Sparke's attention to the differential ordering of citizenship, the networks used to ship containers across boundaries foster the movement of goods while also being intertwined with a politics of security that targets stowaways and other migrants.

The attention to the precise ways in which different institutions of state power treat populations and places differently also leaves us with a complicated account of state sovereignty.

Although sovereignty is primarily treated in relation to state practices of differentiation, with states defining some as proper and appropriate for political life and excluding others, scholars are also clear that it is through these differential classifications that states produce themselves and their institutional power. Building on Timothy Mitchell's (1999) argument that states are the effects of diverse institutional practices and only retroactively recognized as coherent objects, Alison Mountz advocates an ethnographic approach to the state that renders it "knowable through its daily interactions with citizens and others" (2010: xxxii). Mathew Coleman (2005), likewise, urges us to recognize that states are not monolithic actors and that the practices of controlling populations and territory are often incoherent. Moreover, this incoherence is magnified as policies move from centers of state power to local areas of enforcement. Although state agencies exercise profound powers over individuals' lives, these studies also suggest strategic possibilities in recognizing the provisional nature of state sovereignty.

Political theology and the crisis of sovereignty

The precariousness of state sovereignty is also central to Wendy Brown's (2010) account of the proliferation of walls and other partitions of political space, which she argues are not an expression of the continuing importance of sovereignty, but rather a symptom of its crisis in relation to the same global flows of migrants and capital. For Brown, sovereignty is a specific power that goes beyond social and spatial demarcation and differentiation, and is fundamentally theological at its root. Brown suggests that almost every major theorist of sovereign power has conceptualized sovereignty on the model of God. As she puts it, "political sovereignty is never without theological structure and overtones, whether it is impersonating, dispelling, killing, rivaling, or serving God" (Brown 2010: 61). The increased use of walls as a strategy for controlling populations and territory represents an aggressive political-theological reaction as state sovereignty wanes in relation to transnational movements of both capital and religious ideologies.

One hears echoes in this argument of Carl Schmitt's (1985 [1922]: 36) claim that "all significant concepts of the modern theory of the state are secularized theological concepts." Like Brown, Schmitt viewed sovereignty as under threat. In his 1922 book *Political Theology*, the threat was from the instability of liberalism during the tumultuous years of the Weimar Republic. Schmitt characterized liberalism as a kind of interminable discussion that was incapable of taking decisive action – up to and including the power over life and death – to save the state from existential threats. For Schmitt, it was this power – to suspend the law to save the state – that best captured sovereignty's divine quality as within and beyond the existing legal order. Later, writing after the Second World War, Schmitt (2003) would come to relocate the threat to sovereignty in the global ambitions of capitalism. Precisely because legal systems, and thus sovereignty, were always defined by a determinant localization, Schmitt argued that the tendencies of capital to transgress political boundaries threatened existing spatio-legal orders. For this reason he advocated the declaration of juridical exceptions, which was "analogous to the miracle in theology" (1985 [1922]: 36), in defense of state sovereignty (see also Galli 2010).

Michel Foucault also gave primacy to the theological power over life and death in his critique of sovereignty as a specific modality of power that was, if not in total decline, at least in significant transformation since the nineteenth century. Foucault treated this power as a "characteristic privilege" (1990: 135) and a "basic attribute" (2003: 240) of the sovereign that was linked to the ancient *patria potesta*; a right that allowed the heads of Roman households to "'dispose' of the life of his children and his slaves" (1990: 135). For Foucault, then, sovereign power was a juridical power (a right) to kill or let live that was subsequently applied to state sovereignty – although in a "diminished form" (1990: 135) – through the early modern period. Its primary characteristic remained the ability to take life, calling on subjects to defend the realm; it also legitimated the appropriation of "a portion of the wealth, a tax of products, goods and services, labor and blood" (1990: 136) in the name of the health, welfare, and security of the state.

Yet, Foucault noted a relative decline in sovereignty as the ability to deduct and seize life slowly shifted into modern forms of disciplinary and governmental reason. Instead of killing or letting live, these new forms of power focused on expanding the capacities of humans "to make live and let die" (2003: 241). Foucault chronicled this transition in his studies of discipline, madness, sexuality, and medicine, and would influentially examine the powers aimed at intensifying life under the term "biopolitics." Nevertheless, the relationship between the two forms of power – the "old right" (2003: 241) of sovereignty and the newer forms of discipline, government, and biopolitics – remained complicated. Although Foucault charted a general shift "from a regime dominated by structures of sovereignty to a regime dominated by techniques of government" (2004: 106), he also argued that sovereignty continued to exist as a problem – "The problem of sovereignty was never more sharply posed than at this moment" (2004: 106) – as well as a set of relations that intermixed with newer modes of power.

In terms of understanding sovereignty, the precise relation between different modes of power or the historical veracity of his account is less important than the link between sovereignty and the power over life and death. On this reading, sovereign power is a juridical relation incorporated in diverse institutional forms. This shift was important for Foucault's own thinking about the emergence of modern states. Conceptualizing sovereignty not as a property of the state created by contract but a specific right, Foucault was able to situate sovereign power in relation to other modes of governing life. As such, his work directs attention to the multiplicity of forces operative both within and beyond the state, and has been critical for redirecting political analysis away from its obsession with central state authority and toward the capillary movement of power in diverse institutional forms.

The political ontology of sovereign power

The problem of the relationship between sovereignty and life is also at stake in Giorgio Agamben's work on sovereignty, which remains a touchstone in contemporary debates. His now well-known argument builds on the work of Schmitt and Foucault and characterizes sovereignty as a paradoxical power in which the domain of law is established through its legally authorized suspension. Agamben follows Schmitt in identifying this paradox with the exception. Whereas the exception for Schmitt signifies an existential crisis in which the state can legally suspend the law to save itself, for Agamben the exception demarcates a threshold between lives that are worthy of being qualified for political existence and simple, biological being that has no place in the political order. Politics in the West, at least since Plato, have thus been characterized by the exclusion of this bare life, which Agamben links with the figure of *homo sacer* – the sacred man who can be put to death but not sacrificed. However, Agamben also notes that the biopolitical link between sovereignty and bare life persists into the present, including in democratic regimes, and for this reason claims that the paradigmatic space of modern law is not the nation-state, as conventionally understood, but the camp in which bare life is confined and managed such that the political sphere can come into being (Agamben 1998, 2000, 2005).

With his extensive use of spatial concepts and metaphors, Agamben has been read widely in the discipline, yet geographers have also identified profound limits to his thought. Agamben has been variously characterized as glossing over the different ways in which bodies are abandoned by law and, therefore, inattentive to the complexities of political space. His totalizing account of sovereign power has also been criticized for not recognizing the negotiated nature of sovereignty and government or the vital powers that even the most thoroughly dominated, exploited, and terrorized individuals continue to exercise (for a range of these critiques see Mitchell 2006; Coleman & Grove 2009; Mountz 2013; Dunn & Cons 2014). These critiques certainly have validity if we understand Agamben's project from the perspective of critical social science, but they are much more difficult to sustain if we take Agamben at his word when he characterizes his studies as "paradigmatic ontology" (Agamben 2009: 32), implying that his key concepts refer not to empirical events, but to metaphysical categories (for this argument, see Lee et al. 2014).

Of course, invocation of that term alone scarcely clarifies what paradigmatic ontology means or how it relates to the history of metaphysics or the dynamics of contemporary sovereignty. Although ontology refers to the study of being and is often treated in the social sciences as something akin to descriptive taxonomies of things, Agamben's reflections on ontology engage with the problem that Martin Heidegger identified concerning the inability of Western thought directly to address not representations of being, but being in general. On this reading, as Mathew Abbott (2012) and Timothy Campbell (2011) have argued, the key distinction in Agamben's thought is the fissure between lives that are proper for the political sphere (*bios*) and those that are simply bare, biological beings (*zoe*). Read as an ontological rather than sociological or historical category, however, bare life does not index any empirical location, individual, or event. Rather, bare life is the metaphysical presupposition that haunts philosophical systems in which being has been abandoned. What makes the exception "the original – if concealed – nucleus of sovereign power" (Agamben 1998: 6) is not because sovereignty is everywhere and at all times organized in the same manner and on the model of the Ancient Greek *polis*. Rather, as Abbott (2012: 27) explains, abandonment is "the unthought ground of the metaphysics underpinning our political systems." Thus, it is the sheer impossibility of separating politically proper life from the impropriety of bare existence that drives the historically distinct apparatuses of government toward death (see Abbott 2012: 28; Campbell 2011).

The implication of Agamben's argument that the diverse geographical and historical processes producing sovereignty remain beset by metaphysical problems has only haltingly been incorporated in the geographical literature. Nonetheless, there are important precedents. Early engagements with Agamben in the discipline (Belcher et al. 2008) focused on the topological and non-localizable character of paradigmatic spaces of exception, most notably "the camp" that Agamben presented as the *nomos* (and therefore the determinant localization) of modern politics (Agamben 1998; on *nomos* see Schmitt 2003). Read ontologically, the camp – as with all of Agamben's historico-political concepts – is more properly understood as a specific, modern apparatus marked by the abandonment of being. In *Homo Sacer*, Agamben (1998: 44–48) deals with the form of this marking by suggesting that the political categories of constituting and constituted power (precisely the elements topologically linked in the figure of the camp) are political forms of the fundamental ontological categories of potentiality and actuality. Moreover, the relationship between these two categories is blurred, as potentiality includes both the virtual form of being that will be realized in action, and also an inexhaustible potential not to be. For this reason, potentiality and actuality exist in the same topological relationship of indistinction to one another as sovereignty and life. As Agamben argues:

> Sovereignty is always double because Being, as potentiality, suspends itself, maintaining itself in a relation of ban (or abandonment) with itself in order to realize itself as absolute actuality (which thus presupposes nothing other than its own potentiality). (Agamben 1998: 47)

In terms of geographical analysis, this has led some geographers (Wainwright 2008; Lee et al. 2014) to focus on an explicitly postcolonial "spacing of sovereignty," a strategy of reading the ways in which colonial texts, themselves inhabited by the ontological problems at the heart of sovereign power, have written the world, including the divisions between the West and the non-West. In this regard, Agamben's political ontology is instructive for pointing out the ways in which even progressive projects are formulated within the biopolitical horizon of sovereignty, thereby "clarifying what is *not* to be done" (Lee et al. 2014). Alternatively, the possibility of a potential-not-to-be has encouraged others (see, for instance, Braun 2014; or my own attempt to deal with this issue in Barkan 2013) to focus on the potentialities that are immanent to, but unrealized in, various regimes of sovereignty and government.

In the wake of these discussions, it is clear that sovereignty is both less and more than it seems. In terms of understanding the policies and practices of states in an era that is variously characterized as globalized or post-Westphalian, sovereignty is potentially far too metaphysical a concept to ground empirical analysis. Instead, we might leave off questions of sovereignty for more detailed accounts of the legal politics of states and capital as they work through and constitute spaces and subjects. This, of course, has long been a strength of critical human geography, and there is certainly an argument that the scholarly vogue of "sovereignty" only muddies the discipline's main questions. For precisely the same reason, however, the metaphysical dynamics of sovereignty provoke questions about how sovereign power not only shapes political regimes, but also inhabits the implicit spatiality by which we think about and orient ourselves in the world.

In this more limited sense, there remain ample questions concerning the geographies of sovereign power. Certainly, there is no reason to believe that the search for immanent potentials within sovereignty will abate any time soon. In addition, if ontological readings of sovereignty are important for considering "what is not to be done," it is also worth asking how discourses of sovereignty are mobilized by those engaged in political projects that explicitly seek to challenge, destabilize, and contest the colonial order of things. Lee et al. (2014), in their study of sovereignty in South Korea, provide one avenue of analysis, where the particular strategies of the decolonizing state reiterate the abandonment of being. However, other lines of research might consider more challenging cases where communities try to redirect sovereign power against the organizing logics that are attempting to govern life in the interests of state and capital.

Tyler McCreary and Richard Milligan (2014) present a useful example of this type of "ontological politics" in their account of the Carrier Sekani people's resistance to the Enbridge Northern Gateway Project. Doing so draws attention to the different ways in which indigeneity is both incorporated in but also unassimilable to various practices of Canadian state agencies and companies engaged in resource extraction. For instance, they describe the extensive practices of encoding Aboriginal tribal knowledge into the permitting process for the proposed pipeline project, a practice legally mandated by the Canadian

National Energy Board. But they also contrast this with dynamic practices of indigeneity by which the Carrier Sekani leaders and activists assert their ongoing power to determine the use of the land, including in ways that run counter to those of the Canadian state. Moreover, as McCreary and Milligan make clear, the issue is not only that the state has failed "to translate Indigenous geographies with sufficient detail" (2014: 121). Rather, it is that the state cannot recognize other ways of "being on the land that makes that land something different, that renders that land subject to other modes of not just use but also government" (2014: 121). Ontological politics, then, does not refer to an essentialized indigenous identity. As with Agamben's political ontology, it names the diverse strategies by which the state attempts to govern being-in-the-world and the profound limit that state sovereignty encounters as being becomes otherwise.

References

Abbott, M. (2012). No life is bare, the ordinary is exceptional: Giorgio Agamben and the question of political ontology. *Parrhesia* 14: 23–36.

Agamben, G. (1998). *Homo Sacer: Sovereign Power and Bare Life*, trans. D. Heller Roazen. Stanford, CA: Stanford University Press.

Agamben, G. (2000). *Means without End: Notes on Politics*, trans. V. Bineti & C. Casarino. Minneapolis, MN: University of Minnesota Press.

Agamben, G. (2005). *State of Exception*, trans. K. Attell. Chicago, IL: University of Chicago Press.

Agamben, G. (2009). *The Signature of All Things: On Method*, trans. L. D'Isanto with K. Attell. New York: Zone Books.

Agnew, J. (1994). The territorial trap: The geographical assumptions of international relations theory. *Review of International Political Economy* 1: 53–80. doi:10.1080/09692299408434268

Agnew, J. (2005). Sovereignty regimes: Territoriality and state authority in contemporary world politics. *Annals of the Association of American Geographers* 95: 437–461. doi:10.1111/j.1467-8306.2005.00468.x

Agnew, J. (2009). *Globalization and Sovereignty*. Lanham, MD: Rowman and Littlefield.

Barkan, J. (2013). *Corporate Sovereignty: Law and Government under Capitalism*. Minneapolis, MN: University of Minnesota Press.

Belcher, O., Martin, L., Secor, A., Simon, S., & Wilson, T. (2008). Everywhere and nowhere: The exception and the topological challenge to geography. *Antipode* 40: 499–503. doi:10.1111/j.1467-8330.2008.00620.x

Benton, L. (2009). Not just a concept: Institutions and the "rule of law." *Journal of Asian Studies* 68: 117–122. doi:10.1017/S0021911809000126

Benton, L. (2010). *A Search for Sovereignty: Law and Geography in European Empires, 1400–1900*. Cambridge: Cambridge University Press.

Bodin, J. (1992). *On Sovereignty: Four Chapters from The Six Books of the Commonwealth*, ed. & trans. J. Franklin. Cambridge: Cambridge University Press.

Braun, B. (2014). A new urban dispositif? Governing life in an age of climate change. *Environment and Planning D: Society and Space* 32: 49–64. doi:10.1068/d4313

Brenner, N. (1999). Beyond state-centrism? Space, territoriality, and geographical scale in globalization studies. *Theory and Society* 28: 39–78. doi:10.1023/A:1006996806674

Brenner, N. (2004). *New State Spaces: Urban Governance and the Rescaling of Statehood*. Oxford: Oxford University Press.

Brown, W. (2010). *Walled States, Waning Sovereignty*. New York: Zone Books.

Campbell, T. (2011). *Improper Life: Technology and Biopolitics from Heidegger to Agamben*. Minneapolis, MN: University of Minnesota Press.

Coleman, M. (2005). U.S. statecraft and the U.S.–Mexico border as security/economy nexus. *Political Geography* 24: 185–209. doi:10.1016/j.polgeo.2004.09.016

Coleman, M. (2007). Immigration geopolitics beyond the Mexico–U.S. border. *Antipode* 39: 54–76. doi:10.1111/j.1467-8330.2007.00506.x

Coleman, M. (2009). What counts as the politics and practice of security, and where? Devolution and immigrant insecurity after 9/11. *Annals of the Association of American Geographers* 99: 904–913. doi:10.1080/00045600903245888

Coleman, M., & Grove, K. (2009). Biopolitics, biopower, and the return of sovereignty. *Environment and Planning D: Society and Space* 27: 489–507. doi:10.1068/d3508

Coleman, M., and Kocher, A. (2011). Detention, deportation, devolution and immigrant incapacitation in the U.S., post 9/11. *Geographical Journal* 177: 228–237. doi:10.1111/j.1475-4959.2011.00424.x

Dunn, E., & Cons, J. (2014). Aleatory sovereignty and the rule of sensitive spaces. *Antipode* 46: 92–109. doi:10.1111/anti.12028

Elden, S. (2009). *Terror and Territory: The Spatial Extent of Sovereignty*. Minneapolis, MN: University of Minnesota Press.

Fluri, J. (2009). "Foreign passports only": Geographies of (post)conflict work in Kabul, Afghanistan. *Annals of the Association of American Geographers* 99: 986–994. doi:10.1080/00045600903253353

Fluri, J. (2012). Capitalizing on bare life: Sovereignty, exception, and gender politics. *Antipode* 44: 31–50. doi:10.1111/j.1467-8330.2010.00835.x

Foucault, M. (1990). *The History of Sexuality: An Introduction, Vol. 1*. New York: Vintage.

Foucault, M. (2003). *"Society Must Be Defended": Lectures at the Collège de France, 1975–1976*, ed. M. Bertani & A. Fontana, trans. D. Macey. New York: Picador.

Foucault, M. (2004). *Security, Territory, Population: Lectures at the Collège de France, 1977–1978*, ed. M. Senellart, trans. G. Burchell. New York: Palgrave Macmillan.

Galli, C. (2010). *Political Spaces and Global War*, ed. A. Sitze, trans. E. Fay. Minneapolis, MN: University of Minnesota Press.

Glassman, J. (1999). State power beyond the "territorial trap": The internationalization of the state. *Political Geography* 18: 669–696. doi:10.1016/S0962-6298(99)00013-X

Gregory, D. (2006). The black flag: Guantánamo Bay and the space of exception. *Geografiska Annaler, Series B* 88: 405–427. doi:10.1111/j.0435-3684.2006.00230.x

Gregory, D. (2007). Vanishing points: Law, violence, and exception in the global war prison. In Gregory, D., & Pred, A. (eds), *Violent Geographies: Fear, Terror, and Political Violence*, New York: Routledge, 205–236.

Hannah, M. (2006). Torture and the ticking bomb: The war on terrorism as a geographical imagination of power/knowledge. *Annals of the Association of American Geographers* 96: 622–640. doi:10.1111/j.1467-8306.2006.00709.x

Hyndman, J. (2012). The geopolitics of migration and mobility. *Geopolitics* 17: 243–255. doi:10.1080/14650045.2011.569321

Kaplan, A. (2005). Where is Guantánamo? *American Quarterly* 57: 831–858. doi:10.1353/aq.2005.0048

Krasner, S. (1999). *Sovereignty: Organized Hypocrisy*. Princeton, NJ: Princeton University Press.

Lee, S.-O., Jan, N., & Wainwright, J. (2014). Agamben, postcoloniality, and sovereignty in South Korea. *Antipode* 46(3): 650–668. doi:10.1111/anti.12070

Lloyd, J., Mitchelson, M., & Burridge, A. (2012). *Beyond Walls and Cages: Prisons, Borders, and Global Crisis*. Athens, GA: University of Georgia Press.

Mann, M. (1984). The autonomous power of the state: Its origins, mechanisms and results. *European Journal of Sociology* 25: 185–213. doi:10.1017/S0003975600004239

Martin, C. (2012). Desperate mobilities: Logistics, security and the extra-logistical knowledge of "appropriation." *Geopolitics* 17: 355–376. doi:10.1080/14650045.2011.562941

Martin, L. (2012). "Catch and remove": Detention, deterrence, and discipline in U.S. noncitizen family detention practices. *Geopolitics* 17: 312–334. doi:10.1080/14650045.2011.554463

Martin, L., & Mitchelson, M. (2009). Geographies of detention and imprisonment: Interrogating spatial practices of confinement, discipline, law, and state power. *Geography Compass* 3: 459–477. doi:10.1111/j.1749-8198.2008.00196.x

McCreary, T., & Milligan, R. (2014). Pipelines, permits, and protests: Carrier Sekani encounters with the Enbridge Northern Gateway Project. *Cultural Geographies* 21: 115–129.

Mitchell, K. (2006). Geographies of identity: The new exceptionalism. *Progress in Human Geography* 30: 95–106. doi:10.1191/0309132506ph594pr

Mitchell, T. (1999). Society, economy, and the state effect. In Steinmetz, G. (ed.), *State/Culture: State-Formation after the Cultural Turn*, Ithaca, NY: Cornell University Press, 76–97.

Mountz, A. (2010). *Seeking Asylum: Human Smuggling and Bureaucracy at the Border*. Minneapolis, MN: University of Minnesota Press.

Mountz, A. (2011). The enforcement archipelago: Detention, haunting, and asylum on islands. *Political Geography* 30: 118–128. doi:10.1016/j.polgeo.2011.01.005

Mountz, A. (2013). Political geography I: Reconfiguring geographies of sovereignty. *Progress in Human Geography* 37: 829–841. doi:10.1177/0309132513479076

Ong, A. (2000). Graduated sovereignty in South-East Asia. *Theory, Culture, and Society* 17: 55–75. doi:10.1177/02632760022051310

Paglen, T., & Thompson, A.C. (2006). *Torture Taxi: On the Trail of the CIA's Rendition Flights*. Hoboken, NJ: Melville House.

Reid-Henry, S. (2007). Exceptional sovereignty? Guantánamo Bay and the re-colonial present. *Antipode* 39: 627–648. doi:10.1111/j.1467-8330.2007.00544.x

Schmitt, C. (1985 [1922]). *Political Theology: Four Chapters on the Concept of Sovereignty*, trans. G. Schwab. Chicago, IL: University of Chicago Press.

Schmitt, C. (2003). *The Nomos of the Earth in the International Law of the Jus Publicum Europaeum*, trans. G.L. Ulmen. New York: Telos Press.

Sitze, A. (2010). Editor's introduction. In Galli, C. (ed.), *Political Spaces and Global War*, Minneapolis, MN: University of Minnesota Press, xi–lxxxv.

Sparke, M. (2006). A neoliberal nexus: Economy, security and the biopolitics of citizenship on the border. *Political Geography* 25: 151–180. doi:10.1016/j.polgeo.2005.10.002

Varsanyi, M. (2008). Immigration policing through the backdoor: City ordinances, the "rights to the city," and the exclusion of undocumented day laborers. *Urban Geography* 29: 29–52.

Wainwright, J. (2008). *Decolonizing Development: Colonial Power and the Maya*. Malden, MA: Blackwell.

Chapter 6

The State

Alex Jeffrey

University of Cambridge, Northern Ireland, UK

In his landmark paper "Notes on the difficulty of studying the state" (1988), Philip Abrams outlines a key problem in writing about "the state." He argues that focusing on the state grants "spurious concreteness and reality to that which has a merely abstract and formal existence" (1988: 58). Emphasizing this point, later in the paper he seeks to highlight the ideological nature of this political concept, arguing that the state is "not an object akin to a human ear. Nor is it even an object akin to a human marriage. It is a third-order object, an ideological project" (1988: 76). Just as Benedict Anderson (1991) argued regarding *Imagined Communities*, the ideological nature of the state does not render this an illusory entity, but it should orient attention toward how the state is imagined rather than an exploration of its fundamental characteristics. In these terms, we undertake some political labor on behalf of the state by rendering it a single political-geographical apparatus; in doing so we can grant this concept imagined clarity and immutability. It is for this reason that Agnew's (1994) warning of the "territorial trap," in which political processes are read off through the lens of state territoriality, remains so prescient (see also Bilgin & Morton 2002). One of the first obligations, then, when writing about the state is to assume a dual perspective: on the one hand recognizing and exploring the existence of a large body of work that focuses on the characteristics of the state, while on the other remaining careful not to reify this apparatus as a timeless ontological truth.

Reflecting this precarious authorial position, a discussion of the state demands a succinct definition of precisely what is meant by this term, while also remaining skeptical of its universal application. As Peter Bratsis (2006: 9) asserts, arguments concerning the nature of the state have been undermined by assuming the existence of the state, proceeding "as if" the state "was indeed a universal *a priori* predicate to our social existence rather than the product of our social existence." Yet, while acknowledging such caveats, there exists a set of common reference points around which discussion of the state coheres. Often understood as a multidimensional phenomenon, definitions of the state generally coalesce around a

The Wiley Blackwell Companion to Political Geography, First Edition.
Edited by John Agnew, Virginie Mamadouh, Anna J. Secor, and Joanne Sharp.
© 2015 John Wiley & Sons Ltd. Published 2017 by John Wiley & Sons Ltd.

notion of a relationship between power (sovereignty) and space (territory). This may be expressed in law; for example, Held (1983: 1) suggests that the state is an "impersonal and privileged legal or constitutional order with the capability of controlling or administering a given territory." Equally, definitions can focus on the centrality of violence to the maintenance of state authority, exemplified in Weber's (1958: 78) definition of the state as a "human community that (successfully) claims the monopoly of legitimate use of physical force within a given territory." As Bratsis (2006) recognizes, the key area of debate within this definition is not necessarily the violence, but the centrality of the more ambiguous concept of legitimacy (which is also present in Held's definition through the use of the term "privileged"). As we will see, establishing the legitimacy of rule has both internal and external dimensions: both those within and those beyond the borders of the state need to regard a particular set of state arrangements as appropriate. Without internal sovereignty a state struggles to govern its people; without external recognition a state cannot participate in international affairs.

Both Weber's and (to a lesser extent) Held's definitions introduce another significant element of studies of the state: that this is both a geopolitical and a biopolitical project, where the governance, control, and maintenance of territories and human communities intersect. One common route through this material and embodied geography is to stratify the state into different varieties: for example, the medieval state, the authoritarian state, the imperial state, and the nation-state (see Giddens 1985). The "nation-state" has become the most common shorthand for the fusion between human community and state sovereignty (Gilmartin 2009: 20). This term is a helpful reminder that the projection of power is not simply a technocratic process, but requires the consent of resident populations. Political scientists from Jean-Jacques Rosseau onward have spoken of the "social contract" that is forged between state and individual, where certain freedoms are surrendered in return for political and social rights. While this has historically been a matter for discussion for scholars of citizenship (see Dickinson et al. 2008), emerging work studying the embodied and everyday geographies of the state has begun to reflect on the forms of practice that convey the legitimacy and authority of prevailing state systems (Painter 2006; Jeffrey 2013).

Taking Abrams's long-standing concerns seriously, this chapter seeks to pluralize understandings of the state and trace these through four sections, each overlapping and intertwined. The first explores the imagined foundations of the state as a natural or organic expression of human territoriality. In some of the bolder pronouncements in this work the state becomes a reflection of natural law, a form of territorialization that is a human quality rather than an expression of power. If the first section outlines some of the ways in which internal sovereignty has been secured, the second looks at the international nature of statehood through consideration of the state system. Orienting attention to the most influential systemic account, Marxist theories of the state, the section considers how these approaches have embedded a sense of an enduring distinction between state and society. The third section examines how feminist, anthropological, and poststructural approaches to the state have reoriented attention away from formal state institutions and toward the more socially embedded processes through which ideas of the state are reproduced. The final section examines debates concerning the transformation of state sovereignty in the face of more global and networked forms of spatiality. As the state endures, the chapter traces the forms of securitization and power projection that have characterized this era of reconfigured territoriality.

Tracing the state

When a concept such as the state does so much to defy precise definition, it is no surprise that quests are undertaken for its origin or foundation. So pervasive is the state in analysis of the dynamics of contemporary political geography that we may be seduced into thinking that this is a timeless expression of the human desire to territorialize. The sense that the state is somehow prepolitical stalks writings on its foundation and characteristics, where the establishment of states reflects the return of humanity to a natural order of things. If we look to historical accounts, we can see both premodern and modern expressions of this tendency. For example, Plato's *Republic* considers the ideal state as the *polis* (or Greek city-state). One of the key contentions that organized the *polis* is that of autochthony, that its inhabitants "sprang fully formed, born of the earth" (Elden 2013: 22). In his exploration of the emergence of conceptions of territory in Western political thought, Elden turns to Greek myth and tragedy to examine the contention of autochthony, a term that makes explicit "the close and organic link between the people, the land (*khora*), and the *polis*" (2013: 23). Drawing on the founding myths of Athens, Alexandria, and Thebes, Elden's account traces the interplay between the human body, land, and deity, each playing a part in constructing autochthonous origins. Elden, drawing on Saxonhouse (1986), summarizes the purpose behind such intricate autochthonic myth-making:

> First … it provides a unity to the *polis*. Second, the boundaries of the *polis* are set by nature rather than by human agreements. The *polis* is natural, rather than set in opposition to nature. Third, the land is seen to belong to the people by right, by birth. There was no need for conquest and forced movement of previous inhabitants. (Elden 2013: 25–26)

It is perhaps no surprise that those conveying a certain form of rule turn to the materials of nature (be that the body or the soil) in order to provide legitimacy for practices of coercion. The era of imperial geopolitics characterized by the work of Ratzel, Kjellén, and Mackinder conveyed a similar sense of the enfolding of natural science and human geography. The work of Ratzel (1896) has risen to prominence for his attempts to provide a neo-Lamarckian (often referred to as social Darwinist) interpretation of the emergence of states. Just as any consideration of the Greek *polis* cannot be extracted from the poetry and politics of that era, so Ratzel's invocation of the states as living organisms struggling against one another must be inserted into the imperial maneuvers of nineteenth-century Europe. In particular, his account provided a justification for the incursion of powerful states on less organized or powerful neighbors. However, we must be careful in extrapolating from this that Ratzel was merely a scholarly echo chamber for expansionist forms of politics. His work has, of course, cast a long political shadow, famously enrolled within Nazi expansionist politics in 1930s Germany and its anarchistic elements feeding into realist approaches to international relations in the twentieth century. Yet, perhaps less frequently addressed are facets of Ratzel's writings that are suggestive of cooperation. On neighboring states, Ratzel was keen to point to the forms of cooperation that are required in the frontier zones between great powers:

> Naturally the efforts of nations are not confined to extension of territory. Even large states come into close contact at length. Neighbouring states share in the advantages of position and natural resources, and hence arises a uniformity of interests and modes of activity. Beside the lines of communication between the Atlantic and Pacific in the United States, Canada has constructed her

Canadian Pacific railway, and the navigation of the great lakes is made available by canals on both sides. (Ratzel 1896: 360)

In addition, while Ratzel's account may be seen as narrating the nature of conflict, it was also a significant argument for the dynamism of states:

The territory of a state is no definite area fixed for all time – for a state is a living organism, and therefore cannot be contained within rigid limits – being dependent for its form and greatness on its inhabitants, in whose movements, outwardly exhibited especially in territorial growth or contraction, it participates. (Ratzel 1896: 351)

Thus, we must be careful about conflating a series of organic approaches to the state (such as autochthonic or neo-Lamarckian) as if each is conveying a similar understanding of state form. Instead, nature exists as a resource deployed to stabilize assertions of rule and challenge the unnatural characteristics of alternatives. While it is straightforward to recognize the limitations to such monolithic and essentialist ideas of statehood, we need to take seriously the legacy of such perspectives: that the state is imagined as a neutral backdrop against which other forms of political action and spatialization are analyzed.

The state system

Since the state cannot be understood as an immutable characteristic of the human experience, scholarly energies have turned to investigating the emergence of the *state system*. This is both an exercise in understanding the significance of external recognition to the existence of states and also a historicization of the process through which states emerged as the preeminent territorialization of human political life. In some senses this entrenches the point that states only exist in the plural, since they require other states to enact and underscore their legitimacy as sovereign actors of specific territories and populations.

The technological, administrative, and ideological processes through which state power may be established and reproduced have long historical roots. In Europe, Charles Tilly (1992) traces these back to the centralization of coercion and violence, as infrastructural and fiscal technologies improve so that state authority may be conveyed over larger areas (see also Ogborn 1998). When trying to pinpoint a moment of consolidation for the state system in Europe, many commentators look to the Peace of Westphalia, a cluster of diplomatic treaties signed between 1643 and 1648, as the moment heralding ceasefire arrangements that allowed the establishment of a new diplomatic order of sovereign territories. However, this origin-point account does not bear scrutiny, and revisionist scholars have been keen to refer to the role of these accords in codifying absolutist sovereignty rather than acting as the handmaiden to modern state sovereignty, which Teschke (2003), among others, views as a nineteenth-century phenomenon (see Elden 2013: 310).

Yet, the state system did not simply summon itself into existence. It may have been a response to consolidating military and infrastructural technologies, but a larger field of critical scholars has pointed to the role of the rise of capitalism in shaping the emergence of state structures (Harvey 1976; Giddens 1985; Jessop 1990). From this perspective (or more properly, these perspectives), the state system exists as a means through which class advantage may be reproduced. But perhaps just as importantly, the capture of state power by a particular social class allows class struggle to be masked behind a system of disinterested domination that appears separate from the realms of economic or corporate life. Étienne Balibar explains the significance of political territories to the capitalist economic system:

> [A] "world-economy" that has developed in the form of a universal market cannot (contrary to liberal myth) form a *homogenous* whole without boundaries; it must be divided into a plurality of *political unities* that allow for a concentration of economic power and the defense of positions of unearned income (*rente*) or "monopoly" by extraeconomic means. There is no market without monopolies, no monopolies without instruments of political (or juridical) compulsion, which in practice means national states. (Balibar 2004: 18, emphasis in original)

One of the consequences of a Marxist account of the state system is a rather ambiguous position on the relationship between state and society. As a focal point of a transformative politics, Marxist accounts separate the state as a set of institutions that shape social relations in a regressive and counter-revolutionary fashion. However, this sits uneasily with a sense of the state as an abstract entity that enters into the reproduction of relations of production through its socially embedded nature (see Abrams 1988: 118).

This paradoxical placement is evident in the Marxist political theory of Nicos Poulantzas (1978). His work crackles with political fomentation, written as it was at a time when the poststructural approaches of Gilles Deleuze and Michel Foucault were rising to scholarly prominence. In a sense, this work stands as a key articulation of the tension between the political theory of the state and the state as the institutionalization of class relations. In a direct challenge to Foucault's invocation of the capillary-like nature of power, Poulantzas (1978: 28) shuns a view of the state that operates exclusively "through repression, force or 'naked' violence," and instead "calls upon ideology to legitimize violence and contribute to a consensus of those classes and fractions which are dominated from the point of view of political power."

The paradox remains, then, between a critical stance that challenges the ontology of the state as a coherent set of institutions and continued reliance on the distinction between state (as political actor) and society (as subjects of state violence). One of the most powerful attempts to transcend this dichotomy has been provided in the work of Bob Jessop. He has pioneered the translation of "regulation theory" into English-medium debates, work that has traced the role of the state in normalizing and reproducing capitalist relations over time. More recently, Jessop has sought to trace the range of social actors that are enrolled within this process through what he terms the strategic-relational approach (1990, 2005, 2008; Jessop, Brenner, & Jones 2008). This, he argues, is a process of subject formation; recalling Althusser's (2001) notion of interpellation, he argues that the "emergence of relatively stable structural ensembles involves not only the conduct of agents and their conditions of action but also the very constitution of agents, identities, interests, and strategies" (Jessop 2005: 53). Despite its theoretical roots, the strategic-relational perspective seeks to historicize and ground the evolution of strategic elements of state practice, emphasizing how the changing nature of the state reflects changes in capitalist systems of production.

The state as experience

In recent years, geographers have sought to reorient attention away from attempts to capture the nature of the state, to look instead at the implications of the *idea* of the state on a variety of political geographies. From this viewpoint, states do not simply exist; rather, they are accomplishments reified and reformulated through prosaic and quotidian activities (Radcliffe 2001; Painter 2006; Jeffrey 2013). From the 1980s onward – although these techniques have a longer scholarly lineage – scholars have sought to understand how ideas of the state reflect certain uneven landscapes of power, and how in turn "state effects" are distributed through society. Capturing this more decentered notion of state scholarship, Kuus and Agnew (2008: 96) emphasize the need to study states "not as autonomous subjects but as processes of sub-ject-making" and urge "more attention to the spatiality of power beyond that of territoriality." Refocusing on a wide variety of material and embodied state practices has seen the applica-tion of feminist and anthropological techniques and perspectives to trace in ethnographic detail the everyday geographies of assertions of the state.

In order to challenge this perspective, while remaining engaged with the formation and reproduction of state power, scholars have moved to an exploration of state effects outside the institutions and spaces that are generally associated with formal state power. As with other areas of political geography, scholars have been drawn to Michel Foucault's concept of gov-ernmentality to illuminate the diverse political rationalities of government, its "technologies," and the considerable intellectual labor involved in bringing into being the things, people, and processes to be governed (Painter 2002: 116; see also Mitchell 1991, 2006; Weber 1998; Jessop 2006, 2008; Painter 2006). In doing so, governmentality has become shorthand for explaining how state power operates to shape the conduct of conduct, as tactics and technol-ogies are deployed to control, subdue, and oppress the citizenry (Butler 2004; Jeffrey 2013). By examining the practice of rule in these terms, the state no longer stands as an unquestioned *source* of power, but, rather, as its *effect* (Marston 2004: 4; Mitchell 1991). For Timothy Mitchell (1991: 81), such effects are diffuse and difficult to grasp, but it is precisely this impre-cision that has been the "source of its political strength as a mythic or ideological construct."

Feminist geographers have been at the forefront of work examining state effects (see Secor 2001; Fincher 2004; Dyck 2005). Across a series of landmark empirical and theoretical inter-ventions, feminist scholarship has challenged the simple binary rendered between state and society, exploring instead the embodied and situated moments in everyday life through which ideas of the state are conceived and reproduced. As Ruth Fincher (2004: 49) explains:

> Everyday places like the domestic home, the playground, and the community center exhibit power relations that are differentiated and fractured by relations of gender, ethnicity, age, ability, and class. They are no different from the famously "big-P" political sites of public space, the parlia-ment, the city council and the large unionised workplace.

In order to capture this diverse array of political practices, geographers have required new techniques. Using qualitative and anthropological techniques, the ambiguities of the state and its embodiment have begun to emerge. Scholars such as Trouillot (2001), Ferguson and Gupta (2002), Sharma and Gupta (2006), Navaro-Yashin (2002), Taussig (1997), and Corbridge et al. (2005) have sought to use ethnographic techniques in order to understand the social and cultural processes through which state ideas are reproduced. The shifting geometry of this form of scholarship is captured in Corbridge et al.'s project of exploring how the state is apprehended in north India by those who are subject to its rule and decision-making. Entitling this *Seeing the State* is a deliberate retort to James C. Scott's influential *Seeing like a State* (1998), a study that traced how state bureaucracies render subjects legible and governable.

There are perhaps two processes under investigation in such political anthropologies of the state. The first examines how subjects negotiate the materials, forces, and affects through which state authority is conveyed. For example, anthropologist Yael Navaro-Yashin (2002, 2012) has pioneered a form of ethnography that seeks to explore the ways in which the state is experienced, lived, and embodied. Through ethnographic fieldwork in both Turkey and the Turkish Republic of Northern Cyprus, Navaro-Yashin traces what she terms in the later work an "affective geography" of sovereignty, one that commences from a study of the documents, sites, and materials through which sovereign power circulates. Indebted to Latour's Actor-Network Theory, where sovereignty is not a "top-down act of political will or event," it is "a worked-on terrain of relationality between human actors, material land and property, and tools or devices of measurement, numeration, and allocation" (Navaro-Yashin 2012: 44). Anna Secor (2007) uses such techniques to trace how the state interpellates – using Althusser's (2001) term – the citizen-subject into particular regimes of power through micro-situations (for example, the visit to the hospital or the meeting with state officials in the street). Instead of a stable administrative form, Secor argues that the Turkish state "coheres in the *resonance* (between sites, agents, rationalities, and techniques) that is discursively produced through the circulation and arrest of people, documents, information, money, and influence" (Secor 2007: 49, emphasis in original).

The second element of anthropological work on the state has centered on the mechanisms through which state authority is reproduced. For example, Ferguson and Gupta (2002) explore the reification of the state across two spatial maneuvers: verticality and encompassment. State verticality refers to the "central and persuasive idea of the state as an institution somehow 'above' civil society, community and the family" (Ferguson & Gupta 2002: 982). This topography of power is reproduced through the rhetoric of "top-down" policies, layers of bureaucracy, or even the notion of the "head" of state. Ferguson and Gupta (2002: 982) go on to use the term "encompassment" to refer to the significance of the state idea in an "ever widening series of circles that begins with the family and local community and ends with a system of nation-states."

Joe Painter (2006) has developed this work through theoretical engagement with the notion of prosaics, in particular examining the ways in which certain articulations of the state are made meaningful within everyday life. One of Painter's key conclusions is that we should not understand the state as a static achievement, but instead use the word "statization" to indicate the perpetual process through which the state comes into being. I have tried to explore this ethnographic and quotidian understanding of the state through a consideration of state-building practices after conflict (see Jeffrey 2007, 2013). The optic of international intervention provides a useful vantage point for the dynamic process through which a new state secures both internal and external sovereignty (see Agnew 2005; Allen & Cochrane 2010; Toal & Dahlman 2012). Drawing on research conducted in Bosnia and Herzegovina, I have sought to examine the materials, discourses, and practices through which the concept of a postconflict Bosnian state is constructed and reproduced, while being constantly attentive to alternative ideas of the state that circulate in political institutions, civil society, and everyday life. In order to theorize the repeated attempts to assert or accomplish the state, my work has emphasized the *improvised* nature of statehood, where improvisation is understood as a combination of *performance* and *resourcefulness*. The theatrical notion of improvisation is a useful analytical lens through which to explore such legitimating practices, since it orients attention to the relationship between what is possible and what is desired. As I suggest:

> moments of political breakdown and reconstruction provide an opportunity to explore how new ideas of the state are communicated, remaining attentive to the circulation of alternative narratives

of political community. The key point here is to understand how certain state ideas rise to prominence while others are cast out as illegitimate and illegal. (Jeffrey 2013: 171)

This sense of the theatricality of the state, and its entwining with questions of legitimacy, has been explored in McConnell's study of the exiled Tibetan state, as it has developed forms of bureaucracy and governance that set it out as a state-in-waiting, even as it currently exists in a deterritorialized form (McConnell 2009). This sense of a "rehearsal state" teases apart elements of internal and external sovereignty, pointing to the ways in which unrecognized states continue to seek a social contract, even in the absence of external recognition (McConnell forthcoming).

The waning state

While some scholars are turning to more embodied and situated accounts of the expression of state power, a parallel debate has taken place as to whether the state itself is losing relevance in an era of purported global flows. Sovereignty is undoubtedly transforming in the face of a series of material, ideological, and infrastructural shifts that have allowed the reorganization of authority and solidarity. Nevertheless, these movements are not in a single direction, as certain forms of statehood have been identified as contingent and appropriate sites of correction, and others bolstered as models of the correct articulation of the state. While elements of these sovereign forms are not new, proclamations of the deficiency of the sovereignty or governance of others has long been justification for forms of interaction under the banners of colonialism, international development, market forces, or global security.

Security discourses have reshaped understandings of state sovereignty both after the end of the Cold War (1989 onward) and, more significantly, in the wake of the terror attacks of September 11, 2001. These political events have prompted a new rubric of sovereignty, in which scholars have reflected on the complex relationship between territoriality and the assertion of authority. Here, the work of Agnew (2005) is again instructive. Building on Murphy's (1996) concepts of de facto and de jure sovereignty, Agnew proposes the notion of *effective sovereignty* to trace unruly geographies of contemporary territoriality. In many ways underscoring the significance of processes and performance to the assertion of state power, effective sovereignty points attention to the ways in which the territoriality of power does not align with the borders of states. Global economic processes and international terrorism are two forces that Agnew uses to illustrate the ways in which certain states project their sovereignty beyond their borders, while certain state-like actors manage to convey influence without centralized state authority. In order to underscore this plural geography of the power of sovereignty, Agnew advocates speaking of "sovereignty regimes," whereby different strategies of territoriality may be illuminated, beyond "classic" accounts of the Westphalian state system.

The sense of differentiated territorial practices of sovereignty is also at the heart of recent scholarship by Stuart Elden (2009, 2013). He uses the concept of "contingent sovereignty" (2009) to narrate the mechanisms through which powerful actors have suspended state sovereignty in places such as Afghanistan, Somalia, and Iraq in order to confront imagined security threats. Elden explores the ways in which security discourses of state elites, in particular the United States, are reconfiguring the traditional alignment of territory and sovereignty within understandings of the state. Crucially, this is not a wholly deterritorialized picture, as new geographies of threat within national security strategies and military doctrine project danger to particular sites and spaces, perhaps most explicitly through the rubric of "failed" or "rogue" states (see Jeffrey 2009). In doing so, the networks and connections within which imagined danger has emerged are erased, replaced by a more straightforward spatialization of threat contained within specific state boundaries. This, of course, does not equate to the erosion or replacement of the state, but rather a set of reterritorializations whereby certain state practices are emboldened while others have their sovereignty transgressed and suspended.

Nevertheless, it is not security discourses alone that are challenging the primacy of the state and the integrity of the international state system. Recent work exploring the rise of financial processes and institutions that defy the governance of single states or multilateral institutions has challenged the primacy of the sovereign state. For example, Agnew (forthcoming) has explored the role of credit rating agencies in exerting influence over the decision-making processes of democratic states. In terms that recall the concept of "conditional conquest" introduced by nineteenth- and twentieth-century US geographer Isaiah Bowman (see Smith 2003), Agnew explains that it is not the appropriation of land or territory that such financial agencies desire:

> Little is known about how credit-rating agencies operate. What is known is that they have been heavily involved in the ranking and management of the financial products that are at the center of the world's financial economy today. They represent a specifically privatized source of authority. Yet, they also illuminate several aspects of an emerging geopolitical order in which states can no longer be seriously regarded as single unified actors ..., key concepts such as market and state, private and public have taken on distinctively novel meanings, and the expropriation of land no longer lies at the center of geopolitical relations but is being replaced by control over financial products and flows. (Agnew forthcoming: 10–11)

Perhaps the most materially tangible evidence of the waning of states is, paradoxically, the forms of border securitization that have proliferated across the globe. As flows and networks beyond the state expand and cast a greater force over state decision-making, states themselves have attempted to perform their exclusive territoriality. Wendy Brown (2010) has explored what she has termed "late-modern walling" as a phenomenon that demonstrates the severing of sovereignty and the nation-state. The solidity and stability of the wall grant an imagined timelessness to the assertions of territoriality as effected by the state, but all the time – Brown asserts – this is merely a palliative act that demonstrates the limitations of state power, acting as "monuments to the fading strength or importance of nation-state sovereignty" (Brown 2010: 32). As Kuus and Agnew (2008: 99) emphasize, imagined security threats perform an important consolidating function:

> as the state's ability to control territory declines in response to an array of material forces, an outside threat becomes even more important for the reproduction and consolidation for the putative national community.

For Brown, the wall is more than a prosaic materiality, it acts as a form of political theology – and she widely cites the work of Carl Schmitt – that acts to demarcate the ordinary from the sacred. The act of walling is not simply an act of bounding, but a spiritual means through which state power is consecrated.

These final examples do not point to a necessarily waning state, but rather to forms of securitization and new hierarchies of power that can be masked by the imagined universalism of the state system. A critical geography of the state is attentive to such occlusions: it does not necessarily mean transcending the state as a category of political life, but rather engaging more substantively with the concept as a structuring device of political thought. The state has always been an assertion and accomplishment; it is both fragile in its need for reification and durable through its embedding in social and political life. Rejecting the state as a form of political geography has its seductions. Its constituent concepts (sovereignty, territory, or even legitimacy) seem to carry more precision and avoid falling into a territorial trap, or merely reproducing the embedded statism that characterizes affiliated disciplines such as international relations. However, this carries both intellectual and political risks. There is an urgent scholarly need to trace the ways in which state power is secured and reproduced, the ways in which sovereignty is legitimized, and the forms of oppression and exclusion that this can foster (see Fincher 2004). It risks assuming that states are transhistorical, and could overlook the ways in which changes in the duties and responsibilities of the state can harm the most vulnerable both within and beyond the territory of a particular state (Jessop et al. 2008; Elden 2009). As shown in McConnell's (2009) work, overlooking the state could also risk ignoring the significance invested in the state as a political project, as a means of securing national survival and making claims to territory. The challenge for political geographers is not to trace the origins or essences of the state, but rather to continue to trace how ideas and discourses of the state are mobilized to legitimize certain transgressions, exclusions, and enclosures.

References

Abrams, P. (1988). Notes on the difficulty of studying the state. *Journal of Historical Sociology* 1(1): 58–89.

Agnew, J. (1994). The territorial trap: The geographical assumptions of international relations theory. *Review of International Political Economy* 1(1): 53–80.

Agnew, J. (2005). Sovereignty regimes: Territoriality and state authority in contemporary world politics. *Annals of the Association of American Geographers* 95(2):437–461.doi:10.1111/j.1467-8306.2005.00468.x

Agnew, J. (forthcoming). Low geopolitics: Credit-rating agencies, the privatization of authority and the new sovereignty. *Geopolitica(s)*.

Allen, J., & Cochrane, A. (2010). Assemblages of state power: Topological shifts in the organization of government and politics. *Antipode* 42: 1071–1089. doi:10.1111/j.1467-8330.2010.00794.x

Althusser, L. (2001) *Lenin and Philosophy and Other Essays*. New York: Monthly Review Press.

Anderson, B. (1991). *Imagined Communities*. London, Verso.

Balibar, E. (2004). *We, the People of Europe? Reflections on Transnational Citizenship*. Princeton, NJ: Princeton University Press.

Bilgin, P., & Morton, D. (2002). Historicising representation of "failed states": Beyond the cold-war annexation of the social sciences? *Third World Quarterly* 23(1):55–80.doi:10.1080/01436590220108172

Bratsis, P. (2006). *Everyday Life and the State*. Boulder, CO: Paradigm.

Brown, W. (2010). *Walled States, Waning Sovereignty*. New York: Zone Books.

Butler, J. (2004). *Precarious Life: The Powers of Mourning and Violence*. London: Verso.

Corbridge, S., Williams, G., Srivastava, M., & Véron, R. (2005). *Seeing the State: Governance and Governmentality in India*. Cambridge: Cambridge University Press.

Dickinson, J., Andrucki, M., Rawlins, E., Hale, D., & Cook, V. (2008). Introduction: Geographies of everyday citizenship. *Acme* 7. http://www.acme-journal.org/vol7/JDetal.pdf, accessed 11 January 2014.

Dyck, I. (2005). Feminist geography, the "everyday", and local–global relations: Hidden spaces of place-making. *The Canadian Geographer/Le géographe canadien* 49(3): 233–243. doi:10.1111/j.0008-3658.2005.00092.x

Elden, S. (2009). *Territory and Terror: The Spatial Extent of Sovereignty*. Minneapolis, MN: University of Minnesota Press.

Elden, S. (2013). *The Birth of Territory*. Chicago, IL: University of Chicago Press.

Ferguson, J., & Gupta, A. (2002). Spatializing states: Toward an ethnography of neoliberal governmentality. *American Ethnologist* 29(4): 981–1002.

Fincher, R. (2004). From dualisms to multiplicities: Gendered political practices. In Staeheli, L., Kofman, E., & Peake, L. (eds), *Mapping Women, Making Politics: Feminist Perspectives on Political Geography*, London: Routledge, 49–70.

Giddens, A. (1985). *Contemporary Critique of Historical Materialism: The Nation-State and Violence*. Los Angeles, CA: University of California Press.

Gilmartin, M. (2009). Nation-states. In Gallaher, C., Dahlman, C., Gilmartin, M., Mountz, A., & Shirlow, P. (eds), *Key Concepts in Political Geography*, London: Sage, 19–27.

Harvey, D. (1976). A Marxian theory of the state. *Antipode* 8(2): 80–89.

Held, D. (1983). *States and Society*. New York: New York University Press.

Jeffrey, A. (2007). The politics of "democratization": Lessons from Bosnia and Iraq. *Review of International Political Economy* 14(3): 444–466.

Jeffrey, A. (2009). Containers of fate: Labelling states in the "war on terror." In Dodds, K., & Ingram, A. (eds), *Spaces of Security and Insecurity: Geographies of the War on Terror*, Aldershot: Ashgate, 43–64.

Jeffrey, A. (2013). *The Improvised State: Sovereignty, Performance and Agency in Dayton Bosnia*. Oxford: Wiley-Blackwell.

Jessop, B. (1990). *State Theory: Putting the Capitalist State in Its Place*. University Park, PA: Penn State University Press.

Jessop, B. (2006). From micro-powers to governmentality: Foucault's work on statehood, state formation, statecraft and state power. *Political Geography* 26(1): 34–40. doi:10.1016/j.polgeo.2006.08.002

Jessop, B. (2005). Critical realism and the strategic-relational approach. *New Formations* 56: 40–53.

Jessop, B. (2008). *State Power*. Cambridge: Polity.

Jessop, B., Brenner, N., & Jones, M. (2008). Theorizing socio-spatial relations. *Environment and Planning D: Society and Space* 26(3): 389–401. doi:10.1068/d9107

Kuus, M., & Agnew, J. (2008). Theorising the state geographically. In Cox, K., Low, M., & Robinson, J. (eds), *Sage Handbook of Political Geography*, London: Sage, 95–106.

Marston, S. (2004). Space, culture, state: Uneven developments in political geography. *Political Geography* 23(1): 1–16.

McConnell, F. (2009). De facto, displaced, tacit: The sovereign articulations of the Tibetan Government-in-Exile. *Political Geography* 28(6): 343–352. doi:10.1016/j.polgeo.2009.04.001

McConnell, F. (forthcoming) *Rehearsing the State: The Governance Practices of the Tibetan Government-in-Exile*. Oxford: Wiley-Blackwell.

Mitchell, T. (1991). The limits of the state – beyond statist approaches and their critics. *American Political Science Review* 85(1): 77–96.

Mitchell, T. (2006). Society, economy and the state effect. In Sharma, A., & Gupta, A. (eds), *The Anthropology of the State: A Reader*, Oxford: Blackwell, 169–186.

Murphy, A.B. (1996). The sovereign state system as political-territorial ideal: Historical and contemporary considerations. In Biersteker, T.J., & Weber, C. (eds), *State Sovereignty as Social Construct*, Cambridge: Cambridge University Press, 81–120.

Navaro-Yashin, Y. (2002). *Faces of the State: Secularism and Public Life in Turkey*. Princeton, NJ: Princeton University Press.

Navaro-Yashin, Y. (2012). *The Make-Believe Space: Affective Geography in a Post War Polity*. Durham, NC: Duke University Press.

Ogborn, M. (1998). The capacities of the state: Charles Davenant and the management of the Excise, 1683–1698. *Journal of Historical Geography* 24(3): 289–312.

Painter, J. (2002). Governmentality and regional economic strategies. In Hillier, J., & Rooksby, E. (eds), *Habitus: A Sense of Place*, Aldershot: Ashgate, 115–139.

Painter, J. (2006). Prosaic geographies of stateness. *Political Geography* 25(7): 752–774.

Poulantzas, N. (1978). *State, Power, Socialism*. London: Verso.

Radcliffe, S. (2001). Imagining the state as space: Territoriality and the formation of the state in Ecuador. In Hansen, T.B., & Stepputat, F. (eds), *States of Imagination: Ethnographic Explorations of the Postcolonial State*, Durham, NC: Duke University Press, 122–145.

Ratzel, F. (1896). The territorial growth of states. *Scottish Geographical Magazine* 12(7): 351–361.

Saxonhouse, A. (1986). Myths and the origins of cities: Reflections on the autochthony theme in Euripides' Ion. In Euben, J.P. (ed.), *Greek Tragedy and Political Theory*, Berkeley, CA: University of California Press, 252–273.

Scott, J.C. (1998). *Seeing Like a State: How Certain Schemes to Improve the Human Condition Have Failed*. New Haven, CT: Yale University Press.

Secor, A. (2001). Towards a Feminist counter-geopolitics: Gender, space and Islamist politics in Istanbul. *Space and Polity* 5(3): 191–211.

Secor, A.J. (2007). Between longing and despair: State, space and subjectivity in Turkey. *Environment and Planning D: Society and Space* 25(1): 33–52.

Sharma, A., & Gupta, A. (2006). Introduction: Rethinking theories of the state in an age of globalisation. In Sharma, A., & Gupta, A. (eds), *The Anthropology of the State: A Reader*, Oxford: Blackwell, 1–42.

Smith, N. (2003). *American Empire: Roosevelt's Geographer and the Prelude to Globalization*. California, CA: University of California Press.

Taussig, M. (1997). *The Magic of the State*. London: Routledge.

Teschke, B. (2003). *The Myth of 1648: Class, Geopolitics and the Making of Modern International elations*. London: Verso.

Tilly, C. (1992). *Coercion, Capital, and European States, AD 990–1992*. Oxford: Blackwell.

Toal, G., & Dahlman, C.T. (2011). *Bosnia Remade: Ethnic Cleansing and Its Reversal*. Oxford: Oxford University Press.

Trouillot, M.-R. (2001). The anthropology of the state in the age of globalization. *Current Anthropology* 42(1): 125–138.

Weber, C. (1998). Performative states. *Millennium: Journal of International Studies* 27: 77–95.

Weber, M. (1958). Politics as a vocation. In Gerth, H.H., & Wright Mills, C. (eds), *Max Weber: Essays in Sociology*, New York: Oxford University Press, 77–128.

Chapter 7

Federalism and Multilevel Governance

Herman van der Wusten

University of Amsterdam, The Netherlands

This chapter is about two organizing frames for understanding the division of authority in large-scale governing institutions: federalism and multilevel governance. An instructive entry for the discussion is the governing saga of the European Union (EU) and its predecessors, that ongoing aberration within the world of states and long-time laboratory for governing experimentation (Murphy 2008; Mamadouh & Van der Wusten 2008).

Started in the early 1950s as a sectoral cooperative venture of six European states, a traditional intergovernmental organization (IGO) with some regulatory innovations, the European Coal and Steel Community morphed into the current European Union: a 28-member governing institution far beyond the traditional IGO profile, but clearly not a state in its own right and still evolving in uncertain directions. In the course of time, steps in these processes have been hailed and bitterly criticized as moves toward a federal United States of Europe. For opponents of the general drift, the F-word was the final destination to avoid at all costs.

Simultaneously, older regionalisms (based on language, religion, and shared histories of common rule) reasserted themselves and new ones (based on the metropolization of ever-growing functionally related urban areas) came to life. These developments were only very partially EU related. However, as the EU engaged with various forms of regional policy, it found partners not only in national states, but particularly also in regional governments. At the same time, the EU adapted to the new governing philosophies of the time by emphasizing the possibilities of governing through the use of market rules and the involvement of private actors. These trends were reflected in the notion of multilevel governance, widely adopted as a new formula to capture the polycentricity and variably patterned spatial reaches of the resulting governing arrangements.

The frames of federalism and multilevel governance are applied in the case of the EU and elsewhere. They have different intellectual pedigrees. Political geographers have not been very active in this field in recent years, and much of the literature in this chapter stems from political science, public administration, and economic geography. I trace the conceptualization

The Wiley Blackwell Companion to Political Geography, First Edition.
Edited by John Agnew, Virginie Mamadouh, Anna J. Secor, and Joanne Sharp.
© 2015 John Wiley & Sons Ltd. Published 2017 by John Wiley & Sons Ltd.

of both terms in the second and third sections of this chapter. The federal idea functions in public discourse and is a presumed clearcut aim to reach or avoid in practical politics. Multilevel governance is too recent a concept to have resulted in reportable life courses of realized projects. The notion has in practical politics notably been embraced by EU officials as a descriptive device and a political ideal (Prodi, president of the Commission in European Commission 2001; Van den Brande, chair of the Committee of the Regions in Hooghe, Marks, & Schakel 2010: IX–XI).

Federalism and multilevel governance were originally developed as different alternatives for unitary states. In research, the two now are increasingly mixed up. The sharp differences between their respective profiles have subsided and hybrids have been recognized. These hybrids also encompass cases of seemingly unitary states that turn out also to have been taking up elements of the two earlier alternatives. There are now widespread tendencies to strengthen multipurpose governing institutions of intermediate (regional) size between the local and the national, and also to erect single-purpose institutions of all possible sizes with varying relations to existing governing complexes. These are the subjects of the fourth section.

The question then arises of how we can account for all these hybrids and spinoffs of federalism and multilevel governance – stronger regional multipurpose authorities and a plethora of single-purpose bodies of all possible sizes. For an answer we look particularly at functional (notably economic) and cultural identity considerations. Both are articulated in claim-making; thus in politics, through various organizational channels, both rest on different bases of legitimacy. Politicization and legitimacy structure the fifth section, followed by the conclusion.

The federal idea and its realizations

Modern political federalism stems from the aftermath of the American Revolution, first resulting in a confederation and then a federation. It had some inspirations and predecessors: the Swiss Confederation and the Dutch Republic of the United Provinces.

The federal idea combines three notions: a set of collectivities with corporate identities that should be preserved and recognized; those collectivities brought under one roof provided with its own corporate identity; the governing institutions of the whole constellation to be based on a constitution. These constitutions contain safeguards for the autonomy of every collectivity in the set and prescribe the governing competences of the overarching institution. They combine self-rule and shared rule elements in a balance maintained by carefully crafted conflict resolution mechanisms, and theoretically only to be altered with the approval of all concerned.

As these three notions can be realized in different shapes, the resulting formats may change over time and can be disputed at any time by opponents defending a "purer" version of the federal idea or defended by supporters proclaiming their realized version as the purest of all. Realizations of the federal idea lack the clear embodiment of sovereignty to be found in the unitary nation-state; the ultimate authority is in the constitution that links different basic entities at two levels. In federations the rule of law and the supporting legal institutions tend to be extra-prestigious governing institutions. This practically implies a separation of powers and for many also a democratic political system. Nonetheless, nominal federations in autocratically governed polities do exist. Federalism, despite its strong orientation to the law, remains an essentially contested concept in the quality of its different versions and even in its demarcation with respect to other systems of government. A recent astute commentator (Burgess 2013) has referred to the penumbra of federalism – one concept shades into another.

As states still are the basic units of the political order, political federations can usefully be distinguished in federations as states (collectivities with corporate identities at federal state level and in multiples below) and federations as a kind of enriched intergovernmental organization consisting of a number of member states plus an overarching institution. In most IGOs the shared rule component is weak, often so weak that an overarching corporate identity does not come alive. They have only a functional permanent link created between states with a technical competence allocated in a single policy sector by means of an international treaty. Constitutionalization of competences at the upper level is absent, no overarching identity emerges, and no federation appears.

In fact, the federal idea has given rise to a range of more or less complete realizations. Elazar (1998: 8) has distinguished eight different political arrangements that he calls federal in nature (Watts's 2013 typology covers roughly the same ground). Some are more marginal than others. Apart from the full-fledged federation (strong self-government constituent units linked within strong but limited overarching government), Elazar has two versions with – in his judgment – relatively weak lower-level government (union – referring to the erstwhile merger of crowns, consociation), three with relatively weak upper-level government (confederation, league, interjurisdictional functional authorities – like the functional IGOs mentioned earlier), and two with a divergent model of upper-level authority (federacy, condominium). The discussion of federalism is generally focused on the distinction of federal and confederal arrangements. This stems obviously from the American experience of the 1780s, but it is also related to the strong urge to determine whether upper and lower levels together form the unit that will be a member of the club of states, or if in the end the overarching institution is deemed too weak and all the lower-level units will separately enter the state system. The real cutoff point between federation and confederation is not easy to determine (the Swiss still officially inhabit a *confederatio helvetica*, even after they provided themselves with – in their own and everybody else's view – a federal constitution in 1848). Even if a little more distant, the other different realizations of the federal idea show family resemblances to these two central versions.

In all but one of these eight federal varieties, the "natural" units of government are territorially circumscribed. The exception is the consociation, where "non-territorial constituent units share power concentrated in common overarching government" (Elazar 1998: 8). The term stems from Althusius, a protestant theologian/lawyer who wrote an influential treatise on politics in the early seventeenth century, opposing the emerging absolutism. Althusius (Carney 1964) considered different, ever larger aggregations of people as the basis of symbiotically living together in a series of consociations. Then sovereignty in the encompassing commonwealth had to be shared by these different levels so that each could live according to its own rules. Laws were the instrument to bring this about. Althusius is considered as an early protagonist of the federal idea. In his book there is no special attention to the territorial nature of the different entities that he has in mind. He refers inside the commonwealth to cities and provinces, but also to families, religious communities, and occupational associations like guilds. Althusius's perspective is related to the principle of subsidiarity (higher-level governments should refrain from doing tasks that lower ones can do better), written in EU law and derived from Catholic social doctrine. There is also a rephrasing of Althusius in the nineteenth-century political Calvinist teaching of "sovereignty for all different social circles" (Kuyper 1880).

The use of "consociational" for specifically non-territorial entities inside a polity in particular is derived from Lijphart (1969: 211–212), who extracted a part of Althusius's analysis for his own purposes. Here – notably in the Netherlands, Belgium, Switzerland, and

Austria, his initial cases – consociationalism refers to the political institutionalization of societies, strongly segmented along cleavage lines of religion, language, or class. Each segment is largely enclosed in its own world of associations backed up, for instance, by public subsidies for schooling proportionate to population strength and united in strong political parties. National politics is run by broad coalitions through an "elite cartel." These different population segments may in principle be scattered across the country; the segments can therefore be called non-territorial. However, this is Elazar's reading, who thought that such non-territorial constructions could not survive more than two generations (Elazar 1998: 207–212). In fact, even if population segments in Lijphart's cases were not completely concentrated in separate territories, they were far from randomly scattered.

Earlier attempts to deal with comparable governing problems concerned the different nationalities under the common roof of the Austrian part of the Austro-Hungarian empire. They lived increasingly in nationally mixed areas due to rural–urban migration. Bauer's (1907) famous analysis of the problem from the perspective of the socialist movement appeared a few years before the collapse of this multinational polity. He proposed a partial system of government to be based on the *Personalitätsprinzip*, the individual affiliation with a nationality giving a person rights in relation to the state administration at all levels (notably as regards education, cultural facilities, and legal matters). The *Personalitätsprinzip* was to be implemented on the basis of semi-voluntary, locally specific official registrations of national affiliation. It was distinguished from and to be added to the *Territorialitätsprinzip*, in which a democratic federation was to be formed with separate territories self-ruled by different nationalities and topped by an overarching government with shared rule. The addition of the *Personalitätsprinzip* was to end growing minority situations in the separate national territories.

In the current collection of 193 states worldwide, about 15 percent call themselves or are considered federations; designations vary. The largest states in terms of land surface nearly all have federal governments, with China as most significant exception. At the suprastate level, efforts to federalize at the global scale via the United Nations have stalled. The European Union has often been described in terms of a mix of federal and confederal elements.

Multilevel governance: The concept

The concept of multilevel governance was born around 1990 to describe new developments within the European Community, soon to become the EU. "Structural policy," aimed at overcoming problems of disadvantaged areas, became a more prominent European concern. Projects should henceforth be administered by partnerships in member states consisting of supranational, national, and subnational actors based on mixed funding. In multilevel governance, "multilevel" referred to the increased interconnection of governing bodies operating at different territorial levels (at higher and lower levels apart from the state level). In his earliest articles on the subject (using the concept in a title for the first time in 1993), Gary Marks drew on the policy networks approach so far used in the context of domestic politics and now added the supranational level (Bache & Flinders 2004: 2–3; Marks 1993). "Governance" referred to a more inclusive and practice-oriented view of governing that also encompassed non-governmental actors, a notion taken up by the World Bank at the time (IEG 2011: 1–11). Within geography, MacLeod (1999) and Boyle (2000) soon thereafter presented elaborate case studies with a similar perspective.

Marks – in collaboration with Hooghe – further developed the multilevel governance concept, making it seemingly much more widely applicable than merely for European studies. Their often cited typology (Hooghe & Marks 2001, 2003) is meant to encompass all concepts of governing that are "radical departures from unitary government" (Hooghe & Marks 2003: 235, 241). The transition from government to governance does not get much attention in this effort. The basic element of analysis is consistently called a "jurisdiction." If there is room for contributions from outside the traditional public sphere of the state, it is to be found in what are called Type II systems. The differences between the two types are described in terms of individual jurisdictions with different characteristics and of systemic variations. I read them here first as separate systems.

Type I multilevel governance has a restricted number of governmental levels (the units operating at a level cover larger or smaller territories referred to as local, regional, national, and so on). These governing units all possess multipurpose decision-making powers and their areas of jurisdiction do not intersect at any level or across levels. Units at smaller partitioned levels fit seamlessly into the larger units of the next degree of partition.

Type II systems are formed by the separate producers of single public goods. They serve common "customers" in geographical arenas (either territorially delimited or in partial networks), but individual providers can be active beyond a specific arena or only serve in parts. Their domains of activity may well intersect. While the units of Type II systems do not split into levels, the arenas of Type II roughly coincide with the jurisdictions in Type I (e.g., utilities, education provided at local or regional level; climate change mitigation, arms limitation at global level). In fact, Type II systems are perceived as consuming markets.

Both types result in polycentricity. Apparently, the polycentricity of Type I, stretching across levels, will tend to provoke questions of hierarchy and control, while Type II may evolve in all directions driven by supplier strategies and market demand. Hooghe and Marks indicate a number of additional differences between Types I and II. Type I systems are based on (imagined) communities, while Type II systems are founded on policy problems. Type I systems are much more sticky, more difficult to change, while Type II systems are nimble and flexible. Conflict resolution in Type I is more through the use of "voice" (in Hirschman's terms), while in Type II systems "exit" is more common. The structural problems in both systems concern coordination among units. Both are designed to diminish such problems: in Type I through compressing the number of autonomous actors (clearly internal coordination problems are thus created, but perhaps not to the same extent), in Type II through limiting interaction by the splicing of competences (this can obviously merely mitigate the unavoidably arising coordination problems).

Three further features of the typology are indicated. The choices to be made in the design of units are similar with regard to national government, government below and above that level. They come back in political arenas of different sizes. This does not mean that the results in terms of frequencies of both types and their mutual relations are necessarily the same, however. In fact, the two types are complementary and therefore they tend to coexist in some way. It is suggested that this complementarity normally shows Type II to be embedded in Type I systems. This opens the question of whether it is really fruitful to consider Type II as a system of a different kind, or perhaps more as a complementary way to add single, separate governing institutions to an established scaffold provided by Type I. With regard to Type I systems, the question has increasingly arisen since the typology was first published of to what extent they are really different in kind from unitary government systems. The perspective has changed in terms of a growing appreciation of gradual transitions and piecemeal alterations.

The introduction of the concept of multilevel governance in European studies has given new inspiration to the ongoing debate concerning the nature of the process of European integration (George 2004). The specification of the two types of governance leaves much to be desired, however. In addition, as one critical contribution on multilevel governance complains, the concept of "[g]overnance is drastically conceptually underspecified" (Welch & Kennedy-Pipe 2004: 141). Nonetheless, the terminology has been in use very broadly and has assisted in sensitizing research to the loosening coupling of territory, autonomy, and sovereignty in the state system (Caporaso 2000), the overriding polycentric nature of all governing arrangements, and the wide variety in types of polycentricity.

Regional authorities and governance

In recent years federalist and multilevel governance frames in the literature have been repositioned, encompassing on the one hand a growing intertwining of both and a renewed interest in the meso level of states, the regional authorities; and on the other, a further elaboration of Type II systems, or better their units, within the frame of multilateral governance. These are focused on the light, flexible regulatory devices, more informal and partly private, that have emerged, thereby giving more substance to the idea of governance in the array of governing institutions as a whole and frequently transgressing state boundaries, notably by way of networks.

In organizing the *Handbook of Regionalism and Federalism*, the general editors suggested that their contributors use the concept of hybridity as a possibly illuminating way to organize their material, based on the apparent breakdown of the earlier categorical distinction between unitary and federal state models (see for further clarification Loughlin 2013). This perceptual change is also reflected in the relevant scientific journals in the field.

Publius started in 1970. It was established by Daniel Elazar, long-standing student and vocal supporter of federalism. The journal was named after the pseudonym collectively used by the three authors of *The Federalist Papers*. It was meant to be a platform for the discussion of federal issues, but is now also open to papers that conceive of the federalist perspective as too restricted, not sufficiently taking into account the gradual transitions of federal and unitary government (Hooghe & Marks 2013: 179–204). That argument was made by the main protagonists of the multilevel governance frame. This also suggested a less sharp contrast between unitary types of government and multilevel governance than earlier suggested.

Regional and Federal Studies started in 1990 as *Regional Politics and Policy*, initiated by researchers in Ireland on both sides of the border from the fields of political science and public administration (among whom was John Loughlin) with an obvious interest in regional questions. In 1995 it was renamed under the impact of the new rush of integration in Europe and its federalizing potential.

Finally, in 2013 the Regional Studies Association, a global forum on regional and urban governance, launched *Territory, Governance, Politics*, edited by John Agnew, which brings together the renovated classic interest of geography in space, place, and territory with the sphere of politics and a widened sense of governing. The journal encourages contributions from different disciplinary origins. In one of the first issues there is an exploration of new single-purpose regulatory bodies with private actors and informal rules breaking out of the confines of the territorial state system as we know it (Sassen 2013: 21–45), thus once more underlining the trend toward Type II units.

In Hooghe, Marks, and Schakel (2010), a new Regional Authority Index (RAI) is introduced to identify the presence and measure the "weight" of one or more layers of regional government and of special cases of autonomous regions. Scores for 42 democracies and semi-democracies during 1950–2006 have been determined annually. They encompass 35 European states plus Turkey and the Russian Federation, several only incorporated after they had transitioned from highly authoritarian regimes; and also Australia, Canada, Japan, New Zealand, and the United States.

The study is constructed on the basis of Type I governing models in nation-states without any reference to non-state actors. The RAI consists of four components indicating aspects of self-rule and four components of shared rule (thus following the federal perspective). The components of both dimensions are tightly related, with the indexes of self-rule and shared rule a little more loosely coupled.

The 42 countries studied show a large variation in terms of absent, weak, and strong regional authorities in 1950; a variation that is still present in 2006. There is a general trend toward more and more substantive regional authorities over time. In 29 countries regional authority increased, in 11 the RAI remained stable, in 2 countries it decreased. In a number of cases this came about as the result of either the installing of new levels of regional authority (15 times) or the granting of special autonomous status to a specific region (14 regions in 8 countries). The overall increase in RAI can also be expressed in the changes of scores on the eight components of regional authority resulting from reforms. All in all there were 394 cases of such changes due to reforms, in 86 percent increasing regional authority (Hooghe et al. 2010: 52–61).

How can we account for these differences? In the 11 countries where RAI remained stable, this was either due to their smallness (regional authority was lacking; it would clearly be superfluous and inefficient) or to an apparent ceiling effect where a high RAI had already been achieved at the start of the period. In the 29 countries where the RAI increased, this was supposedly the fruit of two basic logics: efficiency in the provision of public goods and the expression of communal identities. The two cases of declining RAIs were singular anomalies.

The main task is to elucidate further the factors that cause the general increase in RAI. Efficiency is the driver that pushes the organization of the provision of public goods in certain molds of different size. The proposed argument here is that while the need for military capabilities strengthened the national level for a long time, the expanding portfolio of the welfare state encouraged the lower levels of state government, in particular the regional one. The resulting functional regions tend to be more or less equal in size, with more or less equal RAIs across the national territory ("symmetrical regionalization"). This provokes a question about the timing of the shift. High growth in the RAI notably occurred in the 1970s, 1980s, and 1990s, when the high tide of welfare state expansion had subsided and trimming efforts started. An alternative explanation foregrounds the shift in dominant beliefs in economic policy inducing further support for decentralization of authority to accommodate intranational diversity, and possibly in fact resulting in increased regional disparities (Rodriguez-Pose & Gill 2004).

The accommodation of communal identities, itself a long-time concern, was supposedly facilitated during this period by two different processes: Democratization lowered the threshold for this type of reform and the emergence of a European governing level encouraged stronger regional government. Many of the 42 countries studied had realized civil and political rights and the concomitant political institutions of democracy long before the 1970s, but democratization can perhaps also be read as the renewed rise of movement politics from the end of the 1960s. In addition, there was a major shift toward less authoritarian polities from the 1970s and this clearly enabled the institution of regional authorities.

European integration stimulated regional government in various ways. The regulation of the market at the European level excluded a priori the possibility for eventual new regional governments to erect new internal trading barriers, thus implying a risk reduction of regionalization for national governments; more intense competition at regional level through open European markets induced more authoritative governments at that level to exploit their comparative advantages; the European Commission strengthened regional government through its structural funds/cohesion policies; and the emergence of a European governing level induced emulation by regional hopefuls. This last bundle of arguments obviously recaptures the early stages of the conceptualization of multilevel governance.

The logic of the expression of communal identities gives rise to different forms of self-rule as the efficiency logic. Sizes of units as well as their levels of authority (asymmetric regionalization) may well vary. Communal identities do not respect minimal thresholds and can therefore, depending on power relations, also be realized in small countries. Communal self-rule tends to be concentrated at one level for each case and not to be split among several levels as in the efficiency logic.

Although self-rule and shared rule tend to move in the same direction, there are differences. Three-quarters of all recorded shifts have been in forms of self-rule. Shared rule progresses in a sequence of rare leaps considered as fundamental transformations; self-rule increase is a far more incremental process. The most frequent growth component in self-rule during this period has been in the field of representation (elected regional assemblies). According to Hooghe et al. (2010: 62), it shows the strengthening of a liberal democratic norm: the obligation to have elected representatives in the case of general-purpose jurisdictions.

In the meantime there has been a tremendous increase in the number of individual governing units shaped like those of the proposed Type II system in multilevel governance, but not necessarily part of that kind of system, not necessarily jurisdictions, and at least partly outside the sphere of the state, thus underlining the transition in governance. A shift is underway in modes of regulatory institutionalization aimed at global issues. Since 1990 the increase in IGOs has diminished, while the growth in private transnational organizations (PTOs) has accelerated (Abbott, Green, & Keohane 2013). In terms of organizational ecology, the available space for regulatory activity is now more frequently filled by PTOs than by IGOs as a result of differences in organizational character and strategic flexibility. In climate change mitigation, PTOs effectively build and use networks with large companies to induce common regulation in order, for instance, to influence the graphs of carbon dioxide emissions. This is considered part of an emerging "global administrative space" that is being explored by legal scholars (Kingsbury, Krisch, & Stewart 2005). These initiatives do not merely operate at the international level, they are also visible in traditionally domestic policy domains. In Latin America a host of independent regulatory agencies has been formed to accommodate apparent regulatory voids in fields such as waste disposal and health care, strongly encouraged and financed by the World Bank.

These regulatory institutions of single-sector public goods eschew attribution to existing governing levels in any orderly fashion. They can link up with all sorts of other regulatory mechanisms and definitely transgress strict distinctions in a domestic field where constitution and derived public law reign supreme, and in an international sphere regulated by treaty-based international law, for example in the field of international investment law and arbitration (Vadi & Gruszczynski 2013), where the domestic legal space in particular is unavoidably compromised. EU practice has for decades made deep inroads into the binary of domestic and international law, but this happened under the legal eye of the Court of Justice in Luxembourg.

In the new world of global administrative space, regulatory practices take shape without an overarching legal basis, yet are firmly founded. Eventually, this is where an integral but obviously highly differentiated Type II system with a more distinctive governance character might emerge.

From a geographer's perspective, it is interesting that deeply geographical terms like ecology and space are used in this connection to indicate metaphorically the field that political scientists and legal scholars are exploring. Geographers might like to enter this emerging field by mapping these spatial metaphors onto the real world. Are the carrying capacities for regulatory organization evenly distributed across the world, can the presumed niche positions be indicated in real-world terms, or are their functional traits completely independent from the actual locations of organizations?

Politicization and legitimacy

The formation of a governmental system is the result of political decision-making. Many decisions date from the past, but the recently installed more weighty layers of government at the regional level of nation-states have been the result of political decisions made since 1950. The constellation that gave rise to the concept of multilevel governance, and the Type II governance units that came with it, were enabled by political decisions in the European Community during the late 1980s and the encompassing ideology of market-based governance redesign supported by leading policy-makers. Such political decision-making is preceded by politicization (mobilization and deliberation). Once implemented, the new governance structure gains a level of legitimacy and in its turn induces actors to review their substantive positions and organization in light of the new circumstances.

The recent growth of regional authority has been largely an incremental process. According to Michael Keating (2013), this process of rescaling government is largely driven by regional parties and, once established, regional governments. Strong regional identities play a major role in some cases, but there is also strong territorialization at the meso level based on interest groups forming territorial policy communities. They engage in continuous competitive regionalism. Comprehensive reform is difficult and a real shift in number of levels and boundaries of units nearly impossible. Key issues in the relevant policy processes are symbolic recognition of the importance of the meso level, tighter definition of competences, and redistribution of fiscal powers. There is no stable end in sight, it is all part of "normal" politics, and no final, federal arrangements can be expected. "Normal" politics implies a high degree of legitimacy for the multilevel political arena in which it occurs.

Keating (2013) also emphasizes that the importance of territorial policy communities is often based on material interests, intercompany networks, and specialized concentrations of knowledge, but adds the existence of strong regional identities as a forceful incentive for politicization. Agnew (2013) has summarized the literature on new economic regions and insisted on the necessity of politicization in order to have them emerge as more than ephemeral entities. For this politicization to happen broadly, strong regional identities (not necessarily in existence for ever, but with sufficiently persuasive past references to have a chance to emerge) and the organizational vehicles to express them seem indispensable.

Politicization of cultural identities translated into national sentiment has very frequently disturbed existing political borderlines. Oriented to existing states, it has produced unsettling diaspora movements; otherwise, it has also resulted in complicated issues to redraw existing borderlines, occasionally involving a number of existing states – once labeled the "Macedonian syndrome" (Weiner 1970) – or adding new and long-disputed ones, as in the Irish case with that extra border written with a capital letter. Moynihan (1994) memorably characterized the problem in general as pandaemonium. However, it does not have to go that way. Politicized cultural identities translated into national, ethnic, or regional movements can be accommodated as part of "normal" politics. This is Keating's verdict for the current politicization of the meso level in Europe. The growing preponderance of that meso level (which still may include a number of formats) would then also be the result of economic processes articulating that level of aggregation (Brenner 2004), which would in their turn stimulate the emergence of policy networks, further encouraged by the development of the EU as a multilevel polity.

Keating's conclusion on the general moderation of mobilized ethnic sentiment in Europe's current meso-level constellation is underwritten by Schrijver's (2006) study on the political and cultural consequences of the introduction of more autonomous regional governments in Galicia, Brittany, and Wales in recent decades. This shows the measured consequences of increased autonomy for regionalist sentiment. It does not decline in any of these settings, but certainly also does not invigorate widespread secession-oriented nationalism.

Various efforts have been made to enumerate the general conditions that accommodate ethnonational claims for autonomy in variants of federalism and result in stable governing institutions over the longer term. McGarry and O'Leary (2007) consider democratic institutions and the rule of law not so much as definitional traits, but as necessary conditions for successful multinational federations. In addition (and significantly related), high levels of prosperity help, as do a voluntary agreement among all participants at the start, a generous and prompt offer of self-rule in case a unitary state transforms into a federation, practices known from consociationalism like large majority coalitions and decisions, and proportional representation electoral systems. For a number of completed, stable federations in democracies (Austria, Belgium, Canada, Germany, Switzerland), Erk (2008) maintains that their different versions of federalism largely reflect their diverse ethnolinguistic structures, and their composition of cultural identities.

Recent papers concentrate on performance and parties' adaptive capacities in a situation of ongoing decentralization of governing. Hepburn and Detterbeck (2013: 77) provide a general overview mentioning the shifting context in which existing state-wide parties have to operate and the large recent increase of "stateless nationalist and regionalist parties" of at least some significance in Western Europe: from 29 to 93 during the period 1982–2009. Hopkin and Van Houten (2009) edited a special issue of *Party Politics* on the relations of state-wide operating parties and the decentralization of government. The contributions focus on party organization, party positions on decentralization, and voter alignments. In their introduction, Hopkin and Van Houten underline the sequential nature of the parties' situation: first as participants in legislating and then as actors that need to redefine their role in the new constellation of decentralized government (Hopkin & Van Houten 2009: 131). One contribution (Hopkin 2009) deals specifically with that sequence in comparing the British Labour Party and Spanish PSOE (Socialist Workers' Party) in their shifting appreciation of decentralization and laborious adaptation to the new governing constellation. In the longer term, statewide operating parties tend increasingly to reflect the forms of decentralized governments in which they are active, but the differences in historical organizational formats across political families remain visible (Thorlakson 2009). As to turnout and voter alignments (Jeffery & Hough 2009), elections for regional authorities tend to follow the model of second-order elections developed for European elections (low turnout, losses for the nationally governing party/ies), but there is extra support for parties based on regional cultural identities, while in the regions where this applies, not only national governing parties but all national parties lose in these regional elections.

The Type II governance institutions that are now increasing in number grow under completely different conditions of politicization and legitimacy. They are often the fruit of expert knowledge, brought together in personal networks and designed by professional organizers, occasionally supported by targeted popular campaigns (about the inherent tensions in that combination, see Choudry & Kapoor 2013). There is no basis of shared encompassing cultural attachment. Large-scale politicization is therefore piecemeal and, if it exists at all, targeted and temporal. Most politicization is personal, informal, and lobbying in style among stakeholders. Legitimacy is a huge problem. It must either be based on accepted claims of specialized expertise, or it should be part of a more generally recognized global administrative space after other specialists (legal scholars) have drawn its contours and recognized the claims under this label as a new extension of the rule of law. However, that could only be the mere beginning of a more general recognition and legitimation process (Peters & Pierre 2004; Verkuil 2007).

Conclusion

Polycentricity in governing systems has so far either been studied in the context of intrastate relations, federal in contrast to unitary; or of interstate relations, different patterns with the hyperpower image of the United States around the millennium as an extreme contrast. As increasing transnational interaction and growing interdepencies have diminished state autonomy and undercut the sharpness of the distinction between intra- and interstate spheres, the discussion of polycentricity in governing stands in a changed perspective.

In this chapter I have discussed two frames for considering matters of polycentricity, federalism, and multilevel governance, taking into account the state and other possibly emergent polities like the EU. Particularly in connection with the EU, "federation" has become an uncertain signpost referring to a seemingly obvious future or to a sharply delineated dystopia, ignoring all the nuances and variations that the federal idea has in store. The concept of multilevel governance was born in the context of the EU when a suprastate political level and a renewed regional level seemed to challenge the preponderance of the member states to some extent, and extrastate institutions appeared to be incorporated in the governing architecture.

It turned out that the federal and multilevel governance literatures have recently become increasingly intertwined, giving rise to a renewed interest in the incremental formation of regional authorities within states. On the other hand, there has also been increasing attention to new governing devices, often more lightly institutionalized, frequently outside the traditional spheres of domestic public law and international law, and transgressing state borders in different ways.

The newly strengthened regional authorities in democracies were the fruit of continuous political conflict and thus came about as the result of waves of politicization. These were to be read as "normal" politics, not as a repeated attempt to overturn the existing political system. In that way, politicization underlined the legitimacy of the polity. In the case of new regulatory devices, often in a still rudimentary global administrative space, their politicization was mostly limited to restricted circles of specialists and their legitimacy was perhaps often assumed as a result of claimed expertise, as well as being studied by legal scholars. Yet, it often leaves room for concern in terms of the accountability and transparency that are thought to be necessary mainstays of the rule of law.

References

Abbott, K.W., Green, J.F., & Keohane, R.O. (2013). Organizational ecology and organizational strategies in world politics. *Discussion paper 13-57*, The Harvard Project on Climate Agreements, 1–35. Cambridge, MA: Harvard Kennedy School.

Agnew, J. (2013). The "new federalism" and the politics of the regional question. In Kincaid, J., & Swenden, W. (eds), *Routledge Handbook of Regionalism and Federalism*, New York: Routledge, 130–139.

Bache, I., & Flinders, M. (2004). Themes and issues in multi-level governance. In Bache, I., & Flinders, M. (eds), *Multi-level Governance*, Oxford: Oxford University Press, 1–11.

Bauer, O. (1907). *Die Nationalitätenfrage und die Sozialdemokratie*. Vienna: Verlag der Wiener Volksbuchhandlung Ignaz Brand.

Boyle, M. (2000). Euro-regionalism and struggles over scales of governance. *Political Geography* 19: 737–769.

Brenner, N. (2004). *New State Spaces: Urban Governance and the Rescaling of Statehood*. New York: Oxford University Press.

Burgess, M. (2013). The penumbra of federalism: A conceptual reappraisal of federalism, federation, confederation and federal political systems. In Kincaid, J., & Swenden, W. (eds), *Routledge Handbook of Regionalism and Federalism*, New York: Routledge, 45–60.

Caporaso, J. (2000). Changes in the Westphalian order: Territory, public authority, and sovereignty. *International Studies Review* 2(2): 1–28.

Carney, F.S. (ed.) (1964). *The Politics of Johannes Althusius: Abridged Translation of Althusius' Main Text of 1614*. Boston, MA: Beacon Press.

Choudry, A., & Kapoor, D. (eds) (2013). *NGOization: Complicity, Contradictions and Prospects*. London: Zed Books.

Elazar, D. (1998). *Constitutionalizing Globalization: The Postmodern Revival of Confederal Arrangements*. Lanham, MD: Rowman and Littlefield.

Erk, J. (2008). *Explaining Federalism: State, Society and Congruence in Austria, Belgium, Canada, Germany and Switzerland*. London: Routledge.

European Commission (2001). *Enhancing Democracy. A White Paper on Governance in the European Union*. Brussels: European Commission.

George, S. (2004). Multi-level governance and the European Union. In Bache, I., & Flinders, M. (eds), *Multi-level Governance*, Oxford: Oxford University Press, 107–126.

Hepburn, E., & Detterbeck, K. (2013). Federalism, regionalism and party politics. In Loughlin, J., Kincaid, J., & Swenden, W. (eds), *Routledge Handbook of Regionalism and Federalism*, New York: Routledge, 76–92.

Hooghe, L., & Marks, G. (2001). Types of multi-level governance. *European Integration Online Papers* 5(11), http://eiop.or.at/eiop/texte/2001-011a.htm, accessed 15 March 2015.

Hooghe, L., & Marks, G. (2003). Unraveling the central state, but how? Types of multi-level governance. *American Political Science Review* 97(2): 233–243.

Hooghe, L., & Marks, G. (2013). Beyond federalism: Estimating and explaining the territorial structure of government. *Publius: The Journal of Federalism* 43(2): 179–204.

Hooghe, L., Marks, G., & Schakel, A.H. (2010). *The Rise of Regional Authority: A Comparative Study of 42 Democracies*. London: Routledge.

Hopkin, J. (2009). Party matters. Devolution and party politics in Britain and Spain. *Party Politics* 15(2): 179–198.

Hopkin, J., & van Houten, P. (2009). Introduction special issue: Decentralization and state-wide parties. *Party Politics* 15(2): 131–135.

IEG (Independent Evaluation Group) (2011). *World Bank Country-Level Engagement on Governance and Anticorruption: An Evaluation of the 2007 Strategy and Implementation Plan*. Washington, DC: Independent Evaluation Group/World Bank Group.

Jeffery, C., & Hough, D. (2009). Understanding post-devolution elections in Scotland and Wales in comparative perspective. *Party Politics* 15(2): 219–240. doi:10.1177/1354068808099982

Keating, M. (2013). *Rescaling the European State: The Making of Territory and the Rise of the Meso.* Oxford: Oxford University Press.

Kingsbury, B., Krisch, N., & Stewart, R.B. (2005). The emergence of global administrative law. *Law and Contemporary Problems* 68: 15–61.

Kuyper, A. (1880). *Souvereiniteit in eigen kring.* Rede ter inwijding van de Vrije Universiteit. Amsterdam: J.H. Kruyt.

Lijphart, A. (1969). Consociational democracy. *World Politics* 21(2): 207–225.

Loughlin, J. (2013). Reconfiguring the nation-state. Hybridity versus uniformity. In Loughlin, J., Kincaid, J., & Swenden, W. (eds), *Routledge Handbook of Regionalism and Federalism*, New York: Routledge, 3–18.

MacLeod, G. (1999). Place, politics and "scale dependence": Exploring the structuration of euro-regionalism. *European Urban and Regional Studies* 6(3): 231–253.

Mamadouh, V., & van der Wusten, H. (2008). The European level in EU governance: Territory, authority and trans-scalar networks. *GeoJournal* 72: 19–31.

Marks, G. (1993). Structural policy and multilevel governance in the EC. In Cafruny, A., & Rosenthal, G. (eds), *The State of the European Community*, Boulder, CO: Lynne Rienner, 391–410.

McGarry, J., & O'Leary, B. (2007). Federation and managing nations. In Burgess, M., & Pinder, J. (eds), *Multinational Federations*, Abingdon: Routledge, 180–211.

Moynihan, D. (1994). *Pandaemonium: Ethnicity in International Politics.* New York: Oxford University Press.

Murphy, A. (2008). Rethinking multilevel governance in a changing European Union: Why metageography and territoriality matter. *GeoJournal* 72: 7–18.

Peters, G., & Pierre, J. (2004). Multilevel governance and democracy: A Faustian bargain? In Bache, I., & Flinders, M. (eds), *Multi-level Governance*, Oxford: Oxford University Press, 75–89.

Rodriguez-Pose, A., & Gill, N. (2004). Is there a global link between regional disparities and global devolution? *Environment and Planning A* 36: 2097–2117.

Sassen, S. (2013). When territory deborders territoriality. *Territory, Politics, Governance* 1(1): 21–45.

Schrijver, F. (2006). Regionalism after Regionalisation: Spain, France and the United Kingdom. PhD thesis. Amsterdam: Vossiuspers.

Thorlakson, L. (2009). Patterns of party integration, influence and autonomy in seven federations. *Party Politics* 15(2): 157–177.

Vadi, V., & Gruszczynski, L. (2013). Standards of review in international investment law and arbitration: Multilevel governance and the commonweal. *Journal of International Economic Law* 16(3): 613–633. doi: 10.1093/jiel/jgt022

Verkuil, P.R. (2007). *Outsourcing Sovereignty: Why Privatization of Government Functions Threatens Democracy and What We Can Do about It.* Cambridge: Cambridge University Press.

Watts, R. (2013). Typologies of federalism. In Loughlin, J., Kincaid, J., & Swenden, W. (eds), *Routledge Handbook of Regionalism and Federalism*, New York: Routledge, 19–33.

Weiner, M. (1970). The Macedonian syndrome: An historical model of international relations and political development. *World Politics* 23: 665–683.

Welch, S., & Kennedy-Pipe, C. (2004). Multi level governance and international relations. In Bache, I., & Flinders, M. (eds), *Multi-level Governance*, Oxford: Oxford University Press, 127–144.

Chapter 8

Geographies of Conflict

Clionadh Raleigh

University of Sussex, England, UK

This chapter reflects on conflict research that takes a geographical perspective, both within and outside the discipline of geography and the subarea of political geography. Geographers bring a distinct interpretation and framing of violence to these discussions: In particular, they alternate the scale, mode, measures, and interpretation of conflict to produce a body of work that largely interrogates how conflict and contestation are produced within places, and how these, in turn, produce dynamics within space.

The role of geography within the study of conflict is increasingly popular, and "space" is taken seriously as an essential component of understanding conflict. Geographers contribute to distinct literatures on the spatial patterns of conflict and to studies of conflict and peace that interrogate risk and vulnerability structures from local to global scales. In this chapter, how geography and geographers engage with political conflict is discussed in reference to these two seemingly separate "factions" of research, noted hereafter as "conflicted" and "contested environments" literatures, respectively. These "factions" and associated communities of scholars have divergent agendas, audiences, and perceptions on how geography is conceptualized and problematized.

A number of central debates in conflict studies rest on the fact that war is waged in and over space, and that the qualities of locations and settings for social interaction influence conflict outcomes. Yet, despite the interest in the spatiality of violence across social science disciplines, geographers are rarely involved in debates concerning geography in conflict studies. Reflecting bluntly on this gap, Kofron (2012) claims that "geographers largely ignore" space and conflict; the vacuum has been filled by others. This chapter suggests that the rather large literature on the geography of war is separated by theoretical, methodological, and conceptual differences across several research factions. The existing disconnects in the focus, tenor, and implications of geography in conflict research are partially a function of disciplinary institutionalism, methodologies, or the limitations of alternative analytical paradigms. Different vocabularies, narratives, and concepts are central to the existing subfields; in

The Wiley Blackwell Companion to Political Geography, First Edition.
Edited by John Agnew, Virginie Mamadouh, Anna J. Secor, and Joanne Sharp.
© 2015 John Wiley & Sons Ltd. Published 2017 by John Wiley & Sons Ltd.

particular, the work on conflicted environments by multiple disciplines often takes a quantitative approach, and employs spatially disaggregated research to model the diffusion of war, the likelihood of violence, conflict determinants, and the dynamics of heterogeneous identities in space. The contested environment approach develops poststructuralist and anthropological narratives on hegemonic dominance within domestic politics, the banality of violence, the "body," neoliberalism, neocolonialism, migration, or other consequences of violence (see the 2009 special issue on conflict by the *Annals of the Association of American Geographers*). Hence, studies that may be deeply geographical (e.g., in a geography journal like *Political Geography*) may not fall within the radius of respective conflict researchers because of differences in goals, terminology, and styles. In short, researchers working on the geographies of war and peace are not reading each other, and therefore not building bridges within political geography.

This chapter examines the contributions of several strands of geographical literature on conflict, acknowledging the divide between perspectives and arguing that the disparate research factions are best served through developing tenets of political geography, including theories on the state, governmentality and territoriality in conflict environments, risk production, and collective action.

In the following sections, the areas of research on political violence in the conflicted and contested environments research is reviewed, each revealing how differently "geography" has been applied to the study of conflict.

Conflicted environments

The conflicted environment literature aims to discern patterns and trends, capture trigger mechanisms, and support or refute causal theories across conflicts. It is firmly located within the field of conflict studies and its association with international relations.[1] Key spatial themes in the conflicted environment literature include: Why do some areas see conflict while others are peaceful? How are groups motivated to rebel? What governance practices encourage or discourage conflict? The use of geography within these studies is designed to capture how the "where" contributes to the "why." Places and hotspots can suggest the relative reasons for conflict over unaffected areas, and spatial movement as a proxy for strategy and imperatives. The patterns of conflict themselves tell us about the underlying "surfaces" of social forces.

There are three general ways in which political violence research has incorporated geographical variables, aspects, and theory into the larger studies of conflict environments: Geography can be treated as the object of study, its subject, or the context.

Geography as object

This literature uses geographical areas as containers through which to compare the occurrence and attributes of political violence, including potential causes and correlates. Conflict patterns are revealed through disaggregated and "event"-based data collections on international conflict or domestic political violence (e.g., the ACLED of Raleigh et al. 2010). The insights from this literature have reversed several long-held assumptions about violent locations. Hegre, Østby, and Raleigh (2009) find that the poorest areas are less violent than wealthier areas; Buhaug (2006), Buhaug, Gates, and Lujala (2009), and Raleigh and Hegre (2009) argue that in poor and poorly connected states, political violence tends to cluster in strategic and target areas (e.g., large towns and cities; areas with high road mass and dense populations) where opposing forces can openly contest each other. "Location" literature also queries the differences between urban and rural sites of violence: Conflicts have become increasingly based in and around urban locations (Buhaug & Urdal 2013), possibly due to population movements in developing states. This shift also indicates that new spaces are the result of new agendas, actors (e.g., gangs, militias, etc.), and modes through which citizens experience political violence (Rodgers 2009).

Researchers ask whether political violence tends to differ within active locations. For example, Boyle (2009) observes how violence against civilians is clustered both at particular time and spatial points within civil wars, revealing a pattern suggesting that the level of competition between government and opposition groups, and intra-opposition hierarchies, structures dynamics. Verpooten (2012) observes the spatial distribution of excess mortality during the Rwandan genocide; Do and Iyer (2010) analyze the spread of conflict in Nepal, and the different location attributes that characterizes the violence; onset and diffusion factors indicate that a war evolves for different reasons than its onset, and little is organized, regimented, or expected from its initial to terminal stages. Of late, the "environmental security" literature has probed different forms of conflict as being more or less likely due to climate shifts (see the special issue of the *Journal of Peace Research* 2012).

The more empirically sophisticated aspects of this subfield look at diffusion and contagion in violence, observing whether theories of conflict patterns, interventions, and terminations stand up to scrutiny once observed across real time. Several diffusion studies in Russia (O'Loughlin, Holland, & Witmer 2011; O'Loughlin & Witmer 2012), Afghanistan, and Pakistan (O'Loughlin, Witmer, & Linke 2010) suggest how new technologies, national and international geopolitics influence local patterns of violence significantly. Schutte and Weidmann (2011) find that conflict escalation is largely driven by diffusion, indicating that interventions or changes in conflict must occur at particular stages of violence if they are to be effective.

Geography as subject

A related field of analysis interrogates how spatial factors and relations can explain political violence. Scale and space, physical factors, clustering, and networks[2] are frequent subjects of analysis, where the approach is generally to observe events through different spatial prisms (across scales of action or stages), or as "contained" by particular – often static – environmental attributes.

A particularly robust debate across this literature is the role of physical factors and attributes of areas, whether resources (see Ross 2004; Le Billon 2008; Korf 2011); terrain (Rustad et al. 2008; Raleigh, 2010), or environmental change as measured by changes in rainfall, temperature, land yields, and so on. The environmental security literature is rapidly advancing to include multiple scales, physical threats, and forms of violence, but it remains contested and open to charges of environmental determinism (see Raleigh, Linke, & O'Loughlin 2014). Some examples of subnational, time-varying work that generally refute a direct connection between environmental change and conflict include Hendrix and Glaser (2007), Buhaug (2010), and Raleigh and Kniveton (2012).

Others have used scale and space more directly. Chi and Flint (2013) observe interstate conflict for how different stages and periods, locations, and themes can be represented by their spatiality. Rustad et al. (2011) consider global temporal and spatial variation in risk, continuities, and change across scales of violence. On a more micro level, Mesev, Shirlow, and Downs (2009) review the geography of conflict and death in Belfast. Whether in case studies or large empirical work, a significant body of literature exists across scales and spaces to emphasize what is generalizable and what is unique about conflicts.

A related field attempts to associate why certain groups/states experience violence over others with their spatial characteristics. The "clustering and networks" approach questions how collective action is helped or hindered by territorial locations, ethnic geographies, and physical positioning. Collective action is required for organized political violence, but not all groups have the ability to organize collectively for political change through violence (despite the motivation to do so). Therefore, it is vital that we understand which conditions allow for particular groups to recruit and operate violently. Such studies are particularly important in locations where ethno-regional identities are the basis of political action and agency (see Posner 2004; Mozaffer & Scarritt 2005; Chabal & Daloz 1999). Weidmann (2009) studies how group geography, including the concentration of a social group over a territory and size in physical space, influences the capacity to fight and organize collective action given the right political and economic conditions. In this vein, Bhavnani and Choi (2012) account for the civil violence in Afghanistan through ethnic geography; and Radil, Flint, and Chi (2013) extend identity and actions into the interstate domain through which power and alliance networks, and embeddedness within the international system, can explain the sequencing of states joining wars. This is extended in the concept of ConflictSpace (Flint et al. 2009) that again uses spatial embeddedness and state networks to observe agency within internationalized wars. Across Africa, Raleigh (2014) investigates how local power structures, national–local relationships, and elite competition affect the production and shape the practices of political violence.

Employing geographical concepts

Geographical concepts are often employed to explain the spatiality, dynamics, and patterns of conflict. Here, political geography tenets are most obvious, and often suggest that conflict is not something irrational or abnormal, separate from "normal" life, but is a reflection of these processes (Duffield 1998). Furthermore, violence takes multiple forms depending on which community is engaging in it, and on how groups exist and operate under often inconsistent and possibly illegitimate power.

In line with the growing interest in how physical factors affect conflict risk, political ecology has been forwarded as a theoretical frame through which to interpret different underlying vulnerability patterns across a population (see *Journal of Peace Research* special issue 2012). The political ecology framework is extended to resources more generally, and to the structural inequalities of economies built on resource management and wealth (see Le Billon & Cevantes 2009). In focusing directly on jungle environments and their appropriation by states through physical force, rhetoric, and discourse, Peluso and Vandergeest (2011) provide an appropriate comparison between how geographers frame physical entities as representing – both physically and theoretically – contested spaces in war. In general, political ecology has much to offer conflict studies by highlighting and deepening how political narratives engage, shape, and frame physical terrains, livelihoods, and resources.

Reconceptualizing the state in, and as a cause of, conflict is a consistent theme in geography-motivated literature. Within this research strand, the state is presented as the sum of its parts, and "governance geography" is investigated as well as representative institutions. A thorough treatment comes from Catherine Boone (2003), who applies political geography/political economy concepts to studies of subnational power distribution in Africa to understand the "political topography" of states. This approach on the uneven production of state power is increasingly popular, fueled by constantly evolving "relations of control and consent, power and authority" (Munro 1996: 148). Raeymaekers (2010, 2012) is also active in applying a political geography lens to ongoing intractable African crises and the shaping and reshaping of states at their margins.

Weberian state concepts are frequently used as a straw man in these debates, and with them the poverty of the "failed state" concept. Hagmann and Péclard (2011) and Ferguson (2006) both consider the forms and site of power and authority that emerge at the intersection of state–society/state–international levels, and how power in many conflict-affected states is a function of ongoing negotiation between elite players on multiple governance levels. Relatedly, Korf, Engeler, and Hagmann (2010), in acknowledging the different, competing systems of power and authority that emerge in any conflict environment, probe how the agency of civilians, elites, conflict contestants, and so on in civil wars is formed within dynamic conflicts. They link this view of conflict spaces to the possibility of multiple narratives and Watts's "governable orders" (2004). This discussion reinforces the reality that power is alternately built and performed in multiple ways. In applying this concept of pervasive "micro politics" in conflict environments, Hasbullah and Korf (2013) consider the specifically "Muslim geographies," violence, and communities within Sri Lanka.

Additional studies situate governance and conflict within the hostile politics of identity and territorial contests and containers. Conflict is a common outcome and producer of hostile group interactions, particularly in spaces where ethno-regional and religious identities are "politically privileged" in that they structure political position access and roles over other forms of allegiance. Central Africa has generated multiple discussions about identity formation in conflict (Chrétien & Banégas 2011; Fujii 2009). Prunier's (2009) masterful discussion of Africa's world war considers the micro, meso, national, and transnational politics of one of the most conflicted areas on the globe, where the production of ethnic antagonism is embedded in accounts of conflict strategy, institutional incompetence, narratives of greed, and marginalization. Cederman, Weidmann, and Gleditsch (2011) test how identity underscores political positioning and promotes collective marginalization, which creates the spatial patterns of political exclusion of groups used as a motivating narrative for conflict. Kalyvas (2005) employs a "micro-logic" to interrogate the likelihood of targeted violence against civilians within civil war "zones" defined by distance and the nature of events therein. This work is perhaps a direct advance on Robert McColl's original research on the "territorial bases of revolution" (1969) in an attempt to understand the spatial imperative of fighting forces to reach their respective conflict-specific goals.

Contested environments

The contested environments literature presents a critical conceptual framework of conflict and political violence within geography. This research is considerably different in perspective from previously reviewed work, as its definitions of political violence, main research questions, and themes are concerned with exposing power differentials within societies, and conceptualizing the ways in which conflict creates and reinforces inequality and vulnerability. This work can be characterized by a rather narrow interpretation of political conflict, but a very wide interpretation of "structural" violence and the ways in which political power is manifest through other media. Across this work is a questioning of the accepted "containers" of power – primarily the state – and the actions and aims of those who wield power within these. In particular, a focus is on how the consequences of war-making and political conflict are creating norms around force, logics of violence, and invisible and visible victims.

Furthermore, this literature seeks to uncover the global, hegemonic, political, and economic relationships and networks that create and sustain political violence as it occurs, primarily across the global South. The work often interrogates the hegemonic narrative and action in conflict, and how "landscapes of violence" are produced. Within it, common tropes include "everydayness," "banal interactions with power," and power's "subversive and violent agendas." Two bodies of work are examined extensively as they represent how the field has embraced the subject matter, and the manner in which conflict is examined. These include *Violent Geographies: Fear, Terror and Political Violence* (Gregory & Pred 2007) and an *Annals of American Geographers* special issue dedicated to conflict in 2009.

Hegemonic narrative and actions

The more recent research (Flint 2005; Gregory & Pred 2007) often focuses on the security environment post–9/11 and the US/hegemonic/neocolonial codes that structure the new security narratives; this same theme is repeated in Dalby (2007), Katz (2007), Pred (2007), Glassman (2007), and Watts (2007). The research revolves around the projection of power and the intended/unintended consequences of violence for localities as a result of hegemonic competition and strategy. Wider geopolitical strategies are responsible for "terrorizing" poorer, frequently developing locations (Oslender 2007). Other contributions include expanding on the geopolitical state of other (often "rogue") states and detailing the ways in which the United States is responsible for creating its own enemies: Mercille and Jones (2009) use "radical geopolitics" to explicate the logics of power in Iran and the "crisis" as a product of American interest.

Conflict is thus presented as both a product of the neoliberal order and a producer of inequalities and marginalization; as examples, acts of political violence are presented as a new form of colonialism (Gregory 2010b on Afghanistan), or militant Islam is associated with neoliberal changes (Watts 2007), despite other lines of argument claiming that it is a manifestation of local power disparities and elite competition within states (Dowd 2015; Basedau & Pfeiffer 2014) and emerging from multiple contests for local power (Dowd & Raleigh 2013). Others expand on conflict's indirect manifestations: Hyndman (2003) details how conflict structures masculine and feminine forms; Coleman (2009) and Hyndman and Mountz (2007) consider the ways in which people are excluded in real and legal terms from space or belonging as a result of conflict.

Landscapes of violence

There is a robust discussion in the contested environments literature of "narratives of violence" and their reflection within landscapes (Ross 2011). Gregory (2010b) describes and codifies two modern dominant modes of conflict – one as surgical, sensitive, and scrupulous; and the second, "populist" form as indiscriminate, callous, and predatory. Violent landscapes are created from "removed" or "embedded" conflict, where spaces of advanced military violence create an abstract and alien space above legal consideration, whereas the other, more localized and brutal form is shaped by countering these landscapes. Both are "waged by the global North," the former presumably with local partnerships. In this type of analysis within the contested environment research, conflict is removed from the domestic arena in which it arose, and situated as a reaction to hegemonic forces. This is coupled with a seeming reluctance to address domestic politics and governance outside of militarily aggressive states. From within this framework, conflict is not borne of marginalization, strategy, greed, or local competition. Instead, many developing countries' political systems, choices, and actions are reduced to reactions, as are citizen responses to the state. The critical lens of hegemonic power practices tends to view action only as subservient responses, and, in doing so, strips wars, soldiers, governments, and victims of their local, national, and regional politics, their agency, attitudes, and choices to engage in and fight within conflicts.

By dissecting the spatiality, experiences, and realities of power through the state, research by Kearns (2006), Secor (2007), Gregory (2007), and others serves to present a theoretical frame by which to examine the influence of violence on its space and citizens, while implicitly questioning the notion of the state itself. Expanding on the points of state action and threats, Holmqvist (2012) interrogates three narratives of war (policing, risk management, and biopolitical empire) through their spatiality in flux; in the case of Lebanon, Fregonese (2012) accounts for the hybrid sovereignties that are in play despite the narrative of weak statehood; Harker (2011) describes the geographies of dealing with power and occupation in Palestine; and Lunstrum (2009) considers the local interpretations of violence and territory of "landscapes of terror and unmaking state power" within the Mozambique civil war. Here, conflict is examined as a component of multiple processes, including migration, identity, unjust social relations (Cowen & Gilbert 2007), and social change (Herb 2005).

Critical reflections and bridging divides

Albeit from starkly different factions, the combined literature using geography as a lens or means through which to understand war and peace is extensive. Approaches differ on the conceptualization and problematizing of the role of the state, the overall definition of political violence and war, and the goal of research. While the conflicted environment camp often begins with an acceptance of the state, its territory and sovereignty, the contested environment camp actively questions these assumptions, problematizing state capacity, governance, regimes, and domestic politics.

A further point of difference between conflicted and contested literatures is in the fundamental grounds of inquiry. As demonstrated by John Agnew's review piece on the geography of war and peace in the *Annals of the Association of American Geographers*' special issue (2009), there is a tendency across conflicted environment research to see war as an "institutionalized phenomenon, organized bureaucratically, and not simply scaled up interpersonal aggression"; in short, it is a specialized and separate social interaction from other forms of force. Yet, in the contested environment literature, conflict has personal, legal, social, transgenerational, economic, moral, and political impacts that transcend scales and identities.

Conflicted environments work is often guilty of limited understandings of socially constructed power, place, and position, and too much emphasis placed on the influence of "static" geographies (Peluso & Vadergeest 2011). Recent work on the links between climate and conflict tends to recycle the tenets of environmental determinism (see Raleigh et al. 2014 for a discussion) and to use "variable" and reductive measures of geometry (e.g., distance) as a replacement for geography. Of particular concern are the ways in which groups and identities are treated in studies of conflict: The agency of violent actors engaged in conflict is reduced to tactical moves, and there exists a tendency for "groupism" (Brubaker 1996), where researchers take groups as "basic" and "discrete," internally homogenous and externally bounded. An equally problematic perspective is from those who present developing states and their citizens as following a codified response to conflict, without reference to the ways in which state action, regime responses, and the domestic practices of power give rise to reactions and force from below. Unfortunately, both conflicted and contested research suffers from a weak understanding of domestic politics in conflict-prone states, and instead relies on remote academic frameworks, a frequent lack of regional knowledge, and an absence of comparative frameworks and counterfactuals. With few exceptions, the work on governance patterns, mechanisms, and possible consequences of conflict has been the purview of social scientists with an interest in geographical concepts, as opposed to political geographers (see Herbst 2000; Boone 2003; Posner 2004, 2007; Kalyvas 2006).

Many of these critiques are valid, but they can be based on a misunderstanding of the modes of research. In particular, quantitative work has embraced ever more sophisticated empirical methods to discern patterns, differences, trends, clusters, impacts, and so on. These bring geography into the study of conflict in a direct, usable, and innovative manner, and answer questions about how landscapes of peace and violence emerge, atrocities occur, and peace-building and development initiatives work. Yet, it is often true that "ordinary people are often written out of the story" (Korf 2011). Individual or "place" scales of study are often not the intent of research on the intersections of geography and conflict, nor perhaps useful in the search for patterns, generalizable triggers, mechanisms, and spaces of conflict. Specifically, conflict environment studies are not developed to "deeply" or "thickly" explain individual acts or sites of violence, but rather to identify similarities in mechanisms or contexts that can provide insights into variation and landscapes of conflict. They often purposefully take a removed perspective, and researchers are aware of the implications of this framework, as embedded researchers surely are of the problems of their methodological and epistemological choices.

One of the main differences in the conflicted and contested environment literatures is the way in which research and advocacy are posited as goals. Kobayashi (2009: 819) notes that a common theme across geographers' work is a "commitment to play a role in creating a world based on peace, not war." This is presented as an advance from the recent past, where "few spoke out to support a role for the discipline in achieving peace and most of the discipline evinced at best a morally neutral, functionalist perspective" (2009: 819). Because of such reliance on and adherence to a critical/radical/geopolitical frame, contested environments can be accused of a fundamental bias in how conflict and violence are approached. In particular, the "advocacy as research" discussion within geography can stifle innovation, producing intradisciplinary elitism and leading to low levels of cross-pollination with other disciplines. Such a perspective on peace (versus war) and the role of geography is presented by Megoran (2011), who lambasts the discipline for being overly focused on war, going so far as to accuse geographers engaged in the study of conflict as lacking a "commitment to peace." Megoran advocates a study of peace equal to the focus on conflict, yet Ross (2011) counters that peace-making and peace-keeping involve distinct forms of subjugation and repression. Holland (2011) registers his concern for the "study of peace" and advocacy as implicitly rejecting the goal of unbiased and empirically rigorous work.

It could also be argued that using the goal of peace advocacy to dismiss the work of others is a disturbing development in geography literature. Quantitative empirical studies and conflict environment insights are reduced to results from a "military-industrial-academic complex," a phrase used by Barnes and Farish (2006) in this regard, cited in Kobayashi (2009). Again, Kobayashi summarizes the internal debate and divide within the discipline over the use of geographical knowledge and skills in public policy and actions, especially in regard to military action:

> the range of opinions among geographers internationally over this precarious relationship surely represents one of the most urgent discourses in the history of our discipline. We expect continued controversy as the result over the role of geographical technologies, the place of *geographers* in formulating public policy, the relationship between *geographers* and the state, and the potential for *geographers* to contribute to peace. (Kobayashi 2009: 821, emphasis added)

Like Gregory and Pred (2007), Crampton, Roberts, and Poorthuis (2014) worry about the use of predictive spatial analysis in geographical intelligence, its new "political economy," and the "violent surveillant state" that it supports. Mason (2013: 298) refers to the "spatial analytics of securitization" that reinforce processes of military-political domination, and ignore "vertical" forms of power. Outside of the analytical excoriation, the "positive peace" literature implies that the study of conflict through detailing the characteristics of states, institutions, landscapes, economics, and societies "hides an economic liberalization agenda that links peace to capitalist development" (Koopman 2011: 194). The implication is that researchers are unaware of how their work can underscore and support narratives that reinforce power hierarchies and subjugation. However, determining the causal processes and spatiality of conflict is pivotal for acting on them, and quantitative spatial research is intended to confirm such pathways to peace empirically.

More importantly for the field and discipline, each researcher and research agenda should complement wider academic and public aims, and be accessible to other geographers and those interested in violence. There is a middle ground between quantified, seemingly reductive external interpretations of violence and idiosyncratic applications of deep theory. For our purposes, the link is a renewed dedication to the tenets of political geography: Contributions could be substantially improved through more and deeper investigations into the "topography" of state power (Boone 2003); a nuanced understanding of developing "statehood" (see Hagmann & Péclard 2010); collective action by territorially defined actors (Weidmann 2013), political ecology, and the transnational behavior of groups. Our ongoing dilemma can be solved by a reinvestment in the very roots of our discipline.

In conclusion, although considerable work is being produced on the "geography of war," it is divided into two factions that rarely reference or learn from each other. A crude distinction might suggest that non-geographers who use concepts of space and political geography within their discipline-specific approaches undertake much of the conflicted environment work, and that the contested environment literature is borne of a poststructuralist tradition firmly rooted in political geography. Seeing past the disciplinary differences would benefit both factions of research; turning a blind eye to the other's work reinforces harmful divisions that limit the study of political violence in general. The ramifications of this divide between conflicted and contested environment studies are quite serious for the direction and future of the geography of conflict: Non-geographers have led the charge in applying geographical concepts to modern conflict studies, and will be central to changing the interpretation and application of those concepts without particularly loud protest from geographers.

Notes

1 Conflict studies is a growing field that emanated from international relations and increasingly concentrates on domestic conflict and its determinants. The field is dominated by research from political science and economics, and is subject to noisy debates on conflict categorization, temporal trends, and measurement issues, regardless of the specific topical area of study (e.g., combatant reintegration, elections, or climate change). A focus on domestic conflict has pushed this field into the study of developing state institutional structures, political regimes, economic development, identity formation, and physical landscapes.

2 Diffusion studies (Zhukov 2012) are designed to measure how signals of conflict are related to proximity or networks.

References

Agnew, J. (2009). Killing for cause? Geographies of war and peace. *Annals of the Association of American Geographers* 99(5): 1054–1061.

Barnes, T.J., & Farish, M. (2006). Between regions: Science, militarism, and American geography from world war to Cold War. *Annals of the Association of American Geographers* 96: 807–826. doi:10.1111/j.1467-8306.2006.00516.x

Basedau, M., & Pfeiffer, B. (2014). Bad religion? Religion, collective action, and the onset of armed conflict in developing countries. *Journal of Conflict Resolution*, online first July 23. doi:10.1177/0022002714541853

Bhavnani, R., & Choi, H.J. (2012). Modelling civil violence in Afghanistan: Ethnic geography, control, and collaboration. *Complexity* 17: 42–51. doi:10.1002/cplx.21399

Boone, C. (2003). *Political Topographies of the African State: Territorial Authority and Institutional Choice*. Cambridge: Cambridge University Press.

Boyle, M. (2009). Bargaining, fear and denial: Explaining violence against civilians in Iraq 2004–2007. *Terrorism and Political Violence* 21(2): 261–287. doi:10.1080/09546550902765565

Brubaker, R. (1996). *Nationalism Reframed: Nationhood and the National Question on the New Europe*. Cambridge: Cambridge University Press.

Buhaug, H. (2006). Relative capability and rebel objective in civil war. *Journal of Peace Research* 43: 691–708. doi:10.1177/0022343306069255

Buhaug, H. (2010). Dude, where's my conflict? LSG, relative strength, and the location of civil war. *Conflict Management and Peace Science* 27(2): 107–128. doi:10.1177/0738894209343974

Buhaug, H., & Urdal, H. (2013). An urbanization bomb? Population growth and social disorder in cities. *Global Environmental Change* 23: 1–10. doi:10.1016/j.gloenvcha.2012.10.016

Buhaug, H., Gates, S., & Lujala, P. (2009). Geography, rebel capability, and the duration of civil conflict. *Journal of Conflict Resolution* 53: 544–556. doi:10.1177/0022002709336457

Cederman, L.-E., Weidmann, N.B., & Gleditsch, K.S. (2011). Horizontal inequalities and ethno-nationalist civil war: A global comparison. *American Political Science Review* 105: 478–495.

Chabal, P., & Daloz, J.-P. (1999). *Africa Works: Disorder as Political Instrument*. Oxford: International African Institute.

Chi, S.-H., & Flint, C. (2013). Standing different ground: The spatial heterogeneity of territorial dispute. *Geojournal* 31: 160–199. doi:10.1007/s10708-012-9451-0

Chrétien, J.-P., & Banégas, R. (eds) (2008). *The Recurring Great Lakes Crisis: Identity, Violence and Power*. London: C. Hurst & Co.

Chretien, J.P., & Banegas, R. (2011). *The Recurring Great Lakes Crisis: Identity, Violence, Power*. New York: Colombia University Press.

Coleman, M. (2009). What counts as the politics and practice of security, and where? Devolution and immigrant insecurity after 9-11. *Annals of the Association of American Geographers* 99(5): 904–913.

Cowen, D., & Gilbert, E. (2007). Citizenship in the "homeland": Families at war. In Cowen, D., & Gilbert, E. (eds), *War, Citizenship, Territory*, London: Routledge, 261–280.

Crampton, J.W., Roberts, S.M., & Poorthuis, A. (2014). The new political economy of geographical intelligence. *Annals of the Association of American Geographers* 104: 196–214.

Dalby, S. (2007). The Pentagon's new imperial map. In Gregory, D., & Pred, A. (eds), *Violent Geographies: Fear, Terror and Political Violence*, New York: Routledge, 295–209.

Do, Q.-T., & Iyer, L. (2010). Geography, poverty and conflict in Nepal. *Journal of Peace Research* 47: 735–748. doi:10.1177/0022343310386175

Dowd, C. (2015). Islamist violence determinants across Africa. *Political Geography, forthcoming*.

Dowd, C., & Raleigh, C. (2013). The myth of global Islamic terrorism and local conflict in Mali and the Sahel, *African Affairs* 112(448): 498–509. doi:10.1093/afraf/adt039

Duffield, M. (1998). Post-modern conflict: Warlords, post-adjustment states and private protection. *Civil Wars* 1(1): 65–102.

Ferguson, J. (2006). *Global Shadows: Africa in the Neoliberal World Order*. Durham, NC: Duke University Press.

Flint, C. (ed.) (2005). *The Geography of War and Peace: From Death Camps to Diplomats*, New York: Oxford University Press.

Flint, C., Diehl, P., Scheffran, J., Vasquez, J., & Chi, S.-H. (2009). Conceptualising ConflictSpace: Toward a geography of relational power and embeddedness in the analysis of interstate conflict. *Annals of the Association of American Geographers* 99: 827–835. doi:10.1080/00045600903253312

Fregonese, S. (2012). Beyond the "weak state": Hybrid sovereignties in Beirut. *Environment and Planning D: Society and Space* 30: 655–674. doi:10.1068/d11410

Fujii, L. A. (2009). *Killing Neighbors: Webs of Violence in Rwanda*. Ithaca, NY: Cornell University Press.

Glassman, J. (2007). Imperialism imposed and invited: The War on Terror comes to Southeast Asia. In Gregory, D., & Pred, A. (eds), *Violent Geographies: Fear, Terror and Political Violence*, New York: Routledge, 93–109.

Gregory, D. (2007). Vanishing points: Law, violence and exception in the global war prison. In Gregory, D., & Pred, A. (eds), *Violent Geographies: Fear, Terror and Political Violence*, New York: Routledge, 150–165.

Gregory, D. (2010a). The everywhere war. *Geographical Journal* 177(3): 238–250.

Gregory, D. (2010b). War and peace. *Transactions of the Institute of British Geographers* 35: 154–186. doi:10.1111/j.1475-5661.2010.00381.x

Gregory, D., & Pred, A. (2007). Introduction. In Gregory, D., & Pred, A. (eds), *Violent Geographies: Fear, Terror and Political Violence*, New York: Routledge, 1–6.

Hagmann, T., & Péclard, D. (2010). Negotiating statehood: Dynamics of power and domination in Africa. *Development and Change* 41: 539–562. doi:10.1111/j.1467-7660.2010.01656.x

Harker, C. (2011). Geopolitics and family in Palestine. *Geoforum* 42: 306–315. doi:10.1016/j.geoforum.2010.06.007

Hasbullah, S., & Korf, B. (2013). Muslim geographies, violence and the politics of community in eastern Sri Lanka. *Geographical Journal* 179: 32–43. doi:10.1111/j.1475-4959.2012.00470.x

Hegre, H., Østby, G., & Raleigh, C. (2009). Poverty and civil war events: A disaggregated study of Liberia. *Journal of Conflict Resolution* 53: 598–623. doi:10.1177/0022002709336459

Hendrix, C.S., & Glaser, S.M. (2007). Trends and triggers: Climate, climate change and civil conflict in Sub-Saharan Africa. *Political Geography* 26: 695–715. doi:10.1016/j.polgeo.2007.06.006

Herb, G. (2005). The geography of peace movements. In Flint, C. (ed.), *The Geography of War and Peace: From Death Camps to Diplomats*, New York: Oxford University Press, 347–368.

Herbst, J. (2000). *States and Power in Africa: Comparative Lessons in Authority and Control*. Princeton, NJ: Princeton University Press.

Holland, E. (2011). Barack Obama's foreign policy, just war, and the irony of political geography. *Political Geography* 30: 59–60. doi:10.1016/polgeo.2010.02.006

Holmqvist, C. (2012). War/space: Shifting spatialities and the absence of politics in contemporary accounts of war. *Global Crime* 13(4): 219–234. doi:10.1080/17440572.2012.715401

Hyndman, J. (2003). Aid, conflict and migration: The Canadian–Sri Lanka Connection. *Canadian Geographer* 47: 251–268. doi:10.1111/1541-0064.00021

Hyndman, J., & Mountz, A. (2007). Refuge or refusal. In Gregory, D., & Pred, A. (eds), *Violent Geographies: Fear, Terror and Political Violence*, New York: Routledge, 37–54.

Journal of Peace Research (2012). Special Issue on Climate Change and Conflict. 49(1).

Kalyvas, S. (2005). Warfare in civil wars. In Duyvesteyn, I., & Angtrom, J. (eds), *Rethinking the Nature of War*, Abingdon: Frank Cass, 88–108.

Kalyvas, S. (2006). *The Logic of Violence in Civil War*. New York: Cambridge University Press.

Katz, C. (2007). Banal terrorism. In Gregory, D., & Pred, A. (eds), *Violent Geographies: Fear, Terror and Political Violence*, New York: Routledge, 349–362.

Kearns, G. (2006). Bare life, political violence, and the territorial structure of Britain and Ireland. In Gregory, D., & Pred, A. (eds), *Violent Geographies: Fear, Terror and Political Violence*, New York: Routledge, 7–36.

Kobayashi, A. (2009). Geographies of peace and armed conflict: Introduction. *Annals of the Association of American Geographers* 99(5): 819–826. doi:10.1080/00045600903279358

Kofron, J. (2012). Space and war, constants and change from a historical perspective. *Geographie* 117(2): 234–252.

Koopman, S. (2011). Alter-geopolitics: Other securities are happening. *Geoforum* 42: 274–284. doi:10.1016/jgeoforum.2011.01.007

Korf, B. (2011). Resources, violence and the telluric geographies of small wars. *Progress in Human Geography* 35, 733–756. doi:10.1177/0309132510394120

Korf, B., Engeler, M., & Hagmann, T. (2010). The geography of warscape. *Third World Quarterly* 31: 385–399. doi:10.1080/01436597.2010.488466

Le Billon, P. (2008). Diamond wars? Conflict diamonds and geographies of resource wars. *Annals of the Association of American Geographers* 98(2): 345–372.

Le Billon, P., & Cervantes, A. (2009). Oil prices, scarcity, and geographies of war. *Annals of the Association of American Geographers* 99: 836–844.

Lunstrum, E. (2009). Terror, territory, and deterritorialization: Landscapes of terror and the unmaking of state power in the Mozambican "civil" war. *Annals of the Association of American Geographers* 99(5): 884–892.

Mason, M. (2013). Climate change, securitization and the Israeli–Palestinian conflict. *Geographical Journal* 179: 298–308. doi:10.1111/geoj.12007

McColl, R. (1969). The insurgent state: Territorial bases of revolution. *Annals of the Association of American Geographers* 59(4): 613–631.

Megoran, N. (2011). War and peace? An agenda for peace research and practice in geography. *Political Geography* 30: 178–189. doi:10.1016/j.polgeo.2010.12.003

Mercille, J., & Jones, A. (2009). Practicing radical geopolitics: Logics of power in the Iranian missile crisis. *Annals of the Association of American Geographers* 99(5): 856–862.

Mesev, V., Shirlow, P., & Downs, J. (2009). The geography of conflict and death in Belfast, Northern Ireland. *Annals of the American Association of Geographers* 99: 893–903. doi:10.1080/00045600903260556

Mozaffar, S., & Scarritt James, R. (2005). The puzzle of African party systems. *Party Politics* 11(July): 399–421. doi:10.1177/1354068805053210

Munro, W. (1996). Power, peasants, and political development: Reconsidering state construction in Africa. *Comparative Studies in Society and History* 38(1): 112–148.

O'Loughlin, J., & Witmer, F.D.W. (2012). The diffusion of violence in the North Caucasus of Russia, 1999–2010. *Environment and Planning A* 44: 2379–2396. doi:10.1068/a44366

O'Loughlin, J., Holland, E.C., & Witmer, F.D.W. (2011). The changing geography of violence in Russia's North Caucasus, 1999–2011: Regional trends and local dynamics in Dagestan, Ingushetia, and Kabardino-Balkaria. *Eurasian Geography and Economics* 52: 596–630. doi:10.2747/ 1539-7216.52.5.596

O'Loughlin, J., Witmer, F.D.W., & Linke, A.M. (2010). The Afghanistan–Pakistan Wars, 2008–2009: Micro-geographies, conflict diffusion, and clusters of violence. *Eurasian Geography and Economics* 51: 437–471. doi:10.2747/1539-7216.51.4.437

Oslender, U. (2007). Violence in development: The logic of forced displacement on Colombia's Pacific coast. *Development in Practice* 17(6): 752–764.

Peluso, N.L., & Vandergeest, P. (2011). Political ecologies of war and forests: Counterinsurgencies and the making of national natures. *Annals of the Association of American Geographers* 101: 587–608. doi:10.1080/00045608.2011.560064

Posner, D.N. (2004). Measuring ethnic fractionalization in Africa. *American Journal of Political Science* 48: 849–886. doi:10.1111/j.0092-5853.2004.00105.x

Posner, D. (2007). *Institutions and Ethnic Politics in Africa*. Cambridge: Cambridge University Press.

Pred, A. (2007). Situated ignorance and state terrorism: Silences, W.M.D., collective amnesia and the manufacture of fear. In Gregory, D., & Pred, A. (eds), *Violent Geographies: Fear, Terror and Political Violence*, New York: Routledge, 246–267.

Prunier, G. (2009). *From Genocide to Continental War: The "Congolese" Conflict and the Crisis of Contemporary Africa*. London: Hurst.

Radil, S.M., Flint, C., & Chi, S.-H. (2013). A relational geography of war: Actor–context interaction and the spread of World War I. *Annals of the Association of American Geographers* 103: 1468–1484. doi: 10.1080/00045608.2013.832107

Raeymaekers, T. (2010). Protection for sale: War and the transformation of regulation on the Congo–Ugandan border. *Development and Change* 41(4): 563–587. doi:10.1111/j.1467-7660.2010.01655

Raeymaekers, T. (2012). Reshaping the state in its margins: The state, the market and the subaltern on a Central African frontier. *Critique of Anthropology* 32(3): 334–350.

Raleigh, C. (2010). Seeing the forest for the trees: Does physical geography affect a state's conflict risk? *International Interactions* 36: 384–410. doi:10.1080/03050629.2010.524524

Raleigh, C. (2014). Political hierarchies and landscapes of conflict across Africa. *Political Geography* 42: 92–103. doi:10.1016/j.polgeo.2014.07.002

Raleigh, C., & Hegre, H. (2009). Population size, concentration, and civil war: A geographical disaggregated analysis. *Political Geography* 28: 224–238. doi:10.1016/j.polgeo.2009.05.007

Raleigh, C., & Kniveton, D. (2012). Come rain or shine: An analysis of conflict and climate variability in East Africa. *Journal of Peace Research* 49: 51–64. doi:10.1177/0022343311427754

Raleigh, C., Linke, A.M., & O'Loughlin, J. (2014). Extreme temperatures and violence. *Nature Climate Change* 4: 76–77. doi:10.1038/nclimate2101

Raleigh, C., Linke, A.M., Hegre, H. & Karlsen, J. (2010). Introducing ACLED: An Armed Conflict Location and Event Dataset. *Journal of Peace Research* 47: 651–660. doi:10.1177/0022343310378914

Rodgers, D. (2009). Slums wars of the 21st century: Gangs, Mano Dura and the new urban geography of conflict in Central America. *Development and Change* 40: 949–976. doi:10.1111/ j.1467-7660.2009.01590.x

Ross, A. (2011). Geographies of war and the putative peace. *Political Geography* 30: 197–199. doi:10.1016/j.polgeo.2011.04.008

Ross, M.L. (2004). What do we know about natural resources and civil war? *Journal of Peace Research* 41: 337–356. doi:10.1177/0022343304043773

Rustad, S.C.A., Buhaug, H., Falch, A., & Gates, S. (2011). All conflict is local modelling sub-national variation in civil conflict risk. *Conflict Management and Peace Science* 28: 15–40. doi:10.1177/0738894210388122

Rustad, S.C.A., Rød, J.K., Larsen, W., & Gleditsch, N.P. (2008). Foliage and fighting: Forest resources and the onset, duration, and location of civil war. *Political Geography* 27: 761–782. doi:10.1016/j. polgeo.2008.09.004

Schutte, S., & Weidmann, N.B. (2011). Diffusion patterns of violence in civilwars. *Political Geography* 30: 143–152. doi:10.1016/j.polgeo.2011.03.005

Secor, A. (2007). An unrecognizable condition has arrived. In Gregory, D., & Pred, A. (eds), *Violent Geographies: Fear, Terror and Political Violence*, New York: Routledge, 37–54.

Verpoorten, M. (2012). Detecting hidden violence: The spatial distribution of excess mortality in Rwanda. *Political Geography* 31: 44–56. doi:10.1016/j.polgeo.2011.09.004

Watts, M. (2004). Resource curse? Governmentality, oil and power in the Niger Delta, Nigeria. *Geopolitics* 9(1): 50–80. doi:10.1080/14650040412331307832

Watts, M. (2007). Revolutionary Islam. In Gregory, D., & Pred, A. (eds), *Violent Geographies: Fear, Terror and Political Violence*, New York: Routledge, 175–204.

Weidmann, N.B. (2009). Geography as motivation and opportunity. *Journal of Conflict Resolution* 53: 526–543. doi:10.1177/0022002709336456

Weidmann, N.B. (2013). The higher the better? The limits of analytical resolution in conflict event datasets. *Cooperation and Conflict* 48: 567–576. doi:10.1177/0010836713507670

Zhukov, Y.M. (2012). Roads and the diffusion of insurgent violence: The logistics of conflict in Russia's North Caucasus. *Political Geography* 31: 144–156. doi:10.1016/j.polgeo.2011.12.002

Chapter 9

Security

Lauren Martin

University of Oulu, Finland

What is security? The term can refer to a state of being, psychological stability, postconflict peace, the attenuation of threats to life and personal liberty, stable livelihoods, predictable financial futures, physical safety, enforced territorial borders, or emancipation from oppression, but it is a term through which actors seek to create certain kinds of political possibilities (Neocleous 2008; Ingram & Dodds 2009; Zedner 2009; Philo 2012). For state security agencies, achieving security requires surveillance, data analysis, and free rein to inspect, detain, and exclude suspect persons from national territory. For those contesting state security practices, security means defending privacy, ending discriminatory rhetoric, or defining alternative political norms. For those fighting domestic violence, security requires state and community protection and spaces of recovery and autonomy. For those fighting starvation, security means consistent access to affordable sustenance and employment. Depending on the context, security can refer to a feeling, an ontological state of being, a field of policy-making, a technical project, or a source of political legitimacy. Security implies stability, predictability, physical safety, and collective cohesion. It is a technical term and a democratic ideal, a rationale for the suspension of law, and the goal of humanitarian aid. Security is, in short, a floating signifier imbued with the power of possibility and mastery of the future.

The September 11, 2001 attacks on New York City's World Trade Center and Washington, DC's Pentagon seemed to reinvent the world and define an era. The event provoked a remapping of global space, governmental practices, and political discourses among North American and European security agencies, and heightened attention to transnational networks and "failed states." And yet in 2001, interdisciplinary scholarship on human security was legion, reflecting both growing scholarly concern with post–Cold War security politics and a shift from security as geopolitics to security as human development. In the intervening years, many scholars have shown that most security practices periodized as "post-9/11" were proposed, planned, or attempted prior to 2001. Still, the date stands as a political catalyst that opened spaces of action and public–private collaboration previously considered either utopian or

The Wiley Blackwell Companion to Political Geography, First Edition.
Edited by John Agnew, Virginie Mamadouh, Anna J. Secor, and Joanne Sharp.
© 2015 John Wiley & Sons Ltd. Published 2017 by John Wiley & Sons Ltd.

distopian, depending on one's political persuasion. If nothing else, 9/11 retrenched the idea that security belongs to nation-states, to which it recentered policy lenses. New agencies in North America and Europe refocused research funding on security-related issues (Bigo & Jeandesboz 2010), particularly geospatial technologies (Crampton, Roberts, & Poorthius 2014). And so, more than 10 years on from the attacks, political geographers and interdisciplinary security studies scholars have tempered their assertions of the era's novelty, and instead contextualize the raft of executive measures authorizing war, rendition, torture, extrajudicial killings, detention, and mass surveillance of both US citizens and others. Thus, 9/11 may not have reinvented the world, but it enabled an unprecedented expansion of state, private, and supranational security practices, along with an amplification of the insecurities that accompany them.

When the previous *Companion to Political Geography* was published in 2003, "security" was mostly absent from the volume's index and chapter titles. Subheadings including "security" pertained to environmental security (Castree 2003: 429) and sustainable security (Dalby 2003: 450); in both chapters, security refers to the creation of predictable futures, not the governmental discourses, practices, and assemblages described in this chapter. While it was recognized, security was simply one of many geo/political terms that political geographers problematized. The current volume's emphasis on security reflects the rise of security as a research object and a powerful political force. This chapter will review current approaches to security in political geography and related disciplines, concentrating on theoretical and methodological approaches and the sites on which security research has focused. As the sections herein will show, security research is diverse, revealing security's power to insert life-and-death urgency into mundane everyday practices and enable new governmental practices.

Theoretical and methodological approaches to security

While 9/11 marked a clear uptick in research on security, the Cold War framed security studies in the post–Second World War era. The relationship between security and state power has, however, a much longer genealogy, particularly acute in Hobbesian and liberal conceptions of the relationship between liberty and security (Foucault 2010). In brief, classical liberal theorists debated the responsibility of governments to create stable futures for their citizens, futures in which citizens could pursue their self-interests in commerce or labor. This concern to provide security to citizens *as populations* came to define nineteenth-century progressive disciplinary institutions seeking to make proper moral subjects, and eventually twentieth-century social security and welfare state institutions. Following the Second World War, state security efforts focused primarily on *national* security, defined by a bipolar East–West ordering of global power. Cold War security studies understood states as the primary global actors, either as threats to or purveyors of security. Nuclear proliferation and its containment raised the specter of species-wide destruction, and critics of proliferation responded by advocating the creation of an international order, ruled by international laws constraining state actions during conflict. Realist and positivist approaches to security, presuming that states act according to objectively discernible rules of self-interest, benefit maximization, and cost minimization, dominated the field of strategic security studies and international relations (Zedner 2009). During this time, military geography primarily concerned itself with applying geographical methods to military problems, an alliance that met with vehement criticism with the rise of critical geography in the 1970s (Flint 2005; Mamadouh 2005; Woodward 2005). Responding in part to the myopia of bipolar Cold War geopolitics and in part to the increasing importance of non-state actors and ethno-nationalism in post–Cold War conflict, mainstream security studies now includes a wider array of actors, problems, and phenomena under the rubric of security.

Since 2001, however, political geographical research on security has built on critical approaches to geopolitics, international relations, and development. Drawing on the work of Michel Foucault and Jacques Derrida, critical geopolitics extended the critical projects of critical security studies, international relations, and feminist geography to the spatial categories of the international state system (Campbell 1992; Tickner 1992; Dillon 1996; Ó Tuathail 1996; Ó Tuathail & Dalby 1996; Enloe 2000). For example, the categorization of "rogue" and "failed" states divides the world into zones of order and disorder, enabling particular forms of intervention, such as humanitarian missions or preemptive military operations. In humanitarian discourses, these terms link "good governance," development, and security in ways that justify breaching territorial integrity (Duffield 2007; Elden 2009). Diagnosing states' failures to achieve particular forms of biopolitical governance and social control, these categorizations are knowledge practices that rely on intertwined neoliberal visions of globalized economic interdependence and security. "Failed" and "rogue" states are considered incapable of guaranteeing the physical security of citizens and of fostering a suitable economy in which to make a livelihood. Thus, neoliberal development discourses (deregulation, privatization, accountability) feed into geopolitical mappings of order and disorder and justify military interventions.

While traditional approaches to security debate the ability of these categories to explain state behavior, the accuracy of the denominations is not the point. These maps visualize security threats by dividing and categorizing global space according to neoliberal conventions of good governance that enable the creation of different rules for "us" and "them" (Mitchell 2010). These mappings exemplify "geostrategic thinking," a style of political reasoning that visualizes claims about danger, threat, and security through reductive categories (Ó Tuathail 2003). In some contexts, those maps come to serve as both explanations of contemporary events and predictions of future patterns of deviance (Roberts, Secor, & Sparke 2003); in other contexts, refugee narratives of violence serve to justify both their claims for state protection and the state's need to exclude them (Martin 2011). To critique the space–times of these geographical imaginaries, however, we need to understand their impact, power, and circulation, especially their ability to naturalize categories, spatial imaginaries, and interventions. Four approaches – constructivism, performativity, governmentality, and embodiment – have dominated studies of security in political geography and international relations, and each holds different implications for the study of security.

Most famously articulated as "securitization theory" by the Copenhagen School (Buzan, Waever, & Wilde 1998; Buzan & Waever 2003), constructivist approaches argue that people, places, or issues are securitized when elite actors make claims about security through illocutionary speech acts and these claims are accepted by public consensus. Ackleson (2005), for example, argues that pre–9/11 US border practices were securitized through elite constructions of border space as chaotic, and that they were then claimed to be potential avenues for terrorists in post–9/11 border policy-making. By coding the US–Mexico border as a disorderly space, border policy elites successfully translated border enforcement into a matter of national security. In the context of the War on Terror, claims that chaotic, uncontrolled borderlands presented an imminent threat to US national security resulted in massive funding infusions, wall-building projects, technological surveillance, and higher staffing levels between 2003 and 2011.

Taking a *performative* approach to securitization, Bialasiewicz et al. (2007) argue that the elites' claims are more than linguistic speech acts, and that securitization requires more than agreement among elite power networks; certain forms of constructivism – and its critiques – rely on a problematic distinction between linguistic speech acts and representations, on the one hand, and material practices and policy implementation, on the other. That is, both constructivists who focus on the content of policy texts and critics who argue for on-the-ground analysis of policy impacts rely on a problematic separation between an ideal, discursive realm and a real, material realm. Theories of performativity, however, begin from a different understanding of what language is and how it works. Language is not separate from the material world, useful for describing things, places, or situations that exist independently from their representations. Rather, the things we take for granted are called into being through discourse, so that ideal and real, discourse and materiality are inextricably bound up with each other. As Bialasiewicz et al. write, "discourse is then not something that subjects use in order to describe objects; it is that which constitutes both subjects and objects" (2007: 407). From this perspective, security discourses produce the worlds they represent, and these discourses gain currency not because certain actors hold power to stipulate policies, but because they reiterate shared mappings of threat and danger, circulate those shared understandings among a variety of institutions, and rework familiar geopolitical tropes.

Picking up on a wider interdisciplinary effort to understand both the endurance of global inequalities and the flexibility of contemporary political economic relations in terms of *assemblages* (Ong & Collier 2005; Anderson & McFarlane 2011) and *global governmentality* (Larner & Walters 2004), Foucault-inflected approaches ask how imagining future catastrophes and dangerous spaces produces particular modes of governing (Dillon 2007; Martin & Simon 2008; De Goede 2012; Grove 2012; Amoore 2013). Understanding security as a mode of governing, some scholars prefer to focus on an *assemblage* of actors, institutions, embodied practices, texts, narratives, infrastructure, and relations that includes but far exceeds the state. Assemblages are not coherent or closed institutions, but fields of power relations that overlap, conflict, change, and break down. Thinking of security in terms of assemblages rather than institutions or elite actors acknowledges how non-state actors, such as consultants (Amoore 2013), contractors (Crampton et al. 2014), think tanks (Bialasiewicz et al. 2007), surveillance technology firms, international organizations (Andrijasevic & Walters 2010), municipal police (Coleman 2009), photographers (Simon 2014), and formal and informal migrant organizations (Staeheli & Nagel 2008; Gill 2009; Martin 2012) play key roles. As De Goede writes,

> An assemblage is understood to exercise power at multiple sites and through diverse elements that work in conjunction but may also encounter friction … It makes it possible to think about the practices of cultural representation, historically constituted modes of calculation, and novel conjunctions of governmental, business, and technological actors at work. It takes into account diverse actants and spaces: both the risk analysts and the models they build, both the security experts and the imagination they deploy, both the midlevel bureaucrats and the law they exceed. (2012: 28–29)

And in so doing, this approach refuses accounts of intentional individual actors making the world according to policy directives. Working from Foucault and Bourdieu, Didier Bigo's influential research on the securitization of migration argues that the "governmentality of unease" emerges from a field of security professionals with expertise in managing a series of social threats, such as criminals, migrants, and terrorists (Bigo 2002). Thus, legitimizing security professionals' expertise requires the active production of insecurity and a generalized sense of "unease" that can be fixed to new threats as they emerge. Generating trust in security experts, the governmentality of unease widens security professionals' discretionary authority, changing the role of policing and surveillance in social life. These approaches require a different explanatory mode from realist and liberal accounts that explain security policy development in terms of intentional acts and individuals.

What is important is not whether security discourses accurately represent the world, or the gap between policy and practice, or whether security technologies adequately calculate risks and threats. What are important are the interventions, actions, and political rationalities that these knowledge practices enable; that is, what do the geographical imaginations of integration, disconnection, failed states, and gray economies *make possible*? What conditions of possibility do they engender, and how do they work through risk calculations, definitions of safety/threat, alignments of economic governance and national security, visual surveillance technologies, and media representations (Coleman 2005; De Goede 2012; Amoore 2013)? For Amoore, analyzing security policies across seemingly distinct spheres of governance, foreign policy, and economy reveals "resonances" between different practices, an "infiltration of each one into the other, such that a moving complex emerges" (2013: 5). There remains some ambiguity in the literature, however, about how to make sense of these quasi-causal relationships. For some, the discursive production of security and insecurity happens simultaneously; insecurity is the necessary other to security's self-image. For others, insecurity's production emerges from the unintended consequences of security practices, and so insecurity cannot be read off textual and visual representations of security. Thus, theoretical and methodological choices about how and where to study security are important to theorizing its spatial politics.

For those working with people targeted as threats, insecurity is not confined to the populations defined by those discourses. Feminist political geographers have been irreverent toward traditional scales of political geographical analysis, drawing important connections between sites and across scales (Katz 2001; Mountz 2011), employing a "connective" rather than comparative approach to security politics (Staeheli & Kofman 2004; Lee & Pratt 2012). Feminist geopolitical approaches to security have, therefore, anchored their research in communities directly affected by securitization practices or in arenas marginalized by state-focused research (Pain 2014). Picking up on calls to *embody* the state (Mountz 2010), this research analyzes the corporeal, banal, and site-specific effects of security practices and traces out forms of insecurity that cannot be deduced from security policy texts (Fluri 2011). Focusing on everyday experiences of and struggles against marginalization, this work marks out a broader field of political struggle over security. Hiemstra (2012), for example, analyzed the US immigration system by working in Ecuador with families of detained migrants. She connects the emotional, physical, and financial insecurities of Ecuadorian families to specific US immigration detention practices, showing how those practices worked to create insecurity abroad. In her research on in/security in Afghanistan, Fluri (2011) demonstrates how the fusion of civilian development aid and military operations values bodies as worthy of particular forms of security, and that civilians actively negotiate these operations in their own strategies for survival in conflict zones. Staeheli and Nagel (2008) have shown how Arab communities in the United States and United Kingdom developed and mobilized their own concepts of security in response to the insecurities produced by post–9/11 policies. Very often, these responses were based on strategies to protect physical safety, exercise rights, claim formal citizenship, and create new spaces of engagement within and beyond their communities. Examining these political technologies refocuses our attention on the political decisions underlying security policies, and also on the politics that activists enact. Analytically, embodied approaches to security prioritize bodies, everyday mobilities, and practices that state security practices render invisible.

The approaches to security described here are not easily separable, especially where Foucauldian and feminist scholars seek to understand how security's politics are enframed, circulated, embodied, disrupted, and challenged. Studies of state security practices include close analysis of the production of difference and securitization of everyday life that have so concerned feminist scholars. Theorizing security not as the exclusive purview of state agencies or sovereign power but as a diffuse field of power relations allows for rich insights into understanding security's spatial politics. This approach has allowed scholars to trace how security practices meet up with political issues outside traditional boundaries of geopolitical decision-making: economic policy, development programs, climate change policy-making, crime policing, health care, and immigration policy-making. It also means that political geographers do not approach security from a fixed methodological approach, but may choose to ground their research in a policy such as immigration detention (Mountz 2010; Martin 2012) or policing (Varsanyi 2008; Coleman 2009); sites such as borderland conservation areas (Sundberg 2011), islands (Mountz 2011), or specific conflict zones (Fluri 2011); technologies such as body scanning (Adey 2009; Amoore & Hall 2009); practices such as the policing of urban space (Graham 2011; Simon 2012); a political rationale such as preemption (Amoore 2013) or good governance (Duffield 2007); a process such as the development of disaster management policy (Grove 2012); or genealogies of concepts such as terror (Elden 2009). This conceptual flexibility and methodological diversity have yielded important contributions to our understandings of the spatial politics of specific sites, processes, and practices, which I review in the next section.

Sites of security

Migration

Migration and mobility have been key loci for studying the extension of security practices to new policy arenas, populations, and spaces. Migration and asylum have received amplified attention from security agencies since 9/11, but problematizing migration as a security issue began long before 2001 (Huysmans 2000). However, contemporary migration management relies on discourses of chaos and crisis, enabling the proliferation of exceptional practices (Mountz & Hiemstra 2014) and an unprecedented linking of domestic policing and external border control (Bigo 2002; Coleman 2007). The securitization of migration has included both the externalization of migration controls and a devolution of immigration policing within states (Hyndman & Mountz 2008; Varsanyi 2008; Gill 2009). The respatialization of immigration inspections has led many scholars to argue that migration's securitization has respatialized the border, multiplying the possibilities for exclusion and fundamentally changing the relationship between states, subjects, and mobile bodies (Balibar 2003; Amoore 2006; Mountz et al. 2013). Migration's threshold status as domestic/foreign policy creates the conditions of possibility for exceptional practices in banal, bureaucratic procedures (Amoore & De Goede 2008) and for de/reterritorializations of the sovereign right to exclude (Agamben 2005; Coleman 2007; Coutin 2010; Mountz 2011). Thus, migration management's shifting of domestic and foreign policy-making not only reconfigures the relationship between the state and mobile, border-crossing bodies, it is also a key factor in the rescaling of state power. For those scholars interested in normative critiques of security, the expansion and entrenchment of immigration policing bear significant implications for how, where, and through which means we can be political (Mountz 2010; Amoore 2013).

Technologies

Casting mobility as a security problem is aided and justified by new technologies, such as biometric data collection, new forms of financial surveillance, massive banking and mining of data, and algorithmic risk analysis. These practices are, however, increasingly aimed at citizen populations in general, as the 2013 disclosures of American NSA electronic data collection have revealed. For geographers interested in how state knowledge practices "think" about their subjects, security technologies have been a particularly important site for analyzing the changing relationship between state and body. These practices range from immigration raids to airport security (Adey 2009; Amoore & Hall 2009), from the designing of public spaces for surveillance (Simon 2012) to financial data mining (De Goede 2012; Amoore 2013). Airports use a combination of identity verification, prescreening, and visual surveillance to compile a "data double" of passengers as they travel (Haggerty & Ericson 2000; Amoore 2006). These data-based surveillance practices are coupled with airport security officers' visual evaluation of threat and reading of travelers' bodily affective performances (Adey 2009; Martin 2010). Far more extensive, however, is the surveillance of financial transactions. This surveillance of "the banal face of counter-terrorism" (Amoore & De Goede 2008) casts suspicion on networks of association and on cash economies that do not conform to particular surveillance practices' data needs (such as migrant remittances, De Goede 2012). Less studied, however, are the racial and colonial implications of these practices. Casting religious and/or "cultural" practices as suspicious, these security practices rely on and circulate difference as a matter of security intervention (Puar 2007; De Goede 2012), while the biopolitical character of biometrics relies on particular norms of whiteness (Pugliese 2012). As postcolonial historians have long argued, the colonies served as laboratories for many policing, surveillance, disciplinary, and biopolitical knowledge practices that have since been brought "back home" (Mitchell 1991; McCoy 2009), and more remains to be done to understand the colonial legacies of security technologies (Loyd, Mitchelson, & Burridge 2012).

Economy

The relationship between economy and security has been framed in a number of ways, but in political geography and critical security studies it has crystallized around open/closed territorial borders and the development–security nexus. At their core, these two literatures revolve around liberal norms of "balancing" state sovereignty and economic governance. Borders, however, have tended to be a flash point for conversations about the "costs and benefits of globalization" in wealthier, migrant-receiving countries, while development, "good governance," and humanitarian intervention have dominated research in postcolonial, poorer, and migrant-sending regions.

Balancing the need for state security against the need for transnational mobility, state governments increasingly employ programs that differentiate between risky and safe travelers, and between desirable and undesirable labor migrants (Amoore 2006; Sparke 2006). For some commentators, heightened border security in North America, Australia, and Europe represents a contradiction between national security and neoliberal demands for a borderless, deregulated global market (Purcell & Nevins 2005). Others, however, do not see these programs as a compromise between competing demands, but as the enabling of capital's mobility through state regulation (Mezzadra & Nielson 2013). From this perspective, securitizing finance, migration, and borders creates opportunities for the temporary spatial fixation – and expansion – of capital, by creating new "dark pools," precarious working classes, and competitive jurisdictions. For other scholars, tensions between economy and security crystallize around liberal norms of individual liberty and state sovereignty (Wilsher 2011) or civil rights and surveillance (Lyon 2003). For still others, this discursive multitude does not cohere into a stable narrative, but signals a heterogeneous assemblage of storylines, rationalities, knowledge practices, and proposed solutions (Coleman 2005). Those working from Foucault's writings on liberalism and neoliberalism (Foucault 2010) note that liberal security norms mobilize very different rationalities for surveillance and interventions than rationalities based on theories of state sovereignty, authorizing state violence within and beyond state territories in new ways (Duffield 2007; Dillon & Reid 2009; Evans 2010).

Liberal approaches to security-based interventions have been particularly salient in international development discourses that link "human security" to specific governance practices and economic opportunities. Duffield (2007), for example, traces the ways in which security became an overarching rationale for specific forms of international aid and militarized interventions in the domestic affairs of aid-receiving countries. This discursive assemblage links individual livelihoods, food security, and entrepreneurship to state policy-making to hold state governments accountable for the socio-economic welfare of their citizens. Thus, the "development–security" nexus has become a taken-for-granted assumption among aid professionals, with a familiar raft of problems identified as the causes of state failure and justifications for donor interventions. There is a self-serving aspect to these discourses, however, that links donor countries' *domestic security* from terrorism to the governability of urban slums (Graham 2011) and economically disconnected areas (Roberts, Secor, & Sparke 2003; De Goede 2012). Foreign interventions are thereby justified as protecting the vulnerable abroad and the population at home. The merging of humanitarian and security mandates is also apparent in Iraq and Afghanistan, where the US military employs biometrics, property rights, agricultural training, and cultural sensitivity to "win the hearts and minds" of the local population (Gregory 2008; Belcher 2012). These practices produce site-specific forms of corporeal insecurity that are experienced quite differently by civilians and international NGO workers (Fluri 2011). Across these different contexts, development and security discourses "resonate" in their common concern for wellbeing and by separating human life into normative categories requiring specific forms of intervention to achieve – or maintain – that wellbeing. These discourses are highly particular, though universalizing, in their claims about humanity, and work to delegitimize other ways of living and being political.

Environment

In many forms of geopolitical thought, environmental processes have served an important role in explaining resource conflicts, ethnic difference, and economic development patterns. Beginning in the 1970s, however, environmental changes threatened by acid rain, nuclear war, and drought came to be understood as man-made and emergent. The environment was no longer a set of background conditions, but a changing – and increasingly unpredictable – landscape. Recently, however, the relationship between environment and security has connected *climate change* to the framings of migration, development, and risk already described (Dalby 2002; Barnett 2003; Nordås & Gleditsch 2007). Changing climatic conditions increase, then, the precarity of human life and pose threats to global order (Barnett & Adger 2007). Scholars and policy-makers make the link between environment, climate change, and security in a variety of ways, based on different assumptions about who and what is at risk; these different positions also imply very different policy solutions to climate problems (e.g., adaptation, mitigation, social change, McDonald 2013). Marketization, insurance, and financial instruments have been especially important for addressing the unpredictability of climate change (Robertson 2006; Grove 2012; Dalby 2013). Among critical security scholars, there has been much debate about the value of "securitizing" the environment (Critical Approaches to Security in Europe Collective 2006) over fears that national security responses would likely rely on military and border control measures over mitigation and adaptation policies (Barnett 2003). Intensifying exclusionary responses to migration would only further undermine livelihood options – and magnify insecurity – for those most directly affected by climate change.

Conclusion

The sites discussed in this chapter are the most salient in political geography at the moment, but by no means the only ones; biosecurity (Hinchcliffe & Lavau 2013), food security, and cybersecurity offer rich potential for future research. While the work reviewed here reveals an exciting diversity of approaches, I would like to conclude by drawing out some broad agreements between these literatures. First, security implies a predictable or pacified future, in which environmental, social, economic, and political conditions remain (or are made to remain) within an acceptable bandwidth of variation. At the very least, a secure future is a knowable future. Second, securitization produces insecurity, as a category and as a condition of everyday life for communities and populations targeted by security practices. Formulating security, whether for economic development or national security, requires the active imagination, identification, and calculation of threats, risks, and interventions. The will to knowledge underlying security knowledge practices has led to unprecedented data collection of banal activities, so much so that living, working, or sending money through unofficial channels becomes suspect in itself. Third, security knowledge practices are increasingly intertwined with economic practices, such as transparency regulations, accounting, financial risk analysis, market integration, and labor migration policies. Political rationalities vary widely between, for example, financial accounting and labor market policies, but neoliberal governance norms of marketization, competition, and risk analysis do create important resonances between seemingly distinct political spheres. In addition, the will to know and master the future relies on universalizing ontologies that marginalize other ways of being, knowing, and relating. Examining how apparently apolitical economic governance practices encode racial, ethnic, and religious difference as suspicious remains a rich but understudied element of security research.

Fourth, in/security invokes a sense of existential threat and impending disaster that constricts political deliberation in favor of technocratic solutions. This means that security practices can simultaneously produce political marginalization and close down spaces in which that marginalization can be contested. These exceptional politics have long been a subject of international law and geopolitics, but the imbrication of those spheres with domestic policing and the proliferation of surveillance actively reconfigure the space and time of the political. Liberal and neoliberal governance norms define security in ways that blur domestic and foreign policy, so that biopolitical population management is a standard component of international policy-making and discretionary, extrajudicial practices are a standard part of domestic policy-making. For minorities, postcolonies, and occupied peoples, these are not new revelations, and the historical geographies of exceptional securitized spaces remain a rich, but understudied, field for political geographers (Sharp 2011). Much work remains to be done to examine security politics beyond the liberal democratic states of North America, the European Union, and Australia, as security has also been a key political discourse of authoritarian, socialist, and colonial regimes. Regardless, political geography's emphasis on spatial politics, governmentality, and geographical imaginations will contribute to geographers' understandings of space and interdisciplinary understandings of security. Security mobilizes political norms, produces difference, circulates discourses of threat to human livelihood, and provokes the imagination of multiple futures, threats, and desires. An understanding of security's specific – and multiple – power relations will continue to be one of political geography's key themes.

References

Ackleson, J. (2005). Constructing security on the U.S.–Mexico Border. *Political Geography* 24(2): 165–184. doi:10.1016/j.polgeo.2004.09.017

Adey, P. (2009). Facing airport security: Affect, biopolitics, and the preemptive securitisation of the mobile body. *Environment and Planning D: Society and Space* 27(2): 274–295. doi:10.1068/d0208

Agamben, G. (2005). *State of Exception*. Chicago, IL: University of Chicago Press.

Amoore, L. (2006). Biometric borders: Governing mobilities in the War on Terror. *Political Geography* 25(3): 336–351. doi:10.1016/j.polgeo.2006.02.001

Amoore, L. (2013). *The Politics of Possibility*. Durham, NC: Duke University Press.

Amoore, L., & De Goede, M. (2008). Transactions after 9/11: The banal face of the preemptive strike. *Transactions of the Institute of British Geographers* 33: 173–185. doi:10.1111/j.1475-5661.2008.00291.x

Amoore, L., & Hall, A. (2009). Taking people apart: Digitised dissection and the body at the border. *Environment and Planning D: Society and Space* 27(3): 444–464. doi:10.1068/d1208

Anderson, B., & McFarlane, C. (2011). Assemblage and geography. *Area* 43(2): 124–127. doi:10.1111/j.1475-4762.2011.01004.x

Andrijasevic, R., & Walters, W. (2010). The international organization for migration and the international government of borders. *Environment and Planning D: Society and Space* 28(6): 977–999. doi: 10.1068/d1509

Balibar, É. (2003). *We, the People of Europe? Reflections on Transnational Citizenship*. Princeton, NJ: Princeton University Press.

Barnett, J. (2003). Security and climate change. *Global Environmental Change* 13(1): 7–17. doi:10.1016/S0959-3780(02)00080-8

Barnett, J., & Adger, W.N. (2007). Climate change, human security and violent conflict. *Political Geography* 26(6): 639–655. doi:10.1016/j.polgeo.2007.03.003 DOI:10.1016/j.polgeo.2007.03.003#doilink

Belcher, O. (2012). The best-laid schemes: Postcolonialism, military social science, and the making of US counterinsurgency doctrine, 1947–2009. *Antipode* 44(1): 258–263. doi:10.1111/j.1467-8330.2011.00948.x

Bialasiewicz, L., Campbell, D., Elden, S., Graham, S., Jeffrey, A., & Williams, A.J. (2007). Performing security: The imaginative geographies of current US strategy. *Political Geography* 26(4): 405–422. doi:10.1016/j.polgeo.2006.12.002 DOI:10.1016/j.polgeo.2006.12.002#doilink

Bigo, D. (2002). Security and immigration: Toward a critique of the governmentality of unease. *Alternatives: Global, Local, Political* 27: 63–92. doi:10.1177/03043754020270S105

Bigo, D., & Jeandesboz, J. (2010). The EU and the European security industry: Questioning the "public–private dialogue." INEX Policy Brief No. 5. Centre for European Policy Studies. http://aei.pitt.edu/14989/1/INEX_PB5_e-version.pdf, accessed January 4, 2014.

Buzan, B., & Wæver, O. (2003). *Regions and Powers: The Structure of International Security*. Cambridge: Cambridge University Press.

Buzan, B., Wæver, O., & de Wilde, J. (1998). *Security: A New Framework for Analysis*. Boulder, CO: Lynne Rienner.

Campbell, D. (1992). *Writing Security: United States Foreign Policy and the Politics of Identity*. Minneapolis, MN: University of Minnesota Press.

Castree, N. (2003). The geopolitics of nature. In Agnew, J., & Mitchell, K. (eds), *Companion to Political Geography*, Malden, MA: Blackwell, 423–439.

Coleman, M. (2005). U.S. statecraft and the U.S.–Mexico border as security/economy nexus. *Political Geography* 24(2): 185–209. doi:10.1016/j.polgeo.2004.09.016 DOI:10.1016/j.polgeo.2004.09.016#doilink

Coleman, M. (2007). Immigration geopolitics beyond the Mexico–US border. *Antipode* 39(1): 54–76. doi:10.1111/j.1467-8330.2007.00506.x

Coleman, M. (2009). What counts as the politics and practice of security, and where? Devolution and immigrant insecurity after 9/11. *Annals of the Association of American Geographers* 99(5): 904–913. doi:10.1080/00045600903245888

Coutin, S.B. (2010). Confined within: National territories as zones of confinement. *Political Geography* 29(4): 200–208. doi:10.1016/j.polgeo.2010.03.005

Crampton, J.W., Roberts, S.M., & Poorthuis, A. (2014). The new political economy of geographical intelligence. *Annals of the Association of American Geographers* 104(1): 196–214. doi:10.1080/00045608.2013.843436

Critical Approaches to Security in Europe Collective (2006). Critical approaches to security in Europe: A networked manifesto. *Security Dialogue* 37(4): 443–487. doi:10.1177/0967010606073085

Dalby, S. (2002). *Environmental Security*. Minneapolis, MN: University of Minnesota Press.

Dalby, S. (2003). Green geopolitics. In Agnew, J., & Mitchell, K. (eds), *Companion to Political Geography*, Malden, MA: Blackwell, 440–454.

Dalby, S. (2013). The geopolitics of climate change. *Political Geography* 37(1): 38–47. doi:10.1016/j.polgeo.2013.09.004 DOI:10.1016/j.polgeo.2013.09.004#doilink

De Goede, M. (2012). *Speculative Security: The Politics of Pursuing Terrorist Monies*. Minneapolis, MN: University of Minnesota Press.

Dillon, M. (1996). *Politics of Security: Towards a Political Philosophy of Continental Thought*. London: Routledge.

Dillon, M. (2007). Governing through contingency: The security of biopolitical governance. *Political Geography* 26(1): 41–47. doi:10.1016/j.polgeo.2006.08.003

Dillon, M., & Reid, J. (2009). *The Liberal Way of War: Killing to Make Life Live*. London: Routledge.

Duffield, M. (2007). *Development, Security and Unending War: Governing the World of Peoples*. Cambridge: Polity Press.

Elden, S. (2009). *Terror and Territory*. Minneapolis, MN: University of Minnesota Press.

Enloe, C.H. (2000). *Bananas, Beaches and Bases: Making Feminist Sense of International Politics*. Los Angeles, CA: University of California Press.

Evans, B. (2010). Foucault's legacy: Security, war and violence in the 21st century. *Security Dialogue* 41(4): 413–433. doi:10.1177/0967010610374313

Flint, C. (2005). *The Geography of War and Peace: From Death Camps to Diplomats*. Oxford: Oxford University Press.

Fluri, J.L. (2011). Bodies, bombs and barricades: Geographies of conflict and civilian (in)security. *Transactions of the Institute of British Geographers* 36(2): 280–296. doi:10.1111/j.1475-5661.2010.00422.x

Foucault, M. (2010). *The Birth of Biopolitics: Lectures at the Collège de France, 1978–1979*. New York: Palgrave Macmillan.

Gill, N. (2009). Governmental mobility: The power effects of the movement of detained asylum seekers around Britain's detention estate. *Political Geography* 28(3): 186–196. doi:10.1016/j.polgeo.2009.05.003

Graham, S. (2011). *Cities under Siege: The New Military Urbanism*. New York: Verso.

Gregory, D. (2008). "Rush to the intimate": Counterinsurgency and the cultural turn in late modern warfare. *Radical Philosophy* 150: 8–23.

Grove, K. (2012). Preempting the next disaster: Catastrophe insurance and the financialization of disaster management. *Security Dialogue* 43(2): 139–155. doi:10.1177/0967010612438434

Haggerty, K.D., & Ericson, R.V. (2000). The surveillant assemblage. *British Journal of Sociology* 51(4): 605–622. doi:10.1080/00071310020015280

Hiemstra, N. (2012). Geopolitical reverberations of US migrant detention and deportation: The view from Ecuador. *Geopolitics* 17(2): 293–311. doi:10.1080/14650045.2011.562942

Hinchcliffe, S., & Lavau, S. (2013). Differentiated circuits: The ecologies of knowing and securing life. *Environment and Planning D: Society and Space* 31(2): 259–274. doi:10.1068/d6611

Hyndman, J., & Mountz, A. (2008). Another brick in the wall? Neo-refoulement and the externalization of asylum by Australia and Europe. *Government and Opposition* 43(2): 249–269. doi:10.1111/j.1477-7053.2007.00251.x

Huysmans, J. (2000). The European Union and the securitization of migration. *Journal of Common Market Studies* 38(5): 751–777. doi:10.1111/1468-5965.00263

Ingram, A., & Dodds, K. (2009). *Spaces of Security and Insecurity: Geographies of the War on Terror*. Farnham: Ashgate.

Katz, C. (2001). On the grounds of globalization: A topography for feminist political engagement. *SIGNS: Journal of Women in Culture and Society*, 26(4): 1213–1234.

Larner, W., & Walters, W. (2006). *Global Governmentality: Governing International Spaces*. London: Routledge.

Lee, E., & Pratt, G. (2012). The spectacular and the mundane: Racialised state violence, Filipino migrant workers, and their families. *Environment and Planning A* 44(4): 889–904. doi:10.1068/a4448

Loyd, J.M., Mitchelson, M., & Burridge, A. (2012). *Beyond Walls and Cages: Prisons, Borders, and Global Crisis*. Athens, GA: University of Georgia Press.

Lyon, D. (2003). *Surveillance as Social Sorting: Privacy, Risk, and Digital Discrimination*. Abingdon: Taylor & Francis.

Mamadouh, V. (2005). Geography and war, geographers and peace. In Flint, C. (ed.), *The Geography of War and Peace: From Death Camps to Diplomats*, Oxford: Oxford University Press, 26–60.

Martin, L. (2010). Bombs, bodies, and biopolitics: Securitizing the subject at the airport security checkpoint. *Social and Cultural Geography* 11(1): 17–34. doi:10.1080/14649360903414585

Martin, L. (2011). The geopolitics of vulnerability: Children's legal subjectivity, immigrant family detention and US immigration law and enforcement policy. *Gender, Place and Culture* 18(4): 477–498. doi:10.1080/0966369X.2011.583345

Martin, L.L. (2012). Governing through the family: Struggles over US noncitizen family detention policy. *Environment and Planning A* 44(4): 866–888. doi:10.1068/a4477a

Martin, L., & Simon, S. (2008). A formula for disaster: The Department of Homeland Security's virtual ontology. *Space and Polity* 12(3): 281–296. doi:10.1080/13562570802515127

McCoy, A.W. (2009). *Policing America's Empire: The United States, the Philippines, and the Rise of the Surveillance State*. Madison, WI: University of Wisconsin Press.

McDonald, M. (2013). Discourses of climate security. *Political Geography* 33: 42–51. doi:10.1016/j.polgeo.2013.01.002

Mezzadra, S., & Neilson, B. (2013). *Border as Method, Or, the Multiplication of Labor*. Durham, NC: Duke University Press.

Mitchell, K. (2010). Ungoverned space: Global security and the geopolitics of broken windows. *Political Geography* 29(5): 289–297. doi:10.1016/j.polgeo.2010.03.004

Mitchell, T. (1991). *Colonising Egypt*. Los Angeles, CA: University of California Press.

Mountz, A. (2010). *Seeking Asylum: Human Smuggling and Bureaucracy at the Border*. Minneapolis, MN: University of Minnesota Press.

Mountz, A. (2011). The enforcement archipelago: Detention, haunting, and asylum on islands. *Political Geography* 30(3): 118–128.

Mountz, A., & Hiemstra, N. (2014). Chaos and crisis: Dissecting the spatiotemporal logics of contemporary migrations and state practices. Annals of the Association of American Geographers 104(2): 382–390. doi:10.1080/00045608.2013.857547

Mountz, A., Coddington, K., Catania, C.T., & Loyd, J.M. (2013). Conceptualizing detention: Mobility, containment, bordering, and exclusion. *Progress in Human Geography* 37(4): 522–541. doi:10.1177/0309132512460903

Neocleous, M. (2008). *Critique of Security*. Edinburgh: Edinburgh University Press.

Nordås, R., & Gleditsch, N.P. (2007). Climate change and conflict. *Political Geography* 26(6): 627–638. doi:10.1016/j.polgeo.2007.06.003

Ó Tuathail, G. (1996). *Critical Geopolitics*. London: Routledge.

Ó Tuathail, G. (2003). Geopolitical structures and geopolitical cultures: Towards conceptual clarity in the critical study of geopolitics. In Tchantouridze, L. (ed.), *Geopolitical Perspectives on World Politics*, Bison Paper 4, Winnipeg, Manitoba: Centre for Defence and Security Studies, 75–102.

Ó Tuathail, G., & Dalby, S. (1998). *Rethinking Geopolitics*. London: Routledge.

Ong, A., & Collier, S.J. (2005). *Global Assemblages: Technology, Politics, and Ethics as Anthropological Problems*. Malden, MA: Blackwell.

Pain, R. (2014). Everyday terrorism: Connecting domestic violence and global terrorism. *Progress in Human Geography*, 38(4): 531–550. doi:10.1177/0309132513512231

Philo, C. (2012). Security of geography/geography of security. *Transactions of the Institute of British Geographers* 37(1): 1–7. doi:10.1111/j.1475-5661.2011.00488.x

Puar, J. (2007). *Terrorist Assemblages: Homonationalism in Queer Times*. Durham, NC: Duke University Press.

Pugliese, J. (2012). *Biometrics: Bodies, Technologies, Biopolitics*. London: Routledge.

Purcell, M., & Nevins, J. (2005). Pushing the boundary: State restructuring, state theory, and the case of U.S.–Mexico border enforcement in the 1990s. *Political Geography* 24(2): 211–235. doi:10.1016/j.polgeo.2004.09.015

Roberts, S., Secor, A., & Sparke, M. (2003). Neoliberal geopolitics. *Antipode* 35(5): 886–897. doi:10.1111/j.1467-8330.2003.00363.x

Robertson, M.M. (2006). The nature that capital can see: Science, state and market in the commodification of ecosystem services. *Environment and Planning D: Society and Space* 24(3): 367–387. doi:10.1068/d3304

Sharp, J. (2011). A subaltern critical geopolitics of the War on Terror: Postcolonial security in Tanzania. *Geoforum* 42(3): 297–305. doi:10.1016/j.geoforum.2011.04.005

Simon, S. (2012). Suspicious encounters: Ordinary preemption and the securitization of photography. *Security Dialogue* 43(2): 157–173. doi:10.1177/0967010612438433

Simon, S. (2014). Disaster, pre-emptive security and urban space in the post-9/11 New York City of Cloverfield and The Visitor. *Journal of Urban Cultural Studies* 1(1): 19–23. doi:10.1386/jucs.1.1.19_1

Sparke, M.B. (2006). A neoliberal nexus: Economy, security and the biopolitics of citizenship on the border. *Political Geography* 25(2): 151–180. doi:10.1016/j.polgeo.2005.10.002

Staeheli, L.A., & Kofman, E. (2004). Mapping gender, making politics: Toward feminist political geographies. In Staeheli, L.A., Kofman, E., & Peake, L. (eds), *Mapping Women, Making Politics: Feminist Perspectives on Political Geography*, New York: Routledge, 1–14.

Staeheli, L.A., & Nagel, C.R. (2008). Rethinking security: Perspectives from Arab-American and British Arab Activists. *Antipode* 40(5): 780–801. doi:10.1111/j.1467-8330.2008.00637.x

Sundberg, J. (2011). Diabolic caminos in the desert and cat fights on the Río: A posthumanist political ecology of boundary enforcement in the United States–Mexico borderlands. *Annals of the Association of American Geographers* 101(2): 318–336. doi:10.1080/00045608.2010.538323

Tickner, J.A. (1992). *Gender in International Relations: Feminist Perspectives on Achieving Global Security*. Vol. 6. New York: Columbia University Press.

Varsanyi, M.W. (2008). Immigration policing through the backdoor: City ordinances, the "right to the city," and the exclusion of undocumented day laborers. *Urban Geography* 29(1): 29–52. doi:10.2747/0272-3638.29.1.29

Wilsher, D. (2011). *Immigration Detention: Law, History, Politics*. Cambridge: Cambridge University Press.

Woodward, R. (2005). From military geography to militarism's geographies: Disciplinary engagements with the geographies of militarism and military activities. *Progress in Human Geography* 29(6): 718–740. doi:10.1191/0309132505ph579oa

Zedner, L. (2009). *Security*. Abingdon: Routledge.

Chapter 10

Violence

James Tyner

Kent State University, Kent, Ohio, USA

Violence is seemingly self-evident. When someone is hit, stabbed, shot, or raped, we *know* that violence has occurred. We *know* also that armed conflict, wars, and genocide are violent. However, when we think for a little bit longer, we quickly realize that violence is actually a very difficult concept to pin down. Tornadoes and earthquakes, for example, are routinely described as violent acts; so too are car accidents, hits on the football field, and verbal arguments. Is it appropriate or useful to equate such disparate events under the singular term "violence"? Richard Mizen (2003: 285) captures this problem when he writes that the "word 'violence' tends to be used in such a loose and broadly defined way that its use as a precise term of description or as a clear concept is severely limited." In this chapter I argue that violence must be understood not as having a universal quality or transhistorical essence, but instead as something that is produced. Indeed, it is necessary to consider the *political and geographical construction* of violence.

Direct and structural violence: An outdated binary

It is common for scholars of violence to begin with the apparent binary of "direct" and "structural" violence. In 1969, for example, Johan Galtung forwarded the position that direct violence occurs when there is an identifiable actor who commits an act of violence. Here, violence refers to any action that impinges on or reduces human potential. The World Health Organization subsequently defined (direct) violence as

> the intentional use of physical force or power, threatened or actual, against oneself, another person, or against a group or community, that either results in or has a high likelihood of resulting in injury, death, psychological harm, maldevelopment or deprivation. (Krug et al. 2002: 1084)

The Wiley Blackwell Companion to Political Geography, First Edition.
Edited by John Agnew, Virginie Mamadouh, Anna J. Secor, and Joanne Sharp.
© 2015 John Wiley & Sons Ltd. Published 2017 by John Wiley & Sons Ltd.

Following this definition, three "forms" of direct violence are identified: self-inflicted, interpersonal, and collective. Self-inflicted violence includes suicide and self-mutilation (e.g., auto-cutting and scarring); interpersonal violence includes violence committed by one person toward another, with two subtypes identified (family/partner and stranger-on-stranger); and collective violence includes group forms of violence, such as gang violence, mob violence, and war.

Structural violence, conversely, is premised to occur when no such actor is identifiable. Galtung (1969: 170–171) elaborates that

> whereas in the first case these consequences can be traced back to concrete persons or actors, in the second case this is no longer meaningful. There may not be any person who directly harms another person in the structure. The violence is built into the structure and shows up as unequal power and consequently unequal life chances.

As Susan Opotow (2001: 151) explains, structural violence is manifest

> as inequalities structured into a society so that some have access to social resources that foster individual and community well-being – high quality education and health care, social status, wealth, comfortable and adequate housing, and efficient civic services – while others do not.

In the United States, for example, an estimated 26,100 people between 25 and 64 years of age die prematurely due to a lack of health coverage (Families USA 2012). For this reason, following Jenna Loyd (2009: 865), it is necessary "to theorize specific economic, political, and social relations of oppression and domination and how they articulate (or intersect) in particular historical, geographic moments."

Galtung's efforts to disentangle the various forms of violence have been especially valuable, but also somewhat limited. A promising avenue has been that pursued by geographers and other social theorists who have challenged the ontological separation of "direct" and "structural" violence (Anglin 1998; Confortini 2006). For these scholars, what superficially appears as a binary between two different and oppositional "forms" of violence is in actuality a dialectics of violence (see also Tyner & Inwood 2014). Graeber (2012), for example, explains that "structural" violence is often presumed to consist of structures that have violent effects: racism, patriarchy, and so on. However, it is more appropriate to consider these "as structures *of* violence – since it is only the constant fear of physical violence that makes them possible" (Graeber 2012: 113). In other words, direct forms of violence must be understood within the context of broader political, social, and economic structures of violence; likewise, it is necessary to understand the imposition and continuance of oppressive structures within the context of the threat or actual use of direct forms of violence.

Contested definitions of violence

Galtung's writings are part of an ongoing effort to expand our awareness and understanding of violence: to move beyond what is often viewed as an overly narrow and restrictive understanding of violence that neglects many practices that harm, injure, or kill people. Edwin Sutherland (1940), for example, early on forwarded the concept of "white-collar criminality"; his work ultimately led to the theorization of "corporate violence." Newton Garver (1968) likewise advanced a more expansive definition of violence. Equating violence more with "violation" than with "force," Garver argued that it is insufficient to focus exclusively on murder, beatings, and rape; instead, it is necessary to address other actions in which a person could be violated.

Galtung, Garver, and other social scientists who promoted more expansive definitions of violence were (and continue to be) met with stiff resistance. Indeed, the debate between those who champion "minimalist" or "restrictive" definitions and those who favor broader definitions remains as vibrant today as it did in the 1960s and 1970s. In an early critique of Garver, for example, Joseph Betz (1977: 341) cautioned:

> if violence is violating a person or a person's rights, then every social wrong is a violent one ... If violence is whatever violates a person and his [sic] rights of body, dignity, or autonomy, then lying to or about another, embezzling, locking one out of his house, insulting, and gossiping are all violent acts.

Betz (1977: 341) concluded that the "enlargement of the extension of the term comes at considerable cost, for there is simply no extension left for the term 'nonviolent social wrong.'"

C.A.J. Coady (1986) also guards against overly capacious definitions of violence. In so doing, he highlights the political contestation of violence. Broad terms such as "structural" violence, Coady (1986: 4) argues, "tend to serve the interests of the political left by including within the extension of the term 'violence' a great range of social injustices and inequalities." This poses a potential danger, he warns, because

> this not only allows reformers to say that they are working to eliminate violence when they oppose, say, a government measure to redistribute income in favor of the already rich, but allows revolutionaries to offer, in justification of their resort to violence, even where it is terrorist, the claim that they are merely meeting violence with violence.

Advocating a narrow understanding predicated on direct, intentional force, Coady (1986: 4) concludes (erroneously, I believe) that the "use of the wide definition seems likely to encourage the cosy but ultimately stultifying belief that there is one problem, the problem of (wide) violence, and hence it must be solved as a whole with one set of techniques." Here, he misses the point, for the argument in favor of expanded definitions is just the opposite. Galtung, Sutherland, and Garver, for example, argue that because violence assumes so many forms, it requires a multiplicity of solutions. Policies designed to address homicide, for instance, will not necessarily reduce social inequalities that lead to illness and death from malnutrition and lack of health care.

Coady is correct, however, in suggesting that definitions, theories, and even statistics of violence are political. To this end, Johanna Oksala (2012) puts forward the argument that violence must be understood as a set of contingent, specific practices; that we must not simply accept its eternal existence; and that we should understand the meaning and specific forms that violence assumes in the particular historical and geographical contexts that produce and sustain it. Geographers Tyner and Inwood (2014) make a similar argument. Thus, arguing against the existence of a pregiven, prediscursive reality, they propose a materially grounded, dialectical approach to violence, stressing that what counts as violence must be theorized as being produced by, and producing, socio-spatially contingent modes of production. Capitalism, for example, is a specific organizational form of production based on the separation of direct producers from the means of production; consequently, the *relations* and structures *of* violence systemic to capitalism are different from those of other forms of production (e.g., feudalism or socialism).

Tyner and Inwood's dialectic approach aligns with the work of another geographer, Simon Springer. In a series of papers, Springer (2011, 2012) also challenges the ontological foundation of violence. He notes (2011: 91) that "while writing about violence directly in empirical terms is a worthwhile endeavor" it is especially problematic if violence itself is not adequately theorized. For him (2012: 138), "to treat the material expression of violence only through its directly observable manifestation is a reductionist appraisal" and it "ignores the complexity of infinite entanglements of social relations." Violence, in other words, is best approached as an assemblage or coalescence of disparate relations and practices. To this end, Springer (2011: 91) argues that *violence sits in places*. This phrase is meant to capture the idea that how we perceive the manifestation of violence is localized and embodied; in other words, to paraphrase Springer, violence is abstracted from specific concrete actions and relations that concern particular places. Violence, therefore, is "no longer confined to its material expression as an isolated 'event' or localized 'thing'" but instead may "more appropriately be understood as an unfolding process, arising from the broader geographical phenomena and temporal patterns of the social world" (Springer 2011: 91). As such, violence is always geographical and political.

Political violence

Violence is frequently divided into "criminal" and "political" forms. Here, the distinction is based on intended goals and objectives. Criminal violence, on the one hand, consists of those actions used to achieve personal gain. Political violence, on the other hand, is used to achieve or effect change at a level far beyond the individual agent, such as particular laws or even governments themselves. Political violence, moreover, may be perpetrated by states, groups, or individuals; and its target may be states, groups, or individuals. A geography of political violence, accordingly, directs attention to the interplay of space and power; in other words, to how the geographical and political organization of societies produces and is produced by violence.

Three types of political violence are readily identifiable: political assassinations, terrorism, and torture. Assassinations can be very broadly likened to homicide: An individual (or group of individuals) is killed by another individual or group. The difference lies in the targeting and motivation for the killing, in that assassinations are deliberate, intentional acts to eliminate a political rival, while homicide is not. Recently, a distinction has been proposed between "assassinations" and "targeted killings." By definition, political assassinations are acts of murder and therefore, according to international law, are considered illegal. Targeted killings, however, are ostensibly conducted in self-defense and therefore are considered (by supporters, that is) as legal state action. The separation of assassination and targeted killing has generated an intense moral and political debate (cf. Guiora 2004; Byman 2006; Gross 2006). Crucial to these debates is the difference between "combatant" and "non-combatant." If those individuals who are targeted, for example, are declared "non-combatants," then the conventions of war and laws of armed conflict make targeted killings illegal.

Recently, the widespread use of drones (unmanned aerial vehicles) by the United States has blurred the distinction – and legality – of assassinations, targeted killings, and, indeed, war itself (Shaw 2013). As Shaw and Akhter (2012: 1495) explain, drones are "heralded by the US military as the apex of targeting logic – accurate, efficient, and deadly." Critics, however, worry about the indiscriminate killing of civilians associated with the use of drones and the ever-expanding geography of "battle-space"; that is, those spaces disciplined, if not destroyed, by remotely delivered violence (Gregory 2011).

A second form of political violence is terrorism, defined as "an act of politics by those who declare that no other channels of political engagement exist" (Flint 2003: 161). In other words, terrorist acts are a form of asymmetric warfare, whereby certain groups – lacking adequate military or political power – resort to acts of violence to effect political change. As historians of the subject have documented, terrorism has a lengthy history and assumes many different forms, characterized in part by its overall objective. Robert Pape (2005) identifies three different forms of terrorism and, while not mutually exclusive, these can be conceived as different means to different goals. *Demonstrative* terrorism is as much political theater as it is violence; the point is less to injure and kill as to be spectacular. It is, in short, a means of gaining publicity, to make one's grievances known to outside parties; to help recruit members to one's cause; and to intimidate and frighten members of the opposition. Fear – but really, the power of terrorism – is augmented by the seeming randomness of terrorist acts: They could happen any time, anywhere. Demonstrative terrorism may also be used to disrupt economic systems, such as the bombing of a railway line. *Destructive* terrorism, conversely, is meant to bring about injury and death; both "military" and "civilian" personnel may be targeted. *Suicide* terrorism, lastly, is often considered the most aggressive form of terrorism. It differs from destructive terrorism in that the terrorist does not expect to survive the mission. Suicide terrorism may be used for demonstrative purposes, but is most often employed as a means to carry out selective assassinations or to kill large numbers of people.

A final form of political violence is torture. While the term brings to mind dark dungeons and medieval instruments of pain and suffering, torture is very much present in contemporary politics (Gregory 2006; Falah 2008). A simple definition holds that torture is "the deliberate infliction of severe pain or suffering"; such an action may be committed "by state agents, or similar acts by private individuals for which the state bears responsibility through consent, acquiescence or inaction" as well as "members of armed political groups" (Amnesty International 2001: 5). Torture has many functions. As a form of corporal punishment, it serves (ostensibly) as a form of deterrence. Likewise, the public spectacle of torture has long been used as a means of reestablishing the authority of the sovereign. Torture is also widely employed as a means of extracting information.

Both the definition and legality of torture are greatly informed by broader political machinations. In the "War on Terror," for example, many states have reworked the definition of torture so that certain actions conform to, or bypass, international law. The Universal Declaration of Human Rights and the covenants and treaties that give it the force of law in the United Nations are designed to protect people from the excesses of the state – including torture (Nagengast 1994: 126). However, under the "ticking time-bomb" scenario, proponents argue that it is only through interrogation and torture of suspected terrorists that it is possible to prevent catastrophic acts of violence (Hannah 2006). Critics counter that no information obtained via torture has prevented any such act.

Political assassination, terrorism, and torture all highlight the political use of violence, but also the political definition and legality of violence. As seen in the example of "targeted killings," states often manipulate the legal system to justify their actions while criminalizing those of opposition (and especially non-state) parties. For instance, states often couch their use of torture as a means of obtaining information necessary for national security. It is also worthwhile to note that discussions of political violence almost exclusively focus on direct forms of violence and neglect the imposition of those *structures* of violence that "let die" certain groups of people.

State violence

Having considered political violence, it is instructive to consider more explicitly the concept of state violence. This is a vast literature, in that most political philosophers, from classical writers such as Hobbes, Locke, and Rousseau to more contemporary theorists like Giorgio Agamben, Jacques Derrida, and Slavoj Žižek have had something to say about violence. Many discussions begin with the observation forwarded by Max Weber that "a state is a human community that (successfully) claims the monopoly of the legitimate use of physical force within a given territory … The state is considered the sole source of the right to use violence." This well-known quote has generated substantial discussion and debate. On the one hand, Weber directs attention to the political contestation over the definition – the abstraction – and legality of violence. Rather than exhibiting an enduring, self-evident quality, violence is defined and justified according to one's access to power. Thus, governments engage not in "political assassinations" but in "targeted killings"; and the use of drones constitutes a practice not of terrorism but of security. On the other hand, Weber calls attention to the complex relationship between violence and other equally contested concepts, including sovereignty, power, and the law. This has generated a vast literature that may be subsumed under the term "administrative violence" (cf. Spade 2011; Gupta 2012; Tyner 2014a).

In large part, the concept of administrative violence originates in the writings of Walter Benjamin and, especially, his 1921 essay "Critique of Violence." From Agamben to Žižek, countless theorists have attempted to decipher the crux of this text. The difficulties in reading Benjamin are twofold. First, definitions in his essay are slippery at best. Indeed, many of his most prominent ideas remain ill defined and thus the source of much confusion. A second, and related, point is that Benjamin was unable to complete a proposed book on politics that presumably would have provided much clarification. Scholars are left, therefore, in a position of reconstructing their own interpretations; this has, according to Bernstein (2013), become something of a Rorschach test.

Benjamin differentiates "law-making" and "law-preserving" violence. The former, also understood as *foundational* violence, is associated with the moments of the investiture of law; the latter, conversely, constitutes an administratively enforced violence, as exemplified by the establishment of judicial and police apparatuses. This raises a fundamental tension (Correia 2013), in that violence appears not only as a method of enforcement – manifest, for example, in the use of armed conflict and police repression – but also as a means of enforcing political legitimacy, as exhibited in the establishment of torture and execution. Later in his essay Benjamin introduces the concept of "mythic violence," which includes both law-making and law-preserving violence. His ideas are complex, but in essence he calls attention to the "mythic" aspects of law and violence whereby mythic violence is "the creator and the protector of the prevailing political and legal order" (Springer 2011: 95). Divine violence, conversely, is "pure" violence in that it challenges and "destroys" mythic violence. How divine violence *appears* has also generated considerable controversy. For Judith Butler (2006), divine violence is equated with non-coercive and non-violent action; for Slavoj Žižek (2008) it is the opposite. Regardless of how one ultimately reads Benjamin, his essay is informative in drawing attention to the complexities of sovereignty, the law, and violence.

Hannah Arendt (1970) similarly works through the idea of state violence. She begins with a critique of Mao Zedong's aphorism that power comes out of the barrel of a gun. Arendt's conceptualization of violence is considerably more nuanced, in that she discerns "power" from "violence." For her, power is not something that is possessed, but rather an expression of a collective will. This counters more conventional approaches – including that of Weber – that conceive of power as "power over" someone. Violence, on the other hand, is instrumental. Thus defined, for Arendt, power and violence are antithetical concepts: Violence may lead to change, but it can never lead to power. Arendt's forwarding of power as opposed to violence complements the thinking of Michel Foucault. For Foucault, power is not some*thing* that is held by any one person, group, or state; rather, power circulates – power is relational. Discipline is a type of power, and while it is not simply a matter of consent, it is also not a renunciation of freedom or a transfer of rights from the powerless to the powerful. Violence, conversely, is according to Foucault totalizing. When violence is engaged it removes the possibilities for active subjects (agents) to act autonomously.

As these all too brief synopses of Benjamin, Arendt, and Foucault illustrate, there is a diverse range of political philosophies when it comes to violence. Geographers, among others, have developed these ideas, especially in the context of transgression and resistance. For Tim Cresswell (1996), a transgression is an action that is not intentional, whereas acts of resistance imply intentionality. An act of resistance, in other words, is purposeful action directed against some disliked entity or rule. The "lunch counter sit-ins" in the United States during the Civil Rights movement of the 1960s were acts of resistance. They were forms of (non-violent) political protest that challenged segregation ordinances (e.g., racial structures of violence) that prohibited the equal participation of American Americans in public space. It is worthwhile noting, also, that these forms of resistance were often met with direct, interpersonal violence in the form of physical beatings.

Indeed, both transgressions and acts of resistance are frequently met with acts of direct violence. Those individuals and/or groups who feel threatened often respond with the actual or threatened use of violence to maintain or reclaim the status quo. The activities of vigilante groups, for example, may be viewed as group behavior directed against perceived acts of resistance or transgression. Etymologically, the term "vigilante" is derived from the same Latin roots as the English words "vigil" and "vigilance"; the word therefore connotes an act of "watching" or "guarding." There is, moreover, an inherent geography to vigilantism in that it is generally the case that what one watches over is a specific place.

When confronted with a threat to their community, however defined, vigilantes often exhibit a willingness to take extraordinary measures, including physical violence. This is seen, for example, in the work of Roxanne Doty (2009), who has documented the various militias that "police" the international border between the United States and Mexico (see also Nevins 2008). Of significance is that vigilante groups are generally not revolutionary in the sense that they seek to abolish or overthrow the state. In fact, many are formed to prevent *other* groups from disrupting or overthrowing the state. As such, vigilante groups are best understood, politically, as being reactionary in their attempts to promote and protect their own values and ways of life that are believed to be under attack (Abrahams 1998: 78). Thus, in defense of their extralegal and often violent actions, community members claim that the state has failed "the people"; they are in effect demanding that the state more forcefully exercises its monopoly on the legitimate use of violence.

Gender violence

Gender violence is a broad umbrella term used to designate a wide range of violations, from rape during wartime to physical and other forms of abuse within the household (Merry 2009). At the most basic level, gender violence refers to violence committed against an individual or group predominantly (or exclusively) because of the victim's gender. In the 1970s and 1980s, for example, terms such as "wife abuse," "battered woman," and "domestic violence" gained familiarity among both academics and those working in the criminal justice system. The deployment of these terms signaled an awareness – *not discovery* – of many forms of violence that were too often ignored or phrased as acts other than violence. As indicated by the terminology, attention focused primarily on married women who suffered physical injury literally from the hands of their husbands. Gradually the definition of domestic violence was expanded to include other, non-physical forms, such as emotional and psychological abuse. Still later, as awareness of violent acts committed in other, non-heterosexual relations was documented, other theorizations and conceptualizations were forwarded: violence within lesbian and gay relationships; violence among unmarried couples, such as cohabitating and dating couples; and violence toward men and women in the process of separation and divorce (Pain 1991, 2014; Cuomo 2013). Currently, the term "intimate partner violence" is the main conceptual tool with which to characterize these forms of violence (cf. Hattery 2009).

In the late 1960s and 1970s, feminists also focused attention on rape as a political problem. As Bumiller (2008: 2–3) explains, many activist groups

> named the problem as the failure of the state to recognize and protect women. In fact, the flagrant denial of violence against women was characterized as state-sanctioned violence and was seen as complicit with other forms of patriarchal control that oppressed women.

Consequently, feminists and other activists helped found shelters and crisis centers; the movement was also effective in challenging laws that neglected or downplayed the seriousness of rape.

Rape as a form of political violence is also widespread in armed conflict. However, as Patricia Hynes (2004: 437) finds, "until recently, little has been known about the prevalence and scale of sexual abuse by men in war." Only within the past three decades has "a growing number of women journalists, lawyers, physicians, and human rights activists … uncovered and exposed the war crimes against women, namely rape, abduction, sexual torture, and trafficking for prostitution." Recent studies, for example, have documented the existence of widespread rape and sexual abuse during both the First and Second World Wars, the Vietnam War, and Bangladesh's war of independence. These forms of violence were widespread in the violence that gripped the former Yugoslavia in the 1980s as well as during the genocide in Rwanda in 1994.

Historically, rape and other forms of sexual and gendered violence have long been (mis)characterized and dismissed by military and political leaders as a "private" act or the result of a few renegade soldiers. Rape in warfare, for example, becomes naturalized – and neutralized – as "just part of war" (Nowrojee 1996; Weaver 2010). War-rape, however, must be viewed as a deliberate instrument of both direct and structural violence; it is a political technique of terror that inflicts both physical and psychological trauma. Moreover, sexual and gendered violence is not limited to the battlefield. Lorraine Dowler (2011), for example, writes of the "hidden war" of female soldiers' increased vulnerability in the US military from state legislation that restricts women from participation in combat units but not men from perpetrating violent attacks. Here, she notes the irony that women are to be protected from the violence of the enemy in combat while subjected to violence within their own ranks. Other research has documented the sexual and gendered violence associated with the often rampant prostitution industry associated with military bases (Sturdevant & Stoltzfus 1992; Moon 1997).

Sexual and gendered violence embodies crucial symbolic components. In many societies, for example, women are not equal to the nation but instead symbolic of it (Domosh & Seager 2001). In part, this symbolism derives from the so-called natural capacity of women to bear children (Yuval-Davis 1996). As Eisenstein (2007: 28) explains, interpersonal forms of violence, such as rape, are directed not solely at the woman but at the "enemy society":

> [the] violated female is no longer a woman that a man wishes to lay claim to. With war-rape, females are reduced to their patriarchal definition as a body vessel and also denied the status of privileged womanhood. In war-rape the woman is totally occupied.

In other words, women's bodies become battlegrounds on which discourses of moral inclusion and exclusion are contested; as a form of political/military violence, rape is intended not just to degrade the individual but to strip humanity from a larger group. It is an extreme form of direct violence that serves another purpose, namely further to marginalize and dehumanize a population (Tyner 2009).

Killing and letting die

Political geographers are becoming increasingly aware of more "hidden" forms of violence; that is, actions and inactions that, on the surface, are not always considered violent and thus are rarely questioned. This is most apparent in the long-standing distinction between "killing" and "letting die." Of all forms of direct, interpersonal violence, killing is often considered the gravest. In part this is because the act of killing takes away that which is most valuable and irreplaceable: a life. Recent work, however, has expanded the discussion to consider the equally grave threats that result from the disallowal of life (Li 2009; Laurie 2014; Tyner 2014b).

For many readers the act of killing is intuitively considered to be morally worse than "letting die." Such a presumption hinges on a particular understanding of agency: To "kill" is considered an action, whereas to "let die" is perceived as an omission, or lack of action. In criminal law, for example, death that results from omission is defined legally as "negligence" and, all else being equal, punished less severely than, say, homicide. At this point, notice that this distinction parallels the binary between "direct" and "structural" violence; and that many restrictive accounts of violence downplay the salience of structural forms of violence (e.g., corporate violence).

The moral partition between "action" and "inaction" – between "killing" and "letting die" – parallels the political concepts of duties and obligations. On the one hand, we have duties not to harm others; these generally require restraint. On the other hand, we have positive duties; that is, obligations to help others. All else being equal, the obligation to not harm people – at least in most Western societies – is more stringent than the obligation to help others. The implications of such logic are profound. Infringements on the (negative) duty *not* to kill assume prominence because "killing" is presumed to exemplify the extreme of two morally relevant factors: certainty that one's action will result in harm (death) and the minimal sacrifice needed to refrain (Lichtenberg 1994: 219). From this perspective, practices that *intentionally* injure or kill – with the exception of both capital punishment and declarations of "just" war – are viewed as more serious than the state's failure to enact positive practices that aid its citizens. Indeed, it is this distinction that designates "political assassination" and "war-rape" as *crimes against humanity*, while the lack of provision of adequate health care (i.e., a structural inequality, or a structure *of* violence) is not considered a crime. In the former situation a state may be found guilty through its perpetuation of specific practices that kill people; in the latter situation, the lack of action is considered to be neither morally nor legally wrong.

A *political reading* of the moral distinction between killing and letting die offers a more nuanced way to understand the legal and spatial separation of direct and structural violence. This is seen, for example, in the often heated political debates surrounding health care in the United States. For example, the Patient Protection and Affordable Care Act (PPACA) was signed into law by President Barack Obama on March 23, 2010 (the following account is based on Dickman et al. 2014). Highly controversial and mired in partisan politics, the intent of the PPACA was to increase access to health insurance for approximately 47 million unin-sured Americans. This was to be accomplished, in part, by requiring states to expand Medicaid eligibility to people with income less than 138 percent of the Federal Poverty Level. In June 2012, however, the US Supreme Court ruled that states may opt out of Medicaid expansion. This, in effect, confirmed a geographical component to the accessibility of health care. As of November 2013, 25 states did in fact opt out, leaving millions of Americans uninsured who otherwise would have been covered by Medicaid. Sam Dickman and his colleagues (2014) estimate that approximately 8 million men, women, and children will be uninsured because of the "opt-out" decision. As a result, they calculate that Medicaid expansion in those states that opted out

> would have resulted in 422,553 more diabetics receiving medication for their illness, 195,492 more mammograms among women aged 50–64 years and 443,677 more pap smears among women age 21–64. Expansion would have resulted in an additional 658,888 women in need of mammograms gaining insurance, as well as 3.1 million women who should receive regular pap smears.

They conclude the number of potential deaths attributable to the lack of Medicaid expansion in opt-out states at between 7,115 and 17,104. In other words, upward of 17,000 people in the United States may die as a result of geographically specific governmental decision-making.

Conclusion

Violence is often separated into "direct" and "structural" forms. Thus, it is appropriate for political geography to consider how these various forms of violence are manifested and interact and how violence is given meaning in particular political and geographical contexts. For instance, political violence (such as assassinations, terrorism, and torture), as distinct from criminal violence, draws attention to the use of violence to effect change beyond the immediate level of the individual. Political geographers' recent engagement with violence has shown that it may result from both action and inaction. A political reading of the moral distinction bet-ween "killing" and "letting die" provides a more nuanced way to understand the legal – and spatial – separation of direct and structural violence.

References

Abrahams, R. (1998). *Vigilante Citizens: Vigilantism and the State*. Malden, MA: Polity Press.
Amnesty International (2001). *End Impunity: Justice for the Victims of Torture*. London: Amnesty International Publications.

Anglin, M.K. (1998). Feminist perspectives on structural violence. *Identities* 5: 145–151.

Arendt, H. (1970). *On Violence*. New York: Harcourt.

Benjamin, W. (1978 [1921]). Critique of violence. In Demetz, P. (ed.), *Reflections: Essays, Aphorisms, Autobiographical Writings,* New York: Schocken, 277–300.

Bernstein, R.J. (2013). *Violence: Thinking without Banisters*. Malden, MA: Polity Press.

Betz, J. (1977). Violence: Garver's definition and a Deweyan correction. *Ethics* 87: 339–351.

Bumiller, K. (2008). *In an Abusive State: How Neoliberalism Appropriated the Feminist Movement against Sexual Violence*. Durham, NC: Duke University Press.

Butler, J. (2006). *Precarious Life: The Powers of Mourning and Violence*. London: Verso.

Byman, D. (2006). Do targeted killings work? *Foreign Affairs* 85: 95–111.

Coady, C.A.J. (1986). The idea of violence. *Journal of Applied Philosophy* 3: 3–19.

Confortini, C.C. (2006). Galtung, violence, and gender: The case for a peace studies/feminism alliance. *Peace Change* 31: 333–367.

Correia, D. (2013). *Properties of Violence: Law and Land-Grant Struggle in Northern New Mexico*. Athens, CA: University of Georgia Press.

Cresswell, T. (1996). *In Place-Out of Place: Geography, Ideology, and Transgression*. Minneapolis, MN: University of Minnesota Press.

Cuomo, D. (2013). Security and fear: The geopolitics of intimate partner violence policing. *Geopolitics* 18: 856–874.

Dickman, S., Himmelstein, D., McCormick, D., & Woolhandler, S. (2014). Opting out of Medicaid expansion: The health and financial impacts. HealthAffairs blog, January 30. http://healthaffairs.org/blog/2014/01/30/opting-out-of-medicaid-expansion-the-health-and-financial-impacts/, accessed January 30, 2014.

Domosh, M. & Seager, J. (2001). *Putting Women in Place: Feminist Geographers Make Sense of the World*. New York: Guilford Press.

Doty, R.L. (2009). *The Law into Their Own Hands: Immigration and the Politics of Exceptionalism*. Tucson, AZ: University of Arizona Press.

Dowler, L. (2011). The hidden war: The "risk" to female soldiers in the US military. In Kirsch, S., & Flint, C. (eds), *Reconstructing Conflict: Integrating War and Post-War Geographies*, Aldershot: Ashgate, 295–314.

Eisenstein, Z. (2007). *Sexual Decoys: Gender, Race and War*. New York: Zed Books.

Falah, G.-W. (2008). Geography in ominous intersection with interrogation and torture: Reflections on detention in Israel. *Third World Quarterly* 29: 749–766.

Families USA (2012). *Dying for Coverage: The Deadly Consequences of Being Uninsured*. Washington, DC: Families USA.

Flint, C. (2003). Terrorism and counterterrorism: Geographic research questions and agendas. *Professional Geographer* 55: 161–169.

Galtung, J. (1969). Violence, peace, and peace research. *Journal of Peace Research* 6: 167–191.

Garver, N. (1968). What violence is. *The Nation*, June 24: 817–822.

Graeber, D. (2012). Dead zones of the imagination: On violence, bureaucracy, and interpretive labor. *Journal of Ethnographic Theory* 2: 105–128.

Gregory, D. (2006). The black flag: Guantanamo Bay and the space of exception. *Geografiska Annaler: Series B* 88: 405–427.

Gregory, D. (2011). From a view to kill: Drones and late modern war. *Theory, Culture and Society* 28: 188–215.

Gross, M.L. (2006). Assassination and targeted killing: Law enforcement, execution or self-defense? *Journal of Applied Philosophy* 23: 323–335.

Guiora, A. (2004). Targeted killing as active self-defense. *Case Western Reserve Journal of International Law* 36: 319–334.

Gupta, A. (2012). *Red Tape: Bureaucracy, Structural Violence, and Poverty in India*. Durham, NC: Duke University Press.

Hannah, M. (2006). Torture and the ticking bomb: The war on terrorism as a geographical imagination of power/knowledge. *Annals of the Association of American Geographers* 96: 622–640.

Hattery, A.J. (2009). *Intimate Partner Violence*. Lanham, MD: Rowman & Littlefield.

Hynes, P.H. (2004). On the battlefield of women's bodies: An overview of the harm of war to women. *Women's Studies International Forum* 27: 431–445.

Krug, E.G., Mercy, J.A., Dahlberg, L.L. & Zwi, A.B. (2002). The World Report on Violence and Health. *The Lancet* 360: 1083–1088.

Laurie, E.W. (2014). Who lives, who dies, who cares? Valuing life through the disability-adjusted life year measurement. *Transactions of the Institute of British Geographers* 40(1): 75–87. doi:10.1111/tran.12055

Li, T.M. (2009). To make live or let die? Rural dispossession and the protection of surplus populations. *Antipode* 41: 66–93.

Lichtenberg, J. (1994). The moral equivalence of action and omission. In Steinbock, B., & Norcross, A. (eds), *Killing and Letting Die*, 2nd edn, New York: Fordham University Press, 210–229.

Loyd, J. (2009). "A microscopic insurgent": Militarization, health, and critical geographies of violence. *Annals of the Association of American Geographers* 99: 863–866.

Merry, S.E. (2009). *Gender Violence: A Cultural Perspective*. Malden, MA: John Wiley & Sons.

Mizen, R. (2003). A contribution towards an analytic theory of violence. *Journal of Analytical Psychology* 48: 285–305.

Moon, K.H.S. (1997). *Sex among Allies: Military Prostitution in U.S.–Korea Relations*. New York: Columbia University Press.

Nagengast, C. (1994). Violence, terror, and the crisis of the state. *Annual Review of Anthropology* 23: 109–136.

Nevins, J. (2008). *Dying to Live: A Story of U.S. Immigration in an Age of Global Apartheid*. San Francisco, CA: Open Media/City Lights Books.

Nowrojee, B. (1996). *Shattered Lives: Sexual Violence during the Rwandan Genocide and Its Aftermath*. New York: Human Rights Watch.

Oksala, J. (2012). *Foucault, Politics, and Violence*. Evanston, IL: Northwestern University Press.

Opotow, S. (2001). Reconciliation in a time of impunity: Challenges for social justice. *Social Justice Research* 14: 149–170.

Pain, R. (1991). Space, sexual violence and social control: Integrating geographical and feminist analyses of women's fear of crime. *Progress in Human Geography* 15: 415–431.

Pain, R. (2014). Everyday terrorism: Connecting domestic violence and global terrorism. *Progress in Human Geography* 38(4): 531–550. doi:10.1177/0309132513512231

Pape, R.A. (2005). *Dying to Win: The Strategic Logic of Suicide Terrorism*. New York: Random House.

Shaw, I.G.R. (2013). Predator empire: The geopolitics of US drone warfare. *Geopolitics* 18: 536–559.

Shaw, I.G.R., & Akhter, M. (2012). The unbearable humanness of drone warfare in FATA, Pakistan. *Antipode* 44: 1490–1509.

Spade, D. (2011). *Normal Life: Administrative Violence, Critical Trans Politics, and the Limits of Law*. Brooklyn, NY: South End Press.

Springer, S. (2011). Violence sits in places? Cultural practice, neoliberal rationalism, and virulent imaginative geographies. *Political Geography* 30: 90–98.

Springer, S. (2012). Neoliberalizing violence: Of the exceptional and the exemplary in coalescing moments. *Area* 44: 136–143.

Sturdevant, S.P., & Stoltzfus, B. (1992). *Let the Good Times Roll: Prostitution and the U.S. Military in Asia*. New York: New Press.

Sutherland, E.H. (1940). White-collar criminality. *American Sociological Review* 5: 1–12.

Tyner, J.A. (2009). *War, Violence, and Population: Making the Body Count*. New York: Guilford Press.

Tyner, J.A. (2014a). Dead labor, landscapes, and mass graves: Administrative violence during the Cambodian genocide. *Geoforum* 52: 70–77.

Tyner, J.A. (2014b). Dead labor, Homo sacer, and letting die in the labor market. *Human Geography* 7: 35–48.

Tyner, J.A., & Inwood, J. (2014). Violence as fetish: Geography, Marxism, and dialectics. *Progress in Human Geography*. Online before print, January 28. doi:10.1177/0309132513516177

Weaver, G.M. (2010). *Ideologies of Forgetting: Rape in the Vietnam War*. Albany, NY: SUNY Press.

Yuval-Davis, N. (1996). Women and the biological reproduction of "the nation." *Women's Studies International Forum* 19: 17–24.

Žižek, S. (2008). *Violence*. New York: Picador.

Chapter 11

Justice

Farhana Sultana

Syracuse University, Syracuse, New York, USA

Injustice anywhere is a threat to justice everywhere. (Martin Luther King, Jr.)

Notions of rightness, fairness and justice are so firmly entrenched in our vocabularies, that we seem powerless to make any political decision without appealing to them. (Harvey 1996: 332)

Justice is an ideal, a contested term, and a fluid and open concept. Justice invokes notions of fairness, of equity, and of doing the "right" thing. Justice also evokes principles by which the benefits and burdens of society should be distributed among people. Social justice crusaders such as Martin Luther King, Jr. and Mahatma Gandhi are well known for their tireless efforts to bring justice to oppressed groups in their respective countries. Both also have global appeal to social justice advocates across the world. Numerous disciplines have approached theorizing and discussing justice, ranging from philosophical debates to activist-oriented work. Justice is an interdisciplinary topic and scholars from across the spectrum have contributed to debating and enriching it ontologically, epistemologically, and methodologically in recent years. The goals and definitions of equality and a just life have thus been critically debated across academic disciplines (Kymlicka 1990). While justice is often associated with the criminal justice system, most social sciences tend to focus on social justice.

Generally viewed as a universalized notion that is supposed to exist above the fray, justice is expected to be a guiding principle for a just social order and a humane society. However, the word is more a signifier than an absolute concept. Justice risks being an all-encompassing term for progressive politics without clear content or meaning. As a normative ideal, justice is not placeless or timeless, but is rather produced through social processes, historical legacies, and political overtures. Distributive justice is often invoked to underscore the spatial, political, and social distribution of resources, rights, and opportunities. Processes that produce unfairness, inequities, and injustices are often identified and deconstructed in an attempt to produce more meaningful understandings of justice that are contextual and realizable. Human sameness in needs and rights is seen as central to overcoming the unequal, arbitrary distribution of

The Wiley Blackwell Companion to Political Geography, First Edition.
Edited by John Agnew, Virginie Mamadouh, Anna J. Secor, and Joanne Sharp.
© 2015 John Wiley & Sons Ltd. Published 2017 by John Wiley & Sons Ltd.

the earth's resources as well as socially constructed enactments of difference that create inequities (Smith 2000). Social and political imaginaries of a universalized ideal of justice thus can drive research and the framing of analysis as well as expose contradictions in different contexts. Issues of power, democracy, political subjectivity, citizenship, role of the state and other institutions, and social struggles are all thus opened up for further critique.

This chapter will engage with the various notions and theorizations of justice across disciplines and with how geographers have contributed to existing debates and can continue to do so. After an overview of the scope and definitional challenges of the notion of social justice, the ways in which rights and democracy are linked to justice are presented. Historical legacies of colonialism and current politics of international development and geopolitics are explored to globalize, spatialize, and temporalize justice. Geography's relevance and contributions to justice are further investigated through the debates around space, place, scale, neoliberalism, and capitalism, whereby the tensions between universalities and particularities become evident. Environmental justice and climate justice are then examined as key advances that geographers have made. Similarly, gender justice and race justice are explored, through which feminist and critical scholars have been paving the way to understanding justice in more nuanced and intersectional ways. The chapter concludes with possible avenues of further engagement by geographers in theory and practice.

Scope and definitional challenges

John Rawls's *A Theory of Justice* (1971) is generally considered to be the foundational text on liberal notions of justice. Rawls argued that rational people would have a common understanding of a "sense of justice," even if it is difficult to define precisely how just institutions and systems are configured. He posited that justice basically means fairness in a society, whereby rules and regulations are agreed on to enact fair treatment and redress injustices in that society. Principles that guide such a social contract are based on criteria that Rawls identified as important for a just society, such as the "original position," where a distanced view of justice is used to define parameters such that, since no one knows how they individually will benefit from the rules (i.e., as they are operating under a "veil of ignorance"), people are more inclined to set a system in place that maximizes fairness and equality. The notion of equal rights to a system of liberties is central to Rawls's idea of justice as fairness. This is supposed to lead to the greatest benefits to the most disadvantaged. Often, liberal discourses of justice tend to focus on the distribution of income and class inequalities within a society. However, what is to be distributed and how it should be done, and whether it should be viewed individually or not, remain contentious in liberal notions of justice.

The emphasis on distribution of income and equality as espoused by liberal discourses of justice has been demonstrated to be inadequate by scholars who focus on a broader range of issues that encompass justice, as well as a more comprehensive set of social differences (e.g., class, gender, race, sexuality, environment). Such scholars generally argue that simple redistribution is insufficient to address justice. Marxist scholars point to structural inequities created by the class system, whereby class is seen to be a source of injustice in society (Peffer 1990; Smith 1994). Oppression is not viewed as merely a problem of redistribution but rather as a symptom of capitalistic logic and a market economy that exploits labor. In contrast, Iris Marion Young (1990) advances a conceptualization of justice as plural and respecting of difference and multiplicity, contrasting with Rawls's ideas of justice that focus more on redistribution. She places a greater emphasis on social structures, institutions, and relations that systemically create and perpetuate injustices, and social justice is theorized and understood more pragmatically and less ideationally. Young stressed the importance of deconstructing and heeding complex issues around exploitation, marginalization, powerlessness, cultural imperialism, and violence. More recently, she focused further on responsibility and social processes that lead to collective action and shared responsibility to address social justice beyond a politics of difference (Young 2011). Understanding processes of marginalization becomes important for excavating the causes of injustice along a range of intersecting social positions (e.g., children, the elderly, disabled, racial groups). Similarly to Young, Nancy Fraser (1997) has also argued that a focus on the misdistribution of benefits and burdens needs to be viewed simultaneously with "recognition" (i.e., redressing the imposition of the norms of dominant groups on others. By engaging with insights from a range of philosophical debates, such articulations of justice open up the possibility of exploring the meanings and practices of in/justice more broadly and in more nuanced ways.

Justice is often understood as equal status or a recognition of equality, so that different groups can expect the same opportunities and treatment as any other group. A sense of justice or feeling of justice is a prevalent way to describe the goals of achieving solidarity and equality around a common humanity (Barnett 2010). It would be a morally appropriate action to treat groups in the right or ethical way. Differential treatment can thus raise calls for addressing injustice, both under the law and in society (e.g., rights to same-sex marriage). However, since notions of what is "right" or "ethical" are often derived from socially held norms (e.g., religious belief) and not merely an abstract notion of justice, such calls often conflict with different notions of morality and ethics, which are also then claimed to disallow equal status. Justice as a process can thus become conflictual and contradictory. Scholars have further posited that injustice may be necessary for justice to be imagined and acted on (Nagel 2005).

Since there is no real metric for justice, it is more comprehensible in its negation; that is, then we are able to identify and measure injustice more readily and widely (e.g., systemic ostracization of groups of people, violation of rights of individuals, structural violence that deprives some groups based on race; Barnett 2010). The multitudinous ways in which justice could be achieved in each instance is far more difficult to agree on, let alone enact. While a reduction in overall cases of injustice or inequality can be viewed as an accomplishment, there will always be complexities and challenges to any articulation of having achieved social justice, as other axes of injustice may exist simultaneously. The complexities contained in any definition of justice thus require critical theoretical analysis as well as grounded empirical work (Hay 1995).

Fears of absolutism or authoritarianism make it nearly impossible to have one concretized definition of justice. Justice must always be held in a constant state of flux, with a general embeddedness in ideals and desires of fairness and rightness. An example of a universalized sense of social processes of justice is enshrined in the United Nations (UN) charter on human rights. However, any notion of justice must be understood and enacted within particular contexts where the universal signifiers come to bear meaning. Even then such meanings of justice may be partial, contested, and heterogeneous. The broader political economy, historical legacies, cultural practices, and social processes are all imbricated in formulating any understanding of justice, especially social justice. The roles of place and scale are equally important, especially for geographers, as justice operates across places and scales in different ways and can bear different meanings and outcomes (Harvey 1996; Waterstone 2010).

Harvey (1996: 330) defines justice as:

> a socially constituted set of beliefs, discourses, and institutionalizations expressive of social rela-
> tions and contested configurations of power that have everything to do with regulating and
> ordering material social practices within places for a time. Once constituted, the trace of a
> particular discursive conception of justice across all moments of the social process becomes an
> objective fact that embraces everyone within its compass. Once institutionalized, a system of jus-
> tice becomes a "permanence" with which all facts of the social process have to contend.

Since justice must thus be seen to be contextual and situated, rather than only universal and
abstract, tension between particularities and universalism creates challenges for the way in
which justice as a concept is defined, interpreted, and contested (Waterstone 2010). Justice
thus often remains a vague ideal that should shape political and social society, but is difficult
to articulate in specifics outside of its context and relational meaning to other discourses (e.g.,
rights). The fluidity and malleability of this esoteric term can propel political mobilization and
action as much as it can confound them. While universal notions of justice can drive goals of
seeking fairness and equality, they can also be challenged by multiple interpretations and practices of what constitutes such normative goals in each location and society. Given a general
lack of what could count as a universal metric for measuring justice, it becomes imperative to
focus on the processes and structures that create injustice and to configure what could be considered to redress such situations.

The dialectical relationship between universality and particularity is mediated through
institutions and practices. Harvey (2000: 242) argues that

> the notion of justice ... acquires universality through a process of abstraction from particular
> instances and circumstances, but becomes particular again as it is actualized in the real world
> through social practices.

Struggles for social justice in a range of places can inform broader notions of justice. Justice is
very much a social process and thus this social and context contingency necessitates that the
meanings and practices of justice anywhere must be open and flexible (Waterstone 2010).
Given this, cultural differences and political variances must be accounted for without falling
into relativism, which can be counter-productive. The uneven geographies of injustice must be
seen not only in relation to universal norms but also within its socio-historical geography, and
this tension – the balancing act between universals and particulars – remains a challenge in
geography. Smith (2000: 1157) thus argues:

> if the definition is grounded in a particular culture, or "thick" conception of the good, this under-
> mines its universality and the possibility of considering justice in distribution at a broad, even
> global, scale.

While the core content of justice is simultaneously contextual and controversial, a common
theme has emerged in geographical debates about justice. The capitalist system's inherent production of uneven geographies of development and equity has become a topic of exploration
and analysis since Harvey's groundbreaking work in *Social Justice and the City* (1973). An
understanding of the scalar and spatial processes that produce and reproduce injustices and
difference through the logic of capital has influenced geographers to advance conceptualizations on a range of issues, from the urban sphere to broader political geographies. Smith's

seminal work *Geography and Social Justice* (1994) further elaborated on how Marxian geography can inform different theories of justice. Marxian notions of class inequality, in arguing for justice as a way to address issues of poverty and discrimination, have been profound in geography. However, as further theoretical work emerged on the various axes of difference across scales and spaces (e.g., critiques of sexism, racism, heteronormativity, ableism), the field of inquiry expanded to account for and accommodate a range of philosophical positionings and empirical analyses (as detailed later in this chapter).

Rights and democracy

Advances in legal and political rights are often seen to be foundational to social justice, but they do not necessarily lead to more progress toward just societies. Individual rights may enable certain claims, but can exist in tension with collective rights and claims. Social justice may be thwarted by competing claims and goals. Moreover, rights discourses can be instrumentalist, utilitarian, technocratic, and limiting (Arendt 1994). They can also be coopted into neoliberal discourses and politics. The constraints of rights discourses are important to heed, while at the same time exploring their liberating opportunities. Rights discourses can enable legal instruments to be used to protect the vulnerable and pursue social justice. Furthermore, rights can be morally claimed, and are not necessarily enforceable or legal (Chatterjee 2004). The following insight captures this conundrum (Smith 2000: 1154):

> If certain things are needed to live a human life, it might be argued that all people everywhere should have them by right. If social justice is to prevail, the moral imperative often associated with rights can give strength to particular entitlements. However, the notion of rights raises difficult issues, with respect to what they are, how they should be prioritised, who bears them (and where), and who have the consequent obligations to ensure that the rights are fulfilled.

Another aspect of rights in justice debates is the role of actors with various agendas. The role of the state in its policies and approaches is critical in enabling or disabling rights, but so are the roles of private corporations, civil society, and international agencies. When neoliberal policies allow accumulation by dispossession, social justice narratives can be used to challenge such processes through exposure of the denial of basic rights. Given that poverty and violence have increased globally with a concomitant growth in extreme wealth and power, such contradictions often give rise to greater calls for social justice. Such pleas open possibilities for the repudiation or examination of political control and disenfranchisement, as well as economic policies that foster injustices that are social, economic, political, and ecological. Thus, while a desire to focus on rights and democracy may lead to a broad-based consensus, it does not necessarily produce social justice (see also Merrett 2004).

International development and global politics

Thinking about justice internationally raises a host of interrelated concerns. Justice is desired by almost all global institutions and nation-states, even if these very entities are often implicated historically and spatially in various forms of injustice. Frequently, addressing justice via alleviating poverty is seen as a critical component of achieving development goals in the global South. However, the specifics are always bound to be contentious, as various stakeholders will vie to define what the best form of development is, how to achieve it, how to measure it, and how to ensure its sustainability. While a vast percentage of humanity lives in dire conditions and faces chronic starvation and political strife, the global consensus on the need to pull people out of suffering and thereby achieve social justice remains trapped in contradictory actions and policies. The rise of poverty and homelessness in the global North and the increasing impoverishment and marginalization of people in the global South are subject to similar but different sets of political and economic factors that make achieving social justice a complex and controversial task, which is not only context based but also subject to a range of political posturing and policy-making internationally (Escobar 1995). Given the political implications of economic globalization, and uneven geographies of distribution of gain and loss, increasingly rapid globalization can exacerbate social inequality and injustice across the globe through a combination of the policies and practices of a range of actors and institutions (Fraser 2008; Kerner 2010). Instances of corporate control, labor rights violations, unfair trade practices, and state-sanctioned violence abound in the news and in academic literature, highlighting the infractions of notions of justice through processes of economic globalization. Demands for justice, reparation, and due process are becoming increasingly common. Similarly, historical injustices are often invoked to highlight current social conditions (e.g., the African slave trade and its legacy in racial discrimination and structural violence in the contemporary Americas and Europe). Colonial legacies of persecution, expropriation, genocide, and war continue to affect postcolonial societies to this day.

One of the most important arenas of international justice is perhaps connected to the relations of power between nation-states and the international development industry (Pogge 2008). Current postcolonial nation-states that are "developing nations," historically colonies of European imperial control, are now subject to international development policies and aid politics. International institutions such as the United Nations, World Bank, International Monetary Fund, and World Trade Organization, which were established in the post–Second World War era, are enormously powerful players in global aid, trade, and the policies of developing nations. While development assistance can be seen as morally imperative and ethically correct to make reparations for colonial exploitation, existing practices of international development have been the subject of much debate (Corbridge 1998). The goal of achieving social justice through loans and interventions may seem laudable, but such control is often critiqued as unfair and even deemed to be neocolonial. More importantly, threats to territorial control and the sovereignty of the nation-state are brought to the fore, thereby raising concerns about whether international interventions in the name of development can undermine international social justice or fracture it in unknown ways. While some scholars have argued for a focus on enhanced capabilities and rights to ensure socially just development processes, these remain controversial (Sen 1999; Nussbaum 2003).

To political geographers and related scholars, global injustices also occur through discourses of terrorism, whereby some countries/peoples are marked as threats to specific interests and emergent problematic discourses of global democracy and peace, and by which military intervention is deemed to be justified (Flint & Radil 2009). Post–9/11, the United States and its allies have carried out extensive assaults and warfare in Afghanistan and Iraq as part of the War on Terror. Similarly, a rise in Islamophobia globally has resulted in discrimination, violence, and death among Muslims who are tainted as "Other" and thereby deemed to be a threat (Esposito & Kalin 2011). Hegemonic constructions of who is worthy or valid or who counts in global politics thus place many communities and individuals in perpetual harm through the rhetoric and tactics of "liberation" or "democracy" on the part of countries and entities that claim political and moral superiority. However, the broader linkages to colonialism and Cold War geopolitics are often erased in such discourses, thereby posing grave injustices to the lives and voices of people as violence is enacted materially and discursively. As such, these historical conjunctures must be seen against a broader canvas of Orientalism and colonial anxiety (Said 1978; Chatterjee 2004). Ongoing conflicts in the Middle East, Southeast Asia, and parts of Africa are closely linked to these concerns. Injustices of never-ending wars, warfare, and occupation thus continue to confound international deliberations about fairer adjudications and resolutions.

Time, space, scale

Other philosophies of justice that are relevant for geographers and social theorists investigate and explore not only institutions and social structures, but also the processes of achieving justice over time and space. Amartya Sen (2009) argues that justice has to be achieved incrementally, that it is not absolute or dichotomous (i.e., there is never a binary whereby there is the full presence or full absence of justice). Rather, he points out that the gradual decrease of injustice is significant to achieving social justice over time, and that a comparative approach is important. For geographers, this insistence on time may seem to overlook the importance of space (discussed later in this section), but the salience of the assertion that justice "takes time" cannot be overlooked, especially given ever-increasing complex political-economic conditions. Barnett (2010) argues that such a notion of justice also unties it from certain normative containments of achieving a perfect theory of justice (see also Nussbaum 2006). In a similar vein, temporal aspects of justice are captured in concerns about intergenerational equity and justice across generations (Meyer 2004).

For geographers, spatial justice has recently emerged as a body of scholarship that focuses on links between social justice and space. Geographers have focused on the spatiality or spatial formation of various injustices. The spatiality of injustices and justice movements is often the key focus of such research and conceptualizations (Soja 2010). Bringing a spatial perspective to understanding and conceptualizing social justice and political life is central to this, as are recent counter-mapping and participatory projects that seek to highlight spatial injustices. Spatial justice is often linked to broad bodies of scholarship on the right to the city as well as radical justice movements in particular places (e.g., Harvey 1973; Dikec 2001; Marcuse et al. 2009; Fincher & Iveson 2012). The focus of spatial justice is related to debates around territorial justice and geographies of injustice, but it focuses more on the importance of the production of space and spatiality in theorizing justice, whereby political objectives are not merely to address spatial fixes for injustices but to think about social justice differently vis-à-vis conflict, difference, and politics. Philippopoulos-Mihalopoulos (2010) argues that spatial justice should not be seen as "add space and stir" in relation to distributive justice, but rather as a more radically informed notion of emplacement and justice. Geographers have thus also studied the relationship between law and space in the spatiality of social justice (e.g., Blomley, Delaney, & Ford 2001). It has been stressed that legal and juridical justice is different from social justice, and that procedural justice is therefore not the same as the various notions of social justice. While the two are related, it becomes imperative to engage with and understand the ways in which social justice is imbricated with the complexities of social systems across scales and sites. While advances may be made in official political and legal rights, there may not be a commensurate enhancement of social justice across all places.

More recently, postcolonial and feminist scholars have broadened the field of analysis to address issues of identity politics, culture, and various forms of difference (e.g., Young 2011;

Fraser 2001). Greater intersectional understandings of oppression and injustice are brought into the discussion on justice, rather than territorially bound or class-based analyses (both of which are dominant and important modes of analysis in political geography, but can also thwart more nuanced comprehensive analysis). Social justice is more broadly debated now and more specifically considered in issues related to gender, race, sexuality, and environment. Ideologically divergent groups have rallied around the vague concept of justice in order to rescue it from relativism by grounding it in a non-oppressive and inclusive dialogue. This is most developed in the fields of environmental justice and gender justice, which are discussed in turn in the following.

Environmental justice and ethics

Environmental justice has emerged as a way to bring nature and society under the rubric of social justice. The distribution of environmental harm has historically been skewed, such that poor and minority communities have been disproportionately exposed to toxic dumps and environmental pollution (Cutter 1995, 2006). Tactics such as NIMBY (not in my backyard), which enabled communities with class and race privilege (i.e., wealthy white neighborhoods) to shift pollution on to more marginalized groups (by class and race), led to a systematic process of discrimination across the United States and Canada. Such geographical and social inequities gave rise to environmental justice as a movement and as a conceptual framework to assess and understand environmental inequities and the spatiality of harm (Pulido 2000). The production of vulnerability, exacerbation of health impacts, and inequities of decision-making processes that unevenly distribute harm are central concerns for environmental justice advocates (Gleeson 1996; Holifield, Porter, & Walker 2010). Scholars have advanced a multiplicity of notions of environmental justice and environmental racism, from scalar politics to urbanization and the splintering of environmental harm (e.g., Low & Gleeson 1998; Swyngedouw & Heynen 2003; Forsyth 2008; Schroeder et al. 2008). Increasing attention is now being given to environmental justice at a global scale and to the differences connecting scales of analysis and the issues at stake, as well as to impacts for future generations (e.g., Clark 2010).

Linked to, but separate from, the discussion above is the increasing attention being paid to an environmental ethic that heeds the needs of non-human others (Whatmore 2002). The impacts of human actions on species and ecosystems are becoming important to scholars and policy-makers concerned with sustainability. Declining biodiversity and species loss raise important questions of social justice, in that the uneven distribution of the globe's resources means that different places face different impacts. For instance, the demand for monocultures of cash crops can displace people from agricultural communities and reduce the growth of food crops, thereby increasing food insecurity and poverty in the area. This form of affecting "distant others" is increasingly being considered with "others" that are animals and plants in accounting for interspecies justice (Low & Gleeson 1998). Calls for engaging in a discourse of justice invoke the need to recognize and address such spatial destruction of habitats (of humans and non-humans), as well as to link these to the broader injustices of rapacious capitalism. Similarly, given the unequal allocation of important natural resources, the increasing neoliberalization of the governance of such critical nature is a growing concern and area of study for geographers. While some places have different endowments of different critical resources (oil, forests, water bodies), the entire globe is increasingly governed through a capitalist logic that seeks to capture and commoditize resources for global consumption. The privatization and capture of water constitute perhaps the most glaring injustice that exists, as the poor are pushed out of the market when water becomes a commodity for market purchase linked to private gain. Given water's non-substitutable and life-giving nature, growing conflicts have arisen as a result of injustices resulting from its being transformed from a public good to a private commodity (Bakker 2010). As a result, to reverse such trends to inequity, a global call for the right to water emerged, articulating that water should be held in the commons as it is necessary for survival. Global water justice movements have been demanding more just and equitable governance of this necessary resource (Sultana & Loftus 2012).

Related to such global efforts, climate justice is a recent development, and is related to environmental justice more broadly. Whereas environmental justice discourses and activism focus on environmental racism and injustices in the ways in which pollution and ecological degradation affect communities of color and other disenfranchised groups, emergent climate justice discourses and activism highlight environmental harm across the globe and the unevenness of the benefits and burdens of climate change. Countries of the global South (or developing nations) have pointed out the historical responsibility in regard to greenhouse gas emissions of countries of the global North (or developed nations) stemming from industrialization and economic growth, whereas the deleterious impacts of climate change are largely experienced in the countries of the global South. Climate justice thus draws attention not only to spatial injustice, but also to historical injustice in the production of wealth in the global North, often at the expense of the global South through colonialism, imperialism, and exploitation for centuries. Ironically, the dramatic impacts of global climate change make geographical areas of the global South particularly vulnerable, through more intense and uncertain weather-related events (violent storms, sea surges, erratic rainfall, etc.) as well as the worsening of agriculture, food security, water availability, territorial integrity (i.e., loss of land to the sea with sea level rises), migration, loss of biodiversity, disease outbreaks, and a host of interrelated socio-ecological concerns (Adger 2001; Burnham et al. 2013). As a result, discourses of climate justice have entered policy debates and no longer remain only in the realm of academic activist work. Geographical and historical injustices are exacerbated through climate impacts that affect different groups of people differentially and thereby worsen social justice and sustainability concerns across and within countries. While concrete actions and efforts are harder to delineate, the overarching point of climate justice remains to highlight inequities and differences that exist socio-spatially and are likely to worsen over time on a global scale (Clark, Chhotray, & Few 2013). Climate justice scholarship has also investigated various scalar, social, and place-based differences to enhance debates that often get reduced to North–South frameworks. For instance, gendering the impacts of climate change demonstrates the connection between gender justice and climate justice, whereby an intersectional understanding of gender informs the variegations of climate injustices on the ground (e.g., Sultana 2014).

Gender justice

Drawing from a broad range of feminist scholarship, gender justice has become common parlance among feminist scholars and activists who highlight the inequities and inequalities that exist across genders around the world. Thus, feminist scholars have sought to interpret and explain justice through a gender framework (e.g., Okin 2004; Fraser 2007; Young 2011; Seguino 2013). Young (1990) argued for two important interrelated dimensions of gender justice: first, a distribution dimension that involved equal access to material distribution of resources, goods, and services; second, an institutional dimension that focused on equal access to participation in decision-making institutions that define and deliver this distribution. The roles of the state, law-making institutions, and judicial processes are brought into the picture, as are cultural practices, social organizations, and international influence. Building on Young's work, Mukhopadhyay and Singh (2007, 4) posit the following:

> Seeing gender justice as outcome and as process helps differentiate between what is to be achieved and how it is to be achieved. Gender justice, as an outcome, implies access to and control over resources, combined with agency (the ability to make choices). Gender justice as a process brings an additional essential element: accountability, which implies the responsibility and answerability of precisely those social institutions set up to dispense justice.

Scholars have also pointed out that while recognition and identity politics are important in drawing attention to injustices, it is critical to include more complex and intersectional understandings of gender as it operates in relation to other social axes of difference (e.g., race, ethnicity, class, age, sexuality). Differently located men and women face various injustices that are mediated through a complex system of interlocking inequities. For instance, concerns about safety, violence, and mobility may be direr for poor women of color working in urban spaces, who are exposed to street violence, unreliable public transportation, and work in insecure low-paying jobs, in comparison to wealthy white women who may have access to personal vehicles and higher-paying employment options in safer locations. Poor women are often made more vulnerable due to their lack of secure housing, low income, and the general urban blight in many cities across the world. This is not to undermine patriarchal concerns that exist across social categories and the manifestations of various forms of gender oppression in the household, economy, and polity. Nevertheless, an understanding of gender injustice complicates any facile notions of female solidarity that do not account for difference across space, scale, and place. It underscores the need to understand and address gender justice through a simultaneous accounting for common gender oppression and systems of marginalization in patriarchal societies as well as the specificities of each context (Mohanty 2003). Feminist geographers have expounded on a range of issues in which the interpretation and understanding of justice are complicated and various emancipatory options and transformative goals are pursued (e.g., Peake & Rieker 2013). Thus, the goal is not simply to identify injustices and expose them, but also to take normative stances on the transformations and justices that are necessary in a politicized notion of justice (Wright 2010). Scholars have demonstrated myriad ways in which it is important to address gender justice across a wide range of issues, such as city planning, transportation, workplace conditions, wages, reproductive rights, political participation, educational opportunities, environmental hazards and exposure, sexual slavery, and the care economy.

Patriarchal systems can be seen as entrenched forms of injustice, in that there is a skewing of power and control over labor, resources, and decision-making in favor of men over women. Addressing the specific and context-based norms, stereotypes, and practices of patriarchal injustices can be a starting point to addressing broader social justice issues. Inequitable relations of power manifest themselves across numerous aspects of life. For instance, gender-biased inheritance rights, limited financial opportunities and restricted property ownership, discriminatory wages, curtailed access to education and health care, and a diminished voice and input in decision-making forums are some of the many ways in which women face inequities around the world. Incidences of normalization of domestic violence, rape culture, and denial of "place" can also be attributed to patriarchal practices that perpetuate violence on women and girls. Indeed, an international collaborative effort around gender injustice resulted in the UN-sponsored CEDAW (Convention to Eliminate All Forms of Discrimination Against Women). A range of issues have thus become important to scholar activists who are interested in addressing concerns of gender justice not only locally but also globally, especially in relation to capitalist patriarchy (Seguino 2013).

Differences across places and spaces are particularly poignant in the context of developing countries, where gender-based discrimination, exploitation, and violence continue to defy the overarching goals of human rights and equality as espoused in development discourses (Molyneux & Razavi 2002). This is not to be reductionist in reifying injustices in postcolonial societies, or to trivialize gender injustices in developed societies, but to bring to the fore the commonalities and differences that exist across places, as well as to highlight the gendered impacts of colonialism, capitalist neoliberal globalization, power politics at international scales, and the impacts of various development interventions (e.g., structural adjustment programs, SAPs). Insofar as these factors play forceful roles in the lives and livelihoods of women and men in marginal places, it becomes important not to analyze gender justice in isolation or in contained ways, but to trace and identify the ways in which local issues are incredibly connected to global issues, both temporally and spatially (O'Neill 1990). Simultaneously, engagement with issues of representation, sexuality, difference, identity, and belonging has been significant in broadening and nuancing debates around gender justice. Given the proliferation of lip service to gender issues globally, feminist scholarship thus also attempts to expose the normative concepts and meanings attached to any reification of gender justice and its appropriation in neoliberal discourses (Mukhopadhyay & Singh 2007). Bell and O'Rourke (2007: 44) argue for more "substantive and material justice for women" as a way to approach transnational justice that takes into account various forms of feminist struggles and unequal power relations. Thus, the debates around gender justice have embodied greater concerns about citizenship and belonging, and the various ways (formal and informal) in which these come to have a bearing on gendered well-being (e.g., Sultana, Mohanty, & Miraglia 2013).

Conclusion

This discussion has highlighted some important strands in the plurality of theorizations and practices of justice. Overall, the notion of justice has to be grounded in knowledge and experience and to develop out of context. No idealized definition is truly possible. This vague yet alluring term has been used in a variety of ways (as noted in the chapter) and has been subject to theorizing from a range of perspectives. Thus, the debates around the meaning and content of an abstract term such as justice are multistranded and complex. Perhaps the growing calls for social justice can be interpreted as a shorthand way of critiquing neoliberal globalization, rapacious capitalism, and the unjust use of power. The injustices and inequities arising from social, cultural, and environmental degradation that are linked to a global economy are increasingly receiving attention from scholars within and outside of geography. Making connections across places and scales, and engaging with a range of theorizations of justice, thus make for more robust critiques of existing realities and gesture at possible alternatives that are more just and fair. Invocations of social justice can also be seen as a call for action and solidarity. In exposing the roots of oppression and marginalization, and articulating the ways in which a range of injustices are linked or related, geographers are able to expound on different responsibilities and options for profound political and social change. Appeals to justice thus can be political moments that foster the envisioning and acting on of democratic and radical alternatives. This is increasingly so in political and social movements that are fighting for democracy and rights in countries across the world. Calls for justice are galvanizing those who are marginalized and oppressed, whether it is related to voter suppression in the United States, dictator rule in the Middle East, or gender-based violence anywhere.

Geographers thus have enormous opportunities to enrich and advance existing debates and scholarship on social justice as well as to contribute to ongoing struggles and praxis. In the age of the Anthropocene, when humans control the fate of the earth unlike any other time in history, it behooves geographers to engage in this area diligently and meaningfully. Emerging research topics can address persistent challenges regarding the ongoing exclusion and marginalization of most of the world's poor, transnational injustices in global geopolitics, existing injustices across gender, race, and class in local landscapes, and scalar connections across ecological injustices. Geographers can thereby continue to contribute to spatializing and placing justice debates, and to demonstrate the connectivities across universalities and particularities. Scholarship and activism on the various forms of environmental injustice and the politics of climate justice, across sites and scales, are also arenas that can benefit from greater geographical analysis. Critical race geographers and feminist geographers can further advance nuanced understandings of how inter-sectionalities and power operate across injustices and spaces. The ever-increasing complexities of geopolitical crises globally, with links to neoliberalism, capitalism, and notions of development, continue to be topics with which political geographers need to engage to reach better explanations of the ways in which justice is challenged and reconfigured. Geographers can thus make great contributions to the existing debates around justice, both theoretically and empirically.

Acknowledgements

I am grateful to the editors for inviting me to write this chapter, particularly to Virginie Mamadouh and Jo Sharp. I also thank the reviewers for their comments and feedback on the chapter. Any errors of interpretation and presentation remain mine.

References

Adger, N. (2001). Scales of governance and environmental justice for adaptation and mitigation of climate change. *Journal of International Development* 13(7): 921–931.

Arendt, H. (1994). *The Origins of Totalitarianism*. New York: Harcourt Books.

Bakker, K. (2010). *Privatizing Water: Governance Failure and the World's Water Crisis*. Ithaca, NY: Cornell University Press.

Barnett, C. (2010). Geography and ethics: Justice unbound. *Progress in Human Geography* 35(2): 246–255.

Bell, C., & O'Rourke, C. (2007). Does feminism need a theory of transitional justice? An introductory essay. *International Journal of Transitional Justice* 1: 23–44.

Blomley, N., Delaney, D., & Ford, R.T. (2001). *The Legal Geographies Reader: Law, Power, and Space*. Oxford: Blackwell.

Burnham, M., Radel, C., Ma, Z., & Laudati, A. (2013). Extending a geographic lens towards climate justice, part 1: Climate change characterization and impacts. *Geography Compass* 7(3): 239–248.

Chatterjee, P. (2004). *The Politics of the Governed: Popular Politics in Most of the World*. New York: Columbia University Press.

Clark, N. (2010). Volatile worlds, vulnerable bodies: Confronting abrupt climate change. *Theory, Culture and Society* 27(2–3): 31–53.

Clark, N., Chhotray, V. and Few, R. (2013). Global justice and disasters. *The Geographical Journal*, 179: 105–113.

Corbridge, S. (1998). Development ethics: Distance, difference, plausibility. *Ethics, Place and the Environment* 1(1): 35–53.

Cutter, S. (1995). Race, class and environmental justice. *Progress in Human Geography* 19(1): 111–122.

Cutter, S. (2006). *Hazards, Vulnerability and Environmental Justice: Risk, Society and Policy*. London: Earthscan.

Dikec, M. (2001). Justice and the spatial imagination. *Environment and Planning A* 33(10): 1785–1806.

Escobar, A. (1995). *Encountering Development: The Making and Unmaking of the Third World*. Princeton, NJ: Princeton University Press.

Esposito, J., & Kalin, I. (2011). *Islamophobia: The Challenge of Pluralism in the 21st Century*. New York: Oxford University Press.

Fincher, R., & Iveson, K. (2012). Justice and injustice in the city. *Geographical Research* 50: 231–241.

Flint, C., & Radil, S. (2009). Terrorism and counter-terrorism: Situating al Qaeda and the global War on Terror within geopolitical trends and structures. *Eurasian Geography and Economics* 50(2): 150–171.

Forsyth, T. (2008). Political ecology and the epistemology of social justice. *Geoforum* 39(2): 756–764.

Fraser, N. (1997). *Justice Interruptus: Critical Reflections on the "Postsocialist" Condition*. New York: Routledge.

Fraser, N. (2001). Recognition without ethics? *Theory, Culture and Society* 18(2–3): 21–24.

Fraser, N. (2007). Feminist politics in the age of recognition: A two-dimensional approach to gender justice. *Studies in Social Justice* 1(1): 23–35.

Fraser, N. (2008). *Scales of Justice: Reimagining Political Space in a Globalizing World*. Cambridge: Polity Press.

Gleeson, B. (1996). Justifying justice. *Area* 28(2): 229–234.

Harvey, D. (1973). *Social Justice and the City*. Oxford: Blackwell.

Harvey, D. (1996). *Justice, Nature and the Geography of Difference*. Oxford: Basil Blackwell.

Harvey, D. (2000). *Spaces of Hope*. Edinburgh: Edinburgh University Press.

Hay, A. (1995). Concepts of equity, fairness and justice in geographical studies. *Transactions of the Institute of British Geographers* 20(4): 500–508.

Holifield, R., Porter, M., & Walker, G. (2010). *Spaces of Environmental Justice*. Chichester: Wiley-Blackwell.

Kerner, I. (2010). "Scales of justice" and the challenges of global governmentality. *Public Reason* 2(2): 40–50.

Kymlicka, W. (1990). *Contemporary Political Philosophy: An Introduction*. Oxford: Clarendon Press.

Low, N., & Gleeson, B. (1998). *Justice, Society and Nature: An Exploration of Political Ecology*. London: Routledge.

Marcuse, P., Connolly, J., Novy, J., Olivio, I., Potter, C., & Steil, J. (2009). *Searching for the Just City: Debates in Urban Theory and Practice*. Abingdon: Routledge.

Merrett, C. (2004). Social justice: What is it? Why teach it? *Journal of Geography* 103(3): 93–101.

Meyer, L. (2004). *Justice in Time: Responding to Historical Injustice*. Baden-Baden: Nomos.

Mohanty, C.T. (2003). *Feminism without Borders: Decolonizing Theory, Practicing Solidarity*. Durham, NC: Duke University Press.

Molyneux, M., & Razavi, S. (2002). *Gender Justice, Development and Rights*. Oxford: Oxford University Press.

Mukhopadhyay, M., & Singh, N. (2007). *Gender Justice, Citizenship, and Development*. Ottawa: IDRC.

Nagel, T. (2005). The problem of global justice. *Philosophy and Public Affairs* 33(2): 113–147.

Nussbaum, M. (2003). Capabilities as fundamental entitlements: Sen and social justice. *Feminist Economics* 9(2–3): 33–59.

Nussbaum, M. (2006). *Frontiers of Justice: Disability, Nationality and Species Membership*. Cambridge, MA: Harvard University Press.

Okin, S. (2004). Gender, justice and gender: An unfinished debate. *Fordham Law Review* 72(5): 1537–1567.

O'Neill, O. (1990). Justice, gender, and international boundaries. *British Journal of Political Science* 20(4): 439–459.

Peake, L., & Rieker, M. (2013). *Rethinking Feminist Interventions into the Urban*. London: Routledge.

Peffer, R.G. (1990). *Marxism, Morality, and Social Justice*. Princeton, NJ: Princeton University Press.

Philippopoulos-Mihalopoulos, A. (2010). Spatial justice: Law and the geography of withdrawal. *International Journal of Law in Context* 6(3): 201–216.

Pogge, T. (2008). *World Poverty and Human Rights: Cosmopolitan Responsibilities and Reforms*, 2nd edn. Cambridge: Polity Press.

Pulido, L. (2000). Rethinking environmental racism: White privilege and urban development in Southern California. *Annals of the Association of American Geographers* 90(1): 12–40.

Rawls, J. (1971). *A Theory of Justice*. Cambridge, MA: Harvard University Press.

Said, E. (1978). *Orientalism*. New York: Vintage.

Schroeder, R., St. Martin, K., Wilson, B., & Sen, D. (2008). Third world environmental justice. *Society and Natural Resources* 21(7): 547–555.

Seguino, S. (2013). Toward gender justice: Confronting stratification and unequal power. *Multidisciplinary Journal of Gender Studies* 2(1): 1–36.

Sen, A. (1999). *Development as Freedom*. New York: Knopf.

Sen, A. (2009). *The Idea of Justice*. Cambridge, MA: Harvard University Press.

Smith, D. (1994). *Geography and Social Justice*. Oxford: Blackwell.

Smith, D. (2000). Social justice revisited. *Environment and Planning A* 32(7): 1149–1162.

Soja, E. (2010). *Seeking Spatial Justice*. Minneapolis, MN: University of Minnesota Press.

Sultana, F. (2014) Gendering climate change: Geographical insights. *Professional Geographer* 66(3): 372–381.

Sultana, F., & Loftus, A. (2012). *The Right to Water: Politics, Governance and Social Struggles*. London: Routledge.

Sultana, F., Mohanty, C.T., & Miraglia, S. (2013) Gender justice and public water for all: Insights from Dhaka, Bangladesh. Municipal Services Project (MSP) Occasional Paper No. 18. Kingston, Canada.

Swyngedouw, E., & Heynen, N. (2003). Urban political ecology, justice and the politics of scale. *Antipode* 35(5): 898–918.

Waterstone, M. (2010). Geography and social justice. In Smith, S.J., Pain, R., Marston, S., & Jones, J.P. (eds), *The SAGE Handbook of Social Geographies*, London: Sage, 419–434.

Whatmore, S. (2002). *Hybrid Geographies*. London: Sage.

Wright, M. (2010). Geography and gender: Feminism and a feeling of justice. *Progress in Human Geography* 34(6): 818–827.

Young, I.M. (2011). *Responsibility for Justice*. New York: Oxford University Press.

Young, I.M. (1990). *Justice and the Politics of Difference*. Princeton, NJ: Princeton University Press.

Chapter 12

Power

Joe Painter

Durham University, England, UK

Politics is about power and power lies at the heart of political geography. Indeed, Mark Haugaard and Stewart Clegg identify power as *the* central concept of all the social sciences (2009: 1). At first glance, power seems straightforward – in everyday language someone is powerful if they can achieve their goals despite opposition. Yet, dig a little deeper and power turns out to be rather more difficult to grasp. Is it a thing, a substance, a quality, an idea, or a relationship? Can it be won and lost? Does it exist when it is not being used? Is it only a negative force, or can it do good? And how does it affect, and how is it affected by, geography? Power certainly shapes political geographies, and yet, while political geographers have long studied in depth the geographical basis and effects of power on their subject matter, their discussions of the nature of power itself have been much rarer (Low 2005). Indeed, for most of the history of the discipline power has been an unexamined concept, whose meaning has been assumed to be sufficiently obvious not to warrant detailed investigation. There are already numerous books surveying theories of power in the social sciences in general (e.g., Clegg 1989; Clegg & Haugaard 2009; Scott 2001), so rather than rehearse the wider debate on power, this chapter will focus on how the concept has been understood in political geography, broadly conceived.

Taking power seriously

It is often said that power is the stuff of politics and that political geography originated in the study of the power-laden conflicts between the most powerful states of the nineteenth and early twentieth centuries. Such talk contains an implicit theory of power; namely, that power is a substance ("stuff"), which is present in political conflicts and of which states can be full or, by implication, empty. Until relatively recently, political geography has been resolutely state-centric, and power meant state power, and in particular the control of territory. State, power, and territory formed a kind of Holy Trinity for political geography. The state was

The Wiley Blackwell Companion to Political Geography, First Edition.
Edited by John Agnew, Virginie Mamadouh, Anna J. Secor, and Joanne Sharp.
© 2015 John Wiley & Sons Ltd. Published 2017 by John Wiley & Sons Ltd.

understood as an institution, territory as bounded space, and power as control. The state "used" power to control territory and sometimes to expand it, usually at the expense of other states. Power as control; power as tool; power as possession; power as substance; power as power *over* territory – thus did political geographers understand one of their central concepts. This understanding was mostly implicit. Although power was a key concept for the subject, it was largely taken for granted. It seemed self-evident that the "Great Powers" would "wield" power (or seek to), that the power of a state could increase or decrease, or that power could be employed to obtain political objectives. And yet, even within this seemingly straightforward discourse, untroubled by theoretical introspection, there are the seeds of something more complex and even contradictory.

Within this power-politics language are several different understandings of power. Thus, we can distinguish the idea of power as a substance that can be possessed from the idea of power as a relationship, such as control or domination. We can also identify an instrumental conception of power – power as something that can be "used" or "wielded," a tool that can be brought into action at will. We might contrast the instrumental conception with the idea of power as an attribute of the state – the idea that the state is the state by virtue of its power over territory, for example. We might also contrast the idea of power as control with that of power as capacity, albeit still the capacity of the state to achieve its objectives. Finally, the notion that power is an instrument, by implication held and wielded centrally, seems different from the idea that power is spread out over territory – the idea that the power of the state is somehow mysteriously everywhere.

While we can spot these seeds of a more developed and reflective understanding of power in the traditional canon of political geography, it has to be admitted that they have taken some time to germinate. Alongside many other concepts in the lexicon of political geography, including "state" and "territory," power has moved from being a term that is largely taken for granted to one that has been the object of considerable theoretical reflection. Like most interesting words, over the centuries "power" has acquired many meanings and nuances of meaning. Difficulties arise not from the multiple meanings, which are inevitable and potentially productive, but from lack of clarity over the meanings and unremarked slippages between them. Geographical scholarship has sought not only to tease out and clarify differing conceptions of power, but also to develop ideas of power in new directions. This broadening of political geographers' engagement with one of their most important concepts has happened for two reasons. First, the field of political geography has itself become much broader, both empirically and theoretically, as the contents of this *Companion* attest. Political geography is markedly less state-centric than it was, and has drawn into its remit a wide range of new topics, from gender relations and sexuality to political ecology and from citizenship to financial crisis. Secondly, the traditional core concepts of political geography, such as the state and territory, have themselves been subject to more rigorous critique and conceptualization than was previously the case.

The theoretical and empirical diversification of political geography makes it not only impossible, but also undesirable, to come up with a single concept of power (Flint & Taylor 2011: 28–43). Steven Lukes, a leading theorist of power, argued that power is an "essentially contested" concept (Lukes 1974). In other words, it will never be possible to reconcile competing notions of power with reference to empirical evidence, because different ideas about power reflect their proponents' normative world views. There is no consensus about what it is that essentially defines power and little purpose in trying to rule on which definitions are correct (Haugaard & Clegg 2009: 4). Not every approach to understanding power is equally present in the field of political geography. Some approaches, such as those drawing on the writings of Michel Foucault, have been much more heavily represented in the geographical literature than others.

Moreover, power itself takes many forms and has many modes (Allen 2003). A key distinction is often drawn between power *over* and power *to*. A master has power over a slave, while an orchestra has the power to produce inspiring music. "Power over" speaks of restriction, domination, limitation, and constraint; "power to" of creation, expression, production, and possibility. Power to is the older meaning, reflecting the term's origins in the Latin verb "posse" (to be able to). Yet, it is not possible to separate entirely power over from power to. The master also wants the slave to be productive, while the orchestra's music arises in part because the conductor has power over the musicians. And in both cases we are talking about power *relations*. To understand the value and importance of a *relational* view of power, we need to return briefly to the earlier understanding of power as substance.

The stuff of politics

We often talk of individuals, organizations, and countries as "having" power – of the United States as having more power than Afghanistan, for example, or of Vladimir Putin as having more power than the Ukrainian president Petro Poroshenko. Similarly, some political positions are said to have more power than others. Thus, a constitutional monarch has less power than an elected prime minister, and a city mayor has less power than a national president. Political actors and states can also be said to "gain" and "lose" power. The so-called BRICS countries (Brazil, Russia, India, China, and South Africa) are said to be gaining power in the global economic and political system. The language of "having," "gaining," and "losing" power suggests that power is some kind of substance that can accumulate or dissipate, be won or lost, and be moved around from one country or leader to another. This can give rise to the assumption that politics is a zero-sum activity – the idea that somehow there is a fixed amount of power in the world and that if the power of one actor increases then the power of another must decline. Yet, power is evidently *not* a physical substance. It is not something that we can touch or store up.

Early political geographers argued that geographical factors contribute to differences in power between states. According to the "heartland thesis" developed by Halford Mackinder (1861–1947), for example, epochal shifts in the balance of power at the world scale were associated with changes in the benefits accruing from being a land-based or a maritime state, which in turn derived from the changing relative effectiveness of land versus sea transport (Mackinder 1904). Richard Hartshorne's attempt to systematize political geography in the mid-twentieth century acknowledged Mackinder, but identified a number of other factors contributing to "national power," including mineral resources (1954: 175). In fact, there have been numerous attempts to identify the factors that affect "national power." Karl Höhn's (2011) comprehensive and systematic study of the literature in geopolitics, political geography, and international relations found no fewer than 69 different multivariate formulas and indexes designed specifically to measure national power and to rank countries in terms of their power relative to each other. The earliest of the formulas appeared in Germany in 1741, the other 68 between 1936 and 2010 (Höhn 2011: 258). This large number of different attempts to measure national power is testament to the fact that power cannot be quantified directly. Instead, those seeking to measure power must infer or derive it from other national attributes (such as population, energy resources, or level of economic development). This implies that those other attributes, whether individually or in combination, are either proxies for power or the sources of power.

Höhn examined the variables used to quantify national power in the 69 cases. Those appearing between 1936 and 1993 used an average of 13 variables each, whereas those appearing since 1993 used an average of 28, reflecting the greater availability of data since the advent of the internet. The variables used typically include land area, population, economic indicators such as GDP, military expenditure and personnel, reserves of raw materials, and so on. It is notable that efforts to quantify differences in national power continue up to the present day in many academic disciplines. However, the basic principles of the approach can be clearly seen in a classic work of political geography from the 1960s. In *Geography and Politics in a World Divided* (1963: 8), Saul B. Cohen sets out five categories that are pertinent to the "power analysis approach" to political geography: the physical environment; movement (flows of goods, people, and ideas); raw materials, semi-finished and finished goods (both employed and potential); population (including its qualitative and ideological characteristics); and the body politic (administrative form, ideas, and spatial organization). To these Cohen adds a sixth category – space – including the "location, shape, and boundaries of political entities" and "the impact of space upon the[ir] internal character and external relations" (1963: 9). In this approach, power is equated with, or understood to stem from, other characteristics of geographical areas such as population and raw materials. Power itself may not be a substance, but is seen to depend on, or be the product of, material things and their spatial arrangement.

Cohen notes several shortcomings of the power inventory approach and its national power indices. For example, one of the factors in his inventory of national power is urbanization, on the grounds that "urbanization usually reflects greater national cohesiveness, more effectively centralized authority, and higher productivity" such that "highly urbanized societies are now more stable politically" (1963: 10–11). Yet, Cohen immediately concedes that some large cities (he gives the example of Naples) are economically weak and politically unstable, and that political revolt in less urbanized countries often begins in, or is led from, the cities. However, the real problem with the power inventory approach is more fundamental. Power inventories and indexes of national power are not really measures of power at all. Instead, for the most part they are measures of different kinds of resources (whether available or potential). Power, then, does not inhere in the resources themselves, but is by implication an attribute or capacity that is generated by them.

Power as capacity

One of the difficulties with power inventory approaches is that they seek to express national power on a unitary scale. While they appear to be attentive to multiple *sources* of power (economic, military, cultural, etc.), they distill them to produce a single measure of power. They thus neglect the diverse forms and modes that power can take, a diversity that becomes apparent when power is understood in relational terms.

The idea of power as capacity or capability is common in the social sciences and appears consistent with the notion of "power to." Specifically, power is often thought of as the capacity to produce effects and transformations, although it is important to remember that the capacity to prevent transformations that would otherwise occur is also an effect of power. The view of power as capacity still leaves us with the sense that power can be held and possessed by one party and used to affect another. However, the notion of power as capacity has been particularly important to feminist understandings of power. For example, Nancy Hartsock argues that

> women's stress on power not as domination but as capacity, on power as a capacity of the community as a whole, suggests that women's experience of connection and relation have consequences for understandings of power and may hold resources for a more liberatory understanding. (Hartsock 1983: 253; see also Miller 1992)

John Scott (2001) contrasts the notion of power as a dispositional capacity with the mainstream view of power inaugurated by the early twentieth-century work of Max Weber (1968), in which power is understood as the episodic imposition of the will of one actor over another. In the dispositional view, capacities provide the disposition to act and affect things, even if they are not used.

John Allen points out that the notion of power as capacity ("power to") is also consistent with thinking of power as control ("power over"): "the idea that individuals or institutions possess specific capacities to secure the compliance of others can be thought about as just so many ways of achieving dominance over others" (Allen 2003: 16). Yet, the capacity of one party to produce effects depends on the liabilities, susceptibilities, and affordances of the other. A conductor only has the power to produce the effect of music to the extent that the orchestra comprises people who know how to play. A government only has the power to control an international border to the extent that it has sufficient personnel with appropriate training and equipment. These examples hint at some of the complexities of power – it is not simply the exertion of force.

Power in concert

The political thinker Hannah Arendt drew a sharp distinction between power and force. Writing in the middle years of the twentieth century, she produced dynamic and innovative political theory that has so far had relatively limited purchase in geography (but see Cloke 2002; Allen 2008; Dikeç 2012a, 2012b, 2013), despite the fact that space is integral to her understanding of politics and the possibility of political action (Dikeç 2013). Arendt's ideas prioritize human action and her theory of power emphasizes human beings acting in concert to "establish relations and create new realities" (Arendt 1958: 200). The word politics derives ultimately from *polis*, the Greek word for city. For Arendt, however, the *polis* "is not the city-state in its physical location; it is the organization of the people as it arises out of acting and speaking together" (1958: 198). The *polis* in this sense is what Arendt calls the "space of appearance." The space of appearance is not permanent, but "comes into being wherever men are together in the manner of speech and action ... Wherever people gather together, it is potentially there, but only potentially, not necessarily and not forever' (1958: 199). Power, according to Arendt, "is what keeps the public realm, the potential space of appearance between acting and speaking me, in existence." The space of appearance therefore depends on power, since 'what first undermines and then kills political communities is loss of power ...; and power cannot be stored up and kept in reserve for emergencies, like the instruments of violence, but exists on in its actualization' (1958: 200). She thus distinguishes sharply between power and force or strength.

For Arendt, power is not dependent on, or derived from, resources, but comes about from communication between human beings and their development of common purposes. However, this does not mean that they act as a single unit. Arendt emphasizes the plurality of human being – the fact that while we are all of one species and thus able to communicate and cooperate, we are also all unique individuals. The geographer Mustafa Dikeç argues that this means that, in Arendt's conceptualization of politics, space both brings us into relation but also separates us by revealing our distinctiveness vis-à-vis one another (Dikeç 2012b). Another implication of Arendt's approach is that public space is not a given, but has to be brought into being through political action (Lee 2009; see also McEwan 2005). Equally, as Lee suggests, the fact that it *can* be brought into being when people act in concert undermines arguments that we necessarily are seeing the end of public space through privatization. The popular uprisings associated with the so-called Arab Spring in the early 2010s, which turned numerous public spaces, such as Cairo's Tahrir Square, into sites of protest, might be seen as examples of Arendtian power.

Discipline and biopower

Arendt's work offers political geographers a positive conception of power that takes space seriously. However, she restricts the exercise of power to the public realm and to positive political action, which she sees as both a precondition and a product of freedom. She excludes violence, strength, and force from her definition of power, as well as the social relations of the private sphere and influences on human thought and behavior that may be hidden from view – influences that are subconscious or involve deceit, for example. For a broader under-standing of power as something that suffuses all kinds of social relations, whether nominally public or private, and whose operation may not be obvious to those that it affects, we must turn elsewhere.

No theorist has had a greater influence on geographers' thinking about power than Michel Foucault. While the impact of his ideas was felt first and probably to the greatest extent in social and cultural geography, political geography has also experienced its own Foucauldian turn. One of the most important features of Foucault's work is the emphasis that he placed on historical understanding. He showed that power itself has a history – in other words, that the nature of power has changed over time. This approach challenges general definitions of power that, by implication, are intended to be equally applicable in all times and places. Foucault identified three forms of power that have evolved over time. Sovereign power – epitomised by the ritual of public execution – involves submitting to the law of the king or other central authority. From the seventeenth century onward, a new form of power emerges in Europe – disciplinary power – whose techniques

> included all devices that were used to ensure the spatial distribution of individual bodies (their separation, their alignment, their serialization, and their surveillance) and the organization, around those individuals, of a whole field of visibility. They were also techniques that could be used to take control over bodies. Attempts were made to increase their productive force through exercise, drill, and so on. (Foucault 2004: 242)

Unlike the highly visible exercise of sovereign power, disciplinary power operates in subtle and less obvious ways through the inculcation of bodily routines and habits. These are developed through institutions such as schools, prisons, hospitals, factories, and barracks and result in bodies that eventually comply with institutional norms and requirements without constant recourse to force. Disciplinary power thus permeates society, rather than being directed or wielded from a central point. The third type of power, biopower, emerges from the eighteenth century onward, and is directed toward populations rather than individual bodies. Biopower is associated more closely with the state than is disciplinary power (Foucault 2004: 250) and uses technologies such as statistics, epidemiology, public health programs, eugenics, and, increasingly, environmental interventions and risk management to improve the health, longevity, composition, and size of the population.

Biopower and its associated political practices (biopolitics) have received considerable attention from geographers. The securitization of territory and borders during the so-called War on Terror has been a particular focus of concern (e.g., Hannah 2006; Sparke 2006; Adey 2009; Coleman & Grove 2009; Martin 2010), but biopower has also been analyzed in relation to the environment and non-human nature (e.g., Alatout 2006; Braun 2007), colonial administration (Legg 2005, 2008), migration (Bailey 2013), and regional policy (Painter 2013), among other topics.

Geographies of power

With Arendt's idea of power as human action in concert and Foucault's theorization of power as operating through a dispersed series of institutions and practices, we have moved well beyond and away from the simple notion that power involves the direction, willful domination, or control of one person or institution by another. Geography is integral to both. Arendt's notion of power depends on the space of appearance, while Foucault proposed that the analysis of power should be concerned with the "points where it becomes capillary" rather than centralized. It is thus perhaps surprising that geographers have paid relatively little attention to the difference that space makes to power (in comparison with the numerous studies of the difference that power makes to space). A notable exception is the work of John Allen. In *Lost Geographies of Power* (2003), he begins by emphasizing that power is not something tangible that can be "wielded or clung on to or flaunted" (2003: 4), nor is it equivalent to resources (as implied by the power inventory approach) nor even to capabilities, which are effectively just another kind of resource. Instead, Allen suggests that power is a "relational effect" of social interactions. Moreover, the geographies of those interactions – their situatedness in time and space and whether they involve geographical proximity or reach – are integral to and constitutive of the different forms of power to which they give effect.

Allen argues that existing theories of power often involve certain spatial understandings – which may be implicit – of how power is realized in different situations (see also Griffin 2012). If power is understood as centralized authority, then it may be thought to radiate out from center to periphery. If power is understood in Foucault's capillary terms, then it may be thought of as dispersed and immanently everywhere. And Allen suggests that Arendt's view of power as action in concert gives rise to what he terms a transverse networked notion of power, involving lateral associations between potentially diverse social actors (2003: 56–58). While these often implicit geographies of power are revealing, for Allen they do not go far enough. In particular, they are insufficiently attentive to the diverse ways in which space is implicated in the exercise of different modalities of power.

Allen distinguishes between eight modes of power: domination, coercion, manipulation, seduction, authority, inducement, negotiation, and persuasion. Space is integral to all of these modes, but in different ways in each case, and typically in more complex ways than conventional notions of power as either centralized, networked, or immanently everywhere. Grasping that complexity requires a move from seeing space in purely topographical terms to understanding it topologically, with different modes of power involving different forms of proximity and reach. For example, authority, according to Allen, is most effective in proximity, where co-presence enables it to be recognized. By contrast, seduction is a form of power exercised in advertising. It is weaker than authority (advertising can be ignored) but has much greater geographical reach, as can be seen from the appearance of brands on hoardings in cities far from the places in which the ads, and their featured products, are designed and produced. More coercive forms of power can also reach into particular distant places – witness the US drone strikes in Pakistan and Afghanistan, controlled in real time by operators working in control rooms thousands of miles away in the United States.

Through persuasion, *Lost Geographies of Power* has gained considerable authority in subsequent geographical research on power – to the extent that it was featured in the "Classics in human geography revisited" section of *Progress in Human Geography* just ten years after it was published (Agnew 2013; Bulkeley 2013; Haugaard 2013). Allen's subsequent work has developed its ideas further (Allen 2009, 2011; Allen & Cochrane 2010), including a consideration of power as always practiced – often tenuously – in particular contexts, drawing critically on the ideas of pragmatist philosophers John Dewey, William James, and Richard Rorty (Allen 2008).

Rethinking sovereign and state power

The turn toward practice-focused approaches in human geography – whether drawn from pragmatist philosophy or other sources such as non-representational theory (e.g., Anderson & Harrison 2010) or Pierre Bourdieu's social theory (e.g., Hillier & Rooksby 2005) – has begun to have a significant impact in political geography. At the same time, global political events since 2001 have drawn political geographers back to the topic of sovereign power, particularly as Western governments have reacted politically and militarily to terrorist attacks and perceived threats and as the "rising powers" of China, Brazil, India, and Russia have sought greater influence in international affairs. These twin influences have combined to generate an extensive body of work on the practices of sovereignty, rather than a reassertion of older understandings of sovereign power as simple domination within a clearly demarcated state territory. As Alison Mountz (2013) has shown, political geographers have been actively rethinking the meaning of sovereignty and its geographies. This reconceptualization has also been driven by the changing nature of sovereignty (Mountz 2013: 830) as governments have reconfigured their relations with their own and others' territories and populations, and engaged in new forms of military violence.

Conventionally, sovereignty refers to the ultimate authority in a legal order such as a state, although in modern societies the constraints on even the highest forms of authority are such that sovereignty is often as much a rhetorical claim as a practical reality. The writings of Giorgio Agamben have been particularly influential in recent geographical work on sovereignty. Drawing on the ideas of Carl Schmitt, Agamben defines the sovereign as "he who decides the state of exception" (Agamben 1998: 15). In other words, sovereignty is the capacity to suspend the normal rules that apply to all but the sovereign. For Agamben, the paradox of sovereignty is that the sovereign is both the pinnacle of the legal order, and yet simultaneously outside it. For political geographers, and many others, Agamben's work has been used to identify and investigate "spaces of exception" – enclaves or zones where the usual constitutional or legal limitations on the exercise of state power do not apply. Epitomised in Agamben's writings by the figure of the detention or concentration camp, spaces of exception are populated by those (such as undocumented migrants, foreign terrorist suspects, and so-called enemy combatants) who are beyond the normal legal protections afforded to citizens and lawful residents, and sometimes by those who have had such protections withdrawn. Geographers have examined a range of spaces of exception, including the Abu Ghraib prison in Iraq during the US-led occupation between 2003 and 2006, the detention facility at the US naval base in Guantánamo Bay, Cuba, and offshore migrant reception/detention centers. Because these spaces are outside the territory of the state, those held there are outside the state's legal jurisdiction, but still under its authority. This creates a legal no-man's land, created by the exercise of sovereign power and within which there are no restrictions on sovereign power.

As Mountz notes, some political geographers have argued that sovereignty increasingly exhibits spatial blurring of "onshore and offshore, internal and external, inside and out." At the same time, Agambenian notions of sovereign power also involve sharp distinctions. Andrew Neal argues that Agamben sets out the problem of sovereignty in binary terms – either sovereign power involves creating exceptions (by definition limited) to the normal or it exceeds all limits. Neal regards this framing of the problem as "apolitical" and unable to do justice to the "extremely complex principle and practice that is sovereignty" (Neal 2004: 375). Neal prefers Foucault's historical account of sovereignty in his 1976 lectures, published as *Society Must Be Defended*. According to Neal, Foucault initially rejects the juridical model of sovereignty as a single point from which all forms of power derive, in favor of an account of power relations that recognizes their multiplicity, differences, and specificity. Rejecting a juridical account of sovereignty, however, does not mean that the problem of sovereignty has gone away. Instead, it returns with a vengeance, allied to the nation as a form of collective identification, so that those deemed to be enemies of the state are simultaneously constructed as enemies of the nation, society, or the people (Neal 2004: 394–395). In spatial terms this combines centered and dispersed forms of power.

A related set of arguments are advanced by geographers Mathew Coleman and Kevin Grove (2009), who claim that Agamben writes geography out of his account of sovereignty:

> On the one hand, we contend that Agamben's use of biopolitics works … to erase the unevenness of political, economic, and social space. We argue that the result is a perverse reconstitution of sovereign (bio)power in a dazzling, all-encompassing, and totalizing spatial form – not unlike the mapping of state power that Foucault found necessary to unpack via his discussion of the biopolitical. (Coleman & Grove 2009: 490–491)

In Foucault, by contrast, we find a concept of government that points "to the 'abidingness' and yet polyvalent and protean character of ensembles of practices and knowledges referred to nominally as the 'state'" (Coleman & Grove 2009: 492). As the work of contemporary political geographers that Mountz (2013) discusses reveals, state and sovereign power – even when expressed in warfare – is still a matter of the disaggregated and situated exercise of power in myriad spaces and institutions.

References

Adey, P. (2009). Facing airport security: Affect, biopolitics, and the preemptive securitisation of the mobile body. *Environment and Planning D* 27(2): 274–295.

Agamben, G. (1998). *Homo Sacer: Sovereign Power and Bare Life.* Stanford, CA: Stanford University Press.

Agnew, J. (2013). Commentary 1: Hidden or lost geographies of power? *Progress in Human Geography* 37(3): 452–454.

Alatout, S. (2006). Towards a bio-territorial conception of power: Territory, population, and environmental narratives in Palestine and Israel. *Political Geography* 25: 601–621.

Allen, J. (2003). *Lost Geographies of Power*. Oxford: Blackwell.

Allen, J. (2008). Pragmatism and power, or the power to make a difference in a radically contingent world. *Geoforum* 39(4): 1613–1624.

Allen, J. (2009). Three spaces of power: Territory, networks, plus a topological twist in the tale of domination and authority. *Journal of Power* 2(2): 197–212.

Allen, J. (2011). Topological twists: Power's shifting geographies. *Dialogues in Human Geography* 1(3): 283–298.

Allen, J., & Cochrane, A. (2010). Assemblages of state power: Topological shifts in the organization of government and politics. *Antipode* 42(5): 1071–1089.

Anderson, B., & Harrison, P. (eds). (2010). *Taking-Place: Non-representational Theories and Geography*. Farnham: Ashgate.

Arendt, H. (1958). *The Human Condition*. Chicago, IL: University of Chicago Press.

Bailey, A.J. (2013). Migration, recession and an emerging transnational biopolitics across Europe. *Geoforum* 44: 202–210.

Braun, B. (2007). Biopolitics and the molecularization of life. *Cultural Geographies* 14: 6–28.

Bulkeley, H. (2013). Commentary 2: Making space for power? *Progress in Human Geography* 37(3): 454–456.

Clegg, S.R. (1989). *Frameworks of Power*. London: Sage.

Clegg, S.R., & Haugaard, M. (eds) (2009). *The SAGE Handbook of Power*. London: Sage.

Cloke, P. (2002). Deliver us from evil? Prospects for living ethically and acting political in human geography. *Progress in Human Geography* 26: 587–904.

Cohen, S.B. (1963). *Geography and Politics in a World Divided*. New York: Random House.

Coleman, M., & Grove, K. (2009). Biopolitics, biopower, and the return of sovereignty. *Environment and Planning D* 27: 489–507.

Dikeç, M. (2012a). Politics is sublime. *Environment and Planning D: Society and Space* 30(2): 262–279.

Dikeç, M. (2012b). Space as a mode of political thinking. *Geoforum* 43(4): 669–676.

Dikeç, M. (2013). Beginners and equals: Political subjectivity in Arendt and Rancière. *Transactions of the Institute of British Geographers* 38(1): 78–90.

Flint, C., & Taylor, P.J. (2011). *Political Geography: World-Economy, Nation-State and Locality*, 6th edn. London: Routledge.

Foucault, M. (2004). *Society Must Be Defended: Lectures at the Collège de France, 1975–76*. London: Penguin.

Griffin, L. (2012). Where is the power in governance? Why geography matters in the theory of governance. *Political Studies Review* 10: 208–220.

Hannah, M. (2006). Torture and the ticking bomb: The war on terrorism as a geographical imagination of power/knowledge. *Annals of the Association of American Geographers* 96(3): 622–640.

Hartshorne, R. (1954). Political geography. In James, P.E., & Jones, D.F. (eds), *American Geography: Inventory and Prospect*, Syracuse, NY: Syracuse University Press, 167–225.

Hartsock, N. (1983). *Money, Sex, and Power: Toward a Feminist Historical Materialism*. Boston, MA: Northeastern University Press.

Haugaard, M. (2013). Commentary 3: The prescience of *Lost Geographies of Power*. *Progress in Human Geography* 37(3): 456–458.

Haugaard, M., & Clegg, S.R. (2009). Introduction: Why power is the central concept of the social sciences. In Clegg, S.R., & Haugaard, M. (eds), *The SAGE Handbook of Power*, London: Sage, 1–24.

Hillier, J., & Rooksby, E. (eds) (2005). *Habitus: A Sense of Place*. Farnham: Ashgate.

Höhn, K.H. (2011). Geopolitics and the measurement of national power. PhD thesis, Universität Hamburg.

Lee, N.K. (2009). How is a political public space made? The birth of Tiananmen Square and the May Fourth Movement. *Political Geography* 28(1): 32–43.

Legg, S. (2005). Foucault's population geographies: Classifications, biopolitics and governmental spaces. *Population Space and Place* 11(3): 137–156.

Legg, S. (2008). Ambivalent improvements: Biography, biopolitics, and colonial Delhi. *Environment and Planning A* 40(1): 37–56.

Low, M. (2005) "Power" and politics in human geography. *Geografiska Annaler Series B: Human Geography* 87B(1): 81–88.

Lukes, S. (1974). *Power: A Radical View*. London: Macmillan.

Mackinder, H.J. (1904). The geographical pivot of history. *Geographical Journal* 23: 421–444.

Martin, L.L. (2010). Bombs, bodies, and biopolitics: Securitizing the subject at the airport security checkpoint. *Social and Cultural Geography* 11(1): 17–34.

McEwan, C. (2005). New spaces of citizenship? Rethinking gendered partcipation and empowerment in South Africa. *Political Geography* 24(8): 969–991.

Miller, J.B. (1992). Women and power. In Wartenberg, T. (ed.), *Rethinking Power*, Albany, NY: State University of New York Press, 240–248.

Mountz, A. (2013). Reconfiguring geographies of sovereignty. *Progress in Human Geography* 37(6): 373–398.

Neal, A.W. (2004). Cutting off the king's head: Foucault's *Society Must Be Defended* and the problem of sovereignty. *Alternatives: Global, Local, Political* 29: 373–398.

Painter, J. (2013). Regional biopolitics. *Regional Studies* 47(8): 1235–1248.

Scott, J. (2001). *Power*. Cambridge: Polity Press.

Sparke, M.B. (2006). A neoliberal nexus: Economy, security and the biopolitics of citizenship on the border. *Political Geography* 25(2): 151–180.

Weber, M. (1968). *Economy and Society: An Outline of Interpretive Sociology*. New York: Bedminster Press.

Chapter 13

Citizenship

Patricia Ehrkamp and Malene H. Jacobsen

University of Kentucky, Lexington, Kentucky, USA

When it comes to citizenship, there is much to debate (Staeheli 2008, 2011; McNevin 2011; Van Doorn 2013; Bauder 2014). In our reading of these scholarly debates, two central issues stand out: the question of rights and the question of political community, items whose complex relationship to one another was central to Hannah Arendt's (1994 [1948]) famous observation that "the right to have rights" is contingent on membership in a political community. Going further, Kesby (2012) suggested that membership in a political community is a way of having a place in the world, which points to a key concern of political geographers: that "far from a neutral background to politics, space is political in itself" (McNevin 2011: 37). A core contribution of political geographers to citizenship studies has been not only to point out that citizenship "is not everywhere the same" (Staeheli et al. 2012: 641), but also to examine the relational nature and uneven spatiality of citizenship – its spatial logics, its application, and the struggles to expand or reduce access to citizenship and rights across space.

Questions of rights and political community are of central concern to political geographers (Kofman 2002; Staeheli 2011), because rights and community are intricately related to space, territory, borders and boundaries, subjectivity, and social practices, and to the broader questions of what counts as politics and how the political is to be conceptualized. As an abundance of scholarly special issues shows, geographers have provided lively contributions to broader debates by identifying central issues in citizenship, such as the spaces of citizenship (Painter & Philo 1995), expanding conversations across the social sciences about borderwork and citizenship (Rumford 2008), cities and citizenship (Staeheli 2003), international migration and migrant transnationalism (Ehrkamp & Leitner 2006), and developing more broadly the geographies of citizenship (Kurtz & Hankins 2005). Together, these works demonstrate a deep and broad engagement with citizenship as a concept, as a state practice, as a social practice, and as lived reality. Without denying the power of the state, scholars have pointed out the importance of examining lived experiences and the practices of citizenship beyond the state (Benhabib 1999, 2004; Ehrkamp & Leitner 2003; Veronis 2006; Chauvin & Garcés-Mascareñas 2012;

The Wiley Blackwell Companion to Political Geography, First Edition.
Edited by John Agnew, Virginie Mamadouh, Anna J. Secor, and Joanne Sharp.
© 2015 John Wiley & Sons Ltd. Published 2017 by John Wiley & Sons Ltd.

Nicholls 2013b) and in various spatialities such as the city (Varsanyi 2006; Van Doorn 2013; Mitchell, Attoh, & Staeheli 2014), a bus (Sziarto & Leitner 2010), across transnational space (Preston, Kobayashi, & Man 2006; Ho 2010), and in the less tangible spaces of international conservation volunteering (Lorimer 2010).

In the following, we examine the debates surrounding citizenship, focusing our discussion around the themes of rights and political community as they are stretched across space, unevenly distributed, contested, constrained, and occasionally expanded. We emphasize in our discussion recent scholarship that has pointed to the importance of finding new vocabularies to understand practices of citizenship and political subjectivity. The next section provides a brief overview of how citizenship has been conceived in the social sciences, how such conceptualizations have been reworked, in large part, and highlights how geographical research has contributed to such changes. This discussion is followed by a more detailed examination of the spaces of citizenship. Here, we show how uneven spatiality and various types of spaces, such as those of social reproduction, are involved in constructions and experiences of citizenship. We then turn to the question of how to establish membership in political communities, discussing both birthright citizenship and morality as entry points into such communities. The final section takes on the question of whether democratic citizenship is in crisis, and illustrates avenues of conceptualizing contemporary transformations of citizenship as new frontiers of the political, challenging political geographers to develop new terminologies and understandings of how space and politics intersect.

Rethinking citizenship

Citizenship is often defined as the formal, legal link between nation-states and their populations, signified by a passport or a birth certificate (Ehrkamp & Leitner 2003). Formal citizenship affords political rights such as the right to vote, freedom of speech, and the right to run for public office (Nyers 2008). Despite originating with the Greek *polis* – that is, as an urban concept of governance, alterity, and inclusion (Isin 2002) – citizenship has since come to be associated largely with the territory of the nation-state (Bauböck 1994), while the nation remains the dominant political community (Isin 2012a). Liberal democratic citizenship is tied to Western notions of democracy that assume such ideals as equality, liberty, and universalism (Marshall 1950). Citizenship is given meaning through definitions of rights and responsibilities (or duties) that delineate substantive citizenship, and through norms and values that are articulated by various communities and groups (Staeheli et al. 2012). Based on T.H. Marshall's classic 1950 piece, scholars frequently distinguish social, civic, and political rights that allow members of the polity to participate in the welfare state, engage in such collective action as demonstrations, and vote or stand in an election (Marshall 1950; see also Shafir 1998 for a summary of debates). A basic assumption of citizenship is that everyone who is part of a polity enjoys the same rights and is held to the same duties.

While much political theory conceives of citizenship and citizenship rights and duties as universally applicable to members of the same polity, this universality has been questioned (Lister 1997a, b; Chatterjee 2011). The struggles for women's suffrage (Ryan 1992) and of sexual and racial minorities to enact rights (Alexander 2012; Hubbard 2013; Van Doorn 2013) have shown that the ideals of equality and universality are rarely realized without contestation. Accordingly, feminist and postcolonial scholars (among others) have maintained that difference – rather than equality – is central to definitions and experiences of citizenship. Scholarship has examined in detail how race, ethnicity, gender, sexuality, or immigration status affects access to membership, rights, and belonging (Kymlicka 1995; Lister 1997b; Yuval-Davis 1997; Benhabib 1999, 2004; Bell & Binnie 2000; Young 2000; Lister 2002; M. Brown 2006; Chauvin & Garcés-Mascareñas 2012; Hubbard 2013; Nicholls 2013a, b).

In addition to these social differentiations, difference across space matters greatly. For example, the *hukou* system in China allocates certain citizenship rights based on citizens' place of residence. The movement of people from rural to urban areas means that they give up certain rights, to health coverage or housing (Fan 2008). This two-class citizenship enables the state to limit and direct rural-to-urban migration in order to regulate its needs for labor in the rapidly developing industrial cities. In another context, Linda Bosniak (2006) usefully illustrates the differentiation of citizenship within a territorially defined polity in her discussion of alienage as a "hybrid legal construct" with regard to immigration and immigrant citizenship. Alienage, on the one hand, refers to legal membership in a national community (Bosniak 2006: 38). On the other, alienage defines the "social relationships among territorially present persons" (2006: 38) who may or may not be eligible for equal rights and equal protection under the law, precisely because they are not members of the legal realm that defines membership in the political community. Hence, the access to rights and protections varies and is often contradictory in that, for example, undocumented school-aged children may not have the legal right to stay in the United States, but still have a constitutionally guaranteed right to a public school education. Here, different legal realms and rights compete within the same territorially defined sphere (see also Kanstroom 2012), making it clear that the relationship between political community and citizenship rights remains complex and contradictory.

Beyond the realm of immigration, LGBT (lesbian, gay, bisexual and transgender) rights are an important current site of struggle. The recent abolition of the Defense of Marriage Act (the federal marriage law positing that marriage was between one man and one woman) in the United States eliminated barriers to same-sex marriage (and tax penalties) at the federal level. However, the recognition of and ongoing legal battles over the constitutionality of same-sex marriage in various states and in front of the US Supreme Court demonstrate that equality is far from achieved for sexual minorities, even as the United States tends to pride itself on being the bastion of liberal democracy (Wolin 2008). This leads to such paradoxes as gay marriage being legal in over 30 states to date, while being gay may still be a cause for getting fired – even in such states as Pennsylvania, Utah, and Virginia where same-sex marriage is now legal, but where there are no discrimination protections for gay workers (Egan 2014). Although national rights to sexual orientation equality are guaranteed in the United Kingdom, displaying such a mundane intimate act as a same-sex kiss may call into question such guarantees of sexual citizenship and highlight that "rights and responsibilities are only partly determined by national laws" (Hubbard 2013: 230). Far from simply taking intimate citizenship to mean the public discussion of intimate relations (Lister 2002), Van Doorn (2013) argues that defining intimate citizenship as limited to talking about intimacy in public spaces reifies problematic distinctions between public and private spaces, and privileges, yet again, the public as the space of politics. For Van Doorn (2013), as for others (Berlant 2011; Hubbard 2013), intimate citizenship and queering citizenship mean to recognize and theorize the intricate connections across such spaces, to take seriously the embodied nature of citizenship, and to recognize the question of intimacy as deeply political and as determined in multiple spatialities.

Focusing solely on rights, then, would obscure the many facets of citizenship as lived experience in daily life, and as social practice (Berlant 2011; Chatterjee 2011; McNevin 2011). While Painter (2006) has rightly emphasized the "prosaic geographies of stateness" – that is, the deep entanglements of the state with people's everyday lives – geographers and other scholars have usefully expanded conceptions of citizenship beyond state-centered notions and status-based understandings of citizenship, otherwise known as passive citizenship. Rather than simply "being a citizen" (cf. Lister 1997b: 36), the notion of acting as a citizen is increasingly central to thinking about citizenship (see also Isin 2008 for a different definition of active citizenship). Consequently, citizenship is not just a status that one holds (Lister 1997b; see also Benhabib 1999; Ehrkamp 2010). This definition of citizenship enacted by active citizens (Lister 1997b; Isin 2002) expands the concept to include also social practices, thereby shifting the focus from the state's influence on people's everyday lives to the ways in which people themselves become political and frequently challenge the state. Reworking conceptions of citizenship to include social practices such as collective action on behalf of undocumented immigrants (Sziarto & Leitner 2010), and demonstrations in public space that both lay claim to and express belonging (Veronis 2006), recent research shows that social practices of citizenship need to be carefully considered when examining the everyday life experiences of ordinary people in a polity. Similarly, struggles over "the right to the city" demonstrate how the collective action of occupying a place, making claims to space by inhabiting it or occupying it, and using – in this case urban – space to advance political goals gives further meaning to citizenship (Varsanyi 2006; Inwood & Bonds 2013; Purcell 2013). Holston (2008: 48) advances the notion of "insurgent citizenship" to describe such everyday practices that disrupt or subvert dominant agendas. He highlights networks of migration, squatter settlements, and migrant labor camps as examples of "sites of insurgence because they introduce … new identities and practices that disturb established histories." Notions of insurgent citizenship have inspired geographers working on the rights claims of immigrants (Leitner & Strunk 2014) or show how university campuses become sites of citizenship and struggle through anti-sweatshop activism (Silvey 2004). These enactments of citizenship and struggles over rights, importantly, are not limited to those who already enjoy citizenship rights (Varsanyi 2006). In fact, recent struggles over immigrant rights in the United States show that practicing citizenship is not necessarily tied to having formal citizenship (Glenn 2011; Nicholls 2013a). This latter point is even more important given some recent claims of the era of postpolitics, to which we return at the end of this chapter.

However, notions of active citizenship also emerge in a less progressive fashion in the contemporary neoliberal era, where personal responsibility and a focus on flexibility and self-governance undermine solidarity, group rights, and the responsibility of the (welfare) state to guarantee such rights. As Erickson and Faria (2011) point out in their study of the South Sudan Women's Empowerment Network (SSWEN) and its transnational feminist activism and peace-building efforts on postconflict Sudan, "an emphasis on the self and the spaces of the home and body divert responsibility from the state for postconflict reconstruction, care, and the dismantling of patriarchal systems" (2011: 627). Similarly, Mitchell (2006) criticizes the new forms of education for immigrant integration in the European Union for their focus on creating flexible laborers rather than on educating citizens who are able to enact social belonging in multicultural societies. Such citizenship education, as Staeheli and Hammett (2013) critically examine for postconflict South Africa, creates self-sufficient citizens who will place fewer demands on the state rather than those who will actively claim rights and membership.

The spaces of citizenship

Geographers have emphasized the spatial complexity of citizenship and have questioned the methodological nationalism and simple location of citizenship at the level of the nation-state. Taking space seriously, this research shows that citizenship "varies across place, across time, and for different people. It is inseparable from the geographies of communities and the networks and relationships that link them, with their attendant inequalities, imperfections, and opportunities" (Staeheli et al. 2012: 641). The street and other public spaces such as Gezi Park in Istanbul remain important sites for protest, opposition politics, claims to rights, and political expressions more broadly (Bayat 2010; Eken 2014). However, recent scholarship has shown that the sites of citizenship politics are often more ordinary, less overt, and even hidden (Staeheli et al. 2012). For example, rural Muslims in Britain use subterranean spaces of citizenship to establish belonging beyond the formal spaces of politics and beyond the public square (Jones 2010). Similarly, Askins (forthcoming: 7) emphasizes the quiet politics of friendship between long-term residents and asylum seekers in Britain that takes place "in homes, neighbourhoods, cafes, going for walks in the local park, and to local shops." That is, it is through spaces of social reproduction that broader notions of belonging in a polity are established. As spaces of sustained encounter, in this case, ordinary spaces of citizenship are implicated in the construction of citizenship norms, values, and belonging. Private and intimate spaces of everyday life are also central to the ways in which individuals may congregate and engage in collective action and political activism, or establish norms and values that imbue citizenship with meaning (Ehrkamp & Nagel 2012). Churches, schools, and community organizations all are sites where norms and values are articulated that, in turn, enter into the politics of citizenship and community. Ehrkamp and Nagel's work on the precarious hospitality offered to undocumented migrants in churches in the US South illustrates this point (Ehrkamp & Nagel 2014). In these churches, newcomers were welcomed as fellow Christians, racialized as "other," and simultaneously stigmatized as breaking secular law in an era of rampant anti-immigrant legislation. In the contradictions of faith-based welcome and law-and-order–based exclusion, the importance of churches as sites for the construction of citizenship becomes obvious: Their members are engaged in policing the borders not only of church membership, but of societal belonging more broadly. Schools, as recent examples of claims to recognition and legalization by the DREAMers movement show, are equally important for networking and for articulating belonging (Glenn 2011). Walter Nicholls (2013a) describes how the DREAMers movement in California solidified, in part, through networking in social media, and in private and intimate spaces. The latter are particularly important. As Nicholls (2013a: 68) writes:

> Interactions in intimate spaces of daily life ... have functioned as moments where activists share their fears with one another, celebrate accomplishments, reinforce their belief that their cause is just and right, and express doubts about their situations and concerns over the movement's directions. These face-to-face interactions foster feelings of trust in other DREAMers and emotional commitment to their general cause.

Intimate spaces, here, are part and parcel of a larger political project, because they accommodate emotional and affective interactions that strengthen political convictions and galvanize energy for political organizing and the continued practice of citizenship.

Membership and (political) community

Thinking about these multiple and multiplying spaces of citizenship also raises the question of how the community of citizens is defined. Lynn Staeheli (2008: 7) has suggested that "community is where contests are waged over membership and the political subjects and subjectivities that 'belong.'" Membership in a political community is often established via birthright citizenship; that is, one is born into citizenship. Citizens acquire citizenship either via the principle of *jus soli* (birth in a territory) or via the principle of *jus sanguinis*, an ethnic definition of citizenship (to whom one is born). Much writing on citizenship takes for granted this birthright citizenship as the entry point into a polity. However, recent scholarship has pointed out that this birthright affords uneven rights and creates inequalities across the globe by the accident of birth in a polity. Among others, Shachar (2003: 347) seeks to unsettle this taken-for-granted idea of the "connection between birth and political membership in any given state," because it affirms certain statuses and rights simply based on place of birth (or birth to a parent who belongs to a political community) without making anyone earn membership. In Shachar's analysis, the birthright citizenship lottery produces and perpetuates global unevenness because "where" and "to whom" one is born take precedence over acts of citizenship, enactment of duties, taking on responsibilities, and so forth (Shachar 2009).

While Shachar (2003, 2009) and others (Stevens 2010) seek to move away from birthright as the condition for belonging in a political community, they do not fully challenge the nation-state as the territorial entity of the polity. Numerous scholars have argued for the allocation of rights based on the place where one lives rather than the place where one is born (Bauböck 1994, 2003; Benhabib 2004; Bauder 2014). Bauböck (2003) proposes expanding formal citizenship for migrants as local/urban citizenship, a call more recently echoed by Bauder (2014), who suggests a move toward *jus domicilus* – that is, residence-based citizenship – in an age of human mobility. Bauböck's urban citizenship, like the *jus domicilus* that Bauder envisions, affords rights derived from presence within a territory, often a spatiality within a nation-state such as a city or a locale. Much like "the right to the city" (Lefebvre 1996), this scholarship seeks to expand citizenship by adding rights and entitlements for those who are not already included in the (national) polity, but are poised to gain access to it through participation and life in cities. Others have sought to move away more clearly from the nation-state as the site of political community and citizenship, arguing that citizenship is increasingly deterritorialized (Soysal 2000) or trying to work toward cosmopolitan ideas of citizenship where national boundaries do not prevent democratic participation in world politics (Archibugi 2008). Soysal (1994) stirred debate with her book *The Limits of Citizenship* over what she called the emergence of postnational citizenship (see also Habermas 2001), arguing that membership in a political community and enjoying citizenship rights are no longer necessarily tied to the same territory because citizenship rights are human rights. This recasting of rights, however, is still problematic as long as the nation-state has to remain the guarantor and enforcer of rights and protections (Russell 2005; Kesby 2012). This work shows that communities of citizens are not necessarily congruent with the territory of nation-states, despite assertions by some that they ought to be (Miller 2008) or that such congruence has been practical and persistent (Calhoun 2007). Several issues remain debatable. One central question when considering political community is the spatial extent of this community and whether or not, and in what ways, it ought to be tied to the territory of a nation-state. Another, related question is how to establish membership in such a political community if not via birthright.

Geographers have also usefully highlighted that conceptions of and membership in a political community rely on a moral community. Michael Brown, for example, urges us to view citizenship as a social contract that includes "the obligations of the sexual citizen and the notion of membership in a political community or polity" (M. Brown 2006: 875; see also Bell & Binnie 2000). Furthermore, "[e]very entitlement is freighted with a duty" (Bell & Binnie 2000: 2–3, cited in M. Brown 2006: 877). That is, having rights is conditional on taking on responsibilities that, in turn, are determined through ideas of morality, norms, and values (Staeheli et al. 2012) as the basis of membership in local and national communities.

How does one become a member of such a moral community of citizens, then? For Honig (2009: 122), a "universalist morality" that derives from abstract understandings of cosmopolitanism or human rights as a universally shared good is insufficient to establish community membership. She argues that it "matters whether we relate to those near or distant under the sign of universality or under the sign of the neighbor," positing that the latter might lead to solidarity while the former may well undermine such solidarity, because its abstraction is too far removed from daily life and relationships between people. Similarly, Staeheli et al. (2012) have identified different models of inclusion. One of these models is based on social justice, which is based on the conviction that people should be treated equally and fairly. The other model is rooted in an ethic of care and explicitly recognizes that not everyone in a polity is equal. In this latter model of an ethic of care, achieving the wellbeing of the citizenry in a democracy requires the "satisfaction of situated and particular needs that arise from the recognition of difference and the practices of accommodation" (Staeheli et al. 2012: 635). These different ethical models, in turn, afford different ways of entry into a moral community, and thereby membership in a polity. The struggles of undocumented immigrants show that achieving membership in the polity through entering moral communities also becomes possible through establishing deservingness. Unauthorized migrants have, in various instances, succeeded at becoming recognized as legitimate political subjects by becoming "good migrants" (and/or "good illegal immigrants"; cf. Chauvin & Garcés-Mascareñas 2012: 249), whose good moral character makes them appear less illegal because it conforms to the dominant moralities and values of the country. Similarly, Nicholls (2013b: 84) argues that if unauthorized migrants "are to gain recognition as legitimate 'voices' and avoid being dismissed as impossible 'noises', they must construct representations of immigrants and their cause in ways that cohere with the core normative and moral values of the nation."

This task is becoming more and more difficult in the globalizing world, however. Mezzadra and Neilson (2013: 164) suggest that there is an "overlapping of multiple lines of inclusion and exclusion, blurring the boundary between them and destabilizing the existence of a unified and homogeneous point of reference against which the position of migrants can be ascertained." Their work shows that the concept of citizenship, with its ties to Western-style liberal democracy and to the territory of nation-states, needs to be critically examined in the twenty-first century. It is perhaps not surprising that critical research on citizenship has made useful advances by "moving from citizenship as membership to citizenship as political subjectivity" (Isin 2012a: 567). Such a move, according to Isin (2012b), helps to decolonize and deorientalize citizenship by exposing its close ties to European notions of modernity and nationhood. These kinds of conceptual moves are even more important at a time when scholars studying Western democracies are pessimistic about citizenship and its political potential, as we discuss next.

A crisis of citizenship or new frontiers of the political?

Etienne Balibar (2012: 437) has diagnosed a "crisis of democratic citizenship" in Western liberal democracies. The "de-democratization" of public life (W. Brown 2006a: 692) has created a society where antagonism has disappeared from political discussions and politics has become a matter of techno-managerial governance and consensus politics (Swyngedouw 2011). These developments are brought about in large part by relegating controversial issues to the private realm (W. Brown 2006b) and by increasingly technocratic solutions to political struggles

(Swyngedouw 2011). Established Western democracies are suffering from a "pervasive mistrust of elected representatives" and "voter apathy" (Chatterjee 2011: 25). The crisis of citizenship, Balibar and others argue, is also due to the increased neoliberalization of governance that undermines citizen participation and collective action by emphasizing individual responsibility (see also Berlant 2011; Swyngedouw 2011). Indeed, Berlant (2011: 222) argues that "neoliberal interests are well served by the displacement of so many historical forms of social reciprocity onto emotional registers, especially when they dramatize experiences of freedom to come that have no social world for them yet." Importantly, these diagnoses of a crisis of citizenship appear largely in those Western democracies that conceive of themselves as the epicenters of democracy. Rarely does this scholarship consider democratic practices in other contexts.

By contrast, others (Honig 2009; Staeheli 2011) consider such crises as integral to the restructuring and transformation of citizenship and democracy. Declarations of citizenship in crisis and of a postpolitical or postdemocratic period have been challenged. Being political is not limited to one's citizenship status or participation in formal and electoral politics. Rather, being political "occurs in a broad range of relations between people and groups wherein norms, practices, ideas, and ways of organizing material life are challenged, questioned, and potentially reconstructed" (Cowen & Gilbert 2008: 1). Many of these challenges come from scholarship on marginalized populations within (or largely excluded from) Western democracies (McNevin 2011; Conlon 2013; Mitchell et al. 2014), and from scholars analyzing how democracy and democratic practices are being reworked, for example in the Middle East and in India (Wedeen 2007; Bayat 2010; Chatterjee 2011). Mitchell et al. (2014) dispute claims of a postpolitical period and insist instead that notions of consensus between stakeholders (cf. Swyngedouw 2011) are not simply postpolitical, but rather highly political themselves as urban dwellers thrash out whose city Boulder, Colorado is. Considering the struggles of homeless people for their right to "sleeping rough" and to challenge city ordinances that criminalize such practices in urban public space, Mitchell et al. (2014: 5) show how homeless activism goes beyond stakeholder politics as the homeless move from the streets to the city council to courtrooms. This analysis affirms Honig's (2009: 1) claim:

> When we treat sovereignty as if it is top down and yet governable by norms we affirm, we help marginalize rather than empower important alternatives, such as forms of popular sovereignty in which action in concert rather than institutional governance is the mark of democratic power and legitimacy.

In a different context, Asef Bayat's (2010: 56) analysis of politics in the Middle East shows that more than overt political action, a "quiet encroachment of the ordinary" – that is, "non-collective but prolonged direct action" – serves to enhance the situation of disempowered and poor people through non-overt challenges of the state and the powerful. As a social practice, this quiet encroachment moves beyond direct notions of resistance or overt confrontation. It involves taking (sometimes protracted) action as it aims to fulfill immediate needs. And yet, over time this practice chips away at state and capitalist power to determine people's lives, because they are taking up more and more space and gaining political ground that way. These quiet mobilizations stand in stark contrast to the assumption that political space necessitates

> a political act that stages collectively the presumption of equality and affirms the ability of "the People" to self-manage and organise their affairs. It is an active process of intervention through which (public) space is reconfigured and through which – if successful – a new socio-spatial order is inaugurated. (Swyngedouw 2014: 131)

Bayat's work, along with that of others who identify democratic practices even within authoritarian regimes (Wedeen 2007), is useful in pointing out the limits of focusing on organized politics, which misses the political efficacy of practices that do not appear to be overtly political or collective.

Similarly, McNevin (2011) argues that new political practices and acts emerge that may not yet be recognizable as political or as enactments of citizenship, especially when they are read exclusively through existing frameworks of citizenship. Hunger strikes by asylum seekers are one such form of political action through which asylum seekers challenge the current system and seek to disrupt the normal order (Conlon 2013). The important aspect here is not the outcome – that is, whether these hunger strikes change the existing order – but the question of how new subjectivities surface in contestations over rights and belonging. This scholarship is interested in how solidarities are being created. Berlant's (2011: 262) conception of "lateral politics" and her focus on the historical present emphasize affective notions of citizenship, since she calls for understanding simple acts of solidarity as politics rather than defining them as being outside the political. Lateral politics entails "becoming a political subject whose solidarities and commitments are neither to ends nor to imagining the pragmatics of a consensual community, but to embodied processes of making solidarity itself" (Berlant 2011: 260). In doing so, Berlant (2011: 262) challenges us to consider "valuing political action as the action of not being worn out by politics."

Even fleeting expressions of solidarity and temporary collective action may serve to shape new political subjectivities. Measuring these practices solely by their outcome misses the point of understanding new ways of becoming political, of fugitive democracy (Wolin 2008), and of quiet encroachment that create the potential for solidarity among ordinary people and may (or may not) form the basis of collective action. Analyzing postcolonial India, Chatterjee (2011: 207) challenges political theory to produce

> a different conceptualization of the subject of political practice – neither as abstract and unencumbered individual selves nor as manipulable objects of governmental policy, but rather as concrete selves necessarily acting within multiple networks of collective obligations and solidarities to work out strategies of coping with, resisting, or using to their advantage the vast array of technologies of power deployed by the modern state.

We contend that political geographers are well positioned to contribute to such new theorizations and to study the "new frontier of the political – that moment of confrontation and destabilization when one account of justice competes with another to shape what we think of as 'common sense' justifications for particular status hierarchies" (McNevin 2011: 5). Much of this recent scholarship shows that "existing lenses" (McNevin 2011: 155) may be insufficient for recognizing and understanding political acts – or "emerging practices" (Chatterjee 2011: 207) – in the contemporary moment. Instead, the "urgency is to reinvent … new idioms of the political, and of belonging itself" (Berlant 2011: 262). These arguments challenge political geographers to examine carefully our own existing ideas and taken-for-granted notions of rights, political community, and what counts as politics in order to pursue new avenues for thinking about citizenship, starting from the recognition that citizenship is enacted by and through people, across multiple and uneven spatialities, and subject to debate. It is these aspects that continue to make the legal concept of citizenship, struggles over rights, political community, and the lived experiences of citizenship across space exciting, timely, and significant topics for contemporary political geographical research.

References

Alexander, M. (2012). *The New Jim Crow: Mass Incarceration in the Age of Colorblindness.* New York: New Press.

Archibugi, D. (2008). *The Global Commonwealth of Citizens: Toward Cosmopolitan Democracy.* Princeton, NJ: Princeton University Press.

Arendt, H. (1994 [1948]). *The Origins of Totalitarianism.* Fort Washington, PA: Harvest Books.

Askins, K. (forthcoming). Being together: Everyday geographies and the quiet politics of belonging. *ACME: An International E-Journal for Critical Geographies.*

Balibar, E. (2012). The "impossible" community of citizens: Past and present problems. *Environment and Planning D: Society and Space* 30(3): 437–449.

Bauböck, R. (1994). *Transnational Citizenship: Membership and Rights in International Migration.* Aldershot: Edward Elgar.

Bauböck, R. (2003). Reinventing urban citizenship. *Citizenship Studies* 7(2): 139–160.

Bauder, H. (2014). Domicile citizenship, human mobility and territoriality. *Progress in Human Geography* 38(1): 91–106.

Bayat, A. (2010). *Life as Politics: How Ordinary People Change the Middle East.* Stanford, CA: Stanford University Press.

Bell, D., & Binnie, J. (2000). *The Sexual Citizen.* London: Polity Press.

Benhabib, S. (1999). Citizens, residents, and aliens in a changing world: Political membership in the global era. *Social Research* 66(3): 709–731.

Benhabib, S. (2004). *The Rights of Others: Aliens, Residents and Citizens.* Cambridge: Cambridge University Press.

Berlant, L. (2011). *Cruel Optimism.* Durham, NC: Duke University Press.

Bosniak, L. (2006). *The Citizen and the Alien: Dilemmas of Contemporary Membership.* Princeton, NJ: Princeton University Press.

Brown, M. (2006). Sexual citizenship, political obligation and disease ecology in gay Seattle. *Political Geography* 25(8): 874–898.

Brown, W. (2006a). American nightmare: Neoliberalism, neoconservatism, and de-democratization. *Political Theory* 34: 690–715.

Brown, W. (2006b). *Regulating Aversion: Tolerance in the Age of Identity and Empire.* Princeton, NJ: Princeton University Press.

Calhoun, C. (2007). Nationalism and cultures of democracy. *Public Culture* 19(1): 151–173.

Chatterjee, P. (2011). *Lineages of Political Society: Studies in Postcolonial Democracy, Cultures of History.* New York: Columbia University Press.

Chauvin, S., & Garcés-Mascareñas, B. (2012). Beyond informal citizenship: The new moral economy of migrant illegality. *International Political Sociology* 6: 241–259.

Conlon, D. (2013). Hungering for freedom: Asylum seekers' hunger strikes – rethinking resistance as counter-conduct. In Moran, D., Gill, N., & Conlon, D. (eds), *Carceral Spaces: Mobility and Agency in Imprisonment and Migrant Detention*, Farnham: Ashgate, 133–148.

Cowen, D., and Gilbert, E. (eds) (2008). *War, Citizenship, and Territory.* New York: Taylor & Francis.

Egan, P.J. (2014). More gay people can now get legally married. They can still be legally fired. *Monkey Cage Blog*, http://www.washingtonpost.com/blogs/monkey-cage/wp/2014/10/06/more-gay-people-can-now-get-legally-married-they-can-still-be-legally-fired/, accessed October 7, 2014.

Ehrkamp, P. (2010). The limits of multicultural tolerance? Liberal democracy and media portrayals of Muslim migrant women in Germany. *Space and Polity* 14(1): 13–32.

Ehrkamp, P., & Leitner, H. (2003). Beyond national citizenship: Turkish immigrants and the (re) construction of citizenship in Germany. *Urban Geography* 24(2): 127–146.

Ehrkamp, P., & Leitner, H. (2006). Guest editorial: Rethinking immigration and citizenship: New spaces of migrant transnationalism and belonging. *Environment and Planning A* 38(9): 1591–1597.

Ehrkamp, P., & Nagel, C. (2012). Immigration, places of worship and the politics of citizenship in the US South. *Transactions of the Institute of British Geographers* 37(4): 624–638.

Ehrkamp, P., & Nagel, C. (2014). "Under the radar": Undocumented immigrants, Christian faith communities, and the precarious spaces of welcome in the U.S. South. *Annals of the Association of American Geographers* 104(2): 319–328.

Eken, B. (2014). The politics of the Gezi Park resistance: Against memory and identity. *South Atlantic Quarterly* 113(2): 427–436.

Erickson, J., & Faria, C. (2011). "We want empowerment for our women": Transnational feminism, neoliberal citizenship, and the gendering of women's political subjectivity in postconflict South Sudan. *Signs* 36(3): 627–652.

Fan, C. (2008). *China on the Move: Migration, the State, and the Household.* New York: Routledge.

Glenn, E.N. (2011). Constructing citizenship: Exclusion, subordination, and resistance. *American Sociological Review* 76(1): 1–24.

Habermas, J. (2001). The postnational constellation and the future of democracy. In Pensky, M. (ed.), *The Postnational Constellation: Political Essays,* Cambridge: Polity Press, 58–112.

Ho, E. (2010). Constituting citizenship through the emotions: Singaporean transmigrants in London. *Annals of the American Association of Geographers* 99(4): 788–804.

Holston, J. (2008). *Insurgent Citizenship: Disjunctions of Democracy and Modernity in Brazil.* Princeton, NJ: Princeton University Press.

Honig, B. (2009). *Emergency Politics: Paradox, Law, Democracy.* Princeton, NJ: Princeton University Press.

Hubbard, P. (2013). Kissing is not a universal right: Sexuality, law and the scales of citizenship. *Geoforum* 49: 224–232.

Inwood, J., & Bonds, A. (2013). On racial difference and revolution. *Antipode* 45(3): 517–520.

Isin, E.F. (2002). *Being Political: Genealogies of Citizenship.* Minneapolis, MN: University of Minnesota Press.

Isin, E.F. (2008). Theorizing acts of citizenship. In Isin, E.F., & Nielson, G.M. (eds), *Acts of Citizenship,* New York: Zed Books, 15–43.

Isin, E.F. (2012a). Citizens without nations. *Environment and Planning D: Society and Space* 30(4): 450–467.

Isin, E.F. (2012b). Citizenship after orientalism: An unfinished project. *Citizenship Studies* 16(5–6): 563–572.

Jones, R.D. (2010). Islam and the rural landscape: Discourses of absence in west Wales. *Social and Cultural Geography* 11(8): 751–768.

Kanstroom, D. (2012). *Aftermath: Deportation Law and the New American Diaspora.* Oxford: Oxford University Press.

Kesby, A. (2012). *The Right to Have Rights: Citizenship, Humanity, and International Law.* Oxford: Oxford University Press.

Kofman, E. (2002). Contemporary European migrations, civic stratification and citizenship. *Political Geography* 21(8): 1035–1054.

Kurtz, H., & Hankins, K. (2005). Guest editorial: Geographies of citizenship. *Space and Polity* 9(1): 1–8.

Kymlicka, W. (1995). *Multicultural Citizenship: A Liberal Theory of Minority Rights.* New York: Oxford University Press.

Lefebvre, H. (1996). *Writings on Cities,* trans. E. Kofman & E. Lebas. Malden, MA: Blackwell.

Leitner, H., & Strunk, C. (2014). Assembling insurgent citizenship: Immigrant advocacy struggles in the Washington DC metropolitan area. *Urban Geography* 35(7): 943–964.

Lister, R. (1997a). *Citizenship: Feminist Perspectives.* New York: New York University Press.

Lister, R. (1997b). Citizenship: Towards a feminist synthesis. *Feminist Review* 57: 28–48.

Lister, R. (2002). Sexual citizenship. In Isin, E.F., & Turner, B.S. (eds), *Handbook of Citizenship Studies,* London: Sage, 191–207.

Lorimer, J. (2010). International conservation "volunteering" and the geographies of global environmental citizenship. *Political Geography* 29(6): 311–322.

Marshall, T.H. (1950). *Citizenship and Social Class, and Other Essays*. Cambridge: Cambridge University Press.

McNevin, A. (2011). *Contesting Citizenship: Irregular Migrants and New Frontiers of the Political*. New York: Columbia University Press.

Mezzadra, S., & Neilson, B. (2013). *Border as Method: Or the Multiplication of Labor*. Durham, NC: Duke University Press.

Miller, D. (2008). Immigrants, nations, and citizenship. *Journal of Political Philosophy* 16(4): 371–390.

Mitchell, D., Attoh, K., & Staeheli, L. (2014). Whose city? What politics? Contentious and non-contentious spaces on Colorado's Front Range. *Urban Studies* online before print: 1–16. doi:10.1177/0042098014550460

Mitchell, K. (2006). Neoliberal governmentality in the European Union: Education, training, and technologies of citizenship. *Environment and Planning D: Society and Space* 24(3): 389-407.

Nicholls, W.J. (2013a). *The DREAMers: How the Undocumented Youth Movement Transformed the Immigrant Rights Debate*. Stanford, CA: Stanford University Press.

Nicholls, W.J. (2013b). Making undocumented immigrants into a legitimate political subject: Theoretical observations from the United States and France. *Theory, Culture and Society* 30(3): 82–107.

Nyers, P. (2008). No one is illegal between city and nation. In Isin, E.F., & Nielsen, G.M. (eds), *Acts of Citizenship*, New York: Zed Books, 160–181.

Painter, J. (2006). Prosaic geographies of stateness. *Political Geography* 25(7): 752–774.

Painter, J., & Philo, C. (1995). Spaces of citizenship: An introduction. *Political Geography* 14(2): 107–120.

Preston, V., Kobayashi, A., & Man, G. (2006). Transnationalism, gender, and civic participation: Canadian case studies of Hong Kong immigrants. *Environment and Planning A* 38(9): 1633–1651.

Purcell, M. (2013). Possible worlds: Henri Lefebvre and the right to the city. *Journal of Urban Affairs* 36(1): 141–154.

Rumford, C. (2008). Introduction: Citizens and borderwork in Europe. *Space and Polity* 12(1): 1–12.

Russell, J. (2005). Rethinking post-national citizenship: The relationship between state territory and international human rights law. *Space and Polity* 9(1): 29–39.

Ryan, M.P. (1992). *Women in Public: Between Banners and Ballots, 1825–1880*. Baltimore, MD: Johns Hopkins University Press.

Shachar, A. (2003). Children of a lesser state: Sustaining global inequality through citizenship law. In Macedo, S., & Young, I.M. (eds), *Child, Family, and State*, New York: New York University Press, 345–397.

Shachar, A. (2009). *The Birthright Lottery: Citizenship and Global Inequality*. Cambridge, MA: Harvard University Press.

Shafir, G. (1998). *The Citizenship Debates: A Reader*. Minneapolis, MN: University of Minnesota Press.

Silvey, R. (2004). A wrench in the global works: Anti-sweatshop activism on campus. *Antipode* 36(2): 191–197.

Soysal, Y.N. (1994). *Limits of Citizenship: Migrants and Postnational Membership in Europe*. Chicago, IL: University of Chicago Press.

Soysal, Y.N. (2000). Citizenship and identity: Living in diasporas in post-war Europe? *Ethnic and Racial Studies* 23(1): 1–16.

Staeheli, L.A. (2003). Introduction: Cities and citizenship. *Urban Geography* 24(2): 97–102.

Staeheli, L.A. (2008). Citizenship and the problem of community. *Political Geography* 27(1): 5–21.

Staeheli, L.A. (2011). Political geography: Where's citizenship? *Progress in Human Geography* 35(3): 393–400.

Staeheli, L.A., Ehrkamp, P., Leitner, H., & Nagel, C.R. (2012). Dreaming the ordinary: Daily life and the complex geographies of citizenship. *Progress in Human Geography* 36(5): 628–644.

Staeheli, L.A., & Hammett, D. (2013). "For the future of the nation": Citizenship, nation, and education in South Africa. *Political Geography* 32(1): 32–41.

Stevens, J. (2010). *States without Nations: Citizenship for Mortals*. New York: Columbia University Press.

Swyngedouw, E. (2011). Interrogating post-democratization: Reclaiming egalitarian political spaces. *Political Geography* 30(3): 370–380.

Swyngedouw, E. (2014). Where is the political? Insurgent mobilisations and the incipient "return of the political." *Space and Polity* 18(2): 122–136.

Sziarto, K.M., & Leitner, H. (2010). Immigrants riding for justice: Space-time and emotions in the construction of a counterpublic. *Political Geography* 29(7): 381–391.

van Doorn, N. (2013). Architectures of the good life: Queer assemblages and the composition of intimate citizenship. *Environment and Planning D: Society and Space* 31(1): 157–173.

Varsanyi, M.W. (2006). Interrogating "urban citizenship" vis-à-vis undocumented migration. *Citizenship Studies* 10(2): 229–249.

Veronis, L. (2006). The Canadian Hispanic Day Parade, or how Latin American immigrants practise (sub) urban citizenship in Toronto. *Environment and Planning A* 38(9): 1653–1671.

Wedeen, L. (2007). The politics of deliberation: Qāt chews as public spheres in Yemen. *Public Culture* 19(1): 59–84.

Wolin, S. (2008). *Democracy Incorporated: Managed Democracy and the Specter of Inverted Totalitarianism*. Princeton, NJ: Princeton University Press.

Young, I.M. (2000). *Inclusion and Democracy*. Oxford: Oxford University Press.

Yuval-Davis, N. (1997). Women, citizenship and difference. *Feminist Review* 57: 4–27.

Chapter 14

The Biopolitical Imperative

Claudio Minca

Wageningen University, The Netherlands

It seems to me that one of the basic phenomena of the nineteenth century was what might be called power's hold over life. ... a certain tendency that leads to what might be termed State control of the biological. ... What does having the right of life and death actually mean? In one sense, to say that the sovereign has a right of life and death means that he can, basically, either have people put to death or let them live, or in any case that life and death are not natural or immediate phenomena which are primal or radical, and which fall outside the field of power.

... after a first seizure of power over the body in an individualizing mode, we have a second seizure of power that is not individualizing but ... massifying, that is directed not at man-as-body [sic] but at man-as-species. After the anatomo-politics of the human body established in the course of the eighteenth century, we have, at the end of that century, the emergence of something that is no longer an anatomo-politics of the human body, but what I would call a "biopolitics" of the human race. (Foucault 2003 [1976])

The notion of biopolitics has become a buzzword, claims Thomas Lemke in opening his *Biopolitics: An Advanced Introduction* (2010). Biopolitics seems to be everywhere, and everything runs the risk of being considered as somehow implicated with the "biopolitical." In the introduction to their key *Biopolitics: A Reader*, Timothy Campbell and Adam Sitze (2013a) caution against inflating the term to the point that it would be rendered useless. From global migrations and refugee mobility to the War on Terror, from biosecurity to public health and the welfare state, from the new states of exception and population management to reconceptualizations of the human body associated with new biotechnologies, all of these questions are directly related to biopolitics and the implementation of biopolitical regimes.

The validity and the practical use of the term biopolitics, and of the presumed "biopolitical turn" in the social sciences and the humanities, are therefore under the close scrutiny of many of its advocates and detractors alike. Biopolitics is indeed often adopted

The Wiley Blackwell Companion to Political Geography, First Edition.
Edited by John Agnew, Virginie Mamadouh, Anna J. Secor, and Joanne Sharp.
© 2015 John Wiley & Sons Ltd. Published 2017 by John Wiley & Sons Ltd.

in different and sometimes even conflicting terms. It is employed by writers of the New Left as well as by representatives of the Right, including some supporting explicitly racist arguments. According to Laura Bazzicalupo (2010), it is therefore necessary to draw a clear distinction between the often unreflexive use of the term biopolitics on the part of political representatives and positivist scientists, and the critical social theory that has engaged with this notion. A positivist reading of the political in merely biological terms indeed stands in conflict to work that has attacked and deconstructed these approaches, especially with reference to emergent themes concerning the impact of biotechnology, but also contemporary manifestations of racism and populism. Questions of privacy – also, but not only, in relation to the presence of terrorist threats – have been central to these divided and radically opposed understandings of biopolitics as well. In this context, human geography, and in particular political geography, has largely embraced the critical stance on biopolitics, and especially the work inspired by Michel Foucault and Giorgio Agamben, while the approaches supporting a calculative and positivist approach to the "bio"political have largely remained unnoticed.

A clear turning point in the critical understanding of biopolitics has indeed been provoked by the genealogical reconstruction offered by Foucault, recognized by many as the initiator of philosophical reflection on the biopolitical. Foucault, in fact, famously associated the origin of this form of non-disciplinary political power with the Christian pastoral, but also with modern state interventions in arenas of public health, policing, and government of the population.

> Unlike discipline, which is addressed to bodies, the new nondisciplinary power is applied not to man-as-body but to the living man, to man-as-having-being; ultimately, if you like, to man-as-species. … discipline tries to rule a multiplicity of men [sic] to the extent that their multiplicity can and must be dissolved into individual bodies … kept under surveillance, trained, used, and, if need be, punished. … a global mass that is affected by overall processes characteristic of birth, death, production, illness, and so on. (Foucault 2003 [1976]: 242)

Work incorporating a biopolitical perspective thus crucially includes research on the "politics of life," often related to debates over family values, abortion, and euthanasia, and the definition of life and death in light of the new horizons of biomedical intervention. More generally, while scholars inspired by Agamben's path-breaking rereading of Foucault in his *Homo Sacer* project (1998, 2002) tend to highlight the production of "bare life" and the possibility that biopolitics may turn into thanapolitics, others inspired by different sources emphasize how biopower may potentially translate into an affirmative bios – for instance, in the form of the cybor politics proposed by Donna Haraway (1989), or via the understanding of the metamorphosis of the body described by Rosi Braidotti (2013), or even the "multitude approach" to biocapitalism advanced by Michael Hardt and Antonio Negri (2000), to mention a few. In this context, Roberto Esposito's (2008, 2011, 2012) affirmative biopolitics based on his radical reconceptualization of immunity is becoming particularly relevant in posthuman debates on the new "bios."

However, the actual origins of the term and how these may have affected its present use remain open questions. In Anglophone critical social theory, the emergence of the concept is often associated with Foucault's first analysis of biopolitics that appeared in his short essay "Right of Death and Power over Life," part of his 1976 book *La volonté de savoir*. That brief intervention would "launch its own share of articles and books was not at all clear in 1978, when the text first appeared in English as Part III of *The History of Sexuality*" (Campbell & Sitze 2013b: 3). However, different genealogical reconstructions of the term exist, born out of different national and disciplinary settings. This is particularly true for interpretations provided not only by Italian contemporary political philosophy, from which this chapter draws extensively, but also by German and French debates in the philosophy of science, sociology, and anthropology. Fundamentally, these different accounts agree on one point: the history of biopolitics is determined by new forms of biopower and state governance that found their utmost expressions in nineteenth-century Europe – and in the colonies, of course – to continue in the manifestations of eugenics as implemented by most Western governments in the first half of the twentieth century and in the realization of the first form of *biocracy* by the Third Reich. The ghost of Adolf Hitler's genocidal biopolitics even today haunts all conversations and policies concerning questions of life and their determinations on the part of public and private institutions, especially when biology, life sciences, and political power seem to be conceived as one and the same.

If biopolitics is key to how we think of life today, and accordingly of politics, it is therefore crucial to understand how geography as a discipline engages with biopolitics, and how geographical practice may imply a biopolitical dimension. Human geography has extensively engaged with Foucault's work in past decades (see, among others, Philo 1992; Driver 1997; and, more recently, Crampton & Elden 2007) and these studies have been crucial in introducing biopolitics among geographers. To gain a better understanding of the origin of the biopolitical, and to link it to the geographical literature, one route is thus tracing back Foucault's legacy and incorporating Agamben's most recent readings of biopolitics as part of that "tradition," so influential in the discipline. An alternative route – curiously left largely unexplored by geographers – is associating the origin of the term and its early implementation with the *Geopolitik* tradition, the writings on *Lebensraum* and the organic state of Rudolf Kjellén and, indirectly, of Friedrich Ratzel. This genealogical account is offered by both Lemke (from a broadly German perspective) and the reconstructions endorsed by several Italian scholars of biopolitics. The biopolitical imperative in geography is thus discussed here mainly via the first route, although in the conclusion I briefly engage with the second as well, proposing it as a viable, alternative way of understanding the ontological relationship between the "biopolitical" and the "geopolitical."

The biopolitical imperative

Roberto Esposito in *Bios* states:

> [A]t the moment in which on one side the modern distinctions between public and private, state and society, local and global collapse, and on the other all other sources of legitimacy dry up, life becomes encamped in the center of every political procedure. No other politics is conceivable other than a politics of life, in the objective and subjective sense of the term. (2008: 15)

But how are *we* to comprehend, asks Esposito, a political government of life? In what sense does life govern politics or in what sense does politics govern life? Does it concern a governing *of* or *over* life?

To address some of these fundamental questions and briefly explore different genealogical reconstructions of the concept of biopolitics, this chapter draws from three mainstream accounts already mentioned: the histories of biopolitics endorsed by Italian political philosophy, including but not limited to Agamben's and Esposito's readings; Lemke's socio-logical introduction to the concept; and Campbell and Sitze's *Reader*. Despite being differently inspired by their respective academic contexts, these narratives nonetheless share some crucial points. First, they all highlight the growing importance of biopolitics in understanding contemporary society and politics. Second, they recognize the biological origin of the term and reflect on the role of science in shaping "the social" during the nineteenth century and beyond. Third, they make reference to the War on Terror and to debates on today's "politics of life." Fourth, they identify Foucault and Agamben as the two scholars whose work has crucially contributed to a "biopolitical turn" in critical social theory and the humanities. Finally, they take a critical stance against all positivist approaches to the biopolitical, and the assumption that culture and politics may be explained merely via biological determinations.

A few lines on how these different approaches introduce biopolitics and its complicated histories may be of help. Campbell and Sitze's reader is indeed an excellent collection of well-established references on biopolitics, complemented by an equally excellent introduction to mainstream related debates. Including key essays from Foucault and Agamben, it proposes also relatively less cited work in geography, like that of Achille Mbembe, Donna Haraway, and Peter Sloterdjik, but also that of Roberto Esposito, Paolo Virno, and Hardt and Negri. However, despite this diversity, the volume is focused on texts that in English-speaking academia are normally considered "must-reads" on biopolitics, while it entirely neglects work marked by a positivistic stance (i.e., positivist political science) and earlier critical references to biopolitics, for example in the French tradition.

Campbell and Sitze clearly state the centrality of Foucault's work in setting the biopolitical agenda of the past decades and in preparing the ground for the presumed "biopolitical turn." Also for Lemke, "biopolitics" in Foucault's work signals a break in the order of politics, by assuming:

> the dissociation and abstraction of life from its concrete physical bearers and referring to the emergence of a specific political knowledge and new disciplines such as statistics, demography, epidemiology, and biology. The ambivalent political figure "population" plays a decisive role in this process. (2010: 5)

Foucault is therefore commonly identified as the founding father of present-day debates on biopolitics, in particular, but not solely, because of the central role that it played in his analysis of governmentality and the institutionalization of pathology in modern society. His path-breaking investigation of the effects of the new "politics of life" is therefore key in reflecting on the biopolitical imperative in geography and in "our" understanding of new biopolitical regimes centered around the care (and the custody) of the population. Population here is intended as a new demo-spatial concept *and* a key economic and political subject in the new formulations of power over life identified by the French philosopher (see Philo 2005 on this):

> one of the greatest transformations the political right underwent in the nineteenth century was precisely that ... sovereignty's old right – to take life or let live – ... came to be complemented by a new right which does not erase the old right but which does penetrate it, permeate it. This is the right, or rather precisely the opposite right. It is the power to "make" live and "let" die. (Foucault 2003 [1976]: 273)

Foucault is adamant about the fact that the sovereign cannot grant life in the same way that he can inflict death:

> The right of life and death is always exercised in an unbalanced way: the balance is always tipped in favor of death. Sovereign power's effect on life is exercised only when the sovereign can kill. (2003 [1976]: 246)

Key to the incorporation of biopolitics in the work of geographers is therefore the fact that "security mechanisms" were established and legitimized around the concept of population as a living organism in order "to optimize a state of life" (246).

A different and perhaps more comprehensive genealogical account of biopolitics is offered by Italian political philosophers Roberto Esposito (2008), Antonella Cutro (2004), and Laura Bazzicalupo (2010), who all focused on the contours and trajectories of the biopolitical, in relation to Foucault, but also before and beyond Foucault. These "Italian" reconstructions are indeed essential to reflect on the travels of the concept from other disciplines to geography, but also on the contribution putatively given by geography to the birth of biopolitical thinking, something discussed later in the chapter. These accounts will be confronted with and complemented by Lemke's broader analysis of the biopolitical today, especially where they converge in highlighting historically the role played by geography in the emergence of biopolitics.

Cutro (2004) and Esposito (2008) suggest that the term biopolitics is used for the first time at the beginning of the twentieth century by Swedish political scientist and geopolitics advocate Rudolf Kjellén, an origin endorsed also by Lemke. However, while the "Italian reconstructions" do recognize the "geographical" origin of the organic approach to the state and its biopolitics, and in particular Ratzel's influence on Kjellén's geopolitical visions, they trace back the origin of biopolitical thinking not to the *Lebenphilosophie* recalled by Lemke, but to French and Anglophone positivistic roots. In particular, for Bazzicalupo (2010: 23), the origin of biopolitical thinking must be put in relation to nineteenth-century research aimed at the direct application of the life sciences, and biology in particular, to society and the understanding of politics. According to the "Italian" genealogies, while the term was adopted only in the twentieth century, its fundamental theoretical roots are linked to the emergence of positivism in the previous century and its inclination to apply scientific thought to all fields of life and knowledge. Cutro (2004) suggests that, while the technocratic view at the origin of the positivistic approach to biopolitics was fueled by evolutionary science, and found in social Darwinism its utmost expression, it was already present in the concept of *biocratie* adopted by Auguste Comte in his *System of Positive Polity, or Treatise on Sociology, Instituting the Religion of Humanity* (1851–54). Biocracy, for Comte, was the form of government and self-discipline of individuals necessary to realize a sociocracy; that is, a society in which a purely biological understanding of life would coincide, perfectly, with modes of life, the social and the cultural (Cutro 2004). The term biocracy reappeared at the very beginning of the 1900s in the work of French hygienist Edward Toulouse, to define the political role of public hygiene, within which psychiatry and mental hygiene were assigned a fundamental role in shaping society (Cutro 2004). Bazzicalupo (2010) notes that while this contribution was indeed key in linking the "bio" to the political within an institutional framework (Toulouse was responsible for the organization of French public psychiatry), even more important is how the roots of biopolitics may be identified within the domain of positivist pedagogy – something that would clearly resonate a few decades later in the eugenic and racial programs launched by many Western governments to produce a purified and healthier body politic (see Werner 2003).

Another key episode in the emergence of the concept identified by Italian philosophy is Marley Roberts's 1938 essay "Bio-politics: An Essay in the Physiology, Pathology and Politics of the Social and Somatic Organism," where the underlying assumption is the implicit connection between politics and biology, and particularly medicine (Esposito 2008). In his organicist perspective on society, Roberts resorts to an interpretation of normality by studying pathology, since the task of politics essentially corresponds to the identification and government of normality (see Canguilhem 1991 on pathology as a political technology). This emphasis on the pathological would later allow Foucault to demonstrate the impossibility of defining what is normal, and that the "government of the living" developed by modern state institutions was fundamentally based on the identification of pathologies in need of normalizing interventions (Bazzicalupo 2010).

From the 1960s onward, what was until then essentially a series of (only partially connected) episodes becomes a more continuous line of research that takes two distinct trajectories, also in light of the emergence of new biotechnologies forging novel approaches to life and politics. The first was based in Anglophone academia and largely founded on a purely technical and biological reading of the political; the second, which was rooted in French academia, was more inclined to host humanistic pedagogic interpretations of human nature, also in relation to the pervasive role played by capitalism in shaping the social and in exploiting the human body (see also Esposito 2008, 2011).

For example, positivistic interpretations of biopolitics experienced a new momentum in the 1970s when the International Political Science Association (IPSA) established a research committee on biology and politics and organized in Paris in (1975) the first congress focused on biopolitics, followed by a series of other similar events worldwide. Biopolitics was presented once again as a way to read human nature and political behavior through the prisms offered by neurology, pharmacology, molecular biology, and even ethology (Cutro 2004). In the early 1980s, the Association for Politics and the Life Sciences was founded, together with the journal *Politics and Life* and the series *Research in Biopolitics*. For Esposito, two distinguished matrices are at the origin of this approach: "Darwinian evolution (or more precisely social Darwinism), and … the ethological research, developed principally in Germany at the end of the 1930s" (2008: 22). However, to locate the beginning of this line of research, Esposito suggests that we should look to the middle of the 1960s when two texts first elaborated the biopolitical lexicon:

> If Lynton Caldwell was the first to adopt the term in question in his 1964 article "Biopolitics: Science, Ethics, and Public Policy", the two polarities within which is inscribed the general sense of this new biopolitical thematization can be traced to the previous year's *Human Nature in Politics* by James C. Davies. (2008: 21)

The second wave of interest in biopolitics was registered in France in the 1960s. The difference from the first, despite the French roots of positivism and social Darwinism previously mentioned, is, for Esposito,

> all too obvious and it couldn't be otherwise in a historical frame that was profoundly modified by the epochal defeat of Nazi biocracy. The new biopolitical theory appeared to be conscious of the necessity of a semantic reformulation even at the cost of weakening the specificity of the category in favor of a more domesticated neohumanistic declination, with respect not only to Nazi biocracy, but also to organicistic theories that had in some way anticipated their themes and accents. (2008: 19)

The volume that in 1960 virtually opened up this new stage of study is Aroon Starobinski's *La biopolitique: Essai d'interprétation de l'histoire de l'humanité et des civilisations*. Despite society here is still explained with reference to the most elementary forms of biological life, a more humanistic tone clearly emerges. However, perhaps the most significant episode in the (rather fragmented) French emergence of the biopolitical is Edgar Morin's 1965 *Introduction à la politique de l'homme*, where he fundamentally calls for a more humane politics, a *bio*politics capable of putting the essential needs of human beings at the center of the political agenda in a moment affected by global capitalism's reconceptualizations of the human. Despite these humanistic leanings, in the French tradition we can identify, from the 1960s onward, numerous attempts on the part of the life sciences to enter the field of politics and, in line with the positivist tradition, to focus on the relationship between biology and society.

Lemke (2010: 3) proposes a rather critical reading of these traditions, which cover "organicist concepts of the state in the first decades of the 20th century through racist modes of reasoning during National Socialism to biologistic ideas in contemporary political science." He suggests that since the 1960s this approach has taken two different forms:

> [F]irst, as an ecological biopolitics that pursues conservative and defensive objectives and seeks to bind politics to the preservation and protection of the natural environment and, second, in a technical reading of biopolitics whose advocates are more interested in dynamic development and productivist expansion than in preservation and protection. ... This interpretation is especially popular nowadays, and is regularly cited in political discussions and media debates to describe the social and political implications and potential of biotechnological innovations. (Lemke 2010: 3)

Largely in line with the "Italian" accounts, he explains that all these approaches fail to capture the essential dimensions of biopolitical processes, since they take for granted the existence

> of a stable hierarchy and an external relationship between life and politics ... The advocates of naturalism regard life as being "beneath" politics, directing and explaining political reasoning and action. The politicist conception sees politics as being "above" life processes; here, politics is more than "pure" biology, going beyond the necessities of natural existence. (Lemke 2010: 3–4)

These positivist readings of the political via the biological have been fundamentally neglected by mainstream critical social theory on biopolitics, including human geography. They are nonetheless relevant for the understanding of the biopolitical imperative in geography, not only because they are indirectly criticized by geographers preoccupied with the implementation of biopolitical policies, but also perhaps because they may be tracked down in several expressions of present-day positivistic geography, especially when uncritically focused on broader questions of nature and population management and distribution.

In light of these reconstructions and the critique of these approaches as developed by the Foucault/Agamben tradition, how can we qualify the role of biopolitics in geography and the contribution that geography may be able to offer to the biopolitics debate? In the next sections, I start from where geographers normally begin, looking at "the spatial" in Foucault and Agamben's work on biopolitics, and conclude with an exploration of the presumed geographical origin of the concept, linking the "bio" to the "geo" via the historical concept of "living/vital space."

Geo-biopolitics

With absolutely no pretense to providing a comprehensive review of the geographical literature on biopolitics, I suggest starting from Foucault once again. Esposito (2008: 26) tellingly asks:

> [W]hat "political space" is it that stirs within Foucault's remarks on biopolitics? Is it the European city, into which and out of which grain flows? The territory of the sovereign European state, as Foucault himself seems to suppose? The concentration camp, as Agamben proposes? The "milieu" of Jean-Baptiste Lamarck and Jakob von Uexküll? Or is Foucault's thinking on biopolitics most notable for the way it seems to abandon any thought of space whatsoever?

These are important questions for understanding the relationship between the "geo" (in biopolitics) and the "bio" (in geography and elsewhere). Perhaps it is useful to recall that while Foucault was introduced to geography more than two decades ago – adopting work, for example, on mental health (see Parr 1997; Philo 2006) or population management (see Philo 2005), to mention two typically Foucauldian topics – the present proliferation of geographical work focused on aspects of biopolitics is only partially related to this initial engagement with the French philosopher. Indeed, it was the popularization of Giorgio Agamben's interpretation of the Foucauldian project on biopolitics that arguably produced the conditions for what could perhaps be identified as a "biopolitical turn" in the discipline. This appears to be in line with what has happened in other fields as well. As Campbell and Sitze (2013b: 4) rightly note:

> [I]t was not until 1998, with the English translation of Giorgio Agamben's provocative rereading of Foucault's "Right of Death and Power over Life" in *Homo Sacer: Sovereign Power and Bare Life*, that Foucault's long-dormant text on biopolitics was reactivated in its current form … [After the controversial but path-breaking Agambenian reading of Foucault, followed in 2000] by the very different but equally controversial appropriation of Foucault by Michael Hardt and Antonio Negri in their book *Empire*, the concept of "biopolitics" began to migrate from philosophy to … the fields of anthropology, geography, sociology, political science, theology, legal studies, bioethics, digital media, art history, and architecture.

For Lemke also, Agamben's and Hardt and Negri's writings are among the most prominent contributions to a reformulation of Foucault's notion of biopolitics. While Agamben argues that the constitution of sovereign power requires the production of a biopolitical regime and of "bare life," the authors of *Empire* describe a new stage of capitalism marked by the merging of economy and politics, production and reproduction. Lemke (2010: 6) identifies two main lines of engagement with Foucault's work on biopolitics. The first is centered on the *mode of politics* and asks how biopolitics is to be distinguished historically and analytically from "classical" forms of political representation and articulation. He includes here the work of Agnes Heller and Ferenc Fehér, but also Anthony Giddens's concept of life politics and Didier Fassin's idea of biolegitimacy. The second is instead focused on the *substance of life* and investigates "how the foundations, means, and objectives of biopolitical interventions have been transformed by a biotechnologically enhanced access to the processes of life and the human body" (Lemke 2010: 7). Under this umbrella he places work on molecular politics, thanatopolitics, anthropolitics, ideas of "biosociality" inspired by Paul Rabinow, and "etho-politics" by Nikolas Rose. While mentioning "vital politics" as promoted by German liberals after the Second World War and the theory of human capital developed by the Chicago School,

Lemke also highlights new visions of "bioeconomy" in contemporary political action plans related to biotechnological innovations and transformations in capitalism.

The "Italian" reconstructions also acknowledge Agamben for having brought biopolitics to the core of the philosophical discussion, and helping it to go beyond Foucault. They identify a number of other further developments in Jacques Rancière's work on the relationship between politics and police, Bruno Latour's stance on biopower, and Isabelle Stengers's *Cosmopolitiques*, all critical of Foucault's understanding of biopolitics when referring to recent biotechnological and institutional innovations. Haraway's classic 1989 essay, "The Biopolitics of Postmodern Bodies," is also discussed in detail, together with Rabinow's proposal for a new anthropology capable of incorporating the effects of the politics of life produced by the new biotechnologies, and Mbembe's notion of *necropolis* as a distressing account of a different dimension of contemporary biopower. Finally, Esposito's work (2011) on affirmative biopolitics and his analysis of the *dispositif* of immunity are also considered key to the present reinterpretations of biopolitics. The result of these interventions is perhaps a broader "biopolitical turn":

> In addition to bioethics, biotechnology, biopower, and biohistory – "bio-"terms ... already in circulation prior to the biopolitical turn – scholars now proposed to study bioculture, biomedia, biolegitimacy, bioart, biocapital, biolabor, bioscience, biohorror, bioeconomics, bioinformatics, biovalue, biodesire, biocomputing, biotheology, biosociety, and biocentrism, among others. (Campbell & Sitze 2013b: 4–5)

Geo/bio turns

Many of these perspectives have been, albeit in different ways, incorporated in the work of geographers. As noted above, Agamben's *Homo Sacer* project has been crucial in determining the presumed biopolitical turn in geography, something extensively discussed in the past decade or so and in no need of further detailed analysis here (see, among others, Ek 2006; Minca 2007; Schlosser 2008). Agamben's reflections on the spatialities of the camp, on "bare life," and on the enactments of sovereign power in modern politics in particular, including his critical appraisal of humanitarianism, seem indeed to have taken center stage in the work of many geographers focused on war, detention, borders, mobility, migrations, the refugee problem, as well as questions of population and the body politic.

Important work has, for example, engaged with the spaces of exception both conceptually (see, among others, Coleman 2007; Belcher et al. 2008; Coleman and Grove 2009) and empirically, mainly but not limited to the geographies produced by the War on Terror launched by the Bush administration in the wake of 9/11 (on the biopolitics and related geopolitics of the War on Terror, see Gregory 2004, 2006a, 2010a, 2010b; Hannah 2006, 2010; Kearns 2008; MacLeavy & Peoples 2010; on the biopolitics of affect related to the War on Terror, see Anderson 2010, 2011). The analysis of the geographies of exception in a biopolitical perspective includes interventions on the effects of preemptive strikes (Amoore & de Goede 2008), drones (Shaw & Akhter 2012), financial surveillance (Atia 2007), and the relationship between terror and territory (Elden 2007a). Drawing directly or indirectly (sometimes rather critically) on Agamben's speculations on "the camp" as a biopolitical space of exception (see Minca 2005, 2006, 2007; Diken & Laustsen 2006; Ek 2006; Gregory 2006b, 2007), geographers have engaged with actual camps, from Auschwitz (Giaccaria & Minca 2011a) and Sobibor (Carter-White 2013), to Cambodia (Cunha et al. 2012) and the Middle East (Ramadan 2009,

2013a; Martin 2015). Work on Guantánamo as a space of "exceptional sovereignty" (Reid-Henry 2007) has also been important in understanding the bio-geopolitical agenda implemented by the US administration and its agencies. Bare life and sovereign exceptions have been linked to gender geographies by Geraldine Pratt (2005) and Jennifer Fluri (2012), and to agency in human geography by Craig Young and Duncan Light (2013), while urban governance has been analyzed as part of the new spatialities of exception (see for example Belina 2007; Schinkel & van den Berg 2011; Boano & Martén 2013).

Further geographical work on biopolitics and the spaces of sovereign exception has taken three distinct additional trajectories concerning new studies on carceral geographies; reconceptualizations of borders and bordering in the age of biometrics; and the related geographies of migration control together with the spatialities of refugees and asylum seekers. Most recent work on carceral geographies, largely supported by a broader biopolitical perspective, includes research on what Alison Mountz (2011) has defined as the "enforcement archipelago," on deportation and removal centers (McGregor 2012), on family detention policies (Martin 2012), and on punishment in the US detention system (Morin 2013). Questions of women's privacy and disciplined mobility in relation to the geographies of detention have been analyzed as well (see Moran, Piacentini, & Pallot 2012; Moran, Pallot, & Piacentini 2013), together with broader issues of mobility, containment, bordering, and exclusion (Mountz et al. 2013).

The new geographies of the border and bordering have also been read using a broader biopolitical analytical framework. Immigration policies, in this perspective, have been analyzed with reference to the United States (Coleman 2007; Belcher & Martin 2013), the United Kingdom (with the border intended as "a space of the ban," see Vaughan-Williams 2010), and the European Union (Van Houtum 2010; for "other" biopolitical borderings see, for example, Artman 2013). In more general terms, new conceptualizations and practices of border management and "bordering" have been studied by looking at questions of sovereign exception (Jones 2009a, 2009b, 2012a), as well as at the new geographies of exception produced by measures aimed at responding to the global terrorist threat (Doty 2011; also Jones 2009c). Airport security is discussed in biopolitical terms by Peter Adey (2009) and Lauren Martin (2010), while biometrics in relation to borders and migrants is analyzed respectively by Louise Amoore (2006), Rebecca Pero and Harrison Smith (2014), and Adam Warren (2013). The biopolitics of migration management has been presented as "migration wars" (Hyndman 2005), as assemblages of global surveillance (Murakami Wood 2013), and as the result of recession in the European context (Bailey 2013).

The redefinition of the principles of citizenship and of the right to mobility has been examined by geographers as part of broader interdisciplinary debates on the biopolitical regimes imposed on the growing masses of refugees and asylum seekers fleeing situations of danger and emergency. The production of bare life in relation to asylum, hospitality, and the politics of encampment has thus been closely interrogated (Darling 2009; also Gill 2009), often associated with what has been described as "domopolitics" (Ingram 2008; Darling 2011, 2014). Issues of mobility in relation to domicile citizenship (Bauder 2014), biological citizenship (Sparke 2006; Kurts, Trauger, & Passidomo 2013), and even post-citizenship (Garmany 2012) have also been discussed in recent geographical writings.

Global health and biosecurity have taken center stage in recent debates as well, incorporating the spatial implications of the biopolitical imperative. Bruce Braun's work has been particularly important in illustrating the workings of the new urban biopolitical dispositif in relation to climate change (Braun 2014), the governance of the "resilient city" (Wakefield &

Braun 2014), and the biopolitics of disease management (Braun 2008), as well as in relation to the new invention and molecularization of life (Braun 2007, 2008). The same can be said for Alan Ingram's contribution to the understanding of biosecurity in relation to HIV/AIDS epidemics (2008, 2010a, 2010b, 2013) and Michael Brown's analysis of the biopolitics of sexuality in urban governance and the birth of the "gay clinic" (Brown 2009; Brown & Knopp 2010, 2014). Biosecurity issues have been discussed with a focus on biosecure citizenship and biological threat (Barker 2010), food security (Bingham & Lavau 2012; Mansfield 2012a, 2012b), biopolitical concerns for neoliberal reconceptualizations of public health (Evered & Evered 2012; related to gender, see Mansfield 2012b; to the welfare state, see Cameron 2006), and the problem of infectious disease linked to urban governance (Keil & Ali 2007), air travel (Budd, Belli, & Brown 2009), and migrations (Warren 2013). The biopolitics of state interventions in medical and family "care and management" has been recently scrutinized in relation to drug courts (Moreno & Curti 2012), organ transfer (Sothern & Dickinson 2011), reproduction and religion (Smith 2012), and the practice of medicine under totalitarian regimes (see Tyner 2012 on the Khmer Rouge).

Three additional areas of geographical investigation influenced by "biopolitical analytics," distinct from but at the same time tightly connected to the above themes, may be tentatively identified: the spaces of humanitarian intervention; the question of affect in biopolitics; and the calculative rationalities traditionally underpinning state biopolitical population regimes. The biopolitics of humanitarian spatialities has been discussed in detail by Stuart Elden (2006) and Simon Reid-Henry (2013) and put in relation to broader geopolitical considerations and the workings of the neoliberal state. Affective and care geographies have been instead closely linked to biopower and, more generally, to the contemporary management of life and population emergency (see Anderson 2010, 2011, 2012; on voluntarism, care, and homelessness, see Evans 2011, 2012; on immunological politics, see Grove 2014). The calculative rationalities associated with the geography of population management have been analyzed by Jeremy Crampton and Stuart Elden (2006, 2007), and linked to National Socialism (Elden 2006) and cartography (Crampton 2010). State biopolitics has also been examined with a focus on bioethics and sovereignty (Tyner 2012), censuses (Hannah 2009), and "population surplus" (Li 2010; Tyner 2013).

Finally, the "politics of life" has entered theoretical debates in geography, largely based on reinterpretations of the Foucauldian/Agambenian legacy in the discipline, but also incorporating the work of other scholars such as Hardt and Negri (2000), Mbembe (2003), Rose (2006), and Esposito (2011). These debates have focused on present-day "vital geographies" (Kearns & Reid-Henry 2009) and, again, affect (Anderson 2012), but also ethics (Barnett 2012) and the "value of life" (and death – see Rose 2009; Gilbert & Ponder 2014), in what Louisa Cadman (2009) has defined as "our posthuman times." This reference to the posthuman brings us to the realms of more-than-human geographies and, more broadly, of animal geographies as well as new (biopolitical) ways to incorporate major environmental issues. Jamie Lorimer and Clemens Driessen, for example, have proposed a biopolitical perspective on past and contemporary policies of "rewilding" nature (2013, 2014; see also Hennessy 2013), while Irus Braverman (2011) has discussed American zoos as spaces/states of exemption. The (biopolitical) subjectivity of animals was analyzed by Lewis Holloway and Carol Morris (2012, 2014) in relation to livestock breeding (see also Bingham 2006; Holloway 2007; Lorimer & Driessen 2014), opening up an entirely new field of investigation on ideas of nature and the environment for geographers (see Asher & Ojeda 2009; Yeh 2012). Biopolitics and conservation biology are examined by Christine Biermann and Becky Mansfield (2014), something of concern also for work on environmental risk (Baldwin & Stanley 2013) and safety (Collard 2012; see Mori 2008 on nuclear biopolitics) together with the realization of "biopolitical environmental citizens" (Baldwin 2010, 2013). More generally, the new ecologies of global capitalism have been studied with biopolitical lenses in reference to questions of water (Bakker 2013; Meehan 2013; Hellberg 2014), climate change (Cupples 2012; Wainwright & Mann 2013), the Anthropocene (Dalby 2013), environmental subjectivities and affective labor (Singh 2013), and even necropolitics (Cavanagh & Himmelfarb 2015).

Research on the biopolitical effects of neoliberalism has focused on "laboring bodies" (Kanngieser 2013; Labban 2014), liberal "care" as expressed by humanitarianism (Reid-Henry 2013), neoliberal violence (Springer 2012), and biological and financial citizenship (Kear 2013; Kurtz et al. 2013; Sparke 2013). Recent engagements with the "new Foucault" (Philo 2012) have continued to highlight the importance of understanding the working of governmentality (see in particular Elden 2007b; also Cadman 2010; Ettlinger 2011), and showing interest in the workings of the biopolitical imperative (see for example Legg's work on India, 2006, 2008, 2009). Along these lines, Reuben Rose-Redwood (2006) has stigmatized the links between governmentality, geography, and the new "geo-coded world," while others have explored the "negatives" of governance (Rose 2014; on illiberal governance see Corva 2008) and its limited and contingent nature (Dillon 2007; Griffin 2010).

The biopolitical turn in geography has also opened the floor to historical investigations on the relationship between the "bio" and the "geo," either by engaging with Nazi bio-geopolitical technologies (see Giaccaria & Minca 2011a, 2011b, 2015; Minca & Rory 2015) or with colonialism and its postcolonialism consequences. James Duncan (2007), for example, has analyzed how climate, race, and biopower were interrelated in nineteenth-century Ceylon, while Steven Legg, as noted above, has investigated the biopolitics of colonial Delhi. The biopolitics of "food provisioning" in the colonial and postcolonial context is illustrated by David Nally (2008, 2011), while postcolonial legacies and some of their biopolitical effects in relation to the "colonial present" (see Gregory 2006a on Iraq, Palestine, and Afghanistan) are observed in Sri Lanka (Korf 2006), Afghanistan (Fluri 2009, 2011), South Korea (Lee, Jan, & Wainwright 2014), Ogaden (Hagmann & Korf 2012), Cambodia (Tyner 2012), Egypt (Rose 2014), and Palestine (Martin 2015), to mention only a few.

Finally, it is perhaps worthwhile recalling the significant amount of theoretical work that the biopolitical turn has sparked in the discipline. From the geographical readings of Agamben's work (see, among others, Ek 2006; Minca 2007), to reflections on the ethico-aesthetics of life and bioethics in Félix Guattari (Hynes 2013), to the biopolitical reading of Jacques Derrida's "the beast and the sovereign" (Rasmussen 2013), to Fluri's (2012) analysis of bare life according to gender politics – and the list could continue – the biopolitical and its spatial projections seem to stimulate an increasing number of interventions populating today's key geographical fora. However, the question of the presumed geographical origin of the notion of biopolitics remains an open one, something that may potentially problematize the relationship between the "bio" and the "geo" as illustrated in this section. It is to this question that we now turn in the concluding paragraphs of this chapter.

Geo-biopolitical origins

Is geography inherently biopolitical? Or, perhaps we should ask: is biopolitics inherently geographical? As noted above, the "Italian" genealogies not only claim that the term bio-politics was adopted for the first time by Kjellén, they also hint at Ratzel's influence on the work of the Swedish initiator of the *Geopolitik* tradition. Indeed, philosophers Giorgio Agamben (2004) and Giorgio Cavalletti (2005) identify a clear trajectory linking Ratzel's bio-geographies to a specific understanding of organic "living space," of *Lebensraum*. Cavalletti (2005) in particular presents the Ratzelian concept of population as a spatial calculative political technology typical of the modern organic state, implicitly suggesting that the "geo" and the "bio" were inherently co-implicated in the politicization of the related living/vital spatialities (see Giaccaria & Minca 2011b, 2015; Minca 2011a; Abrahamsson 2013).

This interpretive line is also endorsed by Lemke, who puts forward the following claim about the appearance of the concept of biopolitics nearly a hundred years ago:

> [It] was as part of a general historical and theoretical constellation. By the second half of the 19th century, *Lebensphilosophie* (the philosophy of life) had already emerged as an independent philosophical tendency; its founders were Arthur Schopenhauer and Friedrich Nietzsche in Germany and Henri Bergson in France. (Lemke 2010: 9)

According to this account, for these *Lebensphilosophen* the re-evaluation of "life" served as a standard by which "processes of rationalization, civilization, mechanization, and technologization," perceived as adversarial to life, ought to be subjected to critical examination. As Lemke (2010: 9) remarked, "The concept of biopolitics emerged in this intellectual setting at the beginning of the 20th century."

In his 1916 *Staten som lifsform* (The State as Form of Life), Kjellén links his geopolitical analysis for the first time to an organicist concept of the state as a "living form" *(Lebensform* in German). According to Esposito, in this idea of the state, as "a whole that is integrated by men and which behaves as a single individual both spiritual and corporeal, we can trace the originary nucleus of biopolitical semantics" (2008: 16):

> In view of this tension typical of life itself … the inclination arose in me to baptize this discipline after the special science of biology as biopolitics; … in the civil war between social groups one recognizes all too clearly the ruthlessness of the life struggle for existence and growth, while at the same time one can detect within the groups a powerful cooperation for the purposes of existence. (Kjellén 1920: 93–94, quoted and translated in Lemke 2010: 10)

Kjellén was not alone in presenting the state as a "living organism," but was accompanied by many social scientists and life scientists of his time. Esposito notes that if

> this process of the naturalization of politics in Kjellén remains inscribed within a historical-cultural apparatus, it experiences a decisive acceleration in the essay that is destined to become famous precisely in the field of comparative biology. I am referring to *Staatsbiologie*, also published in 1920 by Baron Jakob von Uexküll, in which the German state is presented with its peculiar characteristics and vital demands but also with the special emphasis placed on pathology compared to anatomy and physiology. Here we can already spot the harbinger of a theoretical weaving – that of the degenerative syndrome and the consequent regenerative program – fated to reach its macabre splendors in the following decades. (Esposito 2008: 16)

Even more relevant, insists Esposito, "is the biopolitical reference to those 'parasites' which, having penetrated the political body, organize themselves to the disadvantage of other citizens" (2008: 18).

Geographer Karl Haushofer takes this approach further with a clearly racist perspective, in line with the spatial ideology at the foundation of the Nazi conceptualization of *Lebensraum* (see Bassin 1987; Giaccaria & Minca 2015). This organicist state immediately becomes a racial state under Hitler's rule, driven by the millennial task of realizing a purified national community via the implementation of a specific bio-geopolitics. For Lemke, two general characteristics decisively marked Nazi biopolitics: first, the foundation of a program of racial hygiene and "hereditary biology" (*Erbbiologie*); second, "the combination of these ideas with geopolitical considerations" and in particular with the doctrine of *Lebensraum*:

> The concept of *Lebensraum*, which was by 1938 at the latest a central element of National Socialist foreign policy, goes back to scientific ideas that had been worked out earlier in the 20th century. The "father" of geopolitics was the German geographer Friedrich Ratzel, who coined the word *Lebensraum* around the turn of the century. His "anthropogeography" examined the relationship between the motionless Earth and the movements of peoples, in which two geographical factors play a central role: space and position. (2010: 13)

The link between Ratzel and Nazi *Lebensraum* has already been scrutinized by geographers and remains somehow a contested terrain, but this is no place to engage with that debate in full (see Bassin 1987; Farinelli 2000; Giaccaria & Minca 2015). What matters here is that Lemke insists on the geographical origin of biopolitics by recalling the role played by Haushofer in shaping the Geopolitik agenda and in founding the popular journal *Zeitschrift für Geopolitik*. In one article appearing in this journal,

> Louis von Kohl explained that biopolitics and geopolitics were together "the basis for a natural science of the state" (1933: 306) ... "When we observe a people or a state we can place greater emphasis either on temporal or spatial observations. Respectively, we will have to speak of either biopolitics or geopolitics. Biopolitics is thus concerned with historical development in time, geopolitics with actual distribution in space or with the actual interplay between people and space" (1933: 308 in Lemke 2010: 14)

According to historian Boaz Neumann (2002), the deeper biopolitical nature of the link between *Lebens* and *Raum*, between the "bio" and the "geo," as established by Nazi spatial ideology and practice, was reflected in the relationship between *Lebensraum* intended as functional "living space" and *Lebens-Raum* as ontological "life-world", as world view. The Nazi formulation of *Lebensraum* thus consisted of the coming together of two bio-geopolitical visions: the first produced and enacted by German imperialism; the second based on a millennial, ontological interpretation of the term/concept (Giaccaria & Minca 2015). With the emergence of Hitler's biocracy, the first state ever organized mainly around a biopolitical imperative – that is, to realize a racially purified spatial formation – the "bio" and the "geo" became one and the same, indistinguishable and entirely fused in the millennial horizon of the Third Reich.

The "biologization of politics," argues Lemke, "is nevertheless neither a German idiosyncrasy nor limited to the period of National Socialism. The state's 'gardening-breeding-surgical ambitions' ... can be traced back at least to the 18th century." In the interwar period, "these fantasies ... emerged in the projects of the 'new Soviet man' under Stalin's dictatorship but also in the eugenic practices of liberal democracies" (Lemke 2010: 14). Arguably, Lemke concludes, "even if racist biopolitics no longer had any serious scientific or political standing after the end of the Third Reich," it continues to be used by right-wing movements "to complain about the ignorance of the *Zeitgeist*" toward the 'question of race'" (2010: 15).

Curiously, this geographical root of biopolitics and its fundamental relationship with Nazism have been almost entirely neglected by contemporary geographical work on "the politics of life." Such forgetting may be due to two main factors: first, the almost entirely Foucauldian root of the geographical interpretations of biopolitics (despite the fact that Foucault in his Lectures at the Collège de France directly addressed the biopolitics of Hitlerism); second, a longstanding relative lack of interest in Nazi spatial theory on the part of geographers (see Barnes & Minca 2013). An alternative history of the geographical roots of biopolitics thus remains to be written. This indeed appears to be an urgent task for geographers if, as claimed by both Agamben and Esposito, it may prove impossible to unravel the workings of modern sovereign power and their incorporation of life without a full engagement with the bio-geopolitical ontological implications of Hitler's biocratic regime and its contemporary legacies.

References

Abrahamsson, C. (2013). On the genealogy of Lebensraum. *Geographica Helvetica* 68: 37–44.

Adey, P. (2009). Facing airport security: Affect, biopolitics, and the preemptive securitisation of the mobile body. *Environment and Planning D: Society and Space*, 27(2): 274–295.

Agamben, G. (1998). *Homo Sacer*. Stanford, CA: Stanford University Press.

Agamben, G. (2002). *Remnants of Auschwitz*. London: Zone Books.

Agamben, G. (2004). *The Open*. Stanford, CA: Stanford University Press.

Amoore, L. (2006). Biometric borders: Governing mobilities in the War on Terror. *Political Geography* 25(3): 336–351.

Amoore, L., and de Goede, M. (2008). Transactions after 9/11: The banal face of the preemptive strike. *Transactions of the Institute of British Geographers* 33(2): 173–185.

Anderson, B. (2010). Morale and the affective geographies of the "War on Terror." *Cultural Geographies* 17(2): 219–236.

Anderson, B. (2011). Population and affective perception: Biopolitics and anticipatory action in US counterinsurgency doctrine. *Antipode* 43(2): 205–236.

Anderson, B. (2012). Affect and biopower: Towards a politics of life. *Transactions of the Institute of British Geographers* 37(1): 28–43.

Artman, V.M. (2013). Documenting territory: Passportisation, territory, and exception in Abkhazia and South Ossetia. *Geopolitics* 18(3): 682–704.

Asher, K., and Ojeda, D. (2009). Producing nature and making the state: Ordenamiento territorial in the Pacific lowlands of Colombia. *Geoforum* 40(3): 292–302.

Atia, M. (2007). In whose interest? Financial surveillance and the circuits of exception in the War on Terror. *Environment and Planning D: Society and Space* 25(3): 447–475.

Bailey, A.J. (2013). Migration, recession and an emerging transnational biopolitics across Europe. *Geoforum* 44: 202–210.

Bakker, K. (2013). Constructing "public" water: The World Bank, urban water supply, and the biopolitics of development. *Environment and Planning D: Society and Space* 31(2): 280–300.

Baldwin, A. (2010). Wilderness and tolerance in Flora MacDonald Denison: Towards a biopolitics of whiteness. *Social and Cultural Geography* 11(8): 883–901.

Baldwin, A. (2013). Vital ecosystem security: Emergence, circulation, and the biopolitical environmental citizen. *Geoforum* 45: 52–61.

Baldwin, A., & Stanley, A. (2013). Risky natures, natures of risk. *Geoforum* 45: 2–4.

Barker, K. (2010). Biosecure citizenship: Politicising symbiotic associations and the construction of biological threat. *Transactions of the Institute of British Geographers* 35(3): 350–363.

Barnes, T., & Minca, C. (2013). Nazi spatial theory: The Dark Geographies of Carl Schmitt and Walter Christaller." *Annals of the Association of American Geographers*, 103(3): 669–687.

Barnett, C. (2012). Geography and ethics: Placing life in the space of reasons. *Progress in Human Geography* 36(3): 379–388.

Bassin, M. (1987). Race contra space: The conflict between German *Geopolitik* and National Socialism. *Political Geography* 6(2):115–134.

Bauder, H. (2014). Domicile citizenship, human mobility and territoriality. *Progress in Human Geography* 38(1): 91–106.

Bazzicalupo, L. (2010). *Biopolitica: Una Mappa Concettuale*. Bari: Carocci.

Belcher, O., & Martin, L. (2013). Ethnographies of closed doors: Conceptualising openness and closure in US immigration and military institutions. *Area* 45(4): 403–410.

Belcher, O., Martin, L., Secor, A., Simon, S., & Wilson, T. (2008). Everywhere and nowhere: The exception and the topological challenge to geography. *Antipode* 40(4): 499–503.

Belina, B. (2007). From disciplining to dislocation: Area bans in recent urban policing in Germany. *European Urban and Regional Studies* 14(4): 321–336.

Biermann, C., & Mansfield, B. (2014). Biodiversity, purity, and death: Conservation biology as biopolitics. *Environment and Planning D: Society and Space* 32(2): 257–273.

Bingham, N. (2006). Bees, butterflies, and bacteria: Biotechnology and the politics of nonhuman friendship. *Environment and Planning A* 38(3): 483–498.

Bingham, N., & Lavau, S. (2012). The object of regulation: Tending the tensions of food safety. *Environment and Planning A* 44(7): 1589–1606.

Boano, C., & Martén, R. (2013). Agamben's urbanism of exception: Jerusalem's border mechanics and biopolitical strongholds. *Cities* 34: 6–17.

Braidotti, R. (2013) *The Posthuman*. Cambridge: Polity Press.

Braun, B. (2007). Biopolitics and the molecularization of life. *Cultural Geographies* 14: 6–28.

Braun, B. (2008). Inventive life. *Progress in Human Geography* 32: 667–679.

Braun, B. (2014). A new urban dispositif? Governing life in the age of climate change. *Environment and Planning D: Society and Space* 32: 49–64.

Braverman, I. (2011). States of exemption: The legal and animal geographies of American zoos. *Environment and Planning A* 43(7): 1693–1706.

Brown, M. (2009). 2008 Urban Geography Plenary Lecture: Public health as urban politics, urban geography: Venereal biopower in Seattle, 1943–1983. *Urban Geography* 30(1): 1–29.

Brown, M., & Knopp, L. (2010). Between anatamo- and bio-politics: Geographies of sexual health in wartime Seattle. *Political Geography* 29(7): 392–403.

Brown, M., & Knopp, L. (2014). The birth of the (gay) clinic. *Health and Place* 28: 99–108.

Budd, L., Bell, M., & Brown, T. (2009). Of plagues, planes and politics: Controlling the global spread of infectious diseases by air. *Political Geography* 28(7): 426–435.

Cadman, L. (2009). Life and death decisions in our posthuman(ist) times. *Antipode* 41(1): 133–158.

Cadman, L. (2010). How (not) to be governed: Foucault, critique, and the political. *Environment and Planning D: Society and Space* 28(3): 539–556.

Cameron, A. (2006). Geographies of welfare and exclusion: Social inclusion and cxception. *Progress in Human Geography* 30(3): 396–404.

Campbell, T., & Sitze, A. (eds) (2013a). *Biopolitics: A Reader*. Durham, NC: Duke University Press.

Campbell, T., & Sitze, A. (2013b). Biopolitics: An encounter. In Campbell, T., & Sitze, A. (eds), *Biopolitics: A Reader*, Durham, NC: Duke University Press, 1–40.

Canguilhem, G. (1991). *The Normal and the Pathological*. New York: Zone Books.

Carter-White, R. (2013). Towards a spatial historiography of the Holocaust: Resistance, film, and the prisoner uprising at Sobibor death camp. *Political Geography* 33: 21–30.

Cavalletti, G. (2005). *La città biopolitica*. Milan: Mondadori.

Cavanagh, C.J., & Himmelfarb, D. (2015). "Much in blood and money": Necropolitical ecology on the margins of the Uganda Protectorate. *Antipode* 47(1): 55–73.

Coleman, M. (2007). Reviews: State of exception. *Environment and Planning D: Society and Space* 25(1): 187–190.

Coleman, M., & Grove, K. (2009). Biopolitics, biopower, and the return of sovereignty. *Environment and Planning D: Society and Space* 27(3): 489–507.

Collard, R.-C. (2012). Cougar–human entanglements and the biopolitical un/making of safe space. *Environment and Planning D: Society and Space* 30(1): 23–42.

Corva, D. (2008). Neoliberal globalization and the war on drugs: Transnationalizing illiberal governance in the Americas. *Political Geography* 27(2): 176–193.

Crampton, J. (2010). Cartographic calculations of territory. *Progress in Human Geography* 35(1): 92–103.

Crampton, J., & Elden, S. (2006). Space, politics, calculation: An introduction. *Social & Cultural Geography* 7(5): 681–685.

Crampton, J., & Elden, S. (2007). Space, Knowledge and Power: Foucault and Geography. London: Ashgate.

Cunha, M.P., Clegg, S., Rego, A., & Lancione, M. (2012). The organization (Ângkar) as a state of exception: The case of the S-21 extermination camp, Phnom Penh. *Journal of Political Power* 5(2): 279–299.

Cupples, J. (2012). Wild globalization: The biopolitics of climate change and global capitalism on Nicaragua's Mosquito Coast. *Antipode* 44(1): 10–30.

Cutro, A. (2004). *Biopolitica: Storia e Attualità di un Concetto*. Verona: Ombre Corte.

Darling, J. (2009). Becoming bare life: Asylum, hospitality, and the politics of encampment. *Environment and Planning D: Society and Space* 27(4): 649–665.

Darling, J. (2011). Domopolitics, governmentality and the regulation of asylum accommodation. *Political Geography* 30(5): 263–271.

Darling, J. (2014). Asylum and the post-political: Domopolitics, depoliticisation and acts of citizenship. *Antipode* 46(1): 72–91.

Diken, B., & Laustsen, C. (2006). The camp. *Geografiska Annaler: Series B, Human Geography* 88(4): 443–452.

Dillon, M. (2007). Governing through contingency: The security of biopolitical governance. *Political Geography* 26(1): 41–47.

Doty, R.L. (2011). Bare life: Border-crossing deaths and spaces of moral alibi. *Environment and Planning D: Society and Space* 29(4): 599–612.

Driver, F. (1997). Bodies in space: Foucault's account of disciplinary power. In Barnes, T., & Gregory, D. (eds), *Reading Human Geography: The Poetics and Politics of Inquiry*, London: Arnold, 279–289.

Duncan, J.S. (2007). *In the Shadow of the Tropics: Climate, Race and Biopower in Nineteenth Century Ceylon*. Aldershot: Ashgate.

Ek, R. (2006). Giorgio Agamben and the spatialities of the camp: An introduction. *Geografiska Annaler Series B, Human Geography* 88(4): 363–386.

Elden, S. (2006). National Socialism and the politics of calculation. *Social and Cultural Geography* 7(5): 753–769.

Elden, S. (2007a). Rethinking governmentality. *Political Geography* 26(1): 29–33.

Elden, S. (2007b). Terror and territory. *Antipode* 39(5): 821–845.

Esposito, R. (2008). *Bios: Biopolitics and Philosophy*. Minneapolis, MN: University of Minnesota Press.

Esposito, R. (2011). *Immunitas: The Protection and Negation of Life*. London: Polity Press.

Esposito, R. (2012). *Living Thought*. Stanford, CA: Stanford University Press.

Ettlinger, N. (2011). Governmentality as epistemology. *Annals of the Association of American Geographers* 101(3): 537–560.

Evans, J. (2011). Exploring the (bio)political dimensions of voluntarism and care in the city: The case of a "low barrier" emergency shelter. *Health and Place* 17(1): 24–32.

Evans, J. (2012). Supportive measures, enabling restraint: Governing homeless "street drinkers" in Hamilton, Canada. *Social and Cultural Geography* 13(2): 185–200.

Evered, K.T., & Evered, E.Ö. (2012). State, peasant, mosquito: The biopolitics of public health education and malaria in early republican Turkey. *Political Geography* 31(5): 311–323.

Farinelli, F. (2000). Ratzel and the nature of (political) geography. *Political Geography* 19(8): 943–955.

Fluri, J. (2009). "'Foreign Passports Only': Geographies of(Post)conflict Work in Kabul, Afghanistan." *Annals of the Association of American Geographers*, 99(5): 986–94.

Fluri, J. (2011). "Bodies, Bombs and Barricades: Geographies of Conflict and Civilian(in)security." *Transactions of the Institute of British Geographers*, 36: 280–96.

Fluri, J. (2012). "Capitalizing on Bare Life: Sovereignty, Exception, and Gender Politics." *Antipode*, 44(1): 31–50.

Foucault, M. (2003 [1976]). Society must be defended, 17 March 1976. In *Society Must Be Defended: Lectures at the Collège de France: 1975–1976*, trans. D. Macey, New York: Picador, 239–265.

Garmany, J. (2012). "Spaces of Displacement and the Potentialities of(Post)citizenship." *Political Geography*, 31(1): 17–19.

Giaccaria, P., & Minca, C. (2011a). Topographies/topologies of the camp: Auschwitz as a spatial threshold. *Political Geography* 30(1): 3–12.

Giaccaria, P., & Minca, C. (2011b). Nazi biopolitics and the dark geographies of the *Selva. Journal of Genocide Research* 13(1–2): 67–84.

Giaccaria, P., & Minca, C. (eds) (2015). *Hitler's Geographies*. Chicago, IL: University of Chicago Press.

Gilbert, E., & Ponder, C. (2014). Between tragedy and farce: 9/11 compensation and the value of life and death. *Antipode* 46(2): 404–425.

Gill, N. (2009). Presentational state power: Temporal and spatial influences over asylum sector decision-makers. *Transactions of the Institute of British Geographers* 34(2): 215–233.

Gregory, D. (2004). The angel of Iraq. *Environment and Planning D: Society and Space* 22(3): 317–324.

Gregory, D. (2006a). *The Colonial Present*. Oxford: Blackwell.

Gregory, D. (2006b). The black flag. Guantanamo Bay and the space of exception. *Geografiska Annaler: Series B, Human Geography* 88(4): 405–427.

Gregory, D. (2007). Vanishing points: Law, violence and exception in the global war prison. In Gregory, D., & Pred, A. (eds), *Violent Geographies: Fear, Terror and Political Violence, New York*: Routledge, 205–236.

Gregory, D. (2010a). Seeing red: Baghdad and the event-ful city. *Political Geography* 29(5): 266–279.

Gregory, D. (2010b). War and peace. *Transactions of the Institute of British Geographers* 35(2): 154–186.

Griffin, L. (2010). The limits to good governance and the state of exception: A case study of North Sea fisheries. *Geoforum* 41(2): 282–292.

Grove, K. (2014). Agency, affect, and the immunological politics of disaster resilience. *Environment and Planning D: Society and Space* 32(2): 240–256.

Hagmann, T., & Korf, B. (2012). Agamben in the Ogaden: Violence and sovereignty in the Ethiopian–Somali frontier. *Political Geography* 31(4): 205–214.

Hannah, M. (2006). Torture and the ticking bomb: The war on terrorism as a geographical imagination of power/knowledge. *Annals of the Association of American Geographers* 96(3): 622–640.

Hannah, M. (2009). Calculable territory and the West German census boycott movements of the 1980s. *Political Geography* 28(1): 66–75.

Hannah, M. (2010). Learning from the "War on Terror." *Political Geography* 29(2): 119–122.

Haraway, D. (1989). The biopolitics of postmodern bodies: Determinations of self in immune system discourse. *Differences: A Journal of Feminist Cultural Studies* 1(1): 3–43.

Hardt, M., & Negri, A. (eds) (2000). *Empire*. Cambridge, MA: Harvard University Press.

Hellberg, S. (2014). Water, life and politics: Exploring the contested case of eThekwini municipality through a governmentality lens. *Geoforum* 56: 226–236.

Hennessy, E. (2013). Producing "prehistoric" life: Conservation breeding and the remaking of wildlife genealogies. *Geoforum* 49: 71–80.

Holloway, L. (2007). Subjecting cows to robots: Farming technologies and the making of animal subjects. *Environment and Planning D: Society and Space* 25(6): 1041–1060.

Holloway, L., & Morris, C. (2012). Contesting genetic knowledge-practices in livestock breeding: Biopower, biosocial collectivities, and heterogeneous resistances. *Environment and Planning D: Society and Space* 30(1): 60–77.

Holloway, L., & Morris, C. (2014). Viewing animal bodies: Truths, practical aesthetics and ethical considerability in UK livestock breeding. *Social and Cultural Geography* 15(1): 1–22.

Hyndman, J. (2005). Migration wars: Refuge or refusal? *Geoforum* 36(1): 3–6.

Hynes, M. (2013). The ethico-aesthetics of life: Guattari and the problem of bioethics. *Environment and Planning A* 45(8): 1929–1943.

Ingram, A. (2008). Domopolitics and disease: HIV/AIDS, immigration, and asylum in the UK. *Environment and Planning D: Society and Space* 26(5): 875–894.

Ingram, A. (2010a). Biosecurity and the international response to HIV/AIDS: Governmentality, globalisation and security. *Area* 42(3): 293–301.

Ingram, A. (2010b). Governmentality and security in the US President's Emergency Plan for AIDS Relief (PEPFAR). *Geoforum* 41(4): 607–616.

Ingram, A. (2013). After the exception: HIV/AIDS beyond salvation and scarcity. *Antipode* 45(2): 436–454.

Jones, R. (2009a). Agents of exception: Border security and the marginalization of Muslims in India. *Environment and Planning D: Society and Space* 27(5): 879–897.

Jones, R. (2009b). Categories, borders and boundaries. *Progress in Human Geography* 33(2): 174–189.

Jones, R. (2009c). Geopolitical boundary narratives, the global War on Terror and border fencing in India. *Transactions of the Institute of British Geographers* 34(3): 290–304.

Jones, R. (2012a). *Border Walls: Security and the War on Terror in the United States, India, and Israel.* New York: Zed Books.

Jones, R. (2012b). Spaces of refusal: Rethinking sovereign power and resistance at the border. *Annals of the Association of American Geographers* 102(3): 685–699.

Kanngieser, A. (2013). Tracking and rracing: Geographies of logistical governance and labouring bodies. *Environment and Planning D: Society and Space* 31(4): 594–610.

Kear, M. (2013). Governing homo subprimicus: Beyond financial citizenship, exclusion, and rights. *Antipode* 45(4): 926–946.

Kearns, G. (2008). The geography of terror. *Political Geography* 27(3): 360–364.

Kearns, G., & Reid-Henry, S. (2009). Vital geographies: Life, luck, and the human condition. *Annals of the Association of American Geographers* 99(3): 554–574.

Keil, R., & Ali, H. (2007). Governing the sick city: Urban governance in the age of emerging infectious disease. *Antipode* 39(5): 846–873.

Kjellén, R. (1920). *Grundriß zu einem System der Politik.* Leipzig: S. Hirzel Verlag.

Korf, B. (2006). Who is the rogue? Discourse, power and spatial politics in post-war Sri Lanka. *Political Geography* 25(3): 279–297.

Kurtz, H., Trauger, A., & Passidomo, C. (2013). The contested terrain of biological citizenship in the seizure of raw milk in Athens, Georgia. *Geoforum* 48: 136–144.

Labban, M. (2014). Against shareholder value: Accumulation in the oil industry and the biopolitics of labour under finance. *Antipode* 46(2): 477–496.

Lee, S.-O., Jan, N., & Wainwright, J. (2014). Agamben, postcoloniality, and sovereignty in South Korea. *Antipode* 46(3): 650–668.

Legg, S. (2006). Governmentality, congestion and calculation in colonial Delhi. *Social and Cultural Geography* 7(5): 709–729.

Legg, S. (2008). Ambivalent improvements: Biography, biopolitics, and colonial Delhi. *Environment and Planning A* 40(1): 37–56.

Legg, S. (2009). Of scales, networks and assemblages: The League of Nations apparatus and the scalar sovereignty of the Government of India. *Transactions of the Institute of British Geographers* 34(2): 234–253.

Lemke, T. (2010). *Bio-politics: An Advanced Introduction.* New York: New York University Press.

Li, T.M. (2010). To make live or let die? Rural dispossession and the protection of surplus populations. *Antipode* 41: 66–93.

Lorimer, J., & Driessen, C. (2013). Bovine biopolitics and the promise of monsters in the rewilding of heck cattle. *Geoforum* 48: 249–259.

Lorimer, J., & Driessen, C. (2014). Wild experiments at the Oostvaardersplassen: Rethinking environmentalism in the Anthropocene. *Transactions of the Institute of British Geographers* 39(2): 169–181.

MacLeavy, J., & Peoples, C. (2010). War on Terror, work in progress: Security, surveillance and the configuration of the US workfare state. *GeoJournal* 75(4): 339–346.

Mansfield, B. (2012a). Environmental health as biosecurity: "Seafood choices," risk, and the pregnant woman as threshold. *Annals of the Association of American Geographers* 102(5): 969–976.

Mansfield, B. (2012b). Gendered biopolitics of public health: Regulation and discipline in seafood consumption advisories. *Environment and Planning D: Society and Space* 30(4): 588–602.

Martin, D. (2015). From spaces of exception to "campscapes": Palestinian refugee camps and informal settlements in Beirut. *Political Geography* 44: 9–18.

Martin, L. (2010). Bombs, bodies, and biopolitics: Securitizing the subject at the airport security checkpoint. *Social and Cultural Geography* 11(1): 17–34.

Martin, L. (2012). Governing through the family: Struggles over US noncitizen family detention policy. *Environment and Planning A* 44(4): 866–888.

Mbembe, A. (2003). Necropolitics. *Public Culture* 15: 11–40.

McGregor, J. (2012). Rethinking detention and deportability: Removal centres as spaces of religious revival. *Political Geography* 31(4): 236–246.

Meehan, K. (2013). Disciplining de facto development: Water theft and hydrosocial order in Tijuana. *Environment and Planning D: Society and Space* 31(2): 319–336.

Minca, C. (2005). The return of the camp. *Progress in Human Geography* 29(4), 405–412.

Minca, C. (2006). Giorgio Agamben and the new biopolitical nomos. *Geografiska Annaler Series B, Human Geography* 88: 387–403.

Minca, C. (2007). Agamben's geographies of modernity. *Political Geography* 26: 78–97.

Minca, C. (2011a). Carl Schmitt and the question of spatial ontology. In Legg, S. (eds), *Spatiality, Sovereignty and Carl Schmitt: Geographies of the Nomos*, London: Routledge, 163–181.

Minca, C. (2011b). No country for old men. In Minca, C., & Oakes, T. (eds), *Real Tourism*, London: Routledge, 12–37.

Minca, C., & Rory, R. (2015). *On Schmitt and Space.* London: Routledge.

Moran, D., Pallot, J., & Piacentini, L. (2013). Privacy in penal space: Women's imprisonment in Russia. *Geoforum* 47: 138–146.

Moran, D., Piacentini, L., & Pallot, J. (2012). Disciplined mobility and carceral geography: Prisoner transport in Russia. *Transactions of the Institute of British Geographers* 37(3): 446–460.

Moreno, C., & Curti, G. (2012). Recovery spaces and therapeutic jurisprudence: A case study of the family treatment drug courts. *Social and Cultural Geography* 13(2): 161–183.

Mori, M. (2008). Environmental pollution and bio-politics: The epistemological constitution in Japan's 1960s. *Geoforum* 39(3): 1466–1479.

Morin, K. (2013). "Security here is not safe": Violence, punishment, and space in the contemporary US penitentiary. *Environment and Planning D: Society and Space* 31(3): 381–399.

Mountz, A. (2011). The enforcement archipelago: Detention, haunting, and asylum on islands. *Political Geography* 30(3): 118–128.

Mountz, A., Coddington, K., Catania, T., & Loyd, J. (2013). Conceptualizing detention: Mobility, containment, bordering, and exclusion. *Progress in Human Geography* 37(4): 522–541.

Murakami Wood, D. (2013). What is global surveillance? Towards a relational political economy of the global surveillant assemblage. *Geoforum* 49: 317–326.

Nally, D. (2008). "That coming storm": The Irish Poor Law, colonial biopolitics, and the Great Famine. *Annals of the Association of American Geographers* 98(3): 714–741.

Nally, D. (2011). The biopolitics of food provisioning. *Transactions of the Institute of British Geographers* 36(1): 37–53.

Neumann, B. (2002). The National Socialist politics of life. *New German Critique* 85: 107–130.

Parr, H. (1997). Mental health, public space, and the city: Questions of individual and collective access. *Environment and Planning D: Society and Space* 15(4): 435–454.

Pero, R., & Smith, H. (2014). In the "service" of migrants: The temporary resident biometrics project and the economization of migrant labor in Canada. *Annals of the Association of American Geographers* 104(2): 401–411.

Philo, C. (1992). Foucault's geography. *Environment and Planning D: Society and Space* 10(2): 137–161.

Philo, C. (2005). The geographies that wound. *Population, Space and Place* 11(6): 441–454.

Philo, C. (2006). Madness, memory, time, and space: The eminent psychological physician and the unnamed artist-patient. *Environment and Planning D: Society and Space* 24: 891–917.

Philo, C. (2012). A "new Foucault" with lively implications – or "the crawfish advances sideways." *Transactions of the Institute of British Geographers* 37(4): 496–514.

Pratt, G. (2005). Abandoned women and spaces of the exception. *Antipode* 37(5): 1052–1078.

Ramadan, A. (2009). Destroying Nahr el-Bared: Sovereignty and urbicide in the space of exception. *Political Geography* 28(3): 153–163.

Ramadan, A. (2013a). Spatialising the refugee camp. *Transactions of the Institute of British Geographers* 38(1): 65–77.

Ramadan, A. (2013b). From Tahrir to the world: The camp as a political public space. *European Urban and Regional Studies* 20(1): 145–149.

Rasmussen, C. (2013). The beast and the sovereign, biopolitics and Derrida's menagerie. *Environment and Planning D: Society and Space* 31(6): 1125–1133.

Reid-Henry, S. (2007). Exceptional sovereignty? Guantánamo Bay and the re-colonial present. *Antipode* 39(4): 627–648.

Reid-Henry, S. (2013). Humanitarianism as liberal diagnostic: Humanitarian reason and the political rationalities of the liberal will-to-care. *Transactions of the Institute of British Geographers* 39(3): 418–431.

Rose, G. (2009). Who cares for which dead and how? British newspaper reporting of the bombings in London, July 2005. *Geoforum* 40(1): 46–54.

Rose, M. (2014). Negative governance: Vulnerability, biopolitics and the origins of government. *Transactions of the Institute of British Geographers* 39(2): 209–223.

Rose, N. (2006). *The Politics of Life Itself: Biomedicine, Power, and Subjectivity in the Twenty-First Century*. Princeton, NJ: Princeton University Press.

Rose-Redwood, R.S. (2006). Governmentality, geography, and the geo-coded world. *Progress in Human Geography* 30(4): 469–486.

Schinkel, W., & van den Berg, M. (2011). City of exception: The Dutch revanchist city and the urban homo sacer. *Antipode* 43(5): 1911–1938.

Schlosser, K. (2008). Bio-political geographies. *Geography Compass* 2(5): 1621–1634.

Shaw, I.G.R., & Akhter, M. (2012). The unbearable humanness of drone warfare in FATA, Pakistan. *Antipode* 44(4): 1490–1509.

Singh, N. (2013). The affective labor of growing forests and the becoming of environmental subjects: Rethinking environmentality in Odisha, India. *Geoforum* 47: 189–198.

Smith, S. (2012). Intimate geopolitics: Religion, marriage, and reproductive bodies in Leh, Ladakh. *Annals of the Association of American Geographers* 102(6): 1511–1528.

Sothern, M., & Dickinson, J. (2011). Repaying the gift of life: Self-help, organ transfer and the debt of care. *Social and Cultural Geography* 12(8): 889–903.

Sparke, M. (2006). A neoliberal nexus: Economy, security and the biopolitics of citizenship on the border. *Political Geography* 25(2): 151–180.

Sparke, M. (2013). From global dispossession to local repossession. In Johnson, N., Schein, R., & Winders, J. (eds), *The New Companion to Cultural Geography*, Oxford and New York: Wiley-Blackwell, 387–408.

Springer, S. (2012). Neoliberalising violence: Of the exceptional and the exemplary in coalescing moments. *Area* 44(2): 136–143.

Tyner, J. (2012). State sovereignty, bioethics, and political geographies: The practice of medicine under the Khmer Rouge. *Environment and Planning D: Society and Space* 30(5): 842–860.

Tyner, J. (2013). Population geography I: Surplus populations. *Progress in Human Geography* 37(5): 701–711.

Van Houtum, H. (2010). Human blacklisting: The global apartheid of the EU's external border regime. *Environment and Planning D: Society and Space* 28(6): 957–976.

Vaughan-Williams, N. (2010). The UK border security continuum: Virtual biopolitics and the simulation of the sovereign ban. *Environment and Planning D: Society and Space* 28(6): 1071–1083.

Wakefield, S., & Braun, B. (2014). Governing the resilient city. *Environment and Planning D: Society and Space* 32(1): 4–11.

Wainwright, J., & Mann, G. (2013). Climate Leviathan. *Antipode* 45(1): 1–22.

Warren, A. (2013). (Re)locating the border: Pre-entry tuberculosis (TB) screening of migrants to the UK. *Geoforum* 48: 156–164.

Werner, A. (ed.) (2003). *Landscaping the Human Garden*. Stanford, CA: Stanford University Press.

Yeh, E.T. (2012). Transnational environmentalism and entanglements of sovereignty: The tiger campaign across the Himalayas. *Political Geography* 31(7): 408–418.

Young, C., & Light, D. (2013). Corpses, dead body politics and agency in human geography: Following the corpse of Dr Petru Groza. *Transactions of the Institute of British Geographers* 38(1): 135–148.

Theorizing Political Geography

Chapter 15

Spatial Analysis

Andrew M. Linke[1] and John O'Loughlin[2]

[1]*University of Utah, Salt Lake City, Utah, USA*
[2]*University of Colorado-Boulder, Boulder, Colorado, USA*

In contemporary spatial and quantitative study of political geographical phenomena, two fundamental and perennial concepts have maintained their significance. Theoretical and empirical understanding of context (places) and non-stationarity (spatial autocorrelation) continue to bedevil spatial analysis. Fundamental questions that relate to issues in defining and measuring "neighborhood," the social setting in which political interactions occur, remain only partly answered despite further work on these topics since the earlier version of this chapter appeared a decade ago. Relatedly, key uncertainties around the handling of spatial autocorrelation in geographical data are as important now as in early work within the revival of political geography.

Three developments in spatial analysis over the last decade have dramatically changed its nature and research outlook. Vast improvements in data availability, data analysis, and new methodological tools have made the technologies and techniques more accessible to a wider audience in geography and related social sciences. After reviewing the two primary tenets of spatial analysis and outlining the three recent changes that we observe, in this chapter we integrate these discussions into a stylized empirical example of understanding beliefs about violence in the North Caucasus region of Russia, an area that has seen continuous but localized violence since the collapse of the Soviet Union in the early 1990s. Our geographical information systems (GIS) analysis connects foundational concepts of political geography with novel analytical tools using both individual- and aggregate-level data.

Timeless and constant fundamental concepts

Context and place

Political geographers, despite their diversity of methodologies and topical foci, share the belief that the social settings of events influence individual- and group-level behaviors and attitudes. Political behaviors such as voting decisions, conflict about territorial control, political

The Wiley Blackwell Companion to Political Geography, First Edition.
Edited by John Agnew, Virginie Mamadouh, Anna J. Secor, and Joanne Sharp.
© 2015 John Wiley & Sons Ltd. Published 2017 by John Wiley & Sons Ltd.

boundary delineation and demarcation, and public goods provision and allocation are a function of a constellation of influences that mix aggregate and individual factors at scales ranging from the locality to the national and international. Unlike other social sciences where individuals tend to be "atomized" and interactions between them and their social settings unrecognized and unexamined, political geographers have highlighted the contextual effects of politics for over a century, since Vidal de la Blache (1903) first documented and described the complex "milieux" (geographical settings) of France.

In this chapter, we illustrate possible extensions for classical approaches to identifying and measuring the effects of "place" in quantitative political geography, which we understand to be the "settings and scenes of everyday life" (Agnew 1987: 5). A "structurationist" approach (the two-way interaction between individuals and their *milieux* that together lead to social outcomes) was the root of much foundational research pointing toward the importance of social context in human geography (e.g., Pred 1983). Johnston and Pattie (1992), for example, showed that a "spatial polarization" of voting trends (the link between class and party support) had developed in the British electorate and that it pointed toward the importance of social context in shaping individual choices. Like many other studies of elections, Pattie and Johnston (2000) provided evidence that "people who talk together vote together," and therefore that personal interactions in social settings condition political behaviors (in Germany, see also Flint, 1998). Relatedly, in political science, Braybeck and Huckfeldt (2002) explicitly identified neighborhood contexts and measured the diffusion of information about candidates and issues in dispersed and condensed social networks that influence voting behavior. In a field experiment, Enos (2014) shows that increased interethnic group contact in a baseline homogenous US context can result in exclusionary social attitudes.

While place-focused quantitative analysis has concentrated on elections or civic beliefs, it can of course be considered for other political phenomena such as violent conflict. As Agnew (1987: 60) asks:

> Why is it, for example, that political violence characterizes the political histories of some places, but not others? Often this may have been the product of place-specific repression, or the absence of other alternatives such as electoral politics.

For most political geographers, understandings of place and context have moved away from what O'Loughlin (2000) decried as a geometric or "Cartesian" view; this move now considers relational understandings of the settings in which people live their lives (Castree 2004). Relative interactive measures of places, locales, and regions can be defined by links that are related to social identity, political economy, or migration. In this sense, place or context may be defined by group membership or network ties *in addition to* physical location (Massey 1994). While relative and absolute concepts are often presented as mutually exclusive, and we acknowledge that our example in this chapter incorporates distance-based metrics, we believe that the concepts and their measures can simultaneously be accommodated in quantitative spatial analysis.

Although few other social scientists study similar topics, the focus of spatial analytical work in political geography is usually the elaboration of the nuanced effects of social forces expressed in localities at various spatial scales. This theme formed the centerpiece of the spatial analysis chapter in the previous edition of this book. The usual political science study tends to be deductively developed from a formal model (often a rational choice model) and oriented toward testing a single observable implication of a theory. Unfortunately, many non-geographers still rely on methods choices that "control away" the influence of contexts on human behavior, or, in Gould's (1970) phrase, they are "throwing out the baby with the bathwater." For these researchers, a geographical analysis is a means to an end, and not the end itself. There is still a tendency in political science research to narrowly view space and place through proximity and/or contiguity lenses, essentially as geometric measures. Although there have been improvements in spatial analysis outside of geography (we highlight several examples in what follows), this has not changed dramatically from the previous version of this *Companion* chapter and the critique a decade before by the prominent political scientist Gary King (King 1996, in response to Agnew 1996). Of course, political geographers are not atheoretical, but, differing from political scientists, they approach hypothesis-testing with more emphasis on probing the empirics of data before predictive modeling. Disciplinary boundaries still sometimes represent silos of distinct conceptual frameworks for research.

The contextual theme is also commonly employed in other geographically sensitive research fields. In studies of health outcomes, educational achievement, crime rates, and social mobility, it is common practice to assess directly the link between contextual qualities (such as neighborhood levels of poverty, racial or ethnic composition, political representation) and individual characteristics in nested data structures. For example, students learn within classrooms that are not identical because teachers are different, classrooms are in schools that may be different in terms of facilities and resources, and those schools are located in school districts that also vary dramatically due to tax base variations and other factors. A student's achievement partly depends on his/her home location, because observable traits like taxes earned from property values and dedicated to education vary widely in American metropolitan areas. Defining the boundary for capturing such contextual effects can influence results vary greatly, but these varying spatial scales have qualities that can be measured and included in multilevel analysis.

Following standard econometric practices, most scholars in fact accept that geographical differences can be important and, consequently, they use statistical methods that incorporate a "fixed effects" approach. In this method, binary dummy variables are defined for each spatial unit, such as voting districts in a city, and thus they can be numerous. If the researcher is probing whether voters' incomes influence electoral participation, s/he would ideally have an accurate estimate for the effect of poverty after ensuring that other personal and contextual characteristics are not also influencing participation. The accessibility of polling stations on a constituency level could influence participation. If there are no data available for the number of polling station locations, comparing individuals statistically in one constituency only to individuals in that same constituency in a fixed effects model eliminates the possibility that differences between constituencies are also responsible for the changes in participation rates. The key problem with the dummy variable approach is that while the model controls for baseline differences between spatial units, the results of the analysis cannot tell us *how* the differences between units of observations matter.

Regardless of the chosen statistical model, the delineation of the areal dimensions of the spatial units matters greatly for the conclusions of spatial analysis. The Modifiable Areal Unit Problem (MAUP) is well known but rarely examined methodically. While Gehlke and Biehl outlined MAUP in 1934, it is most commonly associated with Openshaw (1983). The key MAUP issue is that the size and direction of a statistical association found at one scale of analysis (e.g., census tract) may not hold at others (e.g., county; Openshaw & Taylor 1979). For example, if we correlate socio-economic class and voting for the Republican party, the coefficient varies across hierarchical scales as a result of the number of data points and the geographical configuration of the districts, and therefore we cannot be sure which coefficient is correct (Openshaw 1996). MAUP dilemmas are different than, but related to, the more conceptual dilemma of the "ecological fallacy." Researchers risk an ecological fallacy when they assume that aggregate-level data represent individual-level processes or phenomena (Robinson 1950; Selvin 1958). In essence, an ecological fallacy is MAUP at the finest resolution (individual level), and confounds explanations and outcomes across aggregate and individual scales.

As accurate measurement of context is central to spatial analysis in political geography and differing neighborhood bounds can influence results, more flexibility for researchers in defining the relevant neighborhood is warranted. Many empirically derived metrics for selecting neighborhood dimensions exist (Root, Meyer, & Emch 2009; Spielman, Yoo, & Linkletter 2013; Spielman & Logan 2013), but fine-resolution data are required for these analytical tools to be effective. It is impossible to know whether a larger unit of analysis is more appropriate than smaller units if one cannot examine and compare results at different levels. For analysts who view the world through the lens of spatial scales, decisions about the proper aggregations of information thus become very important (Nelson 2001; Maclaurin, Leyk, & Hunter 2015). While it is possible to aggregate up from fine resolutions (location-based data) to coarse (large areal-unit data), disaggregating from coarse to fine resolutions is impossible (or at least very difficult, requiring calculations of uncertainty). If one wants to know how best to understand political violence in Afghanistan, knowing only the provincial-level violence rates limits our ability effectively to measure the local ebb and flow of conflict as it occurs relative to the border with Pakistan. We would also have no understanding of how violence is geographically distributed within ethno-regional enclaves at localized scales. With location-level data, however, it is possible to aggregate points (and all of the attributes of individual violent events) to a zone defined by any absolute spatial reference (e.g., border) or relative spatial dimension (e.g., ethnic community dominance in a region; O'Loughlin et al. 2010).

Nonstationary data and spatial autocorrelation

In addition to considerations of context (*milieux*), another fundamental and timeless quality of spatial analysis in political geography is its non-stationarity. Data are non-stationary when the relationships between variables vary across locations within the dataset and are not consistent in different regions. As Tobler (1970) famously wrote, "everything is related to everything else, but near things are more related than distant things." Spatial dependencies (a term that reflects this distance effect) among units of analysis are important observable artifacts of largely unobserved social processes. Spatial analysts are interested in how, where, and why data are geographically related. If human behavioral patterns cluster in space (and they almost always do), geographers strive to understand how and why, rather than simply controlling for the trend as a nuisance that sullies an otherwise non-spatial model.

Questions about spatially non-stationary data are answered inductively. The calculation of Local Indicators of Spatial Association (LISAs) has been the workhorse of spatial analysts for two decades and identifies where a locally specific measure of autocorrelation may not match a global or overall trend (Anselin 1995). LISA statistics are based on a straightforward comparison between observed and simulated (expected uniform or random) spatial distributions. The Getis–Ord Gi* ("hot spot") statistic is one example of a LISA indicator (Ord & Getis 1995). For a study of insurgency in Afghanistan between 2004 and 2009, using daily military reports of "significant activity" (SIGACTS), O'Loughlin and colleagues (2010), for small 25 km^2 gridcells, were able to identify changes in the locations and sizes of clusters of events initiated by both insurgent and coalition forces. As one would expect, some general overlap between the clusters can be observed and they also display differing locational trends. This is an example of using LISA-type statistics inductively, but such procedures can also be employed to identify spatial trends that distort the results of non-spatial econometric analysis. Using a local clustering statistic for the residuals of a regression (difference between observed and predicted values), for example, can reveal clusters of unexplained variance in a dataset, suggesting that assumptions about the independence of observations (explained later in this chapter) are violated.

Point pattern process identification, related to LISA indicators because it compares an observed spatial distribution to an expected (simulated) distribution, can be used to uncover local nuances and trends in political data (O'Loughlin 2002). Recent advances in this style of analysis are extremely powerful, including the routines implemented in SaTScan, software initially designed to identify space–time clusters for epidemiological and other spatially referenced health data (Kulldorff et al. 2005; Kulldorff and Information Management Services 2009). The benefit of using SaTScan is that the temporal and spatial dimensions of the window that defines clustering are allowed to vary and the output is – at least conceptually – a cylinder with vertical and horizontal dimensions that can be mapped by location coordinates. For the study of violence in Afghanistan, this method delivers even more detail about the spatio-temporal dimensions of conflict than just a LISA indicator (O'Loughlin et al. 2010: 490).

With geographically dependent data, "knowing one value on the surface provides the observer with a better than random chance of predicting nearby values" (Gould 1970: 444), thus violating the assumption that observation units are independently and identically distributed (IID). IID assumption violations can give researchers faulty assurances that the results of their quantitative analyses are valid, because standard errors for testing the significance of a coefficient point estimate are artificially small (Anselin 1988). After McCarty's (1954) analysis of geographical patterns of the vote for Wisconsin's right-wing senator Joseph McCarthy – one of the first studies explicitly promoting a spatial approach to statistical analysis – a steady stream of quantitative political geography methods textbooks and articles emerged that provide technical overviews of the confounding influences of spatial dependencies and the many possible solutions (Cliff & Ord 1973; Anselin 1988, 2002).

One of the most straightforward treatments of spatial dependency is introducing an autoregressive (also called "spatial lag") term into a regression model. The classical estimator in regression analysis can be represented in simplistic form as $Y_i = \beta X_i + \varepsilon_i$, in which, for observation i, Y represents the outcome variable, X the main independent variable, and ε the unobserved variation (error) in the relationship. The value of β quantifies the association between X and Y. More advanced models would include temporal dimensions and control variables that might also contribute to the outcome. A spatial autoregressive model is represented by $Y_i = \rho W_i Y + \beta X_i + \varepsilon_i$, with spatial weights matrix, W, used to define the presence of Y within a neighborhood surrounding observation i. W may be defined by proximity (distance) or contiguity (shared border). The estimate of ρ in the model can be interpreted in substantial terms. Interpreting the autoregressive term of the equation reveals a high level of predictability within a model (e.g., O'Loughlin et al. 2012).

In contrast to lag models that accept the role of geographical affinities in the data and try to use them to understand human behavior, the "spatial error" solution to the problems posed by spatial dependency is an adjustment of the standard error of each coefficient estimate of the independent variables (β in the equation above), so that their influence on the outcome of interest is not overstated. In effect, a spatial error approach is similar to common adjustments to the error structure of a regression model (e.g., clustered errors at some spatial scale), but is specifically based on geographical distances or contiguities, defining the connections among the units (as is the case for W in the earlier equation). Because the spatial lag model introduces a new term into the model, consideration of the spatial contiguity effects has been explicitly incorporated into regression models of electoral choices for the past 35 years (O'Loughlin 1981; O'Loughlin, Flint, & Anselin 1994).

Trends and changes in spatial analysis over the last decade

Data availability

While the foundational concepts remain perennially important, there have been major changes in the landscape of spatial analysis over the last decade. Primary among the changes helping to spur more spatial analysis is a dramatic increase in the availability of diverse types of data that are readily available for political geographical research. Following closely on the publication of the earlier version of this chapter – on the cusp of important changes in data and analysis for human geography – Johnston et al. (2004: 367) wrote:

> there are many hypotheses regarding neighborhood effects in the geographical and related litera-
> tures, but their successful testing has been hampered by the absence of data. In particular, analysts
> have lacked data on both individuals and their neighborhood milieux, which allow the interactions
> of different types of people in different types of local context to be explored.

In their article, Johnston et al. (2004) combined survey and electoral outcome data in "bespoke neighborhoods" in a manner that facilitated the creative identification of multiple scales at which social forces operate, thus effectively avoiding criticisms related to MAUP.

Quantitative political geography research had classically relied mostly on administrative unit polygon data, usually in electoral studies. However, quantitative analysis of political geographical subjects is increasingly going beyond only areal unit data to include a variety of formats and types (e.g., surveys in Linke 2013; Secor & O'Loughlin 2005). Examples of the various formats include network (Radil & Flint 2013), census small area (Verpoorten 2012), point pattern location (Linke, Witmer, & O'Loughlin 2012), satellite remote sensing (Henderson, Storeygard, & Weil 2012), roads (Zhukov 2012), land use change detection (Witmer & O'Loughlin 2009), and even mobile phone service data (Pierskalla & Hollenbach 2013). Some research uses non-uniform polygons at a global scale to capture spatial demo-graphic qualities of territories such as ethnic groupings (e.g., Wucherpfennig et al. 2011). Individual research teams most often collect these data, but they are increasingly available from the United Nations, international non-governmental organizations, and even national governmental agencies in the developing world.

The push toward sharing data among academics has also picked up pace alongside the increasingly mandated expectation of journals that data for quantitative analysis be made public for replication by other researchers in the field (King 1995). The Dataverse Network Project at Harvard University (http://www.thedata.org), for example, is "free and open to all researchers worldwide to share, cite, reuse and archive research data." However, concerns regarding the confidential nature of individual data that are anchored to a specific location are growing at a time when social media, governmental, and private information collection is on the rise (VanWey et al. 2005).

A major "neogeography" shift in the availability of geospatial data has also seen volunteered private (e.g., individual location) data used for a number of academic and non-academic applications, in addition to the non-expert use of GIS to further the aims of a given community. The original rise of such data uses accompanied a revolution in the accessibility of web-based mapping applications, such as Google Maps' API and even the many extensions of the basic Google Maps (and Earth) interface(s). Where non-traditional cartographic tools exist (i.e., those that do not rely on proprietary software like ESRI's ArcMap) and are accompanied by spatial data, the potential for "citizen science" emerges. Activist or otherwise community-based activities developed on these platforms can have important real-world consequences. One example of such a technical neogeography within the realm of political spatial analysis would be the role of social media and cartographic mapping within the Syrian conflict. According to the *Washington Post*, for instance, activists and those sympathetic to the anti-Assad forces used Google extension Map Maker to rename the streets and locations in key Syrian cities to reflect opposition movement historical sympathies (Lynch 2012). As a kind of hybrid academic–popular data collection and analysis effort, Voix des Kivus (directed by Peter van der Windt and Macarten Humphries at Columbia University) was a project that introduced truly innovative methods of data collection in the Democratic Republic of Congo (DRC). It distributed mobile phones to communities across the war-torn eastern regions of DRC, with the goal of capturing daily reports of violent incidents that occur below the radar of major private media, governmental, or non-governmental reporting. Voix des Kivus was based on a top-down organizational structure, but the principal operating premise was the acceptance that everyday citizens play profoundly important roles in gathering the geographical data that researchers in academic fields regularly use.

Data use

A second recent trend in spatial analysis relates to the compilation and integration of multiple data formats. In traditional electoral geography, for example, analysis was often conducted with data available at a single scale (vote totals and predictors based on census data). In spatial analysis limited to a single scale and format of data, important "place influences" or contextual-level effects may not be captured effectively by distance and contiguity scores. To advance the measurement of possible local-level influences on outcomes of interest in political geography, hybrid units of analysis and mixed data structures are now more commonly used.

An analysis of conflict in Iraq can serve as an example of diverse data merging into a single GIS platform, allowing for the discovery of social relationships that are hidden when analysis is bound by a single source, analytical unit, or data format (Linke et al. 2012). Four different data formats and dimensions were merged: socio-economic status of districts (survey data aggregated into vector polygons), ethnic group distributions (scanned paper map converted to vector polygons), violent event location data (latitude and longitude point coordinates), and satellite night-time lights to measure urbanization (raster image). Using a Granger causal effects estimator, a tit-for-tat dynamics of insurgent–regime forces reciprocity emerged where the actions of one side of the combat strongly predicted a timely and local response from the other. These associations varied across different spatio-temporal thresholds and across ethno-sectarian, income, and population density regions. A similar compilation of survey, violent events, and population data into a common subnational unit of analysis for a conflict diffusion study of 16 countries in sub-Saharan Africa is found in Linke, Schutte, and Buhaug (2015).

This kind of data integration on a specific spatial scale with complementary spatially sensitive methods is developing rapidly. PRIO-GRID (Tollefson, Strand, & Buhaug 2012) is one example of a major effort to merge multiple freely available data sources for governmental, population, socio-economic status, terrain, and ecological-climatological data into a ½ degree gridcell unified structure for all world regions. While the basic dataset has a static spatial resolution (thus not automatically resolving MAUP issues), it represents a substantial advance in how conflict inquiry in political science is carried out, with a move away from the "territorial trap" of the nation-state in favor of geographically disaggregated research.

Analytical tools

As well as the important changes in data availability and data use, recent years have also seen a dramatic rise in the number of statistical and graphical tools available for spatial analysts. Free platforms for statistical analysis supporting hundreds of procedures for spatial data management, mapping, and regression can now help researchers in innumerable ways after data are correctly formatted and organized (Bivand, Pebesma, & Gomez-Rubio 2008; Griffith & Paelinck 2011; Plant 2012; Brunsdon & Comber 2015). Many free stand-alone software platforms exist for this kind of spatial analysis, including GeoDa (Anselin, Syabri, & Kho 2006) for Exploratory Spatial Data Analysis (ESDA) or LISA calculations (see also SatScan, noted earlier), as well as regression. Dedicated spatial analysis packages exist in the software platform *R* for managing classes of spatial data (**sp**), reading and writing shapefiles (**maptools**), creating and using spatial weights matrices (**spdep**), geostatistics and anisotropy (**gstat**), survey sampling points based on population distributions (**spsurvey**), and point-pattern processes (**spatial, spatstat, splancs,** and **spatialkernel**), multilevel modeling (**nlme4**), geographically weighted regression (**spgwr**), and advanced mapping and visualization (**ggplot2**). Researchers with coding experience can assemble a GIS dataset, execute their preferred statistical estimation, and display graphical results in a single program. This is a change from the past when software was mainly proprietary, and often GIS/mapping and statistical analysis were completed separately. While the learning curve for using command line programming interfaces in *R* can be steep, the payoffs are substantial in the long term.

Spatial analysis of violence and public opinion in the North Caucasus of Russia

In this section we illustrate some of the principles and methods of spatial analysis in political geography research. We use survey data for 2,000 individuals from the North Caucasus of Russia were collected in December 2005. The survey was part of a comparative project with Bosnia-Herzegovina and was designed as a study of postconflict attitudes toward group reconciliation and prospects for peaceful relations. The North Caucasus conflicts between 1994 and 1996, and restarting in 1999, were marked in the later years by guerrilla warfare, terrorist attacks, reprisals by the Russian forces, and a diffusion outward from their original core in Chechnya. Over time, the militancy took on a more Islamist character as its leaders declared a "shariat" (state run by Islamic law) for the Muslim republics across the region. By 2005, the conflict had waned significantly, but violent events still occurred on a daily basis in Chechnya and neighboring republics. Our research question concerns the possible effects of the violence in the immediate area on survey respondent attitudes.

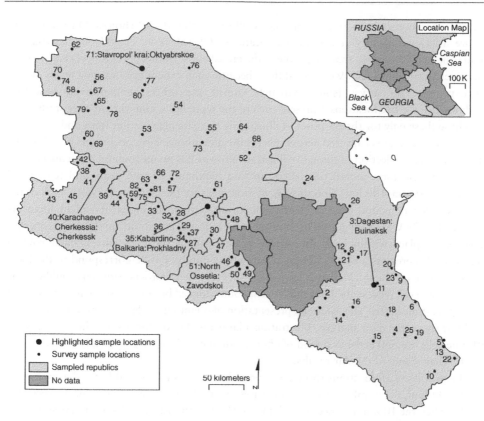

Figure 15.1 The 82 primary sampling locations for the 2005 survey in five republics of the North Caucasus. To ease comparison, the sample sites identified in the figure are also highlighted on the graphic in Figure 15.2. This map also appears in Linke and O'Loughlin (2015), an article that uses the same data for a more comprehensive spatial analysis of multiple survey questions.

The survey was designed to be geographically stratified and was conducted in 82 sampling points within the republics of North Ossetia, Dagestan, Karachevo-Cherkassia, Kabardino-Balkaria, and the territory of Stavropol' (Figure 15.1). Chechnya and Ingushetia were excluded from the study because high levels of violence at the time of the survey made including them impossible. We joined the location of survey respondents with conflict data for the two years before the December 2005 survey. These violent events were coded from newspapers and the data included the exact location, day of event, perpetrators and targets, and estimates of casualties. Coders reviewed thousands of Lexis-Nexis stories to identify events with enough information to ensure reliability in geographical precision. The violent event file was precisely georeferenced by UTM Zone X and Y coordinates after the violent event locations were identified, allowing for precise distance calculations in a projected coordinate system. The spatial non-stationarity of the violent events data has been illustrated by kernel density surfaces and conditional probabilities of reciprocal violence in O'Loughlin and Witmer (2012).

Aggregating the data formats

With the goal of measuring multiple settings or neighborhoods of violence, we allow the spatial boundaries around survey sampling points to vary in making violent event aggregations. Using a range of dimensions around a location allows us to address the uncertainty of the relevant context for a respondent and to illustrate the MAUP choices. On the map in Figure 15.1, we identify five sampling locations (Oktyabrskoe, Prokhladny, Zavodskoi, Buinaksk, and Cherkessk) that we profile in the tabled event counts of neighborhood violence measures (see Figure 15.2). We selected these because each experienced varying levels of violence prior to the survey (some extremely violent and others comparatively peaceful). Additionally, the five locations represent each of the republics within the broader study area. We chose these only for the purposes of illustrating the data aggregation steps and the variation of violence that is captured across space–time dimensions. For conclusions based on the later models, the selection of these five illustrative towns has no meaningful consequence.

We use eight three-month time slices up to a maximum of two years and ten spatial thresholds (10 km–100 km). Specifically, we first measure the distance between each of the 82 survey locations and every violent event location (total locations of violent incidents number 1,367). We use the *pointDistance* function in *R* and count the number of events that have taken place within each time–space buffer around those locations. Each row in Figure 15.2 represents a survey sample point, whose numbers 1–82 correspond to the locations on the map. The thicker black horizontal lines represent borders between republics. The shade of each cell represents the conflict event count (logged because of some high values) of each space–time buffer. This heatmap presentation of quantitative data can be very helpful in exploratory spatial data analysis. Illustrating a large number of dimensions (here 82 space × 80 time cells = 6,560 cells) as a kind of choropleth matrix is more graphically arresting and helpful than standard tabular results.

The profiles of all locations show the expected trends. As the spatial buffer becomes very large (100 km), the number of events increases. Many locations have no violence recorded nearby when the 10 km distance is used, but at 100 km almost every location has some violence. Secondarily, as the temporal range used to define our pairing of violence data with locations expands from 3 months to 24 months, a clear trend toward higher conflict event counts is also visible.

Five illustrative locations demonstrate different overall trends in violence. Zavodskoi near Ingushetia (number 51) is quite violent, with conflict taking place across nearly all temporal and spatial dimensions (and with very high rates at coarse boundary definitions). In dramatic contrast, Oktyabroskoe (76) in northern Stavropol' territory is far from the core of violence in Chechnya and, thus, relatively peaceful. Cherkessk (40) represents a medium level of violence with no conflict found at very fine temporal and spatial resolutions, but accumulating as distances increase. From this graphical display, we make the straightforward conclusion that the "violence neighborhood" surrounding locations varies dramatically between localities and according to threshold delimitation. The more important consideration relates to our earlier discussion of MAUP and it remains unclear what is the correct range for delimiting the violence context.

Varying intercepts in statistical models

The multilevel modeling approach is one that allows the evaluation of individual and contextual effects to be measured and compared, since it is now almost axiomatic in political geography that "context matters" in influencing individual choices and behaviors. Multilevel modeling (MLM) has not achieved prominence in political geography compared to other social sciences, especially epidemiology, criminology, and other public policy research. By allowing the basic relationship between the outcome of interest and a key predictor to vary

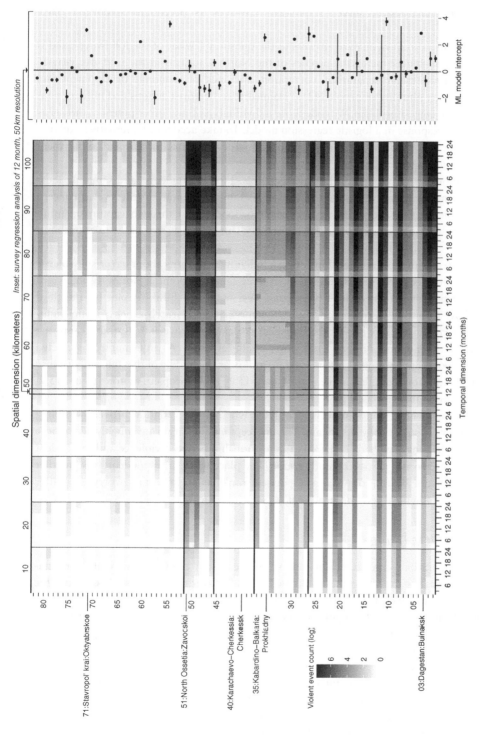

Figure 15.2 The number of violent events aggregated within 80 spatio-temporal dimensions (3 months breaks by 10 km boundaries) for each of 82 survey sample locations in the North Caucasus. Survey locations (vertical axis) corresponds to the map locations in Figure 15.1. Thicker black horizontal lines represent borders of the republics. For legibility, not all labels are shown. The smaller graphical figure (right) is an illustration of the multilevel/hierarchical regression model intercept for a single space–time dimension (here, 12 months and 50 km). Similar variation exists for the quantitative modeling of survey responses in each column representing a space–time dimension. This figure also appears in Linke and O'Loughlin (2015), an article that uses the same data for a more comprehensive spatial analysis of multiple survey questions.

across settings/contexts – which are known to be qualitatively different in unobserved, unmeasurable, and unknown ways – we allow for the influence of contextual factors to be incorporated into the analysis of survey respondent attitudes as potentially important place influences.

For our illustration, we analyze answers to the question: "What of the following listed is the most serious danger facing the peoples of the North Caucasus in the next five years?" More than crime, ethnic separatism, corruption, and lack of economic development and unemployment, a plurality of people (42%) claimed that "terrorist actions and military conflicts" were the biggest problem. We code those individuals as 1 and other potential threats as 0 for a binary response in a logistic regression model. To take account of alternative explanations for the fear of terrorism, we control for individual-level factors (income and employment status, education, age, and gender).

The relationship we model can be represented as, , for respondents and sampling locations. The term represents random intercepts across sampling points j. is the level of concern for new conflict, is the effect of the violence neighborhood (), and through capture the influences of variables through individual-level controls. Unexplained error in the relationship is captured in.

To indicate whether the relationship between individual concerns about terrorism/violence and level of violence in the locality changes by survey sampling point, a plot of the intercepts of the regression model for each of the second-level units is helpful. (A plot for the third republic level would be warranted in a more comprehensive research article.) In the right-hand side of Figure 15.2, we present the statistical association between the variable capturing rates of local violence and a respondent's worry about terrorism. As an example, we selected a spatio-temporal dimension (12 months and 50 km) near the middle of the time and distance ranges. Each location identified in the map (Figure 15.1) and cloropleth matrix (Figure 15.2, left) is included in this graphical plot.

A key component of our analysis is identifying whether the variation between sampling point intercepts is statistically significant. The intercept values vary from roughly –2 to more than +3.5, indicating a wide divergence in the relationship between violence and respondent worries about terrorism across the study region. The place where a respondent lives, in other words, has a major bearing on what a respondent ranks as an important worry, even after controlling for individual-level influences such as age or education.

Illustrating MLM results

Careful consideration of geographical context and scale can change substantial interpretations of statistical analysis. For the results in Table 15.1, we use the **lme4** package and the generalized linear model (for binary outcomes) function *glmer*. Using a logistical regression functional form, a statistically significant influence of contextual violence on worries about terrorism emerges (columns b–d). For every additional violent event that took place within 100 km of the survey location in the preceding year, the odds of respondent worry about terrorism increase marginally. Since the number of events within the space–time buffer ranges from 2–1,319, 0–454, and 0–135 for 100 km, 50 km, and 20 km ranges, respectively, this is a noteworthy effect even though the coefficient estimate is small. Drawing conclusions from the basic non-hierarchical model, one would conclude that a context of violence and insecurity influences worries about terrorism, and that this relationship is robust across definitions of spatial neighborhood. However, out multilevel model results show that the basic relationship

Table 15.1 The influence of violence taking place in various space–time scales on survey respondent concerns about terrorism and war in their area. Statistical estimator is a random intercept multilevel binary outcome logistical regression model, with intercepts varying by the 82 survey locations. Bold table values are highlighted as the main estimates of interest (effect of violence neighborhood on attitudes). Est, StdEr, and OR columns in the table are coefficient estimates, standard errors, and odds ratios, respectively.

	a) Controls only			b) 20km			c) 50km			d) 100km			e) RE 20km			f) RE 50km			g) RE 100km		
	Est	StdEr	OR	Est	StdEr	OR	Est	StdEr	OR	Est	StdEr	OR	Est	StdEr	OR	Est	StdEr	OR	Est	StdEr	OR
(Intercept)	−.29	.10*	0.748	−.37	.10*	0.688	−.38	.10*	0.685	−.39	.10*	0.677	−.30	.19	0.738	−.32	.19*	0.729	−.40	.19*	0.670
12mo 20km total				**.00**	**.00***	**1.003**							**.00**	**.00**	**1.002**						
12mo 50km total							**.00**	**.00***	**1.002**							**.00**	**.00**	**1.001**			
12mo 100km total										**.00**	**.00***	**1.001**							**.00**	**.00***	**1.001**
Education	−.09	.10	0.917	−.09	.10	0.918	−.09	.10	0.911	−.10	.10	0.908	−.15	.11	0.859	−.15	.11	0.860	−.15	.11	0.861
Male	−.10	.09	0.906	−.10	.09	0.904	−.10	.09	0.903	−.10	.09	0.903	−.12	.10	0.885	−.12	.10	0.886	−.12	.10	0.885
Poor	.31	.10*	1.357	.32	.10*	1.371	.29	.10*	1.334	.28	.10*	1.328	.12	.11	1.130	.12	.11	1.132	.12	.11	1.129
Rural	−.04	.09	0.961	.00	.09	0.997	−.02	.09	0.980	−.04	.09	0.959	.03	.20	1.034	.00	.21	0.997	.00	.20	1.004
Tense interview	−.20	.10*	0.819	−.21	.10*	0.808	−.22	.10*	0.804	−.22	.10*	0.802	−.20	.11*	0.819	−.20	.11*	0.816	−.21	.11*	0.809
AIC	2719.51			2715.52			2711.64			2707.29			2620.86			2622.95			2621.78		
RE significant	N/A			N/A			N/A			N/A			Yes			Yes			Yes		

RE = Random intercept model; N/A = not applicable; * = Statistically significant at 10%

between regional violence and worries about terrorism varies widely across survey points, as indicated by the intercepts plotted in Figure 15.2.

The multilevel model results for place-based statistical intercept variations tell a modified but important story (columns e–g of Table 15.1). In contrast to our initial conclusion, the link between regional violence and fear of terrorism is statistically significant only at the 100 km threshold. In other words, a small spatial range (20 km) does not yield significant results because the number of events at such a local level is small. It is the broader, more regional concentration of violence that influences respondent perceptions. In political geography, such place-specific influences are important and consistently appear in different locales and settings, and we should investigate whether our models remain consistent and robust after taking such effects into account. While our example is specific to the violent North Caucasus in 2005 and the distance effect would be different in other settings, we cannot know a priori what the relevant thresholds are until we measure a variety of them and show the variation in the model of contextual effects.

Conclusion

Political geography is a changing and dynamic discipline, as it should be. However, certain thematic continuities are threaded from the post–Second World War resurgence of the discipline to the present. Spatial analysis in human (including political) geography must wholeheartedly embrace a nuanced approach to understanding the contexts, places, and social settings of human behavior. It must also fully recognize the spatial dependencies measured by different metrics between locales and administrative units, including the linkages among individuals living in those places. While we reiterate the importance of these perennial and timeless issues for political geography – we have referred to early recognition of contextual effects in 1903 and quantitative work from 1934 – there have also been remarkable changes over the past decade in the availability and use of data for spatial social science. These empowering changes can foster robust and valuable improvements to quantitative research, but must be accompanied by thoughtful and meticulous training in the assumptions that underlie statistical methods. A firm understanding of theory and concepts in political geography can effectively be merged with cautious and careful quantitative approaches to our discipline; the two are not mutually exclusive and hybrid mixed methods research agendas will hopefully continue to illustrate this combination in the future.

Acknowledgements

We are grateful for National Science Foundation Human and Social Dynamics Program funding (grants 0433927 and 0827016). The survey was designed and organized in collaboration with Gerard Toal (Virginia Tech University) and Vladimir Kolosov (Russian Academy of Sciences) and was implemented by the Levada Center Moscow under the direction of Alexei Grazhdankin. Nancy Thorwardson prepared Figures 15.1 and 15.2 for publication.

References

Agnew, J. (1987). *Place and Politics: The Geographical Mediation of State and Society*. London: Allen and Unwin.
Agnew, J. (1996). Mapping politics: How context counts in electoral geography. *Political Geography* 15(2): 129–146.

Anselin, L. (1988). *Spatial Econometrics: Methods and Models*. Dordrecht: Kluwer.

Anselin, L. (1995). Local indicators of spatial association – LISA. *Geographical Analysis* 27(2): 93–115.

Anselin, L. (2002). Under the hood: Issues in the specification and interpretation of spatial regression models. *Agricultural Economics* 27: 247–267.

Anselin, L., Syabri, I., & Kho, Y. (2006). GeoDa: An introduction to spatial data analysis. *Geographical Analysis* 38(1): 5–22.

Bivand, R.S., Pebesma, E.J., & Gomez-Rubio, V. (2008). *Applied Spatial Data Analysis in R*. New York: Springer.

Braybeck, B., & Huckfeldt, R. (2002). Urban contexts, spatially dispersed networks, and the diffusion of political information. *Political Geography* 21(2): 195–220.

Brunsdon, C., & Comber, L. (2015). *An Introduction to R for Spatial Analysis and Mapping*. London: Sage.

Castree, N. (2004). Differential geographies: Place, indigenous rights and "local" resources. *Political Geography* 23(2): 133–167.

Cliff, A.D., & Ord, J.K. (1973). *Spatial Autocorrelation*. London: Pion.

Enos, R.D. (2014). Causal effect of prolonged intergroup contact on exclusionary attitudes: A test using public transportation in homogenous communities. *Proceedings of the National Academy of Sciences* 111(1): 3699–3704.

Flint, C. (1998). Forming electorates, forging spaces: the Nazi party vote and the social construction of space. *American Behavioral Scientist* 41(9): 1282–1303.

Gehlke, C.E., & Biehl, K. (1934). Certain effects of grouping upon the size of the correlation coefficient census tract material. *Journal of the American Statistical Association* 29(185): 169–170.

Gould, P. (1970). Is statistix inferens the geographical name for a wild goose? *Economic Geography* 46(Special Issue): 439–448.

Griffith, D.A., & Paelinck, J.H.P. (2011). *Non-standard Spatial Statistics and Spatial Econometrics*. New York: Springer.

Henderson, J.V., Storeygard, A., & Weil, D.N. (2012). Measuring economic growth from outer space. *American Economic Review* 102(2): 994–1028.

Johnston, R.J., & Pattie, C. (1992). Is the seesaw tipping back? The end of Thatcherism and changing voting patterns in Great Britain 1979–92. *Environment and Planning A* 24(10): 1491–1505.

Johnston, R., Jones, K., Burgess, S., Propper, C., Sarker, R., & Bolster, A. (2004). Scale, factor analyses, and neighborhood effects. *Geographical Analysis* 36(4): 350–368.

King, G. (1995). Replication, replication. *PS: Political Science and Politics* 28: 443–499.

King, G. (1996). Why context should not count. *Political Geography* 15(2): 159–164.

Kulldorff, M., Heffernan, R., Hartman, J., Assuncao, R., & Mostashari, F. (2005). A space–time permutation scan statistic for disease outbreak detection. *PLoS Medicine* 2(3): 216–224.

Kulldorff, M., & Information Management Services (2009). *SaTScan: Software for the Spatial and Space-Time Scan Statistics, v 9.0.1*. http://www.satscan.org/

Linke, A.M. (2013). The aftermath of an election crisis: Kenyan attitudes and the influences of individual-level and locality violence. *Political Geography* 37: 5–17.

Linke, A.M., & O'Loughlin, J. (2015). Reconceptualizing, measuring, and evaluating distance and context in the study of conflicts: Using survey data from the North Caucasus of Russia. *International Studies Review* 17(1): 107–125.

Linke, A.M., Schutte, S., & Buhaug, H. (2015). Population attitudes and the spread of political violence In Africa. *International Studies Review* 17(1): 26–45.

Linke, A.M., Witmer, F., & O'Loughlin, J. (2012). Space–time Granger Analysis of the war in Iraq: A study of coalition and insurgent action-reaction. *International Interactions* 38: 402–425.

Lynch, C. (2012). Syrian opposition seeks to wipe the Assad name off of the map – via Google. *The Washington Post* , 14 February.

Maclaurin, G., Leyk, S., & Hunter, L. (2015). Understanding the combined impacts of aggregation and spatial non-stationarity: the case of migration-environment associations in rural South Africa. *Transactions in GIS*, in press, doi: 10.1111/tgis.12134

Massey, D. (1994). *Space, Place, and Gender*. Minneapolis, MN: University of Minnesota Press.

McCarty, H.H. (1952). McCarty on McCarthy: The spatial distribution of the Wisconsin vote, 1952. Unpublished manuscript.

Nelson, A. (2001). Analysing data across geographic scales in Honduras: detecting levels of organization within systems. *Agriculture Ecosystems and Environment* 85(1–3): 107–131.

O'Loughlin, J. (1981). The neighborhood effect in urban voting surfaces: An international comparison. In Burnett, A.D., & Taylor, P.J. (eds), *Political Studies from Spatial Perspectives: Anglo-American Essays in Political Geography*. Chichester: John Wiley & Sons, 357–388.

O'Loughlin, J. (2000). Geography as space and geography as place: The divide between political science and political geography continues. *Geopolitics* 5(3): 126–137.

O'Loughlin, J. (2002). The electoral geography of Weimar Germany: Exploratory Spatial Data Analysis (ESDA) of Protestant support for the Nazi Party. *Political Analysis* 10(3): 217–243.

O'Loughlin, J. & Witmer, F. (2012). The diffusion of conflicts in the North Caucasus of Russia, 1999–2010. *Environment and Planning A* 44(10): 2379–2396.

O'Loughlin, J., Flint, C., & Anselin, L. (1994). The geography of the Nazi vote: Context, confession and class in the Reichstag election of 1930. *Annals of the Association of American Geographers* 84(3): 351–380.

O'Loughlin, J., Witmer, F., Linke, A.M., & Thorwardson, N. (2010). Peering into the fog of war: The geography of the Wikileaks Afghanistan war logs 2004–2009. *Eurasian Geography and Economics* 51(4): 437–471.

O'Louglin, J., Witmer, F., Linke, A.M., Laing, A., Gettleman, A., & Dudhia, J. (2012). Climate variability and conflict risk in East Africa, 1990–2009. *Proceedings of the National Academy of Sciences* 109(45): 18344–18394.

Openshaw, S. (1983). *The Modifiable Areal Unit Problem*. Norwich: Geo Books.

Openshaw, S. (1996). Developing GIS-relevant zone-based spatial analysis methods. In Longley, P., & Batty, M. (eds), *Spatial Analysis: Modeling in a GIS Environment*. New York: John Wiley & Sons, 55–73.

Openshaw, S., & Taylor, P. (1979). A million or so correlation coefficients: Three experiments on the modifiable areal unit problem. In Wrigley, N. (ed.), *Statistical Applications in the Spatial Sciences*, London: Pion, 127–144.

Ord, J.K., & Getis, A. (1995). Local spatial autocorrelation statistics: Distributional issues and application. *Geographical Analysis* 27(4): 286–306.

Pattie, C., & Johnston, R. (2000). "People who talk together vote together": An exploration of contextual effects in Great Britain. *Annals of the Association of American Geographers* 90(1): 41–66.

Pierskalla, J., & Hollenbach, F. (2013). Technology and collective action: The effect of cell phone coverage on political violence in Africa. *American Political Science Review* 107(2): 207–224.

Plant, R.E. (2012). *Spatial Data Analysis in Ecology and Agriculture Using R*. Boca Raton, FL: CRC Press.

Pred, A. (1983). Structuration and place: On the becoming of sense and place and structure of feeling. *Journal for the Theory of Social Behavior* 13(1): 45–68.

Radil, S., & Flint, C. (2013). Exiles and arms: The territorial practices of state making and war diffusion in post-Cold War Africa. *Territory, Politics, Governance* 1(2): 183–202.

Robinson, W.S. (1950). Ecological correlations and the behavior of individuals. *American Sociological Review* 15(3): 351–357.

Root, E.D., Meyer, R., & Emch, M. (2009). Evidence of localized clustering of gastroschisis in North Carolina, 1999–2004. *Social Science and Medicine* 68: 1361–1367.

Secor, A.J., & O'Loughlin, J. (2005). Social and political trust in Istanbul and Moscow: A comparison of individual and neighborhood effects. *Transactions of the Institute of British Geographers* 30(1): 66–82.

Selvin, H.C. (1958). Durkheim's suicide and problems of empirical research. *American Journal of Sociology* 63(6): 607–619.

Spielman, S.E., & Logan, J. R.. (2013). Using high-resolution population data to identify neighborhoods and establish their boundaries. *Annals of the Association of American Geographers* 103(1): 67–84.

Spielman, S.E., Yoo, E.-H., & Linkletter, C. (2013). Neighborhood contexts, health, and behavior: Understanding the role of scale and residential sorting. *Environment and Planning B* 40(3): 489–506.

Tobler, W. (1970). A computer movie simulating urban growth in the Detroit region. *Economic Geography* 46(2): 234–240.

Tollefson, A.F., Strand, H., & Buhaug, H. (2012). PRIO-GRID: A unified spatial data structure. *Journal of Peace Research* 49(2): 363–374.

VanWey, L.K., Rindfuss, R.R., Gutmann, M.P., Entwisle, B., & Balk, D.L. (2005). Confidentiality and spatially explicit data: Concerns and challenges. *Proceedings of the National Academy of Sciences* 102(43): 15337–15342.

Vidal de La Blache, P. (1903). *Tableau de la geographie de la France*. Paris: Hachette (reprinted 1979, Paris: Tallandier).

Verpoorten, M. (2012). Detecting hidden violence: The spatial distribution of excess mortality in Rwanda. *Political Geography* 31(1): 44–56.

Witmer, F., & O'Loughlin, J. (2009). Satellite data methods and application in the evaluation of war outcomes: Abandoned agricultural land in Bosnia-Herzegovina after the 1992–1995 conflict. *Annals of the Association of American Geographers* 99(5): 1033–1044.

Wucherpfennig, J., Weidmann, N.B., Girardin, L., Cederman, L.-E., & Wimmer, A. (2011). Politically relevant ethnic groups across space and time: Introducing the geoEPR dataset. *Conflict Management and Peace Science* 28(5): 423–437.

Zhukov, Y.M. (2012). Roads and the diffusion of insurgent violence: The logistics of conflict in Russia's North Caucasus. *Political Geography* 31(3): 144–156.

Chapter 16

Radical Political Geographies

Simon Springer

University of Victoria, Greater Victoria, British Columbia, Canada

Political geography as a subfield of geographical thought has done a great deal to support the interests of the powerful. There was and continues to be a desire among many political geographers to service the status quo, rather than seeking to use the field as a means to communicate alternatives and undermine existent structures of authority. To a significant extent this speaks to the wider historical positioning of geography as a discipline vis-à-vis the machinations of imperialism (Godlewska & Smith 1994), but in particular, political geography has had a tendency to focus its energies on promoting the articulation of realpolitik, the practice of statecraft, and the apotheosis of war, none of which lends itself well to a radical trajectory. The result is that for much of the twentieth century, political geography was viewed as a pugnacious domain, receiving a very chilly reception from other practitioners within the wider discipline of human geography (Johnston 1981). Peter Taylor (2003: 47) acknowledged these exact patterns and deficiencies in an earlier edition of this volume, where in considering the radical turn of the late 1960s, he argued that political geography was ignored by radicals and that "there has not been an identifiable radical tradition in political geography." While there is a lot to be said for the lack of engagement and the potential reasons behind this, I want to focus my efforts on demonstrating how radical critique can be and has been sutured together with political geography. In doing so, I actually want to challenge Taylor's interpretation and suggest that while his view of the last few decades is correct vis-à-vis the development of radical geography as a Marxist undertaking, had he looked further into the past and beyond Marxism, he would have discovered that there was indeed an identifiable radical tradition that showed significant potential, which continues to resonate in the present.

In understanding where to look, we need to unpack the etymology of "radical," coming from the Latin *radix*, meaning "root." When we dig down to the source by going back to the emergence of political geography as a field, we of course come face to face with the environmental determinism of Friedrich Ratzel and Ellen Churchill Semple, the *geopolitik* of Karl Haushofer, and the explicitly imperialist geography of Halford Mackinder (Driver 2001;

The Wiley Blackwell Companion to Political Geography, First Edition.
Edited by John Agnew, Virginie Mamadouh, Anna J. Secor, and Joanne Sharp.
© 2015 John Wiley & Sons Ltd. Published 2017 by John Wiley & Sons Ltd.

Kearns 2009a). Yet, within this historical milieu, where empire, geography, and naturalized assumptions all marched hand in hand as an ostensibly seamless corporation, we also find the anti-colonialist geographies of Peter Kropotkin and Élisée Reclus, who rejected both the machinations of imperialism and the class-centric outlook of Marxism, instead arguing for political patterns that reflected voluntary association, mutual aid, and the evasion or abolition of the state (Ferretti 2013; Springer 2013a). While the threads of their radical project were sewn into understandings of social ecology, evolutionary theory, environmental studies, and bioregionalism, Kropotkin and Reclus were first and foremost geographers whose concerns were decidedly political (Clark 2009; Kearns 2009b). Indeed, today these two men are remembered primarily for their activism and contributions to the political philosophy of anarchism, and it is here, in the emancipatory geographies of direct action, reciprocity, and prefigurative politics, that we can find a very different tradition of thinking through just what exactly political geographies might look like through a radical lens.

While an anarchist antecedent to contemporary radical political geographies certainly existed, we should also recognize that the word "anarchism" has been subjected to a century-long project of vilification at the hands of political elites. What this means for the radical political geographies that have evolved, particularly in the aftermath of the countercultural movement, is that very often what are anarchistic tendencies have not been recognized as such. For example, while feminist and postcolonial political geographies certainly offer something new, if we look again at their political impulses and underlying motivations, we may just find precursors in the anarchist movement, which was not simply a rejection of the authority of the state, but a manifold critique of all forms of domination, ranging from capitalism to sexism to racism and beyond. An important part of the history of radical political geographies, then, is the silence that has been afforded to anarchism with respect to explicit engagements with this idea in the literature, even if its aims and intentions have been unknowingly adopted and advanced within the frameworks of both the academic practice of postcolonial and feminist political geographies, and the domain of everyday organizing and resistance. It is also notable how the most progressive filaments of the Marxist tradition have been creeping ever closer toward an autonomist line over the last decade, suggesting that all of the contemporary domains of radical political geographies are starting to recognize and appreciate a common ground that is highly anarchistic in its outlook.

In this chapter, I explore three branches of radical political geographies, examining the major developments and strains of inquiry within each category, and suggesting how these relate to or diverge from the early expressions of Kropotkin and Reclus. I begin with Marxist political geographies, which have been the most popular variant of radical geographical thought for the past three decades, particularly owing to the influence of David Harvey. Here, I speak of a boomerang effect, where Marxism's divergence from anarchism during the middle part of the nineteenth century, particularly over differences around the time of the First International, started to loop back into an increasing coalescence with the rise of autonomist Marxism. I then turn to postcolonial political geographies and the critiques that have been mounted against colonialism, drawing on Edward Said's intrepid work and its uptake in articulating a critical geopolitics. Before becoming known as "anarchism," the anarchists of the day referred to their engagements as anti-colonialism, and so the synergies are not hard to find within the postcolonial project. Next, I focus my attention on anarchist political geographies in their explicit formulation through the prefigurative politics of direct action. Anarchism has returned to the geographical literature in recent years following periods of long silence, even if its spatial practice beyond the academy never actually went away (Springer 2013a). Finally, it is important to note that although feminist political geographies deserve attention as being decidedly radical, owing to space restrictions I have not included a discussion of them here. This terrain is covered in depth by Fluri in Chapter 18 of this volume, where she makes a strong case for their ongoing importance in thinking through politics both geographically and radically.

Marxist geographies: Proletarian politics

One of the earliest expressions of Marxian critique within political geography was articulated by Karl Wittfogel, who set his sights on German *geopolitik*, arguing that this was a form of "bourgeois" materialism (Ó Tuathail 1996b). The conservative character of geography at the time ensured that his ideas were largely ignored, and it really was not until the founding of the journal *Antipode* in 1969 that Marxist political geographies started to gather momentum. The increasing radicalism of geography was part of the countercultural movement going on outside of the academy, yet much of the work that was being done at the time, while intensively political, was not considered to be "political geography" proper. In many ways it was a case of purposeful disassociation, as the subdiscipline carried the stigma of being considered a "moribund backwater" (Berry 1969), owing to the continuing influence of environmental determinism and in particular Mackinder's heartland theory, which reduced all politics to the expression of a tautological geography. While a degree of introspection began to emerge among political geographers, where Richard Muir (1976), for example, asked if the subdiscipline was a "dead duck" or a "phoenix," Marxist geographers paid little attention. Instead, they set about tilling the soil of what they considered to be the greener pastures of economic, social, and urban geography. Here, they carved out space for critiques of poverty, racism, labor exploitation, and exclusion, wherein David Harvey (1972, 1973) and Manuel Castells (1977) began to spatialize Marxist theory by focusing attention on the city, while Richard Peet (1975, 1977) sought to draw attention to what Marxist ideas could tell us about the social and economic issues of the day. These articulations of Marxism, which were political but not considered political geography, set the tone for the bulk of the radical geography that followed. Aside from a special issue on anarchism edited by Myrna Breitbart (1978), the radical trajectory of geography beginning in the late 1960s onward was almost exclusively Marxist.

With the publication of *The Limits to Capital*, Harvey (2007 [1982]) established himself as the leading figure of the emergent radical geography. He brought Marxism to the forefront of geographical theory, which had ripple effects across the discipline and inspired others to take up a Marxian analysis in ways that spoke more specifically to the more "traditional" concerns of political geography. Ron Johnston (1982, 1984) incorporated Marxist ideas into his critique of the state and the ways in which its formation was intercalated with processes of capital accumulation, which brought a new dimension to political geography insofar as it provoked critical reflection on the capitalist superstructure that forms the basis of state power. Digging deeper into the workings of the state, Gordon Clark and Michael Dear (1984) were also quite explicit in their radical approach. Citing capitalism as the lifeblood of the state apparatus, they developed a taxonomy to address systematically the intersections between these enterprises, and in doing so, they brought geography to center stage in theorizations of the state by infusing it with a political economy perspective that acknowledged class. Outside of the Anglo-American sphere, francophone geographer Yves Lacoste (2000) was similarly inspired to bring Marxian critiques into conversation with political geography through the expression of a radical geopolitics that sought to undermine the long-held assumptions of realpolitik, which again provoked critical questions about the state and its strategic outlook. While these critiques of the state were welcome antidotes to the celebrations of state power that had preceded them within political geography, there was still room to push further in transcending the state-centric views that continued to dominate political economy approaches across the social sciences. The work of Peter Taylor (1985, 1994) stepped in to fill this void in political geography by looking to Immanuel Wallerstein's (1974, 1979) work on world systems, which insisted that in understanding social change we must look to the global rather than to the state. This had a profound effect on geographical understandings of politics, and hence on political geography. Taylor (1991a), for his part, saw much to be gained from this move away from a focus on the state as the primary power holder in society, recognizing that there was a potentially emancipatory politics in thinking beyond it. He flirted with anarchism (Taylor 1991b) and critiqued Marxist politics for their scalar focus on the state (Taylor 1987), and yet, like most other radical geographers at the time, he remained committed to the idea of a welfare state in ways that anarchists would reject. Notably though, Fabrizio Eva (1998) did work against the state-centric grain, questioning the "functional fiction" of geopolitical discourse with respect to international boundaries, arguing for the importance of anarchist thought in breaking the statist hold on our political imaginations.

More recently, the character of Marxist political geography has started to shift away from both the class-centric outlook that came to dominate radical geography and the state-centric view of early political geography, toward a more autonomist position that shares a number of similarities with the earlier anarchism of Kropotkin and Reclus. Marxism has had many appendages that have been added on to and extended from the original iteration of Marx's ideas, where one of the most powerful and influential is autonomist Marxism. Michael Hardt and Antonio Negri (2000), although not geographers themselves, are the best-known proponents of this alternative reading of Marxism, and they have provoked considerable reflection and debate among geographers (Corbridge 2003; Sparke 2005). Invoking the idea of a multitude, Hardt and Negri (2000: 15) transcend the class centrism that traditional Marxism maintains, suggesting that the collective creative forces that sustain empire "are also capable of autonomously constructing a counter-Empire, an alternative political organization of global flows and exchanges." While political geographers have found both "intriguing possibilities and problematics deserving further reflection" within this work (Sidaway 2005: 68), particularly with respect to the geography of sovereignty and the geopolitics of world order, it is the possibility of political agency that has the most powerful resonance with anarchism, evoking a "boomerang effect" that brings radical political geography back to its earliest proponents, Reclus and Kropotkin. This (re)turn comes by way of the idea that the populace exudes a spontaneous power that foments cracks in state hegemony (Hardt & Negri 2005), which in many ways mirrors anarchist notions of direct action and prefigurative politics. In particular, the analyses of autonomist Marxists, who actively seek to create the future by fostering alternative social relations and new forms of being in their everyday lives (Katsiaficas 1997; Pickerill & Chatterton 2006; Marks 2012), find significant correspondence with the anarcho-geography of Kropotkin (2008 [1902]), wherein the potential for a new society is seen to exist already within the materiality of capitalism. Through this we start to see a picture where what has come to define radical political geography as set apart from the moribund backwater of traditional political geography is its willingness to look beyond and outside of the state-centric sphere in the construction of its politics (Agnew & Corbridge 2002).

Postcolonial geographies: Discursive politics

With increasing militarism in the United States, including violent excursions into Iraq and Afghanistan, critiques of colonialism have been gathering momentum in radical political geography in recent years (Agnew 2005; Harvey 2003). Derek Gregory's (2004) *The Colonial Present* stands out as a watershed text that reinvigorated political geographers' interest in the postcolonial project inaugurated by Edward Said (2003 [1978], 1993), Gayatri Chakravorty Spivak (1999), and Homi K. Bhabha (1994). Although there are important critical questions to be asked about the meaning of "postcolonialism" (Sidaway 2000), the general thrust against colonialism within contemporary political geography represents a sea change from the earlier position of the subdiscipline as a colonial vanguard, disrupting and contesting the "imperiality of knowledge" that has been sustained for far too long (Slater 2004: vii). This trajectory has been taken up in multiple ways, including closer attention to culture (Escobar 2001), race (Pred 2000), citizenship (Staeheli 2011), and violence (Springer 2011b, 2012b), and particularly the ways in which "othering" is shot through each of these processes. The other key implication of an anti-colonial imperative is that in paying attention to the ways in which justifications for colonialism have been mounted in the past, political geographers are better equipped to critique the attempts at its legitimation in the present. The work of Michel Foucault (2002 [1972], 1990 [1978]) has been instrumental in this regard, where his insights on discourse have found a prominent home among radical political geographers who have applied a critical discourse analysis to a range of political phenomena, including climate change (Boykoff 2008), ethnic conflict (Korf 2006), good governance (Springer 2010), and questions of sovereignty (Kuus 2002). Extensions of Foucault's arguments have also been applied to considerations of biopolitics (Coleman & Grove 2009; Sparke 2006) and governmentality (Luke 1996; Watts 2004), which invoke a more nuanced sense of thinking through the "post"colonial insofar as sovereign rule still functions, but operates through a less overt power interface that relies on the rationalities and techniques of producing governable subjects through the "conduct of conduct" rather than via brute force.

One of the key developments in political geography where discourse has been taken up is the domain of geopolitics. Although long associated with a governmentalized center of power over geographical knowledge, wherein judgment is meted out across space in an effort to organize and discipline experience into ideological, racial, and nationalistic truths, geopolitics has more recently been invigorated with a radical outlook in the form of "critical geopolitics." Ó Tuathail (1996b) spearheaded this movement of dissident geographers, exploring what he calls the convenient and inconvenient "fictions" that geopolitics presents as "regimes of truth" by looking to how power is evoked through the construction of particular discursive formations that are intended to *geo-graph* the global political scene. To Ó Tuathail (1996b: 14), geopolitics "is not a concept that is immanently meaningful and fully present to itself but a discursive 'event' that poses questions to us whenever it is evoked and rhetorically deployed." Rather than a fixed conceptual category or fully demarcated field, the intention of critical geopolitics is for it to be read as a work in progress, wherein multiple paths are able to evolve along undefined trajectories that call power into question in a variety of domains, including cultural studies, actor networks, psychoanalysis, and gender identities, so that the assumptions of practitioners may be challenged and rethought (Ó Tuathail & Dalby 2002). In taking up this notion of critical geopolitics being a domain for the exploration of new critiques against hierarchical political arrangements, the concept of "anti-geopolitics" introduced by Ó Tuathail (1996a), but later developed by Paul Routledge (1998: 245), served to identify a specific focus on counter-hegemonic struggle, which included the "assertion of permanent independence from the state." In a similar vein, but with greater attention to grassroots practice and the ways in which groups engage in geopolitics at the level of the everyday and beyond the gaze of authority, Sara Koopman (2011: 277) proposed the term "alter-geopolitics" as a radical extension of both anti- and feminist geopolitics, wherein "[g]rassroots groups are not waiting for (or trusting) the state, but coming together on their own, nonviolently, for safety." Likewise, David Featherstone's (2008, 2012) work on counter-global solidarity networks and Jo Sharp's (2009, 2013) postcolonial, subaltern geopolitics from the margins equally offer important interventions that move us away from state-centricity, and it is not difficult to see how these engagements resonate strongly with anarchist ideas.

While the turns toward critical, anti-, postcolonial, and alter-geopolitics are welcome antidotes to the centralizing logics of power that have been the traditional domain of geopolitics in particular, and political geography more generally, the connections between state and colonial power remain understated and largely unexamined. A precursor to these anti-colonial trajectories of contemporary radical political geography can be found in anarchism, and specifically the work that was done by Kropotkin and Reclus over a century ago, wherein linkages between state and colonial projects became overt. With exceptions like Gerry Kearns's (2004) attention to Kropotkin's liberatory geographical imagination, and Kevin Hewitt's (2001) discussion of what geography, through the contrasting lenses of Kropotkin and Pinochet, can tell us about state terror and human rights, few political geographers have engaged the anarchist literature for insights into thinking about anti-colonialism today. Recent developments in anarchist geographies have attempted to redress this absence, where Simon Springer (2012a: 1607) has argued that

> there is no fundamental difference between colonization and state-making other than the scale upon which these parallel projects operate, meaning that any substantively "post-colonial" positionality must also be "post-statist" or anarchic.

Although state power and colonial power differ in their diffusion and distribution across space, they nonetheless both attempt to map a singular identity on antecedent ways of imagining belonging, and therein represent the same violent principles of a privileged few wielding influence over others (Springer 2013b). In light of these observations, coupled with the decolonizing epistemologies, ontologies, and methodologies that have been highlighted by indigenous scholars (Shaw, Herman, & Dobbs 2006; Coombes 2007; Smith 1999), it would seem that the time is ripe for political geography to develop a robust "anti-colonial imagination" that specifically raises a poststatist or anarchist challenge (Anderson 2005). Such a critical encounter with anarchism might provide indigenous activists and progressive Settler peoples with more common ground in their political struggles (Barker & Pickerill 2012). Indeed, Reclus and Kropotkin demonstrated long ago that geography lends itself well to the communication of emancipatory ideas (Clark & Martin 2004; Pelletier 2009; Ferretti 2011), and if "it was no accident that two of the major anarchists of the late Nineteenth Century were also geographers" (Ward 2010: 209), then it is time to take stock of what anarchist political geographies might contribute in anticipation of an anarchist (re)turn.

Anarchist geographies: Prefigurative politics

The political philosophy of anarchism has often been misrecognized as simply advancing an anti-state outlook. While concerns for the state and its violence have indeed formed the core of anarchist thought, it is incorrect to assume that this is its only or even primary locus. Anarchism has long held that the devastating effects of capitalism are just as important as the tyranny of the state (Bakunin 2002 [1873]; Goldman 1969 [1917]). Indeed, given that Pierre-Joseph Proudhon was the first person ever to declare himself an "anarchist," it is only through caricature that the state can be considered the sole concern of anarchism. In *What Is Property? or, An Inquiry into the Principle of Right and of Government*, Proudhon (2008 [1840]) answered that "property is theft," which subsequently became a rallying cry against capitalism and an early defining feature of the anarchist movement. In this vein, Kropotkin (1995 [1908]: 94) argued that

> the State ... and capitalism are facts and conceptions which we cannot separate from each other.
> In the course of history these institutions have developed, supporting and reinforcing each other.

Part of the reason for the confusion stems from Marxists' attempts to position their philosophy as the only viable anti-capitalist alternative, where beginning with Friedrich Engels, who "distorts Bakunin's argument, which also held capital to be an evil necessary to abolish" (Gouldner 1982: 864), the myth that anarchists consider the state as the only enemy has been repeated again and again. This fiction has hopefully been shattered once and for all by the recent "reanimation" of anarchist geographies (Clough & Blumberg 2012; Springer et al. 2012), where a new crop of radical political geographers have sought to advance an aggregate understanding of anarchism. Anthony Ince (2010: 294) contends that "anarchism's holism – its recognition of the many different factors that influence and feed off each other as interrelated and inseparable in capitalist systems – means that it is ideally suited to an analysis of capitalism's contested geographical terrain," while Simon Springer (2014b) argues that anarchist geographies go "beyond simply crossing interdisciplinary boundaries to initiate a process of 'undisciplining' by engaging both critical geopolitics and geopoetics to suture together all the ostensibly separate pieces in articulating an 'integral anarchism.'" The promise of anarchism for radical political geography rests with its refusal to assign priority to any one of the multiple dominating apparatuses that we face in the current conjuncture, as all are considered as irreducible to one another. In short, while political geography has been state-centric, anarchism does not repeat this mistake by holding a mirror to it that invokes a singular anti-state-centric position.

Springer (2012a: 1607) defines anarchist geographies as

> kaleidoscopic spatialities that allow for multiple, non-hierarchical, and protean connections between autonomous entities, wherein solidarities, bonds, and affinities are voluntarily assembled in opposition to and free from the presence of sovereign violence, predetermined norms, and assigned categories of belonging.

Given the inherently integral character of such geographies, it is not unfair to suggest that anarchism offers greater possibilities for inclusion than even the most progressive variants of Marxism, precisely because it does not prioritize class identity over any other category of belonging or exclusion (Springer 2014c). The Marxian obsession with class has produced a political outlook that imagines stages of history, wherein the ultimate goal is a revolution that ends with an ostensibly temporary proletarian dictatorship and a promise for the eventual withering away of the state. Unlike the end-state politics of Marxism, which envisions a moment wherein history ends and a perfect polity is instantiated, the prefigurative politics of anarchism are considered as an infinitely demanding struggle (Critchley 2008). In this way, anarchism does not advance a revolutionary imperative like Marxism (Smith 2010), but an insurrectionary imperative, which is to be understood as a processual *means without end* (Springer 2011a). The geographical implications of this outlook suggest that anarchism fully appreciates the processual nature of space, where the politics of waiting – for the revolution, for the withering away of the state, for the stages of history to pass – are all rejected in favor of the realism that comes with acknowledging that the everyday is the only moment and space in which we have any tangible control over our lives (Springer 2012a). Thus, far from being utopian, anarchism offers an antidote to Marxian political deferral by embracing the *here* and *now* of the everyday. Such a position represents a deeper appreciation of space–time as a constantly folding, unfolding, and refolding story, where direct action, radical democracy, and mutual aid allow us instantaneously to reconfigure its parameters.

Prefigurative politics refers to the idea that we breathe life into anarchism through effecting social relationships and organizing principles in the present that attempt to reflect the future society that we seek. The idea of prefiguration is thus not to be confused with predetermination, as it is about the active and ongoing process of building a new society in the shell of the old (Ince 2012). Prefiguration is thus synonymous with direct action, which, in contrast to civil disobedience and its grand gesture of defiance, proceeds with no consideration of authority whatsoever, considering all authority to be illegitimate (Graeber 2009). Such a politics offers a very different political orientation to the vanguardism of Marxism (Dean 2012), as the spectacular moment of revolution is replaced with the ongoing process of insurrection, or "rising up" above government, religion, and other hierarchies, not necessarily to overthrow them, but simply to disregard these structures by taking control of one's own individual life and creating alternatives on the ground (Stirner 1993 [1845]). Anarchist political geographies accordingly call attention to the "diverse economies" and "other worlds" that already exist and are constantly being remade through experimentation beyond capitalism (Gibson-Graham 2008; White & Williams 2012). The political geographies of prefigurative direct action demand our consideration precisely because we stare them in the face on a daily basis, but scarcely recognize them for what they are. Every time you have ever lent your neighbor a screwdriver, watched your sister's kids, borrowed a friend's bike, helped a stranger load groceries into their car, or had friends over for dinner, you have – perhaps unknowingly – engaged in an anarchist politics of reciprocity, which Kropotkin (2008 [1902]) dubbed "mutual aid." As a decentralized practice, the prefigurative politics of direct action and mutual aid evokes a geography that spontaneously liberates a particular area (of land, time, or imagination) to produce a temporary autonomous zone (TAZ), which exists outside of the gaze of authority, only to dissolve and reform elsewhere before capitalism and sovereignty can crush or co-opt it (Bey 1991). A TAZ is driven by the confluence of action, being, and rebellion, wherein there is no separation of theory and practice. Organizational form is not subordinated to future events as in the Marxist frame of understanding. Instead, it is continually negotiated as an unfolding rhizome that arises though the actual course of *process* (Springer 2014a).

Conclusion

In this chapter I have attempted to show how anarchism has been an "absent presence" (Derrida 1997) within political geography for as long as the field of inquiry has existed. While often obscured behind other radical movements within the subdiscipline, an anarchist basis nonetheless forms the root of radical geographical thought, including its application to "the political." I have argued that while Marxist geographies have been the main trajectory of radical geography since its resurgence in the late 1960s, there was an earlier tradition of radical political geography to be found in the work of Kropotkin and Reclus during the latter part of the nineteenth century. While radical geographers writing in the last few decades of the twentieth century largely overlooked these early contributions, there has been a more recent turn toward autonomous Marxism that finds many synergies with anarchism. I have also looked to postcolonial geographies as a particularly productive branch of radical political geography, specifically in terms of the latter's contributions to recasting how we might think about and apply geopolitics. The anti-colonial implications of this work also found strong resonances with anarchist critique, particularly when we examine how state-making and colonial projects overlap in terms of their strategies and outlook. Finally, I look to the (re)turn to anarchist geographies among radical political geographers and the contributions that have been made particularly in terms of thinking through a prefigurative politics and a geography of direct action. Here we find a radical political geography that finally removes the state-centric frame that has dominated political geographical thought throughout its history, exploring the geographies of autonomy, mutual aid, and voluntary association, wherein politics operates as an expression of horizontalism rather than through hierarchical modes of power that have been assumed for far too long.

For all of the radical gains that have been made within political geography, there is still significant room for exploration into more emancipatory domains. Perhaps the biggest obstacle to a more thoroughly radicalized political geography is the continuing fixity of the state in our collective geographical imaginations. Although recent developments in anarchist geographies have shown that even Leviathan has an Achilles' heel, where cracks in the monolith grow deeper with every passing day, the state by and large remains a sacred cow within the discipline. Most geographers continue to take its presence for granted and thereby forward state-centric analyses that assume a certain geographical form, which limits what can be said about the radical possibilities of the *here* and *now* in articulating new spatial realities. Instead of opening the state up to interrogations that undermine its dominance in ways that question its actual basis, as Giorgio Agamben (1998) and Michel Foucault (2003) ask us to do with respect to sovereignty, the typical line of critique runs along a reformist axis that challenges particular iterations of state violence, suggesting that certain states can and should do better. To be "radical" is not to embrace reticence by giving in to notions of reform that produce mere variations on a theme, but, rather, it is to be bold and intrepid in outlook by showing a willingness to return to the roots. What this means for political geography is a recognition that in human experience, the root of all power rests within the ability of human beings, and the ways in which we choose to organize our activities and relationships through and across space. The phenomenon of the state is but a blip on the radar screen in the long march of history, and it has been the insistent focus of political geography on the ebbs and flows of civilizational power that has obscured our vision of what exists beyond such a myopic view of humanity's collective possibilities.

A radical political geography would do well to explore more thoroughly the domain of politics beyond the state, a provocation that does not mean that the state should be ignored. It remains critical to contest the ongoing violence of sovereignty, the divisive processes of othering that are made possible through discourses of "race," "ethnicity," and "nation," and the seemingly endless wars that are fought under state banners. Yet, for a more emancipatory strain of radical political geography to bear fruit, there must be a greater willingness to explore prefigurative politics, to embrace an ethos of "do it yourself" that looks to direct action as a mode of change, and to coordinate with activist struggles on the ground in building greater solidarities beyond the academy (Koopman 2008). Anarchism, as a continually unfolding process that is perpetually being prefigured through mutual aid, voluntary association, and self-organization, offers significant hope for a reinvigoration of the "radical" in political geography. Whereas Marxian political geographies with their stages of history, revolutionary imperatives, and end-state politics have seemingly hit a wall, thereby necessitating a turn toward autonomism, in contrast anarchism is a political philosophy that already appreciates the essential dynamism of the social world. By abandoning any pretext of achieving a completely free and harmonious society in the future, anarchism instead focuses on the immediacies of a prefigurative politics of direct action in the present. Radical political geography thus requires the rise of a greater sense of entitlement, not to "power over," as is the traditional domain of political geography, but to "prefigurative power," where the desired outcomes are lived through the embodiment of the *here* and *now*. Such a movement would give radical political geography a greater consonance with what is already happening on the street, where recent political movements use space as the very expression of ideology. When commentators dismiss contemporary social movements as lacking a coherent ideology owing to their decentralized forms and lack of leadership, what they fail to realize is that this mode of organizing *is* their ideology (Graeber 2004), a political *geo-graphing* that is radical through and through.

References

Agamben, G. (1998). *Homo Sacer: Sovereign Power and Bare Life*. Stanford, CA: Stanford University Press.

Agnew, J.A. (2005). *Hegemony: The New Shape of Global Power*. Philadelphia, PA: Temple University Press.

Agnew, J., & Corbridge, S. (2002). *Mastering Space: Hegemony, Territory and International Political Economy*. London: Routledge.

Anderson, B. (2005). *Under Three Flags: Anarchism and the Anti-Colonial Imagination*. London: Verso.

Bakunin, M. (2002 [1873]). *Statism and Anarchy*. Cambridge: Cambridge University Press.

Barker, A.J., & Pickerill, J. (2012). Radicalizing relationships to and through shared geographies: Why anarchists need to understand indigenous connections to land and place. *Antipode* 44: 1705–1725.

Berry, B.J.L. (1969). "International Regions and the International System", Book review. *The Geographical Review* 59: 450-451.

Bey, H. (1991). *T.A.Z.: The Temporary Autonomous Zone, Ontological Anarchy, Poetic Terrorism*. Brooklyn, NY: Autonomedia.

Bhabha, H.K. (1994). *The Location of Culture*. London: Routledge.

Boykoff, M.T. (2008). The cultural politics of climate change discourse in UK tabloids. *Political Geography* 27: 549–569.

Breitbart, M. (1978). Introduction. *Antipode* 10: 1–5.

Castells, M. (1977). *The Urban Question: A Marxist Approach*. London: Edward Arnold.

Clark, G.L., & Dear, M.J. (1984). *State Apparatus: Structures and Language of Legitimacy*. Boston, MA: Allen & Unwin.

Clark, J.P. (2009). Reclus, E. In Kitchen, R., & Thrift, N. (editors-in-chief), *International Encyclopedia of Human Geography*, Amsterdam: Elsevier, 107–110.

Clark, J.P., & Martin, C. (eds) (2004). *Anarchy, Geography, Modernity: The Radical Social Thought of Elisée Reclus*. Oxford: Lexington.

Clough, N., & Blumberg, R. (2012). Toward anarchist and autonomist Marxist geographies. *ACME: An International E-Journal for Critical Geographies* 11: 335–351.

Coleman, M., & Grove, K. (2009). Biopolitics, biopower, and the return of sovereignty. *Environment and Planning D: Society and Space* 27: 489–507.

Coombes, B. (2007). Postcolonial conservation and kiekie harvests at Morere New Zealand – abstracting indigenous knowledge from indigenous polities. *Geographical Research* 45: 186–193.

Corbridge, S. (2003). Countering empire. *Antipode* 35: 184–190.

Critchley, S. (2008). *Infinitely Demanding: Ethics of Commitment, Politics of Resistance.* London: Verso.

Dean, J. (2012). *The Communist Horizon.* London: Verso.

Derrida, J. (1997). *Of Grammatology.* Baltimore, MD: Johns Hopkins University Press.

Driver, F. (2001). *Geography Militant: Cultures of Exploration and Empire.* Oxford: Blackwell.

Escobar, A. (2001). Culture sits in places: Reflections on globalism and subaltern strategies of localization. *Political Geography* 20: 139–174.

Eva, F. (1998). International boundaries, geopolitics and the (post)modern territorial discourse: The functional fiction. *Geopolitics* 3: 32–52.

Featherstone, D. (2008). *Resistance, Space and Political Identities: The Making of Counter-Global Networks.* Chichester: Wiley-Blackwell.

Featherstone, D. (2012). *Solidarity: Hidden Histories and Geographies of Internationalism.* London: Zed Books.

Ferretti, F. (2011). The correspondence between Élisée Reclus and Petör Kropotkin as a source for the history of geography. *Journal of Historical Geography* 37: 216–222.

Ferretti, F. (2013). "They have the right to throw us out": Élisée Reclus' new universal geography. *Antipode* 45: 1337–1355.

Foucault, M. (2002 [1972]). *The Archeology of Knowledge.* London: Routledge.

Foucault, M. (1990 [1978]). *The History of Sexuality: An Introduction.* New York: Vintage.

Foucault, M. (2003). *Society Must Be Defended: Lectures at the Collège de France, 1975–76.* New York: Picador.

Gibson-Graham, J.K. (2008). Diverse economies: Performative practices for "other worlds." *Progress in Human Geography* 32: 613–632.

Godlewska, A., & Smith, N. (eds) (1994). *Geography and Empire.* Oxford: Blackwell.

Goldman, E. (1969 [1917]). *Anarchism and Other Essays.* New York: Dover.

Gouldner, A.W. (1982). Marx's last battle: Bakunin and the First International. *Theory and Society* 11: 853–884.

Graeber, D. (2004). *Fragments of an Anarchist Anthropology.* Chicago, IL: Prickely Paradigm.

Graeber, D. (2009). *Direct Action: An Ethnography.* Oakland, CA: AK Press.

Gregory, D. (2004). *The Colonial Present: Afghanistan, Palestine, Iraq.* Malden, MA: Blackwell.

Hardt, M., & Negri, A. (2000). *Empire.* Cambridge, MA: Harvard University Press.

Hardt, M., & Negri, A. (2005). *Multitude: War and Democracy in the Age of Empire.* New York: Penguin.

Harvey, D. (1972). Revolutionary and counter revolutionary theory in geography and the problem of ghetto formation. *Antipode* 4: 1–13.

Harvey, D. (1973). *Social Justice and the City.* London: Edward Arnold.

Harvey, D. (2003). *The New Imperialism.* Oxford: Oxford University Press.

Harvey, D. (2007 [1982]) *Limits to Capital.* New York: Verso.

Hewitt, K. (2001). Between Pinochet and Kropotkin: State terror, human rights and the geographers. *Canadian Geographer* 45: 338–355.

Ince, A. (2010). Whither anarchist geography? In Jun, N., & Wahl, S. (eds), *New Perspectives on Anarchism.* Lanham, MD: Lexington, 281–302.

Ince, A. (2012). In the shell of the old: Anarchist geographies of territorialisation. *Antipode* 44: 1645–1666.

Johnston, R.J. (1981). British political geography since Mackinder: A critical review. In Burnet, A.D., & Taylor, P.J. (eds), *Political Studies from Spatial Perspectives*, Chichester: John Wiley & Sons, 11–32.

Johnston, R.J. (1982). *Geography and the State: An Essay in Political Geography*. London: Macmillan.

Johnston, R. (1984). Marxist political economy, the state and political geography. *Progress in Human Geography* 8(4): 473-492.

Katsiaficas, G. (1997). *The Subversion of Politics: European Autonomous Movements and the Decolonization of Everyday Life*. Atlantic Highlands, NJ: Humanities Press.

Kearns, G. (2004). The political pivot of geography. *Geographical Journal* 170: 337–346.

Kearns, G. (2009a). *Geopolitics and Empire: The Legacy of Halford Mackinder*. Oxford: Oxford University Press.

Kearns, G. (2009b). Kropotkin, P. In Kitchen, R., & Thrift, N. (editors-in-chief), *International Encyclopedia of Human Geography*. Amsterdam: Elsevier, 56–58.

Koopman, S. (2008). Imperialism within: Can the master's tools bring down empire? *ACME: An International E-Journal for Critical Geographies* 7: 283–307.

Koopman, S. (2011). Alter-geopolitics: Other securities are happening. *Geoforum* 42: 274–284.

Korf, B. (2006). Who is the rogue? Discourse, power and spatial politics in post-war Sri Lanka. *Political Geography* 25: 279–297.

Kropotkin, P. (2008 [1902]). *Mutual Aid: A Factor in Evolution*. Charleston, SC: Forgotten.

Kropotkin, P. (1995 [1908]). Modern science and anarchism. In Woodcock, G. (ed.), *Peter Kropotkin, Evolution and Environment*. Montreal: Black Rose, 15–110.

Kuus, M. (2002). Sovereignty for security? The discourse of sovereignty in Estonia. *Political Geography* 21: 393–412.

Lacoste, Y. (2000). Rivalries for territory. *Geopolitics* 5: 120–158.

Luke, T.W. (1996). Governmentality and contragovernmentality: Rethinking sovereignty and territoriality after the Cold War. *Political Geography* 15: 491–507.

Marks, B. (2012). Autonomist Marxist theory and practice in the current crisis. *ACME: An International E-Journal for Critical Geographies* 11: 467–491.

Muir, R. (1976). Political geography: Dead duck or phoenix? *Area* 8: 195–200.

Ó Tuathail, G. (1996a). An anti-geopolitical eye: Maggie O'Kane in Bosnia, 1992–93. *Gender, Place and Culture* 3: 171–186.

Ó Tuathail, G. (1996b). *Critical Geopolitics: The Politics of Writing Global Space*. Minneapolis, MN: University of Minnesota Press.

Ó Tuathail, G., & Dalby, S. (eds) (2002). *Rethinking Geopolitics*. London: Routledge.

Peet, R. (1975). Inequality and poverty: A Marxist-geographic theory. *Annals of the Association of American Geographers* 65: 564–571.

Peet, R. (1977). *Radical Geography: Alternative Viewpoints on Contemporary Social Issues*. Chicago, IL: Maaroufa Press.

Pelletier, P. (2009). *Élisée Reclus: Géographie et Anarchie*. Paris: Les éditions du monde libertaire.

Pickerill, J., & Chatterton, P. (2006). Notes towards autonomous geographies: Creation, resistance and self-management as survival tactics. *Progress in Human Geography* 30: 730–746.

Pred, A. (2000). *Even in Sweden: Racisms, Racialized Spaces and the Popular Geographical Imagination*. Berkeley, CA: University of California Press.

Proudhon, P.-J. (2008 [1840]). *What Is Property? An Inquiry into the Right and Principle of Government*. Charleston, SC: Forgotten.

Routledge, P. (1998). Anti-geopolitics: Introduction. In Ó Tuathail, G., Dalby, S., & Routledge, P. (eds), *The Geopolitics Reader*, London: Routledge, 245–255.

Said, E. (1993). *Culture and Imperialism*. New York: Knopf.

Said, E. (2003 [1978]). *Orientalism*, 25th anniversary edn. New York: Vintage Books.

Sharp, J. (2009). *Geographies of Postcolonialism: Spaces of Power and Representation*. London: Sage.

Sharp, J. (2013). Geopolitics at the margins? Reconsidering genealogies of critical geopolitics. *Political Geography* 37: 20–29.

Shaw, W.S., Herman, R.D.K., & Dobbs, G.R. (2006). Encountering indigeneity: Re-imagining and decolonizing geography. *Geografiska Annaler: Series B* 88: 267–276.

Sidaway, J.D. (2000). Postcolonial geographies: An exploratory essay. *Progress in Human Geography* 24: 591–612.

Sidaway, J.D. (2005). Empire's geographies. *ACME: An International E-Journal for Critical Geographies* 3: 63–78.

Slater, D. (2004). *Geopolitics and the Post-colonial: Rethinking North–South Relations*. Malden, MA: Blackwell.

Smith, L.T. (1999). *Decolonizing Methodologies: Research and Indigenous Peoples*. London: Zed Books.

Smith, N. (2010). The revolutionary imperative. *Antipode* 41(1): 50–65.

Sparke, M. (2005). *In the Space of Theory: Postfoundational Geographies of the Nation-State*. Minneapolis, MN: University of Minnesota Press.

Sparke, M.B. (2006). A neoliberal nexus: Economy, security and the biopolitics of citizenship on the border. *Political Geography* 25: 151–180.

Spivak, G.C. (1999). *A Critique of Postcolonial Reason: Toward a History of the Vanishing Present*. Cambridge, MA: Harvard University Press.

Springer, S. (2010). Neoliberal discursive formations: On the contours of subjectivation, good governance, and symbolic violence in posttransitional Cambodia. *Environment and Planning D: Society and Space* 28: 931–950.

Springer, S. (2011a). Public space as emancipation: Meditations on anarchism, radical democracy, neoliberalism and violence. *Antipode* 43: 525–562.

Springer, S. (2011b). Violence sits in places? Cultural practice, neoliberal rationalism, and virulent imaginative geographies. *Political Geography* 30: 90–98.

Springer, S. (2012a). Anarchism! What geography still ought to be. *Antipode* 44: 1605–1624.

Springer, S. (2012b). Neoliberalising violence: Of the exceptional and the exemplary in coalescing moments. *Area* 44: 136–143.

Springer, S. (2013a). Anarchism and geography: A brief genealogy of anarchist geographies. *Geography Compass* 7: 46–60.

Springer, S. (2013b). Violent accumulation: A postanarchist critique of property, dispossession, and the state of exception in neoliberalizing Cambodia. *Annals of the Association of American Geographers* 103: 608–626.

Springer, S. (2014a). Human geography without hierarchy. *Progress in Human Geography* 38(3): 402–419. doi:10.1177/0309132513508208

Springer, S. (2014b). War and pieces. *Space and Polity* 18(1): 85-96. doi:10.1080/13562576.2013.878430

Springer, S. (2014c). Why a radical geography must be anarchist. *Dialogues in Human Geography* 4(3): 249–270. doi:10.1177/2043820614540851

Springer, S., Ince, A., Pickerill, J., Brown, G., & Barker, A. (2012). Reanimating anarchist geographies: A new burst of colour. *Antipode* 44: 1591–1604.

Staeheli, L.A. (2011). Political geography: Where's citizenship? *Progress in Human Geography* 35: 393–400.

Stirner, M. (1993 [1845]) *The Ego and Its Own: The Case of the Individual Against Authority*. London: Rebel.

Taylor, P.J. (1985). *Political Geography: World-Economy, Nation-State and Locality*. London: Longman.

Taylor, P.J. (1987). The paradox of geographical scale in Marx's politics. *Antipode* 19: 287–306.

Taylor, P.J. (1991a). The crisis of the movements: The enabling state as quisling. *Antipode* 23: 214–228.

Taylor, P.J. (1991b). "The Geography of Freedom: The Odyssey of Élisée Reclus", Book Review. *Urban Studies* 28: 658–660.

Taylor, P.J. (1994). The state as container: Territoriality in the modern world-system. *Progress in Human Geography* 18: 151–162.

Taylor, P.J. (2003). Radical political geographies. In Agnew, J., Mitchell, K., & Toal, G. (eds), *A Companion to Political Geography*, Malden, MA: Blackwell, 47–58.

Wallerstein, I. (1974). *The Modern World-System*. New York: Academic Press.

Wallerstein, I. (1979). *The Capitalist World-Economy*. Cambridge: Cambridge University Press.

Ward, D. (2010). Alchemy in Clarens: Kropotkin and Reclus, 1977–1881. In Jun, N., & Wahl, S. (eds), *New Perspectives on Anarchism*, Lanham, MD: Lexington, 209–226.

Watts, M. (2004). Resource curse? Governmentality, oil and power in the Niger Delta, Nigeria. *Geopolitics* 9: 50–80.

White, R.J., & Williams, C.C. (2012). The pervasive nature of heterodox economic spaces at a time of neo-liberal crisis: Towards a "post-neoliberal" anarchist future. *Antipode* 44: 1625–1644.

Chapter 17

Geopolitics/Critical Geopolitics

Sami Moisio

University of Helsinki & Academy of Finland, Helsinki, Finland

The term geopolitics is often used by scholars, journalists, military strategists, and different kinds of management experts, both unintentionally and intentionally, in multiple ways and for various purposes. It is thus a contested term that has been given different meanings in different spatial and temporal contexts and within different kinds of networks of actors. In short, the term geopolitics "poses a question to us every time it is knowingly evoked and used" (Ó Tuathail 1996: 66).

Over the past two decades, the term geopolitics has, yet again, been expropriated for new usages and it seems pointless to try to capture its "correct" meaning (cf. Murphy et al. 2004). Two introductory points may be made, however. First, geopolitics is a scholarly practice, often regarded as a subfield of political geography, political science, and international relations. Geopolitics can also be understood as a continually remade scholarly field that is articulated in (text)books and articles through certain development trajectories and the ideas of key scholars (see, e.g., Parker 1998; Agnew 2003; Flint 2006; Dittmer & Sharp 2014a).

Secondly, geopolitics is a political practice that brings together power, place, "world," and subjects in unique combinations. From a historical perspective, geopolitics must be understood not only as an academic theorizing of politics, but also as the political action of all sorts of actors who have sought to mold political spaces. Moreover, geopolitical scholarship has not only informed political practices, it has also been affected by the world's political ruptures and the emergent and dominant political rationalities of a given time. Because social scientific knowledge production is, albeit often implicitly, tied up with political discourses and ideologies, it is difficult to draw a blunt line between academic geopolitics and geopolitical action.

This chapter scrutinizes the historical roots and recent developments of geopolitics. Throughout, the focus is not restricted to political geographical works that are self-consciously labeled as geopolitical (in the title or otherwise). What follows is a decidedly subjective mapping of trends and studies that have made a contribution to the field. The rest of the chapter

The Wiley Blackwell Companion to Political Geography, First Edition.
Edited by John Agnew, Virginie Mamadouh, Anna J. Secor, and Joanne Sharp.
© 2015 John Wiley & Sons Ltd. Published 2017 by John Wiley & Sons Ltd.

proceeds through five sections. The second section discusses classical geopolitics. Section three inquires into the foundations of critical geopolitics. Section four discusses recent topics in critical geopolitics. Section five suggests that bringing critical geopolitics closer to the cultural political economy would be worthwhile. The final section makes a few remarks on the impact and future of critical geopolitics.

Classical geopolitics

John Agnew (2003) suggests that a particular geopolitical visioning evolved in the sixteenth century and was associated with the age of exploration. The term geopolitics, however, was coined much later. In 1899, Swedish politician and political scientist Rudolf Kjellén used it to denote the territorial aspects of the state. The invention of the term nonetheless "marked only a terminological modification of an existing intellectual agenda, previously labeled 'political geography'" (Heffernan 2000: 28). Kjellén and his German colleague Friedrich Ratzel played a role in developing geopolitics as an organicist theory of the state. Within such a frame, the state is not only bound to earth but also determined by it. The origins of the term geopolitics were thus inspired by theories of evolution. A discernible socio-biological reasoning was not uncommon within European academic circles in the early twentieth century.

The second branch of classical geopolitics is often associated with British conservative politician and geographer Halford Mackinder and with American naval officer Alfred Thayer Mahan. Although Mackinder never used the term geopolitics, his theory on global competition between leading states (the so-called heartland theory) is a telling example of the early tenets of classical geopolitical reasoning (Kearns 2009). Mackinder's political map was constituted through spatial binaries such as sea power/land power and heartland/rimland, and by cultural distinctions between Western individualist philosophy and Eastern (European) communal traditions (Parker 1982). These binaries constituted a grand theory of world affairs. In such a view, geography in international affairs refers to a "big picture": timeless laws of nature, political geographical outcomes of technological developments, factors such as proximity or physical geographical landscapes, and the distribution of land and sea. This type of classical geopolitics seeks to understand forces emanating from both nature and technology, forces that shape international affairs. This kind of geopolitics, which conceives of physical geographical elements, as well as the role of technological development with regard to these elements, as enabling or constraining factors of state strategy is often dubbed geostrategy.

The classical geopolitical theories were of course products of their own time and reflect the competitive ambitions of European states in the early twentieth century (Ó Tuathail 1996; Parker 1998; Agnew 2003; Dittmer & Sharp 2014b), the anxieties over the balance of power in Europe, as well as the related need to enhance state security in the face of perceived external threats. For instance, Mackinder's attempt to develop the study of geography as an aid to statecraft was explicitly linked with these concerns (Parker 1982).

The two traditions discussed indicate that the origins of classical geopolitical scholarship in fin-de-siècle Europe, as well as its later developments in the twentieth century, are entangled with changing modes of political action and engagement with the state, nationalism, colonialism, and imperialism, as well as with justifications for particular interstate rivalry (Ó Tuathail 1996; Heffernan 2000; Agnew 2003; Moisio 2013). As a form of reasoning, classical geopolitics discloses some of the key political characteristics of the twentieth century: Command of territory and natural resources were understood as pivotal dimensions of interstate rivalry and fundamental constituents of state status.

After the Second World War, German geopolitical reasoning (*Geopolitik*), which was based on environmental determinism and thus treated the state as an organic entity, was associated with Adolf Hitler's thinking and politics in particular. Given this historical legacy, geopolitics became a notorious signifier denoting control over space as states vie for power. In the postwar context, geopolitics was a problematic legacy for the discipline of geography. The end of the war marked the demise of geopolitics among academic geographers. It was replaced by a seemingly neutral political geography.

The new political geography, which was developed during the Cold War by a handful of individuals, understood the spatiality of the world political map as a set of geographical factors, forces, and resulting capacities of state actors that could be analyzed by a trained scholar. This form of geopolitical reasoning, which was possible to conflate with balance-of-power thinking, became prominent particularly in the realist tradition of international relations scholarship. Saul Cohen's attempt to divide the world into geopolitical and geostrategic regions is illustrative of the new political geography (see Cohen 1963, 2009). His work "regionalizes" international affairs into geostrategic units and discusses the processes, forces, and substances that characterize these dynamic "regions." During the Cold War, the legacy of classical geopolitics lived on also in the various branches of strategic studies. Colin Gray's *The Geopolitics of Super Power* (1988) is exemplary in this context. It draws explicitly from Mackinder's original conceptual apparatus.

The political significance of the authors of classical geopolitics should not be exaggerated, of course. Agnew (2012) argues that the history of academic geography is marked by the lack of connection to state-building and political power. Accordingly, the discipline was "a marginalized field with limited influence beyond that exercised by a few individuals like Bowman in the United States" (2012: 322). Some of the ideas of classical geopolitics nevertheless had a presence in foreign and security policy practices during the Cold War.

Today, variants of classical geopolitics persist in the ways in which politicians, foreign and security policy experts, military strategists, scholars, and the general public make sense of international affairs (see, e.g., Megoran 2010). Consider, for example, the political crisis in Ukraine that broke out in 2014. On both sides of the Atlantic the crisis has been explained and thus made meaningful through classical geopolitical reasoning. To illustrate, Alexander Dugin, who is one of the most active Russian geopolitical debaters (see Ingram 2001) and whose ties with the Kremlin have been speculated on recently, provides a classical geopolitical reading of the crisis. Accordingly, the crisis is not about Ukraine per se; rather, it is about the new geopolitical order. In such reasoning, the US-dominated unipolar international order becomes challenged in Russian policies toward Ukraine.

Following some of the tenets of classical geopolitics, Dugin perceives Ukraine as a specific "battleground" in a wider geopolitical struggle for power. He legitimizes particular interstate rivalry through a script that essentializes two geopolitical cultures. In such a script, the crisis in Ukraine encloses and discloses geographically determined global contradictions between a liberal Anglo-Saxon civilization, which is premised on sea power and characterized by US hegemony, and a conservative continental civilization based on commanding territories. The latter civilization is represented by Russia. Dugin narrates the world political map as if it were composed of two "value systems." The maritime value tradition is premised on money and trade, whereas the continental tradition is rooted in traditional values associated with conservatism and heroism (Dugin, cited in Heiskanen 2014).

Critical geopolitics: Foundations

After being banished by geographers for decades, geopolitics experienced a revival in the late 1980s. Critical geopolitics, which challenged most of the tenets of classical geopolitics, was not the only approach that characterized this revival (for a good review, see Mamadouh 1998), but it was the most vocal. At the risk of oversimplifying matters slightly, this section maps out five of the founding characteristics and ideas of critical geopolitics.

First, the emergence of critical geopolitics was contextual and was influenced by the development of world political developments and the related rise of leftism in Anglo-American academia in the 1970s and 1980s (Agnew 2013). The structuralist openings in political geography in the 1970s and early 1980s – notably the world-systems approach developed in political geography, most forcefully by Peter J. Taylor – paved the way for critical geopolitics. Structuralism not only provided an example of the potential redeployment of geopolitics in a new fashion (Hepple 1986), it also formed a mirror against which agency-centered poststructuralist scholarship could be cultivated. Critical geopolitics emerged in the 1980s and early 1990s as part of the cultural/linguistic/constructivist turn in the social sciences and human geography. Critical geopolitics thus highlights agency (the capacity to act politically) and the constitutive role of language in geopolitical practice, and conceptualizes geopolitics as culturally embedded spatial practices – both representational and material – of statecraft. Geopolitics is thus understood as a "set of socially constructed, rather than naturally given, practices and ideas through which the international political economy is realized geographically" (Agnew & Corbridge 1995: 4–5).

The origins of critical geopolitics can be traced back to the work of John Agnew, Simon Dalby, and Gearóid Ó Tuathail in particular. There are not only interesting postpositivist intertextualities but also theoretical tensions between the approaches of these scholars – particularly with regard to the central issue of the materiality/textuality of geopolitical discourse and agency. As Agnew (2013: 24) has himself noted in retrospect, this tension "came down to a question of privileging the representational versus emphasizing the socio-economic resources that made some representations more powerful than others." The early tensions in critical geopolitics thus stem from the difference between asking "how" questions and "why" questions in social scientific inquiry.

Secondly, early critical geopolitics builds on postpositivist and postmodern writings whose main goal is epistemological critique. It reconceptualizes geopolitics through drawing from debates on identity and difference, feminism, positionality, and postcolonialism. In such a view, critical geopolitics demonstrates that geography matters in world politics, but not quite in the classical geopolitical sense. Critical geopolitics is not about purported natural forces or constraining/enabling geographical factors in politics. Rather, critical geopolitics understands geography in international politics as socially constructed and bound to larger questions regarding identity/difference, for instance. In order to explicate this form of geography, critical geopolitics extends socio spatial theorizing into the study of international affairs. It inquires into the spatial theories and assumptions that guide action in international affairs. One of the founding ideas of critical geopolitics is thus to problematize the seemingly neutral and objective scientific gaze of classical or "conventional" realist geopolitics through deconstruction. It disturbs the putatively scientific "view from nowhere" by arguing that classical geopolitics as a science that emerged in the late nineteenth century is a significant constituent of the modernist perception of international space made visible by geopolitical experts, as "detached viewing subjects" surveying a worldwide stage from above (Ó Tuathail 1996: 29). In so doing, critical geopolitics argues that the spaces of world politics are "produced" in representations and practices, and thus that this socially constructed spatiality of international affairs becomes visible in different mapping exercises, such as all sorts of strategic plans and related imaginations. Critical geopolitics thus claims that an inquiry into the spatiality of international affairs must be pivotal to the study of politics (Kuus 2010: 683).

Critical geopolitics' emphasis on international affairs is a notable phenomenon. It is nonetheless understandable given that the emergence of critical geopolitics cannot be detached from so-called dissident international relations (IR). The key dissidents introduced poststructuralism into IR in the 1980s and challenged the foundational ideas of neorealism in particular. In so doing, they developed a critique of modernism in IR theory. They sought to unsettle state-related practices through which an "uncertain" world was constantly territorialized (ordered) by the "sovereign man" employing putatively rational and scientific methods (see Ashley & Walker 1990). For them, statecraft was a specific action

> inscribing problems and dangers that can be taken to be exterior to sovereign man and whose exteriority serves to enframe the "domestic population" in which the state can be recognized as a center and can secure its claims to legitimacy. (Ashley 1989: 302)

This approach highlights the socially constructed nature of international affairs and the importance of the practices of statecraft in this process. Critical geopolitics draws on this idea of statecraft. Ó Tuathail and Dalby (1998: 3) propose that "particular cultural mythologies of the state" must be taken into account in critical geopolitical study and that geopolitics is not a distinct school of statecraft, but should rather be understood as the spatial practices (material, representational) of statecraft itself.

The "dissident effect" resulted in growing awareness of the intertwined nature of foreign/ security policy, power, and space. What followed was a set of seminal critical geopolitical investigations on the ways in which the maps of global politics are carved up and simultaneously made public by different kinds of intellectuals of statecraft such as political leaders, pundits, advocates, and advisers (Dalby 1990a; Ó Tuathail 1992). Here, geopolitics is not understood as a somewhat contained action by scholars who explicitly speak the language of classical geopolitics. Rather, it elaborates on how international politics is continually "spatialized" by powerful political actors who depict it as a "world" characterized by specific places, people, and events (Ó Tuathail & Agnew 1992: 192). The representations put together by the powerful influence the ways in which the world is understood, and these understandings, in turn, have effects on subsequent policy practices (Sharp 2008: 191). The emphasis is thus turned more generally to the politics of the spatial specification of politics (Dalby 1991).

Thirdly, critical geopolitics conceptualizes geopolitics as a discursive practice (for comments and critique, see Müller 2008; Müller & Reuber 2008). It is a broad social and cultural phenomenon, so that there is always a particular relationship between geopolitical discourse/text/representation and practice/action (Agnew 2013: 28). All analyses of international politics make spatial assumptions and are necessarily predicated on particular geographical assumptions. Critical geopolitics brings these up and puts them under scrutiny (Kuus 2010) by examining discursive operations, the associated "geopolitical truths" as well as social mechanisms through which geopolitical truth claims and larger geopolitical regimes of truth are produced. Here, critical geopolitics draws on Foucault (1980a), who conceptualized discourse as referring to power-driven knowledge, certain accepted truths that are historically constructed and together form larger regimes of truth. The idea is to trace the function of particular geopolitical expertise and related "geopolicy-relevant" knowledge in international affairs. One of the foundational ideas of a more materially grounded critical geopolitics is, in turn, to examine the conjunction of historically contingent, material geopolitical orders and associated geopolitical discourses (see Agnew & Corbridge 1995). From this angle, even if historically and geographically evolved social practices have an existence outside practical geopolitical discourses, geopolitics must be understood as a discursive practice, too (Agnew 2013).

What follows from the poststructural underpinnings of critical geopolitics is that it does not treat geography as an innocent setting on which international politics is conducted or a structural element that sets the limits for rational policy-making. Critical geopolitics argues that geography and politics are inescapably co-constituted, and that space is power. In such a conceptualization, geography must be understood as earth-writing, the active production of "global political space" by institutionally bounded intellectuals (Ó Tuathail 1996: 185; see also Dodds 1994). This founding idea of critical geopolitics is utilized in a range of studies, which, for instance, encompass the power of the geopolitical discourses of "development" in places like Africa (e.g., Mercer, Mohan, & Power 2003).

Fourthly, geopolitics can be understood as a set of traveling theories that have been introduced and negotiated in different intellectual, spatial, and temporal contexts (Dodds & Atkinson 2002) and with different kinds of consequences. Inquiries into different (national) geopolitical traditions/cultures (for a recent example, see Guzzini 2013) and thinkers typically result from the early critical geopolitical problematizations of classical geopolitical knowledge production. However, more recently critical geopolitics has succinctly demonstrated how, as interpretive cultural practices, widely traveled geopolitical theories (such as the Huntingtonian "clash of civilizations") and powerful geopolitical markers (such as East and West) are constituted and played out in an interplay between the universalizing "geopolitical centers" where theories and markers are invented and the margins where they are consumed and appropriated for particular purposes by local intellectuals of statecraft (Kuus 2007).

Fifthly, critical geopolitics scrutinizes national identity-building and the associated spatial practices of politico-spatial inclusion and exclusion. It points out that political subjectivities, identities, and objects are formed in geopolitical action and within discourses. Following dissident IR, national identities are conceptualized as constituted through powerful narratives of national security, a central constituent of statecraft and sovereign man (see Ashley 1989; Ashley & Walker 1990; Dalby 1990a; Campbell 1992). By following this line of thinking, Dalby (1990b) argues that geopolitics is an inherently ideological process of constructing borders to demarcate the domestic space (here) from threats outside the state, as well as to exclude the threatening "other" (there) and the "otherness" that it represents spatially as well as culturally.

The elaboration of national identity-building (and national interest-building as well) not only resonated with the ideas of those who sought to push border studies in new directions (e.g., Paasi 1996), but also expanded critical geopolitics to diverse geographical contexts (see, e.g., Megoran 2006; Bialasiewicz & Minca 2010). Since the 1990s, "border" has thus become one of the key concepts and empirical subjects in critical geopolitics. The important idea here is to investigate how national identities are narrated and performed in various bordering practices, ranging from the upper echelons of state authority to more mundane events and related "nationalizing" symbols and rituals (see, e.g., Paasi 2013). This does not entail, however, that geopolitical identities can be equated only with seemingly fixed or stable "national" ones (see Herb & Kaplan 1999).

The "Self–Other thematic" in critical geopolitics inspired the development of popular geopolitics, which studies the circulation of geopolitical narratives and the constitution of geopolitical discourses in popular culture. It argues that instead of paying attention only to the formal and practical geopolitical reasoning of the intellectuals of statecraft (for these terms, see Ó Tuathail & Dalby 1998), attention must be devoted to visual and non-visual products such as newspapers and magazines (Sharp 2000), radio shows (Dodds & Pinkerton 2009), films (Carter & Dodds 2011), art (Ingram 2009), comic strips (Dittmer 2013), and cartoons (Dodds 1996; Ridanpää 2009). In the conventional analysis these representations are understood as belonging to the sphere of "low politics," but according to popular geopolitics they are pivotal in enacting and circulating particular popular geopolitics among different "audiences" (Dittmer 2010; see also Holland 2012). Focusing on audiences is important, for it highlights the dual role of audiences as both consumers as well as producers of geopolitical discourses. Through an investigation of less formal arenas of geopolitics, scholars of popular geopolitics have more generally taken the field closer to the geopolitics of mundane and associated identity politics (Sharp 2008).

Critical geopolitics: An unfinished business

During the past two decades critical geopolitics has become a vibrant scholarly field. As a sign of this vibrancy, some of the foundations of critical geopolitics have been critically discussed. To illustrate, the pioneering text by Ó Tuathail (1996) has been criticized for itself representing a perspective of a detached subject (see Smith 2000) and for neglecting the ontological dimension of geopolitics (Agnew 2000). Conceptual inconsistencies and obscurities regarding the important theoretical issues of discourse and agency/actor have also been taken up by those researchers critiquing an epistemology-centered critical geopolitics (e.g., Albert, Reuber, & Wolkersdorfer 2014).

Feminist scholars, in turn, have suggested that the early critical geopolitics not only remasculinizes geopolitics (Sharp 2000), but also neglects the important issue of subjectification inherent in geopolitical practices (Dowler & Sharp 2001). Feminists have thus pointed out that critical geopolitics focuses too much on text and representations and pays too little attention to embodied practices and to the materiality of geopolitical discourses. Accordingly, "its deconstructive impulses are insufficient to generate change for building alternative futures" (Hyndman 2001: 213). Feminist geopolitics has continued to locate human agency in geopolitics and to study related "embodiment" and the role and construction of masculinity in political space-making. It "destabilizes" the state-centered view on geopolitics by scrutinizing marginal voices, by discussing "informal arenas" (the private) of politics, and by bringing the mundane aspects of the geopolitics of the local into closer investigation (e.g., Secor 2001).

These critiques, together with the founding ideas, inform contemporary critical geopolitics (see, e.g., Coleman 2005; Bialasiewicz 2008, 2011; Bachmann & Sidaway 2009; Dittmer et al. 2011; Morrissey 2011). The analysis of how spatial assumptions about the world "figure into the calculi" of political elites and inquiry into the related imaginaries that animate popular "worldviews" have remained pivotal in critical geopolitical studies (Agnew 2013: 22). Critical geopolitics nonetheless remains an unfinished business in which some themes such as immigration, borders, "development," war, the environment, and associated practices of security are thematically central (for a thorough review of current developments, see Dodds, Kuus, & Sharp 2013).

The war–environment–security nexus is one of the classical topics of geopolitics and it has also been scrutinized fruitfully from critical perspectives (e.g., Le Billon 2008). The critical geopolitics of the environment is perhaps the most "policy-relevant" subfield in critical geopolitics. This work challenges some of the basic forms of security thinking, which are based on simplistic assumptions about the relationship between environmental change and conflicts. Simon Dalby's (2013) plea to radically rethink the basic geopolitical premises in security thinking in light of the insights from earth system science is exemplary here. His concern is that climate security is too often understood as a matter of environmental change causing political difficulties, which are in the end reasoned through the lenses of traditional geopolitics. According to Dalby, this view should be turned upside down. The issue of contemporary environmental security should be understood as a matter of today's political difficulties causing accelerating climate change. This new understanding forces one to develop a fresh type of geopolitical reasoning with regard to both environment and security.

The geopolitics of the world after 9/11 and the related territorial and relational spaces of "security and insecurity" and their implications for issues such as human rights, public space, and state sovereignty (Elden 2009; Ingram & Dodds 2009; Mountz 2013; on sovereignty see also Steinberg, Nyman, & Caraccioli 2012) have been examined conceptually as well as empirically. It is also noteworthy that critical scholars are currently opening up new spaces to examine the geographies of peace, a topic that is too often left unnoticed in geopolitical scholarship (see Megoran, McConnell, & Williams 2014).

One of the interesting branches in the critical geopolitical work on war touches on new technologies of warfare and the ways in which these technologies open up new social/cultural spaces within the military, form subjectivities and identities, as well as constitute relational battlefield geographies. Ian Shaw's (2013) work on "the dronified US security strategy" and the broader exercise of biopolitical power in the widening "battlespace" exemplifies the ongoing interest in what Stephen Graham (2004) calls vertical geopolitics. Here, critical geopolitics not only interrogates foreign and security policy discourses, it also inquires into technologies, knowledge, practices, and strategies "arranged around the deployment of drones for targeted killings" (Shaw 2013: 540).

Derek Gregory (2013) has examined "drone geographies" by elaborating on the practices of killing with Predators and Reapers by the US Air Force in Afghanistan and Iraq. Through a creative reading of the many socio-spatial dimensions of the relationship between the different military personnel and people who live under the surveillance and actions of these military personnel, Gregory (2013) argues that drone operations are characterized by "remote splits" and "individuation of killing." Drone warfare is waging war that is simultaneously distant/impersonal yet intimate/personal for the various operators of these drones. Moreover, the use of drones is predicated on the "sanitized battlefield," which includes a subjectification through which the target is constituted as a particular type of individual (criminal) whose spatiality is both local and global. Through an examination of drone geographies, Gregory problematizes many key issues and constitutive elements of modern warfare and military violence. One of the messages is that not only is geographical knowledge "used" in warfare, the deployment of drones for targeted killings is also constitutive of the new (cultural) geographies of war, power, and the everyday.

[I]t is precisely the ways in which armed drones – their technologies, visualities and dispositions – have become part of everyday life that needs the closest scrutiny … Drones have undoubtedly made a difference to the conduct of later modern war – and, in the case of targeted killings, to its transformation into something else altogether – but their use cannot be severed from the matrix of military and paramilitary violence of which they are but a part. (Gregory 2013: 16)

"Drone geopolitics" also touches on the ways in which war and related killing are structured, legitimized, and rationalized in "homelands" and beyond, as well as how they are problematized in the context of military personnel located in American suburbia far away from the battlefield, and those often most vulnerable and defenseless populations who live under the not-so-distant eye of the new war machines.

Dodds (2001) lamented more than a decade ago that critical geopolitics is methodologically disappointing. In particular, he pointed to the lack of careful ethnographic approaches as to how geopolicies emerge in the interaction between institutions, individuals, and textual as well as bodily practices (see also Paasi 2000). A particular geopolitical study of spatial expertise, to mention just one example, has taken this critique seriously (also Megoran 2006). It draws from an idea that geopolicy should not be understood as a ready-made "blueprint," but rather as a dynamic social process, and that approaching this process requires one to draw on ethnographic approaches. Merje Kuus (2014) has developed this perspective in the context of the external relations of the European Union (EU). Her work examines how particular policies get shaped within the EU bureaucracy and how the constitution of these policies involves a particular geopolitical expertise that is negotiated among different kinds of officials who deal with the EU's external affairs. Based on a set of interviews among "Eurocrats," Kuus examines the transnational diplomatic practices by which actors in the European quarter in Brussels communicate particular geopolitical imaginations.

These themes exemplify critical geopolitical scholarship, but the list is not all-encompassing, of course. A quick glance at the pages of the journal *Geopolitics* reveals that the geographical scope of critical geopolitics has been expanding, too. This is significant, given that in the beginning critical geopolitics was to a great extent an Anglo-American endeavor that was eager to scrutinize US foreign policies. Even if it can be argued that some particularly prominent thinkers and associated critical engagements with geopolitics have been overlooked in Anglophone critical geopolitics (see Sharp 2013), one also notices that critical geopolitics is today practiced in a number of contexts outside the Anglophone sphere.

Important also is that scholars have opened up new theoretical spaces by taking into account the affective, emotional, and experiential aspects of geopolitical practices and discourses (Pain 2009), as well as by suggesting that concepts such as assemblage and complexity would provide new openings for critical geopolitical inquiry (Dittmer 2014). More generally, the flirtation with the language of non-representational theory almost inescapably leads to a critique of some of the founding ideas of critical geopolitics. Be that as it may, the flexibility of the critical geopolitics banner, and the lack of attempts to define what precisely is meant by the signifier "critical," is perhaps its tremendous strength.

On the interface of critical geopolitics and cultural political economy

The discussion suggests that bringing critical geopolitical scholarship closer to the cultural political economy (see, e.g., Jones 2008) would be worthwhile. One of the many possible ways of doing this is to focus on the knowledge-based society as a new form of geopolitical governance. One may elaborate the ways in which the state in particular becomes reconstituted as a new kind of geopolitical persona through the discourses of the knowledge-based economy; a persona whose authority is shared between private and public agencies (cf. Cowen & Smith

2009). The knowledge-based society concerns the reterritorializing of certain key assumptions of the deterritorializing global knowledge-based economy, giving rise to a specific understanding of society. The knowledge-based society is thus made possible by the "first concerted global discursive operation of the cultural circuit of capital which involved attempts to describe itself to the world (as the 'knowledge economy')" (Thrift 2005: 12).

The constitution and specification of the state as a knowledge-based society through the deployment of reterritorializing logics form a geopolitical act, which involves specific kinds of spatial practices, both material and representational. The knowledge-based society can be considered as an ensemble of material and discursive elements through which power is exercised (cf. Foucault 1980a: 196). This ensemble thus touches on the spatial structures of the state as well as social and political relations within and beyond state borders.

An inquiry into the functioning of the knowledge-based society ensemble – into how it translates the purportedly fluid and borderless knowledge-based economy into a societal discourse at large – may be elucidated and examined through four imbricated elements. First, imaginaries of the state as the knowledge-based society are entangled with new technical systems that constitute "virtual spaces of comparison" (Larner & Le Heron 2002). If during the Cold War the strength of the state in interstate competition was measured in terms of natural resources, population, "national homogeneity," steel production capacity, and the size of particular industries such as machinery, the world of knowledge-based societies is also pregnant with ways of measuring state capacities. The miscellaneous competitiveness, creativity, innovativeness, and productivity indices and associated statistical tables, diagrams, and charts should be regarded as important geopolitical techniques in the context of the knowledge-based society ensemble. These techniques not only enable interspatial comparison and contrast, they also make the key objects that constitute the knowledge-based society "real." Through these techniques, the state as the knowledge-based society becomes an object that can be known and acted on (cf. Moisio & Luukkonen 2014).

Secondly, one may explicate the ways in which the relational spatial imaginaries – predicated for instance on the view that (global) city-regions are becoming more pivotal than territorial states in the contemporary capitalist world – are valorized in socio-spatial formations of the knowledge-based society. To exemplify, the discourses of the knowledge-based society are almost invariably clustered around the geopolitical imaginary of a network. Here, the network is one of the many discursive geopolitical productions of the world, the operation of which has effects and counter-effects (see Barry 1996; Walters 2012). In the contexts of the knowledge-based society, a network is a normative idea with the aim of organizing the state through flows and nodes.

The knowledge-based society is closely tied up with emerging forms of global governance that operate through all sorts of spatial exceptions and that increase spatial differentiation (Moisio & Paasi 2013). There is already a rich body of geopolitical work that deals with different kinds of spaces of exception (see, e.g., Coleman 2009; Giaccaria & Minca 2011; Martin 2012), certain economized political spaces that "escape the grip of national-state territoriality" and denationalize territories (Sassen 2013: 28). These seem to be constitutive of the knowledge-based society. Particular zoning strategies that graduate state sovereignty (Ong 2006) seem to characterize the spread of neoliberal governance more generally (Kangas 2013).

Thirdly, the geopolitics of the knowledge-based society ensemble interrogates governmental interventions of the state that aim at regulating both the population and spatial forms of the state (cf. Legg 2005). The knowledge-based society is predicated on reasoning that a sort of "war over talent" exists at the heart of interspatial competition and that, in order to cope with such competition, there is a need to give birth to a new citizen with particular skills (see Mitchell 2003). The ensemble of the knowledge-based society thus makes the "skill problem appear like a society-wide 'we' concern" (Jones 2008: 391). The practices of the knowledge-based society are thus "subjectivating by invoking and legitimizing certain images of the self while excluding others" (Bröckling, Krasmann, & Lemke 2011: 12; cf. Foucault 1980b).

Fourthly, the knowledge-based society is geopolitically constituted through particular circuits of knowledge. It would thus be crucial to examine the geopolitical imaginaries of the knowledge-based society as linked to the "cultural circuits of capitalism" (see Thrift 2005) and as produced through what Roberts (2003) calls "global strategic visions." These visions are produced and mobilized by prestigious business schools, think tanks, lobbying organizations, consultants, and management gurus "who reflexively circulate ideas and ideologies on capitalism *for itself*" (Jones 2008: 382) and they should be critically interrogated from a geopolitical perspective.

Conclusion

Critical geopolitics brings together scholars who employ different kinds of theoretical and philosophical perspectives and research materials to study the politics and governance of space within and beyond the state. It continues to examine the ways in which the political world we inhabit – its places, subjects, and dramas – are spatially constituted but not necessarily territorially demarcated in the age of neoliberalism (see Roberts, Secor, & Sparke 2003; Springer 2013). In short, critical geopolitics is a vibrant field in political geography that attracts students both within and beyond Anglo-American academia. Critical geopolitics has made a contribution to human geography, and users of the concept of geopolitics no longer have to worry about someone using the subject's past against them within academia (cf. Hepple 1986). It is easy to agree with Agnew's statement that "[t]here will be a need for critical geopolitics of some sort as long as the established nostrums of how the world is spatialized and divided up for political purposes remain in effect" (Agnew 2013: 30).

The contribution of critical geopolitics within academia seems more significant than its impact on geopolicy formation. This is perhaps due to the fact that scholars of critical geopolitics have been more willing to deconstruct political processes than to imagine and propose alternative strategies (see also Dodds 2001: 475). It must be noted, however, that an explicitly progressive tone seeking to develop a kind of "anti-intellectual poison" is clearly visible, for instance, in Simon Dalby's early works (1990a, 1991). He insisted on a transformation to a more peaceful and democratic form of geopolitical action (more recently, see Megoran 2011), but arguably, critical geopolitics has never represented any sort of political singularity.

It is very likely that critical geopolitics will continue to be expanded both geographically and theoretically. As regards the latter, new political philosophers and spatial theorists will be brought into a dialogue with the founding ideas of critical geopolitics. Arguments have been aired suggesting that it is problematic theoretically and otherwise that geopolitics is often made a synonym for "global politics." Elden (2013a) has suggested that geopolitics could be rethought through an explication of "dimensions of territory," and that the attention of geopolitical scholarship should be refocused from "global politics" to a "politics of the earth" and should

> take into account the power of natural processes, or resources; the dynamics of human and environment; the interrelation of objects outside of human intervention; the relation between the biosphere, atmosphere and lithosphere; and the complex interrelations that produce, continually

transform and rework the question of territory and state spatial strategies. This geopolitics would sit alongside, rather than replace, the attention given to biopolitics in recent years. (Elden 2013b)

These and other suggestions point to the fact that there is always room for reconceptualization in critical geopolitical scholarship.

Acknowledgments

The author would like to thank Anni Kangas, Anssi Paasi, Veit Bachmann, and Jo Sharp for their valuable comments on an earlier version of this chapter.

References

Agnew, J.A. (2000). Global political geography beyond geopolitics. *International Studies Review* 2(1): 91–99.

Agnew, J.A. (2003). *Geopolitics: Revisioning World Politics*, 2nd edn. London: Routledge.

Agnew, J.A. (2012). Of canons and fanons. *Dialogues in Human Geography* 2: 321–323.

Agnew, J.A. (2013). The origins of critical feopolitics. In Dodds, K., Kuus, M., & Sharp, J. (eds), *The Ashgate Research Companion to Critical Geopolitics*, Farnham: Ashgate, 19–32.

Agnew, J.A., & Corbridge, S. (1995). *Mastering Space*. London: Routledge.

Albert, M., Reuber, P., & Wolkersdorfer, G. (2014). Critical geopolitics. In Schieder, S., & Spondler, M. (eds), *Theories of International Relations*, London: Routledge.

Ashley, R.K. (1989). Living on border lines: Man, poststructuralism and war. In Der Derian, J., & Shapiro, M. (eds), *International/Intertextual Relations*, New York: Lexington Books, 259–322.

Ashley, R.K., & Walker, R.B.J. (1990). Speaking the language of exile: Dissident thought in international studies. *International Studies Quarterly* 34(3): 259–268.

Bachmann, V., & Sidaway, J.D. (2009). Zivilmacht Europa: A critical geopolitics of the European Union as a global power. *Transactions of the Institute of British Geographers* 34(1): 94–109.

Barry, A. (1996). The European network. *New Formations* 29: 26–37.

Bialasiewicz, L. (2008). The uncertain state(s) of Europe. *European Urban and Regional Studies* 15(1): 71–82.

Bialasiewicz, L. (ed.) (2011). *Europe in the World: EU Geopolitics and the Making of European Space*. Farnham: Ashgate.

Bialasiewicz, L., & Minca, C. (2010). The "border within": Inhabiting the border in Trieste. *Environment and Planning D* 28(6): 1084–1105.

Bröckling, U., Krasmann, S., & Lemke, T. (2011). From Foucault's lectures at Collège de France to studies of governmentality. In Bröckling, U., Krasmann, S., & Lemke, T. (eds), *Governmentality: Current Issues and Future Challenges*, London: Routledge, 1–32.

Campbell, D. (1992). *Writing Security*. Minneapolis, MN: University of Minnesota Press.

Carter, S., & Dodds, K. (2011). Hollywood and the "War on Terror": Genre-geopolitics and "Jacksonianism" in The Kingdom. *Environment and Planning D* 29(1): 98–113.

Cohen, S.B. (1963). *Geography and Politics in a World Divided*. New York: Random House.

Cohen, S.B. (2009). *Geopolitics: The Geography of International Relations*, 2nd edn. Lanham, MD: Rowman and Littlefield.

Coleman, M. (2005). US statecraft and the US–Mexico border as security–economy nexus. *Political Geography* 24(2): 185–209.

Coleman, M. (2009). What counts as geopolitics, and where? Devolution and immigrant insecurity after 9/11. *Annals of the Association of American Geographers* 99(5): 904–913.

Cowen, D., & Smith, N. (2009). After geopolitics? From the geopolitical social to geoeconomics. *Antipode* 41(1): 22–48.

Dalby, S. (1990a). *Creating the Second Cold War*. London: Pinter.

Dalby, S. (1990b). American security discourse: The persistence of geopolitics. *Political Geography Quarterly* 9(2): 171–188.

Dalby, S. (1991). Critical geopolitics: Discourse, difference, and dissent. *Environment and Planning D* 9(3): 261–283.

Dalby, S. (2013). Rethinking geopolitics: Climate security in the Anthropocene. *Global Policy* 5(1): 1–9.

Dittmer, J. (2010). *Popular Culture, Geopolitics, and Identity*. Lanham, MD: Rowman and Littlefield.

Dittmer, J. (2013). *Captain America and the Nationalist Superhero*. Philadelphia, PA: Temple University Press.

Dittmer, J. (2014). Geopolitical assemblages and complexity. *Progress in Human Geography* doi:10.1177/0309132513501405

Dittmer, J., & Sharp, J. (eds) (2014a). *Geopolitics: An Introductory Reader*. Abingdon: Routledge.

Dittmer, J., & Sharp, J. (2014b). General introduction. In Dittmer, J., & Sharp, J. (eds), *Geopolitics: An Introductory Reader*, Abingdon: Routledge, 1–10.

Dittmer, J., Moisio, S., Ingram, A., & Dodds, K. (2011). Have you heard the one about the disappearing ice: Recasting Arctic geopolitics. *Political Geography* 30(4): 202–214.

Dodds, K. (1994). Geopolitics and foreign policy: Recent developments in Anglo-American political geography and international relations. *Progress in Human Geography* 18(2): 186–208.

Dodds, K. (1996). The 1982 Falklands War and a critical geopolitical eye: Steve Bell and the If… cartoons. *Political Geography* 15(6–7): 571–592.

Dodds, K. (2001). Political geography III: Critical geopolitics after ten years. *Progress in Human Geography* 25(3): 469–484.

Dodds, K., & Atkinson, D. (2002). Introduction: A century of geopolitical thought. In Dodds, K., & Atkinson, D. (eds), *Geopolitical Traditions: A Century of Geopolitical Thought*, London: Routledge, 1–24.

Dodds, K., & Pinkerton, A. (2009). Radio geopolitics: Broadcasting, listening and the struggle for acoustic spaces. *Progress in Human Geography* 33(1): 10–27.

Dodds, K., Kuus, M., & Sharp, J. (eds) (2013). *The Ashgate Research Companion to Critical Geopolitics*. Farnham: Ashgate.

Dowler, L., & Sharp, J. (2001). A feminist geopolitics. *Space and Polity* 5(3): 165–176.

Elden, S. (2009). *Terror and Territory: The Spatial Extent of Sovereignty*. Minnesota, MN: Minnesota University Press.

Elden, S. (2013a). Secure the volume: Vertical geopolitics and the depth of power. *Political Geography* 34: 35–51.

Elden, S. (2013b). Geo-metrics abstract. *Progressive Geographies*. http://progressivegeographies.com/2013/07/15/geo-metrics-abstract/, accessed January 10, 2014.

Flint, C. (2006). *Introduction to Geopolitics*. London: Routledge.

Foucault, M. (1980a). *Power/Knowledge: Selected Interviews and Other Writings 1972–1977*. Brighton: Harvester Press.

Foucault, M. (1980b). *The History of Sexuality. Volume 1: The Introduction*. Brighton: Harvester Press.

Giaccaria, P., & Minca, C. (2011). Topographies/topologies of the camp: Auschwitz as a spatial threshold. *Political Geography* 30(1): 3–12.

Graham, S. (2004). Vertical geopolitics: Baghdad and after. *Antipode* 36(1): 12–23.

Gray, C. (1988). *The Geopolitics of Super Power*. Lexington, KT: University of Kentucky Press.

Gregory, D. (2013). Drone geographies. *Radical Philosophy* 183: 7–19.

Guzzini, S. (ed.) (2013). *The Return of Geopolitics in Europe? Social Mechanisms and Foreign Policy Identity Crises*. Cambridge: Cambridge University Press.

Heffernan, M. (2000). Fin de siècle, fin du monde? On the origins of European geopolitics. In Dodds, J., & Atkinson, D. (eds), *Geopolitical Traditions: A Century of Geopolitical Thought*, London: Routledge, 27–51.

Heiskanen, H. (2014). Russian philosopher: The new world order is being decided in Ukraine. *YLE*. http://yle.fi/uutiset/venalaisfilosofi_ukrainassa_ratkotaan_maailmanjarjestysta/7248248, accessed May 18, 2014.

Hepple, L.W. (1986). The revival of geopolitics. *Political Geography Quarterly* 5(4): 521–536.

Herb, G., & Kaplan, D. (eds) (1999). *Nested Identities: Nationalism, Territory, and Scale*. Lanham, MD: Rowman and Littlefield.

Holland, E.C. (2012). "To think and imagine and see differently": Popular geopolitics, graphic narrative, and Joe Sacco's "Chechen War, Chechen Women." *Geopolitics* 17(1): 105–129.

Hyndman, J. (2001). Towards a feminist geopolitics. *Canadian Geographer* 45(2): 210–222.

Ingram, A. (2001). Alexander Dugin: Geopolitics and neo-fascism in post-Soviet Russia. *Political Geography* 20(8): 1029–1051.

Ingram, A. (2009). Art and the geopolitical: Remapping security at green zone/red zone. In Ingram, A., & Dodds, K. (eds), *Spaces of Security and Insecurity: Geographies of the War on Terror*, Farnham: Ashgate, 257–277.

Ingram, A., & Dodds, K. (eds) (2009). *Spaces of Security and Insecurity: Geographies of the War on Terror*. Farnham: Ashgate.

Jones, M. (2008). Recovering a sense of political economy. *Political Geography* 27(4): 377–399.

Kangas, A. (2013). Governmentalities of Big Moscow: Particularizing neoliberal statecraft. *Geopolitics* 18(2): 299–314.

Kearns, G. (2009). *Geopolitics and Empire: The Legacy of Halford Mackinder*. Oxford: Oxford University Press.

Kuus, M. (2007). *Geopolitics Re-framed: Security and Identity in Europe's Eastern Enlargement*. New York: Palgrave Macmillan.

Kuus, M. (2010). Critical geopolitics. In Denemark, R. (ed.), *The International Studies Encyclopedia Volume II*, Oxford: Blackwell, 683–701.

Kuus, M. (2014). *Geopolitics and Expertise: Knowledge and Authority in European Diplomacy*. Oxford: Wiley-Blackwell.

Larner, W., & Le Heron, P. (2002). The spaces and subjects of a globalizing economy: A situated exploration method. *Environment and Planning D* 20: 753–774.

Le Billon, P. (2008). Diamond wars? Conflict diamonds and geographies of resource wars. *Annals of the Association of American Geographers* 98(2): 345–372.

Legg, S. (2005). Foucault's population geographies: Classifications, biopolitics and governmental spaces. *Population, Space and Place* 11(3): 137–156.

Mamadouh, V. (1998). Geopolitics in the nineties: One flag, many meanings. *GeoJournal* 46: 237–256.

Martin, L. (2012). "Catch and remove": Detention, deterrence, and discipline in US noncitizen family detention practice. *Geopolitics* 17(2): 312–334.

Megoran, N. (2006). For ethnography in political geography: Experiencing and re-imagining Ferghana Valley boundary closures. *Political Geography* 25(6): 622–640.

Megoran, N. (2010). Neoclassical geopolitics. *Political Geography* 29(4): 187–189.

Megoran, N. (2011). War and peace? An agenda for peace research and practice in geography. *Political Geography* 30(4): 178–189.

Megoran, N., McConnell, F., & Williams, P. (eds) (2014). *The Geographies of Peace*. London: I.B. Tauris.

Mercer, C., Mohan, G., & Power, M. (2003). Towards a critical political geography of African development. *Geoforum* 34(4): 419–436.

Mitchell, K. (2003). Educating the national citizen in neoliberal times: From the multicultural self to the strategic cosmopolitan. *Transactions of the Institute of British Geographers* 28(4): 387–403.

Moisio, S. (2013). The state. In Dodds, K., Kuus, M., & Sharp, J. (eds), *The Ashgate Research Companion to Critical Geopolitics*, Farnham: Ashgate, 231–246.

Moisio, S., & Luukkonen, J. (2014). European spatial planning as governmentality: An inquiry into rationalities, techniques and manifestations. *Environment and Planning C: Government and Policy* doi:10.1068/c13158

Moisio, S., & Paasi, A. (2013). Beyond state-centricity: Geopolitics of changing state spaces. *Geopolitics* 18(2): 255–266.

Morrissey, J. (2011). Closing the neoliberal gap: Risk and regulation in the long war of securitization. *Antipode* 43(3): 874–900.

Mountz, A. (2013). Political geography I: Reconfiguring geographies of sovereignty. *Progress in Human Geography* 37(6): 829–841.

Müller, M. (2008). Reconsidering the concept of discourse for the field of critical geopolitics: Towards discourse as language and practice. *Political Geography* 27(3): 322–338.

Müller, M., & Reuber, P. (2008). Empirical verve, conceptual doubts: Looking from the outside in at critical geopolitics. *Geopolitics* 13(3): 458–472.

Murphy, A.B., Bassin, M., Newman, D., Reuber, P., & Agnew, J. (2004). Is there a politics to geopolitics? *Progress in Human Geography* 28(5): 619–640.

Ong, A. (2006). *Neoliberalism as Exception*. Durham, NC: Duke University Press.

Ó Tuathail, G. (1992). The Bush administration and the "end" of the Cold War: A critical geopolitics of US foreign policy in 1989. *Geoforum* 23(4): 437–452.

Ó Tuathail, G. (1996). *Critical Geopolitics*. London: Routledge.

Ó Tuathail, G., & Agnew, J.A. (1992). Geopolitics and discourse: Practical geopolitical reasoning in American foreign policy. *Political Geography* 11(2): 190–204.

Ó Tuathail, G., & Dalby, S. (1998). Rethinking geopolitics: Towards a critical geopolitics. In Ó Tuathail, G., & Dalby, S. (eds), *Rethinking Geopolitics*, London: Routledge, 1-15.

Paasi, A. (1996). *Territories, Boundaries, and Consciousness*. Chichester: John Wiley & Sons.

Paasi, A. (2000). Book review on "Rethinking Geopolitics." *Environment and Planning D* 18(2): 282–284.

Paasi, A. (2013). Borders. In Dodds, K., Kuus, M., & Sharp, J. (eds), *The Ashgate Research Companion to Critical Geopolitics*, Farnham: Ashgate, 213–229.

Pain, R. (2009). Globalized fear? Towards an emotional geopolitics. *Progress in Human Geography* 33(4): 466–486.

Parker, G. (1998). *Geopolitics: Past, Present, Future*. London: Pinter.

Parker, W.H. (1982). *Mackinder: Geography as an Aid to Statecraft*. Oxford: Clarendon Press.

Ridanpää, J. (2009). Geopolitics of humour: The Muhammed cartoon and the Kaltio comic strip episode in Finland. *Geopolitics* 14(4): 729–749.

Roberts, S. (2003). Global strategic vision: Managing the world. In Perry, R.W., & Maurer, B. (eds), *Globalization under Construction*, Minneapolis, MN: University of Minnesota Press, 1–37.

Roberts, S., Secor, A., & Sparke, M. (2003). Neoliberal geopolitics. *Antipode* 35(5): 886–896.

Sassen, S. (2013). When territory deborders territoriality. *Territory, Politics, Governance* 1(1): 21–45.

Secor, A. (2001). Towards a feminist counter-geopolitics: Gender, space and Islamist politics in Istanbul. *Space and Polity* 5(3): 199–219.

Sharp, J. (2000). *Condensing the Cold War: Reader's Digest and American Identity, 1922–1994*. Minneapolis, MN: University of Minnesota Press.

Sharp, J. (2008). Critical geopolitics (1996): Gearóid Ó Tuathail. In Hubbard, P., Kitchin, R., & Valentine, G. (eds), *Key Texts in Human Geography*, Thousand Oaks, CA: Sage, 189–196.

Sharp, J. (2013). Geopolitics at the margins? Reconsidering genealogies of critical geopolitics. *Political Geography* 37: 20–29.

Shaw, I.G.R. (2013). Predator empire: The geopolitics of US drone warfare. *Geopolitics* 18(3): 536–559.

Smith, N. (2000). Is a critical geopolitics possible? Foucault, class and the vision thing. *Political Geography* 19(3): 365–371.

Springer, S. (2013). Neoliberalism. In Dodds, K., Kuus, M., & Sharp, J. (eds), *The Ashgate Research Companion to Critical Geopolitics*, Farnham: Ashgate, 147–163.

Steinberg, P., Nyman, E., & Caraccioli, M.J. (2012). Atlas swam: Freedom, capital and floating sovereignties in the seasteading vision. *Antipode* 44(4): 1532–1550.

Thrift, N. (2005). *Knowing Capitalism*. Thousand Oaks, CA: Sage.

Walters, W. (2012). *Governmentality: Critical Encounters*. London: Routledge.

Chapter 18

Feminist Political Geography

Jennifer L. Fluri

University of Colorado-Boulder, Boulder, Colorado, USA

Feminist geographers have contributed many radical, insightful, and innovative approaches to the study of political geography. Early interventions addressed the invisibility of women as research subjects and worked to expose gendered aspects of political power (Kofman & Peake 1990; Pratt 1993; Staeheli & Cope 1994; Staeheli & Lawson 1995; Sparke 1996; McDowell & Sharp 1997). This research influenced the way in which politics is understood spatially, relationally, and at multiple scales. Feminist scholarship did not simply include gender as a category, but rather explicated the narrow analyses associated with masculinist approaches to critical geopolitical scholarship (Dowler & Sharp 2001). Research that examined the embodied experiences of individuals and groups has brought new insights and valuable contributions to the study of politics and spaces, such as spatial marginalization and politically induced migration and mobility. Static and disembodied analyses of political conflict and war have also been challenged by feminist scholars who attend to the complex and multilayered analyses of the body as a geographical site, corporeal experiences of conflict, and bodies as a geopolitical tool or weapon (Hyndman & De Alwis 2004; Mountz 2004; Fluri 2011; Koopman 2011a).

Feminist geographers have implicitly resisted conventional research methodologies through rich ethnographies and qualitative analyses (Kofman 2008). These methods have illuminated the complexities of the corporeal experience of power geometries, structural violence, resistance movements, and conflict. This research has illustrated a breadth of scholarship from the representational identity politics of nation-states to informal political processes and political resistance. Research on racial, ethnic, religious, and other forms of identity politics illuminates the manifestation of politics through the manipulation of socially constructed gender "norms" and relations (Fenster 1996). Some of this research reveals the tentacles of power through gendered identity politics such as state governmental attempts to change societal structures through biological and social reproduction (Smith 2009, 2012). This research has also questioned conventional notions of gendered political spaces and assumptions that politics only happens through "formal" processes (Secor 2001b).

The Wiley Blackwell Companion to Political Geography, First Edition.
Edited by John Agnew, Virginie Mamadouh, Anna J. Secor, and Joanne Sharp.
© 2015 John Wiley & Sons Ltd. Published 2017 by John Wiley & Sons Ltd.

With its breadth and depth, feminist scholarship has had a great impact on political geography, yet more scholarship is needed to push the discipline further toward integrating aspects of race, sex and sexuality, class, and belief systems at the intersection of gender and politics. This chapter is organized thematically in an effort to provide an overview of the various contributions to the study of political geography by feminist scholars, methods, and methodologies. These themes include gendered analyses of the state and nation; feminist contributions to political analyses of public and private space; scholarship on the intersecting relationship between gender borders, mobility, and security studies; and recent and emerging scholarship on corporeal political geographies.

Gender, state, and nation

Much scholarship by feminist political geographers examines how gender intersects with discursive national ideologies and state power. Feminist interventions have challenged the dichotomous representations of the male state and female nation, which both physically and socially positioned women as reproducers rather than productive or active political participants (Sharp 1996; Mayer 2000). Gendered political analyses critique and confront the material and ideological struggles involved in the making of nations as well as counter or revolutionary groups poised to challenge the state (Cowen & Gilbert 2008). Research that counters masculinized representations of the state also attends to feminist nationalist political movements, and challenges hetero-normative nationalisms and dichotomous gender roles and relations (Mayer 2000; Chen 2003; Paur 2007; Fluri 2008). Examining the embodied experiences of women (and men) and attending to the continual formations of place, boundaries, bodies, and power relations remain central for understanding the dynamic affiliations between gender, state, and nation (Mayer 2004).

Feminist political geographers have also explored women's participation and agency within the mechanisms of statecraft. Gilmartin and Kofman (2004) highlighted the "textual invisibility" of women's historical political labor and influence. Other forms of critical scholarship examine the use of gender as a political tool. This has been exemplified in scholarship that critiques the use of various gendered tropes by states in an attempt to marginalize and situate the distant "other" as inferior. Correspondingly, the symbolic representation of "women as nation" has been incorporated into several independence and nationalistic movements (Fluri 2013). Contemporary geopolitical tools have used gendered tropes in an effort to reinforce or solidify the "need" for military occupations and development interventions in an effort to "save or protect" women in locations such as Afghanistan as part of the Global War on Terror (GWOT; Riley, Bruce-Pratt, & Mohanty 2008; Fluri 2011). In this way, GWOT continues to be framed through a gendered lens that reduces complex social systems to narrow representations, such the framing of Muslim men as abusive patriarchs and Muslim women as helpless victims. Research on borders, public and private space, and mobility by feminist political geographers both questions and reveals the ways in which the gendering of state and nation resonates as part of a particularly and often politically crafted spatial imaginary. The following two sections provide a more detailed review of research on public and private space, borders, mobility, and security by feminist political geographers.

Public and private space

In addition to noting the boundaries between feminist and political geography, feminist scholars were making significant contributions toward destabilizing the implicit division between public (political) and private (apolitical) in much of political geography (Hyndman 2004). Feminist research on the public and private and the ways in which they influence, enact, and reproduce each other elucidates not the divide or boundary between public and private, but rather their interlinkages and the complications associated with political action and security (Secor 2001a, 2002; Nagar 2002; Cope 2004).

Radcliffe and Westwood's (1993) volume on women and popular protests in Latin America addressed several ways in which gendered expectations of private lives and behavior were used to confront politics in public spaces. Communities marked by racial, class, and gendered marginalization have worked to gain political legitimacy by appropriating public space for political action, such as political demonstrations (Clark 1994). Feminist scholarship has argued that political participation includes the involvement of individuals operating in both formal and informal political spaces (Cope 1997; see also Ali & Hopkins 2012). A counter-geopolitics, as argued by Secor (2001b), challenges the historical representations of politics as operating among states by examining "alternative and multiple loci of power" (2001b: 193). Thus, what constitutes power, spatial, and scalar relations remains fluid, geometric, and multifarious rather than static, hierarchical, or dichotomous. In addition to challenging conventional political analyses in geography, feminist scholars have also sought to focus on the use of conventional gender norms by organizations resisting state policy, action, or violence.

Women activists have opportunistically manipulated the predisposition of the home as an apolitical space in order to plan or execute covert political actions against the state (Dowler 1998, 2001). Female activists, such as the Madres de la Plaza de Mayo and the protestors at the Greenham Common Peace Camp, incorporated representations of the home and motherhood as an act of spatial transgression and protest against the state (Cresswell 1996; Bosco 2001). These studies illustrate disparate forms of gendered spaces that challenge the existing dichotomous demarcations of public and private spaces and the use of gendered norms or tropes to counter and challenge the actions of the state and its military.

Domestic spaces have also been associated with women's seclusion, immobility, and abuse, while conversely being associated with comfort, care, and retreat from abuse. Within some domiciles both conflict and its mitigation occur in tandem, which challenges the "expected" role of outside forces such as the state or other agencies in mitigating domestic violence (Fluri 2011; Cuomo 2013). As feminist scholars have shown, the home is a complex and complicated space and one that offers important insights for social, cultural, and political epistemologies. Feminist geographers' examinations of domestic labor provide a nuanced understanding of the fluidity and relational and intersectional linkages between public and private spaces. For example, paid domestic work has historically been seen as less important and outside state regulation, which rendered domestic workers more vulnerable to abuse by employers (Pratt & Yeoh 2003; Pratt 2004). Domestic labor has also been a key area of research for studying the personal and political aspects of gendered migration and mobility.

Borders, mobility, and security

Mobility has in some respects became a "barometer" for the study of gender, class, and other sociopolitical relations (Hyndman 2004: 169). Feminist scholarship examines the intersectionalities between gender and other forms of identity that spatially situate persons as central to, associated with, or at the margins of political machinations and power relations. The complex relationship between ethnicity, gender, and citizenship was evidenced by way of city planning for Ethiopian immigrant women in Israel (Fenster 1998). Jewish Ethiopian migrants were racially othered due to their phenotype and subsequently spatially marginalized, which was used politically to call into question their "Jewishness" (Fenster 1998).

The role and work of migrants working as domestics became integrated into Peruvian nationhood because these predominantly female laborers exemplified their "proper" place within the national imaginary through their embodied representations of gender, ethnicity, and class (Radcliffe 1990). Research on Filipina migrant domestic workers in Canada illustrates a complicated mix of intersecting and at times conflicting social and political requisites from the state, domestic workers, and their employers (Pratt 2002). In other examples the state has been actively involved in planning and regulating remittances from its migratory labor force (Silvey 2004). The Indonesian state labeled female out-migrants working as domestics in other countries "heroes of national development." This was done in an effort to ensure the continuation of this form of migratory labor because of the financial benefits it provided to the state, and despite the negative and at times abusive experiences of many Indonesian women working as domestics abroad (Silvey 2005). Feminist geographers have contributed new insights into the role of migratory laborers in the formation of national identities and ideologies.

Feminist political geographers have made important strides that demonstrate the interweaving of global geopolitical maneuvers with the embodiment of everyday experiences, emotions, and actions (Sharp 2009; Philo 2012). This research examines the ways in which displaced persons experience mobility or containment, along with the ways in which refugee spaces are managed by nations and supranational organizations such as the United Nations (Hyndman 2004; Mountz 2004).

Research on detention has shown the ways in which the state concludes what is both "secure" and in the "best interests of the child" to reinforce family detention practices at the border legally and discursively (Martin 2011). The geopolitics of childhood vulnerability has been used by the state to question parental practices by framing migratory laborers traveling with children as criminals rather than caregivers. Research on prisons and borders by critical geographers challenges the conventional belief that these create social order and security by illustrating how they act as forms of coercive mobility or oppression against the poor or displaced (Loyd, Mitchelson, & Burridge 2012).

Feminist political geographers have significantly shaped the study of mobility as a political process (Silvey 2004). Embodied analysis of mobility and asylum-seeking offers a nuanced understanding of geopolitical and economic modes of security and insecurity, respectively. Mountz (2011) argues that island detention is a form of suppression "where the bodies and identities of asylum-seekers are contained and regulated in the name of border enforcement, national security, and geopolitical imperatives" (2011: 119).

The study of conflict and security by feminist geographers includes macro-scale analyses of security and at the intersection of the global and the intimate (Enloe 1993, 2000; Hyndman 2007). Within locations embroiled in political conflict or war, security is distinctly spatialized and considerably contentious, leading feminist researchers to question "whose security is of concern and to whom" (Hyndman & DeAlwis 2004: 535; see also Kleinfeld 2007; Fluri 2011). The increased presence of gender and feminist research on security was further evidenced by the number of feminist scholars and gender-based research included in the 2009 *Annals of the Association of American Geographers* Special Issue on Peace and Armed Conflict (Blumen & Halevi 2009; Fluri 2009a; Henderson 2009; Kleinfeld 2009; Kobayashi 2009; Loyd 2009; Lunstrum 2009). Research on border, mobility, and security tends to include a multiscalar understanding of power relations, particularly from the perspectives and experiences of the marginalized and disenfranchised. Analyses by feminist political geographers further explore the concepts and use of scale as a mode of inquiry. Scalar interventions in political geography include the Marston–Brenner debates (Marston 2000; Brenner 2001; Marston & Smith 2001) and the political construction of scale (Delany & Leitener 1997; Cox 1998; Herod & Wright 2002; Fluri 2009b).

Research focusing on the interlinkages between the intimate and the global have disrupted grand narratives that dominated hegemonic and imperial political reasoning in order to chart a path outside the normative framework (Wright 2009; Koopman 2011a; Pratt & Rosner 2013). As part of feminist political geographers' engagements with scale and calls for examining private, social, and apolitical spaces and places, the body increasingly resonates within feminist study as both a site for political geographical analysis and a methodology for destabilizing the disembodied macro-scale study of politics and space. Feminist political geographers' analyses of the corporeal have significantly altered the geographical understanding of space by emphasizing the importance of embodied practices, place, and intimate knowledge as integral to feminist political geographical research (McDowell 1999; Dowler & Sharp 2001; Mountz & Hyndman 2006; Sharp 2007).

Corporeal geographies

Feminist and critical examinations of geographical scale combined with gender and critical race analyses opened up spaces for corporeal geographies of the political. Security, conflict, and violence are also key areas in which a distinct focus on bodies, body politics, and place-based examinations presents significant and valuable contributions. Some notable examples include Dowler's (2002) analysis of post–9/11 Ground Zero in New York as a masculine space, which included a print media reimagining of the Twin Towers as the body of a male firefighter and police officer. The ways in which bodies carry particular representations of citizenship and its associated privileges elucidate important linkages among citizenship, identity, transnationalism, and mobility studies. Transnational solidarity groups in Guatemala and Columbia put the bodies of privileged volunteers to work accompanying threatened activists in order to provide secured mobility. This form of "proxy citizenship" attempts to alter conflict dynamics through corporealized protection (Henderson 2009). Koopman's (2011a) research on accompaniment further argues that this form of "alter-geopolitics" should garner more attention from scholars seeking to broaden our definitions and understanding of security in order to bolster the amount of attention paid to these and other forms of peace-building geopolitics.

Proxy bodies have also been incorporated into the research on sex workers and brothels as sites for male sexual aggression within conflict zones (Enloe 1993, 2000). In Afghanistan, where the "saving women" discourse curtails sexual engagements between international men and local Afghan women, various forms of sexual abuse, misconduct, and harassment by international men against international women abound (Fluri 2009b). The illegitimacy and illegality of sex workers place their corporeality as a proxy for the release of male sexual aggression and has been mythologized by some men as a method to "protect" local women and other "legitimized" female laborers (i.e., international aid/development workers) from this aggression. Sex work as argued by Wright (2004) became a discursive strategy for questioning the morality and legitimate position of female victims of murder in Ciudad Juárez:

> Those … who blame the kidnapping and murder victims for the crimes perpetrated against them, claim that these events do not reflect problems in the city but instead, problems within the women/girls who brought about this trouble. (Wright 2004: 377)

Associating women's public labor with sex work effectively devalued women's economic agency and their membership within society. In the above scenarios, feminist researchers illustrate the ways in which women's bodies become flexible nodes for an intimately aggressive geopolitics that diverts attention away from male actors/perpetrators and the structures of aid/development and urban governance, respectively, along with the unwillingness of these institutionalized structures to address or curtail these abuses.

Intimate linkages between social and political conflict and daily life are further explored in Sara Smith's work (2009, 2011). She examines the complicated ways in which geopolitical conflicts in Ladakh, India resonate on bodies and currently construct barriers to the formerly accepted practice of intermarriage between Buddhists and Muslims. As Smith ardently argues, "when population becomes part of a territorial struggle, the body itself becomes a geopolitical site" (Smith 2011: 456–457). In many respects, bodies signify and approximate the fragile scale at which processes and promises of geopolitics and political identities resonate (Dixon & Marston 2011). Situating bodies at the center of research on political conflict is necessary to ensure that the bodies of individuals and collectives count rather than being reduced to aggregate sums associated with the collateral excesses of war (Hyndman 2007; Tyner 2009).

The Muslim headscarf/veil has been and remains a prominent symbolic and gendered representation of, or challenge to, the nation within colonial and postcolonial politics. Gökarıksel and Secor's (2009, 2010, 2012) work on veiling fashion provides an excellent method for addressing the contradictions of politics and the economy around the headscarf as a social and political symbol of piety and religious devotion, which simultaneously functions as a motif of capitalist production and consumption. The capitalist use of the veil incorporates it as "fashion" capitalism, which both subsumes and challenges its utility and marginalization within Turkish society. The politics of the veil and its very intentionality are therefore disrupted, while becoming entrenched within the equally oppressive and problematic mechanisms of neoliberalism (also see Gökarıksel and Secor 2010, 2012; Secor 2002). The aesthetics of gendered bodies and geopolitics, as discussed by feminist political geographers, examine the representative ways in which idealized forms of womanhood become representative of nationalism and citizenship through beauty pageants and related symbolic representations of femininity (Sharp 1996; Oza 2006; Fluri 2009c; Faria 2010, 2013).

In addition to incorporating gender and feminist theories as accepted and integral aspects of analysis within political geography, feminist geographers have also significantly contributed to research methodologies and methods. Feminist methodologies and research methods provide another important (if at times overlooked) contribution to political geographical inquiry. Methodologically, feminist research seeks to critique the epistemological and ontological understandings of power relations (Staeheli & Lawson 1995).

Methodologies and research methods

The extensive amount of fieldwork conducted by feminist political geographers is astounding. Feminist political geography by and large has included qualitative and ethnographic methods, often in sites and situations that offer several challenges for researchers. These qualitative approaches have contributed both empirically sound and theoretically rich research to the study of political geography and critical geopolitics. Feminist geographers' contributions to research methodologies include challenging the assumptions of research objectivity. Thus, the concept of intersubjectivity between researcher and research populations has become an integral research strategy (Rose 1997; McDowell 1999; Moss 2002; Gibson-Graham 2010). This includes attempting to disengage the binary structure between subjects and objects, and acknowledging that knowledge can be constructed from research participants rather than only through existing hierarchies of understanding between researchers and research populations. Many feminist political geographers incorporate the concept of situated knowledges (Haraway 1988) to illustrate that the production of knowledge and understanding of politics must be contextualized across space at multiple scales. Situated knowledges refers to including "knowledges that are explicit about their positioning, sensitive to the structures of power that construct these multiple positions and committed to making visible the claims of the less powerful" (McDowell 1992: 413; see also Rose 1997).

Other theoretical interventions include counter-topographies, counter-geopolitics, and the topological. Counter-topographies, a term coined by Cyndi Katz (2004), provide a metaphorical elaboration of the contour line to track the lineage of theory and practice in the production of knowledge. Katz's topography connects the politics of location and difference by examining the linkages between the particularities of marginalization and political struggle in distinctive locales. Melissa Wright (2008) took up this concept to highlight the counter-topographies of geographical knowledge and activism across the globe. Mountz (2011) examines the counter-topographies of asylum seekers en route between nation-states and argues that "feminist counter-topographies confront directly the fixity of exclusion" in current geometries of power, "seizing precisely on the potentiality of other topological processes of transformation" (2011: 392).

Feminist geographers have also employed topology in addressing the ways in which power can become endlessly twisted formations of abuse and oppression (Koopman 2008). In addition to these methodological interventions, feminist political geographers highlight the importance of field-based empirical research in order to provide rich linkages between theory and praxis, and stress the interconnectedness between the intimate and the global (Mountz & Hyndman 2006).

Most feminist political geographers conduct fieldwork and view this as a necessary if not essential aspect of doing research. This work does not rely on examples or discussions of the mundane, but rather provides a rich and nuanced understanding of small-scale politics (Sharp 2004; Dixon & Marston 2011). Much of feminist geography requires engagements with other populations in various communities in and outside of Anglo-speaking countries in order to inform theory through extensive qualitative, ethnographic, and empirical studies. Feminist methodologies have also influenced research on and use of geographical information systems (GIS) through the incorporation of qualitative methods (Kwan 2002; Cope & Elwood 2009). Fieldwork by feminist scholars contributes nuanced and qualitatively rich research to political geography. Arguably, this is an area that requires more attention from political geographers, particularly those who do not actively engage in fieldwork.

These research methods include many discussions and debates about methods and methodologies along with issues of race, positionality, and critical self-reflexivity (Staeheli & Lawson 1994). The 2002 "Talking across Worlds" special issue in *Gender, Place and Culture* addressed several concerns and debates among feminist scholars on the role of the researcher conducting fieldwork. Several articles argue for feminist geographers to recognize the legitimacy of activists as research participants and political actors, and to reinforce the acceptance of various forms of knowledge (Raju 2002). This includes addressing the tension between institutional demands for theoretically driven scholarship and the political and ethical need for on-the-ground research that is meaningful to the populations with which we work or that we study (Nagar 2002). In addition to the extensive amount of discussion of positionality and reflexivity and their inclusion in feminist geographical research (England 1994; Rose 1997),

Nagar (2002) identifies the continuing problem associated with talking across worlds (particularly on political issues), asking feminist researchers to consider "Who are we writing for, how, and why?" and "[W]hat does it mean to co-produce relevant knowledge across geographical, institutional, and/or cultural borders?" (Nagar 2002: 179). Similar concerns have been taken up by feminist geographers' critiques of critical geopolitics, such as "[C]an there be a more constructive side to critical geopolitics – a more positive politics?" (Dowler & Sharp 2001: 167). Feminist geographers, along with other critical researchers, engage in various forms of participatory or activist research as a transformative process (Moss 2002; Cahill 2007). The ethics associated with participatory research are also regularly discussed and debated among feminist scholars (Cahill, Sultana, & Pain 2007).

Working with grassroots organizations should also include formal institutions and the state, in order to elucidate the ways in which body and identity politics can be a site for countering or resisting hegemonic or institutional forms of oppression (Mountz 2002; Silvey 2002). Additionally, linkages and intersections between macro and micro scales, the universal and the particular, are often identified as a basis for collaborative politics across social, political, and spatial differences (Pratt 2002). Wright's (2009) review article on activism highlights feminist scholarship that attempts to contend with the difficult issues associated with struggles over power, global linkages, and the politics of epistemology.

The future for feminist political geography

Several recent articles and discussions within feminist geography more broadly about the role of emotions or affect have included the work of feminist political geographers (Woon 2013). Sharp (2009: 78) argues that "emphasis on the political manipulation of emotion/affect is key, and indeed offers a necessary line of examination for geography." By linking fears about global terrorism with the lack of attention (and funds) paid to an embodied understanding of the everyday terrors experienced by women in situations of domestic or intimate partner violence, Pain (2009) highlights the need to incorporate emotional geopolitics as part of our analyses. In the recent special issue of *Geopolitics* on "Feminist Geopolitics: Unpacking (In)Security, Animating Social Change," emotional geographies and feminist geopolitics are intersected in the study of security in a number of articles (Casolo & Doshi 2013; Clark 2013; Cuomo 2013; Dowler 2013; Ojeda 2013; Williams & Boyce 2013; Williams & Massaro 2013).

There have indeed been tremendous strides and contributions by feminist political geographers since Brown and Staeheli's (2003) "Are we there yet?" article. Going forward, these geographical inquiries must remain vigilant to ensure that gender-based research continues to engage critically with a diverse array of research populations and collaborators, both within and outside the academy. Cultivating transnational connections and collaborations provides another area for increased attention by feminist geographers. Collaboration and learning both from our research populations and scholars outside our respective interpretive communities remain a growth area for political geography, led predominantly by feminist geographers. Intimate geopolitics and scholarship on the global intimate are an exciting growth area for feminist political geography.

The lack of gendered considerations and the disembodied analysis of political conflict among political geographers demonstrate the continued need for feminist interventions in political geography. Research methodologies such as critical self-reflexivity and positionality remain the purview of feminist researchers, rather than political geography more generally.

It is important to continue to push political geographers to incorporate these methodologies in various forms of critical scholarship. Feminist geographers have had a radical impact on political geography, while more research that challenges conventional epistemologies about gender, space, politics, and scale is necessary to move feminist geography from an interventionist subdiscipline to one that is incorporated throughout the study of political geography.

References

Ali, R., & Hopkins, P. (2012). Everyday making and civic engagement among Muslim women in Scotland. In Ahmad, W., & Saradar, A. (eds), *Muslims in Britain: Making Social and Political Space*, Oxford: Routledge, 141–155.

Blumen, O., & Halevi, S. (2009). Staging peace through a gendered demonstration: Women in black in Haifa, Israel. *Annals of the Association of American Geographers* 99(5): 977–985.

Bosco, F.J. (2001). Place, space, networks, and the sustainability of collective action: The Madres de Plaza de Mayo. *Global Networks* 1(4): 307–329.

Brenner, N. (2001). The limits to scale? Methodological reflection on scale structuration. *Progress in Human Geography* 25(4): 591–614.

Brown, M., & Staeheli, L.A. (2003). "Are we there yet?" Feminist political geographies. *Gender, Place and Culture* 10(3): 247–255.

Cahill, C. (2007). The personal is political: Developing new subjectivities through participatory action research. *Gender, Place and Culture* 14(3): 267–292.

Cahill, C., Sultana, F., & Pain, R. (2007). Participatory ethics: Politics, practices, institutions. *ACME: An International E-Journal for Critical Geographies* 6(3): 304–318.

Casolo, J., & Doshi, S. (2013). Domesticated dispossessions? Towards a transnational feminist geopolitics of development. *Geopolitics* 18(4): 800–834.

Chen, T.M. (2003). Female icons, feminist iconography? Socialist rhetoric and women's agency in 1950s China. *Gender and History* 15(2): 268–295.

Clark, H. (1994). Taking up space: Redefining political legitimacy in New York City. *Environment and Planning A* 26(6): 937–955.

Clark, J.H. (2013). "My life is like a novel": Embodied geographies of security in Southeast Turkey. *Geopolitics* 18(4): 835–855.

Cope, M. (1997). Participation, power, and policy: Developing a gender-sensitive political geography. *Journal of Geography* 96(2): 91–97.

Cope, M. (2004). Placing gendered political acts. In Staeheli, L., Kofman, E., & Peake, L. (eds), *Mapping Women Making Politics: Feminist Perspectives on Political Geography*, New York: Routledge, 71–86.

Cope, M., & Elwood, S. (2009). *Qualitative GIS: A Mixed Methods Approach*. Thousand Oaks, CA: Sage.

Cowen, D., & Gilbert, E. (2008). *War, Citizenship, Territory*. London: Routledge.

Cox, K.R. (1998). Spaces of dependence, spaces of engagement and the politics of scale, or looking for local politics. *Political Geography* 17(1): 1–23.

Cresswell, T. (1996). *In Place/Out of Place: Geography, Ideology, and Transgression*. Minneapolis, MN: University of Minnesota Press.

Cuomo, D. (2013). Security and fear: The geopolitics of intimate partner violence policing. *Geopolitics* 18(4): 856–874.

Delaney, D., & Leitner, H. (1997). The political construction of scale. *Political Geography* 16(2): 93–97.

Dixon, D.P., & Marston, S.A. (2011). Introduction: Feminist engagements with geopolitics. *Gender, Place and Culture* 18(4): 445–453.

Dowler, L. (1998). "And they think I'm just a nice old lady": Women and war in Belfast, Northern Ireland. *Gender, Place and Culture* 5(2): 159–176.

Dowler, L. (2001). No man's land: Transgressing the boundaries of West Belfast, Northern Ireland. *Geopolitics* 6(3): 158–176.

Dowler, L. (2002). Women on the frontlines: Rethinking war narratives post 9/11. *Geojournal* 58(2–3): 159–165.

Dowler, L. (2013). Waging hospitality: Feminist geopolitics and tourism in West Belfast, Northern Ireland. *Geopolitics* 18(4): 779–799.

Dowler, L., & Sharp, J. (2001). A feminist geopolitics? *Space and Polity* 5(3): 165–176.

England, K.V.L. (1994). Getting personal: Reflexivity, positionality, and feminist research. *Professional Geographer* 46(1): 80–89.

Enloe, C.H. (1993). *The Morning After: Sexual Politics at the End of the Cold War.* Berkeley, CA: University of California Press.

Enloe, C.H. (2000). *Bananas, Beaches and Bases: Making Feminist Sense of International Politics.* Berkeley, CA: University of California Press.

Faria, C. (2010). Contesting Miss South Sudan: Gender and nation-building in diasporic discourse. *International Feminist Journal of Politics* 12(2): 222–243.

Faria, C. (2013). Styling the nation: Fear and desire in the South Sudanese beauty trade. *Transactions of the Institute of British Geographers* 20(1): 87–106.

Fenster, T. (1996). Ethnicity and citizen identity in planning and development for minority groups. *Political Geography* 15(5): 405–418.

Fenster, T. (1998). Ethnicity, citizenship, planning and gender: The case of Ethiopian immigrant women in Israel. *Gender, Place and Culture* 5(2): 177–189.

Fluri, J. (2008). Feminist-nation building in Afghanistan: An examination of the Revolutionary Association of the Women of Afghanistan (RAWA). *Feminist Review* 89(1): 34–54.

Fluri, J. (2009a). "Foreign passports only": Geographies of (post) conflict work in Kabul, Afghanistan. *Annals of the Association of American Geographers* 99(5): 986–994.

Fluri, J. (2009b). Geopolitics of gender and violence "from below." *Political Geography* 28(4): 259–265.

Fluri, J. (2009c). The beautiful "other": A critical examination of "western" representations of Afghan feminine corporeal modernity. *Gender, Place and Culture* 16(3): 241–257.

Fluri, J. (2011). Bodies, bombs and barricades: Geographies of conflict and civilian (in)security. *Transactions of the Institute of British Geographers* 36(2): 280–296.

Fluri, J. (2013). Women. In Dodds, K., Kuus, M., & Sharp, J. (eds), *Ashgate Research Companion to Critical Geopolitics*, Farnham: Ashgate, 509–526.

Gibson-Graham, J.K. (2010). Gibson-Graham, J.K. 1996: *The End of Capitalism (as we Knew it): A Feminist Critique of Political Economy.* Oxford: Blackwell. *Progress in Human Geography* 34(1): 117–127.

Gilmartin, M., & Kofman, E. (2004). Critically feminist geopolitics. In Staeheli, L.A., Kofman, E., & Peake, L.J. (eds), *Mapping Women, Making Politics*, New York: Routledge, 113–126.

Gökarıksel, B. (2012). The intimate politics of secularism and the headscarf: The mall, the neighborhood, and the public square in Istanbul. *Gender, Place and Culture* 19(1): 1–20.

Gökarıksel, B., & Secor, A. (2009). New transnational geographies of Islamism, capitalism and subjectivity: The veiling-fashion industry in Turkey. *Area* 41(1): 6–18.

Gökarıksel, B., & Secor, A. (2010). Islamic-ness in the life of a commodity: Veiling-fashion in Turkey. *Transactions of the Institute of British Geographers* 35(3): 313–333.

Gökarıksel, B., & Secor, A. (2012). "Even I was tempted": The moral ambivalence and ethical practice of veiling-fashion in Turkey. *Annals of the Association of American Geographers* 102(4): 847–862.

Haraway, D. (1988). Situated knowledges: The science question in feminism and the privilege of partial perspective. *Feminist Studies* 14(3): 575–599.

Henderson, V.L. (2009). Citizenship in the line of fire: Protective accompaniment, proxy citizenship, and pathways for transnational solidarity in Guatemala. *Annals of the Association of American Geographers* 99(5): 969–976.

Herod, A., & Wright, M.W. (eds) (2002). *Geographies of Power: Placing Scale.* Oxford: Blackwell.

Hyndman, J. (2004). Mind the gap: Bridging feminist and political geography through geopolitics. *Political Geography* 23(3): 307–322.

Hyndman, J. (2007). Feminist geopolitics revisited: Body counts in Iraq. *Professional Geographer* 59(1): 35–46.

Hyndman, J., & De Alwis, M. (2004). Bodies, shrines, and roads: Violence, (im)mobility and displacement in Sri Lanka. *Gender, Place and Culture* 11(4): 535–557.

Katz, C. (2004). *Growing Up Global: Economic Restructuring and Children's Everyday Lives.* Minneapolis, MN: University of Minnesota Press.

Kleinfeld, M. (2007). Misreading the post-tsunami political landscape in Sri Lanka: The myth of humanitarian space. *Space and Polity* 11(2): 169–184.

Kleinfeld, M. (2009). The political utility of the nonpolitical child in Sri Lanka's armed conflict. *Annals of the Association of American Geographers* 99(5): 874–883.

Kobayashi, A. (2009). Geographies of peace and armed conflict: Introduction. *Annals of the Association of American Geographers* 99(5): 819–826.

Kofman, E. (2008). *Feminist Transformations of Political Geography.* Los Angeles, CA: Sage.

Kofman, E., & Peake, L. (1990). Into the 1990s: A gendered agenda for political geography. *Political Geography Quarterly* 9(4): 313–336.

Koopman, S. (2008). Cutting through topologies: Crossing lines at the School of the Americas. *Antipode* 40(5): 825–847.

Koopman, S. (2011a). Alter-geopolitics: Other securities are happening. *Geoforum* 42(3): 274–284.

Koopman, S. (2011b). Let's take peace to pieces. *Political Geography* 30(4): 193–194.

Kwan, M.-P. (2002). Feminist visualization: Re-envisioning GIS as a method in feminist geographic research. *Annals of the Association of American Geographers* 92(4): 645–661.

Loyd, J.M. (2009). "A microscopic insurgent": Militarization, health, and critical geographies of violence. *Annals of the Association of American Geographers* 99(5): 863–873.

Loyd, J.M., Mitchelson, M., & Burridge, A. (2012). *Beyond Walls and Cages: Prisons, Borders, and Global Crisis.* Athens, GA: University of Georgia Press.

Lunstrum, E. (2009). Terror, territory, and deterritorialization: Landscapes of terror and the unmaking of state power in the Mozambican "civil" war. *Annals of the Association of American Geographers* 99(5): 884–892.

Marston, S. (2000). The social construction of scale. *Progress in Human Geography* 24(2): 219–242.

Marston, S., & Smith, N. (2001). States, scales and households: Limits to scale thinking? A response to Brenner. *Progress in Human Geography* 25(4): 615–619.

Martin, L. (2011). The geopolitics of vulnerability: Children's legal subjectivity, immigrant family detention and US immigration law and enforcement policy. *Gender, Place and Culture* 18(4): 477–498.

Mayer, T. (2000). *Gender Ironies of Nationalism: Sexing the Nation.* London: Routledge.

Mayer, T. (2004). Embodied nationalisms. In Staeheli, L., Kofman, E., & Peake, L. (eds), *Mapping Women, Making Politics: Feminist Perspectives on Political Geography*, New York: Routledge, 153–168.

McDowell, L. (1992). Doing gender: Feminism, feminists and research methods in human geography. *Transactions of the Institute of British Geographers* 17: 399–416.

McDowell, L. (1999). *Gender, Identity and Place: Understanding Feminist Geographies.* Minneapolis, MN: University of Minnesota Press.

McDowell, L., & Sharp, J.P. (1997). *Space, Gender, Knowledge: Feminist Readings.* London: Arnold.

Moss, P. (ed.) (2002). *Feminist Geography in Practice: Research and Methods.* Oxford: Blackwell.

Mountz, A. (2002). Feminist politics, immigration, and academic identities. *Gender, Place and Culture* 9(2): 187–198.

Mountz, A. (2004). Embodying the nation-state: Canada's response to human smuggling. *Political Geography* 23(3): 323–345.

Mountz, A. (2011). The enforcement archipelago: Detention, haunting, and asylum on islands. *Political Geography* 30(3): 118–128.

Mountz, A., & Hyndman, J. (2006). Feminist approaches to the global intimate. *Women's Studies Quarterly* 34(1&2): 446–463.

Nagar, R. (2002). Footloose researchers, "traveling" theories, and the politics of transnational feminist praxis. *Gender, Place and Culture* 9(2): 179–186.

Ojeda, D. (2013). War and tourism: The banal geographies of security in Colombia's "retaking." *Geopolitics* 18(4): 759–778.

Oza, R. (2006). *The Making of Neoliberal India: Nationalism, Gender, and the Paradoxes of Globalization.* New York: Routledge.

Pain, R. (2009). Globalized fear? Towards an emotional geopolitics. *Progress in Human Geography* 33(4): 466–486.

Philo, C. (2012). Security of geography/geography of security. *Transactions of the Institute of British Geographers, New Series.* 37(1): 1–7.

Pratt, G. (1993). Reflections on poststructuralism and feminist empirics, theory and practice. *Antipode* 25(1): 51–63.

Pratt, G. (2002). Collaborating across differences. *Gender, Place and Culture* 9(2): 195–200.

Pratt, G. (2004). *Working Feminism.* Philadelphia, PA: Temple University Press.

Pratt, G., & Rosner, V. (eds) (2012). *The Global and the Intimate: Feminism in Our Time.* New York: Columbia University Press.

Pratt, G., & Yeoh, B. (2003). Transnational (counter) topographies. *Gender, Place and Culture* 10(2): 159–166.

Puar, J. (2007). *Terrorist Assemblages: Homonationalism in Queer Times.* Durham, NC: Duke University Press.

Radcliffe, S. (1990). Ethnicity, patriarchy, and incorporation into the nation: Female migrants as domestic servants in Peru. *Environment and Planning D: Society and Space* 8(4): 379–393.

Radcliffe, S., & Westwood, S. (eds) (1993). *Viva: Women and Popular Protest in Latin America.* New York: Routledge.

Raju, S. (2002). We are different but can we talk? *Gender, Place and Culture* 9(2): 173–178.

Riley, R.L., Bruce-Pratt, M., & Mohanty, C.T. (eds) (2008). *Feminism and War: Confronting US Imperialism.* London: Zed Books.

Rose, G. (1997). Situating knowledges: Positionality, reflexivities and other tactics. *Progress in Human Geography* 21(3): 305–320.

Secor, A. (2001a). Ideologies in crisis: Political cleavages and electoral politics in Turkey in the 1990s. *Political Geography* 20(5): 539–560.

Secor, A. (2001b). Toward a feminist counter-geopolitics: Gender, space and Islamist politics in Istanbul. *Space and Polity* 5(3): 191–211.

Secor, A. (2002). The veil and urban space in Istanbul: Women's dress, mobility and Islamic knowledge. *Gender, Place and Culture* 9(1): 5–22.

Sharp, J.P. (1996). A feminist engagement with national identity. In Duncan, N. (ed.), *Bodyspace: Destablising geographies of gender and sexuality,* London: Routledge, 97–108.

Sharp, J. (2004). Doing feminist political geographies. In Staeheli, L., Kofman, E., & Peake, L. (eds), *Mapping Women, Making Politics: Feminist Perspectives on Political Geography.* New York: Routledge, 87–98.

Sharp, J.P. (2007). Geography and gender: Finding feminist political geographies. *Progress in Human Geography* 31(3): 381–387.

Sharp, J.P. (2009). Geography and gender: What belongs to feminist geography? Emotion, power and change. *Progress in Human Geography* 33(1): 74–80.

Silvey, R. (2002). Sweatshops and the corporatization of the university. *Gender, Place and Culture* 9(2): 201–207.

Silvey, R. (2004). Power, difference and mobility: Feminist advances in migration studies. *Progress in Human Geography* 28(4): 490–506.

Silvey, R. (2005). Borders, embodiment, and mobility: Feminist migration studies in geography. In Nelson, L., & Seager, J. (eds), *A Companion to Feminist Geography,* Oxford: Blackwell, 138–149.

Smith, S. (2009). The domestication of geopolitics: Buddhist–Muslim conflict and the policing of marriage and the body in Ladakh, India. *Geopolitics* 14(2): 197–218.

Smith, S. (2011). "She says herself, 'I have no future'": Love, fate and territory in Leh District, India. *Gender, Place and Culture* 18(4): 455–476.

Smith, S. (2012). Intimate geopolitics: Religion, marriage, and reproductive bodies in Leh, Ladakh. *Annals of the Association of American Geographers* 102(6): 1511–1528.

Staeheli, L.A., & Cope, M.S. (1994). Empowering women's citizenship. *Political Geography* 13(5): 443–460.

Staeheli, L.A., & Lawson, V.A. (1994). A discussion of "women in the field": The politics of feminist fieldwork. *Professional Geographer* 46(1): 96–102.

Staeheli, L.A., & Lawson, V.A. (1995). Feminism, praxis, and human geography. *Geographical Analysis* 27(4): 321–338.

Tyner, J.A. (2009). *War, Violence and Population: Making the Body Count.* New York: Guilford Press.

Williams, J., & Boyce, G.A. (2013). Fear, loathing and the everyday geopolitics of encounter in the Arizona borderlands. *Geopolitics* 18(4): 895–916.

Williams, J., & Massaro, V. (2013). Feminist geopolitics: Unpacking (in)security, animating social change. *Geopolitics* 18(4): 751–758.

Woon, C.Y. (2013). For "emotional fieldwork" in critical geopolitical research on violence and terrorism. *Political Geography* 33: 31–41.

Wright, M.W. (2004). From protests to politics: Sex work, women's worth, and Ciudad Juarez modernity. *Annals of the Association of American Geographers* 94(2): 369–386.

Wright, M. (2008). Gender and geography: Knowledge and activism across the intimately global. *Progress in Human Geography* 33(3): 379–386.

Wright, M.W. (2009). Gender and geography: Knowledge and activism across the intimately global. *Progress in Human Geography* 33(3): 379–386.

Chapter 19

Postcolonialism

Chih Yuan Woon

National University of Singapore, Singapore

On February 6, 2014, the Indonesian government announced that it would name a warship after two "national heroes" who had been executed in Singapore some 46 years ago for their involvement in the 1965 terror attack on a civilian building (MacDonald House) located along Singapore's popular Orchard Road.[1] The destructive act initiated by the two Indonesian marines was part of a historical chapter known as the Confrontation or *Konfrantasi* in the 1960s, in which Indonesian radicals infiltrated the streets of Singapore (then part of Malaysia), planting bombs to generate alarm and stir up latent racial tensions. The Confrontation was Indonesian President Sukarno's attempt at disrupting the new state of Malaysia so as to enable the creation of a pan-Malay Muslim region called Maphilindo[2] that would come to symbolize a self-reliant, free, and independent Malay entity in postcolonial Southeast Asia (see Poulgrain 1998; Tan 2008). As the late American scholar George Kahin wrote in 1964, the *Konfrantasi* represented the "powerful, self-righteous thrust of Indonesian nationalism" and the widespread belief that Indonesia, because of its size, had the "moral right" to lead a (resurgent) Muslim-centered Southeast Asian polity in the postcolonial era (1964: 260–261). Given such a scenario, the incident raised fears for Singapore (as well as other Southeast Asian states) of fresh rounds of (geo)political struggles in the region, with Indonesia replacing the retreating Western imperialists as the neocolonial power. Thus, according to Law (2007), it is this possibility of a Malay Muslim geopolitical entity, with all its weighty historical roots, that continues to haunt postcolonial (multicultural) Singapore. It can then hardly be considered surprising that Jakarta's latest decision to name a new navy frigate in honor of two individuals who are remembered in Singapore for their crimes of terrorism has given such fears renewed veracity: It unsettles the boundaries of postcolonial Singapore, reinvigorating concerns about the nation's vulnerability in a largely Muslim region. In the words of Singapore's defense minister, Teo Chee Hean, this "insensitive" move by Indonesia will "re-open wounds" and "revive painful memories of an Indonesia which sought deference from its neighbours and was prepared to use force to implement its desires" (cited in Kausikan 2014).

The Wiley Blackwell Companion to Political Geography, First Edition.
Edited by John Agnew, Virginie Mamadouh, Anna J. Secor, and Joanne Sharp.
© 2015 John Wiley & Sons Ltd. Published 2017 by John Wiley & Sons Ltd.

Extrapolating the postcolonial insights from this opening account, it can be argued that postcolonialism does not simply refer to a linear temporal transition toward a political condition of "after colonialism." Rather than a distinct break from the past, there are "post-colonial hauntings," to borrow Law's (2007: 63) term, whereby Singapore's anxieties as a postcolonial state remain unabated (even today), with Indonesia's "new" imperialist designs on Southeast Asia looming large as the backdrop. Interestingly, this incident attests to Young's (2012) felicitous exhortation that "postcolonial remains." The postcolonial will always be "left over" and it is, in many ways, "an unfinished business" – "the continued project of past conflicts into the experience of the present, the insistence of the afterimages of historical memory that drive the desire to transform the present" (Young 2012: 21). It is under such circumstances that Young urges that more attention be paid to postcolonialism's lingering legacies and how they have been transformed in the present in different and novel ways (e.g., alternative "neocolonial" powers rising from outside the West).

This chapter responds to Young's call by examining the contributions of political geography to understanding these contemporary (re)configurations of the postcolonial. I contend that political geographers' emerging research forays into these new postcolonial formations not only help to locate and expose the hidden rhizomes of colonialism's historical reach (what remains invisible, unseen, silent, or unspoken), they simultaneously go beyond critiques to open up possibilities for projects of postcolonial empowerment – to reconstruct Western knowledge formations, reorient ethical norms, turn the power structures of the world upside down, and refashion the world from below (cf. Young 2012). The aims of the chapter are thus threefold. First, I map out the genealogy and development of postcolonialism as a field of study, in order to elucidate the impacts of postcolonial theory on political geography. This will then provide the basis to illuminate and account for the different research agendas of postcolonial political geography, which in turn demonstrate the subdiscipline's continual sensitivity toward postcolonialism's changing concerns and challenges. The final part of the chapter will reflect on the methodological question of "how" to do postcolonial political geographies in order to chart the future possibilities for the subdiscipline.

Postcolonialism: Origins and developments

Postcolonialism is a complex and contested term and has been deployed in multiple ways. Blunt and McEwan (2002: 3) suggest that this is in part due to the "post" in postcolonialism possessing two distinct meanings. First, it can direct attention to an epoch in time – a period after colonialism or after independence that is linked to a variety of events and processes arising specifically from the decline and fall of European colonialism since the end of the Second World War.[3] However, critics have pointed out that such a conception of postcolonialism is problematic because it seems to connote the contemporary irrelevance of colonialism. Indeed, while (European) colonialism may no longer exist as before, it is far from clear that it has been completely banished into history. Colonies and colonial powers continue to be evident today (Britain and France, for instance, still possess scattered territories around the world), but, more importantly, colonialism has taken on a new form in the era of globalization. Claims abound that neocolonialism has enabled developed nations to continue their domination and exploitation of formerly colonized states through inequitable international trading relations and geopolitical interactions.

In lieu of such circumstances, there is an increasing consensus that postcolonialism is not simply an epochal reference point; rather, it can also refer to ways of criticizing the material and discursive legacies of colonialism and how they continue to shape unequal North–South relations[4] (Jacobs 1996; Sidaway 2000). Such a view has arguably been promulgated and advanced most forcefully by postcolonial theory, whose influence in the social sciences has been attributed to a number of key writers, notably Edward Said, Homi Bhabha, and Gayatri Spivak (see also Sharp 2003).[5] Said is best known for his book *Orientalism* (1978), which is often considered to be the founding text of contemporary postcolonial theory. Orientalism, according to Said, offers a critical pathway to conceptualize the history of relations between what we might commonsensically call the "West" and the "East" or the Occident and the Orient. Rather than accepting the term as one that designates an area of neutral scholarly expertise (be it oriental languages, literatures, or customs), Said argues that Orientalism continues to function as a discourse in which the West's knowledge of the Orient is inextricably bound up with its domination over it. Evoking Michel Foucault's proposition that all forms of knowledge are productive of power (constituting someone/something as an object of knowledge is to assume power over it), Said assesses the implications of the Western construction of the Orient as an object of knowledge during colonial expansion. Because he refuses to accept the innocence of knowledge about and representations of the Orient, Said is able to consider how Orientalism's classification of the East as different and inferior has performative effects insofar as it legitimized Western interventions and rule. Thus, Orientalism hinges on a set of polarities that draws on imaginative geographies of the Orient as irrational, exotic, erotic, despotic, and heathen, thereby securing the West in contrast as rational, familiar, moral, just, and Christian. It is crucial in drawing out the process of "othering" and how power functions to place the European experience at the center, while positing other cultures/peoples in other places as inferior and abnormal.

Like Said, Bhabha is concerned with how Western negative imag(in)ings of postcolonial countries have hinged on classifying the latter group as a homogenous category. However, Bhabha presents a profound challenge to the geography of Orientalism, insofar as he refutes the binary logic inherent in it by redefining the postcolonial positively in terms of uncertainty and ambivalence, a third space of mixing and multiple cultural borders. Indeed, in *The Location of Culture* (1994), Bhahba introduces concepts such as mimicry and hybridity to question critically the colonial production of binary oppositions (center/margin, civilized/savage, superior/inferior), so as to underscore that cultures interact, transgress, and transform each other in much more complex and nuanced ways than dichotomous thinking allows. Hence, hybridity has the potential to intervene and dislocate processes of domination through reinterpreting and redeploying dominant discourse – the spaces where differences meet become important, which "ultimately heralds the end of colonialism because it is not possible to keep Occidentalism pure and separate from Orientalism" (Sharp 2009: 121). In this sense, rather than seeing impurity/mixing as necessarily bad (as colonial administrators did), Bhabha conceptualizes the hybrid as a form of resistant politics that does not simply redraw boundaries, but actively seeks to subvert them. Although Bhabha, in celebrating cultural heterogeneity has been criticized for relying on essentialist notions of original and distinct cultures (see, e.g., Mitchell 1997), his emphasis on the subversive effects of hybridization has been particularly influential for postcolonial theorizations.

Spivak's ethico-political project, on the other hand, is concerned with the least powerful (the impoverished class, women, and other marginalized people), to whom she collectively refers as subalterns. In her widely cited essay "Can the subaltern speak?" (1988), Spivak argues that Western ways of knowing have always been deemed the *only* legitimate way, thereby downplaying or even negating other (non-Western or indigenous) forms of knowledge. She is thus engaging in

> a critique of the self-assured, scientific approach to studying other cultures that has been characteristic of Western scholarship, which makes the Western self the subject of history and the nonwestern other its object. (Sharp 2009: 111)

As a result, in order for their voices to be heard and taken seriously, subalterns have to adopt Western thought, language, and reasoning. In this sense, Spivak questions whether subalterns can truly speak for themselves or whether their views necessarily have to be represented (sometimes even in a distorted manner). The issue here, according to Spivak, does not lie with the inability of the subaltern to speak, but with the unwillingness of the dominant groups to listen. As such, "the subaltern must always be caught in translation, never truly expressing herself, but always already interpreted" (Briggs & Sharp 2004: 665).

Crystallizing the theoretical insights from Said, Bhabha, and Spivak, it can be argued that postcolonialism has two critical agendas. First, it addresses issues such as identity, race, ethnicity, and gender, challenging the ways in which they are intimately bound up with the power/knowledge nexus, both in terms of how colonial powers produce and use the knowledge of colonized peoples for their own agendas and in how these forms of knowledge continue to structure inequitable relations between the colonizers and the formerly colonized. However, a second concern of postcolonialism goes beyond critique to reignite interest in the agency of marginalized/subaltern groups in articulating, reclaiming, and celebrating their cultural identities. Despite the significant uptake of postcolonial theory in a number of disciplines within the social sciences and the humanities, there has been ongoing contemplation of its actual impacts and continued relevance. There have been growing concerns that the postcolonial intellectual agenda of decolonizing knowledge has yet to be fulfilled. Although there has been more than a decade of talk of decolonizing knowledge, it has been argued that postcolonialism represents the interests of Western-based intellectual elites (Bhabha, Said, and Spivak were employed at universities in Europe and/or the United States when they made their most influential interventions, for example) who speak the language of the Western academy, thereby perpetuating the exclusion of the colonized and oppressed. Furthermore, there is the sentiment that postcolonial perspectives seem to have lost some vigor, especially in the context of various disciplines' fast-changing theoretical predilections, and that certain empirical settings may feel past the postcolonial (e.g., Europe; Sidaway, Woon, & Jacobs 2014). However, critics have pointed out that postcolonial thinking remains vital and relevant in many other places, and that scholars should not lose sight of the temporal and geographical variations in postcolonial theory and its relevance (Gilmartin & Berg 2007; see also special issue of *Singapore Journal of Tropical Geography* 2014).

In reflecting on these concerns, Young (2012) argues that thinking through the postcolonial still has yet-to-be-realized potentials. In particular, he suggests that the postcolonial remains insofar as it "lives on, ceaselessly transformed in the present into new social and political configurations" (Young 2012: 22). As such, he is essentially imploring contemporary postcolonial analyses to take on the challenge of understanding how the postcolonial itself has been changed in response to the historical transformations that have been occurring in the last decades. In other words, what conditions and situations have risen to a new visibility on the radar of postcolonialism? However, even if such questions are increasingly important, Young cautions that to truly decolonize knowledge, postcolonial research must avoid the tendency to employ the language of the "other" while trying to deconstruct this exact same process of "othering" in Western discourses. If not, postcolonialism will simply be "accept[ing] the discriminatory gesture of social and political othering that it appears to contest" (Young 2012: 37). Hence, for Young, the critical task for postcolonial researchers is to delineate possible pathways to stop majority groups from othering minorities altogether, "at which point minorities will be able to represent themselves as they are, in their specific forms of difference, rather as they are othered" (Young 2012: 37).

The rest of this chapter will critically discuss political geographers' ventures into the postcolonial. Apart from documenting how the critical agendas of postcolonialism have shaped the developmental trajectory of political geography, I will also offer insights into how political geographers have, in line with Young's critical exposition, avoided the language of the "other" in order to engage with the novel geopolitical, economic, and cultural logics that imperial and neocolonial geographies promulgate, as well as the resistances to them.

Postcolonial political geography

Until recently, there was little dialogue between postcolonialism and political geography, reflecting differences in disciplinary traditions and politics, and divergences in the languages and concepts used to articulate core issues. However, the possibility of producing a truly decolonized, postcolonial knowledge in political geography became a subject of keen interest and considerable debate during the 1990s. Postcolonialism is an increasing challenge to dominant ways of apprehending North–South relations and has begun to pose significant questions for political geography, particularly regarding the material effects of discourse, the (geo)politics of knowledge, and notions of agency and power. Hence, in what follows I critically discuss the ways in which postcolonial inflections of political geography have allowed for the decentering of "Western" knowledges and the initiation of culturally sensitive scholarship across the North–South divide. Specifically, I argue that attempts to move beyond critique toward critical interventions have brought about a renewed interest in area studies, such that productive pathways are generated in the thinking and doing of postcolonial political geographies.

Critiquing the "colonial present"

The previous section has highlighted how postcolonialism forces us to confront the fact that (Western) forms of knowledge continue to be deployed for the structuring of unequal relations between colonizers and colonized. Drawing strength from such a way of thinking, one of the key contributions of postcolonial political geography is to bring into critical view how dominant Western discourses have pervasively shaped asymmetrical geopolitical relationships on the global stage, which in turn have negative impacts on political dynamics in many countries outside of the European and North American contexts. There are various examples that will fit well into this paradigm, but Derek Gregory's acclaimed monograph *The Colonial Present* (2004) is perhaps a reference point. As the title of the book suggests, Gregory is chiefly concerned with the ways in which colonialism continues to manifest itself in the present through the construction of imaginative geographies or "constructions that fold distance into difference through a series of spatializations" (2004: 17). In this formulation, Gregory upholds that the absence of traditional colonies does not imply the absence of colonialism. Specifically, he extends Said's idea of Orientalism to "real" spaces to show how the US-initiated global

"War on Terror" was framed in cultural terms and reminiscent of the colonial past. Indeed, through the case examples of Afghanistan, Iraq, and Palestine, Gregory demonstrates how the War on Terror is premised on a dichotomous relationship between "us" and "them" or "civilization" and "barbarism," ultimately suggesting that the United States was involved in a moral and just battle against evil. Such performances of space, articulated through imaginative geographies, essentially reduce the "other" to the status of *Homo sacer*: one who occupies the space of the exception and has no political voice, and whose death does not matter either. The ultimate goal of *The Colonial Present*, then, is to highlight the physical and corporeal manifestations of contemporary colonialism. In doing so, Gregory exposes the brutality of violence that accompanies imperial (territorial) control in the Middle East, which a complacent, if not complicit, American government and media have tried to shield from its public.

The colonial violence that is unleashed by the United States in its global War on Terror also forms the primary focus of an article by Roberts, Secor, and Sparke (2003). However, these authors go one step further in speculating on the fundamental reasons/ambitions that drive America's contemporary counter-terrorism initiatives in different parts of the world. Elaborating on what they term "neoliberal geopolitics," Roberts et al. (2003: 887) explore "how a certain globalist and economistic view of the world, one associated with neoliberalism, did service in legitimating the war [on terror]." Using the 2003 Iraq war as a case in point, the authors go on to illuminate the ways in which a Western, neoliberal world vision has served to obscure the more "traditional" geopolitical purposes of US interventions in Iraq. Indeed, while the Bush administration insists that the Iraq war is one that seeks to augment the much-desired Western neoliberal values of democracy and freedom in the Middle Eastern state, Roberts et al. echo other geographers in arguing that the war was in fact a traditional national, imperial war aimed at the monopolization of resources (cf. Graham 2004; Le Billon & El Khatib 2004). They say:

> It was ... partly a war about securing American control over Iraqi oil. Russia's Lukoil and France's TotalFinalElf will thereby lose out vis-à-vis Chevron and Exxon; more importantly, the US will now be able to function as what Christian Parenti (2003) calls an "energy gendarme" over key oil supplies to East Asia and Europe. (Roberts et al. 2003: 888)

The quote duly illustrates how resource geopolitics and the military-industrial complex were pertinent in shaping the Bush administration's decision to go to war. In other words, the specific vision of neoliberal geopolitics has resulted in many neoliberals who

> support the war insofar as it helped to ... facilitate the planning and overarching coordination of the violence and insofar as the war showed how the extension of neoliberal practices on a global scale has come to depend on violent interventions by the US. (Roberts et al. 2003: 895)

As such, Roberts et al.'s work can be said to embody the spirit of postcolonial political geography in two ways. First, it exposes how Western neoliberal values have been projected as the universal, governing norm in a variety of (geo)political circumstances. Second, these neoliberal discourses consequently function as a source of power to enable certain Western actors to hide their insidious, imperial agendas, thereby leading to wider socio-economic dispossession of communities in the so-called global South (given that these communities' everyday geographies are defined by existential violence and their country's resources exploited to nourish foreign interests).

Moving away from the spaces and events associated with the War on Terror, a number of political geographers have also utilized a postcolonial lens to look at the (geo)political discourses and practices associated with Western interventions in the African continent (Power 2001, 2010; Power, Mohan, & Mercer 2006; Power & Mohan, 2010). For instance, Sidaway (2003: 157) offers a "postcolonial criticism of a central object of political geography: conceptualizations of sovereignty and the territorial state." He argues that Western caricatures of African sovereignties have tended to employ the language of "weak" or "failed" states to connote their lack or absence of authority and connections such as being excluded from processes of economic liberalization, global market integration, democratization, and so on. Hence, African states are often seen through the lens of exceptionalism, whereby the Western model of statehood is taken to be the reference point, the norm, and African countries are always dichotomously represented as the less-than-ideal "other" (see also Doty 1996; Power 2003; Carmody 2012). However, as Sidaway rightly points out, African sovereignties should not be scripted as "lacking" in certain critical Western elements and qualities; rather, it is the *excess* of these Western influences that are shaping the supposed "abnormalities" of African states. Recounting the case of Angola, he suggests that the geohistorical developments of the African state are profoundly connected to Western normalities such as the transnational flows of oil, gems, weapons, and capital into Angola (especially Swiss bank accounts, plus European and American multinationals). In his words (2003: 164):

> Angola is symptomatic of how malign combinations of imperialism, Cold War, the power of money, minerals (global demand for oil and diamonds) and violence may interact. Angola's situation is a production of these interactions.

In sum, Sidaway has convincingly revealed how the endurance of derogatory (colonial) frames in looking at African statehood has helped to obscure the process whereby the West is fully caught up with (not external to) the production of what are scripted as weak states.

Critiquing "Western" production of (geo)political knowledge

Insofar as critiques have emerged to question the universality of Western knowledges that are invoked for imperial agendas, political geographers have also engaged in introspective reflections to ascertain whether the subdiscipline has been complicit in privileging particular ways of knowing. Indeed, claims abound that many political geographical terms and categories tend to operate in a limited selection of languages and traditions (Dodds & Atkinson 2000; Mamadouh 2003; Agnew 2013; Sidaway et al. 2014). As such, the infusion of postcolonial theory into political geography has culminated in calls for the subdiscipline to be more critically aware of the situated basis of its claims and vantage point.

Responding to such calls, Sidaway (2008) has attempted to map out "the geography of political geography." In doing so, he points out that political geography has always offered a distinctly located assessment of its themes (such as territory, sovereignty, heartland, and shatterbelts). Indeed, the view from the West (from the United States, United Kingdom, and Germany, most commonly) has shaped political geography's theorizations of topics as diverse as geopolitics, conceptions of the state, or feminist interventions in the field of the political. For instance, the writings of Halford Mackinder (1904) and Friedrich Ratzel (1940) exemplified how state-building and competing national and imperial projects provided intellectual space and patronage for political geography in a number of European countries in the early twentieth century. These are assessments to which we are still encouraged to turn today in order to sketch out how and why territories link states to their populations in relations of authority, legitimacy, and surveillance. As such, for anyone trying to explore the field of the political beyond this Euro-American axis, the model of the Western state continues to serve as *the* theory of the state, which leaves vast realms of the world of states beyond the comprehension of this kind of political geography. This has led Finnish political geographer Jouni Häkli (2003) to caution that the concepts associated with political geography as developed in Anglophone literatures might carry the risk of an unwarranted universalism:

> A particular parochiality is thus universalized and made to pass as the best available internationally recognized scholarship. But perhaps there is a market for other parochialities. French, Mediterranean, Nordic, Iberian, South American, African, ones that are poorly known by those who cannot read work done outside Anglo-American circles. If this is the case, then the universalism of Anglophone geography is but an illusion caused by lack of knowledge concerning the richness of the political geographical world. (Häkli 2003: 660)

Häkli acknowledges the existence of other geopolitical traditions outside of the United States–European Union (US–EU) axis (see, e.g., Foresta 1992 and Dodds 1993 for engagements with Latin American (geo)political ideas), but he argues that the limited language capabilities of many Anglophone political geographers inherently create a certain linguistic imperialism on the part of the wider Anglophone academy – interrogations are largely limited to English-language materials (see also Hepple 2000). Häkli's concerns are echoed by a number of political geographers who highlight that the development of political geography as a subdiscipline has been marked by a primary focus on Anglo-American cases and interests. Perry (1987: 6), writing in the 1980s, claimed that "Anglo-American political geography poses and pursues a limited and impoverished version of the discipline, largely ignoring the political concerns of four-fifths of humankind." This situation persisted into the 1990s, with Kofman (1994: 437) noting "the heavily Anglocentric, let alone Eurocentric, bias of political geography writing."

In reflecting on the possibilities of doing political geography in a postcolonial context, Jenny Robinson (2003a) goes one step further to divulge the factors contributing to the US–EU being established as the hegemonic zone of the production of knowledge. She coins the term "knowledge-production complex" (KPC) to "capture the way in which the knowledge produced within the contemporary circuits of intellectual labour is deeply uneven in its structure and profoundly biased against scholars working outside of the heartlands of geography" (Robinson 2003b: 281). She argues that academics in privileged locations have direct access to publishers and markets that ensure the circulation and citation of their work. Publishers function as active gatekeepers to maintain tight surveillance of the contents of texts; prospective sales in these key markets discipline authors, referees, and editors alike so as to ensure the profitability of the sector. For example, Robinson points out that editors and referees of books or journal articles are often asked to consider questions such as: Does the scholarship contribute to original knowledge? Does it draw on relevant literatures and methods? However, assumptions are often made about appropriate forms of scholarship and what constitutes a relevant field of reference for topics of inquiry that simply disadvantage academics from marginal contexts. Robinson relates her own encounters with the KPC, whereby her work on South Africa was deemed too parochial and not general enough to sustain publication in its own right (Robinson 2003a). Hence, through the KPC, Western knowledges that are arguably similarly located and parochial in nature gain universal acceptance; for instance, "western feminist scholarship becomes 'feminist geography', and theories of a few western states produce 'political geographies of the state'" (Robinson 2003a: 648).

Beyond critique: Postcolonial political geography and area studies

As alluded to earlier in this chapter, postcolonial theory refrains from lapsing into mere critiques of the problematic representations and initiatives associated with "non-Western" communities and societies. Instead, attention has been paid simultaneously to charting alternative ways forward in terms of critical interventions. To achieve this, researchers have responded to exhortations for area studies to be reinvigorated for the postcolonializing of political geography (Robinson 2003a; see also Sidaway 2013). For area studies to be used as a tactic of postcolonializing (political) geography might seem strange given that the history of area studies is intimately bound up with the geopolitics of knowledge, arising out of colonial imperatives to compile detailed information about places and regions for the classification and ordering of the world (Gibson-Graham 2004; Roy 2009). However, there has been a rethinking of area studies, such that the emphasis is no longer on "trait geographies" but on "process geographies" (Appadurai 2000); in other words, on the forms of movement, encounters, and exchanges that confound the idea of bounded world regions of immutable traits. As Olds (2001: 129) notes:

> the large regions which dominate the current map for area studies are not permanent geographical facts. They are problematic heuristic devices for the study of global geographic and cultural processes.

It is this reformulated area studies that can help to postcolonialize political geography. This not only goes beyond the global/local divide that has encapsulated debates related to area studies, it also eschews the (imperially driven) hierarchization of societies and economies by highlighting the potentialities of learning from different worlds (McFarlane 2006; Jazeel & McFarlane 2010). In this view, area studies can purposefully "provincialize" the forms of knowledge that are produced in Europe and the United States, and that pass themselves off as universal (Chakrabarty 2000). As Appadurai (1996: 16) puts it, "area studies is a crucial resource for postcolonialising the production of academic knowledge" insofar as it has "provided the major counterpoint to the delusions of the view from nowhere that underwrite much canonical social science."

Adopting such an approach, I will highlight two key ways in which postcolonial political geographers have used area studies to call into question the universality of Western systems of knowledge and positively address the interrelated histories (and geographies) of violence, domination, inequality, and injustice. In particular, I will showcase how these works have sought to interrogate the transformative energies of the postcolonial into new social and political configurations, while materializing Young's (2012) plea not to reify the "other" through postcolonial interventions.

"New" empires?

Recent years have seen significant interest in the economic ascendancy of countries outside of the West, which contributes to a shifting of paradigms of geoeconomic and geopolitical power that have certainly modified the sensibility of colonial dependency. One area that has arguably attracted widespread attention is Asia, as attested by the buzz around the discourse of "emerging Asia" (or what some term "rising Asia") featuring two key "global" players, China and India (Raghuram, Noxolo, & Madge 2014). I am going to focus on the case of China here, not because India is any the less important, but rather because my expertise allows me to speak more convincingly of the (geo)politics surrounding China's purported emergence as a global power (see Woon 2012). China's current status requires us to rethink the logics of imperialism and empire. Rather than being the peripheral and non-Western "other," China has been a focus of extensive debates, with many Western scholars considering whether the Asian giant is set to become the next global hegemon with imperialist ambitions (Horner 2009; Luttwak 2012). China's ventures into Africa have more often than not been used as case studies to augment such a perspective. Despite calls for fuller understandings of the historical and institutional genealogies that shape Sino-African relations (e.g., Mawdsley 2007; Mohan & Power 2008), claims abound that China is walking down the same path toward empire in Africa as the former European powers did a century ago (Alden, Large, & Soares De Oliveria 2008; French 2010). China's role as a new imperial power is similarly explored in Anand's (2012) work, albeit the focus here is on China's "internal" relationship with its minority groups. According to Anand (2012: 83), China acts imperially within its sovereign space, especially near its contested peripheries (e.g., Xinjiang and Tibet), "acknowledging cultural difference but erasing political identities [defined as secessionist]." As such, China's imperial status is not defined by expansion, but is more to do with these internal forms of domination and consolidation.

However, rather than allowing these Western narratives to dictate what can be seen and heard about China's "place in the world," political geographers such as Agnew (2010, 2012), Cartier (2013), and Woon (2012) have drawn inspiration from postcolonial theory to champion the case for more research to be conducted on the ways in which Chinese elites (e.g., scholars and policymakers) envisage their future on the international stage (cf. Callahan 2011). This is not to suggest that Chinese geopolitical narratives are necessarily more authentic, nor is it to uphold an essentialist view that there are fundamental ideas intrinsic to the existing Chinese polity. Rather, we argue that looking at Chinese accounts allows insights into how local elites understand important (geo)political issues and, in the process, construct imag(in) ings that will plausibly provide justifications for China's foreign policy. Such revelations are important, as world leaders and opinion-makers can better address China's global challenge if they see China for what it is, as opposed to what they want it to be.

It is interesting to note that ideas related to a Chinese empire do not emerge purely out of Western geopolitical imaginations. Crucially, there have been voices emanating within China that seek to harness the notion of empire to usher in Chinese visions of world order in the twenty-first century (Qin 2005; Zhao 2005, 2011).[6] Agnew (2012), however, has pointed out that competing accounts do exist that call into question the possibility of an emerging Chinese empire. In particular, he argues that while discourses about China's rise focus on divulging insights about the country's present policies and future orientations, very little attention is devoted to examining China's colonial past and how postcolonial consciousness affects the country. Chinese elites have frequently drawn on China's collective memory of humiliation (国耻) at the hands of foreign powers in the nineteenth and twentieth centuries to remind audiences worldwide of two important "facts." First, China was never a colonial power but has been subjected to the tyranny of different colonizers. As such, it will never reproduce the painful experience of colonialism on others. Second, this historical humiliation of China has now fueled a desire for recognition and respect of the country's sovereign space, both from foreigners as well as from minorities scattered across China's territories (see also Wang 2008). Citing the works of William Callahan (2009, 2010), Agnew highlights the significance of the "cartography of national humiliation," the use of historical and contemporary maps to document both Chinese claims and foreign denials. Through these various humiliation maps, it can be seen that China's shift from premodern unbounded understandings of space and territory (疆域) to bounded understandings of sovereign territory (主权领土) in the twentieth century has been anything but a simple linear progression. Indeed, it is in the interplay of the otherwise contradictory cartographic conventions of imperial domain space and sovereign territorial space that China's geobody actually emerges. However, as Agnew goes on to illustrate, China's geobody is still neither stable nor hegemonic, even after a century of crafting – it faces counter-discursive resistances on multiple fronts. For example, the long-standing conflicts between Taiwan – the "renegade province," the seat of the Kuomingtang government "in exile" since 1949 – and China reflect the increasingly separate identity of Taiwan and the refusal of the Beijing government to entertain any ideas to do with the independence of Taiwan from "China proper." With the rise of the independence movement in Taiwan, however, official maps of the island have shifted from accepting the possibility of incorporating Taiwan into the geobody of China to seeing it as separate (e.g., through portraying Taiwan as a maritime nation facing away from mainland China in a deliberate act of politicized cartography; Agnew 2012: 308). In this sense, is perhaps inaccurate to claim that China has irredentist geopolitical ambitions for the current times, given that it is more concerned with the challenges posed by different factions within its sovereign space (e.g., Taiwan, Tibet), rather than actively seeking out more territory abroad.

Thus, in highlighting these dimensions, (postcolonial) political geographers have gone beyond critiques to advance debates related to empire, colonialism, and imperialism through diversifying understandings and usages of these conceptual parameters beyond Western epistemological traditions. Chinese historian Emma Teng (2004: 7) has called for a corrective to the assumption that imperialism was "essentially a Western phenomenon" – the case of China indicates that when this corrective is made to the focus of scholarship, some novel variants of imperialism, colonialism, and postcolonialism can start to emerge.

Subaltern geopolitics

A second way in which political geographers have positively intervened in the postcolonial through area studies is to look for evidence of counter-hegemonic voices in subaltern spaces that may contest and to varying extents transform the power relations of hegemonic discourse. Young (2012: 27) asserts that the postcolonial question in current times is no longer about a formal colonizer–colonized relation; rather, it is about "how to make the dream of emancipation accessible for all those people who fall outside the needs of contemporary modernity." In other words, there is an urgent need to "reformulate the emancipatory aims of anticolonial struggle outside the parameters of nation-state" (2012: 27). Such an objective similarly informs a handful of political geographical works that operate under the conceptual rubric of "subaltern geopolitics." As Sharp (2011) explains, "subaltern geopolitics" is concerned with subjects that are cast outside of the state, but it does so by not simply categorizing them as the "other," as alternative, as enacting resistance. Instead, it is a

> positioning that recognizes the possibility that political identities can be established through
> geographical representations that are neither fully "inside" nor "outside", and thus seeks a model

of political subjectivity to challenge that perpetuated by dominant western geopolitics that does not rely on otherness. (Sharp 2011: 271–272)

Hence, subaltern geopolitics is in agreement with Young (2012) that postcolonial concerns should not be couched in the language of the "other," given that looking at spaces of geopolitical knowledge production in terms of the dominant/resistant framework can only serve to reproduce this binary geopolitical structure, rather than challenging it.

Indeed, in a special issue on the theme of subaltern geopolitics (*Geoforum* 2011), various scholars have anchored on Sharp's conceptualization to advance progressive (geo)political agendas. These works seek to intervene in what Judith Butler (2009: 25–26) terms "precarity," a "politically induced condition in which certain populations suffer from failing social and economic networks of support and become differentially exposed to injury, violence and death." To do so, the authors acknowledge Butler's call for a common humanity catalyzed by vulnerability as a generalized condition rather than a differential way of marking a cultural identity (see also Woon 2014). Thus, these works can be said to embody a "feminist ethics of cosmopolitanism" whereby "subjects are endlessly (re)constituted through dialectical processes of recognition within multiple networks of power" (Mitchell 2007: 6). For instance, Koopman (2011) has for some time promulgated the notion of "alter-geopolitics" – new proposals and practices that challenge hegemonic geopolitics and create new geopolitics. Although Koopman does not employ the term "subaltern geopolitics," Sharp (2011: 272) argues that Koopman shares the concerns of subaltern geopolitics by recognizing "the dispersion of agency through the political system, to see those outside of the formal circuits of power/knowledge of international relations and statecraft." Koopman draws attention to grassroots movements that build international relations of solidarity in opposition to mainstream geopolitics, arguing that academics might become involved in struggles where principles of feminist geopolitics are already being translated into political reality. Using the case of international accompaniment in Colombia, a strategy that "puts bodies that are less at risk next to bodies that are under threat," Koopman (2011) highlights how "western" volunteers actively rely on their "privileged" bodies to shield and protect precarious lives. This help to resist oppressive powers by "establishing their own performances of security through putting ... bodies together for safety, by establishing ... people to people relations across space" (Koopman 2011: 274). In this sense, the accompaniers are using the fact that their (Western) bodies/lives "count" more to protect the bodies that are deemed outside of state protection, able to be injured – a performance that is perhaps reminiscent of Butler's ideas about forging a community on the basis of shared vulnerability and that ultimately attempts to build a world where everyone's lives "count."

Alternatively, I have been interested for some time now in the efforts of the Philippines rebel group Rebulusyonaryong Partido ng Manggagawa ng Mindanao (RPM-M), and in how this dissident organization has actively repudiated the label of terrorism that has come to be associated with it by emphasizing instead issues of state-induced vulnerability and marginalization (Woon 2011). Through this case study, I argue for a form of security that emphasizes linkages rather than difference – peoples in the Philippines (and beyond) can place themselves in an "extended tapestry of connection and belonging that does not arise from an imagined fear of 'others' nor does it have exclusion as its basis" (Woon 2011: 294). This case of the Philippines thus helps to highlight the possibilities of subaltern agency in promoting the recognition of the shared precariousness of lives, so as to (re)define who counts as legitimate subjects, rather than allowing hegemonic powers to create an unequal distribution of vulnerability among different social groups.

"Doing" postcolonial political geographies

Following this chapter's insistence on being sensitive to the postcolonial's lingering presence, I would like to end not by crafting a "conventional" conclusion, but by reflecting on the different ways of doing postcolonial political geographies in order to signal the open-ended possibilities of the subdiscipline's developmental trajectory. While critique remains a crucial role of postcolonial political geography, this chapter has simultaneously foregrounded the prospects that area studies holds for advancing progressive alternatives. However, how exactly should area studies be initiated such that its full potential for postcolonializing political geography can be realized? In seeking the answer to this pertinent question, I am increasingly convinced by Spivak's (2003) notion of "planetarity" and its implications for area studies. For Spivak (2003: 16), area studies has to be a deterritorialized discipline insofar as it "must always cross borders" as well as "displace globalization into planetarity." Essentially, Spivak is imploring us to respect linguistic and cultural diversity by "imagining ourselves as planetary subjects rather than global agents" (2003: 73). The distinction here between globalization and the planetary is an important one. According to Spivak, while globalization implies the same system of exchange everywhere, the planet is "the species of alterity, belonging to another system, on loan to us" (2003: 72). What this means is that area studies should not be about utilizing the language of one-worldism that globally extends Western categories of thought – to bring others into our universe or to judge others according to our views. Rather, it is for us, together with others, to envision in a collective process working toward the plain of coexistence. Hence, planetarity poses a challenge to decolonize our knowledge of the world by extending an invitation to know it from outside the categories of Western thought. The key point here is recognizing, as Spivak (2003: 73) puts it, that "alterity remains underived from us," not immediately comprehensible by the violent normalizations of a universal claiming to speak for the particular. As such, since planetarity is situated in the domain of uncertainty, the to-come, it urges continual hard work to keep on decentering ourselves in the face of ungraspable otherness and other worldings. In Jazeel's (2011: 88) words, planetarity "looks in fact to continually displace from the West, and realign, the axis of comparison in knowledge production and in extensions towards difference." Put simply, how can I know difference in ways that do not prescribe otherness in my own terms?

If we incorporate such planetary consciousness into doing postcolonial political geographies, the methodological implications are significant. (Postcolonial) political geographers should not reduce the language and communities of the southern hemisphere to mere objects of study; rather, they must take up the position of a learning subject inclined to "learn from below" (Spivak 2003: 15). This type of learning is not about speaking for an individual or group in order to produce something like a notionally more cosmopolitan geographical knowledge. Instead, it seeks to develop new positions through interactions between researchers and people in disparate locations. This hinges on a greater sensitivity to the relationships of power, authority, positionality, and knowledge; an ethic that demands consistent critical reflections in an effort that seek to learn between contexts and constituencies (Jazeel & McFarlane 2010). It requires continual questioning and reflexivity about one's relative position (e.g., nationality, class, and gender) and how that can potentially have negative effects on attempts to create knowledge and to represent others. A good example of this responsible learning process is embodied in the methodological commitment to praxis that refocuses on resistance, agency, and action in developing societies. There are already ongoing efforts by political geographers to move out of the ivory tower and collaborate with the communities and social movements in the South. This can involve employing field techniques that introduce a collaborative ethos such as those offered by participatory action research, so that the dangers of misrepresenting different subaltern groups can be avoided through culturally informed understandings of the agency of research participants in the (re)making of their individual life-worlds. Thus, the key lesson here for political geographers is that they must always be prepared to work with the uncertainties inherent in a postcolonial approach to learning, and come to accept that their knowledges, ideas, or theories may be altered, transformed, or even rejected through responsible engagements with different constituencies.

Notes

1 The warship's proposed name, *Osman-Harun*, is derived from the two Indonesian marines, Osman Mohamed Ali and Harun Said. Their act of terror killed 3 and wounded 33 others (see Sebastian 2014).

2 Maphilindo was the name for the United Malay state consisting of Malaysia, the Philippines, and Indonesia that was proposed in the late 1950s.

3 When the hyphen is used in the term post-colonialism, it normally refers to the "common-sensical definition of post-colonialism as the period following independence from colonialising power" (Sharp 2009: 4).

4 I am aware that the dualistic categories of "North–South" and "East–West" are loaded in meaning and have specific genealogies (see Sheppard & Nagar 2004). However, I am using these categories here to highlight the unequal power relationships that exist between these various spatial blocs, with "North" and "West" widely used to connote the dominant, powerful states on the global stage.

5 Other writings that are primarily based on "real-life" anti-colonial/postcolonial struggles have also shaped the development of postcolonialism – e.g., Franz Fanon and his experiences in Algeria and the Subaltern Studies Group (see Young 2001; McEwan 2009).

6 The work of Zhao Tingyang, a prominent philosopher based at the Chinese Academy of Social Sciences, has been especially influential in this respect. According to Zhao, the Chinese concept of 天下 (Tianxia), literally translated as "all under heaven," can do the conceptual work that is necessary for a stable world. His claim is premised on the fact that that in comparison to other historical empire systems, the Tianxia system is "the most appropriate empire for the twenty-first century" (2005: 102–109), given that it is the only system that thinks through the world, while previous empires have taken a particular nation-state (and its values and standards) as the model.

References

Agnew, J. (2010). Emerging China and critical geopolitics: Between world politics and Chinese particularity. *Eurasian Geography and Economics* 51(5): 569–582.

Agnew, J. (2012). Looking back to looking forward: Chinese geopolitical narratives and China's past. *Eurasian Geography and Economics* 53(3): 301–314

Agnew, J. (2013). The origins of critical geopolitics. In Dodds, K., Kuus, M., & Sharp, J. (eds), *The Ashgate Research Companion to Geopolitics*, Aldershot: Ashgate, 19–32.

Alden, C., Large, D., & Soares De Oliveria, R. (eds) (2008) *China Returns to Africa: A Rising Power and a Continent Embrace*. London: C. Hurst.

Anand, D. (2012). China and India: Postcolonial informal empires in the emerging global order. *Rethinking Marxism: A Journal of Economics, Culture and Society* 24: 68–86.

Appadurai, A. (1996). *Modernity at Large: Cultural Dimensions of Globalization*. Minneapolis, MN: University of Minnesota Press.

Appadurai, A. (2000). Grassroots globalization and research imagination. *Public Culture* 12(1): 1–19.

Bhabha, H. (1994). *The Location of Culture*. London: Routledge.

Blunt, A., & McEwan, C. (2002). Introduction. In Blunt, A., & McEwan, C. (eds), *Postcolonial Geographies*, London: Continuum, 1–8.

Briggs, J., & Sharp, J. (2004). Indigenous knowledges and development: A postcolonial caution. *Third World Quarterly* 25(4): 661–676.

Butler, J. (2009). *Frames of War: When Is Life Grievable?* London: Verso.

Callahan, W. (2009). The cartography of national humiliation and the emergence of China's geobody. *Public Culture* 21: 141–173.

Callahan, W. (2010). *China: The Pessoptimist Nation*. Oxford: Oxford University Press.

Callahan, W. (2011). Tianxia, empire and the world: Chinese visions of world order for the twenty-first century. In Callahan, W., & Barabantseva, E. (eds), *China Orders the World: Normative Soft Power and Foreign Policy*, Washington, DC: Woodrow Wilson Center Press, 91–117.

Carmody, P. (2012). *The New Scramble for Africa*. Cambridge: Polity Press.

Cartier, C. (2013). What's territorial about China? From geopolitical narratives to the "administrative area economy." *Eurasian Geography and Economics* 54(1): 57–77.

Chakrabarty, D. (2000). *Provincialising Europe: Postcolonial Thought and Historical Difference*. Princeton, NJ: Princeton University Press.

Dodds, K. (1993). Geopolitics, cartography and the state in South America. *Political Geography* 12(4): 361–381.

Dodds, K., & Atkinson, D. (eds) (2000). *Geopolitical Traditions: Critical Histories of a Century of Geopolitical Thought*. London: Routledge.

Doty, R. (1996). *Imperial Encounters: The Politics of Representation in North–South Relations*. Minneapolis, MN: University of Minnesota Press.

Foresta, R. (1992). Amazonia and the politics of geopolitics. *Geographical Review* 82(2): 128–142.

French, H. (2010). The next empire. *The Atlantic* April 13. http://www.theatlantic.com/magazine/archive/2010/05/the-next-empire/308018/, accessed March 21, 2014.

Geoforum (2011). Special Issue on Subaltern Geopolitics. 42(3).

Gibson-Graham, J.K. (2004). Area studies after poststructuralism. *Environment and Planning A* 36(3): 405–419.

Gilmartin, M., & Berg, L. (2007). Locating postcolonialism. *Area* 39(1): 120–124.

Graham, S. (ed.) (2004). *Cities, War and Terrorism: Towards an Urban Geopolitics*. Malden, MA: Blackwell.

Gregory, D. (2004). *The Colonial Present: Afghanistan, Palestine, Iraq*. Malden, MA: Blackwell.

Häkli, J. (2003). To discipline or not to discipline, is that the question? *Political Geography* 22(6): 657–661.

Hepple, L. (2000). Geopolitiques de gauche: Yves Lacost, Herodote and French radical geopolitics. In Dodds, K., & Atkinson, D. (eds), *Geopolitical Traditions: A Century of Geopolitical Thought*, London: Routledge, 268–301.

Horner, C. (2009). *Rising China and Its Postmodern Fate: Memories of Empire in a New Global Context*. Athens, GA: University of Georgia Press.

Jacobs, J. (1996). *Edge of Empire*. London: Routledge.

Jazeel, T. (2011). Spatializing difference beyond cosmopolitanism: Rethinking planetary future. *Theory, Culture and Society* 28(5): 75–97.

Jazeel, T., & McFarlane, C. (2010). The limits of responsibility: A postcolonial politics of academic knowledge production. *Transactions of the Institute of British Geographers* NS 35(1): 109–124.

Kahin, G. (1964). Malaysia and Indonesia. *Pacific Affairs* 37(3): 253–270.

Kausikan, B. (2014). Indonesia's naming of navy ship: Sensitivity is a two-way street. *The Straits Times* 13 February.

Kofman, E. (1994). Unfinished agendas: Acting upon minority voices of the past decade. *Geoforum* 25(4): 429–443.

Koopman, S. (2011). Alter-geopolitics: Other securities are happening. *Geoforum* 42(3): 274–284.

Law, L. (2007). Remapping the geopolitics of terror: Uncanny urban spaces in Singapore. In Berkling, L., Frank, S., Frers, L., Low, M., Steets, S., & Stoetzer, S. (eds), *Negotiating Urban Conflicts: Interaction, Space and Control*, Bielefeld: Transcript, 53–65.

Le Billon, P., & El Khatib, F. (2004). From free oil to "freedom oil": Terrorism, war and US geopolitics in the Persian Gulf. *Geopolitics* 9(1): 109–137.

Luttwak, E.W. (2012). *The Rise of China vs. the Logic of Sovereignty*. Cambridge, MA: Belknap Press.

Mackinder, H. (1904). The geographical pivot of history. *Geographical Journal* 23(4): 421–437.

Mamadouh, V. (2003). Some notes on the politics of political geography. *Political Geography* 22(6): 663–675.

Mawdsley, E. (2007). China and Africa: Emerging challenges to the geographies of power. *Geography Compass* 1(3): 405–421.

McEwan, C. (2009). Postcolonialism/postcolonial geographies. In Kitchin, R., & Thrift, N. (eds), *International Encyclopedia of Human Geography*, London: Elsevier, 327–333.

McFarlane, C. (2006). Crossing borders: Development, learning and the North–South divide. *Third World Quarterly* 27(8): 1413–1437.

Mitchell, K. (1997). Different diasporas and the hype of hybridity. *Environment and Planning D: Society and Space* 15(5): 533–553.

Mitchell, K. (2007). Geographies of identity: The intimate cosmopolitan. *Progress in Human Geography* 31(5): 706–720.

Mohan, G., & Power, M. (2008). New African choices? The politics of Chinese engagement. *Review of African Political Economy* 35(1): 23–42.

Olds, K. (2001). Practices for "process geographies": A view from within and outside the periphery. *Environment and Planning D: Society and Space* 19(2): 127–136.

Perry, P. (1987). Political Geography Quarterly: A content (but discontented) review. *Political Geography Quarterly* 6(1): 5–6.

Poulgrain, G. (1998). *The Genesis of Konfrontasi: Malaysia, Indonesia, Brunei, 1945–1965*. London: C. Hurst and Co.

Power, M. (2001). Geo-politics and the representation of Portugal's African colonial wars: Examining the limits of "Vietnam syndrome." *Political Geography* 20(4): 461–491.

Power, M. (2003). Re-imagining postcolonial Africa: A commentary on Michael Watts' "Development and Governmentality." *Singapore Journal of Tropical Geography* 24(1): 49–60.

Power, M. (2010). Geopolitics and development: An introduction. *Geopolitics* 15(3): 433–440.

Power, M., & Mohan, G. (2010). Towards a critical geopolitics of China's engagement with African development. *Geopolitics* 15(3): 462–495.

Power, M., Mohan, G., & Mercer, C. (2006). Post-colonial geographies of development: An introduction. *Singapore Journal of Tropical Geography* 27(3): 231–234.

Qin, Y.Q. (2005). Guoji zhengzhi lilun de hexin wenti yu Zhongguo xuepai de shengcheng (The core problem of international relations theory and the emergence of the Chinese School). *Zhongguo shehui kexue* 3: 165–176.

Raghuram, P., Noxolo, P., & Madge, C. (2014). Rising Asia and postcolonial geography. *Singapore Journal of Tropical Geography* 35(1): 119–135.

Ratzel, F. (1940). *Erdenmacht und Voelkerschicksal*. Stuttgart: Stuttgart University Press.

Roberts, S., Secor, A., & Sparke, M. (2003). Neoliberal geopolitics. *Antipode* 35(5): 886–897.

Robinson, J. (2003a). Political geography in a postcolonial context. *Political Geography* 22(6): 647–651.

Robinson, J. (2003b). Postcolonialising geography: Tactics and pitfalls. *Singapore Journal of Tropical Geography* 24(3): 273–289.

Roy, A. (2009). The 21st century metropolis: New geographies of theory. *Regional Studies* 43(6): 819–830.

Said, E. (1978). *Orientalism*. New York: Random House.

Sebastian, L. (2014). The Usman-Harun issue: Some thoughts for Indonesia to ponder. *RSIS Commentaries*, https://www.rsis.edu.sg/rsis-publication/idss/2154-the-usman-harun-issue-some-th/#.VR6l0VxmnZc, accessed April 3, 2015.

Sharp, J. (2003). Feminist and postcolonial engagements. In Agnew, J., Mitchell, K., & Toal, G. (eds), *A Companion to Political Geography*, Malden, MA: Blackwell, 59–74.

Sharp, J. (2009). *Geographies of Postcolonialism*. London: Sage.

Sharp, J. (2011). Subaltern geopolitics: Introduction. *Geoforum* 42(3): 271–273.

Sheppard, E., & Nagar, R. (2004). From East-West to North-South. *Antipode* 36(4): 557–563.

Sidaway, J. (2003). Sovereign excess? Portraying postcolonial sovereigntyscapes. *Political Geography* 22(2): 157–178.

Sidaway, J. (2008). The geography of political geography. In Cox, K., Low, M., & Robinson, J. (eds), *The Sage Handbook of Political Geography*, London: Sage, 41–56.

Sidaway, J. (2013). Geography, globalization and the problematic of area studies. *Annals of the Association of American Geographers* 103(4): 984–1002.

Sidaway, J., Woon, C.Y., & Jacobs, J. (2014). Planetary postcolonialism. *Singapore Journal of Tropical Geography* 35(1): 4–21.

Singapore Journal of Tropical Geography (2014). Special Issue on Advancing Postcolonial Geographies. 35(1).

Spivak, G. (1988). Can the subaltern speak? In Nelson, C., & Grossberg, L. (eds), *Marxism and the Interpretation of Culture*, Urbana, IL: University of Illinois Press, 271–313.

Spivak, G. (2003). *Death of a Discipline*. New York: Columbia University Press.

Tan, G.S.H. (2008). *Indonesian Confrontation and Sarawak Communist Insurgency, 1963–1966: Experiences of a Local Reporter*. Kuching, Malaysia: Penerbitan Sehati.

Teng, E.J. (2004). *Taiwan's Imagined Geography: Chinese Colonial Travel Writing and Pictures, 1683–1895*. Cambridge, MA: Harvard University Press.

Wang, Z. (2008). National humiliation, history education, and the politics of historical memory: Patriotic education campaign in China. *International Studies Quarterly* 52(4): 783–806.

Woon, C.Y. (2011). Undoing violence, unbounding precarity: Beyond the frames of terror in the Philippines. *Geoforum* 42(3): 285–296.

Woon, C.Y. (2012). Comment on looking back to looking forward: Geopolitics, identity and engaging China. *Eurasian Geography and Economics* 53(3): 331–337.

Woon, C.Y. (2014). Precarious geopolitics and the possibilities of nonviolence. *Progress in Human Geography* 38(5): 654–670.

Young, R. (2001). *Postcolonialism: An Historical Introduction*. Malden, MA: Blackwell.

Young, R. (2012). Postcolonial remains. *New Literary History* 43(1): 19–42.

Zhao, T.Y. (2005). Tianxia tixi: shijie zhidu zhexue daolun (The Tianxia system: A philosophy for the world institution). Nanjing: Jiangsu jiaoyu chubanshe.

Zhao, T.Y. (2011). Rethinking empire from the Chinese concept "all-under-heaven" (Tianxia). In Callahan, W., & Barabantseva, E. (eds), *China Orders the World: Normative Soft Power and Foreign Policy*, Washington, DC: Woodrow Wilson Center Press, 21–36.

Chapter 20

Children's Political Geographies

Kirsi Pauliina Kallio and Jouni Häkli

University of Tampere, Finland

Children's absence from the concerns of the subfield of political geography has drawn some attention during the past decade or so, and several authors have sought to account for this relative invisibility. Some point at children's marginal position and limited capacities to gain visibility in Western societies, which place them in a special position of exclusion (Matthews, Limb, & Taylor 1999). Others have emphasized children and young people's liminal position "in between" competence and incompetence, liability and unaccountability, responsibility and irresponsibility, which tends to obscure children's roles as actors in political analysis (Skelton 2010). Yet another explanation puts the blame on the conventional wisdom according to which children's lives should be safe from adult concerns, placing them rather categorically outside the political (Brocklehurst 2006; Kallio and Häkli 2010).

Children in political geography

While children's matters have certainly been overlooked by political geographers, there are ways in which children and youth have figured in the subfield. In a *Political Geography Quarterly* editorial comment titled "Children and politics," Peter Taylor (1989) notes, maybe a bit tongue in cheek, that the title evokes images of politicians campaigning and garnering votes by "kissing babies" in order to appear caring and attentive to the most vulnerable of human beings. Yet, he goes on to state that "in recent years children have been entering politics in a completely different manner to appear at the heart of our most important debates" (Taylor 1989: 5). What Taylor refers to is the political geography approach that finds children as victims of war, oppressive societal orders, unfavorable socio-economic circumstances, and natural disasters, or as foci of social policies such as schooling, health care, or participatory practices (cf. Katz 1993; Gruffudd 1996; Wood 1996; Kalipeni & Oppong 1998; Cheney 2005; Mitchell 2006; Barker 2012; Yea 2013). Anticipating what would become a growing area of scholarly interest, Taylor (1989) also envisions a political geography concern with

The Wiley Blackwell Companion to Political Geography, First Edition.
Edited by John Agnew, Virginie Mamadouh, Anna J. Secor, and Joanne Sharp.
© 2015 John Wiley & Sons Ltd. Published 2017 by John Wiley & Sons Ltd.

children as active participants in political events and processes, mentioning the uprisings of school children in the West Bank and apartheid South Africa as two examples.

Since Taylor's early observations, a growing political geography literature has emerged focused on children insofar as they are exposed to ideological goals, abuse, armed conflicts, or other grave circumstances. Children have long been seen as an important segment of the population for ideological or biopolitical interventions, with aims that range from spatial socialization into national and/or state subjectivities, to manipulating family and gender structures, building geopolitical and cultural dispositions, or cultivating healthy neoliberal citizens (Bar-Gal 1993; Maddrell 1996; Newman & Paasi 1998; Mitchell 2001; Conlon 2010; Biesta 2012; Martin 2012; Jackson 2013; Mills 2013). The studies of children as implicated in major processes of (geo)political regulation and direction have made it abundantly clear that they are anything but safe from adult concerns. On the contrary, the possibility of influencing children's growth and development toward adulthood makes them prime targets for the manipulation of the shape of future societies, subjecting them to some very powerful political passions (Gruffudd 1996; Gagen 2000; Wainwright & Marandet 2011).

Another important strand of scholarship approaches children's political geographies from the point of view of socio-economic vulnerability and abuse (e.g., Ennew & Swart-Kruger 2003; Young 2004). There are several countries where it is commonplace for children to participate in the labor force, but the conditions in which this occurs vary dramatically from responsible and rewarded contribution to household sustenance to downright slavery in plantations, factories, or sweatshops (Robson 2004; Aitken et al. 2006; Swanson 2009; Jeffrey 2010; Evans 2011). In extreme situations, children are oppressed through practices of human trafficking and sexual abuse (Cream 1993; Kesby, Gwanzura-Ottemoller, & Chizororo 2006). While the latter is certainly always injurious to children, the consequences of trafficking are more dependent on contextual factors, such as age, gender, place, poverty, and traditions that shape the conditions in which children are lured or forced to work away from their parents (Manzo 2005; Van Blerk 2008; Yea 2013).

Yet another literature that has relevance to political geography focuses on the ways in which children are victimized by armed conflicts (e.g., Grundy-Warr & Wong Siew Yin 2002). Children's involvement in war is typically approached as experienced directly in conflict societies or indirectly through forced displacement (Kalipeni & Oppong 1998; Lang & Knudsen 2009). Expanding on the social consequences of violent conflicts, studies have also charted the ways in which war distress is recalled in later life or transmitted intergenerationally from parents to children (McDowell 2004; Kuusisto-Arponen 2009). Some work has also been done on the direct involvement of child soldiers in conflicts, but there are surprisingly few detailed studies targeting the issue (Cheney 2005; Hyndman 2010).

As this discussion indicates, it is possible to pinpoint several examples of scholarship attentive to children's political geographies. Yet, it would be an exaggeration to say that even this literature in itself signals extensive interest. We can only agree with Tracey Skelton (2010), who notes, on the basis of a survey on the contents of the subdiscipline's flagship journal *Political Geography,* that the scholarship could focus some more attention on children and young people. To illustrate just how rare it is to find children as key subjects in political geography analysis, we use the remaining part of this section to discuss the treatment of the 2004 North Ossetian Beslan school hostage crisis in an article by Gearoid Ó Tuathail, published in the first issue of the 2009 volume of *Political Geography.* The article is based on the *Political Geography* Plenary Lecture at the 2008 Annual Meeting of the Association of American Geographers, and sets out to develop "a critical geopolitical account of the ways in which key actors involved in the terrorist incident at School Number 1 in Beslan North Ossetia constructed its meaning and justified their actions" (Ó Tuathail 2009a: 4). In the issue the article is followed by three commentaries and Ó Tuathail's response.

Since the hostage crisis took place in a school, it is not surprising that children figure strongly in the tragedy. The crisis demanded the life of 334 hostages, more than half of whom were children who were attending a ceremony with their parents to mark the beginning of the school year. Ó Tuathail (2009a: 4) examines the event from three perspectives: "the terrorists' Beslan," "the Kremlin's Beslan," and "Beslan among Ossetians and others in the North Caucasus." Echoing Taylor's (1989) concerns about "children's deaths [as] an unfortunate side effect" of the ways in which our system works, the paper aptly introduces children as victims of the tragedy: as members of the community under siege by terrorists, as hostages whose release was negotiated by the President of Ingushetia, and as human beings who suffered injuries or died in the incident. Children are also portrayed as people protected by the Russian Special Forces, as the subject of concern by the Russian and North Ossetian presidents, and as targets of resurrection mobilized by the Mothers of Beslan (Tuathail 2009a: 4, 8, 12–13).

Despite their focal role in both the public media reporting the events and Ó Tuathail's (2009a) assessment of the incident, children remain surprisingly invisible in the subsequent discussions on the paper. In their commentaries, Bakke (2009) and Nicley (2009) refrain from any child-related terminology, whereas Gorenburg (2009) refers to the terrorists as "child killers" (citing Vladimir Putin) and recalls that the attack targeted "innocent schoolchildren." Ó Tuathail's (2009b) response, instead, notes (in passing) that "the life conditions of every Chechen child is constrained by the fact that every Russian child learns Lermontov's poem about a 'wicked Chechen' sharpening his *kinzhal* (dagger)." He makes the point in reference to Åsne Seierstad's (2008) book that places children at the heart of the inquiry into the Chechnyan conflict, thus offering a potential starting point for "bringing children in" also as political subjects, not merely passive victims. Nevertheless, he does not follow this line of thought and, as yet, the dialogue has not prompted further discussion on the Beslan case in the journal or elsewhere within the subdiscipline.

Ó Tuathail's (2009a) analysis of the events that drew hundreds of children violently into the core of a troubled geopolitics is an adept treatment of a complex and multilayered conflict. It is also an apt example of just how remote the idea of children's agency has been and largely still is to political geographical research, including its critical dimensions. Despite Taylor's (1989) early optimism about children participating in politics in new and remarkable ways, it seems that in political geography scholarship this change has been a slow train coming. Unlike other people, children have typically not been appreciated as agents actively *present* in political events, *operating* in particular ways, and *developing* as political subjects, let alone *creating* political settings, dynamics, and practices in their everyday lives, and *involving* other people in these geographies. The traditional political geography approach has afforded children predominantly passive, or at the least non-initiative-taking, roles as members of the political world. Consider again the pages of *Political Geography*. The term "children's agency" appears for the first time in 2010 in Aspasia Theodosiou's review introducing a book co-authored by Yiannis Papadakis, Nicos Peristianis, and Gisela Welz (none of whom is a political geography scholar). Similarly, vocabulary highlighting children's subjectivity and active political roles can only be found in some recent articles discussing the matter explicitly (Kallio & Häkli 2010, 2011a; Bartos 2012; Wood 2012).

The sea change concerning children's political roles that we are now witnessing has found a foothold somewhat earlier in areas of research with a less explicit political geographical focus (see the next sections). In particular, scholarship inspired by feminist and postcolonial theorization has been influential for the study of youthful political agency that has explicitly sought to include in the realm of mundane political agency the hitherto excluded "half of the world's population" (Brocklehurst 2006: 1). In what follows we provide an overview of this development and seek to assess its significance for political geography more broadly.

Politics in children's geographies

The development of political geography research toward acknowledging children and young people as important subjects, agents, and actors was prefaced by more general work in the emerging subfields of children's geographies and geographies of young people since the early 1990s. These interconnected but distinguishable fields have built strong linkages to feminist social and cultural geographies, and many of the key concepts, theoretical perspectives, and methodological approaches of the early scholarship were drawn from literatures concerning the family, the school, welfare institutions, and urban space. Early publications include a discussion series in *Area* with contributions from Sarah James (1990, 1991), David Sibley (1991), and Hilary Winchester (1991); articles in geography journals and books by Cindi Katz (1991, 1993), Stuart Aitken, Joan Wingate, and Thomas Herman (Aitken & Wingate 1993; Aitken & Herman 1997), Teresa Ploszajska (1994), Hugh Matthews (1995a, b), Hilary Winchester and Lauren Costello (1995), David Sibley (1995a, b), and Gill Valentine (1996a, b); as well as monographs by Hugh Matthews (1992) and Stuart Aitken (1994), among others.

This discussion set out to bring to the fore childhood issues and children's matters in geography at large, introducing children and youth as active agents in their lived worlds. The discussion involved perspectives and vocabularies familiar to political geographical research. For instance, ideas from the work of William Bungé (1973), Colin Ward (1977), and Robin Moore (1986) were introduced to draw attention to some fundamental issues largely ignored in the scholarship of the time. Hilary Winchester's comment in *Area* captures the spirit well:

> The socio-spatial relationships of children, their dependence on adults, and the power relationships which circumscribe their lives are certainly the most underdeveloped and potentially fruitful area of geographical research. (Winchester 1991: 359)

The positive thrust did not, however, push the emerging field of children's geographies toward explicitly political inquiry. Since the mid-1990s, discussions on childhood and youth expanded in both size and scope, but politics remained a rather marginal issue. Youthful agency was randomly noticed in the context of policy and children's rights, with Hugh Matthews and Melanie Limb (1999) as the most influential scholars, but any broader assessment of children's place in politics and the place of politics in different kinds of childhoods remained absent. At the same time, feminist geography was establishing its position within political geography, exploring various kinds of political processes, dynamics, and practices, and expanding the notion of "the political" to include individual and collective, official and mundane, rational and affective ways of acting and having an impact politically (for an overview, see Brown & Staeheli 2003). The relative stagnation within children's geographies can partly be explained by the influence of the new childhood studies paradigm that started to direct the scholarly debate from the early 1990s.

Conceptions of childhood have undergone a radical change in the past 20 years. The 1980s was a decade of discursive change in the interlinked yet separate fields of children's rights advocacy and social studies of childhood. During that time, the United Nations Convention on the Rights of the Child was going through an extremely prolonged compilation that resulted in a new agenda of children's human rights (Häkli & Kallio 2014a). In its final form in 1989, the treaty introduced the idea of *voice* as a fundamental right of the child, comparable to *provision* and *protection* that form the traditional twosome of children's rights. Simultaneously, a forceful social theoretical critique was launched against adult-centered notions of children's lived worlds that stressed behavior and development over agency and "being" (e.g., Jenks 1982; Henriques et al. 1984; Adler & Adler 1986; Alanen 1988; Chisholm et al. 1990). This critique led to a paradigmatic change, producing an interdisciplinary discussion generally known as the "new" social studies of childhood. The book *Constructing and Re-Constructing Childhood: Contemporary Issues in the Sociological Study of Childhood*, edited by Allison James and Alan Prout (1990) and published in concert with the Convention, came to manifest the new research agenda.

During the 1990s, the altered human rights frame and the new disciplinary approach were tightly knit together. The emphasis given to the child's right to be heard, combined with the approach stressing children as agents "here and now," generated a concept that has thereafter dominated both research and policy agendas: *children's participation*. Scholars working in multiple disciplinary fields, children's rights advocates and benefit organizations, professionals working with children and youth, policymakers and administrative actors seeking "the best of the child," as well as the media, quickly embraced the concept. Human geographers, too, took a new course, moving from "top-down socialization" toward interest in child-centered participation and children as active agents in the worlds where they are situated in particular ways (e.g., Holloway & Valentine 2000). This work took notice of children's agency in both mundane everyday environments and more official arenas, thus covering children's everyday "political" and formal "Political" geographies (e.g., Owens 1997; Valentine 1997; Jones 1999). Yet, the political aspects of children's agency were not explicitly emphasized, largely because the new childhood studies paradigm affiliated politics chiefly with the adult-led world and thus skirted the theme as contradicting child-centered perspectives.

A new course emerged by the beginning of the twenty-first century. Two publications with long-standing influence in the politically oriented study of children's geographies appeared in 2003. First, *Space and Polity* published a special issue on the "Political Geographies of Children and Young People," edited by Chris Philo and Fiona M. Smith (2003), based on a conference session, "Politicising Child Life," held at the Annual Meeting of the Association of American Geographers (AAG) in 2001. Second, the *Children's Geographies* journal was launched as a specific forum for children and young people's geographies. Little by little, discussions concerning youthful political presence in different geographical settings evolved, first in the context of young adults and late teenagers (e.g., Freeman, Nairn, & Sligo 2003; O'Toole 2003; Wridt 2004; Cahill 2007; Skelton 2007; Hörschelmann 2008; Thomas 2008; Staeheli & Hammett 2010; Kallio & Häkli 2011b; Azmi, Brun, & Lund 2013), but increasingly focusing on younger children as well (e.g., Kelley 2006; Kallio 2007; Lund 2007; Cope 2008; Bosco 2010; Kallio & Häkli 2011a; Bartos 2012; Bordonaro 2012; Mitchell & Elwood 2012; Wood 2012; Marshall 2013).

A productive coupling between the research streams on children and young people has recently resulted in joint conference sessions, edited collections, and collaborative research projects. For example, papers from a session on children and young people's everyday politics at the AAG meeting in 2012 were published as an edited collection (Kallio & Häkli 2013), and a conference session on children, young people, and critical geopolitics at the RGS-IBG meeting in 2013, organized by Peter Hopkins and Matthew Benwell, is leading to another edited volume. A 12-volume major reference work is also being prepared by Tracey Skelton. With these and many other contributions, the discussion on children's politics is moving toward themes and theoretical orientations generally employed in political geography. The concomitant appreciation of children as complete human beings in the political worlds in which they live alongside other people helps to locate "politics" in children's geographies and "children" in political geographies. This opens up further avenues for an inquiry into the "geography" in children's politics.

Geography in children's politics

Explicit theorization of the *spatialities* of children's politics may have been overshadowed by the need to justify the idea of youthful realities as *political*, but the consolidation and further expansion of the research area are likely to change the situation, as some recent works indicate (e.g., Vanderbeck 2008; Ansell 2009; Kallio & Häkli 2013). However, in empirical terms, certain geographies have received more attention than others during the past three decades, providing fruitful starting points for further theorization and methodological work.

Given that economic geography is one of the last terrains in which children's agency is still to gain a foothold, it is interesting that one of the first research streams in children's political geographies has a strong political economic emphasis. Cindi Katz's (1986, 1991, 1993, 2004) long-standing work in rural Sudan and urban New York has paid attention to children's positions, roles, and agencies in the world of economic restructuration. Binding together the mundane spheres of work, play, and education with reference to "knowing" and livelihood as they appear in the multiscalar world, she has shown how the global time–space compression unfolds in children's lived worlds as time–space *expansions*. Katz's approach has inspired many people; in particular, scholarship concerning child labor and children's work in underdeveloped and developing regions has taken a political geographical tone (e.g., Robson 2004; Abebe 2007; Dyson 2008; Ansell 2009; Evans 2011). In parallel, Stuart Aitken (1994, 2001) has studied children's livelihoods from a social geographical perspective at the US–Mexico border, contributing to feminist and postcolonial geographies. This work has been extended to various dimensions in the San Diego research group (e.g., Aitken et al. 2006; Bosco 2010; Bosco, Aitken, & Herman 2011) and elsewhere (e.g., Aitken, Kjørholt, & Lund 2007; Forsberg & Pösö 2011).

Another strong research stream targets children's place in public/private space, in both urban and rural contexts, and in the minority and the majority world. Spanning from homes, streets, and neighborhoods to various natural and built environments, as well as demonstrations and other semi-formal participation venues, these studies have come to ask what are the "right places" for children and youth and why they seem "out of place" in other locations and events. Sarah Holloway, Gill Valentine, and Tracey Skelton have been active in this field at an early stage, emphasizing feminist theoretical perspectives (Valentine 1996b; Skelton & Valentine 1997; Holloway & Valentine 2000, 2001; Valentine & Holloway 2001). Other studies have worked to broaden the approach further (e.g., Punch 2001; Tucker & Matthews 2001; Nairn, Panelli, & McCormack 2003; Christensen & O'Brien 2004; Young 2004). The work that links to the "right to the city" idea is one of the recent openings pertinent to children's political geographies (e.g., Bosco et al. 2011; Elwood & Mitchell 2012; Cele 2013).

Perhaps the most explicit political landscape that childhood scholars have explored is policymaking, especially in relation to democratic practice, political participation, and citizenship. This strongly adult-led arena has been unsettled by the idea of children as full human beings who should have a place in a democratic society alongside others. Here, childhood and youth geographers have largely joined forces, making space for the idea that membership in a political community – be it of any scalar extent – is not a question of age. Tracey Skelton (2010) has written extensively on the matter, with a serious attempt to *spatialize* children's participation between the mundane and official political realities (also Skelton & Valentine 2003). Hugh Matthews and Melanie Limb (1999), Suzie Weller (2003), Barry Percy-Smith (2006), Janet Habashi and Jody Worley (2008, 2014), and Kirsi Pauliina Kallio and Jouni Häkli (2011a) have sought to unpack various aspects of children's societal agency and belonging, and this work has been further developed by scholars such as Ann Bartos (2012, 2013), Bronwyn Wood (2012), and David Marshall (2013).

Research on children's participation is closely linked to the question of the rights of the child, which has received considerable attention among geographers (e.g., Lund 2007; Ruddick 2007a, b; Skelton 2007; Kallio 2012; Häkli & Kallio 2014a), not to mention more explicit discussions concerning citizenship (e.g., Cope 2008; Driskell, Fox, & Kudva 2008; Pykett 2009; Staeheli & Hammett 2010; Kallio & Häkli 2011a). Many of these studies resonate with political geography research that approaches the school as an institution reinforcing particular kinds of citizenship, and as a space of interactive citizenship formation where children's agency also plays a part. In a similar vein, the rights-based research often links with previous work in areas such as development studies, peace and conflict research, and postcolonial and migration studies. Thus, the research themes introduced here are both interconnected and intertwined with more conventional political geography analysis in many ways, showing how developing new insights often draws on the tradition of the subdiscipline.

Conclusions

This chapter has sought to provide an overview of children's political geographies as a hetero-geneous but consolidating area of scholarly activity. While it is feasible to discuss the emerging field in terms of specific research streams, there are, of course, many studies that do not easily fall under those introduced here. There is interesting work, for example, on the politics of mobility (Barker 2003, 2012; Kearns & Collins 2003; Benwell 2009; Kullman 2010; Evans 2011); on identity politics related to race, gender, and class in everyday communities (Hyams 2000; Morris-Roberts 2004; Van Ingen & Halas 2006; Thomas 2008, 2011; Mitchell & Elwood 2012); and on power relations and political subjectivity (e.g., Cahill 2007; Gallagher 2008; Pike 2008; Häkli & Kallio 2014b).

That said, there are some common denominators informing the many differently angled research streams and projects. One broadly accepted view is that, regardless of the empirical focus or the theoretical framework, it is nearly impossible to assess children's lived worlds as neatly structured according to one (or another) scalar logic. The everyday realities of children's lives and the mundane and more formal modes of their politics often appear micro-scalar at first sight, but contemporary scholarship has shown this to be an oversimplification. Chris Philo and Fiona Smith (2003) were among the first to discuss the issue explicitly, and it has since become somewhat common knowledge (e.g., Vanderbeck & Dunkley 2004; Wridt 2004; Kallio 2007; Ansell 2009). To understand the variable meanings of the everyday, it is necessary to retain theoretical open-mindedness toward the many forms, arenas, starting points, and foci of children's political agencies.

We are convinced that the expanding scholarly interest in children's politics will contribute to an enlivened political geography that is able to recognize and discover politics in extraordinary and unexpected places and situations. Rather than contradicting other existing understandings of politics, this approach helps to identify and study events, acts, and contexts that in political analysis are easily bypassed as apolitical. Approaching childhood and children in this way, not as exotic issues marginal to political geography but as considerations at the heart of its debates, will yield conceptual tools that facilitate theoretical work on the limits and borders of politics more generally.

References

Abebe, T. (2007). Changing livelihoods, changing childhoods: Patterns of children's work in rural Southern Ethiopia. *Children's Geographies* 5(1–2): 77–93. doi:10.1080/14733280601108205

Adler, P.A., & Adler, P. (eds) (1986). *Sociological Studies of Child Development*. Vol. 1. Greenwich, CT: JAI Press.

Aitken, S.C. (1994). *Putting Children in Their Place*. Association of American Geographers. Washington, DC: Edwards Bros.

Aitken, S.C. 2001. *Geographies of Young People: The Morally Contested Spaces of Identity*. London: Routledge.

Aitken, S.C., & Herman, T. (1997). Gender, power and crib geography: Transitional spaces and potential places. *Gender, Place and Culture* 4(1): 63–88. doi:10.1080/09663699725503

Aitken, S.C., & Wingate, J. (1993). A preliminary study of the self-directed photography of middle-class, homeless, and mobility-impaired children. *Professional Geographer* 45(1): 65–72. doi:10.1111/j.0033-0124.1993.00065.x

Aitken, S.C., Kjørholt, A.T., & Lund, R. (2007). Why children? Why now? *Children's Geographies* 5(1–2): 3–14. doi:10.1080/14733280601108114

Aitken, S.C., Estrada, S.L., Jennings, J., & Aguirre, L.M. (2006). Reproducing life and labor: Global processes and working children in Tijuana, Mexico. *Childhood* 13(3): 365–387. doi:10.1177/0907568206066356

Alanen, L. (1988). Rethinking childhood. *Acta Sociologica* 31(1): 53–67. doi:10.1177/000169938803100105

Ansell, N. (2009). Childhood and the politics of scale: Descaling children's geographies? *Progress in Human Geography* 33(2): 190–209. doi:10.1177/0309132508090980

Azmi, F., Brun, C., & Lund, R. (2013). Young people's everyday politics in post-conflict Sri Lanka. *Space and Polity* 17(1): 106–122. doi:10.1080/13562576.2013.780716

Bakke, K.M. (2009). Beslan and the study of violence. *Political Geography* 28(1): 16–18.

Bar-Gal, Y. (1993). Boundaries as a topic in geographic education: The case of Israel. *Political Geography* 12(5): 421–435. doi:10.1016/0962-6298(93)90011-U

Barker, J. (2003). Passengers or political actors? Children's participation in transport policy and the micro political geographies of the family. *Space and Polity* 7(2): 135–151. doi:10.1080/1356257032000133900

Barker, J. (2012). A free for all? Scale and young people's participation in UK transport planning. In Kraftl, P., Horton, J., & Tucker, F. (eds), *Critical Geographies of Childhood and Youth*, Bristol: Polity Press, 169–184.

Bartos, A.E. (2012). Children caring for their worlds: The politics of care and childhood. *Political Geography* 31(3): 157–166. doi:10.1016/j.polgeo.2011.12.003

Bartos, A.E. (2013). Friendship and environmental politics in childhood. *Space and Polity* 17(1): 17–32. doi:10.1080/13562576.2013.780711

Benwell, M.C. (2009). Challenging minority world privilege: Children's outdoor mobilities in post-apartheid South Africa. *Mobilities* 4(1): 77–101. doi:10.1080/17450100802657970

Biesta, G. (2012). Becoming public: Public pedagogy, citizenship and the public sphere. *Social and Cultural Geography* 13(7): 683–697. doi:10.1080/14649365.2012.723736

Bordonaro, L.I. (2012). Agency does not mean freedom: Cape Verdean street children and the politics of children's agency. *Children's Geographies* 10(4): 413–426. doi:10.1080/14733285.2012.726068

Bosco, F.J. (2010). Play, work or activism? Broadening the connections between political and children's geographies. *Children's Geographies* 8(4): 381–390. doi:10.1080/14733285.2010.511003

Bosco, F.J., Aitken, S.C., & Herman, T. (2011). Women and children in a neighborhood advocacy group: Engaging community and refashioning citizenship at the United States–Mexico border. *Gender, Place and Culture* 18(2): 155–178. doi:10.1080/0966369X.2010.551652

Brocklehurst, H. (2006). *Who's Afraid of Children? Children, Conflict and International Relations*. Aldershot: Ashgate.

Brown, M., & Staeheli, L.A. (2003). "Are we there yet?" Feminist political geographies. *Gender, Place and Culture* 10(3): 247–255. doi:10.1080/0966369032000114019

Bungé, W. (1973). The geography. *Professional Geographer* 25(4): 331–337. doi:10.1111/j.0033-0124.1973.00331.x

Cahill, C. (2007). The personal is political: Developing new subjectivities through participatory action research. *Gender, Place and Culture* 14(3): 267–292. doi:10.1080/09663690701324904

Cele, S. (2013). Performing the political through public space: Teenage girls' everyday use of a city park. *Space and Polity* 17(1): 74–87. doi:10.1080/13562576.2013.780714

Cheney, K.E. (2005). "Our children have only known war": Children's experiences and the uses of childhood in Northern Uganda. *Children's Geographies* 3(1): 23–45. doi:10.1080/14733280500037133

Chisholm, L., Brown, P., Kruger, H.H., & Buchner, P. (eds) (1990). *Childhood, Youth and Social Change: A Comparative Perspective*. London: Falmer Press.

Christensen, P., & O'Brien, M. (eds) (2004). *Children in the City: Home, Neighbourhood and Community*. London: Routledge.

Conlon, D. (2010). Ties that bind: Governmentality, the state, and asylum in contemporary Ireland. *Environment and Planning D: Society and Space* 28(1): 95–111. doi:10.1068/d11507

Cope, M. (2008). Patchwork neighborhood: Children's urban geographies in Buffalo, New York. *Environment and Planning A* 40(12): 2845–2863. doi:10.1068/a40135

Cream, J. (1993). Child sexual abuse and the symbolic geographies of Cleveland. *Environment and Planning D: Society and Space* 11(2): 231–246. doi:10.1068/d110231

Driskell, D., Fox, C., & Kudva, N. (2008). Growing up in the new New York: Youth space, citizenship, and community change in a hyperglobal city. *Environment and Planning A* 40(12): 2831–2844. doi:10.1068/a40310

Dyson, J. (2008). Harvesting identities: Youth, work, and gender in the Indian Himalayas. *Annals of the Association of American Geographers* 98(1): 160–179. doi:10.1080/00045600701734554

Elwood, S., & Mitchell, K. (2012). Mapping children's politics: Spatial stories, dialogic relations and political formation. *Geografiska Annaler: Series B, Human Geography* 94(1): 1–15. doi:10.1111/j.1468-0467.2012.00392.x

Ennew, J., & Swart-Kruger, J. (eds) (2003). Homes, places and spaces in the construction of street children and street youth. *Children, Youth and Environments* 13(1), special issue.

Evans, R. (2011). Young caregiving and HIV in the UK: Caring relationships and mobilities in African migrant families. *Population, Space and Place* 17(4): 338–360. doi:10.1002/psp.583

Forsberg, H., & Pösö, T. (2011). Childhood homes as moral spaces – new conceptual arena. *Social Work and Society* 9(2): 1–14.

Freeman, C., Nairn, K., & Sligo, J. (2003). "Professionalising" participation: From rhetoric to practice. *Children's Geographies* 1(1): 53–70. doi:10.1080/14733280302182

Gagen, E.A. (2000). An example to us all: Child development and identity construction in early 20th-century playgrounds. *Environment and Planning A* 32(4): 599–616. doi:10.1068/a3237

Gallagher, M. (2008). "Power is not an evil": Rethinking power in participatory methods. *Children's Geographies* 6(2): 137–150. doi:10.1080/14733280801963045

Gorenburg, D. (2009). The causes and consequences of Beslan: A commentary on Gerard Toal's Placing Blame: Making Sense of Beslan. *Political Geography* 28(1): 23–27. doi:10.1016/j.polgeo.2009.01.003

Gruffudd, P. (1996). The countryside as educator: Schools, rurality and citizenship in inter-war Wales. *Journal of Historical Geography* 22(4): 412–423. doi:10.1006/jhge.1996.0028

Grundy-Warr, C., & Wong Siew Yin, E. (2002). Geographies of displacement: The Karenni and the Shan across the Myanmar–Thailand border. *Singapore Journal of Tropical Geography* 23(1): 93–122. doi:10.1111/1467-9493.00120

Habashi, J., & Worley, J. (2008). Child geopolitical agency: A mixed methods case study. *Journal of Mixed Methods Research* 3(1): 42–63. doi:10.1177/1558689808326120

Habashi, J., & Worley, J.A. (2014). Children's projected political preference: Transcending local politics. *Children's Geographies* 12(2): 205–218. doi:10.1080/14733285.2013.812306

Häkli, J., & Kallio, K.P. (2014a). The global as a field: Children's rights advocacy as a transnational practice. *Environment and Planning D: Society and Space* 32(2): 293–309. doi:10.1068/d0613

Häkli, J., & Kallio, K.P. (2014b). Subject, action and polis: Theorizing political agency. *Progress in Human Geography* 38(2): 181–200. doi:10.1177/0309132512473869

Henriques, J., Holloway, W., Urwin, C., Venn, C., & Walkerdine, V. (1984). *Changing the Subject: Psychology, Social Regulation and Subjectivity*. London: Methuen.

Holloway, S.L., & Valentine, G. (2000). Spatiality and the "new" social studies of childhood. *Sociology*, 34(4): 763–783. doi:10.1177/S0038038500000468

Holloway, S.L., & Valentine, G. (2001). Children at home in the wired world: Reshaping and rethinking home in urban geography. *Urban Geography* 22(6): 562–583. doi:10.2747/0272-3638.22.6.562

Hörschelmann, K. (2008). Populating the landscapes of critical geopolitics – young people's responses to the war in Iraq (2003). *Political Geography* 27(5): 587–609. doi:10.1016/j.polgeo.2008.06.004

Hyams, M. (2000). "Pay attention in class… [and] don't get pregnant": A discourse of academic success among adolescent Latinas. *Environment and Planning A* 32(4): 635–654. doi:10.1068/a3239

Hyndman, J. (2010). The question of "the political" in critical geopolitics: Querying the "child soldier" in the "War on Terror." *Political Geography* 29(4): 247–255. doi:10.1016/j.polgeo.2009.10.010

Jackson, P.S.B. (2013). The crisis of the "disadvantaged child": Poverty research, IQ, and muppet diplomacy in the 1960s. *Antipode* 46(1): 190–208. doi:10.1111/anti.12027

James, A., & Prout, A. (eds) (1990). *Constructing and Re-Constructing Childhood: Contemporary Issues in the Sociological Study of Childhood*. London: Falmer Press.

James, S. (1990). Is there a "place" for children in geography? *Area* 22(3): 278–283.

James, S. (1991). Children and geography: A reply to David Sibley. *Area* 22(3): 270–272.

Jeffrey, C. (2010). Geographies of children and youth I: Eroding maps of life. *Progress in Human Geography* 34(4): 496–505. doi:10.1177/0309132509348533

Jenks, C. (ed.) (1982). *The Sociology of Childhood*. London: Batsford Academic and Educational.

Jones, O. (1999). Tomboy tales: The rural, nature and the gender of childhood. *Gender, Place and Culture* 6(2): 117–136. doi:10.1080/09663699925060

Kalipeni, E., & Oppong, J. (1998). The refugee crisis in Africa and implications for health and disease: A political ecology approach. *Social Science and Medicine* 46(12): 1637–1653. doi:10.1016/S0277-9536(97)10129-0

Kallio, K.P. (2007). Performative bodies, tactical agents and political selves: Rethinking the political geographies of childhood. *Space and Polity* 11(2): 121–136. doi:10.1080/13562570701721990

Kallio, K.P. (2012). Desubjugating childhoods by listening to the child's voice and the childhoods at play. *ACME: An International E-Journal for Critical Geographies* 11(1): 81–109.

Kallio, K.P., & Häkli, J. (2010). Political geography in childhood. *Political Geography* 29(7): 357–358. doi:10.1016/j.polgeo.2009.11.001

Kallio, K.P., & Häkli, J. (2011a). Tracing children's politics. *Political Geography* 30(2): 99–109. doi:10.1016/j.polgeo.2011.01.006

Kallio, K.P., & Häkli, J. (2011b). Young people's voiceless politics in the struggle over urban space. *GeoJournal* 76(1): 63–75. doi:10.1007/s10708-010-9402-6

Kallio, K.P., & Häkli, J. (2013). Children and young people's politics in everyday life. *Space and Polity* 17(1): 1–16. doi:10.1080/13562576.2013.780710

Katz, C. (1986). Children and the environment: Work, play and learning in rural Sudan. *Children's Environments Quarterly* 3(4): 43–51.

Katz, C. (1991). Sow what you know: The struggle for social reproduction in rural Sudan. *Annals of the Association of American Geographers* 81(3): 488–514. doi:10.1111/j.1467-8306.1991.tb01706.x

Katz, C. (1993). Growing girls/closing circles: Limits on the spaces of knowing in rural Sudan and US cities. In Katz, C., & Monk, J. (eds), *Full Circles: Geographies of Women over the Life-Course*, London: Routledge, 88–106.

Katz, C. (2004). *Growing Up Global: Economic Restructuring and Children's Everyday Lives.* Minneapolis, MN: University of Minnesota Press.

Kearns, R., & Collins, D. (2003). Crossing roads, crossing boundaries: Empowerment and participation in a child pedestrian safety initiative. *Space and Polity* 7(2): 193–212. doi:10.1080/1356257032000133937

Kelley, N. (2006). Children's involvement in policy formation. *Children's Geographies* 4(1): 37–44. doi:10.1080/14733280600577145

Kesby, M., Gwanzura-Ottemoller, F., & Chizororo, M. (2006). Theorizing other, "other childhoods": Issues emerging from work on HIV in urban and rural Zimbabwe. *Children's Geographies* 4(2): 185–202. doi:10.1080/14733280600807039

Kullman, K. (2010). Transitional geographies: Making mobile children. *Social and Cultural Geography* 11(8): 829–846. doi:10.1080/14649365.2010.523839

Kuusisto-Arponen, A.-K. (2009). The mobilities of forced displacement: Commemorating Karelian evacuation in Finland. *Social and Cultural Geography* 10(5): 545–563. doi:10.1080/14649360902974464

Lang, H., & Knudsen, A. (2009). "Your subject of protection is a dangerous one': Protracted internal conflict and the challenges for humanitarian agencies. *Norsk Geografisk Tidsskrift* 63(1): 35–45. doi:10.1080/00291950802712111

Lund, R. (2007). At the interface of development studies and child research: Rethinking the participating child. *Children's Geographies* 5(1–2): 131–148. doi:10.1080/14733280601108247

Maddrell, A.M.C. (1996). Empire, emigration and school geography: Changing discourses of imperial citizenship, 1880–1925. *Journal of Historical Geography* 22(4): 373–387. doi:10.1006/jhge.1996.0025

Manzo, K. (2005). Exploiting West Africa's children: Trafficking, slavery and uneven development. *Area* 37(4): 393–401. doi:10.1111/j.1475-4762.2005.00644.x

Marshall, D. (2013). All the beautiful things: Trauma, aesthetics and the politics of Palestinian childhood. *Space and Polity* 17(1): 53–75. doi:10.1080/13562576.2013.780713

Martin, L.L. (2012). Governing through the family: Struggles over US noncitizen family detention policy. *Environment and Planning A* 44(4): 866–888. doi:10.1068/a4477a

Matthews, H. (1995a). Culture, environmental experience and environmental awareness: Making sense of young Kenyan children's views of place. *Geographical Journal* 161(3): 285–295.

Matthews, H. (1995b). Living on the edge: Children as "outsiders." *Tijdschrift voor Economische en Sociale Geografie* 86(5): 456–466. doi:10.1111/j 1467-9663.1995.tb01867.x

Matthews, H., & Limb, M. (1999). Defining an agenda for the geography of children: Review and prospect. *Progress in Human Geography* 23(1): 61–90. doi:10.1191/030913299670961492

Matthews, H., Limb, M., & Taylor, M. (1999). Young people's participation and representation in society. *Geoforum* 30(2): 135–144. doi:10.1016/S0016-7185(98)00025-6

Matthews, M.H. (1992). *Making Sense of Place: Children's Understanding of Large-Scale Environments.* Hemel Hempstead: Harvester Wheatsheaf.

McDowell, L. (2004). Cultural memory, gender and age: Young Latvian women's narrative memories of war-time Europe, 1944–1947. *Journal of Historical Geography* 30(4): 701–728. doi:10.1016/j.jhge.2003.08.020

Mills, S. (2013). "An instruction in good citizenship: Scouting and the historical geographies of citizenship education. *Transactions of the Institute of British Geographers* 38(1): 120–134. doi:10.1111/j.1475-5661.2012.00500.x

Mitchell, K. (2001). Education for democratic citizenship: Transnationalism, multiculturalism, and the limits of liberalism. *Harvard Educational Review* 71(1): 51–79.

Mitchell, K. (2006). Neoliberal governmentality in the European Union: Education, training, and technologies of citizenship. *Environment and Planning D: Society and Space* 24(2): 389–407. doi:10.1068/d1804

Mitchell, K., & Elwood, S. (2012). Mapping children's politics: The promise of articulation and the limits of nonrepresentational theory. *Environment and Planning D: Society and Space* 30(5): 788–804. doi:10.1068/d9011

Moore, R.C. (1986). *Childhood's Domain: Play and Place in Child Development*. London: Croom Helm.

Morris-Roberts, K. (2004). Girls' friendships, "distinctive individuality" and socio-spatial practices of (dis)identification. *Children's Geographies* 2(2): 237–255. doi:10.1080/14733280410001720539

Nairn, K., Panelli, R., & McCormack, J. (2003). Destabilizing dualisms: Young people's experiences of rural and urban environments. *Childhood* 10(1): 9–42. doi:10.1177/0907568203010001002

Newman, D., & Paasi, A. (1998). Fences and neighbours in the postmodern world: Boundary narratives in political geography. *Progress in Human Geography* 22(2): 186–207. doi:10.1191/030913298666039113

Nicley, E.P. (2009). Placing blame or blaming place? Embodiment, place and materiality in critical geopolitics. *Political Geography* 28(1): 19–22.

O'Toole, T. (2003). Engaging with young people's conceptions of the political. *Children's Geographies* 1(1): 71–90. doi:10.1080/14733280302179

Ó Tuathail, G. (2009a). Placing blame: Making sense of Beslan. *Political Geography* 28(1): 4–15. doi:10.1016/j.polgeo.2009.01.007

Ó Tuathail, G. (2009b). Displacing blame & counter-terrorist number one: Response to commentaries. *Political Geography* 28(1): 28–31. doi:10.1016/j.polgeo.2009.02.003

Owens, P.E. (1997). Adolescence and the cultural landscape: Public policy, design decisions, and popular press reporting. *Landscape and Urban Planning* 39(2): 153–166. doi:10.1016/S0169-2046(97)00052-2

Percy-Smith, B. (2006). From consultation to social learning in community participation with young people. *Children, Youth and Environments* 16(2): 153–179.

Philo, C., & Smith, F.M. (2003). Guest editorial: Political geographies of children and young people. *Space and Polity* 7(2), 99–115. doi:10.1080/1356257032000133883

Pike, J. (2008). Foucault, space and primary school dining rooms. *Children's Geographies* 6(4): 413–422. doi:10.1080/14733280802338114

Ploszajska, T. (1994). Moral landscapes and manipulated spaces: Gender, class and space in Victorian reformatory schools. *Journal of Historical Geography* 20(4): 413–429. doi:10.1006/jhge.1994.1032

Punch, H. (2001). Negotiating autonomy: Childhoods in rural Bolivia. In Alanen, L., & Mayall, B. (eds), *Conceptualizing Child–Adult Relations*, London: Routledge/Falmer, 23–36.

Pykett, J. (2009). Making citizens in the classroom: An urban geography of citizenship education? *Urban Studies* 46(4):,803–823. doi:10.1177/0042098009102130

Robson, E. (2004). Hidden child workers: Young carers in Zimbabwe. *Antipode* 36(2): 227–248. doi:10.1111/j.1467-8330.2004.00404.x

Ruddick, S. (2007a). At the horizons of the subject: Neo-liberalism, neo-conservatism and the rights of the child. Part one: From "knowing" fetus to "confused" child. *Gender, Place and Culture* 14(5): 513–527. doi:10.1080/09663690701562180

Ruddick, S. (2007b). At the horizons of the subject: Neo-liberalism, neo-conservatism and the rights of the child. Part two: Parent, caregiver, state. *Gender, Place and Culture* 14(6): 513–527. doi:10.1080/09663690701659101

Seierstad, Å. (2008). *The Angel of Grozny: Orphans of a Forgotten War*. New York: Basic Books.

Sibley, D. (1991). Children's geographies: Some problems of representation. *Area* 23(3): 269–270.

Sibley, D. (1995a). Families and domestic routines: Constructing the boundaries of childhood. In Pile, S., & Thrift, N. (eds), *Mapping the Subject: Geographies of Cultural Transformation*, London: Routledge, 123–137.

Sibley, D. (1995b). *Geographies of Exclusion*. London: Routledge.

Skelton, T. (2007). Children, young people, UNICEF and participation. *Children's Geographies* 5(1–2): 165–181. doi:10.1080/14733280601108338

Skelton, T. (2010). Taking young people as political actors seriously: Opening the borders of political geography. *Area* 42(2): 145–151. doi:10.1111/j.1475-4762.2009.00891.x

Skelton, T., & Valentine, G. (eds) (1997). *Cool Places: Geographies of Youth Cultures*. London: Routledge.

Skelton, T., & Valentine, G. (2003). Political participation, political action and political identities: Young D/deaf people's perspectives. *Space and Polity* 7(2): 117–134. doi:10.1080/1356257032000133892

Staeheli, L.A., & Hammett, D. (2010). Educating the new national citizen: Education, political subjectivity and divided societies. *Citizenship Studies* 14(6): 667–680. doi:10.1080/13621025.2010.522353

Swanson, K. (2009). *Begging as a Path to Progress: Indigenous Women and Children and the Struggle for Ecuador's Urban Spaces*. Atlanta, GA: University of Georgia Press.

Taylor, P. (1989). Children and politics. *Political Geography Quarterly* 8(1): 5–6.

Theodosiou, A. (2010). Book review: Divided Cyprus: Modernity, History and an Island in Conflict, edited by Yiannis Papadakis, Nicos Peristianis, and Gisela Welz, 2006. Bloomington: Indiana University Press. *Political Geography* 29(1): 53–55. doi:10.1016/j.polgeo.2010.01.003

Thomas, M.E. (2008). The paradoxes of personhood: Banal multiculturalism and racial–ethnic identification among Latina and Armenian girls at a Los Angeles high school. *Environment and Planning A* 40(12): 2864–2878. doi:10.1068/a40141

Thomas, M.E. (2011). *Multicultural Girlhood: Racism, Sexuality and the Conflicted Spaces of Urban Education*. Philadelphia, PA: Temple University Press.

Tucker, F., & Matthews, H. (2001). "They don't like girls hanging around there": Conflicts over recreational space in rural Northamptonshire. *Area* 33(2): 161–168. doi:10.1111/1475-4762.00019

Valentine, G. (1996a). Angels and devils: Moral landscapes of childhood. *Environment and Planning D: Society and Space* 14(5): 581–599. doi:10.1068/d140581

Valentine, G. (1996b). Children should be seen and not heard: The production and transgression of adults' public space. *Urban Geography* 17(3): 205–220. doi:10.2747/0272-3638.17.3.205

Valentine, G. (1997). "Oh yes I can." "Oh no you can't": Children and parents' understandings of kids' competence to negotiate public space safely. *Antipode* 29(1): 65–89. doi:10.1111/1467-8330.00035

Valentine, G., & Holloway, S.L. (2001). A window on the wider world? Rural children's use of information and communication technologies. *Journal of Rural Studies* 17(4): 383–394. doi:10.1016/S0743-0167(01)00022-5

Van Blerk, L. (2008). Poverty, migration and sex work: Youth transitions in Ethiopia. *Area* 40(2): 245–253. doi:10.1111/j.1475-4762.2008.00799.x

Van Ingen, C., & Halas, J. (2006). Claiming space: Aboriginal students within school landscapes. *Children's Geographies* 4(3): 379–398. doi:10.1080/14733280601005856

Vanderbeck, R.M. (2008). Reaching critical mass? Theory, politics, and the culture of debate in children's geographies. *Area* 40(3): 393–400. doi:10.1111/j.1475-4762.2008.00812.x

Vanderbeck, R.M., & Dunkley, C.M. (2004). Introduction: Geographies of exclusion, inclusion and belonging in young lives. *Children's Geographies* 2(2): 177–183. doi:10.1080/14733280410001720494

Wainwright, E., & Marandet, E. (2011). Geographies of family learning and aspirations of belonging. *Children's Geographies* 9(1): 95–109. doi:10.1080/14733285.2011.540442

Ward, C. (1977). *The Child in the City*. London: Architectural Press.

Weller, S. (2003). "Teach us something useful": Contested spaces of teenagers' citizenship. *Space and Polity* 7(2): 153–171. doi:10.1080/1356257032000133919

Winchester, H.P.M. (1991). The geography of children. *Area* 23(3): 357–360.

Winchester, H.P.M., & Costello, L.N. (1995). Living on the street: Social organisation and gender relations of Australian street kids. *Environment and Planning D: Society and Space* 13(3): 329–348. doi:10.1068/d130329

Wood, B.E. (2012). Crafted within liminal spaces: Young people's everyday politics. *Political Geography* 31(6): 337–346. doi:10.1016/j.polgeo.2012.05.003

Wood, W.B. (1996). From humanitarian relief to humanitarian intervention: Victims, interveners and pillars. *Political Geography* 15(8): 671–695. doi:10.1016/0962-6298(95)00066-6

Wridt, P. (2004). Block politics. *Children's Geographies* 2(2): 199–218. doi:10.1080/1473328041000172 0511

Yea, S. (2013). Mobilising the child victim: The localisation of human trafficking in Singapore through global activism. *Environment and Planning D: Society and Space* 31(6): 988–1003. doi:10.1068/d15411

Young, L. (2004). Journeys to the street: The complex migration geographies of Ugandan street children. *Geoforum* 35(4): 471–488. doi:10.1016/j.geoforum.2003.09.005

Doing Politics

Chapter 21

Electoral Geography in the Twenty-First Century

Michael Shin

University of California, Los Angeles, California, USA

In no other period in history have there been more democracies or elections around the world. More people have the right to vote today than ever before. Between 2010 and 2014 alone, more than 500 presidential, parliamentary, and legislative elections and referenda were held globally (www.electionguide.org). By adding local, city, and mayoral elections, and the range of other popular contests for public office, this number balloons well into the thousands. Keeping a tally of such political contests is one thing, but making sense of them is quite another. By framing elections, electoral outcomes, and political behavior geographically, electoral geography provides important insights into democracy and the democratic process.

Studies within contemporary electoral geography can be categorized into three complementary analytical traditions (Taylor & Johnston 1979; Taylor 1993; Warf & Leib 2011). The first and perhaps most classic tradition in electoral geography is the geography of voting, or using maps to understand and explain election results. Mapping election results and explaining voting patterns dates back well into the twentieth century, but the ubiquity and near real-time dissemination of such analyses, sophisticated electoral geographical predictions, and interactive web mappings of twenty-first-century electoral geography merit further scrutiny. The next tradition concerns identifying and evaluating the geographical influences on voting, such as the "effects" of neighborhoods and local campaigns. Included in this line of work are studies that treat concepts such as place, context, and scale as social constructs through which elections and voting are mediated, rather than mere units of analysis by which voting data are reported. The final analytical tradition is concerned with understanding how votes are translated into representation, party seats, and governments. Like the first tradition, recent methodological and technological advances, in conjunction with outstanding issues such as redistricting and gerrymandering, make explorations into the implications of the geography of representation a compelling area of study.

The Wiley Blackwell Companion to Political Geography, First Edition.
Edited by John Agnew, Virginie Mamadouh, Anna J. Secor, and Joanne Sharp.
© 2015 John Wiley & Sons Ltd. Published 2017 by John Wiley & Sons Ltd.

Despite the stability of the analytical traditions of electoral geography, the landscape of democracy and elections is continually shifting, and merits continued study. One outstanding question concerns the future of democracy and elections around the world in the twenty-first century given their fragility in some regions, and the entrenchment and renewal of authoritarianism in states such as Russia and China (Ignatieff 2014). In established democracies, the overall quality, role, and efficacy of the state and democratic institutions are under scrutiny as voters tune out, demographics change, and party membership declines against the backdrop of increased political polarization and economic inequality. Moreover, as already noted, the role of technology, and in particular geospatial information technology, in both understanding and shaping democracy and elections has dramatically increased in recent years. How such technology is implicated in the democratic process, as well as its downstream consequences and benefits, presents itself as a promising area of inquiry. The three analytical traditions described here are used to structure this chapter, which reviews established and emerging themes within electoral geography and voting studies.

Mapping voters and votes

The mapping of election results, whether for candidates, parties, or referenda, is a common practice. From Andre Siegfried's (1949) work in the Ardèche region of France to mappings of the 2014 Indian election, the largest ever election to date, such mappings serve as accessible and compelling ways to communicate electoral outcomes. Moreover, technological advances in mapping, visualization, and data analysis enable new methods to explore the geography of voting, and also introduce new topics for further inquiry. Despite such technological advances, and the numerous opportunities to map elections around the world, the central question behind the geography of voting remains: What accounts for the geographical patterning of votes? Behind this seemingly simple question are perspectives on electoral behavior that have shaped both the field of electoral geography, and debates within voting studies at large.

One of the central premises of electoral geography is that important insights into democracy and political behavior can indeed be gleaned from the geography of voting and the spatial distribution of voters. In other words, the voting patterns revealed on a map are arguably a reflection of a polity's social composition and structure. This compositional view of voting considers individual and social characteristics such as class, education level, religiosity, ethnicity, and age to be key determinants of voting behavior. Knowing something about the geographical distribution of such traits can arguably lend insights into voting outcomes and, subsequently, the electoral geography of a given democracy. From this perspective, geography is considered the canvas or backdrop for patterns of political participation rather than a fundamental aspect of voting and elections.

A simple map comparison, for instance, between the percentage of native Russian speakers and support for the Party of Regions in the 2012 Ukrainian general election serves to illustrate this approach. The similarity in the geographical patterns and the strong positive relationship between the two variables suggest that the geography of support for the Party of Regions was in fact related to the spatial distribution of native Russian speakers (see Figures 21.1a, 21.1b and 21.1c).

Such map sequences that encourage comparisons are common before voting begins, as well as after the polls close. Although useful to frame and stimulate discussions and conversations about politics, electoral geography as map comparisons and exercises in correlation and

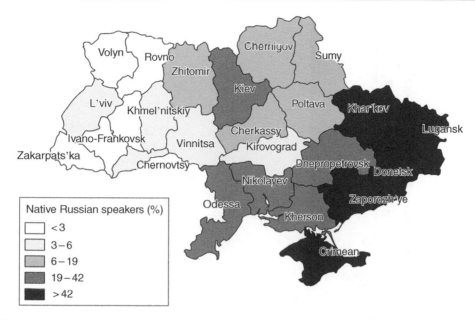

Figure 21.1a Percentage of native Russian speakers in Ukraine. *Source*: Data from State Statistics Service of Ukraine (http://www.ukrstat.gov.ua/)

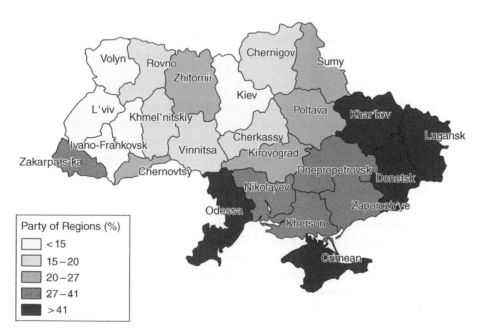

Figure 21.1b Percentage support for Party of Regions, 2012. *Source*: Data from Central Election Commission of Ukraine (http://www.cvk.gov.ua/)

regression has its limits. In particular, correlation is not causation (i.e., by itself, being a native Russian speaker is not a causal mechanism to cast a ballot for a given party), and such maps and analyses often warrant further explication and investigation beyond the interpretation of a particular election result.

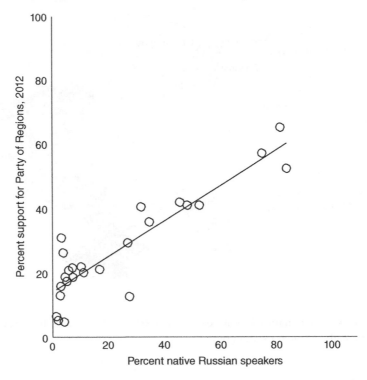

Figure 21.1c The relationship between native Russian speakers and 2012 support for the Party of Regions in Ukraine. *Source*: Data from State Statistics Service of Ukraine (http://www.ukrstat.gov.ua/)

One approach that extends voting studies beyond a single election is the social cleavage framework. Based on the idea that voting behaviors reflect underlying social divisions or cleavages within a democracy, social cleavage approaches frame elections within a set of fundamental struggles and conflicts within a democracy. For example, emerging with modernization and nationhood in nineteenth-century Western Europe were four cleavages – church vs. state, urban vs. rural, center vs. periphery, and owner vs. labor – around which modern European party politics and elections coalesced (Lipset & Rokkan 1967). Such cleavages arguably represent key social conflicts and tensions that continue to define social group membership, the platforms of political parties, and, ultimately, voter preferences. For instance, in many Western European democracies, electoral support for labor and left-of-center parties is considered to be indicative of the owner vs. worker cleavage. In the aftermath of communism's collapse at the end of the twentieth century, Kitschelt (1995) suggested that cleavages around market liberalism vs. economic populism, secular interests vs. religious traditions, and ethnic majority vs. minority identity would figure prominently in postcommunist democracies. The maps in Figure 21.1 arguably illustrate the geographical manifestation of Kitchelt's ethnicity cleavage by using the percentage native Russian-speaking population as a surrogate for such an ethnic division across the Ukraine.

The cleavage framework is relevant to electoral geography for several reasons. First, as articulated by Taylor and Johnston (1979), cleavages provide a way to theorize electoral geography by linking voters and voting patterns to larger societal tensions, concerns, and questions that extend beyond any particular election. Although initially introduced to understand electoral politics in Cold War Western Europe, the cleavage model has proven useful in other contexts, such as Africa (Basedau et al. 2011), Latin America (Bonilla et al. 2011), and in newer democracies like Russia (O'Loughlin, Shin, & Talbot 1996; Perepechko, Kolossov, & ZumBrunnen 2007). Second, some cleavages are inherently geographical in nature. From Lipset and Rokkan's (1967) urban vs. rural and core vs. periphery cleavages to Kitschelt's (1995) ethnic majority cleavage, geography can be central to political geographical divisions, which are often considered *regional* cleavages by themselves (Knutsen 2010). Finally, social cleavages align well with existing sources of data, such as national censuses, which in turn facilitate electoral mappings and related analyses.

Despite the adoption of social cleavage frameworks in electoral geography and voting studies, there are outstanding questions concerning their decline and continued significance (Elff 2007; Evans & Tilley 2012). As the social structure of a democracy changes, for example, due to shifting demographics or economic restructuring, the relevance of cleavages may in fact wane. Moreover, shifting policy preferences and changing socio-economic realities within a democracy, coupled with technological change, arguably promote greater individualization that undermines the fundamental premise of such *social* cleavages (Beck 2014). Regardless of the future of social cleavages, the individual characteristics and social composition of voters continue to be recognized as key determinants of voting behavior and patterns of political participation.

One electoral pattern that receives considerable interest, especially in contemporary American politics, is political polarization. The political polarization construct that pits republican "red" states against democratic "blue" states is itself defined, understood, and reproduced in electoral geographical terms. Determining the causes, consequences, and degree of polarization remains a subject of interest (Mellow & Trubowitz 2005; Morrill, Knopp, & Brown 2007; Tam Cho, Gimpel, & Hui 2013) and to some degree skepticism. For example, notable geographical variations in support for Republicans and Democrats call into question the red/blue dichotomy (Gimpel & Schuknecht 2004; Gelman 2010), and the degree to which the American electorate is truly geographically divided or ideologically split has also been questioned (Glaeser & Ward 2006; Fiorina, Abrams, & Pope 2011).

The geographical framing of political polarization in the United States and elsewhere highlights the continued interest in both the geography of voting and the methods and techniques of electoral geographers (Brunn et al. 2011; Tam Cho & Gimpel 2012). Significant increases in the amount, quality, and accessibility of georeferenced election data, and the rapid and concurrent diffusion of internet and mapping technologies, have led to an explosion of maps, mappings, and electoral geographical analyses by professionals and non-professionals alike. This democratization of electoral geography, and in particular the tools of the trade, is referred to as political "neo-geography" (Shin 2009). Simple, hard-copy, and static maps of election returns have been supplanted by interactive visualizations of sources and amounts of campaign contributions (see www.fec.gov), colorful animations of voting shifts (see http://elections. nytimes.com), and user-defined scenarios of electoral geographies and outcomes that draw from the latest polls (see www.realclearpolitics.com; www.politico.com). Such maps, visualizations, and analyses are not only part of contemporary post-election analysis, but are now expected in the lead-up to an electoral contest. Although the primary concern of electoral geography remains mapping politics, the politics of the map warrant careful consideration as well (Crampton 2010; Wood 2010).

More formal examinations of voting patterns with spatial and geostatistical analysis are also common in contemporary electoral geography (O'Loughlin 2003). One objective of such spatial analyses is to determine whether or not voting patterns are in fact geographically random or statistically substantive, and in need of additional or alternative explanations, above and beyond what purely compositional accounts offer (Tam Cho & Rudolph 2008). From exploratory spatial data analyses (ESDA) that evaluate the geographical clustering of party support, change, and campaign contributions (Tam Cho 2003; Shin & Passarelli 2012) to various spatial regression techniques that attempt to integrate compositional approaches with more contextual ones (Tam Cho & Gimpel 2007; Shin & Agnew 2011), the formal analysis of spatial data provides important insights into electoral geography by incorporating electoral data into explicit geographical frameworks of analysis.

Although characteristics such as income, educational attainment, and age certainly matter with regard to political behavior, the presumption that voter composition trumps all when it comes to understanding the geography of support is open to discussion and debate. The vote is the end result of a multifaceted and complex decision-making process that does not occur in the vacuum of outer space, but in ever-changing geographical contexts that are shaped and mediated by social institutions, practices, and interactions. With a focus on the geographical influences on voting, such contextual approaches in electoral geography are the subject of the next section.

Geographical influences on voting

Complementing compositional approaches to voter behavior are perspectives that recognize the importance of socialization, and, more specifically, how and *where* socialization and political attitude formation occur. While a voter's income, age, and level of education may indeed be related to political attitudes and behaviors, social processes and interactions that occur in particular places are also arguably important shaping and influencing factors. Consider two voters with identical compositional characteristics (i.e., the same age, income, level of education), but voter A lives in a predominantly conservative neighborhood and voter B in a highly liberal one. According to the compositional perspective, both voter A and voter B should cast identical ballots, regardless of where they live. Contextual approaches to electoral behavior, however, contend that place matters and that geography is intrinsic to political behavior. From this perspective, the political attitudes and behaviors of both voters A and B will be influenced by their respective partisan environments. In aggregate terms, this can lead to a party performing better than expected in its electoral strongholds. Contextual approaches to voting behavior differ from compositional ones by emphasizing sociogeographical processes and interactions that are situated in particular places over the ascribed characteristics of voters alone.

Electoral studies that focus on socialization and group effects often make use of the terms "context" and "contextual effects" (Huckfeldt & Sprague 1987; Agnew 1996; Van der Wusten & Mamadouh 2014). Although there is a tacit understanding that voters are socialized and influenced within particular contexts, what constitutes a context is not always clear. Some studies refer to context as the particular competitive environment of an election, but in others the concept of context clearly refers to a specific geographical location or a local unit of analysis such as a neighborhood, voting district, or electoral precinct. Much about voting behavior has certainly been learned from purely local-level analyses, but because the local is a relative concept and cannot be understood without reference to other scales of analysis (e.g., regional, national, global), considering context in broader geographical terms, beyond the neighborhood and the local, is also useful when examining electoral behavior (Agnew 1996, 2002; Shin & Agnew 2008).

Within electoral geography, the context of the neighborhood and its effect on voters has received a considerable amount of attention (Butler & Stokes 1971; Johnston 1986; Cutts et al. 2014). Social processes and interactions that constitute neighborhood effects (Galster 2012), such as informal conversations between residents (Pattie & Johnston 2000) and local and targeted campaign effects (e.g., Carty & Eagles 2005; Johnston & Pattie 2013; Cutts et al. 2014), are believed to be especially relevant to political behavior. Such social exchanges and interactions that take place in a neighborhood setting arguably bias the local environment in such a way that all voters residing within the partisan context are more likely to support the locally dominant party or candidate. Evidence for neighborhood effects, where support for a party or candidate is greater than expected in areas where the party is already well established, has been found in the United States (Dyck, Gaines, & Shaw 2009; Meredith 2013), as well as across Europe (Johnston et al. 2007; David & Van Hamme 2011; van Gent, Jansen, & Smits 2014).

Extending these neighborhood effects are conceptions of context that elevate the importance of scales of analysis and the concept of context-as-place (Agnew 1987, 1996). As a unit of analysis, the neighborhood is frequently associated with the local scale of analysis and referred to as the "local" context. Understanding the local, however, requires reference to other scales, such as the regional, national, and global. By relating the local to other scales of analysis, context-as-place becomes a much richer concept and elevates geography to more than an effect to be controlled in a model or a backdrop for voting. The local no longer only refers to a unit of analysis or the spatial scale of a study, but becomes the place where multiple processes emanating from different scales converge, intersect, and are mediated by voters.

Geographical variations in voting are therefore not artifacts or geographical deviations from a national mean; rather, they are the constituent parts of a democracy's electoral geography. The geographical distribution of an electorate's social composition certainly contributes to such patterns and variations in political participation. How such compositional differences are manifest in differences in political behavior is also a function of how places are shaped, reproduced, and redefined by multiscalar processes like economic restructuring and political campaigns. From this point of view, knowing only the compositional qualities of a voter such as income or ethnicity is not sufficient. Understanding how and why the same compositional categories vary in meaning and significance across a democracy, and how such variations are associated with particular political attitudes and behaviors, highlights the potential of place-as-context approaches.

Contextual investigations into political behavior and electoral geography have delivered important insights into, for example, political participation in the United States (Tam Cho & Rudolph 2008) and electoral change across Italy (e.g., Shin & Agnew 2008). Despite the potential of contextual approaches, the effects of both neighborhood and, more broadly, context are not without their skeptics. McAllister and Studlar (1992) find neighborhood effects in Britain to be more statistical artifact than substantive explanation for regional variations in voting. Similarly, in a response to Agnew's (1996) position that context-as-place matters in electoral geography, King (1996) suggests that contextual effects rarely matter, although it is clear that he defines context as a local "neighborhood effect" rather than a geosociological, place-based construct. Despite such differences in opinion about the nature and impact of geographical influences on voting, increased access to georeferenced data sets and the tools to analyze them are creating new opportunities to explore geographical influences on voting, and electoral and political geography more broadly (Ward & O'Loughlin 2002).

From votes to seats

Regardless of the position one takes concerning the key determinants of electoral behavior, compositional or contextual, the act of casting a ballot is similar for all electors. The translation of votes into seats, parties, coalitions, and governments, however, varies considerably across the world of representative democracy (Norris 2004; Golder 2005). From the type of political and electoral system in place within a democracy to the nature of the election itself, there is more behind a polity's electoral geography than simply the map.

The rules that structure elections within a democracy are often inherently geographical (Johnston 2002; Gudgin & Taylor 2012). Most electoral systems rely on the geographical partitioning of a country into constituencies and electoral districts for which representation is determined. Frequently based on the underlying geographical distribution of a country's population, the creation of electoral districts is an exercise in applied electoral cartography. Since there are numerous ways in which such districts can be drawn, there are arguably several possible electoral outcomes even when the vote is held constant. Consequently, the creation of constituencies and the redrawing of voting districts lead to both biases and abuse.

Malapportionment and gerrymandering represent the most common forms of electoral bias and abuse in representative democracies (Chen & Rodden 2013). Malapportionment occurs when there is a notable discrepancy in the number of electors between constituencies. In an unbiased system, each constituency should have roughly the same number of voters, and the ratio of the most populous constituency to the least should be close to one. Consider the 2014 Indian election, which was not only the largest election to date in the world but featured a difference between the largest and smallest constituencies of over 9.4 million voters and a ratio between the two of nearly 200. It was also among the most biased, with a staggering amount of unequal representation (Wall Street Journal 2014).

Gerrymandering refers to the intentional drawing of electoral district boundaries that favor one party or candidate over another. This is a long-standing practice in US politics, and different gerrymandering strategies have emerged. For instance, to minimize the number of seats a party wins, district boundaries can be drawn in a way that concentrates the supporters of the given party. The alternative "divide-and-conquer" approach does the opposite, by splitting a bloc of voters between several districts to dilute and minimize their impact at the polls. By creating a surplus of unnecessary votes or by ensuring that votes are cast for a losing candidate or party, the practice of gerrymandering in effect "wastes" votes. Despite efforts to eliminate gerrymandering by using non-partisan or bi-partisan redistricting commissions charged with drawing neutral boundaries, overcoming such biases remains problematic (Grofman & Lijphart 2003; Gudgin & Taylor 2012). In the United States, one of the consequences of bi-partisan redistricting is that the interests of voters have been superseded by those of the parties concerned with creating "safe" uncompetitive seats. This has also led to more discussions about whether or not gerrymandering dampens political participation (Hayes & McKee 2009) and causes political polarization (McCarty, Poole, & Rosenthal 2009).

Research on redistricting and gerrymandering is at its core electoral geography, and overlaps with the interests of political science, legal studies, and public policy. Furthermore, redistricting in practice has significant political, geographical, and legal implications. Just as the availability of election data and recent technical innovations have led to developments in election mapping, similar tools and techniques, such as simulation, are being used to reassess the effects of gerrymandering and partisan redistricting (Altman & McDonald 2011; Chen & Rodden 2013). Such approaches are not limited to redistricting in America, but are critical in other democracies as well (Handley & Grofman 2008).

Electoral bias in the form of gerrymandering and malapportionment is most commonly associated with the majoritarian electoral systems of the Anglo-American world. It is important to note that geographical concerns are also implicated in proportional representation (PR) systems that allocate seats according to the percentage of votes that a party or candidate list receives. In PR elections where multiple seats are assigned to constituencies, geographical variations in the number of seats allocated per constituency can lead to inter-constituency differences in the percentage of votes required to secure seats. The possibility of such disproportionate outcomes versus increased representation and the reduction of wasted votes constitute one of the trade-offs between PR and majoritarian systems. The transferable vote system, in use in Irish elections, also attempts to reduce the number of wasted votes by taking into consideration voter rankings and assigning excess votes to lower-ranked parties and candidates. Producing results similar to PR in many cases, the transferable vote system can also promote geographically based voter management strategies by parties to ensure and increase representation (Johnston & Pattie 2011). Finally, mixed or hybrid systems like those used in Germany and Bolivia combine majoritarian and PR system elements to provide direct local representation while also aiming to ensure that the legislature reflects the nationwide vote.

In addition to translating votes to seats, there are various rules and regulations that govern voting and electoral participation. For instance, voting is compulsory in several democracies, such as Brazil, Singapore, and Belgium. By increasing electoral participation, compulsory voting arguably contributes to the legitimacy of an election and the quality of a democracy at large, and also ensures the enfranchisement of as many voters as possible (Birch 2009). Opponents of such regulations cite that mandatory political participation is in itself an infringement on an individual's civic rights and freedoms.

Regardless of whether or not voting is compulsory, with each election cycle the act of voting reinforces and reproduces the key practices and norms of a country's electoral system, as well as democracy in general. Oftentimes, the process of voting is neither consistent across a democracy nor uncontested. In established democracies, the regular disenfranchisement of certain groups of voters, for example through onerous voter registration requirements, undermines the legitimacy and extent of democracy itself. In new and emerging democracies like Afghanistan, democratic norms and practices are not only in their nascent stages, they are also contested and challenged by non-democratic interests. Figure 21.2 maps the number of votes in question at polling stations where electoral irregularities were reported after the 2009 presidential election in Afghanistan.

The irregularities in question occurred when any presidential candidate received over 95 percent of the vote at a polling station, or when the vote totals at a polling station exceeded

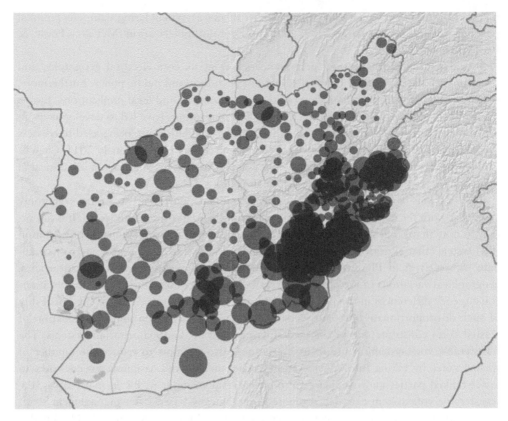

Figure 21.2 The geography of electoral irregularities in the 2009 presidential election in Afghanistan. Each circle refers to the location of a polling station and the size of the circle corresponds to the number of votes in question – which in total exceeded 1.5 million. *Source*: National Democratic Institute Afghanistan, 2014. CC-BY-SA

the maximum capacity of a ballot box (National Democratic Institute Afghanistan 2014; Nordland & Ahmed 2014). Just as the geography of support for parties varies across a democracy, the map of electoral irregularities in Afghanistan in 2009, and, more specifically, the over 1.5 million ballots – or 19 percent of all ballots cast – in question, display notable geographical variations in potential fraud.

The political geographical elements and consequences of electoral systems are fundamental to democracy, and represent compelling opportunities for inquiry. With election and political data from democracies around the world more readily available than ever before, several opportunities exist to examine the formation and shifting of such electoral geographies, and to question and reframe the practices, processes, and patterns behind elections and political behavior. Additionally, through the innovative use and application of technology (e.g., mobile devices, social media, crowdsourcing), questions about political attitudes and behavior in seemingly data-poor areas can now be addressed. For example, Linke's (2013) study of political attitudes after episodes of electoral violence in Kenya frames elections not as the endgame, but as a catalyst for shaping political attitudes, and also reflects increases in access to different types of data. With the diffusion of democracy and advances in information technology, calls to expand the world and scope of electoral geography (Osei-Kwame & Taylor 1984) are not only being met, but embraced.

Whither electoral geography?

Although the primary concerns of electoral geography – namely, the geography of support, geographical influences on voting, and the geography of representation – have remained consistent over time, the ways in which elections and election results are understood, analyzed, and reported have changed considerably in recent years. For example, feminist and poststructuralist perspectives provide alternative ways to frame voting and electoral geography (Secor 2004; Brown, Knopp, & Morrill 2005; Cupples 2009; Schurr 2013), and complement and contrast the analytical traditions discussed in this chapter. What is more, maps of election returns are not only expected immediately following an election, maps of predicted outcomes based on the latest polling data and analysis are also expected in anticipation of voters going to the polls. Several online media outlets also permit readers to create and share their own electoral geographical predictions. The availability of such election prediction and outcome maps is a direct reflection of the diffusion of geospatial and other information technologies, the increasing demand for such types and forms of information, and the creation of such demand.

The increasing visibility of and demand for electoral geography underscores the importance of geographical perspectives and the value of spatial methodologies. As geographical information systems (GIS) and related mapping technologies continue to drive the "spatial turn" across the social sciences and beyond (Logan 2012), several opportunities to extend electoral geography present themselves. For instance, additional efforts to expand technical and quantitative approaches to the study of elections are warranted. As the availability of election, polling, and other types of data and information continues to increase, greater levels of numeracy and technical skills (e.g., GIS, statistics, computer programming, database management) are both required and in demand when examining elections and voting. Elections continue to be an ideal expression of political geography, and, given their frequency, proliferation, and data, they warrant renewed interest, mapping, and analysis by political geographers. Furthermore, there is a growing need to be able to integrate, synthesize, and visualize political data and information from diverse sources, scales, and contests.

The rationale underlying this call for more technical and quantitative approaches is twofold. First, electoral geography is a quantitative endeavor and in order to understand and communicate electoral processes and voting outcomes, a certain degree of numeracy is arguably required. The appeal and insights delivered by pundits and bloggers like Nate Silver (2012) of *The Signal and the Noise* fame, who correctly predicted how each of the states voted in the 2012 US presidential election, are upping the ante for all consumers of political information. Second, new tools, techniques, and technologies are making what were once very technical and complex procedures, such as interactive data visualizations, predictive modeling, and simulations, far more accessible. Prior to applying such techniques, knowledge of various tools and programming languages is often needed to compile, shape, and reshape the necessary data. Moreover, having an understanding of the potential and pitfalls of spatial data is becoming critical as such tools and techniques are more widely adopted.

Although formal, state-level elections remain the primary concern of electoral geography, citizens around the world have opportunities to vote in supra-national contests (e.g., European elections), as well as in local, regional, and other subnational contests, referenda, and initiatives. Are such increased opportunities to cast a vote diluting democracy and contributing to increases in voter fatigue and apathy (Mair 2013)? How are technology and social media altering electoral behavior and civic engagement? It is important to note that the adoption, diffusion, and use of such technology vary between generations as well as geographically, which in turn may accentuate both generational and geographical differences in political behavior (Shin & Agnew forthcoming). Beyond providing the tools for electoral geographers, understanding how technology is implicated with democracy, elections, and electoral geography more broadly are further important topics to consider.

Efforts to expand the world of electoral geography should not simply be limited to the study of recent elections or new representative democracies. As noted at the beginning of this chapter, the state of democracy, democratic institutions, and elections is neither fixed nor assured. Moreover, both new and established democracies face internal pressures and external challenges that are often manifested electorally. Electoral geographers need to continue to address these and other questions, and engage with other disciplines (Johnston 2005), as well as with a public that is both interested in the geospatial and mappings of all sorts. Grounded in three complementary analytical traditions that are well suited to grappling with the proliferation of political data, technology, and innovative methods, electoral geography is well positioned to contribute substantive insights into democratic processes and the democratic enterprise at large.

References

Agnew, J.A. (1987). *Place and Politics: The Geographical Mediation of State and Society*. Boston, MA: Unwin Hyman.

Agnew, J. (1996). Mapping politics: How context counts in electoral geography. *Political Geography* 15(2): 129–146. doi:10.1016/0962-6298(95)00076-3

Agnew, J. (2002). *Place and Politics in Modern Italy*. Chicago, IL: University of Chicago Press.

Altman, M., & McDonald, M.P. (2011). BARD: Better Automated Redistricting. *Journal of Statistical Software* 42(i04). http://econpapers.repec.org/article/jssjstsof/42_3ai04.htm, accessed March 26, 2015.

Basedau, M., Erdmann, G., Lay, J., & Stroh, A. (2011). Ethnicity and party preference in sub-Saharan Africa. *Democratization* 18(2): 462–489. doi:10.1080/13510347.2011.553366

Beck, U. (2014). Individualization is eroding traditions worldwide: A comparison between Europe and China. In Beck, U. (ed.), *Ulrich Beck: Pioneer in Cosmopolitan Sociology and Risk Society*, 90–99. SpringerBriefs on Pioneers in Science and Practice, 18. New York: Springer, 90–99. http://link.springer.com/chapter/10.1007/978-3-319-04990-8_9, accessed March 26, 2015.

Birch, S. (2009). *Full Participation: A Comparative Study of Compulsory Voting*. Manchester: Manchester University Press.

Bonilla, C.A., Carlin, R.E., Love, G.J., & Silva Méndez, E. (2011). Social or political cleavages? A spatial analysis of the party system in post-authoritarian Chile. *Public Choice* 146(1–2): 9–21. doi:10.1007/s11127-009-9580-2

Brown, M., Knopp, L., & Morrill, R. (2005). The culture wars and urban electoral politics: Sexuality, race, and class in Tacoma, Washington. *Political Geography* 24(3): 267–291. doi:10.1016/j.polgeo.2004.09.004

Brunn, S.D., Webster, G.R., Morrill, R.L., Shelley, F.M., Lavin, S.J., & Clark Archer, J. (2011). *Atlas of the 2008 Elections*. Lanham, MD: Rowman and Littlefield.

Butler, D., & Stokes, D.E. (1971). *Political Change in Britain*. New York: St. Martin's Press.

Carty, R.K., & Eagles, M. (2005). *Politics Is Local: National Politics at the Grassroots*. Don Mills, Ontario: Oxford University Press.

Chen, J., & Rodden, J. (2013). Unintentional gerrymandering: Political geography and electoral bias in legislatures. *Quarterly Journal of Political Science* 8(3): 239–269. doi:10.1561/100.00012033

Crampton, J.W. (2010). *Mapping: A Critical Introduction to Cartography and GIS*. Malden, MA: Wiley-Blackwell.

Cupples, J. (2009). Rethinking electoral geography: Spaces and practices of democracy in Nicaragua. *Transactions of the Institute of British Geographers* 34(1): 110–124. doi:10.1111/j.1475-5661.2008.00324.x

Cutts, D., Webber, D., Widdop, P., Johnston, R., & Pattie, C. (2014). With a little help from my neighbours: A apatial analysis of the impact of local campaigns at the 2010 British general election. *Electoral Studies* 34(June): 216–231. doi:10.1016/j.electstud.2013.12.001

David, Q., & Van Hamme, G. (2011). Pillars and electoral behavior in Belgium: The neighborhood effect revisited. *Political Geography* 30(5): 250–262. doi:10.1016/j.polgeo.2011.04.009

Dyck, J.J., Gaines, B.J., & Shaw, D.R. (2009). The effect of local political context on how Americans vote. *American Politics Research* 37(6): 1088–1115. doi:10.1177/1532673X09332932

Elff, M. (2007). Social structure and electoral behavior in comparative perspective: The decline of social cleavages in Western Europe revisited. *Perspectives on Politics* 5(2): 277–294. doi:10.1017/S1537592707070788

Evans, G., & Tilley, J. (2012). How parties shape class politics: Explaining the decline of the class basis of party support. *British Journal of Political Science* 42(1): 137–161. doi:10.1017/S0007123411000202

Fiorina, M.P., Abrams, S.J., & Pope, J. (2011). *Culture War? The Myth of a Polarized America*. Boston, MA: Longman.

Galster, G.C. (2012). The mechanism(s) of neighbourhood effects: Theory, evidence, and policy implications. In van Ham, M., Manley, D., Bailey, N., Simpson, L., & Maclennan, D. (eds), *Neighbourhood Effects Research: New Perspectives*, Amsterdam: Springer, 23–56. http://link.springer.com/chapter/10.1007/978-94-007-2309-2_2, accessed March 26, 2015.

Gelman, A. (2010). *Red State, Blue State, Rich State, Poor State: Why Americans Vote the Way They Do*. Princeton, NJ: Princeton University Press.

Gimpel, J.G., & Schuknecht, J.E. (2004). *Patchwork Nation Sectionalism and Political Change in American Politics*. Ann Arbor, MI: University of Michigan Press.

Glaeser, E.L., & Ward, B.A. (2006). Myths and realities of American political geography. *Journal of Economic Perspectives* 20(2): 119–144.

Golder, M. (2005). Democratic electoral systems around the world, 1946–2000. *Electoral Studies* 24(1): 103–121. doi:10.1016/j.electstud.2004.02.008

Grofman, B., & Lijphart, A. (2003). *Electoral Laws and Their Political Consequences*. New York: Algora Publishing.

Gudgin, G., & Taylor, P.J. (2012). *Seats, Votes, and the Spatial Organisation of Elections*. Colchester: ECPR Press.

Handley, L., & Grofman, B. (2008). *Redistricting in Comparative Perspective*. Oxford: Oxford University Press.

Hayes, D., & McKee, S.C. (2009). The participatory effects of redistricting. *American Journal of Political Science* 53(4): 1006–1023. doi:10.1111/j.1540-5907.2009.00413.x

Huckfeldt, R., & Sprague, J. (1987). Networks in context: The social flow of political information. *American Political Science Review* 81(4): 1197–1216. doi:10.2307/1962585

Ignatieff, M. (2014). Are the authoritarians winning? *New York Review of Books* July 10. http://www.nybooks.com/articles/archives/2014/jul/10/are-authoritarians- winning/, accessed March 26, 2015.

Johnston, R. (2002). Manipulating maps and winning elections: Measuring the impact of malapportionment and gerrymandering. *Political Geography* 21(1): 1–31. doi:10.1016/S0962-6298(01)00070-1

Johnston, R. (2005). Anglo-American electoral geography: Same roots and same goals, but different means and ends? *Professional Geographer* 57(4): 580–587. doi:10.1111/j.1467-9272.2005.00500.x

Johnston, R., & Pattie, C. (2013). Learning electoral geography? Party campaigning, constituency marginality and voting at the 2010 British general election. *Transactions of the Institute of British Geographers* 38(2): 285–298. doi:10.1111/j.1475-5661.2012.00527.x

Johnston, R., Jones, K., Propper, C., & Burgess, S. (2007). Region, local context, and voting at the 1997 general election in England. *American Journal of Political Science* 51(3): 640–654. doi:10.1111/j.1540-5907.2007.00272.x

Johnston, R.J. (1986). The neighbourhood effect revisited: Spatial science or political regionalism? *Environment and Planning D: Society and Space* 4(1): 41–55.

Johnston, R.J., & Pattie, C. (2011). Electoral systems, geography and political behaviour: United Kingdom examples. In Warf, B., & Lieb, J. (eds), *Revitalizing Electoral Geography*, Farnham: Ashgate, 31–57.

King, G. (1996). Why context should not count. *Political Geography* 15(2): 159–164. doi:10.1016/0962-6298(95)00079-8

Kitschelt, H. (1995). Formation of party cleavages in post-communist democracies: Theoretical propositions. *Party Politics* 1(4): 447–472. doi:10.1177/1354068895001004002

Knutsen, O. (2010). The regional cleavage in Western Europe: Can social composition, value orientations and territorial identities explain the impact of region on party choice? *West European Politics* 33(3): 553–585. doi:10.1080/01402381003654577

Linke, A. (2013). The aftermath of an election crisis: Kenyan attitudes and the influence of individual-level and locality violence. *Political Geography* 37: 5–17. doi:10.1016/j.polgeo.2013.08.002

Lipset, S.M., & Rokkan, S. (1967). *Party Systems and Voter Alignments.* Toronto: Free Press.

Logan, J.R. (2012). Making a place for space: Spatial thinking in social science. *Annual Review of Sociology* 38(1): 507–524. doi:10.1146/annurev-soc-071811-145531

Mair, P. (2013). *Ruling the Void: The Hollowing of Western Democracy.* London: Verso.

McAllister, I., & Studlar, D.T. (1992). Region and voting in Britain, 1979–87: Territorial polarization or artifact? *American Journal of Political Science* 36(1): 168–199.

McCarty, N., Poole, K.T., & Rosenthal, H. (2009). Does gerrymandering cause polarization? *American Journal of Political Science* 53(3): 666–680. doi:10.1111/j.1540-5907.2009.00393.x

Mellow, N., & Trubowitz, P. (2005). Red versus blue: American electoral geography and congressional bipartisanship, 1898–2002. *Political Geography* 24(6): 659–677. doi:10.1016/j.polgeo.2005.01.010

Meredith, M. (2013). Exploiting friends-and-neighbors to estimate coattail effects. *American Political Science Review* 107(4): 742–765. doi:10.1017/S0003055413000439

Morrill, R., Knopp, L., & Brown, M. (2007). Anomalies in red and blue: Exceptionalism in American electoral geography. *Political Geography* 26(5): 525–553. doi:10.1016/j.polgeo.2007.03.006

National Democratic Institute Afghanistan (2014). Election irregularities and fraudulent returns. http://afghanistanelectiondata.org/front, accessed March 26, 2015.

Nordland, R., & Ahmed, A. (2014). Voting fraud hangs stubbornly over Afghan elections, with runoff likely. *New York Times*, April 24. http://www.nytimes.com/2014/04/25/world/asia/voting-fraud-stubbornly-hangs-over-afghanistan-election.html, accessed March 26, 2015.

Norris, P. (2004). *Electoral Engineering: Voting Rules and Political Behavior.* Cambridge: Cambridge University Press.

O'Loughlin, J. (2003). Spatial analysis in political geography. In Agnew, J., Mitchell, K., & Toal, G. (eds), *A Companion to Political Geography*, Malden, MA: Blackwell, 30–46. http://onlinelibrary.wiley.com/doi/10.1002/9780470998946.ch3/summary, accessed March 26, 2015.

O'Loughlin, J., Shin, M., & Talbot, P. (1996). Political geographies and cleavages in the Russian parliamentary elections. *Post-Soviet Geography and Economics* 37(6): 355–385. doi:10.1080/10889388.1996.10641023

Osei-Kwame, P., & Taylor, P.J. (1984). A politics of failure: The political geography of Ghanaian elections, 1954–1979. *Annals of the Association of American Geographers* 74(4): 574–589. doi:10.1111/j.1467-8306.1984.tb01475.x

Pattie, C., & Johnston, R. (2000). "People who talk together vote together": An exploration of contextual effects in Great Britain. *Annals of the Association of American Geographers* 90(1): 41–66. doi:10.1111/0004-5608.00183

Perepechko, A.S., Kolossov, V.A., & ZumBrunnen, C. (2007). Remeasuring and rethinking social cleavages in Russia: Continuity and changes in electoral geography 1917–1995. *Political Geography* 26(2): 179–208. doi:10.1016/j.polgeo.2006.09.006

Schurr, C. (2013). Towards an emotional electoral geography: The performativity of emotions in electoral campaigning in Ecuador. *Geoforum* 49(Oct.): 114–126. doi:10.1016/j.geoforum.2013.05.008

Secor, A. (2004). Feminizing electoral geography. In Staeheli, L., Kofman, E., & Peake, L. (eds), *Mapping Women, Mapping Politics: Feminist Perspectives on Political Geography*, New York: Routledge, 261–272.

Shin, M.E. (2009). Democratizing electoral geography: Visualizing votes and political neogeography. *Political Geography* 28(3): 149–152. doi:10.1016/j.polgeo.2009.03.001

Shin, M.E., & Agnew, J.A. (2008). *Berlusconi's Italy: Mapping Contemporary Italian Politics*. Philadelphia, PA: Temple University Press.

Shin, M., & Agnew, J.A. (2011). Spatial regression for electoral studies: The case of the Italian Lega Nord. In Warf, B., & Lieb, J. (eds), *Revitalizing Electoral Geography*, Farnham: Ashgate, 59–74.

Shin, M.E., & Agnew, J. (forthcoming). Demography and democracy: Exploring the linkage between age and voter turnout in Italy with geospatial analysis. In Howell, F.M., Porter, J.R., & Matthews, S.A. (eds), *Recapturing Space*, New York: Springer.

Shin, M.E., & Passarelli, G. (2012). Northern League in national, European and regional elections: A spatial analysis. *Polis* 3. doi:10.1424/38671

Siegfried, A. (1949). *Géographie électorale de l'Ardèche sour la IIIème République*. Paris: A. Colin.

Silver, N. (2012). *The Signal and the Noise: Why So Many Predictions Fail – but Some Don't*. Harmondsworth: Penguin.

Tam Cho, W.K. (2003). Contagion effects and ethnic contribution networks. *American Journal of Political Science* 47(2): 368–387. doi:10.1111/1540-5907.00026

Tam Cho, W.K., & Gimpel, J.G. (2007). Prospecting for (campaign) gold. *American Journal of Political Science* 51(2): 255–268. doi:10.1111/j.1540-5907.2007.00249.x

Tam Cho, W.K., & Gimpel, J.G. (2012). Geographic information systems and the spatial dimensions of American politics. *Annual Review of Political Science* 15(1): 443–460. doi:10.1146/annurev-polisci-031710-112215

Tam Cho, W.K., & Rudolph, T.J. (2008). Emanating political participation: Untangling the spatial structure behind participation. *British Journal of Political Science* 38(2): 273–289. doi:10.1017/S0007123408000148

Tam Cho, W.K., Gimpel, J.G., & Hui, I.S. (2013). Voter migration and the geographic sorting of the American electorate. *Annals of the Association of American Geographers* 103(4): 856–870. doi:10.1080/00045608.2012.720229

Taylor, P.J. (1993). *Political Geography: World-Economy, Nation-State, and Locality*. London: Longman Scientific and Technical.

Taylor, P., & Johnston, R.J. (1979). *Geography of Elections*. Harmondsworth: Penguin.

Van der Wusten, H., & Mamadouh, V. (2014). "It is the context, stupid!" Or is it? British-American contributions to rlectoral geography since the 1960s. *L'Espace politique* 23(July). http://espacepolitique.revues.org/3048, accessed March 26, 2015.

van Gent, W.P.C., Jansen, E.F., & Smits, J.H.F. (2014). Right-wing radical populism in city and suburbs: An electoral geography of the Partij Voor de Vrijheid in the Netherlands. *Urban Studies* 51(9): 1775–1794. doi:10.1177/0042098013505889

Wall Street Journal (2014). In perspective: India's smallest and largest constituencies. *WSJ Blogs – India Real Time*. http://blogs.wsj.com/indiarealtime/2014/03/31/in-perspective-indias-smallest-and-largest-constituencies/, accessed March 26, 2015.

Ward, M.D., & O'Loughlin, J. (2002). Spatial processes and political methodology: Introduction to the special issue. *Political Analysis* 10(3): 211–216. doi:10.1093/pan/10.3.211

Warf, B., & Leib, J. (2011). *Revitalizing Electoral Geography*. Farnham: Ashgate.

Wood, D. (2010). *Rethinking the Power of Maps*. New York: Guilford Press.

Chapter 22

Nation and Nationalism

Marco Antonsich

Loughborough University, England, UK

On a warm Saturday night in September 2013, I stand in front of St. Peter's Basilica, in Vatican City. I am watching the spectacle of about 100,000 people leaving the square at the end of the vigil for peace in Syria led by Pope Francis. As I notice a group of persons carrying a large Colombian flag, I decide to approach them to appease my geopolitical curiosity. A woman in her fifties replies to my question about why they decided to bring the flag along: "We came here to pray for the world, but also for our country – this is our identity." Maybe puzzled by my question, she then asks in return: "And where is your flag?" Playing the card of the host, I reply: "Well, I come from this country, I don't need to wave my flag." Slightly annoyed, she remarks: "So you don't care about your country!"

Catholicism is universal in scope; it knows no (national) boundaries. And yet, when Pope John Paul II died, St. Peter's Square was also inundated by flags from all over the world, proudly wielded by Catholics who came to mourn their pope. Why did people sharing the same common identity (Catholic) and witnessing a very human (i.e., universal) event feel the need to display their national flags? To answer this question is to answer that about the persisting power of nationalism, even in this age of globalization: Why do nations still matter? Not an easy question, indeed. To Calhoun (2007), who wrote an insightful book on this very subject, nations still matter because they offer a meaningful framework for interpreting social, political, and economic life. Echoing Raymond Williams's often cited idea, Calhoun (2007: 171) suggests that nations can be associated with "structures of feeling" that link categories of thought to emotional engagements. Furthermore, Brubaker (2006: 15, 207) acknowledges the important role of nationhood and ethnicity in helping people make sense of their world; and Bhabha (1990: 1) similarly talks of nations as systems of cultural signification. In other words, today, like yesterday, the national register can offer a medium for interpreting and experiencing reality. The thousands of people who gathered in St. Peter's Square were not there simply as Catholics, but as Colombian Catholics, Polish Catholics, Filipino Catholics, and so on. Their universal affiliation was articulated, performed, felt, and visually expressed in their

The Wiley Blackwell Companion to Political Geography, First Edition.
Edited by John Agnew, Virginie Mamadouh, Anna J. Secor, and Joanne Sharp.
© 2015 John Wiley & Sons Ltd. Published 2017 by John Wiley & Sons Ltd.

national vernaculars. As Calhoun (2007: 171) again observes, nations can also provide individuals with a familiar category through which they, as a group, can participate in history – and the death of John Paul II certainly was a historical event. One should also not forget the fact that sharing and experiencing an event as part of a cultural group can be enjoyable, as Bratsis (2006) suggests with his notion of the "libidinal value" of (national) identities.

Although national identities have today become just one of the many forms of collective formations with which people might identify (Scholte 1996), they remain the most basic form of social identity (Bechhofer and McCrone 2009: 1) or, in Goswami's elegant phrasing: "one of the most universally legitimate articulations of group identity and one of the most enduring and pervasive forms of modern particularism" (2002: 775). A quick look at survey data for various forms of territorial attachment would only confirm this point (Antonsich 2009a).

As national identities are far from disappearing, so nationalism – understood in Gellner's (1983: 1) terms as "a political principle which holds that the political and the national unit should be congruent" – continues to be the fundamental organizing principle of the interstate order and the primary source of political legitimacy (Özkırımlı 2010: 2). It is on the basis of the nationalist principle that the world is divided into "us" and "them," that "we" claim a given portion of the earth as ours, and that "we" look into the past to find confirmation of the existence of ourselves and our homeland. These identity, spatial, and temporal claims are, according to Özkırımlı (2010: 208–209), what distinguishes the nationalist discourse from any other similar discourse. Yet, exactly because of these claims, many social scientists tend to regard nationalism as a suspicious notion or as something to be fixed, when not explicitly dismissed as an exclusionary, regressive, and aggressive principle (Nussbaum 1994), intimately connected with war and militarism (Billig 1995: 7). In the opposite camp, there are those who observe instead that nationalism cannot be treated as "a moral mistake" (Calhoun 2007: 1), since it can motivate and sustain civic engagement, responsibility, and solidarity, provide support for redistributive policies, and foster the inclusion of a diverse population (Brubaker 2004a: 121–122). And yet, according to one of its old students, it might in the end be meaningless to distinguish between "good" and "bad" nationalism, since all nationalisms are always morally and politically ambiguous (Nairn 1977: 348).

This chapter will first address the classic debate on nationalism, with its long-standing focus on the origins ("when") and substance ("what") of the nation, before discussing new theoretical and methodological approaches that instead privilege the "how" and "where" of the nation; that is, the ways in which the nation is reproduced and in which socio-spatial contexts it emerges as a salient category. The discussion will then explore one of the most important challenges that nationalism is facing today, namely global migration and its impact on the national idea, and conclude by highlighting those aspects that deserve further empirical investigation.

The classic debate: "When" and "what" is the nation?

Traditionally, nation and nationalism have been the privileged subjects of historians. It is thus not surprising that the question of the origins of these two notions has occupied a central place in their investigations. "When is a nation?" – as Connor (1990) famously put it – is a question that still divides scholars today. As a way of creating order among the answers, Smith (1971) proposed a tripartite division (primordialism, modernism, and ethno-symbolism) that, although it risks glossing over scholarly differences in each category, remains the most widely recognized and used categorization in nationalism studies (Özkırımlı 2010: 201). As detailed descriptions of these three approaches can be found in various textbooks, I will here only briefly present the key aspects of each of them, largely following Özkırımlı's (2010) account.

Primordialism, as the name suggests, maintains that nations have always existed; they are a "natural" part of human life (Ozikirimli 2010: 49). This is often the view espoused by nationalist actors and ordinary people alike. The real primordialists, for Brubaker (2004b: 83), believe in the immemorial and perennial character of the nation (see also Fox & Jones 2013: 388). From this perspective, as Gellner (1983: 6) phrased it, "a man [sic] must have a nationality as he must have a nose and two ears." From cradle to grave, we are all constituted as a sort of "homo nationalis" (Balibar 1990); our national identity is part of our own being (see also Pickel 2004; Özkırımlı and Uyan-Semerci 2011).

Closely related to primordialism is perennialism, which maintains that nations are not natural, but antique (Özkırımlı 2010: 58) – a distinction, however, that does not convince Özkırımlı (2010: 60), who criticizes both approaches for their essentializing treatment of the nation as a given, fixed, static entity. Primordialism is found guilty of "retrospective nationalism"; that is, "the tendency to project modern concepts and categories onto earlier social formations" (Özkırımlı 2010: 69–70). It correctly points to the deeply rooted attachment that people have toward their nation, but it fails to explain the reasons for this attachment and eschews any issues of power behind the primacy and legitimacy of a particular attachment over others. As such, its analytical purchase is rather marginal (Özkırımlı 2010: 202).

In the 1960s, first with Kedourie (1960) and then with Gellner (1964), a new paradigm emerged in nationalism studies: modernism. In its essence, modernism suggests that nations and nationalism are intrinsic features of the modern world and are not primordial or perennial features (Smith 1998: 3). In the words of Gellner (1983: 8), maybe the most influential representative of the modernist approach, "nations, like states, are a contingency, and not a universal necessity." The key historical event that made the idea of nation and nationalism possible was, for Gellner, the passage from agricultural (traditional) to industrial (modern) societies. Industrialization requires mutually substitutable, atomized individuals, and the idea of a homogenous culture superimposed on the place-specific particularities of the agrarian society resonated well with nationalism (Gellner 1983: 57). In other words, nationalism was not simply the product of industrialization, but as a principle of societal organization it perfectly matched the imperatives of industrialization (Gellner 1983: 46). In this sense, contrary to primordialism (and ethno-symbolism), nations did not engender nationalism, but the other way around (Gellner 1983: 8).

The modernist account stresses the role of the modern, centralized, bureaucratic state (Tilly 1990; Mann 1993), also in relation to its communication and transportation networks (Deutsch 1953; Seton-Watson 1977) and its homogenizing practices associated with mass education and public ceremonies and iconographies (Hobsbawm & Ranger 1983). Accordingly, political elites figure centrally in modernist accounts, which highlight the instrumental nature of ethnicity and national identities as tools in the political struggles among competing actors (Brass 1991). It is indeed when elites feel disempowered by the uneven economic development of a given territory that they might decide to resort to nationalism as a way to respond to this perceived injustice (Hechter 1975; Nairn 1977; see also Greenfeld 1992). Regarding nations as the product of political engineering, modernist authors have also observed the artificial character of national traditions, which are indeed "invented" in order to create continuity with the past so to give present legitimacy and glory to the nation (Hobsbawm & Ranger 1983).

Like primordialism, different aspects of modernism have also been criticized. These are well summarized in Özkırımlı (2010: 123 ff), who observes that nationalism is too complex a phenomenon to be explained by a single factor (modernity) and by the exclusive role of elites and their manipulative strategies. In so doing, modernists end up heralding the nationalist experience of eighteenth- to nineteenth-century Western Europe as a universal model and remain blind to (and unable to explain) the emotive power of nationalism.

It was exactly by building on this latter critique that Anthony Smith, a former student of Gellner, developed an alternative view: ethno-symbolism. While Smith (1986) acknowledges that nationalism is a modern doctrine, nations are not. Their roots are in premodern ethnic communities (ethnies), whose myths, symbols, values, and traditions have remained relatively unchanged since the fifteenth and sixteenth centuries – and in some cases (Armenians, Egyptians, and Jews) even earlier (Smith 2008a). Nationalist elites, therefore, do not invent, but merely rediscover or reconstruct traditions of a pre-existing ethnic past (Smith 1991a: 358). This belief in the persistence and durability of the nation makes ethno-symbolism rather similar to perennialism (Özkırımlı 2010: 202–203) and exposes it to the same critique formulated for the latter. According to Connor (2004), nations are a mass phenomenon and it would not make sense to trace their origin back to a past when only a small aristocracy shared a national consciousness. Malešević (2006: 129) accuses ethno-symbolism of determinism, fatalism, and finalism, as ethnies are indeed predestined to become nations.

The question of "When is a nation?" finds a natural complement in "What is a nation?" In its French version, the question was famously asked by Ernest Renan (1997 [1882]), whose answer is clearly in subjectivist language: The nation is "a soul, a spiritual principle." Scholars espousing the modernist approach have instead put forward more objectivist interpretations (Goswami 2002). In an attempt to reach an ideal-typical definition, applicable across all geographical and historical contexts (Özkırımlı 2010: 19), Smith (2005: 98) tried to combine both subjectivist and objectivist elements:

> [A nation is] a named and self-defined community whose members cultivate common myths, memories, symbols and values, possess and disseminate a distinctive public culture, reside in and identify with a historic homeland, and create and disseminate common laws and shared customs.

One of the most debated issues regarding the question of "What is a nation?" is its civic or ethnic character. While a civic nation can be defined as a political-legal community of citizens sharing a common civic culture, an ethnic nation stands for a community of common descent (Smith 1991b: 11). Although "ethnic" is often associated with "cultural" – ethno-cultural is indeed a term frequently used in nationalism studies – some scholars, echoing Meinecke's (1919 [1907]) distinction between *Staatsnation* and *Kulturnation*, prefer instead to keep them separate (Peters 2002), arguing that there is also empirical evidence in people's views supporting this case (Janmaat 2006). Originally, the civic–ethnic distinction was introduced by Hans Kohn (1958 [1944]), who saw the emergence of civic nationalism in the West as a liberal achievement produced by the state, and ethnic nationalism as the typical trait of the non-Western world (frequently identified with the East), trapped in prepolitical ethnic divisions (Calhoun 2007: 128). Thus, while civic nationalism has come to represent a form of liberal, voluntarist, universalist, and inclusive nationalism, its ethnic variant has been downplayed as illiberal, ascriptive, particularist, and exclusive (Brubaker 1999: 56). It is true, however, that almost all nationalisms combine civic and ethnic elements (Xenos 1996; Yack 1996; Calhoun 2007: 42, 145; Özkırımlı 2010: 37), thus Kohn's distinction might not prove very useful analytically (Brubaker 1999: 58). The United States, for instance, like most Western nations, has only recently become civic, at least in rhetorical terms (Kaufmann 2000; Kuzio 2002).

Yet, even so, it still embodies, in its institutional settings, the cultural values and practices of the majoritarian group (Kymlicka 2001; Bader 2005). When operationalized in quantitative studies, the civic/ethnic distinction also returns contradictory results. According to Hjerm (1998), there is in fact a significant relationship between the likelihood of xenophobia and ethno-national feelings, whereas Janmaat (2006) argues that it is the intensity of national identification rather than its civic/ethnic character that correlates positively with xenophobia. Other authors have instead found either small or no differences in the association between West/civic and East/ethnic nationalisms (Jones & Smith 2001; Shulman 2002).

New approaches: "How" and "where" is the nation?

In the 1990s, the debate on nationalism entered a new stage (Smith 2008b: 564; Özkırımlı 2010: 169–170). Influenced by the "cultural turn" in social sciences, scholars – now sociologists, psychologists, and geographers more than historians – started challenging the static, given, coherent notion of culture that had characterized nationalism studies until then. Rather than asking "when" and "what" is a nation – questions that, according to Brubaker (1994), lead necessarily to substantialist and essentialist accounts of the nation – the new approaches focus on "how" and, to a more limited extent, "where" is a nation. How is nationhood as a political and cultural form institutionalized within and among states? And, once established, how does it work as a practical category, as a classificatory scheme, and as a cognitive frame? How do ordinary people reproduce the nation in their discourses and practices (Brubaker 1994: 6; Skey 2009: 333). For Balibar (1990), the central question should not be about a beginning or an end of the nation, but its reproduction. Similarly, Connor dismisses the theoretical relevance of "When is a nation?," since in people's views nations are timeless and what matters are not facts, but perceptions of facts, as these shape attitudes and behavior (Connor 2004). For Brubaker (2004a, 2006), scholars should avoid treating ethnicity and nationness (a term that, together with nationhood, he uses in place of "nation") as categories of analysis, since this approach would reify them as substantial entities. Ethnicity and nationhood should instead be viewed as categories of practice, as ways of talking, seeing, and acting: cognitive, discursive, or pragmatic frames for understanding and interpreting experience (Brubaker 2006: 207). Nationhood should be the object of analysis, not the tool of analysis (Brubaker 2004a: 116). Far from being an entity, nationhood should be conceptualized as an event; it is not a continuous, but an intermittent and contingent phenomenon. Thus, it demands not a developmentalist (as in traditional, substantialist accounts of nationalism) but an eventful analysis, capable of mapping the situational context of its mobilization (Brubaker 1994: 8, 2006: 208, 361). Accordingly – pace Anthony Smith – no general theory of nationalism is possible (Özkırımlı 2010: 195).

The new approaches bring forward the importance of human agency. With the sole exception of Hobsbawm, who was the first to call attention to the views of ordinary people, classic accounts have treated nationalism as an exclusive matter of elites. When present, masses have entered the picture only as passive receptacles of nationalist messages or mere choreographies in public ceremonies. By building on Hobsbawm's "view from below," the new approaches argue instead that individuals are key actors in the reproduction of the nation. Although there is a general consensus that nations are produced in a "top down" fashion, they are continuously reproduced by ordinary people. According to Thompson (2001: 20), despite their abstract, objectified appearance, nations do not exist above and beyond the agency of individuals. Nations and national identities are not given categories that exist "out there," they are not conferred on individuals like a certificate of birth, but are acquired by individuals, who "make them real" in the course of their daily social interactions (Thompson 2001: 24). Each of us is implicated in organizing, categorizing, and invoking ideas of nation and national identities (Thompson 2001: 29), through discursive acts that are themselves constitutive of those very categories (Edwards & Stokoe 2004). This is what Thompson (2001) calls the "local" production of national identity and what Cohen (1996), adopting a more anthropological perspective, calls "personal nationalism." Drawing on Herzfeld's (1997) seminal work on the ways in which ordinary people recast national official narratives in the more familiar terms of their own daily experiences, Cohen shows how these personalized versions of nationalism enter the constitution of a sense of selfhood (see also Hearn 2007).

Liberated from its predominant institutional, historical heritage and reconceptualized as a category of practice activated in daily contexts, the nation thus becomes an object of analysis in discursive terms. Anderson's (1983) famous book *Imagined Communities* opened the way to this discursive, subjectivist approach (Goswami 2002: 773), which Bhabha (1990) and Calhoun (1997) have most proficiently theorized and Brubaker (2006) has most skillfully operationalized for empirical research. Discursive psychologists (Condor 2000) have also extensively contributed knowledge to the field, by carefully tracing the ways in which ordinary people resort to nationally framed utterances in their daily conversations (Skey 2011). Social interactions have accordingly become the primary focus of any discursive analysis of nation, as for instance in the series of publications by the so-called Edinburgh National Identity Group, which has highlighted the importance of "identity markers" and "identity rules" in verbal identity claims made by ordinary people (Bechhofer & McCrone 2009).

With social interaction has also come a focus on the everyday. The new approaches do indeed show an interest in the micro (little, daily things) more than the macro (historical grand narratives) of nationhood. As Fox and Miller-Idriss (2008: 554) observe, "the nation is not simply the product of macro-structural forces; it is simultaneously the practical accomplishment of ordinary people engaging in routine activities." Billig's (1995) *Banal Nationalism* is possibly the most widely known and celebrated contribution to the study of the everyday salience of nationalism. For him, the nation is continuously "flagged" or called to mind for its citizens in very banal and subliminal ways. Although many followers of this approach have later investigated the material aspects (coins, bank notes, stamps, etc.) of the banality of the nation, Billig's focus is not so much on the material, but on the discursive flagging. It is not about the weather map published in a newspaper, but *the* weather: the little words ("the," "here," "no," "we," etc. – what he call deixis) that often go unnoticed, but are more powerful than grand, memorable phrases in making the nation a daily, unconscious presence in people's lives (see also Thrift 2000 for a more geopolitical account of the same perspective).

A possibly less known but equally important contribution to this form of banal nationalism comes from Edensor (2002, 2004, 2006). Here, the focus is not on the official symbols that permeate daily life in unconscious ways, but on the asymbolic material aspects of nationhood that, through their pervasiveness in time and space, make the national landscape a familiar, homely space. House style, traffic lights, street furniture, fencing style, leisure facilities, parks, and petrol stations are only a few of numerous, unremarkable material aspects (to which one can also add referents of the "soundscape," such as ambulance sirens, and "smellscape," such as food), which do not belong to the official national iconography, but do contribute making the nation a visible, tangible presence in people's routine experience of space (Edensor 2002: 51–54; see also Jones & Merriman 2012). It is exactly this familiar landscape that generates "a cognitive, sensual, habitual and affective sense of national identity, providing a common-sense spatial matrix which draws people and places together" (Edensor 2002: 37).

Like Brubaker (2006), Edensor also focuses on nationness as everyday ways of doing and talking. Mundane habits, routines, and social interactions are here explored in terms of not their content, but their performativity. As Löfgren (1989: 15) observes, a national culture is not "what" but "how" its members talk about it: "the styles in which a problem is addressed, an argument is carried on or a conflict resolved (or suppressed)." Although gender, class, age, and other socio-demographic features might influence ways of doing and talking, they are also specific to a given national culture. For instance, ways of queuing, interacting with strangers, and behaving in public spaces might be ordered and organized according to certain national traits. In an insightful study on the nationalization of trivialities, Linde-Laursen (1993) observes how even dishwashing might become an identity marker between Swedes and Danes. In contrast to Butler (1993), who distinguishes performance (self-conscious and deliberate) from performativity (unreflexive), Edensor argues for the impossibility of drawing a clear boundary, as the two dimensions are always imbricated one with another:

> Reflexivity and unreflexivity are not properties that are associated with particular kinds of enaction, but depend upon contexts and the conditions which shape the frequency of performance. (Edensor 2002: 89)

The focus on performance also brings forward the importance of the body in the reproduction of the nation. Habits are indeed embodied practices. Ways of walking, sitting, standing, and moving are inflected by the socio-demographics already mentioned as much as by national traits (Edensor 2002: 72). The repetition of embodied routines (how and when to eat, wash, play, work, etc.) generates synchronic, mundane choreographies, which in turn feed a sense of a collective "we" grounded not in common ethno-cultural features, but in "the sharing of spatial and temporal co-presence" (Edensor 2004: 110; see also Antonsich 2009b for a similar point). Already anticipated by Anderson (1983: 26), the importance of synchronized practices is further illustrated by Edensor (2006) in terms of working hours, school holidays, television schedules, and so on – a banal, routinized, synchronized everyday life that also contributes to generating a sense of security and functional predictability (see also Löfgren 2001).

As already discussed, although the new approaches are mainly concerned with the "how" of the nation, a few scholars have also approached its "where." Geographers in particular have a long tradition of investigating the geography of the nation, with a particular emphasis on its territory (Williams & Smith 1983; Herb & Kaplan 1999; White 2000; Penrose 2002; Murphy 2013). More recently, however, there has been a new interest in the scalar "happening" of the nation. Jones and Fowler (2007: 333), for instance, echoing Appleton's (2002) argument about the plurality of scales implicated in the representation of the nation, have contested "the implicit and unproblematized emphasis placed upon the national scale as the only appropriate scale at which to study nations." Accordingly, they have focused on the local, showing how the particularities of place actively shape the formation and inflection of nationalist ideologies. Similarly, Confino and Skaria (2002) have challenged traditional renditions of the local as either a preexisting repository of subnational sentiments awaiting to be awakened, or an anti-national entity in need of being nationalized, and have instead mapped vernacular translations of the nation (see also Edensor 2002: 186).

Finally, the new approaches have also rectified the gender-blind and Eurocentric perspective of classic accounts, by providing a feminist-informed understanding of nations and nationalism (Yuval Davis & Anthias 1989; Yuval Davis 1997) and by opening up these categories to the postcolonial notion of hybridity, which challenges any national unitary claim (Bhabha 1990; Chatterjee 1993). This points to the important question of "who" is the nation, as various authors (McClintock 1995; Sharp 1996; Dowler 2001) have observed that nationalism is essentially a masculine product, which often denies women any direct relation to national agency.

The remaking of the nation in the age of globalization

It is commonplace to say that one of the transformations associated with globalization is the decline of fixed, territorial identities and the rise of fluid, mobile forms of identification (Appadurai 1996; Urry 2007). Trasnationalism and cosmopolitanism have become the new flags of this mobile world and they are often heralded in opposition to nationalism. The convergence between the nation-state and the idea of society in scholarly accounts has also been questioned under the critique of "methodological nationalism" (Wimmer & Glick Schiller 2002; Chernilo 2007) or the "territorial trap" (Agnew 1994). And yet, both nation and nationalism seem far from disappearing. In a world perceived as uncertain and constantly changing, national identities might offer people a comforting anchorage, a sort of ontological security (Kinnvall 2004; Skey 2011). "Entitativity" is what psychologists call the reification of a social category, which in turn makes people feel secure and protected from external threats (Sacchi, Castano, & Brauer 2009). And even more so today, people perceive the nation as such an entity, also imbued with a clear ethno-cultural jargon that defies the "civic turn" of many Western governments (Jones & Smith 2001; Joppke 2010). Expressions of "paranoid nationalism" (Hage 2003), or white backlash against migrants, are rather widespread (Triandafyllidou 2001); they also populate the internet, which, contrary to some expectations, appears to have strengthened rather than weakened national sentiments (Eriksen 2007), also in the shape of new diasporic forms of nationalism (Mavroudi 2007; Nieswand 2012).

However, if one goes beyond the above commonplace, one can see how globalization and nationalism are actually complementary (Chernilo 2007; Halikiopoulou & Vasilopoulou 2011). Similarly, transnational processes are integral to the dynamics of national belonging (Dragojlovic 2008) and some authors (Delanty 2006; Calhoun 2008; Beck & Levy 2013) suggest that cosmopolitanism can also be written into nationalism. Moreover, heralding transnationalism and cosmopolitanism for their potentially liberating effects risks overlooking the continuing importance of nationalism for subaltern groups in the global South (Yeğenoğlu 2005). One should also not forget that contrasting a new postmodern cosmopolitan present with a modern national past implicitly contributes to reifying the myth of a historically stable, coherent nation-state that has never existed (Chernilo 2007). Historically, the nation-state has always reinvented itself (Baumann 2004). Thus, the point is not to discuss whether today the nation-state is in crisis, but how it is being reworked to remain salient among new sociospatial formations; and how national identities are renegotiated and reconfigured in the age of globalization (Biswas 2002).

In this respect, how to reconcile unity and diversity is the big challenge that nationalism is facing today. Theoretically, this question has been addressed through various lenses, from (neo)assimilation (Alba & Nee 1997; Brubaker 2001; Nagel 2009) to multiculturality (for a review, see Antonsich 2014). Here, I would like to focus on two normative perspectives that

I believe are most relevant to the present study: liberal nationalism and multicultural nationalism. While both emphasize the importance of nationhood as the founding principle around which to bring diverse people together, they show a rather different approach to cultural diversity, which in turn influences their conceptualization of national identity. From a liberal nationalist perspective, cultural diversity is something to be "fixed," either by providing minorities with special rights so to enhance their public participation (Kymlicka 2001) or, with a more civic nationalist tone, by treating ethnicity as a private cultural phenomenon (Miller 2000: 122, 137). As a consequence, the type of national identity envisioned to maintain social unity in a pluralistic state is rather "thin," based on sharing liberal values (Miller 1995), common language, and history (Kymlicka 2001: 312–315).

In contrast, multicultural authors acknowledge the inescapability and desirability of cultural diversity (Parekh 2000: 340). Accordingly, the challenge is to build a plural and inclusive national identity, based on a composite culture constituted through intercultural dialogue (Parekh 2000: 235–236). Contra national liberalism, this project does not "take off" but "adds" ethno-cultural diversity to national identity (Modood & Meer 2012: 52), which should then be understood as an overarching shared identity built on diversity (Uberoi 2008; Modood 2011). All citizens should be involved in rewriting this new collective "we" (Papademetriou 2012), giving shape to forms of "intercultural nationalism" (Blad & Couton 2009) or "cosmopolitan nationalism" (Brett & Moran 2011; Beck & Levy 2013), which incorporate notions of tolerance, openness, and alterity within national identity (Mavroudi 2010).

Conclusion

Nations and nationalism are not relics of the past. Wishing them away and calling for new postnational configurations (Closs Stephens 2013) risks diverting attention from the ways in which the nation(al) is continuously resignified in adjusting to mutating socio-political conditions (Antonsich et al. 2014). While classic accounts have greatly contributed to the study of the historical and institutional development of the nation, they have overlooked its resonance in people's daily lives. A more sustained dialogue between these two perspectives would therefore be welcomed. For instance, banal and everyday nationalisms would benefit from a closer look at the political and economic forces that have produced the phenomenon of nationalism in the first instance. These forces are indeed associated with structured inequalities, which in turn shape the ways in which people talk, feel, and experience the nation (Skey 2011: 168). As Smith (2008b) observes, the new approaches also seem to operate with an undifferentiated "ordinary people," which fails to specify how forms of banal and everyday nationalisms might resonate differently among people who differ in terms of class, gender, age, race, and so on (Skey 2009). A more fruitful dialogue between classic and new approaches could also contribute to better understanding the persistence of nationalism. Recent work that, building on some classic accounts (Connor 1994), has explored the emotional dimension of the nation is certainly a good start (Caluya 2011; Wood 2012) and it can be further supported by psychological investigations (Baldacchino 2011). Responses to "Why is a nation?" should also be accompanied by additional studies on "Where is a nation?" that could map more extensively the sites where the nation(al) "happens." Finally, we know too little about the impact of an increasingly ethno-culturally diverse population on the meanings, materiality, and performance of the nation. Empirical studies in this direction would nicely complement existing normative accounts and help in exploring further the "who" of the nation.

There is still quite a lot of work to be done before we can get a better sense of why one night in September 2013, thousands of Catholics gathered in St. Peter's Square, waving their national flags.

References

Agnew, J. (1994). The territorial trap: The geographical assumptions of international relations theory. *Review of International Political Economy* 1(1): 53–80. doi:10.1080/09692299408434268

Alba, R., & Nee, V. (1997). Rethinking assimilation theory for a new era of immigration. *International Migration Review* 31(4): 826–874.

Anderson, B. (1983). *Imagined Communities*. London: Verso.

Antonsich, M. (2009a). National identities in the age of globalization. *National Identities* 11(3): 281–299. doi:10.1080/14608940903081085

Antonsich, M. (2009b). On territory, the nation-state and the crisis of the hyphen. *Progress in Human Geography* 33(6): 789–806. doi:10.1177/0309132508104996

Antonsich, M. (2014). Living together *in* diversity. *Bollettino Società Geografica Italiana* 13(7): 317–337.

Antonsich, M., Fortier, A.-M., Darling, J., Woods, N., and Closs-Stephens, A. (2014). Reading Angharad Closs Stephens's The Persistence of Nationalism. *Political Geography* 40: 56–63. doi:10.1016/j.polgeo.2014.02.005

Appadurai, A. (1996). *Modernity at Large*. Minneapolis, MN: University of Minnesota Press.

Appleton, L. (2002). Distillations of something larger: The local scale and American national identity. *Cultural Geographies* 9(4): 421–447. doi:10.1191/1474474002eu257oa

Bader, V. (2005). Ethnic and religious state neutrality: Utopia or myth. In Sicakkan, H., & Lithman, Y. (eds), *Changing the Basis of Citizenship in the Modern State*, Lewiston, NY: Mellen Press, 161–198.

Baldacchino, J.-P. (2011). The eidetic of belonging: Towards a phenomenological psychology of affect and ethno-national identity. *Ethnicities* 11(1): 80–106. doi:10.1177/1468796810388702

Balibar, E. (1990). The nation form: History and ideology. *Review* XIII(3): 329–361.

Baumann, G. (2004). Introduction. In Schiffauer, W., Baumann, G., Kastoryano, R., & Vertovec, S. (eds), *Civil Enculturation*, Oxford: Berghahn, 1–18.

Bechhofer, F., & McCrone, D. (eds) (2009). *National Identity, Nationalism and Constitutional Change*. London: Palgrave.

Beck, U., & Levy, D. (2013). Cosmopolitanized nations: Re-imagining collectivity in world risk society. *Theory Culture Society* 30(2): 3–31. doi:10.1177/0263276412457223

Bhabha, H. (ed.) (1990). *Nation and Narration*. London: Routledge.

Billig, M. (1995). *Banal Nationalism*. London: Sage.

Biswas, S. (2002). W(h)ither the nation-state? National and state identity in the face of fragmentation and globalisation. *Global Society* 16(2): 175–198. doi:10.1080/09537320220132910

Blad, C., & Couton, P. (2009). The rise of an intercultural nation. *Journal of Ethnic and Migration Studies* 35(4): 645–667. doi:10.1080/13691830902765277

Brass, P. (1991). *Ethnicity and Nationalism*. London: Sage.

Bratsis, P. (2006). *Everyday Life and the State*. Boulder, CO: Paradigm.

Brett, J., & Moran, A. (2011). Cosmopolitan nationalism: Ordinary people making sense of diversity. *Nations and Nationalism* 17(3): 188–206. doi:10.1111/j.1469-8129.2010.00451.x

Brubaker, R. (1994). Rethinking nationhood. *Contention* 4(1): 1–14.

Brubaker, R. (1999). The Manichean myth: Rethinking the distinction between "civic" and "ethnic" nationalism. In Kriesi, H., Armingeon, K., Siegrist, H., & Wimmer, A. (eds), *Nation and National Identity*, Zürich: Rügger, 55–71.

Brubaker, R. (2001). The return of assimilation? Changing perspectives on immigration and its sequels in France, Germany, and the United States. *Ethnic and Racial Studies* 24(4): 531–548. doi:10.1080/0141987012004977 0

Brubaker, R. (2004a). In the name of the nation. *Citizenship Studies* 8(2): 115–127. doi:10.1080/1362102042000214705

Brubaker, R. (2004b). *Ethnicity without Groups*. Cambridge, MA: Harvard University Press.

Brubaker, R. (2006). *Nationalist Politics and Everyday Ethnicity in a Transylvanian Town*. Princeton, NJ: Princeton University Press.

Butler, J. (1993). *Bodies That Matter*. London: Routledge.

Calhoun, C. (1997). *Nationalism*. Buckingham: Open University Press.

Calhoun, C. (2007). *Nations Matter*. London: Routledge.

Calhoun, C. (2008). Cosmopolitanism and nationalism. *Nations and Nationalism* 14(3): 427–448. doi:10.1111/j.1469-8129.2008.00359.x

Caluya, G. (2011). Domestic belongings: Intimate security and the racial politics of scale. *Emotion, Space and Society* 4(4): 203–210. doi:10.1016/j.emospa.2010.11.001

Chatterjee, P. (1993). *The Nation and Its Fragments*. Princeton, NJ: Princeton University Press.

Chernilo, D. (2007). *A Social Theory of the Nation-State*. London: Routledge.

Closs Stephens, A. (2013). *The Persistence of Nationalism*. London: Routledge.

Cohen, A. (1996). Personal nationalism: A Scottish view of some rites, rights, and wrongs. *American Ethnologist* 23(4): 802–815. doi:10.1525/ae.1996.23.4.02a00070

Condor, S. (2000). Pride and prejudice: Identity management in English people's talk about "this country." *Discourse and Society* 11(2): 175–204. doi:10.1177/0957926500011002003

Confino, A., & Skaria, A. (2002). The local life of nationhood. *National Identities* 4(1): 7–24. doi:10.1080/14608940120115657

Connor, W. (1990). When is a nation? *Ethnic and Racial Studies* 13(1): 92–103. doi:10.1080/01419870.1990.9993663

Connor, W. (1994). *Ethnonationalism*. Princeton, NJ: Princeton University Press.

Connor, W. (2004). The timelessness of nations. *Nations and Nationalism* 10(1/2): 35–47. doi:10.1111/j.1354-5078.2004.00153.x

Delanty, G. (2006). Nationalism and cosmopolitanism: The paradox of modernity. In Delanty, G., & Kumar, K. (eds), *The SAGE Handbook of Nations and Nationalism*, Thousand Oaks, CA: Sage, 357–368.

Deutsch, K. (1953). *Nationalism and Social Communication*. Cambridge, MA: MIT Press.

Dowler, L. (2001). Till death do us part: Masculinity, friendship and nationalism in Belfast Northern Ireland. *Environment and Planning D* 20(1): 53–71. doi:10.1068/d40j

Dragojlovic, A. (2008). Reframing the nation. *Asia Pacific Journal of Anthropology* 9(4): 279–284. doi:10.1080/14442210802509747

Edensor, T. (2002). *National Identity, Popular Culture and Everyday Life*. Oxford: Berg.

Edensor, T. (2004). Automobility and national identity. *Theory, Culture and Society* 21(4–5): 101–120. doi:10.1177/0263276404046063

Edensor, T. (2006). Reconsidering national temporalities. *European Journal of Social Theory* 9(4): 525–545. doi:10.1177/1368431006071996

Edwards, D., & Stokoe, E.H. (2004). Discursive psychology, focus group interviews and participants' categories. *British Journal of Developmental Psychology* 22(4): 499–507.

Eriksen, T.H. (2007). Nationalism and the internet. *Nations and Nationalism* 13(1): 1–17. doi:10.1111/j.1469-8129.2007.00273.x

Fox, J., & Jones, D. (2013). Migration, everyday life and the ethnicity bias. *Ethnicities* 13(4): 385–400. doi:10.1177/1468796813483727

Fox, J., & Miller-Idriss, C. (2008). Everyday nationhood. *Ethnicities* 8(4): 536–576. doi:10.1177/1468796808088925

Gellner, E. (1964). *Thought and Change*. London: Weidenfeld & Nicolson.

Gellner, E. (1983). *Nations and Nationalism*. Oxford: Blackwell.

Goswami, M. (2002). Rethinking the modular nation form. *Comparative Studies in Society and History* 44(4): 770–799. doi:10.1017/S001041750200035X

Greenfeld, L. (1992). *Nationalism*. Cambridge, MA: Harvard University Press.

Hage, G. (2003). *Against Paranoid Nationalism*. Sydney: Pluto.

Halikiopoulou, D., & Vasilopoulou, S. (eds) (2011). *Nationalism and Globalisation*. London: Routledge.

Hearn, J. (2007). National identity: Banal, personal and embedded. *Nations and Nationalism* 13(4): 657–674. doi:10.1111/j.1469-8129.2007.00303.x

Hechter, M. (1975). *Internal Colonialism*. London: Routledge.

Herb, G., & Kaplan, D. (eds) (1999). *Nested Identities*. Lanham, MD: Rowman and Littlefield.

Herzfeld, M. (1997). *Cultural Intimacy*. London: Routledge.

Hjerm, M. (1998). National identities, national pride and xenophobia. *Acta Sociologica* 41(4): 335–347. doi:10.1177/000169939804100403

Hobsbawm, E., & Ranger, T. (eds) (1983). *The Invention of Tradition*. Cambridge: Cambridge University Press.

Janmaat, J. (2006). Popular conceptions of nationhood in old and new European member states. *Ethnic and Racial Studies* 29(1): 50–78. doi:10.1080/01419870500352363

Jones, F., & Smith, P. (2001). Diversity and commonality in national identities. *Journal of Sociology* 37(1): 45–63. doi:10.1177/144078301128756193

Jones, R., & Fowler, C. (2007). Placing and scaling the nation. *Environment and Planning D* 25(2): 332–354. doi:10.1068/d68j

Jones, R., & Merriman, P. (2012). Network nation. *Environment and Planning A* 44(4): 937–953. doi:10.1068/a44159

Joppke, C. (2010). *Citizenship and Immigration*. Oxford: Polity Press.

Kaufmann, E. (2000). Ethnic or civic nation? Theorizing the American case. *Canadian Review of Studies in Nationalism* 27(1/2): 133–155.

Kedourie, E. (1960). *Nationalism*. London: Hutchinson.

Kinnvall, C. (2004). Globalization and religious nationalism. *Political Psychology* 25(5): 741–767. doi:10.1111/j.1467-9221.2004.00396.x

Kohn, H. (1958 [1944]). *The Idea of Nationalism*. New York: Macmillan.

Kuzio, T. (2002). The myth of the civic state. *Ethnic and Racial Studies* 25(1): 20–39. doi:10.1080/01419870120112049

Kymlicka, W. (2001). *Politics in the Vernacular*. Oxford: Oxford University Press.

Linde-Laursen, A. (1993). The nationalization of trivialities. *Ethnos* 58(3–4): 275–293.

Löfgren, O. (1989). The nationalization of culture. *Ethnologia Europæa* 19(1): 5–24.

Löfgren, O. (2001). The nation as home or motel? Metaphors of media and belonging. *Sosiologisk Årbok* 14(1): 1–34.

Malešević, S. (2006). *Identity as Ideology*. Basingstoke: Palgrave.

Mann, M. (1993). *The Sources of Social Power*, Vol. II. Cambridge: Cambridge University Press.

Mavroudi, E. (2007). Learning to be Palestinian in Athens: Constructing national identities in diaspora. *Global Networks* 7(4): 392–411. doi:10.1111/j.1471-0374.2007.00176.x

Mavroudi, E. (2010). Nationalism, the nation and migration. *Space and Polity* 14(3): 219–233. doi:10.1080/13562576.2010.532951

McClintock, A. (1995). *Imperial Leather*. London: Routledge.

Meinecke, F. (1919 [1907]). *Weltburgertum und Nationalstaat*. Munich: Oldenbourg.

Miller, D. (1995). *On Nationality*. Oxford: Oxford University Press.

Miller, D. (2000). *Citizenship and National Identity*. Cambridge: Polity Press.

Modood, T. (2011). *Multiculturalism and Integration: Struggling with Confusions*. Fiesole: EUI.

Modood, T., & Meer, N. (2012). Framing contemporary citizenship and diversity in Europe. In Triandafyllidou, A., Modood, T., & Meer, N. (eds), *European Multiculturalisms*, Edinburgh: Edinburgh University Press, 33–60.

Murphy, A. (2013). Territory's continuing allure. *Annals of the Association of American Geographers* 103(5): 1212–1226. doi:10.1080/00045608.2012.696232

Nagel, C. (2009). Rethinking geographies of assimilation. *Professional Geographer* 61(3): 400–407. doi:10.1080/00330120902941753

Nairn, T. (1977). *The Break-up of Britain*. London: NLB.

Nieswand, B. (2012). Banal diasporic nationalism. *Ethnic and Racial Studies* 35(11): 1874–1892. doi:10.1080/01419870.2011.607505

Nussbaum, M. (1994). Patriotism and cosmopolitanism. *Boston Review* 19(5): 3–16.

Özkırımlı, U. (2010). *Theories of Nationalism*. Basingstoke: Palgrave.

Özkırımlı, U., & Uyan-Semerci, P. (2011). Pater familias and Homo nationalis: Understanding nationalism in the case of Turkey. *Ethnicities* 11(1): 59–79. doi:10.1177/1468796810388701

Papademetriou, D. (2012). *Rethinking National Identity in the Age of Migration*. Washington, DC: MPI.

Parekh, B. (2000). *Rethinking Multiculturalism*. London: Macmillan.

Penrose, J. (2002). Nations, states and homelands: Territory and territoriality in nationalist thought. *Nations and Nationalism* 8(3): 277–297. doi:10.1111/1469-8219.00051

Peters, B. (2002). A new look at "national identity"? *Archives Européennes de Sociologie* 43(1): 3–32. doi:10.1017/S0003975602001005

Pickel, A. (2004). *Homo nationis*: The psycho-social infrastructure of the nation-state order. *Global Society* 18(4): 325–346. doi:10.1080/1360082042000272445

Renan, E. (1997 [1882]). *Qu'est-ce qu'une nation?* Paris: Fayard.

Sacchi, S., Castano, E., and Brauer, M. (2009). Perceiving one's nation: Entitativity, agency and security in the international arena. *International Journal of Psychology* 44(5): 321–332. doi:10.1080/0020 7590802236233

Scholte, J.A. (1996). The geography of collective identities in a globalizing world. *Review of International Political Economy* 3(4): 565–607. doi:10.1080/09692299608434374

Seton-Watson, H. (1977). *Nations and States*. London: Methuen.

Sharp, J. (1996). Gendering nationhood: A feminist engagement with national identity. In Duncan, N. (ed.), *BodySpace*, London: Routledge, 97–107.

Shulman, S. (2002). Challenging the civic/ethnic and West/East dichotomies in the study of nationalism. *Comparative Political Studies* 35(5): 554–585. doi:10.1177/0010414002035005003

Skey, M. (2009). The national in everyday life: A critical engagement with Michael Billig's thesis of *Banal Nationalism*. *Sociological Review* 57(2): 331–346. doi:10.1111/j.1467-954X.2009.01832.x

Skey, M. (2011). *National Belonging and Everyday Life*. Basingstoke: Palgrave.

Smith, A. (1971). *Theories of Nationalism*. London: Duckworth.

Smith, A. (1986). *The Ethnic Origins of Nations*. Oxford: Blackwell.

Smith, A. (1991a). The nation: Invented, imagined, reconstructed? *Millennium* 20(3): 353–368. doi:10.1 177/03058298910200031001

Smith, A. (1991b). *National Identity*. Harmondsworth: Penguin.

Smith, A. (1998). *Nationalism and Modernism*. London: Routledge.

Smith, A. (2005). The genealogy of nations. In Ichijo, A., & Uzelac, G. (eds), *When is the Nation?* London: Routledge, 94–112.

Smith, A. (2008a). *The Cultural Foundations of Nations*. Oxford: Blackwell.

Smith, A. (2008b). The limits of everyday nationhood. *Ethnicities* 8(4): 563–573. doi:10.1177/14687968 080080040102

Thompson, A. (2001). Nations, national identities and human agency: Putting people back into nations. *Sociological Review* 49(1): 18–32. doi:10.1111/1467-954X.00242

Thrift, N. (2000). It's the little things. In Atkinson, D., & Dodds, K. (eds), *Geopolitical Traditions*, London: Routledge, 380–387.

Tilly, C. (1990). *Coercion, Capital and European States*. Oxford: Blackwell.

Triandafyllidou, A. (2001). *Immigrants and National Identity in Europe*. London: Routledge.

Uberoi, V. (2008). Do policies of multiculturalism change national identities? *Political Quarterly* 79(3): 404–417. doi:10.1111/j.1467-923X.2008.00942.x

Urry, J. (2007). *Mobilities*. Cambridge: Polity Press.

White, G. (2000). *Nationalism and Territory*. Lanham, MD: Rowman and Littlefield.

Williams, C., & Smith, A.D. (1983). The national construction of social space. *Progress in Human Geography* 7(4): 502–518. doi:10.1177/030913258300700402

Wimmer, A., & Glick Schiller, N. (2002). Methodological nationalism and beyond. *Global Networks* 2(4): 301–334. doi:10.1111/1471-0374.00043

Wood, N. (2012). Playing with "Scottishness": Musical performance, non-representational thinking and the "doings" of national identity. *Cultural Geographies* 19: 195–215. doi:10.1177/1474474011420543

Xenos, N. (1996). Civic nationalism: Oxymoron? *Critical Review* 10(2): 213–231. doi:10.1080/08913 819608443418

Yack, B. (1996). The myth of the civic nation. *Critical Review* 10(2): 193–211. doi:10.1080/0891 3819608443417

Yeĝenoĝlu, M. (2005). Cosmopolitanism and nationalism in a globalized world. *Ethnic and Racial Studies* 28(1): 103–131. doi:10.1080/0141987042000280030

Yuval-Davis, N. (1997). *Gender and Nation*. London: Sage.

Yuval-Davis, N., & Anthias, F. (eds) (1989). *Woman-Nation-State*. London: Macmillan.

Chapter 23

Regional Institutions

Merje Kuus

University of British Columbia, Vancouver, British Columbia, Canada

The largest economy in the world is not a country but a regional institution – the European Union. This surprises many people, because the clout of the union is rarely mentioned as something over and above the power of its member states. To this day, pundits like to quote Henry Kissinger's quip: "You say Europe but can you tell me which number I should call?"[1] This incongruous framing of the European Union illustrates a bigger puzzle about the role of regional institutions in world affairs today. One the one hand, political bargaining on the international arena increasingly utilizes regional connections; consider the growth of trade blocs in both number and influence in recent decades. In the organizations that are global in reach, such as the United Nations or the World Trade Organization, regional interests are often at play. Beyond formal institutional arrangements, informal assumptions about regional affinities permeate political analysis. Arguments about Europe, East Asia, the Pacific Rim, the Arab world, and so on are omnipresent in media and scholarly analyses alike. On the other hand, regional organizations are seen mostly as alliances of states. In the common-sense view, understanding such organizations requires that we start and end with the interests of states. Regional institutions are then a subcategory of international organizations and regional power relations a footnote to interstate relations. Maps of regional organizations prominently display the borders of states and news reports deal almost exclusively in national-level data. Regional relations and interests as distinct from national ones seem to have limited explanatory power. If Europe as the most powerful regional institution of our time garners dismissive remarks about its lack of a phone number, what chance do the others have?

For political geographers, regional institutions are interesting in part because they cultivate, at least potentially, spatial relations beyond the familiar framework of nation-states. Political maps of regionalism are not the same as the maps of states – although it is by no means clear what the differences are and how they matter. Regional practices and their institutional frames are diverse and ambiguous. The diversity is perhaps obvious. Some ideas and practices of regions are old: The notions of continents and cultural regions, such as the Arab world or

The Wiley Blackwell Companion to Political Geography, First Edition.
Edited by John Agnew, Virginie Mamadouh, Anna J. Secor, and Joanne Sharp.
© 2015 John Wiley & Sons Ltd. Published 2017 by John Wiley & Sons Ltd.

Eastern Europe, date back decades and even centuries. Other concepts and institutional arrangements, such as the Asia-Pacific or Europe's neighborhood, are recent. Some regions, including Europe, the North Atlantic, or the Asia-Pacific, have clear institutional frames: the European Union (EU), the North Atlantic Treaty Organization (NATO), and Asia-Pacific Economic Cooperation (APEC), respectively. Others are highly ambiguous: The Arctic region involves only small parts of Arctic states and Eurasia generates vague incomprehension in most people. There is no one conceptual lens, whether economic interest, cultural identity, historical legacies, or anything else, that can adequately explain all of these arrangements.

The actual or potential bases of regional affinities and institutions are equally unclear. For starters, there is the question of whether regions are the roots or effects of regional institutions; that is, whether regions exist prior to the institutions that operate in their name. Regional claims tend to assume that distinct regional dynamics undergird international politics. To speak of geopolitical regions, such as Europe or the Arab world, is to assert that a regional affinity exists prior to being labeled as such. Yet, even a cursory look at world affairs writing reveals that regions are delineated very differently depending on the writer's own position and interests, and this suggests that the content of regions is far from self-evident. It is likewise not clear whether a regional affinity is necessarily after or below the national kind or how their relationship does and can change. The diversity and ambiguity of regional dynamics therefore highlight the vital importance of our analytical frameworks: What we see cannot be divorced from our conceptual lenses.

This chapter takes stock of the political geographical work on regional institutions so as to illuminate the variety and import of such institutions and geographical analyses of them. My goal is not to offer a comprehensive review of what we know and do not know about regional institutions. Neither do I try to pinpoint the cutting edge of geographical work on the theme. The effort is, rather, to cultivate a better understanding of why regionalism is a difficult and fascinating subject and what political geography has added and can add to our understanding of it. I will argue that regional interests are not the self-evident bases of regional institutions. Rather, these very interests are made in part through institutional practices. There are no obvious regional dynamics separate from the institutions that codify these processes and thereby bring them into being as objects of practice. This makes institutions central to our efforts to analyze the spatiality of power.

The rest of the argument proceeds in three steps. The following section situates political geographical research in its broader scholarly context in human geography and the social sciences. It notes that most geographical writing about regional institutions is carried out by economic rather than political geographers. Much of the work examines regional institutions at subnational or cross-border levels. Put simply, regions are taken to be smaller than states. The realm of the international or regions larger than states has remained the turf of another academic discipline: international relations or IR. We need more work on international and supranational regionalism in geography. In particular, there is ample room for research that can bring the sophisticated geographical thinking on spatiality into the analysis of international affairs. The subsequent section charts some current trends in the study of regional institutions on the international scene. Using European, Arctic, and North Atlantic institutions as examples, I highlight the amorphous nature of regions and the institutions that work in their name. This does not mean that regions can be undone at will, however. Once created, regional institutions can powerfully shape national and local interests in the region. A number of my examples come from the European Union, first because Europe is the most thickly institutionalized region in the world today; and secondly because there is substantial political geographical scholarship on it. My argument is not about Europe as such, though; it is about the making and remaking of regions in international politics. The concluding section summarizes how political geography can help us understand regional institutions and power relations.

Regions and institutions beyond the state

Messy regions

Ever since the emergence of the idea of the world as one coherent unit in the seventeenth and eighteenth centuries, regional representations have been used to make sense of it. The concepts of East and West or Orient and Occident are some of the most enduring tools of geopolitical writing. During the Cold War, global affairs were customarily analyzed in terms of three blocs: the First, Second, and Third Worlds (Agnew 2003b). These categories were delineated ostensibly on ideological grounds – the First World standing for the free West, the Second World for the communist East, and the Third World being cast as the arena of competition between the first two – but each "world" also contained a regional component. The institutions created by the First World were designed to serve the West as a region. Today, claims about America's pivot to Asia, the formation of Europe's neighborhood, or Eurasia as a geopolitical formation are familiar fixtures in current affairs writing (Bialasiewicz et al. 2009; Agnew 2012b; Megoran & Sharapova 2013). After the terrorist attacks of September 11, 2001, the idea of the integrated core and the non-integrated gap served to legitimize American foreign policy around the world (Roberts, Secor, & Sparke 2003; Dalby 2007). In the European Union, regional references to the north and south of Europe have surfaced powerfully in the efforts to explain the sovereign debt crisis (Engelen et al. 2011).

These examples tell us little about how we should define regions, but they tell us that regions can have many meanings. A region can fit into a contemporary state or transcend international borders; it can be small or large; it can be institutionalized thickly or hardly at all. Many regions have historical roots and connotations: Consider the untold numbers of regions that used to be major seats of power but survive mostly as administrative regions of states or as labels on consumer products. Saxony was a power center until well into the modern period, but how many people can locate it on Europe's political map today? Burgundy is now known for its wine, not its geopolitical power. Some claims about regions, such as the rumblings about a north–south divide in Europe, rely on vague historical and cultural imageries. Others draw on and solidify specific institutional arrangements, either visible structures like regional organizations or the largely invisible frames like the geographical desks of foreign ministries. Europe and North America are regions not only in geography books but also in the organizational charts of policy-making institutions; there are Europe and North America departments in foreign ministries around the world.

A region is often a creation of contemporary institutional interests; administrative regions of states are prime examples of this. Regional identities can breed much opposition to state power, but this does not make them any more natural or consequential than national affinities. Few people think of Catalonia, Scotland, or Quebec as international players because, although these places have strong regional identities, they fit into the international system of states. Wales, Galicia, Eastern Europe, the Arctic, and Central Asia are all regions in the common sense, but their institutional frames are very different.

Regions can come into existence and recede again from practical significance. Central Europe was unimportant to most people, other than the region's intellectuals, until the 1990s, but it then became a powerful source of identity in several states' drive for EU and NATO membership. In 1991, Central Europe was institutionalized in terms of the Visegrád Group of Poland, Hungary, and Czechoslovakia (later the Czech Republic and Slovakia; Moisio 2002; Kuus 2007). When the Visegrád states all acceded to the EU, they came to be called simply European and Central Europe as a geopolitical designation lost import again. Eurasia as a regional specification separate from the continent likewise emerged only after the Cold War. It is used for varied political programs, from Russian nationalism to academic criticism of that ideology (Diener 2014). In traditional geopolitical analysis, Eurasia is a code word for a Russian sphere of influence – think of the proposals for, and the critique of, the Eurasian Union. In the more imaginative writings on power and identity, Eurasia signifies something quite different: the areas and ways of demarcating space that do not map conveniently onto our preconceived spatial imageries like Europe, Asia, Central Asia, and so on (Grant 2012). Tell me your Eurasia and I tell you who you are.[2] The last example shows that a regional reference can constrain and stifle debate, but it can also open up discussions of power and spatiality in directions unimaginable from other spatial lenses. Not all of these regions are widely known among the general public, and this is significant: Regionalism is a messy business compared to the seeming clarity and stability of the state-based political map.

Elephants in the room

Regionalism is a key theme in geographical scholarship and one would therefore expect a great deal of writing on regional institutions in political geography. However, most of the existing work focuses on specific cases of subnational and cross-border regionalism and says little about institutions beyond these cases. Geographers study the ideas and practices of regionalism – issues like regional identities or cross-border flows of know-how – rather than the nitty-gritty of institutional structures. A substantial part of the geographical work on regional institutions is penned by economic geographers. It focuses on economic processes on subnational levels: agglomerations, regional development efforts, interregional competition, and so on. Thus, a search for the term "regional institutions" in *Progress in Human Geography*, the principal journal to review contemporary scholarship in human geography, brings up economic geographical analyses of substate regionalism. A search for "international organizations" (with an *s* or a *z*) in that same journal yields no items – *nothing*. There was no chapter on regional institutions in the first edition of this *Companion*. The recent *Companion to Critical Geopolitics* does include a chapter on international organizations (Bachmann 2013), but that essay relies substantially on IR sources and says little about regional dynamics.

True, political geographers have written extensively on regional concepts, such as Europe, the West, the Middle East, and so on (e.g., Ó Tuathail 1996; Heffernan 1998; Agnew 2003b; Gregory 2004; Kuus 2007). We know a great deal about elite and popular views on these concepts. There is far less on the institutions that codify these ideas. Europe is in many ways an exception, partly because the EU has over 26,000 civil servants on its payroll (i.e., not diplomats seconded to it by nation-states, as is typically the case with international organizations) and partly because many geographers study it (data from Kuus 2014: 82). We know little about regional institutions beyond the EU (but see Sidaway 2002). One might say that institutions are a kind of elephant in the room: acknowledged in theory but not analyzed in practice.

The relative paucity of political geographical work on institutions is understandable in some ways. Geographers are interested in social realities rather than their administrative frames (Albert & Reuber 2007). They study economic, political, and cultural patterns as the bases for regional affinity and interest. They examine everyday practices rather than bureaucracies, so as to avoid naively privileging formal organizational structures over informal social ones. The unintended consequence of this set of priorities is that geographical work is oblique about the role of institutions in producing regional dynamics. Institutions are not a layer on top of existing social relations; institutions make these relations. They both reflect existing patterns of communication and shape these patterns. Speaking of the prospects of regional integration in a 1950s Europe traumatized by war and prejudice, Jean Monnet (a founding visionary figure in the integration project) noted: "Nothing is possible without men: nothing is lasting without institutions." If the European Union has been successful in creating an influential regional institution – and in the global scale it clearly has – this is not because of some primordial European identity, but because of the gradual institutionalization of regional cooperation. There is nothing preordained about that process. It was only in 2009, more than half a century after the start of the integration process, that the Treaty of Lisbon consolidated the European Union's legal personality and enabled the union to sign international treaties.[3]

The somewhat casual treatment of institutions in political geographical work is one problem. The other is that most of the existing work investigates substate rather than supranational regions. There are good reasons for this focus on local and substate scales. One of the weak spots in contemporary social science is its methodological nationalism: a tendency to treat the state as a container of society and the basic unit of scholarly analysis. This happens in part because states are treated as equivalent to persons (Agnew 2003a). Just as the individual is viewed as the self-evident unit of scientific analysis, international politics is implicitly framed as an arena of interstate action. That lens produces studies of American, Indian, or Italian societies as more or less coherent entities and downplays the processes and patterns that cross national borders. Political geography counters this by showing that the state is not nearly as coherent as nationalist politicians or central government documents may lead one to believe. To take Italy as an example, any close investigation of that country immediately reveals regional traditions and discourses of national *dis*unity. Even in the 1950s, fewer than one fifth of Italian adults communicated exclusively in the Italian language; the remaining four-fifths may have used Italian on some occasions, but relied on regional dialects for most of their daily lives (Judt 2006: 256; see also Agnew 2003a). The broader point is that we cannot depend on the concept of the nation-state to understand social and political life, even in the continent that invented that concept. And yet, even though social scientists readily assume that the international sphere is one of competition and disunity, they still treat the inside of the state as a coherent national space.

Against this background, political geographers seek to challenge the state-centered assumptions of scholarly analysis (Kuus & Agnew 2008). There are many sophisticated accounts of the intertwining of nationalist and regionalist politics. There is also vibrant work on cross-border regionalism. That research investigates the nexus between region and border: regionalism across borders, borders amid regions, and the mutual constitution of these concepts and practices (Paasi 2003; Sparke et al. 2004; Gilbert 2007). It shows the power of the state in shaping regional institutions at any scale and the inadequacy of analyzing regionalization principally through the lens of the state. As a result of these efforts, we know a great deal about how the state interacts with local and regional entities below it, especially in the sphere of local economic development. By the same token, we know little about international organizations. There is a kind of glass ceiling in geographical scholarship: A great deal of thought goes into substate and cross-border regions, but politics "above" the state – the realm of supranational and interstate bargaining – remains an afterthought. The specificity of diplomatic institutions is a part of the reason – international organizations are more difficult to access and study than the local governments of border regions – but not the whole reason. One might say that the international arena is also an elephant in the room: Its importance is not denied but it is not explicitly analyzed either.

Supranational regional institutions, such as trade blocs, are thus left to scholars in international relations. These scholars have indeed written a great deal on regionalism. There are well-researched and influential accounts of the histories and metageographies of regionalism: concepts like Europe, Asia, and so on (Lewis & Wigen 1997). There is likewise a large body of work on regional institutions as variants of international organizations. Some analysts even claim that our world is best understood in terms of large regional blocs, which are economic and political as well as cultural in their roots and manifestations (e.g., Huntington 1996; Katzenstein 2005). One of the most discussed geopolitical treatises of the post–Cold War era – political scientist Samuel Huntington's *The Clash of Civilizations and the Remaking of World Order* – is a book about supranational regions. For Huntington, cultural regions or what he calls civilizations are the most basic unit of political struggle in the twenty-first century. Peter Katzenstein (2005), another influential political scientist, likewise underscores the need to consider regions alongside states as geopolitical actors in their own right.

Mainstream IR work correctly identifies the import of regional dynamics, but it oversimplifies these processes. Claims about Europe, Asia, East Asia, the North Atlantic, and so on too often create the false impression of stable and essential interests and identities. Underneath the seemingly descriptive language, as in Huntington's narrative about civilizations, are unexamined assumptions about primordial identities and strategic interests. This happens in part because regions are treated as given: not as complicated spatial processes, but as pre-existing territorial blocs that the analyst needs to discover and demarcate. That reified view of regions draws on a simplified view of power as a zero-sum game among territorial blocs. A region is too often treated as a scaled-up state: A strong regional institution is state-like, whereas a weak one is dissimilar from states. This is why the European Union fails to measure up in such accounts: Although not lacking in influence, its power is not fully traceable to the sum of the member states (Bialasiewicz 2008; see also Bachmann & Sidaway 2009).

States are certainly powerful and must be center stage in our efforts to understand regional institutions. Many regional dynamics are created by and for states. The administrative regions of states may have historical roots, but they are also the products of state power: They emerge from the efforts of the state to codify its authority over the national space (Häkli 1998). Regional projects that do not garner the support of states tend to flounder. The pan-Europeanist movement of the interwar years or the pan-Africanism of the 1960s can serve as examples (Heffernan 1998; Sharp 2013). Both cases indicate that visions of cooperation and integration require sustained institutional support to thrive. If a regional institution, such as the Association of Southeast Asian Nations or the African Union, accomplishes something, then this is not because it prevails over states, but because it works with and through the interests of its member states.

The European Union is once again a case in point. The union's institutions are set up explicitly to guard and promote the interests of its member states. Contrary to stereotypes, the weightiest decisions in the EU are not taken by an unelected Brussels bureaucracy. Such decisions are, rather, adopted by an intergovernmental body – the European Council – that consist of the heads of state and government of the member states. The Brussels-based civil service is itself a product of continuous interstate negotiation, down to informal national quotas for personnel. The nationalities of high-ranking officials are informally agreed before individuals are even considered for these posts. This means that state interests are built into the union's ostensibly independent civil service. When looking at the organizational charts of EU institutions, a casual observer sees the administrative units of a bureaucracy. An insider also sees national flags and, more broadly, the European system of states (Kuus 2014: 95). And yet, it would be a grave mistake to stop the analysis at that point. The civil service created by states and tasked with the supranational regulation of space changes the playing field for everyone, including the member states. The meaning and significance of the national and the regional are continuously transformed and national institutions are not the only players in town. In high-level EU appointments, states cannot put too much pressure on EU institutions: Even big states can only prevent unfavorable appointments, they usually cannot impose favorable ones. The creation of supranational European institutions has fueled political dynamics beyond interstate bargaining and thereby enhanced the power of subnational and cross-border regions, such as Scotland or Oresund.[4] The nation-state is transformed – not unmade but remade – through regional pressures from above (the EU) as well as below (substate regions; Jensen & Richardson 2004; Mamadouh & Van der Wusten 2008).

Empirical examples of such processes can become highly context specific quickly, and this is significant. Regional institutions and affinities blend with national, supranational, and local ones in ways that cannot be neatly disentangled. When trying to grasp regional institutions, we must consider the power of states, but not confine our analysis to states. The question of what a region is cannot be closed; it must remain center stage in geographical and political inquiry. The task is to sort out in empirical terms the negotiation and mutation of multiple interests and identities in transnational settings. This necessitates keen attention to the scales and practices of everyday life beyond formal institutions: the seemingly mundane and unremarkable ways in which regions and institutions are enacted on a daily basis.

Performing regions

This discussion underscores that political geographers are interested in the making of regions: the historical and contemporary processes that create regional patterns of political practice. These practices, in turn, shape the conceptual toolbox of analysis. "Performing" here refers to the ways in which daily social practices create our categories of analysis. To say that regional institutions perform regional interests therefore does not mean that their practices are false – theatrical tricks of some sort. Rather, it suggests that these practices do not simply represent but, more precisely, produce regional interests. To illustrate this point and to showcase political geographical thinking on these processes, I give three examples of the ambiguity of regional affinity and the crucial role of institutions in its production.

The first example concerns European identity. The common-sense view is that Europe has been more successful than other regions in advancing distinctly regional interests, because EU institutions can build on a sense of European identity among European citizens. Europe as a political and cultural entity is taken to exist prior to EU institutions: The institutions mostly represent and cultivate this underlying feeling of Europeanness among the population. Yet, even in this thickly institutionalized region, the very meaning of what is represented – Europe – remains profoundly unclear. Some observers argue that any sense of a European regional identity in Europe is (still) too weak to be the basis for institution-building. Others note that Europe cannot be neatly defined in any event. Historian Timothy Garton Ash (2001) pinpointed this presciently some time ago:

> "Will Europe never be Europe because it is becoming Europe?" To most speakers of the English language, the sentence must look like nonsense. But in Brussels, the capital of Europe and the inner temple of the European debate, it is perfectly comprehensible and indeed vitally important. One just needs to insert the different meanings of the word "Europe". The sentence then reads: "Will the current European Union of fifteen states – that is, Europe in sense 1 – never attain the long dreamed-of political unity – that is Europe in sense 2 – because it is now committed to including most other states on the geographical continent of Europe – that is, Europe in sense 3?"

An effort to pin down the content of European identity or to assess whether EU institutions represent it properly is doomed to failure. The question is not what Europe really is, but how the concept functions in political struggles. Once we pose the question in such processual terms, we see that Europe is produced in part by and for the institutions that exist in its name (Jones 2006; *Geopolitics* 2005; Clark & Jones 2008, 2011; Bialasiewicz 2011). There is certainly a discourse of Europeanness outside EU institutions, but it is codified into practical policy measures primarily by the union's institutions. It is no accident that the definition and demarcation of Europe have been closely tied to the European Union throughout the post–Cold War era. Through its institutions, the EU has been able to claim Europe as "its" region. The idea that Europe stops at the eastern border of the European Union may seem logical now, but the Cold War era bounding of the region seemed equally natural to many people a generation ago. The lesson beyond Europe is that, to put it bluntly, regions do not make institutions but the other way around. It is not helpful to ask whether a region is real or whether any institution codifies its meanings well. A more fruitful inquiry concerns the work done in the name of that region, and this includes work by regional institutions (Moisio et al. 2013; Kuus 2014).

The second example relates to the Arctic as a seemingly new region. The Arctic as a social space is not a new phenomenon, of course, but its inscription as a particular kind of region – as a stage of interstate competition for resources and posturing – is relatively recent. During the Cold War, the Arctic was viewed in terms of superpower competition. For practical purposes, there was no Arctic region; there was an Arctic theater of the Cold War. It was only after the dissolution of the Soviet Union that the Arctic became available for interstate competition and meaning-making more broadly (Steinberg 2001; Dodds 2010; Dittmer et al. 2011). The Arctic Council was founded in 1996, when the Arctic came into being as a supranational region.[5] Casting the Arctic in such fluid terms may seem odd. A concept like Europe, one might say, is vague because its eastern border is not fixed. The Arctic, in contrast, is clearly defined in natural terms as the area above the Arctic Circle. A tropical state, such as Singapore, is obviously not in the Arctic. And yet, being an Arctic player is not as self-evident as it may appear at first. The Arctic region is indeed defined in terms of latitude, but it is governed from national capitals much farther south. The southern border of the Arctic is at 66.33 degrees northern latitude, but the borders of the Arctic Council as a regional organization reach far beyond that area. The Arctic Council can, furthermore, grant observer status to non-Arctic states. In late 2013, Singapore had that status, along with India, China, Japan, and eight other non-Arctic states.[6] Beyond the Arctic, this example reminds us that geopolitical claims, including regional references, can be produced much beyond the region in question. To understand such references, we need to consider the geographies of geopolitical knowledge.

The meaning of the Arctic as a supranational region is produced largely outside that region, not always to the benefit of Arctic populations.[7]

The third example focuses on the North Atlantic as a regional marker. That category is a product of the Cold War. In everyday parlance, it means two things: first, relations between the United States and Europe (whatever the latter might mean); and second, the relations and institutions created around NATO. The organization was created in 1949 as the military alliance of Western Europe and the United States against the Soviet Union. At the end of the Cold War, as NATO lost its enemy, the organization became a centerpiece of the US-led Western bloc of influence. As more states have become involved in institutional cooperation with NATO, the North Atlantic or Euro-Atlantic space has become marketed in terms of an expansive region of stability and Western influence. In NATO's own rhetoric, the North Atlantic space reaches "from Vancouver to Vladivostok" (Kuus 2009). Some even talk about a "global NATO," a networked military alliance capable of projecting its power to all corners of the globe (ibid.). The North Atlantic space includes Turkey, Finland, Russia, and Afghanistan, since all are integrated into NATO-based institutional structures, albeit in different ways.[8]

Whether we are dealing with a seemingly established cultural region like Europe, a natural one like the Arctic, or a strategic formation like the North Atlantic space, regional links are cultivated and sometimes invented by institutions in specific historical circumstances. All three examples illustrate that the meanings and borders of regions change continuously and a sustained attention to institutional structures goes a long way to explain these processes. The argument does not deny the existence of regional identities; it simply highlights the role of institutions in their production. True, an institution cannot invent a region and safely expect social realities to follow its design. The European Union has possessed a highly qualified civil service for decades and spent fortunes on promoting European identity, but has little to show for it: Opinion polls consistently indicate that Europeans identify with the nation much more than Europe (Moisio et al. 2013). It does suggest, however, that to understand regions we must investigate processes of institution-building.

Conclusion: Regions as questions

It is perhaps clear by now that this chapter does not tell us how to bound and define regions and what exactly is the role of regional institutions in any such exercise. I hope to have shown that these questions do not have clear answers. Regions and their institutional frames are unstable spatial and political formations. Ironically, the claims about regional affinity and animosity are used so much because they seem to stabilize the world intellectually and politically (Agnew 2012a). In ambiguous and controversial questions, such as the shifting alliances of states or the rise and decline of regions (as in the rise of the Asia-Pacific or the decline of Europe), regional language is convenient. Claims about continents or geographical blocs rather than countries seem to carry more weight analytically and politically. At a time when the authority of the nation-state is being challenged from many directions, a map of larger regions seems to offer something new. When a Scandinavian politician mentions northern and southern Europe in her criticism of Italy or Greece (or the other way around), she frames complex historical and political matters in terms of seemingly clear-cut geographical categories. When a Spanish civil servant speaks of Catalonia as a region rather than a nation, he implicitly molds political struggles into the framework of state sovereignty. In that framework, the ultimate authority in Spain rests with the national rather than the regional government. The reverse is true as well: When Catalan nationalists shun regional language and evoke the Catalan nation instead, they reach for the national as a scale above the regional. Regional claims serve many purposes for many institutions.

Analyses of regional patterns and institutions are in constant danger of making the same mistake that nationalist claims do: of assuming internal homogeneity within a region and external differentiation among regions in situations where neither exists. A region like Southeast Asia or Europe is neither homogenous internally nor clearly distinct from the outside world. Both internal unity and external differentiation must be continually maintained lest cracks appear in them. Regional institutions are not necessarily stronger or more natural than any other institutions. We are not transitioning to the world of regional blocs, but neither are we staying in some "basic" world of states. Regional patterns and institutions are an integral part of the transnationalization of international politics. Attention to these phenomena helps us see the forms of solidarity and competition that are not neatly mappable as German or European, Chinese or Asian, American or North American, and so on. They raise difficult questions about formal and informal politics, the spatiality of power, and the changing role of the state in our world, but these questions vex scholars much beyond geography. Political geography has much to add to our thinking on these issues.

A regional approach is not a magical solution to the difficulties of making sense of global affairs. It can produce impoverished simplifications just as other geopolitical models can. Political geographical work on spatial relations (as distinct from territorial dynamics) in everyday spaces indicates that our best bet is to keep the geographical definition of politics a question and not an answer. The task is to offer the kind of nuanced and thick geographical knowledge that undercuts the thin universalist theorizing of mainstream geopolitical analysis (Toal 2003: 655). At the same time, geographers' scant attention to formal institutions can hamper our understanding of political space. Close examination of institutions can certainly produce narrow and dry accounts of organizational rhetoric, but it can also destabilize trite spatial lenses. European institutions may seem to promote European interest when viewed from afar, but closer up a myriad of other agendas and relations come into view. Sustained study of regional institutions can yield analyses that are richer in detail, more open in their spatial assumptions, and more imaginative about the inherently transnational operation of power today.

Notes

1   Henry Kissinger was a high-ranking and influential American diplomat in 1969–77 and a prominent commentator on international affairs for decades afterward. The quip is quoted in Garton Ash 2001: 61.
2   This paraphrases historian Timothy Garton Ash's (1999: 350) point about Central Europe: "Tell me your Central Europe and I tell you who you are."
3   Prior to the Lisbon Treaty, which was signed in 2007 and came into effect in 2009, the European Union functioned through a complicated system of three legal pillars. It could sign trade deals as a bloc, but did not have a consolidated legal personality in many other spheres of international law.
4   The Oresund is a cross-border area that centers on the metropolitan regions of Copenhagen (Denmark) and Malmö (Sweden).
5   There are many more such examples: The Baltic Sea Region and the Black Sea Region illustrate similar processes of reinvention and re-demarcation. The Baltic and the Black Sea certainly existed before the end of the Cold War and certain regional connections were in place, but the areas surrounding these seas were not viewed as coherent regions until the end of the bipolar world order.

6   These states, and some others, are interested in observer status because of their economic interests (e.g., shipping) in the Arctic and beyond.
7   The Mediterranean illustrates similar dynamics. It perhaps conjures up images of a blue sea and its coasts. Institutionally, the Union for the Mediterranean as a regional organization consists of 43 states, plus EU institutions and the Arab League. See Pickles 2005; Jones 2006; Smith 2013 for geographical analyses of Mediterranean spaces.
8   The North Atlantic is not simply a space of cooperation between the United States and Canada on the one hand and European states on the other. The cooperation in question is usually a specifically NATO-based project of military collaboration. When international affairs professionals speak of the North Atlantic or Euro-Atlantic space, they do not mean cultural exchange; they mean military matters.

References

Agnew, J.A. (2003a). Territoriality and political identity in Europe. In Berezin, M., & Schain, M.A. (eds), *Europe without Borders: Remapping Territory, Citizenship, and Identity in a Transnational Age*, New York: Columbia University Press, 319–342.

Agnew, J.A. (2003b). *Geopolitics: Re-visioning World Politics*, 2nd edn. London: Routledge.

Agnew, J.A. (2012a). Arguing with regions. *Regional Studies* 47: 6–17. doi:10.1080/00343404.2012.676738

Agnew, J.A. (2012b). Is US security policy "pivoting" from the Atlantic to Asia-Pacific? *Dialogue on Globalization*. Berlin: Friedrich-Evert-Stiftung. http://library.fes.de/pdf-files/iez/global/09318.pdf, accessed March 26, 2015.

Albert, M., & Reuber, P. (2007). Introduction: The production of regions in the emerging global order – Perspectives on "strategic regionalization." *Geopolitics* 12: 549–554. doi:10.1080/14650040701546038

Bachmann, V. (2013). International organizations. In Dodds, K., Kuus, M., & Sharp, J.P. (eds), *The Ashgate Research Companion to Critical Geopolitics*, Burlington, VT: Ashgate, 405–420.

Bachmann, V., & Sidaway, J.D. (2009). Zivilmacht Europa: A critical geopolitics of the European Union as a global power. *Transactions of the Institute of British Geographers* 34: 94–109. doi:10.1111/j.1475-5661.2008.00325.x

Bialasiewicz, L. (2008). The uncertain state(s) of Europe? *European Urban and Regional Studies* 15(1): 71–82. doi:10.1177/0969776407081279

Bialasiewicz, L. (ed.) (2011). *Europe in the World: EU Geopolitics and the Making of European Space*. Farnham: Ashgate.

Bialasiewicz, L., Dahlman, C., Apuzzo, G.M., Ciută, F., Jones, A., Rumford, C., Wodak, R., Anderson, J., & Ingram, A. (2009). Interventions in the new political geographies of the European "Neighbourhood." *Political Geography* 28(2): 79–89. doi:10.1016/j.polgeo.2008.12.002

Clark, J., & Jones, A. (2008). The spatialities of Europeanisation: Territory, government and power in "Europe." *Transactions of the Institute of British Geographers* 33(3): 300–318. doi:10.1111/j.1475-5661.2008.00309.x

Clark, J., & Jones, A. (2011). The spatialising politics of European political practice: Transacting "Eastness" in the European Union. *Environment and Planning D: Society and Space* 29(2): 291–308. doi:10.1068/d4609

Dalby, S. (2007). Regions, strategies and empire in the global War on Terror. *Geopolitics* 12(4): 586–606. doi:10.1080/14650040701546079

Diener, A.C. (2014). Russian repositioning: Mobilities and the Eurasian regional concept. In Walcott, S.M., & Johnson, C. (eds), *Eurasian Corridors of Interconnections: From the South China to the Caspian Sea*, London: Routledge, 72–109.

Dittmer, J., Moisio, S., Ingram, A., & Dodds, K. (2011). Have you heard the one about the disappearing ice? Recasting Arctic geopolitics. *Political Geography* 30(4): 202–214. doi:10.1016/j.polgeo.2011.04.002

Dodds, K. (2010). Flag planting and finger pointing: The law of the sea, the Arctic, and the political geographies of the outer continental shelf. *Political Geography* 29(2): 63–73.

Engelen, E., Hendrikse, R., Mamadouh, V., & Sidaway, J. (2011). Commentary: Turmoil in Euroland: The geopolitics of a suboptimal currency area? *Environment and Planning D: Society and Space* 29(3): 571–583. doi:10.1068/d2904com

Garton Ash, T. (1999). Where is Central Europe now? In *History of the Present: Essays, Sketches and Dispatches from Europe in the 1990s*, London: Penguin, 349–361.

Garton Ash, T. (2001). The European orchestra. *New York Review of Books* May 17: 60–67.

Geopolitics (2005). Forum: Multiple Europes: Boundaries and margins in European Union enlargement. 10(3): 567–558.

Gilbert, E. (2007). Leaky borders and solid citizens: Governing security, prosperity and quality of life in a North American partnership. *Antipode* 39(1): 77–98. doi:10.1111/j.1467-8330.2007.00507.x

Grant, B. (2012). We are all Eurasian. *NewsNet: News of the Association of Slavic, East European, and Eurasian Studies* 52(1): 1–6.

Gregory, D. (2004). *The Colonial Present*. Oxford: Blackwell.

Häkli, J. (1998). Manufacturing provinces: Theorizing the encounters between governmental and popular "geographs" in Finland. In Ó Tuathail, G., & Dalby, S. (eds), *Rethinking Geopolitics*, New York: Routledge, 131–151.

Heffernan, M. (1998). *The Meaning of Europe: Geography and Geopolitics*. London: Arnold.

Huntington, S.P. (1996). *The Clash of Civilizations and the Remaking of the World Order*. New York: Simon and Schuster.

Jensen, O., & Richardson, T. (2004). *Making European Space: Mobility, Power and Territorial Identity*. London: Routledge.

Jones, A. (2006). Narrative-based production of state spaces for international region building: Europeanization and the Mediterranean. *Annals of the Association of American Geographers* 96: 415–431. doi:10.1111/j.1467-8306.2006.00484.x

Judt, T. (2006). *Postwar: A History of Europe since 1945*. Harmondsworth: Penguin.

Katzenstein, P. (2005). *A World of Regions: Asia and Europe in the American Imperium*. Ithaca, NY: Cornell University Press.

Kuus, M. (2007). *Geopolitics Reframed: Security and Identity in Europe's Eastern Enlargement*. New York: Palgrave Macmillan.

Kuus, M. (2009). Cosmopolitan militarism? Spaces of NATO expansion. *Environment and Planning A* 41(3): 545–562. doi:10.1068/a40263

Kuus, M. (2014). *Geopolitics and Expertise: Knowledge and Authority in European Diplomacy*. Oxford: Wiley-Blackwell.

Kuus, M., & Agnew, J. (2008). Theorizing the state geographically: Sovereignty, subjectivity, territoriality. In Cox, K., Low, M., & Robinson, J. (eds), *The Handbook of Political Geography*, Thousand Oaks, CA: Sage, 117–132.

Lewis, M.W., & Wigen, K.E. (1997). *The Myth of Continents: A Critique of Metageography*. Berkeley, CA: University of California Press.

Mamadouh, V., & van der Wusten, H. (2008). The European level in EU governance: Territory, authority and trans-scalar networks. *GeoJournal* 72(1–2): 19–31. doi:10.1007/s10708-008-9162-8

Megoran, N., & Sharapova, S. (2013). *Central Asia in International Relations: The Legacies of Halford Mackinder*. London: C. Hurst.

Moisio, S. (2002). EU eligibility, Central Europe and the invention of applicant state narrative. *Geopolitics* 7(3): 89–116. doi:10.1080/714000971

Moisio, S., Bachmann, V., Bialasiewicz, L., dell'Agnese, E., Dittmer, J., & Mamadouh, V. (2013). Mapping the political geographies of Europeanization: National discourses, external perceptions and the question of popular culture. *Progress in Human Geography* 37(6): 737–761. doi:10.1177/0309132512472093

Ó Tuathail, G. (1996). *Critical Geopolitics: The Politics of Writing Global Space*. Minneapolis, MN: University of Minnesota Press.

Paasi, A. (2003). Region and place: Regional identity in question. *Progress in Human Geography* 27(4): 475–485. doi:10.1191/0309132503ph439pr

Pickles, J. (2005). "New cartographies" and the decolonization of European geographies. *Area* 37(4): 355–364. doi:10.1111/j.1475-4762.2005.00645.x

Roberts, S., Secor, A., & Sparke, M. (2003). Neoliberal geopolitics. *Antipode* 35(5): 886–896. doi:10.1111/j.1467-8330.2003.00363.x

Sharp, J. (2013). Geopolitics at the margins? Reconsidering genealogies of critical geopolitics. *Political Geography* 37: 20–29. doi:10.1016/j.polgeo.2013.04.006

Sidaway, J. (2002). *Imagined Regional Communities: Integration and Sovereignty in the Global South.* London: Routledge.

Smith, A. (2013). Europe and an inter-dependent world: Uneven geo-economic and geo-political developments. *European Urban and Regional Studies* 20(1): 3–13. doi:10.1177/0969776412463309

Sparke, M., Sidaway, J., Bunnel, T., & Grundy-Warr, C. (2004). Triangulating the borderless world: Geographies of power in the Indonesia–Malaysia–Singapore growth triangle. *Transactions of the Institute of British Geographers* 29(4): 485–498. doi:10.1111/j.0020-2754.2004.00143

Steinberg, P. (2001). *The Social Construction of the Ocean.* London: Routledge.

Toal, G. (2003). Re-asserting the regional: Political geography and geopolitics in world thinly known. *Political Geography* 22(6): 653–655. doi:10.1016/S0962-6298(03)00073-8

Chapter 24

The Banality of Empire

Luca Muscarà

Università del Molise, Italy

A certain confusion surrounds the concept of "empire" due to the different analytical and metaphorical uses of the term (e.g., Agnew 2003; MacDonald 2009). Analytically, beyond expressing an asymmetric relation of power and control, empire often implicates a territorial dimension, and both these connotations imply geography and politics. A quick reference to etymology and history shows how at its origins empire was a personal attribute, and only later did it become a territorial projection of power.

Empire, imperial, and imperialism derive from the Latin *imperium*, which comes from *imperāre*, "to make preparations + in, into, against, directed at" (Partridge 1959). Thus, *imperium* involves planning, organizing, and conducting a centrally coordinated action toward a specific goal – a military one, since *imperator* was a title attributed to the winner of a military campaign. Originally an Etruscan import into the late Roman monarchy, *imperium* was an individual attribute of the king consisting of *coercitio*, the power to decide life and death, and *auspicium* (Latin *avis spicere*), to interpret the divine will through the observation of birds' flight. Despite the violence and destruction of the Roman wars, it thus implied a degree of moral limitation to the absolute power of the king. In the Republican age (509–27 BCE), *imperium* described a discretional power to exercise the functions of the *consules*, *praetores*, and *dictatores*, so not only military but also executive and legislative power; the *Lex Valeria* allowed any citizen of Rome to oppose it. When, in 23 BCE, Augustus was awarded the *imperium maius et infinitum* (ruling over all military forces), he refused the title of *imperator* and preferred that of *princeps*, although his reorganization of the state, the administration, the military, and its territories further expanded *imperium*'s semantic range.

The historical and geographical range of the term expanded too. If, at least in the West, the Roman empire represents the classic paradigm of military, political, and economic power, through the centuries it was extended to antecedent realms, such as the empire of Alexander the Great, or outside the European tradition, such as the Chinese, Inca, Ottoman, or Mogul empires. The Holy Roman empire, and its tripartition at Verdun (843 CE), exemplifies how the

The Wiley Blackwell Companion to Political Geography, First Edition.
Edited by John Agnew, Virginie Mamadouh, Anna J. Secor, and Joanne Sharp.
© 2015 John Wiley & Sons Ltd. Published 2017 by John Wiley & Sons Ltd.

imperium became the territorial projection of a personal attribute, that of the emperor, thus implying a specific geographical political domain. The relationship between empire and geography re-emerged at the end of the Middle Ages, when the commercial interests of the Italian city-states contributed to the development of geography as a form of descriptive knowledge in the fields of chorography, topography, and cartography, (Gambi 1994: 74). It progressed during the Age of Discoveries, often coupled with the evangelizing purposes of the Christian religion, as in the military conquest and colonization undertaken by the Portuguese and Spanish empires, to be followed by France and England, and by the Low and Scandinavian Countries. In the sixteenth century, under the Tudor dynasty, England began to see itself as an empire, distinct from that of Rome. As the study of geography gained popularity, a first indication of "imperial thinking" was recognized in its attempts at establishing England as a standard of comparison with the rest of the world, viewed as "inferior," which encouraged the English "to see the world as theirs by right of conquest" (Cormack 1994: 19–30).

Such a self-centered and hierarchical view of the world began to take on an explicit nationalistic tone with Napoleon, who established France's First Empire (1804) and liquidated the Holy Roman empire (1806), while geographers contributed to his aggressive international imperialism (Godlewska 1994). While Russia's nationalist imagination and geographical identity fueled its imperial expansion toward the Pacific (Bassin 1999), geography became the science of the empire *par excellence* in the Victorian age with the extraordinary growth of the British empire (Livingstone 1992, 1994). After all, the progress of science that brought about the industrial revolution was transforming the national landscape in both the country and the city, and the eighteenth-century ideals of more political and economic freedom for both individuals and their communities, brought into being by the American and French revolutions, strongly affected two of the major empires of the time, with consequences that went far beyond their respective territories (Gottmann 1973).

The ideologies of the time and rapid changes related to industrialization gave way to a sense of the acceleration of history, which required states to deal urgently with increasing political-economic and technological complexity. The publication of Darwin's theories came to provide a general frame to explain the evolution of species, which was soon arbitrarily extended to explaining the diversity of human societies around the world in an ethno-hierarchical and Eurocentric vision of a world divided into nation-states, whose relative "vitality" justified their territorial and colonial expansion. Perhaps it is not a coincidence that the origins of the term "imperialism" can be traced back to the nineteenth century when it was used in Britain to designate the aggressive foreign policy of France's Second Empire under Napoleon III. In the 1870s, it was employed by the opponents of the allegedly aggressive and ostentatious imperial policies of British prime minister Disraeli (later appropriated by explicit supporters of imperialism such as Chamberlain). The term re-emerges in the Second International and later in the perspectives of John Hobson, Rosa Luxemburg, and Lenin.

Modern political geography develops in this period, the so-called Age of Empires (Hobsbawm 1987), an epoch that only closed with the end of the Second World War and the subsequent decolonization process. The concept of imperialism continued to spread during the Cold War, to designate American power politics (in a Marxist perspective) or the Soviet equivalent (in a North Atlantic perspective), and it is now undergoing a controversial revival, even if today the only existing emperor is in Japan. Meanwhile, it has extended its semantic range beyond the geopolitical and geoeconomic domains, to the extent that there is even a computer algorithm named "imperialist" because of its dominating behavior.

The historic canon

In the Age of Empires, the relationship between geography and imperialism, as a science serving respective nation-states, has been described as "symbiotic" (Heffernan 1994: 93–94). It is not merely a chronological coincidence: Empire itself was "a quintessentially geographical project" (Godlewska & Smith 1994: 2) because of geography's usefulness for imperial purposes, as shown in the active role in support of colonialism played by the many geographical societies founded during the nineteenth century.

By identifying with their nation-state, geographers, as did historians or linguists, also contributed to the prevailing interests of their respective homelands: from nation-building to the conservation or expansion of their territorial sovereignties in the metropole (to unify common ethnic and linguistic groups) and in the colonies (by exploiting the resources seen as vital to the national economy).

Colonialism was not only a matter of the economic advantage of trading and exploiting resources to be used in industrial processes. It was rooted in the perspective of Social Darwinism, which put Europeans and Americans (at least those of European descent) above all others. Imperial expansion was therefore a sort of "natural right." Since they considered themselves at the top of a racial ladder, those nations believed that they were naturally entitled to submit other peoples to their military rule in order to expand territorially. Any eventual sense of guilt was removed under the justification of a "civilizing mission": They were bringing education to the "uncivilized" other(s).

In the Social Darwinist perspective, states could be compared to living organisms, their territorial expansion was a natural fact, and even competition among empires was seen as natural, dictated by the very interpretation of the relative place of different human societies within nature. These notions were part of the conceptual equipment that political geography provided to European imperialism, which has been defined as "naturalized knowledge" (Agnew 2002), since it also had the prestige of being akin to the sciences that found order in nature.

Industrialization produced a demographic boom and an increase in population densities in all the main European capital cities and urban centers, while the speed of commercial exchanges was accelerated by new technologies of transport and communication. By the end of the nineteenth century, there were no more "free" lands to conquer and territorial expansion had become a zero-sum game, creating a sense of a "closing world." Since territory and the colonies of the industrial nations were also a matter of identity, essential in terms of their international status to how each nation-state perceived itself and was perceived by its peers, such premises fueled the growing interimperial rivalries of the time, which, combined with nationalism and militarism, ended up propelling the two world wars.

Such ideas were largely common up to the Second World War and the theories developed in that epoch constitute the so-called historic canon of political geography (Agnew 2002), within which the thought of Ratzel or Mackinder, if different in much of its content, was largely about serving the interests of their respective states.

Even if the huge corpus of Friedrich Ratzel (1844–1904) remains largely to be confronted in its entirety, with this founder "the State took possession of geography and became its supreme object" (Farinelli 2001: 44). His writings have been variously interpreted depending on epoch and theoretical perspective. If some of his ideas migrated to the United States through the work of some of his students, such as Ellen Churchill Semple, in the context of Germany between the two world wars his concept of *Lebensraum*, or his "laws" of the spatial growth of the state, gave way to new interpretations, usually within the thread of *Geopolitik*. Kjellén and Hausofer, Maull or Penck are considered as having differently influenced the territorial expansion of Hitler's Germany and attracted followers also in Japan and Italy. Such interpretations were strongly disputed in France (Demangeon 1939) and, during the Second World War, in the United States (Bowman 1942; Gottmann 1942; Gyorgy 1944). After the war, the condemnation of *Geopolitik* not only resulted in a discrediting of the entire German school, in a rough equation among Ratzel, *Geopolitik*, and Nazism, but also involved the whole subdiscipline of political geography, which, affected by the "original sin" of determinism, faced in those years a true eclipse. Only gradually did a re-evaluation of Ratzel (Gottmann 1947, 1952, 1973; Bassin 1984, 1987a), and of the relationship between Nazism and *Geopolitik* develop, absolving the latter from the accusation of anti-Semitism (Bassin 1987b), while a larger reassessment of Ratzel took place in the 1990s (Antonsich et al. 2001).

Even more relevant are perhaps the ideas of another "founder," Halford Mackinder (1861–1947), already then considered an "Aid to Statecraft" (Teggart 1919) or serving the interests of the British empire, whose maritime primacy he perceived as threatened by the rise of the new industrial powers' development of railroads and modern fleets. In the first version of his global geopolitical model, that of the "pivot area" (Mackinder 1904), those technologies, together with the relative geographical location of empires at a global scale, highlighted a risk for the British empire, seen as peripheral, as opposed to the continental centrality of the Russian empire. Mackinder therefore predicted "the war between Russia and Germany for the control of Eastern Europe" (Blouet 2001), which seems to have returned to geopolitical topicality with the 2014 events in Crimea and Ukraine (or is it Mackinder's enduring influence?). His concept of the pivot area, whose control would have granted world supremacy, was reformulated, with a more Western location, as the theory of the heartland in 1919. Not all his ideas were always listened to at home, such as the proposal for a second *cordon sanitaire*, a second series of buffer states between Germany and Russia, expressed after his mission as British High Commissioner to South Russia, devastated by the civil war (Blouet 1976).

In the United States Mackinder's heartland theory inspired an attempt by Dutch political scientist Spykman (1944) to invert the political map of the world (and its related location of power) by substituting the central location of Russia with that of the United States. This may have inspired the policy of containment of the Soviet Union, the US Cold War's driving idea. And even if Mackinder's 1943 paper could be seen as inspiring the idea of a North Atlantic alliance, it was necessary to wait for the end of the Cold War for a new assessment of his thinking (Blouet 1987) and of the role it may have played in global geopolitics and geostrategy (Blouet 2001, 2005; Kearns 2009).

Indeed, not all political geography can be reduced to Ratzel and Mackinder. The French school, founded by Vidal de la Blache (1845–1918) and developed by his students, was also serving its nation-state. Theoretically not determinist, even in Reclus's anarchic variant it never questioned the colonial empire, legitimized as *mission civilisatrice*. In the United States, the geography of Ratzelian influence was replaced by that developed by Isaiah Bowman (1878–1950). The political geographical doctrine emerging during the Wilson administration broke with the Monroe doctrine, beginning to foresee a global role for the United States. This perspective confronted the world largely in terms of problem solving, with an inventory approach to the critical areas for which "scientific" solutions were proposed to the advantage of the American state. The geography of Vidal and his students and that of Bowman have been critically reinterpreted in the last decades (e.g., Berdoulay 1981; Robic 1994; Claval & Sanguin 1996; Smith 2003a).

As a whole, the historic canon does not seem adequately to have confronted the theme of power theoretically: Power is seen as implicit in the behavior of nation-states, while geographical "laws" are proposed to support the various empires' colonial and territorial policies. Only between the two wars was a notion of such power found in the distinction between *have* and *have not* states, the latter claiming their "natural" right to colonize, without concern for people, social consequences, or environmental damage. In this perspective the two world wars were also failed attempts to curtail the power of the British and French empires, and of the United States in the Pacific, by *have not* states (e.g., Gottmann 1973; Van Valkenburg 1939).

Finally, these developments in the history of ideas, and of political geography, cannot be abstracted from the context of impressive demographic growth intertwined with industrialization at an increasingly global scale. Proceeding with a formidable acceleration, this has contributed to increased political-economic complexity (Muscarà 2001). The next section presents the epochal turn that the end of the Second World War represented for both empires and political geography.

Cold War

With the end of the Second World War, the colonial empires also came to an end; the geopolitical order and the political geographical theories that followed were very different. Among the many explanations for such a significant change, one is certainly related to the invention of the atomic bomb, which changed forever the balance of power, with numerous general implications:

- The supremacy granted by nuclear weapons, combined with the further introduction of intercontinental ballistic missiles, ended territory's traditional function of providing security (Gottmann 1973).
- On the geopolitical plane, the progressive extension of nuclear arsenals to all members of the UN Security Council (followed by India, Pakistan, etc.) established a new international hierarchy of *haves* and *have nots*, based on the possession not of colonies, but nuclear weapons, replacing the idea of an ethnic hierarchy among nations.
- Conceptually, by reversing the order of the relationships between nature and the human species, the nuclear bomb constituted an "argument" that buried any notion of environmental determinism, together with the theories of *Geopolitik*.
- The bomb also changed the relationship between political geography and governments; the former lost its *aid to statecraft* role typical of the previous epoch, replaced by other disciplines, from physics to area studies, and underwent a temporary eclipse.
- When political geography was reinvented beginning in the 1960s, it looked for inspiration in other disciplines; after Foucault, the subject of power became central, replacing that occupied by nature in the historic canon.
- While the process of decolonization was progressively dismantling the colonial empires on the global scale, the focus on power inevitably brought to greater attention the concept of imperialism and the related development of anti-imperialist movements.

Beyond these general implications, it is necessary also to consider another element specific to the historical trajectory of the United States. At the end of the Second World War, the winners on the military, economic, and ideological planes, thanks to the atomic bomb, appeared to be endowed with an enormous power, symbolically comparable almost to a divine one. Still, the bomb was developed in the context of the war on totalitarianism and the Social Darwinist principles by which it was inspired. The defeat of totalitarianism and the victory of the Allied forces obscured the fact that a totalitarian element remained prisoner of the bomb itself, in the culture of the "military-industrial complex" that had developed since the Manhattan Project and its successors in the United States and permeated a part of government culture. Such logic contributed to producing the bipolar ontology of the Cold War: The absolute weapon required an absolute enemy, necessary for the military-industrial complex to perpetuate itself. In this perspective, even if the USSR shared the notion of a fundamental ideological conflict between the two "systems," the Cold War was essentially an American invention (Isaac & Bell 2012).

Such a legacy of totalitarianism embodied in the bomb could explain the temporary bursts of paranoid behavior within the United States (Hofstadter 1964) and the "imperialism" of the two superpowers, as confirmed by a long string of regional wars and more or less classified operations of regime change, with which, beyond the principles of democracy, the logic of force resumed its rule.[1] This development is perhaps also ascribable to the *pathologies of power* associated with nuclear weapons (e.g., Fettweis 2013).

This logic carried strong implications in terms of scale as well. The armaments race, and the related proliferation of nuclear arsenals, required huge financial effort, and a parallel scientific, military, and industrial gigantism (Galison & Hevly 1992) that from *Big Science* and the growth of large-scale research was extended to the cultural arena, and was mirrored also by the increasing gigantism of corporations (Lynn 2010).

It can be said that in this way the end of historical empires gave way to a new "imperial culture," first invisible, then more apparent. That in turn provoked the development of anti-imperial movements in the United States and elsewhere, signaling an inner fracture between the political, military, and industrial apparatuses, on the one hand, and a relevant part of national societies, especially beginning with the generations born during and after the Second World War, on the other. It is in this period that political geography began, at least in the West, to be critical of the establishment and to problematize the issue of power, while simultaneously it transcended the strict national dimension that had characterized the previous period.

Thus, imperialism in its new form engendered a new anti-imperialism and its deconstruction of power, in the form of a vertical fracture within national societies, which then tried to coalesce horizontally with its peers on a transnational basis. This phenomenon would multiply and evolve into the present-day counter-globalization transnational networks (Featherstone 2007, 2008, 2013). This may also be connected to the thread of critical geopolitics inaugurated by Agnew and Ó Tuathail (1992), and others, which introduced analysis of the ways in which hegemonic narratives are established within wider societies, a thread that has been expanded in the direction of subaltern geopolitics, which studies how the practices of pan-Africanism, for example, tried to shape a postcolonial future different from the logic of the Cold War (Sharp 2013).

Meanwhile, in the deep ideological opposition that characterized the Cold War, reciprocal accusations of imperialism were exchanged between the USSR and the United States (and their respective allies), while geography was denounced as having always served the waging of war (Lacoste 1976). With the 1970s, the new political geography discovered a renewed interest in past empires (Hudson 1972). New Left historians contributed a distinction between formal and informal empires to explain the actual reach and influence of the British empire beyond its formal territory (Reynolds 2002).

As the process of decolonization progressed, in the context of cultural and literary studies a new and successful postcolonial thread developed following Said's deconstruction of the language of power (1978, 1993). Furthermore, from an interest in the voices from within formerly colonized territories, the postcolonial suggestion to listen to colonized peoples has extended to their presence in the former metropole, and even a concept of "vernacular imperialism" has been proposed for these ideas and language about the continuing impact of imperialism (Phillips 2011). (See Chapter 19 in this volume for more on postcolonialism.)

Within the postcolonial approach, the first decade of the twenty-first century also saw historiographical studies dedicated to cultural interpretations of the imperial past, especially regarding the British empire and its transformation. As Ballantyne (2010) has argued, while it is necessary to examine closely the relationship between the economic and cultural dimensions of empires, including the significance of precolonial structures, three analytical threads have emerged: "the importance of information and knowledge in empire building, the centrality of cultural difference within imperial social formations, and the place of imperial networks and patterns of cross-cultural exchange in the operation of the empire," the latter emphasizing the relationships between the metropole and the colonies, following the center–periphery model.

Revival of empires

Since 2000, *Ab Imperio*, a new social sciences journal, has been published in Kazan, Russia in five languages (English, Russian, German, French, Ukrainian) dedicated to studies of new imperial history and nationalism in the post-Soviet space. Interrogating the relation between the end of the Russian empire (1917) and that of the "Soviet empire" (1992), Strada (2014) highlighted the importance of the ideal meaning that empires have had in history for their respective people and proposed the term "imperiology" as a new field in historical studies. This could well be applied also to the Western surge in the use of terms such as empire, imperial, and imperialism after the attacks of September 11, 2001, and in particular the use of the "American empire" metaphor (Wainwright 2004; Lester 2008; MacDonald 2009), both positively and negatively. In a positive vein, as a normative model for US foreign policy some authors championed the regressive idea that the United States *should act in a more imperialist way*.

Already with the collapse of the USSR, the necessity had been proposed of an assertive American leadership to address the issue of global governance through the use of "soft power" (Nye 1990). It was followed by various interpretations of "America's unipolar moment" (Krauthammer 1990) and a string of "endist" theories (Fukuyama 1992; O'Brien 1992; Ohmae 1995). After the initial triumphalism, "declinist" concerns appeared about the possibilities of ensuring a new "American century" (Nye 2002). The neoconservative Project for a New American Century (2000), on which the Bush doctrine was based (the right to invade other countries to root out potential terrorists), represented a response to such fears of a US decline, to reassure the right wing about America's continuing primacy, a concern that has continued in the years since (e.g., Friedman 2010).

Following the shock of 9/11, such concerns were subsumed in a wider national identity crisis in the United States, which partly explains the Bush administration's turn to "hard power" and its so-called War on Terror. Terrorism was to be fought anywhere, for an indefinite time, with all possible means, but actually the threat was regressively reterritorialized at the scale of nation-states (Gregory 2011). This happened not only through the hardening of US borders and the process of domestic securitization, but through military campaigns that, by targeting the regimes of independent states, aimed at re-establishing the international status of the United States as a superpower. By reminding everybody of US military superiority, this strategy also aimed to reassure the American people that the terrorist threat was being contained outside the homeland (Caracciolo 2011).

The Bush doctrine found enthusiastic supporters among neoconservatives such as Kagan (1998, 2003), while the media playing a key role in its promotion became an object of study for critical geopolitics (e.g., Ó Tuathail 2003). While Nye was cautioning against the use of unilateralism, some liberal historians advocated a more formal imperial US intervention in a sort of neocolonial civilizing mission (e.g., Ignatieff 2003). The United States was an "empire in denial," and as long as it refused to identify itself as an imperial power and act consequently, global chaos would continue (Ferguson 2003). It was good to be an empire, and the argument was constructed through a historical comparison with the British empire, reinterpreted as a positive force for the world. While some authors called for an American empire in the name of moral and humanitarian imperatives, the overall effect was to support the unilateralism of the Bush administration, particularly in the context of the military invasion of Iraq in 2003.

Certainly, empire was also used as a negative metaphor. Neil Smith (2003a) drew opposite conclusions as to the benevolence of the American empire. In his long-researched biography, Isaiah Bowman became the key to discovering in the "lost geography of the American Century" a blueprint for American globalism. His work for Wilson and Roosevelt, up to and into the post–Second World War period, revealed an imperial project of economic and political dominance spanning the twentieth century. With hindsight, Smith (2009: 736) redefined empire as "a deeply conservative project of economic expansion, power and control bound up with a social civilizational mission."

In the same Marxist tradition, Harvey (2003) interpreted America's contemporary "new imperialism" as focused on maintaining control over the stability of supplies of a strategic resource: it is "all about oil." The imperative of controlling the "oil spigot" was seen as the key to explaining both US imperial policy in the Middle East and the unilateral invasion of Iraq, although Harvey and Smith agree that the latter was "not simply a war to control oil resources but rather to control the global political economy within which the disposition of oil resources will be organized" (Smith 2003b: 265).

From a non-Marxist perspective, historical sociologist Mann (2003) considered four different types of power (military, economic, political, and ideological), concluding that while the "American empire" was not aiming at ruling foreign lands permanently, its imperialism was deeply "incoherent," as the Bush doctrine was asymmetrically tilted toward an "overconfident, hyperactive militarism" and neglected all the other dimensions. Thus, it was doomed, since it would have ended up creating more terrorism and weakening American leadership in the world. In this forecast, Mann followed the line of Johnson (2000) who, in *Blowback*, had already analyzed the "unintended consequences of the American empire," and was denouncing the implications of American militarism for globalization and the US financial deficit (Johnson 2003).

Within the tradition of cultural studies, Gregory's (2004) emphasis on colonialism widened the perspective beyond the United States. The aggressive policies pursued by American, British, and Israeli governments in Iraq, Afghanistan, and Palestine revealed a shared practice of "othering," thus seeing imperialism as a question of territorial identity. By establishing a correspondence between the binaries of "us/them" and "good/bad" – explicit in the Bush doctrine – Gregory explained the reassertion of colonial power in terms of spatializing the "bad," which in turn created the new binary of "occupier/occupied." The latter was exemplified by the internal divisions in Afghanistan, Baghdad's Green and Red Zone, the wall separating Israel–Palestine, or extrastate territories such as Guantánamo.

Underlining the contradiction between US unilateralism and the political identity of the former British colony, Agnew (2003) instead argued that the concept of hegemony is analytically more accurate than that of empire to describe the historical relationship between the United States and the rest of the world, since hegemony is an indirect form of dominance in which the leader rules other states, groups, and so on by means of power, the threat of force, or economic or cultural means. Rather than an American empire, Agnew saw "a hegemonic project intimately connected to the geopolitical calculus of the US government and economic interests during the Cold War and to the incorporation of the entire world into its grip" (Agnew 2005).

From this viewpoint, the aggressive reaction of the United States after 2001 in fact proved more its weakness than its ability to establish an empire, given the instability that US intervention engendered in Afghanistan, Iraq, and neighboring countries. Moreover, the obsessive focus on security and terrorism not only contributed to tripling the fiscal deficit, but diverted US attention from other global challenges, especially in the geoeconomic realm. The rise of China and other global actors, within the ongoing processes of economic globalization, was largely overlooked, until the explosion of the financial crisis in 2007–08 brought the economy back onto the radar. In fact, the economic crisis proved America's vulnerability "to the very forces of globalization that successive US governments and businesses have helped unleash" (Agnew 2009).

Since then, the subject of financial crisis has alternated in the dominant narratives with geopolitical concerns about distant threats. The securitization started by the Bush administration has continued to curtail domestic civil liberties, as well as to undermine the tenets of international law and universal human rights. Increasingly, the digital surveillance activities of US intelligence have involved spying not only on "enemy" countries, but also on traditional allies (Engelhardt 2014). And the War on Terror continues, with President Obama's controversial recourse to drones (Shaw 2013), which from their initial usage in Afghanistan have now extended to Pakistan, Yemen, Somalia, Iraq, Syria, and Libya.

Meanwhile, another anti-imperial thread has been developing, this time focused on economic globalization. In the post-Marxist tradition, Hardt and Negri (2000) have used the metaphor of empire as a substitute for globalization, envisaging a deterritorialized and denationalized empire superseding the nation-state. The empire to which they refer is not the United States; it is instead a consequence of the decline of sovereignty of nation-states in the process of globalization, and the concurrent rise of a new global sovereignty with no territorial center of power or fixed boundaries, which has "progressively incorporated the entire global realm within its open, expanding frontiers" (Hardt & Negri 2000: xii). Abstractly conceived as a regime "outside of history or at the end of history" (2000: xv) and presented as "permanent, eternal, and necessary" (2000: 11), such an empire would eventually "envelop all power relations within its world order" (2000: 20). This all-encompassing perspective was attacked from the perspective of international relations (Barkawi & Laffey 2002), while Hardt and Negri's abstraction from geography and history was criticized as echoing a globalized version of Nagel's "view from nowhere" (Coleman & Agnew 2007).

From a different political-economic perspective, Marxist sociologist Arrighi (2007) argued that the failure of the Project for a New American Century in Iraq and Afghanistan, ending "a sixty-year long struggle of the US to become the organizing center of a world state" (2007: 261) and combined with the extraordinary economic growth of China, could instead realize Adam Smith's intuition of a global market society, based on greater equity among the different world regions. Rather than empire, he called this a "Commonwealth," a product of hybridization between the Western and Chinese traditions.

Zielonka (2012) instead suggested that the present multipolar context could be explained "by four or five vast territorial actors, all resembling empires: the US, China, Russia, and European Union." The advantage of the term empire would be in the possibility of including the "moral and ideological factors that form civilising missions." And even if the present economic crisis in the EU keeps in check its potential as a global actor no less than its incomplete project, its cultural, economic, and territorial expansion (11 new member states since 2004) has been compared to that of an empire (Zielonka 2006; Parker 2011). In a similar vein, an analysis of the EU's cartographic discourse on its websites and in its continental currency argues that the exclusion of "other" Europeans from the maps reflects a subconscious – yet gradually emerging – imagination of empire (Foster 2013).

Seeing imperialism as immanent in speculative international financial capital, Marxist economist Patnaik (2010, 2011, 2014) has attacked it as a manifestation of neoliberal ideology. Its dominant position in the contemporary world economy is achieved through the imposition of neoliberal policies on governments under the threat that otherwise financial capital would move elsewhere, causing a liquidity crisis that would push the state to collapse. Ireland and Greece stand out as examples of the results of these policies on national sovereignty. In allowing the centralization of capital at the global scale in the hands of a global financial community, such policies boost national unemployment and lower economic growth, increasing the distance between the rich and the poor, at different scales, as proved also by Piketty (2013). The neoliberal project aims at turning states into agents of global capital in order to operate over an entire undivided planet, as if the world could be an isotropic financial space, abstracted from geography and history.

In a similar vein among recent contributions from within political geography, the Kilburn manifesto (Hall, Massey, & Rustin 2013–14) may better succeed at creating conceptual tools for resistance to the penetration of neoliberal ideologies in the wider public. In fact, its focus on the United Kingdom does not prevent its deconstruction of economic fundamentalism from being applicable at the global scale. Without falling into the "horizontalism" of Hardt and Negri, according to whom there can be nothing external to their "empire," the Kilburn Manifesto actually aims at reinventing democracy and the state, beginning at the national scale.

The normative reductionism of empire

Why has there been so much recourse to the notion of empire to explain the geopolitical, geoeconomic, or cultural condition of the present? The causes may be many. As a metaphor, emphasizing a power asymmetry that endorses a hierarchical view of the world, empire and imperialisms are used as hyperbole for the concepts of state and hegemony, to capture the attention of the reader. Moreover, mindful of historical experiences such as the Roman, Byzantine, and British empires (e.g., Münkler 2007), the concept provides a reassuring comparison for military interests and political conservatives about the essential similarities between present and past.

Furthermore, by positing a singular agency, the recourse to the metaphor of empire simplifies the epistemological complexity of the world, reducing it to simpler and more manageable terms. It is a powerful and versatile rhetorical expedient of *reductio ad unum* (reduction to one), which, in simplifying complexity and the multiplicity of actors, provides a center to the discourse. Paradoxically, even anti-imperial literature may mirror an analogue logic. In both cases, its analytical value is limited, since it produces narratives that are not easily generalized, as each anti-imperial author seems to compete for his/her own version of empire, thus producing a fragmentation that resembles that produced by imperialism itself. *Divide et impera.*

Finally, all of these points would seem to support Noel Parker's intuition (2010) of the *ordering power* of empire, in view of increasing global complexity, over a world perceived as a disorderly space. In this sense it is then preferable to classic analytical concepts such as the Westphalian state, or "great powers" and "hegemony," especially within international relations. In fact, Parker explicitly rejects those explanations of empire that focus on a single historical force, such as modernization or capitalism, and articulates an analysis of many aspects of empires, including their territorial expansionism, their centers and boundaries, and their unintended consequences. In rapidly changing times such as ours, to give order to the complexity of the world is an enduring problem, and in this sense, after the end of historical empires, recourse to the iconography of empire could be considered, with Gottmann (1952), as a defensive reaction to the identity threats posed by increasing complexity (Muscarà 2005).

After all, the notion of empire as a way of ordering complexity existed already in the Roman empire, if one thinks that Augustus came to power following a period of civil wars, when the city of Rome had close to 1 million inhabitants (Carcopino 1940). Italy then had a population of between 7 and 15 million (Lo Cascio & Malanima 2005); at its height, the Roman empire encompassed about 4 million square kilometers, with an ethnically diverse and polyglot population probably in excess of 60 million (Mattingly 2011). This would support an interpretation of *imperium* as *ordering power* related to growing sizes, varieties, and scales, and therefore complexity, since complexity increases with scale (Mofson 1999), thus also intrinsically linking geography and empire. Empire would be an ideal normative model for the problem of growing complexity, both for its proponents and, as a negative one, for its detractors.

Beyond the hyperbole, the rhetorical expedient, and the ordering power, it is indisputable that, in anti-imperial literature, empire and imperialism are used to stress power asymmetries at any scale, and relationships of violence, coercion, and domination. Their proliferation signals their continuing and increasing presence, and in political geography this points to the continuing nexus between historical empires and Social Darwinism, not only their association during the heyday of the historic canon.

If the teleological power of explanation attributed to imperialism in anti-imperial literature may seem to imply a specular theoretical monism, or a similar monocausal perspective on the world, this may be rooted in the need to confront a culture that has at its root the very dictatorship of the bomb and of the military-industrial complex currently permeating the most powerful governments. That the Cold War discourse may not have ended yet, either in the DNA of the geopolitical culture of the US government and its allies or perhaps in its recent rebirth in Russia, and well beyond the military sphere, is manifested in the fact that a logic based on force has successfully penetrated the economic culture of capitalism in the last decades. Neoliberal determinism and its expanding global ambition in fact seem to aim at shaping globalization through the morphing of state regulations and policies in favor of large-scale global business (Gilman 2014).

In the end, if contemporary imperialism does exist, its forms are not limited to geopolitical, economic, energetic, technological, cultural, or linguistic domains, as imperialism could in fact be seen as an attempt at the colonization of a series of distinctive narratives by reference to a single historical force, under the lure of reducing complexity or under the threat of superior military or economic power. Thus, it can hardly be challenged through a symmetric anti-imperial fundamentalism that would reproduce its Social Darwinian logic.[2] It is instead the penetration of deterministic thinking in daily operations and culture that has to be exposed, as Hannah Arendt's concept of the "banality of evil" (1951) famously implied. It is the seeming autonomy of politics and the economy from ethical considerations that should be debated at large, in view of its potential global political and environmental implications. Talk of empire, imperialism, and so on may thus become a barrier to a real understanding of complexity.

Notes

1 The autonomy of politics from ethics existed already in Machiavelli, but Social Darwinism intertwined it with the industrial revolution, with consequences at a wider scale. The introduction of nuclear weapons extended its potential implications to the prospect of complete annihilation of life on earth, a responsibility that requires ethics to be reintroduced into politics, as so many physicists have asked since 1945.
2 For example, a novel by Hamid (2007) highlighted an interesting equation between Islamist fundamentalism and neoliberal financial fundamentalism.

References

Agnew, J. (2002). *Making Political Geography*. London: Arnold.

Agnew, J. (2003). American hegemony into American empire? Lessons from the invasion of Iraq. *Antipode* 35(5): 871–875.

Agnew, J. (2005). *Hegemony: The New Shape of Global Power*. Philadelphia, PA: Temple University Press.

Agnew, J. (2009). *Globalization and Sovereignty*. Lanham, MD: Rowman and Littlefield.

Antonsich, M., Kolossov, V., & Pagnini, M.P. (2001). *On the Centenary of Ratzel's* Politische Geographie: *Europe Between Political Geography and Geopolitics*. Rome: Memorie della Società Geografica Italiana, Vol. 63, 2 tomes.

Arendt, H. (1951). *The Origins of Totalitarianism*. New York: Schocken Books.

Arrighi, G. (2007). *Adam Smith in Beijing: Lineages of the Twenty-First Century*. London: Verso.

Ballantyne, T. (2010). The changing shape of the modern British empire and its historiography. *Historical Journal* 53(2): 429–452.

Barkawi, T., & Laffey, M. (2002). Retrieving the imperial: Empire and international relations. *Millennium Journal of International Studies* 31(1): 109–127.

Bassin, M. (1984). Friedrich Ratzel's travels in the United States: A study in the genesis of his anthropogeography. *History of Geography Newsletter* 4: 11–22.

Bassin, M. (1987a). Imperialism and the nation state in Friedrich Ratzel's political geography. *Progress in Human Geography* 11(4): 473–495.

Bassin, M. (1987b). Race contra space: The conflict between German Geopolitik and National Socialism. *Political Geography Quarterly* 2: 115–134.

Bassin, M. (1999). *Imperial Visions*. Cambridge: Cambridge University Press.

Berdoulay, V. (1981). *La formation de l'école française de géographie (1870–1914)*. Paris: CTHS.

Blouet, B.W. (1976). Sir Halford Mackinder as British High Commissioner to South Russia 1919–1920. *Geographical Journal* 142(2): 228–236.

Blouet, B.W. (1987). *Halford Mackinder: A Biography*. College Station, TX: Texas A & M University Press.

Blouet, B.W. (2001). *Geopolitics and Globalization in the Twentieth Century*. London: Reaktion Books.

Blouet, B.W. (ed.) (2005). *Global Geostrategy: Mackinder and the Defense of the West*. London: Frank Cass.

Bowman, I. (1942). Geography versus geopolitics. *Geographical Review* 4: 646–658.

Caracciolo, L. (2011). *America vs. America*, Roma-Bari: Laterza.

Carcopino, J. (1940). *Daily Life in Ancient Rome*. New Haven, CT: Yale University Press.

Claval, P., & Sanguin, A.-L. (1996). *La géographie française à l'epoque classique, 1918–1968*. Paris: L'Harmattan.

Coleman, M., & Agnew, J. (2007). The problem with empire. In Elden, S., & Crampton, J. (eds), *Space, Knowledge and Power: Foucault and Geography*, Aldershot: Ashgate, 317–339.

Cormack, L.B. (1994). The fashioning of an empire: Geography and the state in Elizabethan England. In Godlewska, A., & Smith, N. (eds), *Geography and Empire*, Oxford: Blackwell, 15–30.

Demangeon, A. (1939). Géographie politique, à propos de l'Allemagne. *Annales de Géographie* 48(272): 113–119.

Engelhardt, T. (2014). Shadows of a thug state. *Le monde diplomatique*, March. http://mondediplo.com/2014/03/12state#nb3, accessed March 26, 2015.

Farinelli, F. (2001). Friedrich Ratzel and the nature of (political) geography. In Antonsich, M., Kolossov, V., & Pagnini, M.P. (eds), *On the Centenary of Ratzel's* Politische Geographie: *Europe Between Political Geography and Geopolitics*, Rome: Memorie della Società Geografica Italiana, Vol. 63-1. Also with the same title in *Political Geography* 19 (2000): 943–55.

Featherstone, D. (2007). The spatial politics of the past unbound: Transnational networks and the making of political identities. *Global Networks* 7(4): 430–452. doi:10.1111/j.1471-0374.2007.00178.x

Featherstone, D. (2008). *Resistance, Space and Political Identities: The Making of Counter-Global Networks*. Malden, MA: Wiley-Blackwell.

Featherstone, D. (2013). The contested politics of climate change and the crisis of neo-liberalism. *ACME: An International E-Journal for Critical Geographies* 12(1): 44–64.

Ferguson, N. (2003). *Empire: How Britain Made the Modern World*. Harmondsworth: Penguin.

Fettweis, C.J. (2013). *The Pathologies of Power: Fear, Honor, Glory, and Hubris in U.S. Foreign Policy*. Cambridge: Cambridge University Press.

Foster, R. (2013). Tabula Imperii Europae: A cartographic approach to the current debate on the European Union as empire. *Geopolitics* 18: 371–402. doi: 10.1080/14650045.2012.716466

Friedman, G. (2010). *The Next 100 Years: A Forecast for the 21st Century*. New York: Anchor Books.

Fukuyama, F. (1992). *The End of History and the Last Man*. New York: Free Press.

Galison, P., & Hevly, B, (eds) (1992). *Big Science: The Growth of Large Scale Research*. Stanford, CA: Stanford University Press.

Gambi, L. (1994). Geography and imperialism in Italy: From the unity of the nation to the "new" Roman empire. In Godlewska, A., & Smith, N. (eds), *Geography and Empire*, Oxford: Blackwell, 74–91.

Gilman, N. (2014). The twin insurgency. *American Interest* 15 June.

Godlewska, A. (1994). Napoleon's geographers (1797–1815): Imperialists and soldiers of modernity. In Godlewska, A., & Smith, N. (eds), *Geography and Empire*, Oxford: Blackwell, 31–54.

Godlewska, A., & Smith, N. (eds) (1994). *Geography and Empire*. Oxford: Blackwell.

Gottmann, J. (1942). The background of geopolitics. *Military Affairs* 4: 197–206.

Gottmann, J. (1947). Doctrines géographiques en politique. In *Les Doctrines Politiques Modernes*, New York: Brentano, 17–40.

Gottmann, J. (1952). *La politiques des états et leur géographie*. Paris: Armand Colin.

Gottmann, J. (1973). *The Significance of Territory*. Charlottesville, VA: University Press of Virginia.

Gregory, D. (2004). *The Colonial Present: Afghanistan, Palestine, Iraq*. Oxford: Blackwell.

Gregory, D. (2011). The everywhere war. *Geographical Journal* 177: 238–250. doi:10.1111/j.1475-4959.2011.00426.x

Gyorgy, A. (1944). *Geopolitics: The New German Science*. Berkeley, CA: University of California Press.

Hall Stuart, D.M., & Rustin, M. (2013–14). After neoliberalism: The Kilburn Manifesto. *Soundings*. http://www.lwbooks.co.uk/journals/soundings/manifesto.html, accessed March 26, 2015.

Hamid, M. (2007). *The Reluctant Fundamentalist*. New York: Harcourt.

Hardt, M., & Negri, A. (2000). *Empire*. Cambridge, MA: Harvard University Press.

Harvey, D. (2003). *The New Imperialism*. Oxford: Oxford University Press.

Heffernan, M. (1994). The science of empire: The French geographical movement and the forms of French imperialism, 1870–1920. In Godlewska, A., & Smith, N. (eds), *Geography and Empire*, Oxford: Blackwell, 92–114.

Hobsbawm, E. (1987). *The Age of Empire: 1875–1914*. London: Weidenfeld and Nicholson.

Hofstadter, R. (1964). The paranoid style in American politics. *Harper's* November: 77–86.

Hudson, B. (1972). The new geography and the new imperialism 1870–1918. *Antipode* 9(2): 12–19.

Ignatieff, M. (2003). *Empire Lite: Nation-Building in Bosnia, Kosovo and Afghanistan*. Toronto: Penguin.

Isaac, J., & Bell, D. (eds) (2012). *Uncertain Empire, American History and the Idea of the Cold War*. Oxford: Oxford University Press.

Johnson, C. (2000). *Blowback: The Costs and Consequences of American Empire*. New York: Henry Holt.

Johnson, C. (2003). *The Sorrows of Empire: Militarism, Secrecy, and the End of the Republic*. New York: Metropolitan Books.

Kagan, R. (1998). The benevolent empire. *Foreign Policy* 111: 24–35.

Kagan, R. (2003). *Of Paradise and Power: America and Europe in the New World Order*. New York: Knopf.

Kearns, G. (2009). *Geopolitics and Empire: The Legacy of Halford Mackinder*. Oxford: Oxford University Press.

Krauthammer, C. (1990). The unipolar moment. *Foreign Affairs* 70(1): 22–33.

Lacoste, Y. (1976). *La géographie, ça sert, d'abord, à faire la guerre*. Paris: Maspero.

Lester, A. (2008). Empire. In Cox, K.R., Low, M., & Robinson, J. (eds), *The Sage Handbook of Political Geography*, Los Angeles, CA: Sage, 455–469.

Livingstone, D. (1992). *The Geographical Tradition*. Oxford: Blackwell.

Livingstone, D. (1994). Climate's moral economy: Science, race and place in post-Darwinian British and American geography. In Godlewska, A., & Smith, N. (eds), *Geography and Empire*, Oxford: Blackwell, 132–154.

Lo Cascio, E., & Malanima, P. (2005). Cycles and stability: Italian population before the demographic transition (225 B.C.–A.D.1900). *Rivista di Storia Economica* XXI(3): 5–40.

Lynn, B.C. (2010). *Cornered: The New Monopoly Capitalism and the Economics of Destruction*. Hoboken, NJ: John Wiley & Sons.

MacDonald, P. (2009). Those who forget historiography are doomed to republish it: Empire, imperialism and contemporary debates about American power. *Review of International Studies* 35: 45–67.

Mackinder, H. (1904). *The Geographical Pivot of History*. London: Royal Geographical Society.

Mann, M. (2003). *Incoherent Empire*. London: Verso.

Mattingly, D.J. (2011). *Imperialism, Power and Identity: Experiencing the Roman Empire*. Princeton, NJ: Princeton University Press.

Mofson, P. (1999). Global ecopolitics. In Demecki, G.J., & Wood, W.B. (eds), *Reordering the World: Geopolitical Perspectives on the Twenty-first Century*, 2nd edn. Boulder, CO: Westview Press.

Münkler, H. (2007). *Empires: The Logic of World Domination from Ancient Rome to the United States*. Cambridge: Polity Press.

Muscarà, L. (2001). Understanding Ratzel and the challenge of complexity. In Antonisch, M., Pagnini, P., & Kolossov, V. (eds), *Europe between Political Geography and Geopolitics*, Vol. I, Rome: Società Geografica Italiana, 79–91.

Muscarà, L. (2005). Territory as a psychosomatic device. *Geopolitics* 10: 24–49.

Nye, J. (1990). Soft power. *Foreign Policy* 80: 153–171.

Nye, J. (2002). *The Paradox of American Power: Why the World's Only Superpower Can't Go It Alone.* Oxford: Oxford University Press.

O'Brien, R. (1992). *Global Financial Integration: The End of Geography.* London: Chatham House.

Ohmae, K. (1995). *The End of the Nation-State: The Rise of Regional Economies.* New York: McKinsey.

Ó Tuathail, G. (2003). "Just out looking for a fight": American affect and the invasion of Iraq. *Antipode* 35: 857–870.

Ó Tuathail, G., & Agnew, J. (1992). Geopolitics and discourse: Practical geopolitical reasoning in American foreign policy. *Political Geography* 11(2): 190–204.

Parker, N. (2010). Empire as a geopolitical figure. *Geopolitics* 15: 109–132. doi:10.1080/14650040903420412

Parker, N. (2011). Imperialism, territory, and liberation: On the dynamics of empire stemming from Europe. *Journal of Political Power* 4(3): 355–374. doi:10.1080/2158379X.2011.628565

Partridge, E. (1959). *Origins: A Short Etymological Dictionary of Modern English.* New York: Macmillan.

Patnaik, P. (2010). Notes on contemporary imperialism. *Monthly Review* December 20. http://mrzine.monthlyreview.org/2010/patnaik201210.html, accessed March 26, 2015.

Patnaik, P. (2011). Capitalism and imperialism. *People's Democracy*, XXXV(25). http://mrzine.monthlyreview.org/2011/patnaik190611.html, accessed March 26, 2015.

Patnaik, P. (2014). The elusive recovery. *Socialist Project: The Bullet* 925(Jan. 8). http://www.socialistproject.ca/bullet/925.php, accessed March 26, 2015.

Phillips, R. (2011). Vernacular anti-imperialism. *Annals of the Association of American Geographers* 101(5): 1109–1125.

Piketty, T. (2013). *Le capital au XXI siècle.* Paris: Le Seuil.

Project for a New American Century (2000). *Rebuilding America's Defenses: Strategy, Forces and Resources for a New Century.* http://www.informationclearinghouse.info/pdf/RebuildingAmericasDefenses.pdf, accessed March 26, 2015.

Reynolds, D. (2002). American globalism: Mass, motion and the multiplier effect. In Hopkins, A.G. (ed.), *Globalization in World History*, London: Pimlico, 243–260.

Robic, M.-C. (1994). National identity in Vidal's Tableau de Géographie de la France. In Hooson, D. (ed.), *Geography and National Identity*, Oxford: Blackwell, 58–70.

Said, E. (1978). *Orientalism.* New York: Vintage Books.

Said, E. (1993). *Culture and Imperialism.* New York: Vintage Books.

Sharp, J.P. (2013). Geopolitics at the margins? Reconsidering genealogies of critical geopolitics. *Political Geography*, 37: 20–29.

Shaw, I.G.R. (2013). Predator empire: The geopolitics of US drone warfare. *Geopolitics* 8(3): 536–559.

Smith, N. (2003a). *American Empire: Roosevelt's Geographer and the Prelude to Globalization.* Berkeley, CA: California University Press.

Smith, N. (2003b). After the American *Lebensraum*: "Empire", Empire, and Globalization. *Interventions: International Journal of Postcolonial Studies*, 5(2): 249–270.

Smith, N. (2009). The imperial present: Liberalism has always been conservative. *Geopolitics* 13(4): 736–739.

Spykman, N. (1944). *The Geography of the Peace.* New York: Harcourt Brace.

Strada, V. (2014). *Europe: La Russia come frontiera.* Venezia: Marsilio.

Teggart, F.J. (1919). Geography as an aid to statecraft: An appreciation of Mackinder's "Democratic Ideals and Reality." *Geographical Review* 8: 4–5.

Van Valkenburg, S. (1939). *Elements of Political Geography.* New York: Prentice-Hall.

Wainwright, J. (2004). American empire: A review essay. *Environment and Planning D* 22(5): 465–468.

Zielonka, J. (2006). *Europe as Empire: The Nature of the Enlarged European Union.* Oxford: Oxford University Press.

Zielonka, J. (2012). Empires and the modern international system. *Geopolitics* 17(3): 502–525.

Chapter 25

Social Movements

Sara Koopman

York University, Toronto, Ontario, Canada

Political geography has traditionally focused on states and formal politics in and across them. It was not until recently that political geography textbooks even mentioned social movements, and then usually very briefly, in an add-spice-and-stir sort of way (e.g., Agnew 2002; Cox 2002); although Painter (1995) is a notable exception (now in a second edition as Painter & Jeffrey 2009). However, in the last 10 years there has been a broadening of what counts as "doing politics," such that it includes social movements, as part of a reshaping of the discipline and reworking of understandings of the relationship between space and power, the political and the spatial, as argued in recent reviews of the field (Cox & Robinson 2007; Dodds, Kuus, & Sharp 2013), and in the first edition of this *Companion* (Agnew, Mitchell, & Toal 2003).

These changes are in part because feminists have for years pushed for the integration of "small *p*" politics and the everyday, and for a recognition of how public and private spheres are intertwined and processes happen across, and through, both the global and the intimate (Rosner & Pratt 2006; Sharp 2007). Some social movements focus more on the small *p*, some on influencing large *P* politics, and many on both, at times using one or the other. Many think of their work as social or economic more than political. Likewise, some geographers researching movements identify not as political geographers, but more as social, economic, urban, development, or feminist geographers. I include their work here as part of the shift in the field to a broader understanding of what is political.

Much of the work that geographers have done on social movements has actually used categories other than movement, such as contentious politics, collective action, connective action, or network. In this chapter I begin by considering these issues of definition. I then review some of the major contributions that geographers have made to understanding the work of social movements, particularly how various spatialities shape and are shaped by movements, and the spatialities of solidarity. I briefly talk about territoriality as one spatiality of movements that is emphasized by Latin American geographers. I then discuss relationships between geographers and social movements, and point to some of the collaborative ways in

The Wiley Blackwell Companion to Political Geography, First Edition.
Edited by John Agnew, Virginie Mamadouh, Anna J. Secor, and Joanne Sharp.
© 2015 John Wiley & Sons Ltd. Published 2017 by John Wiley & Sons Ltd.

which geographers have gone about thinking with and for movements. Ultimately, I argue that geographers have made and can make important contributions to building the power of movements to make a better world.

What is a movement, or who moves what how?

Social movement is not a clearly defined or well-agreed category for collective action and organizing for social change, and that is true in geography too (Pile & Keith 1997; Routledge & Cumbers 2009; McCarthy 2011; Chatterton, Featherstone, & Routledge 2013). What counts as a social movement, how to categorize such movements, and even what they should be called have been contested, both by academics and by activists – sometimes for theoretical or methodological reasons, sometimes for political or ethical ones. Issues have included: How many groups does it take to count as a movement? Do they have to be connected and/or coordinated? Do they have to be effective at creating social change to count? Can groups have paid staff? Can they be involved in electoral campaigns? Can they take up arms? Can they engage in property destruction? What if groups have no office, or no formal structure?

In the last 15 years it has become more and more possible for groups to self-organize, informally and transnationally. Information and communication technologies (ICTs) have played a key role in this new era of looser linked organizing, with much broader frames (like "we are the 99%") that can encompass many different issues and identities (Bennett 2005; see also Chapter 29 in this volume on social media). ICTs help make this sort of organizing possible, but it is also born out of generational and societal shifts, and the breakdown in all sorts of group membership and institutional loyalties in late modernity.[1]

Bennett and Segerberg (2012) call this new wave "connective action," because the quick resonance and spread of Occupy, for example, happened through a back and forth between online and offline activisms that feed and shape each other. The changing online tools available, such as the recent ubiquity of both social media and smartphones, also change what is possible on the streets, and in turn the actions there continue to alter how the online tools are used. This is quite different than what used to be one-way communication from an organization to its supporters, and it is much more than the internal listservs widely used for organizing in the 1990s and 2000s, or even Indymedia aggregation sites. This is about more than merely the low-input clicktivism of signing a petition or sending a form letter. This online communication goes in many directions, in personalized ways, through action sites where people post pictures, videos, stories, and analysis, comment on posts, share posters for wheatpasting, and add calendar events, say to an interactive map, or do other participatory geovisualization, like open street mapping of Occupy sites. It is about activities off the main action site, such as posting clickthrough action links and widgets, but also, importantly, radicalizing social media with personalized tweets, Facebook and Pinterest posts, perhaps geolocated, and using hashtags. Yet again, the change is not only in virtual space, but is entangled with physical space in ways that Gregory (2013) argues depend on, and create, a particular performance of space.

New ICT tools are yet again changing the look and feel of organizing and what it means to "do" politics. As burgeoning smartphone use makes the internet available to many more people in the global South, these dynamics are growing fast. However, widespread autonomous horizontal organizing, in many ways similar to the recent wave in the global North, also has a longer history in Latin America, on the other side of the digital divide. Notable here are the El Alto uprising in Bolivia in 2005 and the "Que se vayan todos" (throw them all out) uprising in Argentina in 2001. Nevertheless, these were not entirely without ICTs of a sort, as pirate and community radio stations have played an important role in organizing throughout Latin America.

Geographers have not written much on the current dynamics of social movements and ICTs (although see Mamadouh 2004; Meek 2012; and good work is being done by communications scholars, e.g., Costanza-Chock 2012). These dynamics have been changing quickly since Pickerill et al. conducted an in-depth study of how the anti-war movement was shaped by ICTs from 2005–07 (Gillan, Pickerill, & Webster 2008). In general there has been less work by geographers as yet on this new form of organizing (although see Halvorsen 2012; Pickerill & Krinsky 2012; Uitermark & Nicholls 2012; Sparke 2013; Zemni, De Smet, & Bogaert 2013; Adams this volume).

Older forms of collective action still coexist with connective action – and often we see hybrid actions where traditional brick-and-mortar membership organizations also participate in these loose, multi-issue networks, say around particular events or days of action (Bennett & Segerberg 2012). Yet, most of the recent wave of protests (particularly the Indignados) rejected the participation of formally organized groups as such, although some accepted material support.[2] Some activists bemoan professionalization, and say that once a group has paid staff and non-profit status it has bureaucratized and is no longer in a movement. This is a double-edged sword, as Katz details (2005), but some non-governmental organizations (NGOs) do function as part of, rely on, and even seek to build a larger movement, even when their grassroots grow into large sequoias, as Brown puts it (2007). Certainly, there are some that are primarily handmaidens to neoliberal states, but NGOs should not be written out of the social movement story (Townsend, Porter, & Mawdsley 2004).[3] Geographical thinking about social movements could be strengthened by more conversations across these literatures.

The spatialities of social movements

Whether they use the term movement, network, NGO, contentious, collective, or connective action, what geographers have brought to the understanding of the work of movements is a focus on space. Social movement studies had until recently tended to write about movements as if they happened "on the head of a pin" (Nicholls 2007). Geographers have asked: How do movements use space? How does space shape movements? How do movements shape space? We have used different spatial concepts to get at this: place, networks, and scale in particular, although also sometimes territory, city, and region. Nicholls, Miller, and Beaumont (2013) argue that from the 1930s to 1950s it tended to be region; 1960–1970s space; 1980s place; 1990s–2000s scale; and now networks and mobility. They offer a helpful review of how each has been taken up for movement studies.

All of these concepts exist in relation both to movements and to each other. This argument was first made by Leitner et al. (2008), who have been widely cited for arguing that too often a favorite "master spatiality," as they call it, has been plopped down on a movement rather than recognizing that these spatialities are "co-implicated." Scale and territory, for example, not only shape each other, but also, as Neil Smith says, "the scale of struggle and the struggle over scale are two sides of the same coin" (Smith 1992a: 74, cited in Martin 2013: 332). Miller argues in the conclusion of the volume *Spaces of Contention: Spatialities and Social Movements* (Nicholls, Miller, & Beaumont 2013), the most recent and useful collection of geographies of social movements, that this is also true for other spatialities of social movements. That is, movements work to shape these spatialities, and they also shape movements. Although this will always be contextual, Miller offers as a guide a complex grid that, in relation to movements, grafts Jessop et al.'s framework for the relationship between territory, place, scale, and network (TPSN) on to the Lefebvrian triad of material space, conceptual space, and lived space, and points to the different socio-spatial power relations created in each (Miller 2013: 334).

A more fluid and process-oriented take on these relations, and perhaps the wave of the 2010s, is the assemblage approach, which McFarlane (2009) argues for using over networks to understand the spatiality of movements. Rutland, for example, employs actor network theory (ANT) to great effect in detailing how the movement for a new Portland energy policy was made possible and constituted through a concatenation of human and non-human elements associated across time and space. Nevertheless, he warns that Latour's approach ignores processes of *disas-sociation* and exclusion, and attends to minute details at the cost of obscuring the broad systemic conditions, like capitalism and patriarchy, under which associations are formed (Rutland 2013: 255). The assemblage approach is not always strictly ANT, however. Davies (2012) uses the practices of the Tibet Support Group to make a strong argument for using the concept of assemblage to convey how movements use both territorial and relational spatialities.

Urban movements like the one in Portland are more likely to be discussed in urban than in political geography journals. Likewise, geographies of social movements of all sorts are being done in development, feminist, cultural, and economic geography, of which labor geographies is often considered its own subfield (Knutsen, Bergene, & Endresen 2012). Castree, Featherstone, and Herod's review (2007) compares labor geographies to geographies of trans-national organizing, and argues that the former tends to emphasize scale, as opposed to the network focus of the latter. Still, community unionism is all about networks, and Wills in particular has made important contributions to thinking about these (Wills et al. 2009). Again, there is no one dominant, or right, spatiality for understanding movements, nor is there one right subfield for geographers interested in movements.

Within political geography, some work has even been done within the subfield of critical geopolitics, looking at how social movements try to influence, and themselves engage in, geo-politics (Routledge 2003; Koopman 2011). Critical geopolitics has been associated with textual analysis, but from its early conception Dalby argued for it to engage critically with social movements as a way to reconceptualize security and push for peace (1991) and over the years he has taken seriously the discourses of movements (1993, 1999; Mackenzie & Dalby 2003). More recently, feminist geopolitics has pushed the field to more engagement with ten-sions across the material and the discursive, and this has led to work that looks at both the discourses and actions of movements, such as that by Casolo and Doshi (2013), Williams and Boyce (2013), and the in-depth work of Loyd (2014).

Work by geographers to understand the spatialities of social movements has been so broad and varied that it cannot be summed up in any one major contribution. What is clear is that space is a key part of how social movements create change, whether the best way to think of that in the case of a given movement is some combination of scale, territory, network, place, or assemblage.

Spatialities of solidarity

Through their work on, with, and across social movements, geographers have also contrib-uted to broader debates about the spatialities of solidarity. Much of political geography is about the relationship between here and there, us and them. Some of what political geographers have offered, as I will review here, is a perspective on what can shift and bring those categories together, and how movements build connections across distance and difference. This has ranged from analyses of commonalities, to how difference is reworked, to the emotional work involved.

Neil Smith's (1992b) argument that movements can be strengthened when they "jump scale" was highly influential. However, it is not always obvious how to expand resistance in this way. Castree et al. (2007) claim that the most useful contribution geographers can make to movements is to "draw lessons" as to "the most effective basis for otherwise place-based constituencies to unite"; that is, common identities, enemies, or goals. Routledge (2007) wants geographers to draw comparisons and account for why social movements emerge *where* they do. He has long written about place as being able to tell us why movements happen where they do, and how, and about the importance of looking at how movements can take the uneven ways in which state and economic power intersect with people's everyday lives in a place and use them to create "terrains of resistance" – both literally (barricades) and meta-phorically (imaginaries; Routledge 1994). Yet, grand overarching patterns about what sort of places inspire what sort of movements, or what always works best for building solidarity across them, seem difficult to find.

However, geographers are well "placed" to do analyses of who is organizing where, how, and for, or against, what – and how different whos are organizing against similar whats, or how organizing here is or is not like organizing there, and how each is facing similar or related dynamics of not only capitalism, but other systems of oppression. In so doing, geographers have drawn what Katz (2001) has called counter-topographical lines that could help move-ments learn from and connect with each other. Some topics seem particularly ripe for geogra-phers, such as resistance to road expansion (Perz 2012) or land grabs (Baletti, Johnson, & Wolford 2008). Miller (2000) could be seen as drawing short lines between four places in metropolitan Boston with different demographics and ties to the defense industry that shaped their different anti-nuclear organizing. Many of these lines have been short, in the sense that they were drawn within parts of a broader movement, for example alternative currencies (North 2007) or Occupy sites (Uitermark & Nicholls 2012), or between fairly similar move-ments, such as racialized radical activists in Los Angeles (Pulido 2005). However, other lines are much longer. Hart (2006), for example, looks at how struggles around land dispossession in South Africa and East Asia (particularly Taiwan) are constituted in relation to each other. Mackenzie and Dalby (2003) connect two struggles, in Scotland and Canada, around siting superquarries.

It is those surprising connections, à la Katz's lines from New York City to Sudan, that are a particularly useful contribution that geographers can offer to movements. Dalby warns that it is common wishful thinking to assume compatibility between diverse resistance in different places (1999), but that is precisely why it is worth doing careful work to draw specific con-nections. Casolo and Doshi draw such lines between two gendered movements in rural Guatemala and urban India, from which they gain insights into how practices of social reproduction can be key to both geoeconomic processes of accumulation by dispossession and to their resistance (2013: 827).

Routledge (2003) has written about ongoing transnational coalitions like People's Global Action in the late 1990s as "convergence spaces" where, as he puts it, movements find common ground and converge some goals and tactics, but do not act as a unified whole. He writes about this as if these convergences were only between place-based groups, but many broader and issue-based groups use this term themselves. The 2014 US Social Forum was organized at multiple "sites of convergence" (that is, locations) so as to support movement building more effectively, because it was "seeking to move beyond convergence towards greater cohesion in our work for social transformation."[4] Geographers have and can contribute to this coming together into a whole that honors its parts.

Massey (2008) suggests that movements that are working to build solidarity across distance and inequality do so by stressing the ways in which those places are connected, either through similarities between the places, or their political and economic relations, or a common enemy such as the World Trade Organization. Brown and Yaffe (2014) argue that these are often used in combination, and that solidarity can and often does flow in more than one direction, not only back and forth but laterally as well (as, in their example, between groups in various countries working against South African apartheid). As they phrase it, political solidarity campaigns try to change "the connections and flows between places, refusing to participate in the reproduction of inequalities and oppression" (Brown & Yaffe 2014: 39). Unfortunately, even as movements work against uneven power relations, it is easy to reproduce these unintentionally. Then again, as Koopman (2008) shows for Colombia, those with more privilege (particularly geopolitical and racial privilege) can leverage that position to work in solidarity with those with less, without necessarily reinforcing the structures that create those inequalities.

Featherstone writes about solidarity as a "transformative relation," one that is "forged through political struggle which seeks to challenge forms of oppression" and works to create "new ways of relating" (2012: 5). He writes about the various practices of solidarity used by movements across time and place to affect and change uneven power relations between here and there, and the many ways in which imaginaries of us and them have been reworked. Rather than focusing on likeness, or the pre-existing connections that solidarity can activate to which Routledge turns, Featherstone emphasizes the new relationships that are in constant creation through solidarity practices. Brown and Yaffe (2014) extend his arguments by emphasizing that the way in which solidarity is framed cannot be separated from how it is performed, and that micropractices at key sites matter for what sort of alternatives are created. They give moving examples from the four-year non-stop picket in front of the South African embassy in London in the late 1980s. They present daily life on the picket as an assemblage of practices through which solidarity was performed. These included learning and singing South African songs banned in that country loudly enough to disrupt embassy business, as well as singing to cohere the group in the face of police pressure. In so doing, those involved both disrupted and reconfigured spatial relationships. Brown also writes about how these acts of standing in solidarity released powerful emotions (Brown 2013).

Geographers have contributed insights into how the geographies of emotion shape and are shaped by social movements. That singing on the picket, for example, helped overcome a sense of isolation, and made it enjoyable to be "unruly" in a public space (Brown 2013). In an influential piece, Brown and Pickerill (2009) argue that examining the spaces of activism is a useful way to get at the relationship between emotions and activism, and in particular to improve emotional reflexivity in movements and make activism more sustainable.[5] Bosco (2007) has argued that the way in which social movements grow and maintain their networks is shaped by the strategic mobilization of emotions, and by activists' emotional labour to create shared feelings of outrage, love, and solidarity, even across distance.

Solidarity, geographers have argued, is more than sameness. Indeed, solidarity builds connections across difference and distance. The practices, emotions, and imaginaries involved in those connections work to change uneven relationships between here and there, us and them.

Hopefully, geographers will continue to offer analyses to strengthen those ties, build new ones, and be part of the process of building coherence between movements.

The territoriality of movements

To this effect, geographers need to build more connections within our discipline, not only between subfields, but also across linguistic and national borders. Thinking globally is not only about doing fieldwork elsewhere, but also about engaging with other literatures. The ability to do this is limited by linguistic abilities. Mine being limited to Spanish and Portuguese, I briefly review some of the work published in those languages here, in particular the thinking about movements and territory.

There is much work being carried out in Latin America on space and social movements, but it is mostly being done outside of geography, although Bernardo Mançano (2012) and Carlos Gonçalves (2009) are notable exceptions. Both are Brazilian, and this may reflect how much stronger geography is as a discipline in Brazil than in the rest of Latin America. Alternatively, it may be a reflection of how much of political geography in most of Latin America is still based in military universities, as reflected by the 2011 IGU conference, hosted by the Chilean military academy, an actual torture site in the 1970s and where Pinochet himself taught geography (Hirt & Palomino-Schalscha 2011).

Mançano (2000) makes a distinction between what he calls sociospatial movements and socioterritorial movements, the latter being those who use territory as a "*trunfo*" (trump card), like the Movimento dos Trabalhadores Rurais Sem Terra (MST), the landless workers' movement. Such movements, he argues, have to continue "territorializing" to be successful; that is, both organizing in and making their territory their own, and expanding to new territory. Gonçalves (2001) writes about movements establishing "new territorialities" through a process of symbolic material appropriation that he describes as "making the earth" or geo-graphing. His own work is with the *seringueiros*, or rubber tappers, and he is particularly interested in movements' new ways of understanding and enacting territoriality as a way of building sustainability.

The most widely recognized writer in Spanish (and most translated into English) on the spatial dynamics of social movements is not a geographer but a journalist, Raúl Zibechi (2010, 2012), although (or perhaps because) he often slides into romanticizing resistance (Sparke 2008). Circling back to the issue of the different terms used for movements, Zibechi argues that they are so different in Latin America today that they should be called something else, and uses the term "societies in movement." As Mançano and Gonçalves do, Zibechi highlights how autonomous Latin American movements are "territorialized," by which he means rooted physically, culturally, emotionally, and even spiritually in spaces that they have either recuperated or retained through struggle. This is a non-state take on territory that goes beyond the indigenous and Afro-Colombian territories legally recognized by the state (Agnew & Oslender 2013), or even those of Colombian small farmers resisting state displacement in peace communities (Koopman 2011), or plantation takeovers by the landless. Zibechi includes factory takeovers by workers, and invasions that become marginalized urban communities ("slums").

There is a geopolitics to whose knowledge counts and, as Sharp (2013) has powerfully argued, language issues are part of it. We can move toward decolonizing our discipline by taking seriously and translating more of the voices and ideas of thinkers in the global South, such as the forthcoming issue of *Antipode*, "What Can a Territory Do? Societies in Movement and New Territorialities in Contemporary Latin America," co-edited by Alvaro Reyes and Brenda Baletti.

How geographers move with movements

Geographers working on social movements often decide to do so because we feel strongly about the issues on which the movement works, as Askins has compellingly described (2009). Sometimes this means that we work *with* movements, but of course there are also all sorts of reasons we might *not* want to collaborate with the movement we are researching (for more on this, see Gillian & Pickerill 2012). Sometimes this work is actually done to support a counter movement. For example, Williams and Boyce (2013) work to understand the vigilante anti-immigrant movement in the US–Mexico borderlands as a way to support immigrant rights.

It is becoming increasingly common for geographers to collaborate with movements that they support, and of which they are even a part, as engaged scholars or activist academics, committed to contributing to the strength of the movement. Sometimes this means analyzing the processes that movements are opposing, or are working to nurture. At other times geographers help movements strengthen their conjunctural analysis. Often the research is about the movement itself and how it organizes.

The process of doing the research can also be a contribution to the movement. Geographers' discussions with activists about how they understand and use various spatial strategies and tactics for social change and the socio-spatial relations that they create (Martin 2013) may themselves offer insights to activists that change their strategies. Research can also help movements be more accountable for what they do and how they do it by fostering conversations about internal dynamics (Koopman 2008). Movements operate in a world full of systems of oppression, so of course forms of domination show up inside resistance (Sharp et al. 1999).

For this work to be more able to contribute to the work of movements themselves, Castree et al. (2007) point to the importance of writing accessibly, and in venues beyond books and journals.[6] Many geographers do regularly present their research to movements, both in person and in movement newsletters, blogs, listservs, and so on. This can be done as a way to keep the conversation going, more than as simply one-way dissemination. There are now many online ways to do this and be not only more accessible, but also more *accountable* to movements (Fuller & Askins 2010).

Many geographers are having these sorts of conversations, and learning from, thinking with, and acting with activists. There are all kinds of creative ways to do this, from theater to discussions that also serve as blockades (Mason & Askins 2012). For more on ways to think, and act, together, see Kindon, Pain, and Kesby (2007). As guiding lights, three geographers who have done ongoing collaborative research and writing with grassroots groups are Geraldine Pratt, Linda Peake, and Richa Nagar. Pratt (2002) has collaborated for over 20 years with the Philippine Women Centre of BC on issues involved in temporary labor migration, including family separation. Lately they have been using theater, in both Canada and the Philippines, as a way to share this research and stimulate community debate. Peake has collaborated for three decades with the Guyanese women's organization Red Thread on research into issues such as the impact of structural adjustment, domestic violence, and reproductive health. She has written about transnational activist academic praxis in a recent collection edited by Nagar (Peake & de Souza 2010). Nagar herself writes with a collective of seven Indian grassroots activists about how NGOs both foster and stifle solidarity (Nagar & Ali 2003).

Unfortunately, this sort of collaboration, and open discussion in and among movements, is sometimes hampered by the potential that such research conversations could be used against a movement, to repress it. In an era of high military surveillance, exemplified by Edward Snowden's revelations about the National Security Agency (NSA), and now that the US military has expressly shifted its focus from anthropology to geography as a tool for controlling the so-called human terrain (Medina 2014), it is quite possible that research by geographers could also be weaponized; that is, actually used to harm or even kill activists or those with whom they work (Koopman forthcoming). This is of course particularly true in conflict zones, like Colombia, but given that police forces across the United States are increasingly militarized and respond more violently to mobilizations, as seen in Ferguson, Missouri in 2014, it is all the more important for geographers working with US movements to be cautious about how their work might be violently misused.

Geographers have collaborated with social movements in all sorts of ways, from going undercover to research the target of a movement's campaign (Routledge 2002), to investigating the counter movement, to fostering conversations inside a movement about how systems of domination are shaping members' resistance. The challenge now is for us to support each other in being creative about how to do this work in ways that are safe and ethical in the current context of high surveillance.

Conclusion

There are various ways in which geographers contribute to the power of social movements. More and more of them identify as scholar activists, but not all are holding collaborative research conversations in the trenches. Some draw counter-topographical lines from afar, others analyze the spatialities at work by analyzing groups' online discourse. Some geographers focus on the scale of movements, others on how they function as networks, and recently the assemblage approach has become increasingly popular. Geographers have looked not only at how these spatialities shape movements, but also at how the spatialities themselves are shaped by the movements.

Geographers have analyzed the spatialities of solidarity in various ways. Rather than solidarity around sameness, this work has tended to focus on connections across difference and distance and how movements can change the relationships between here and there, us and them. Geographers have contributed to growing connections between movements by drawing counter-topographical lines between not only capitalist forms of violence, but also around how movements resist and create alternatives; that is, between the pieces of a greater peace with justice that is being built around the world. Geographers in Latin America have particularly focused on the territoriality of movements, by which they mean not only physical rootedness, but also cultural, emotional, and spiritual connections to territory gained through struggle.

Some geographers use the term movement, others prefer network, grassroots groups, NGO, contentious, collective, or connective politics. Some consider themselves political geographers; many do not. I have argued here for more solidarity in geography and for making more connections: across geographies that use different terms for social movement, across subdisciplines of geography, across languages of research, and, most of all, with and across various movements. I particularly hope that geographers will pay increasing attention to the collaborative thinking being done through and with the action of movements around the world. Let us honor all of these ways of doing this research and continue to contribute to these movements in our various ways, so that together we can build a better world for us all.

Notes

1 New ICTs are changing not only movements but also how we collaborate as academics, and I invite you to add to the bibliography at www.zotero.org/groups/geographies_of_social_movements.
2 For example, the port-a-potties at Vancouver Occupy were provided by a labor union.
3 For a useful overview of critical geopolitics work on NGOs, see Jeffrey (2013).
4 http://www.ussocialforum.net/node/508, accessed March 26, 2015.
5 That paper is part of a special issue that Brown and Pickerill edited for *Emotion, Space and Society* on emotions and movements; 2(1), 2009. For a fuller review of literature in and around the geography on emotion and protest, see Wright (2010).
6 Activists certainly can and do read academic articles as well, though this is hampered by paywalls and they have to rely on pre-proof versions reposted on academics' own sites and repositories like academia.edu.

References

Agnew, J.A. (2002). *Making Political Geography*. Plymouth: Rowman and Littlefield.

Agnew, J., & Oslender, U. (2013). Overlapping territorialities, sovereignty in dispute: Empirical lessons from Latin America. In Beaumont, J., Miller, B., & Nicholls, W. (eds), *Spaces of Contention: Spatialities and Social Movements*, Aldershot: Ashgate, 121–140

Agnew, J., Mitchell, K., & Toal, G. (eds) (2003). *A Companion to Political Geography*. Malden, MA: Blackwell.

Askins, K. (2009). "That's just what I do": Placing emotion in academic activism. *Emotion, Space and Society* 2(1): 4–13.

Baletti, B., Johnson, T.M., & Wolford, W. (2008). "Late mobilization": Transnational peasant networks and grassroots organizing in Brazil and South Africa. *Journal of Agrarian Change* 8(2–3): 290–314.

Bennett, W.L. (2005). Social movements beyond borders: Understanding two eras of transnational activism. In della Porta, D., & Tarrow, S. (eds), *Transnational Protest and Global Activism*, New York: Rowman and Littlefield, 203–226.

Bennett, W.L., & Segerberg, A. (2012). The logic of connective action. *Information, Communication and Society* 15(5): 739–768.

Bosco, F.J. (2007). Emotions that build networks: Geographies of human rights movements in Argentina and beyond. *Tijdschrift voor Economische en Sociale Geografie* 98(5): 545–563.

Brown, G. (2013). Unruly bodies (standing against apartheid). In Cameron, A., Dickinson, J., & Smith, N. (eds), *Body/State*, Farnham: Ashgate, 145–156.

Brown, G., & Pickerill, J. (2009). Space for emotion in the spaces of activism. *Emotion, Space and Society* 2(1): 24–35.

Brown, G., & Yaffe, H. (2014). Practices of solidarity: Opposing apartheid in the centre of London. *Antipode* 46(1): 34–52.

Brown, M. (2007). Working political geography through social movement theory: The case of gay and lesbian Seattle. In Cox, K.R., Low, M., & Robinson, J. (eds), *The SAGE Handbook of Political Geography*, London: Sage, 353–377.

Casolo, J., & Doshi, S. (2013). Domesticated dispossessions? Towards a transnational feminist geopolitics of development. *Geopolitics* 18(4): 800–834.

Castree, N., Featherstone, D., & Herod, A. (2007). Contrapuntal geographies: The politics of organizing across sociospatial difference. In Cox, K.R., Low, M., & Robinson, J. (eds), *The Sage Handbook of Political Geography*, London: Sage, 305–321.

Chatterton, P., Featherstone, D., & Routledge, P. (2013). Articulating climate justice in Copenhagen: Antagonism, the commons, and solidarity. *Antipode* 45(3): 602–620.

Costanza-Chock, S. (2012). Mic check! Media cultures and the Occupy movement. *Social Movement Studies* 11(3–4): 375–385.

Cox, K. (2002). *Political Geography: Territory, State and Society.* Chichester: John Wiley & Sons.

Cox, K.R., Low, M., & Robinson, J. (eds) (2007). *SAGE Handbook of Political Geography.* London: Sage.

Dalby, S. (1991). Dealignment discourse: Thinking beyond the blocs. *Current Research in Peace and Violence* 13(3): 140–155.

Dalby, S. (1993). The "Kiwi disease": Geopolitical discourse in Aotearoa/New Zealand and the South Pacific. *Political Geography* 12(5): 437–456.

Dalby, S. (1999). Against "globalization from above": Critical geopolitics and the World Order Models project. *Environment and Planning D: Society and Space* 17(2): 181–200.

Davies, A.D. (2012). Assemblage and social movements: Tibet support groups and the spatialities of political organisation. *Transactions of the Institute of British Geographers* 37(2): 273–286.

Dodds, K., Kuus, M., and Sharp, J. (eds) (2013). *The Ashgate Research Companion to Critical Geopolitics.* Aldershot: Ashgate.

Featherstone, D. (2012). *Solidarity: Hidden Histories and Geographies of Internationalism.* London: Zed Books.

Fuller, D., & Askins, K. (2010). Public geographies II: Being organic. *Progress in Human Geography* 34(5): 654–667.

Gillan, K., & Pickerill, J. (2012). The difficult and hopeful Eehics of research on, and with, social movements. *Social Movement Studies* 11(2): 133–143.

Gillan, K., Pickerill, J., & Webster, F. (2008). *Anti-War Activism: New Media and Protest in the Information Age.* Basingstoke: Palgrave Macmillan.

Gonçalves, C.W.P. (2001). *Geo-grafías: Movimientos sociales, nuevas territorialidades y sustentabilidad.* Mexico: Siglo XXI.

Goncalves, C.W.P. (2009). Geo-grafías-movimientos sociales, nuevas territorialidades y sustentabilidad. *GEOgraphia* 5(9). http://www.uff.br/geographia/ojs/index.php/geographia/article/viewArticle/123, accessed March 26, 2015.

Gregory, D. (2013). Tahrir: Politics, publics and performances of space. *Middle East Critique* 22(3): 235–246.

Halvorsen, S. (2012). Beyond the network? Occupy London and the global movement. *Social Movement Studies* 11(3–4): 427–433.

Hart, G. (2006). Denaturalizing dispossession: Critical ethnography in the age of resurgent imperialism. *Antipode* 38(5): 977–1004.

Hirt, I., & Palomino-Schalscha, M. (2011). Geography, the military and critique on the occasion of the 2011 IGU regional meeting in Santiago de Chile. *Political Geography* 30(7): 355–357.

Jeffrey, A. (2013). Non-governmental organizations. In Dodds, K., Kuus, M., & Sharp, J.P. (eds), *The Ashgate Research Companion to Critical Geopolitics*, Aldershot: Ashgate, 387–404.

Katz, C. (2001). Vagabond capitalism and the necessity of social reproduction. *Antipode* 33(4): 709–728.

Katz, C. (2005). Partners in crime? Neoliberalism and the production of new political subjectivities. *Antipode* 37(3): 623–631.

Kindon, S.L., Pain, R., & Kesby, M. (eds) (2007). *Participatory Action Research Approaches and Methods: Connecting People, Participation and Place.* New York: Psychology Press.

Knutsen, I.I.M., Bergene, A.C., & Endresen, S.B. (2012). *Missing Links in Labour Geography.* Aldershot: Ashgate.

Koopman, S. (forthcoming). Beware, your research may be weaponized. *Annals of the Association of American Geographers.*

Koopman, S. (2008). Imperialism within: Can the master's tools bring down empire. *ACME: An International E-Journal for Critical Geographies* 7: 283–307.

Koopman, S. (2011). Alter-geopolitics: Other securities are happening. *Geoforum* 42(3): 274–284.

Leitner, H., Sheppard, E., & Sziarto, K. (2008). The spatialities of contentious politics. *Transactions of the Institute of British Geographers* 33(2): 157–172.

Loyd, J.M. (2014). *Health Rights Are Civil Rights: Peace and Justice Activism in Los Angeles, 1963–1978.* Minneapolis, MN: University of Minnesota Press.

Mackenzie, A.F.D., & Dalby, S. (2003). Moving mountains: Community and resistance in the Isle of Harris, Scotland, and Cape Breton, Canada. *Antipode* 35(2): 309–333.

Mamadouh, V. (2004). Internet, scale and the global grassroots: Geographies of the Indymedia network of independent media centres. *Tijdschrift voor Economische en Sociale Geografie* 95(5): 482–497. doi:10.1111/j.0040-747X.2004.00334.x

Mançano Fernandes, B. (2000). Movimento social como categoria geográfica. *Terra Livre* 15: 59–85.

Mançano Fernandes, B. (2012). Movimentos socioterritoriais e movimentos socioespaciais: Contribuição teórica para uma leitura geográfica dos movimentos sociais. *Revista Nera* 6: 24–34.

Martin, D.G. (2013). Place frames: Analyzing practice and production of place in contentious politics. In Nicholls, W., Beaumont, J., & Miller, B. (eds), *Spaces of Contention: Spatialities and Social Movements,* Farnham: Ashgate, 85–102.

Mason, K., & Askins, K. (2012). "Us and us": Faslane 30 and academic direct action. *Medicine, Conflict and Survival* 28(4): 282–288. doi:10.1080/13623699.2012.739045

Massey, D. (2008). Geographies of solidarities. In Clark, N., Massey, D., & Sarre, P. (eds), *Material Geographies: A World in the Making,* London: Sage, 311–362.McCarthy, J. (2011). Social movements. In Gregory, D., Johnston, J., Pratt, G., Watts, M., & Whatmore, S. (eds), *The Dictionary of Human Geography,* Malden, MA: John Wiley & Sons.McFarlane, C. (2009). Translocal assemblages: Space, power and social movements. *Geoforum* 40(4): 561–567.

Medina, R.M. (2014). From anthropology to human geography: Human terrain and the evolution of operational sociocultural understanding. *Intelligence and National Security, online before print.* doi:10.1080/02684527.2014.945348

Meek, D. (2012). YouTube and social movements: A phenomenological analysis of participation, events and cyberplace. *Antipode* 44(4): 1429–1448. doi:10.1111/j.1467-8330.2011.00942.x

Miller, B.A. (2000). *Geography and Social Movements: Comparing Antinuclear Activism in the Boston Area.* Minneapolis, MN: University of Minnesota Press.

Miller, B. (2013). Spatialities of mobilization: Building and breaking relationships. In Nicholls, W., Miller, B., & Beaumont, J. (eds), *Spaces of Contention: Spatialities and Social Movements,* Aldershot: Ashgate, 285–298.

Nagar, R., & Ali, F. (2003). Collaboration across borders: Moving beyond positionality. *Singapore Journal of Tropical Geography* 24(3): 356–372.

Nicholls, W.J. (2007). The geographies of social movements. *Geography Compass* 1(3): 607–622.

Nicholls, W., Miller, B., & Beaumont, J. (eds) (2013). *Spaces of Contention: Spatialities and Social Movements.* Aldershot: Ashgate.

North, P. (2007). *Money and Liberation: The Micropolitics of Alternative Currency Movements.* Minneapolis, MN: University of Minnesota Press.

Painter, J. (1995). *Politics, Geography and Political Geography: A Critical Perspective.* London: E. Arnold.

Painter, J., & Jeffrey, A. (2009). *Political Geography,* 2nd edn. London: Sage.

Peake, L., & de Souza, K. (2010). Feminist academic and activist praxis in service of the transnational. In Nagar, R., & Lock Swarr, A. (eds), *Critical Transnational Feminist Praxis,* Albany, NY: SUNY Press, 105–123.

Perz, S. (2012). Social mobilization in protest of trans-boundary highway projects: Explaining contrasting implementation outcomes. *Development and Change* 43(3): 797–821.

Pickerill, J., & Krinsky, J. (2012). Why does Occupy matter? *Social Movement Studies* 11(3–4): 279–287.

Pile, S., & Keith, M. (1997). Introduction. In Pile, S., & Keith, M. (eds), *Geographies of Resistance,* London: Routledge, 1–32.

Pratt, G. (2002). Collaborating across our differences. *Gender, Place and Culture* 9(2): 195–200.

Pulido, L. (2005). *Black, Brown, Yellow, and Left: Radical Activism in Los Angeles.* Berkeley, CA: University of California Press.

Rosner, V., & Pratt, G. (2006). *Women Studies Quarterly*, special issue "The global and the intimate." Spring/Summer.

Routledge, P. (1994). Backstreets, barricades, and blackouts: Urban terrains of resistance in Nepal. *Environment and Planning D: Society and Space* 12(5): 559–578.

Routledge, P. (2002). Travelling East as Walter Kurtz: Identity, performance, and collaboration in Goa, India. *Environment and Planning D* 20(4): 477–498.

Routledge, P. (2003). Anti-geopolitics. In Agnew, J., Mitchell, K., & Ó Tuathail, G. (eds), *A Companion to Political Geography*, Malden, MA: Blackwell, 236–248.

Routledge, P. (2007). Transnational political movements. In Cox, K.R., Low, M., & Robinson, J. (eds), *SAGE Handbook of Political Geography*, London: Sage, 335–349.

Routledge, P., & Cumbers, A. (2009). *Global Justice Networks: Geographies of Transnational Solidarity*. Manchester: Manchester University Press.

Rutland, T. (2013). Energizing environmental concern in Portland. In Nicholls, W., Miller, B., & Beaumont, J. (eds), *Spaces of Contention: Spatialities and Social Movements*, Aldershot: Ashgate, 250–271.

Sharp, J.P. (2007). Geography and gender: Finding feminist political geographies. *Progress in Human Geography* 31(3): 381–387. doi:10.1177/0309132507077091

Sharp, J.P. (2013). Geopolitics at the margins? Reconsidering genealogies of critical geopolitics. *Political Geography* 37: 20–29. http://www.sciencedirect.com/science/article/pii/S0962629813000449, accessed March 26, 2015.

Sharp, J.P., Paddeson, R., Philo, C., & Routledge, P. (1999). *Entanglements of Power: Geographies of Domination/Resistance*. London: Routledge.

Smith, N. (1992a). Geography, difference and the politics of scale. In Doherty, J., Graham, E., & Malek, M. (eds), *Postmodernism and the Social Sciences*, New York: St. Martin's Press, 57–79.

Smith, N. (1992b). Contours of a spatialized politics: Homeless vehicles and the production of geographical scale. *Social Text*, 33: 55–81.

Sparke, M. (2008). Political geography: Political geographies of globalization III: Resistance. *Progress in Human Geography* 32(3): 423–440.

Sparke, M. (2013). From global dispossession to local repossession: Towards a worldly cultural geography of Occupy activism. In Johnson, N., Schein, R., & Winders, J. (eds), *The Wiley-Blackwell Companion to Cultural Geography*, Malden, MA: Wiley-Blackwell, 385–408.

Townsend, J.G., Porter, G., & Mawdsley, E. (2004). Creating spaces of resistance: Development NGOs and their clients in Ghana, India and Mexico. *Antipode* 36(5): 871–889.

Uitermark, J., & Nicholls, W. (2012). How local networks shape a global movement: Comparing Occupy in Amsterdam and Los Angeles. *Social Movement Studies* 11(3–4): 295–301.

Williams, J., & Boyce, G.A. (2013). Fear, loathing and the everyday geopolitics of encounter in the Arizona borderlands. *Geopolitics* 18(4): 895–916.

Wills, J., Datta, K., Evans, Y., Herbert, J., May, J., & McIlwaine, C. (2009). Religion at work: The role of faith-based organizations in the London living wage campaign. *Cambridge Journal of Regions, Economy and Society* 2(3): 443–461.

Wright, M.W. (2010). Geography and gender: Feminism and a feeling of justice. *Progress in Human Geography* 34(6): 818–827.

Zemni, S., De Smet, B., & Bogaert, K. (2013). Luxemburg on Tahrir Square: Reading the Arab revolutions with Rosa Luxemburg's The Mass Strike. *Antipode* 45(4): 888–907.

Zibechi, R. (2010). *Dispersing Power: Social Movements as Anti-State Forces*. Oakland, CA: AK Press.

Zibechi, R. (2012). *Territories in Resistance: A Cartography of Latin American Social Movements*, trans. R. Ryan. Oakland, CA: AK Press.

Chapter 26

Religious Movements

Tristan Sturm

Queen's University Belfast, Northern Ireland, UK

Religion has been understudied in the twenty-first century despite its ability to mobilize and abet social movements (see Chapter 25 in this volume). The reason for this lacuna is in large part because of blind adherence to secularization theory, for which political geography has only recently begun to make up. It was not until after the Cold War that Western foreign policymakers were shaken from their secularized security slumber, especially by the events around September 11, 2001. Many reached two major geopolitical presuppositions: That religion just did not matter in other parts of the world, especially after the Iranian revolution and within the developing world (Marty 2005: 15); and, more precipitously damning, that the post–Cold War world might best be drawn through a "clash of [largely religion-based] civilizations" bent on war to shore up their speciously foundational differences that were seen as under attack by American cultural, economic, and political globalization (Huntington 1993; cf. Katzenstein 2010).

The secularized Cold War world was largely about competing capitalist and communist political-economic regimes. But it was about good and evil, between those with a God (the United States) and those without one (the USSR). The decade after the Cold War had various geopolitical theories attached to it, from multipolarity, to clash of civilizations, to perpetual peace. Absent from the literature was that this decade was also religious in tone. Following September 11, 2001, the world was reimagined and reterritorialized by scholars and cultural pundits, to be between those with a God (the Christian West) and those with the wrong God (the Islamic Middle East). This new geopolitical polarity was then transposed by scholars, policymakers, and cultural institutions into a new geopolitics of enmity, no longer based primarily on political-economic differences but, rather, reduced to cultural and religious differences (Debrix 2008). As Bialasiewicz (2006: 720) writes of popular cultural and governmental geopolitical renditions of the world, "international politics is being increasingly scripted in the spatial grammar of a millennial struggle between Good and Evil." The near decade-long rule of George W. Bush, with his religiously tinged pandering to his Evangelical "base" and the rise of

The Wiley Blackwell Companion to Political Geography, First Edition.
Edited by John Agnew, Virginie Mamadouh, Anna J. Secor, and Joanne Sharp.
© 2015 John Wiley & Sons Ltd. Published 2017 by John Wiley & Sons Ltd.

the voices of Middle Eastern Islamists, made it seem like "religion is the emerging political language of the time" (Agnew 2006: 183).

Indeed, religion is returning to the discourses of the public sphere; or, as some research on the "postsecular" suggests, it was always there (Thomas 2005; Hurd 2008). And this imbrication of religion and politics suggests the need for a focused analysis of political geographies of religion (cf. Nyroos 2001; Secor 2001; Agnew 2006; Megoran 2006; Dittmer & Sturm 2010; Sturm 2013). The co-editor of the journal *Geopolitics*, David Newman (2012), echoes this need, saying that religion is the future for geopolitical research: "We don't understand enough about the geopolitical and global underpinnings of religion and, more recently, religious fundamentalism." In this chapter, I attempt to make research on the power-laden manifestations of religion present, applicable, and doable, including its production of geographies of fear, danger, power, exceptionalism, universalism, security, and, ultimately, qualifications of difference.

Thus far, the vast majority of political geography on religious topics, specifically critical geopolitics, has been about fundamentalism and Christian imaginations of an imminent apocalypse (cf. Dittmer & Sturm 2010). This means not only that geopolitical interest in the topic has been sparse, but also that it fluctuates as religiously inspired internationally violent events hit the newspapers and as apocalyptic events are predicted and gain cultural momentum. Examples include the temporally titled events of Y2K, 9/11, and the failed environmental apocalypse of December 21, 2012. I suspect that this fluctuating interest is compounded by geography's relatively small disciplinary standing among specifically North American research institutions. Political geography is perhaps attracted to current issues that may bring interested cross-disciplinary readership from the larger social sciences into geography.

What political geography of religion lacks is a sustained research agenda. Agnew (2010: 44) writes on this point that religious geopolitics has focused on "the theological exotica more than the mundane everyday practices of churches and religions in relation to politics that attract attention." Precisely because we live in an increasingly postsecular world, such everyday religion should be taken seriously, for example by those within foreign policy circles who are influenced not only by popular culture, but also by its religious belief systems, whether explicit or culturally embedded religious significations, beliefs, and rituals (Philpott 2009; Snyder 2011). Beaumont and Baker (2011) make this point relative to religious leaders within urban social movement politics. In this chapter, by giving examples of potential research projects that are not "fundamentalist" or "millennial" movements, I set an agenda for political geographers to study not only extreme religious movements, but also the more banal. Such scholarship would take religious movements seriously, rather than simply reporting on the newest topical geopolitical world view/event in order to debunk its spatial assumptions.

The chapter is split into three major sections. The first defines some key terms, and illustrates how Marxists and poststructuralists often research religion. The second asserts that a distinction should be made between "religious geopolitics" (secular discourses that utilize religious discourse) and "geopolitics of religion" (geopolitics by religious groups and of theologies). It provides an organizational analytical tapestry in which future scholars of religion and geopolitics can find root and inspiration. The third section briefly explores two possible theoretical interventions for future research on religion and political geography given recent interests: affect and mobilities. Following that, I lay out a theoretical foundation for "religious nationalism," which I believe to be the most promising subtopic within this cluster of proposed research possibilities. While I review some of the literature on political geographies of religion, I am most concerned with outlining the debates from several social sciences that are of interest to political geographers. In so doing, I define various terms, identify pitfalls, and attempt to carve out analytically various research possibilities for this still nascent intersection with political geography.

Definitions and caveats

The political geography of religion is relatively inchoate. The category "religion(s)" is too broad a designation, because each religion is made up of divergent sects, interpretations, and performances. Religious variance makes it difficult for scholars to write of "Islam," "Christianity," or other religions as a whole (Asad 1999). We therefore need to take methodological and theoretical care. A definition of religion and thus of what counts as religion is necessary to study what I have termed "religious geopolitics"; that is, secular geopolitical discourses that utilize religious language. If we are to identify symbols, vocabularies, and actions as *religious* among otherwise secular individuals, groups, or institutions, then a definition, however simplistic, is required. In line with mainline sociology of religion, I define religion as a set of beliefs, actions, performances/rituals, and institutions that are founded on the existence of supernatural deities who can interfere with and judge earthly, spiritual, and heavenly scales of being. Religion requires an

> array of symbolic, moral, and organizational resources ... [It] embraces a creed, a cult, a code of conduct, and a confessional community ... [and there are] numerous and diverse social identities, political parties, and legal claims that they underwrite (Appleby 2000: 8–9).

From a poststructuralist perspective, however, "religion" would be defined by the actors and actions of self-identified religious groups. Such a definition, where it is understood that self-identified religious movements outline geopolitical divisions in the world, is more appropriate for what I have termed the "geopolitics of religion."

Religion is often understood in academia as epiphenomenal: It is the socio-cultural place or context from which one's religious beliefs arise and which determines one's actions. A Marxist perspective, for example, suggests that social class and capitalist relations determine social realities, where religion functions to pacify people by giving them a distractive purpose from oppressive capitalist relations. Conversely, a poststructuralist perspective suggests that religion itself cannot be the cause of anything. First, religion is too broadly interpreted and defined, or is simply the multiplicity of other competing, intersectional, and resonating social discourses. Second, culture and politics are co-constitutive: Religions and ideas about religions are constituted through various practices, performances, and associated meanings of political space and the politics of spatial practices. Religion, therefore, cannot be studied from a poststructuralist perspective as a thing apart or insular from other socio-cultural phenomena. It is intimately wrapped into systems and domains and representations, society and economy. It does not operate independently as a separate variable.

Certainly, generalizations or essential truths about religions are problematic. While Buddhists might be presented as peaceful in popular Western culture, some have armies, for example in the formally theocratic Tibet or present-day Nepal, and therefore differ in doctrine and practice by state, region, and temple. As McConnell (2013) notes, Buddhism can be better explained not by cultural context, but rather by whether it is an official state religion or a separate institution. Therefore, only when many various contributing factors come together can we attempt to explain anything. Admittedly, causal linkage explanations are not my goal.

Religion functions through language and can help interpret some geopolitical facets. While religion may be interpreted in many ways by various groups, it cannot be interpreted *ad infinitum*. There are a number of somewhat foundational beliefs. The Catholic Church, for example, uses absolution and the transference of holiness through an institutional hierarchy, whereas most Protestant churches require individual good deeds for someone to be saved. However, more importantly, it is precisely in the messy multiplicity of religious institutions and sects that most political geographers will be interested; that is, how certain religious social movements interpret the politics of the world through a religious lens and discursively impose their own explanatory religious geopolitics on it. Thus, rather than generalize about "religions" as monolithic power structures that impose power from a center onto political regimes, most political geographers will instead be interested in how religions are part of wider normative power structures and in religion's influence in social movements (Gerhardt 2008).

Furthermore, we should take seriously the many discourses that justify actions through religion, especially when an individual or group insists on their religious motivation. Unless compelling evidence to the contrary is provided, we should assume that religious categories of practice are relevant to motives and actions. Without doing so, we inadvertently absolve a group from responsibility for action and can be dismissive of their intentions. Nevertheless, to what extent should we assume that those who claim to be religious actors should be accepted as such? When should we as scholars question others' self-presentation and when would we do better to take it for granted? For example, the libertarian-type Christian identity movements tend to claim a religious foundation and inspiration, but are primarily white supremacists who are not supported by even the most extreme and fundamentalist congregations (Barkun 1997). The US–Mexico border has become their cause célèbre in recent years in terms of "the minutemen" (Doty 2007). These groups justify violent actions through a Christian foil for their political beliefs based in ideas of a racially pure nation-state.

Religious geopolitics and geopolitics of religion

The term critical geopolitics has flourished in the past two decades and has in the last ten years made motions to put religions under its microscope and scalpel. Such geopolitical discourses claim to render world politics transparent and argue that their perspectives are "the only possible real," and are often natural or predetermined results of complex social, political, and cultural contexts (Ó Tuathail 1996a). Briefly defined, critical geopolitics attempts to defamiliarize our presuppositions of space, showing them to be political on the one hand, and containing overt epistemological assumptions about political space on the other (Dalby 1991). This is what Bertolt Brecht called *Verfremdungseffeckt* or the "alienation effect," the act of taking what may seem familiar or obvious and making it strange (see Jameson 1998). Of course, Brecht is allowing for the unpredictability of directing and acting in the theater, but this concept has been foundational for critical geopolitics, especially as space is often treated as a stage by geopoliticians for the theatrics of politics rather than being co-constitutional of politics (Ó Tuathail 1996a; Sturm 2006).

However, can we consider religion – particularly if its "reach is largely confined to the private sphere of home and hearth" (Bruce 2003: 1)? As the religious becomes reinserted into and even conflated with the secular or postsecular in the political and public spheres of life, religious identities, movements, beliefs, and performances cannot claim impunity, especially when engaged in the realm of politics. This said, we have to be critical of our own situated knowledges by asking who is criticizing and who is being criticized. Therefore, what ethno-centric assumptions are built into these critiques in a manner that distorts the subject matter? This is a central problem with islamophobic critiques of Islam, where the religion is often misrepresented as a barbarous Other against a backdrop of a rational West. Nevertheless, simply because religious beliefs are dearly held or often possess a master status for identities does not mean that performances made in the name of religion cannot be criticized. While certain religious actors are minoritized via Western hegemonic or strong religions, religion itself is not a minority requiring protection or an apolitical cultural purveyor of peace and private culture. It is part of the public sphere of legitimate cultural critique, including critical geopolitics. Moreover, precisely because religion is so prevalent and pervasive within popular culture and everyday life, it may better resonate with or limit the flow, perception, and incorporation of geopolitics from popular culture to practical governmental imaginations. This said, religion is not simply an oppressive ideology; it can also provide the tools for transformative, emancipatory, and revolutionary change within the discourse of critical geopolitics (Little 2007; Megoran 2010).

Critical geopolitics of religion should not simply be about criticizing various theologies or practices by religious actors, it should also criticize its own religious targets that are the result of secularization theory and eurocentricism. Various religions have had well-intentioned, pacific, progressive goals, and disruptive revolutionary goals as or within social movements (Smith 1996). While Bruce (2003: 244) says that "religion is oppressive," because of various doctrinal commitments from universal application to millennial thinking, such doctrines, like millennialism, can also potentially be emancipatory (Megoran 2013). There is the potential for religion to counter the oppressive generality of geopolitics and that can be transformative and "anti-geopolitical" (Ó Tuathail 1996b). Emerging interest in peace studies in political geography or what Megoran (2010) has termed "pacific geopolitics" has had its beginning in looking at the emancipatory potential in religious movements like the Reconciliation Walk, which seeks to bring various faiths together to solve geopolitical disputes. Such work on religious peacemakers from Christianity to Islam has had much promise in the last decade and will challenge the radical atheist's perception of religion as being inherently oppressive and even violent (Gopin 2000; Johnston 2003; Little 2007; Funk & Said 2009).

The link between religion and geopolitics has been called "spiritual geopolitics," "religeopolitics," or simply "religious geopolitics" (Ó Tuathail 2000; Nyroos 2001; Dittmer & Sturm 2010). To add some definitional structure, I make a distinction that was hinted at earlier between the terms "geopolitics of religion" and "religious geopolitics" (Sturm 2013). The geopolitics of religion refers to religious actors who are concerned with theologically inspired representations of the borders of the world (or lack thereof in certain utopian proselytizing imaginaries), whether between religions or between the faithful and the secular. Religious geopolitics refers to secular geopolitical discourses and actions that nevertheless can be seen to employ political-theological vocabularies, symbols, and action. The latter definition acknowledges contributions to recent literatures on postsecularism, which recognize the religiosity of everyday life and its embeddedness in cultures.

There are at least four ways to study religious geopolitics and the geopolitics of religion at a broad, sociologically analytical level: Explanation; element; analogous; and typology.

- Religion might be able to *explain* something about geopolitics, particularly its origins. This of course can take many forms, depending on what it is about geopolitics that is being explained and what part of religion is doing that explanation. We might argue that the Westphalian state system has its roots in Christian ideas of division and exclusion that found form in the separate Lutheran and Catholic territorial spaces that ended the Thirty Years' War, for example.

- Religion could be seen as integral to all, or specific to some, geopolitical imaginations; as often an *element* of geopolitical constructions. We could claim using Carl Schmitt's idea that all significant concepts of the modern theory of the state are secularized theological concepts (cf. Asad 1999), and therefore that all of geopolitics is imbricated with religion. However, to be more specific, and as Ó Tuathail (2000) points out, some geopoliticians write of world politics being exasperated by or even foundational to their religion. Religion does not define a certain geopolitical imagination, then, but rather supplies myths that are central to geopolitical imaginations.

- As an *analogous* construction, we might think of religion *as* geopolitics. Geopolitics and religion are both ways of seeing the world. Geopolitics is a perspective on the world, a way of seeing the world, not a thing in the world: A way of expressing interests, categorizing the world, and signifying events. Religion too, analogous to geopolitics, can serve these same functions.

- It could be argued that there is a particular religious *typology* of geopolitics: "Religious geopolitics" and/or "geopolitics of religion." This is analytically separate from the tripartite practical, formal, and popular, and requires new methods and theories. It would focus on religious discourse emanating from texts and sermons. While such a formulation is fraught with concerns regarding what counts as religious or secular, we can assume that a religious actor who claims to speak in the name of their religion is performing for that cause. This said, rather than creating a new typology, various types of religious geopolitics and geopolitics of religion can be slotted in and across the Lefebvrean-based conceptualization of so-called formal, popular, and practical "geopolitical discourses" (Ó Tuathail & Agnew 1992; Sharp 2000).

I provide below an example for each type of geopolitical discourse distinguished in critical geopolitics to illustrate how the concept of religion might be approached:

- Practical (governmental, policy, and bureaucratic) geopolitics is perhaps the most obvious geopolitical influence, but also the most contentious (Dijkink 2009). In this argument, the idea is that a government is itself led and/or influenced by a religious group or religious ideas attempting to make a theocracy, pining for the apocalypse or holy war, or actively interpreting world events from a theological perspective. The popularized phrase within Iran that would see Israel pushed into the sea was a practical form of geopolitics. It could be interpreted as coming from a secular foundation because of Israel's occupation of Palestine or the decolonization of an imagined Ummah Islamiyyah or Muslim Middle East. While Iran is a democracy, it injects religious norms into government as a first principle, thereby limiting choices and constructing a starting point for conservative Shi'ite Islam.

- Popular (cultural influence and culturally fabricated) geopolitics has formative power. It is often how people learn about the world not immediately around them, be this through (tabloid) media sources, books, including comic books and cartoons, and other cultural manifestations. Militant fundamentalisms are much less isolated now and have access through information and communication technologies (ICT) like television, email and the internet, and air travel. Mass media has "evoked so many defensive strategies and provoked offensive tactics from those they attacked" (Marty 2005: 14). More moderate religious groups in the United States, like Roman Catholics, Reform and conservative Jews, and mainstream Protestants, largely failed to adopt radio and television as a dissemination tool as the more fundamentalist-like groups did to win political power. Televangelists, for example, like Oral Roberts, Tim LaHaye, Jerry Falwell, Pat Robertson, and Jimmy Swaggart, reach an Evangelical audience of over 100 million Americans weekly through radio and television programs. Tim LaHaye and Jerry Jenkins have also sold more than 60 million of their Evangelical vision of the apocalypse, *Left Behind*, which has a very particular geopolitical view about the regional make-up of the world as being divided between good and evil armies. This novel presents a fictional account of what many American Christian Zionists believe to be real-world prophecies about the making of Israel and the wider Muslim Middle East (Dittmer & Sturm 2010). There is a large appetite for these books and Evangelicals are one of the largest social and political movements in the United States and Israel. Such prophecy writings helped shape American attitudes toward the Cold War and the so-called War on Terror, and they encourage a systematic view of how to look at the world.

- Formal (intellectual, think tank, or academic-based) geopolitics, which could include, for example, academics with a religious agenda, seminary scholars, and Christian lobbyists in Washington and Jerusalem/Tel Aviv, have a direct and indirect influence on geopolitics. American Evangelicals are the second largest base of support for pro-Israel interests in the United States, and have the most unqualified, unwavering, and uncritical support for militant political agendas, for example the nuclear first strike against Iran. Furthermore, Mearsheimer and Walt argue in *The Israel Lobby* (2007) that Christian lobby groups, through complex, open, and back-door channels, helped justify (and possibly convince) the Bush administration to go to war in Iraq based on their geopolitical assumptions about Iraq's biblical role in world politics.

Given the above definitions, caveats, and discursive clarifications, it is now possible to explore the research possibilities for students of political geographies of religious movements, especially given the current theoretical climate and thrust within critical geopolitics and political geography.

Research possibilities

For the purposes of a political geography of religious movements, we might explore the various historical, contemporary, and analytical themes that lead to religious generalities and how various beliefs and actions might be productive threads for future political geographies. In this section I briefly illustrate three possible topics with examples for analysis in political geography that are by no means exhaustive. Others could have included religious origins and ontologies, gender and sexualities, orientalism, and religion and state stability, among others. Instead, I outline three research possibilities, starting with affect. I follow with election myths, mobilities, and missionaries. I then give the lion's share of attention to the topic of religious nationalism, providing a literature review and a theoretical justification for this seemingly contradictory topic. I believe that religious nationalism has the most potential for sustained future research interest in the political geography of religion, because religious movements have often contributed to nationalisms, because nationalisms are the most territorial of all social systems, and because any national claim is also a geopolitical assertion (Dijkink 2009). I chose these three research possibilities since they overlap with current theoretical trends in political geography generally; because of this overlap, political geographers might feel more comfortable studying religion and therefore expanding the literature; and, furthermore, many of these trends overlap recent work in the sociology of religion and some work in international relations that has recently engaged, analyzed, and critiqued religious social movements. Therefore, the intent here is to provide politico-spatial examples overlaid with a short set of cross-disciplinary citations to entice political geographers to explore religion.

Affect

Despite religion being defined earlier by a system of symbols and codes, affect is an ambiguous force lying just beyond symbolization and representational capture, either as surplus or prior to it. However, for a political geography student taking a more poststructuralist definition of religion as a starting point – that is, that religion consists in practices and performances by self-identified religious actors – then such an affective geopolitics of religion is made possible by escaping the question of whether religion per se is potentiating geopolitical divisions of the world. This is not to advocate a return to structural-functionalist theories that saw religion as irrational and emotional, but rather to acknowledge religion's affective potentiality in geopolitics (Smith 1996: 3).

Within the pervasive smokescreen of secularization theory, religion plays a functional and epiphenomenal role in political geography by justifying otherwise rational calculations of violence at a macro scale, rather than seeing religion and enchantment as an affective force (Hyndman 2004). While much of geography has had sustained engagement with theories of affect in the last 15 years (and, more generally, with non-representational theory), political geography has been slow to contribute, with some recent and exciting exceptions (Carter & McCormack 2006; Sharp 2009).

A sustained discussion beginning with current issues in religious geopolitics makes sense, especially within the analytical category of popular geopolitics. This is because of the affective power of religion in producing, assembling, and maintaining the motivation and commitment to geopolitical violence (Massumi 2002). As an embodied process that troubles individualization and socialization, affect is also a potentially emancipatory source of resistance to the seemingly irresistible forces of capitalism and colonialism, while also being a potential source of repressive politics as well (Barnett 2008). As an ambiguous force lying just beyond representation and ideological capture, affect implies a religious potentiality for resistance to social phenomena and even oppression from religion.

Election myths, mobilities, and missionaries

Religious election myths – that is, the idea that a group is superior because it has been chosen by a deity – are geopolitical assertions that see social groups sharply divided, often territorially, into black and white, us vs. them, geo-packages. Bruce (2003: 225) writes, "[I]f we have the true faith and everyone else is wrong, then clearly we are superior to the heathen." Orientalist othering is a well-covered topic in political geography, but religious justifications that view other religions or even secular states as Others have not been of prime interest to political geographers.

Perhaps this form of geopolitics is further facilitated by proselytizing religions, especially as they become coterminous with aspirations for colonialism, imperialism, and empire. As Agnew (2010) points out, "numbers matter" in the religious geopolitics for souls, and with such proselytizing comes a kind of religious conflict for sovereignty over those souls. Judaism is not currently a proselytizing religion and is therefore not a potentially boundless religion like Christianity or Islam (Sand 2010). Thus, various religious forms of Zionism, for example, concern themselves with the territorial maximalism of Israel to encompass Palestine and even a wider region of the Middle East (Genesis 15:18–21, Numbers 34:1–15 and Ezekiel 47:13–20).

Such missionary activity and universalizing doctrines should be included in the growing canon of "mobilities" and "postcolonial" research in political geography. The nation-state has limited such expansionism, but has certainly not stopped it, especially given globalization and, for example, the global consistency of some religions, such as Islam's use of Arabic that allows it to transcend borders, especially Arab states in the Middle East (Haynes 1994). Such transcendence has come in the way of support for other states with the same religions, as in the Greek support for Milosevic's Orthodox Serbian regime, and also through missionary activity. Many Evangelical Christians from around the world, for example, have since the 1970s focused their missionary efforts on the geopolitical designation between latitudes 10 and 40, commonly known as the "10/40 window," which by their assessment is the least Christian area of the world (Han 2010). At times throughout history such Christian missionary activities have created or exasperated political boundaries, creating ethnic groups where none had existed before, resultantly fostering conflict between religions and between various Christian sects.

Religious nationalism

There is almost a religious devotion among all nationalisms to a territory or homeland. More often than not, there are religious discourses embedded within performances of nationalism. As Agnew (2006: 185) notes, "much nationalism and imperialism have found purpose and justification in religious difference and in proselytizing." Countless sentiments of devotion comprise nationalism, but territory and a geopolitical imagination of it are key, especially in contrast to common enemies and neighboring territories. Because of the nationalist identification of "internal" and "external," "our nation" and "their nation," Agnew argues (2008) that these binaries make nationalism the most territorial of all major ideologies among socialism, liberalism, and nationalism. Herb and Kaplan (1999: 2) argue that "territory becomes a vital constituent of the definition and identification of the group living within it."

The "nation" is often defined as a group of people who feel that they share a set of myths concerning a territory, having common experiences of danger, destiny, historical struggles, and cultural affinity in relation to common places. However, I argue, consistent with theories that the "nation" is a "social text," that the nation does not exist outside of the performance of such binding myths. Instead, these common performances are what can be termed the verb for "nationalism"; that is, the doing of the "nation" (Calhoun 2001; Agnew 2008).

Given that all national claims are also geopolitical assertions and that most nationalisms are inflected with religious discourse, they slot within categories of either "religious geopolitics" (RG) or a "geopolitics of religion" (GoR). For instance:

- Give a moral core (RG).
- Further cement common belief systems among a community of common values (RG or GoR).
- Act as a legitimizing power for state leaders, especially for rallying support in religiously conservative regions (RG).

- Act as a powerful political lobbying body (GoR).
- Use the power of religion to elicit a common heritage and, therefore, to create common socio-cultural ground (largely RG).
- Provide the geopolitical counter discourse to perceived Western-based models of secular nationalism, as we see in the case of Islamic movements in much of the Middle East over the last two decades and especially following the "Arab Spring" (RG and GoR; Secor 2001; Mullin 2010).

Explicit and well-known examples of these functional, instrumental, and facilitative uses of religion are currently employed in Turkey, Ukraine, Russia, and the United States, among many other countries. Such a tally is clearly not exhaustive of the ways in which religion and nationalism are fused. While not all nationalist expressions are religious, almost any religious discourse can be adopted into cultural or governmental politics and be performed as such (that is, as a religious geopolitics perspective), just as any religion (within a geopolitics perspective) can adopt national allegiances as part of its ritualized performance (McAlister 2008).

Much of the work on nationalism has tended to ignore religion or explain it as a function of nationalism. In this view, nationalism is a modernist project that replaces religion by emphasizing socio-economic factors or cultural or political modernity (Gellner 1994). That said, there are many examples of synergy between religion and nationalism, some mobilizing strong nationalisms such as in Ireland and Poland, and other relatively weak movements such as in Russia and Ukraine (Bruce 2003). However, care must be taken in fusing these two terms together as "religious nationalism," because there are often many reasons for, expressions of, and geographically specific types of nationalism, the use of religious signifiers being but one of them that is often of secondary influence, epiphenomenal, or used as a guise for political means (Agnew 2008).

In this respect, Friedland (2001: 138) writes that religion can be used to constitute the "state and territorially bounded population in whose name it speaks." He has suggested that there is an alternative type of nationalism, in form and content, which he calls "religious nationalism," where nationalism is defined as "discursive practice by which the territorial identity of a state and the cultural identity of the people whose collective representation it claims are constituted as a singular institutional fact" (2001: 138). Religion in this conception fuses state, territory, and culture in an analytical way. Friedland's "religious nationalism," however, hinges on the state, as if religious expression can only be funneled through the state's political voice or religious expression matters only if it is sanctioned by the state. Rather than "religious nationalism," Friedland's observation might better be termed "religious statism" (Brubaker 2006).

Friedland's argument is thus underinclusive, ignoring other non-state-centric forms of nationalism. Nevertheless, for the purposes of this scholarship, his conception has relevance to nationalism. There are still shared perceptions of origin and myth that are utilized. A more nuanced and somewhat less state-centric case can be made as well, where the lack of nationalist congruence with the state may bear and nurture a type of religious nationalism.

Although nationalism certainly has a "spiritual principle," as Ernest Renan (1996 [1882]: 52) classically observed, the attempt to formulate a theory for religious nationalism is one fraught with problems. Some have suggested that religious nationalism is an oxymoronic term; as Brubaker (2006: 23) provocatively concluded, "nationalist politics is carried out in the name of the nation, religious politics in the name of God." Despite this, I argue that the fusion between nationalism and religion takes place most convincingly not in how national discourse is inflected by religious language or how religious discourse is inflected with nationalist language, but rather, following Asad's (1999) demand, in understanding the category of religion as one of practice rather than solely one of belief. Therefore, what we assume to be nationalist language is also religious practice through prayers, sermons, and pilgrimages, to mention only a few performances. As McAlister (2008: 875) explains, it is not that "everything is religion, it is just that religion can be virtually anything." Nationalism, too, can be religious practice.

There is then not a natural or inevitable perennial outcome either. Anthony Smith (2003) makes the argument that proto-nationalisms based on religious and ethnic groupings, what he calls "ethnies," pre-existed and predisposed the Western world to modern nationalism. He outlines four kinds of religion-based proto-nationalist cultural resources: The myth of ethnic election; attachment to terrains as sacred; a yearning to recover the spirit of a golden age; and claims to the regenerative powers of sacrifice to ensure a glorious destiny. The feeling of having been Chosen (part of the Elect) to act in a grand play to claim sacred national territory given in a Covenant with God that will result in favor and glory has overlapping religious/ national affinities. That said, most nationalisms imagine themselves in these terms, which suggests that religion is often a legitimating factor in nationalist identifications. Therefore, while often employed to make the argument for a "religious nationalism," Smith's analytical division is not particularly convincing. I would therefore deny that there is a natural continuity as presented by Smith; nationalisms are emergent for a variety of reasons from many different historical experiences.

Thus, Smith's notion of proto-nationalisms based on sacred texts and religious practices is not argued here. Rather, I maintain that religious myths are often symbols of heritage and destiny and are pulled into performances that are co-constitutive of space and national identity. So, while modern nationalism has a vast array of mythical heritage events, religion is often a central legitimating claim. It is only within this vein of practice that is informed by such discourse that Smith's (2003: 254) charge is taken up to serve the period of nationalism with "sacred foundations" and "deep cultural resources" in the nation's commitment to authenticity, however contrived and selective. Nationalism is then constructed from pre-existing culture, but, through its performative making, it also changes that culture.

Understanding the distinction between "civic" and "ethnic" nationalism is essential to grasping the meaning of the ethno-religious nationalism emerging throughout the world. Both Brubaker and Smith make distinctions between civic and ethnic nationalism (Smith 2003). The concept of American civic nationalism – otherwise referred to as the "American creed," which is based on the democratic and enlightenment beliefs on which the American Constitution was founded – is alive and well and continues to bind the American state and people, for example. Nevertheless, it is thought to be under attack by a broad spectrum of culturally conservative Americans, giving way to the so-called culture wars.

Like French civic nationalism, American civic nationalism is assimilationist, not pluralist; therefore, while based demographically on immigration, it has a nativist expectation of conversion to the American creed. This is an especially poignant observation when that creed is imagined to be founded in a mythologized Jewish and Christian synergy. This mix of Protestantism and Enlightenment thought is what has been called "civil religion" in the United States (Bellah 1976). However, what Lieven (2004) terms the American nationalist "Antithesis" or ethnic nationalism, a sometimes competing form of American civic nationalism, he argues co-exists with civic nationalism, but has its roots in ethno-religious beliefs and Jacksonian ideals. While this latter form of nationalism is often subordinate to civic nationalism, it can rise to prominence in times of crisis. Imaginations of an imminent apocalypse, being an embattled minority, and the belief that Israel and Jews are being persecuted by the surrounding world provide such a crisis mentality, particularly when juxtaposed against a common enemy, which has often been Islam (Sturm 2010).

It is this form of ethno-religious nationalism that Benedict Anderson predicted to be "the wave of the future" for nationalism, one consisting of "ethnic and racial stereotypes, xenophobia, sectarian 'multiculturalism' and the more brutal forms of identity politics" (Anderson 1996: 12–13). For Lieven (2004: 6), in his condemnation of America's support for Israel's radical right, it is this ethno-religious nationalism that cements "America's attachment to Israel, [and] ethno-religious factors have become dominant, with extremely dangerous consequences for the war on terror." What distinguishes ethno-religious nationalism from other cultural identities is not only its territorial and political dimensions and ideals, but also its commitment to the pursuit of the authenticity (however mythical) of the heritage and destiny of the imagined national community.

Conclusion

There are many fruitful possibilities for political geographers of religion to engage with affect, mobilities, and religious nationalisms, especially insofar as more ethno-religious forms bubble up in global space. There are many other fruitful possibilities, such as gender, sexualities, and inter/intrastate conflict, among others, and these research possibilities will hopefully expand and sustain political geographies of religion research. Through the funneled and varied structure of this chapter, it is hoped that students of political geography will be inspired by one or more of the many possibilities provided in this burgeoning inclusion of religion in political geography.

References

Agnew, J. (2006). Religion and geopolitics. *Geopolitics* 11(2): 183–191.

Agnew, J. (2008). Nationalism. In Duncan, J., Johnson, N.C., & Schein, R.H. (eds), *A Companion to Cultural Geography*, Oxford: Blackwell, 223–238.

Agnew, J. (2010). Deus vult: The geopolitics of the Catholic Church. *Geopolitics* 15: 39–61.

Anderson, B. (1996). Introduction. In Balakrishnan, G. (ed.), *Mapping the Nation*. New York: Verso, 1–7.

Appleby, S. (2000). *The Ambivalence of the Sacred: Religion, Violence, and Reconciliation*. Lanham, MD: Rowman and Littlefield.

Asad, T. (1999). Religion, nation-state, secularism. In van der Veer, P., & Lehmann, H. (eds), *Nation and Religion: Perspectives on Europe and Asia*, Princeton, NJ: Princeton University Press, 178–196.

Barkun, M. (1997). *Religion and the Racist Right*. Chapel Hill, NC: University of North Carolina Press.

Barnett, C. (2008). Political affects in public space: Normative blind-spots in non35 representational ontologies. *Transactions of the Institute of British Geographers* 33(2): 186–200.

Beaumont, J., & Baker, C. (2011). *Postsecular Cities: Space, Theory and Practice*. London: Continuum.

Bellah, R. (1976). Civil religion in America. *Daedalus* 96(1): 1–21.

Bialasiewicz, L. (2006). The death of the West: Samuel Huntington, Oriana Fallaci and a new "moral" geopolitics of births and bodies. *Geopolitics* 11(4): 701–724.

Brubaker, R. (2006). Religion and nationalism: Four approaches. Paper presented at the *Nation/religion* conference, Konstanz, July 6–8.

Bruce, S. (2003). *Politics and Religion*. Cambridge: Polity Press.

Calhoun, C. (2001). *Nationalism*. Minneapolis, MN: University of Minnesota Press.

Carter, S., & McCormack, D. (2006). Film, geopolitics and the affective logics of intervention. *Political Geography* 25(2): 228–245.

Dalby, S. (1991). Critical geopolitics: Discourse, difference, and dissent. *Environment and Planning D: Society and Space* 9(3): 261–283.

Debrix, F. (2008). *Tabloid Terror: War, Culture, and Geopolitics*. New York: Routledge.

Dijkink, G. (2009). Geopolitics and religion. In Thrift, N., & Kitchen, R. (eds), *International Encyclopedia of Human Geography*, Amsterdam: Elsevier, 453–457.

Dittmer, J., & Sturm, T. (eds) (2010). *Mapping the End Times: American Evangelical Geopolitics and Apocalyptic Visions*. Farnham: Ashgate.

Doty, R.L. (2007). States of exception on the Mexico–U.S. border: "Decisions" and civilian border patrols. *International Political Sociology* 1(2): 113–137.

Friedland, R. (2001). Religious nationalism and the problem of collective representation. *Annual Review of Sociology* 27(1): 125–152.

Funk, N.C., & Said, A.A. (2009). *Islam and Peacemaking in the Middle East*. Boulder, CO: Lynne Rienner.

Gellner, E. (1994). *Encounters with Nationalism*. Oxford: Blackwell.

Gerhardt, H. (2008). Geopolitics, ethics, and the Evangelicals' commitment to Sudan. *Environment and Planning D: Society and Space* 26(5): 911–928.

Gopin, M. (2000). *Between Eden and Armageddon: The Future of World Religions, Violence, and Peacemaking*. New York: Oxford University Press.

Han, J.H.J. (2010). Reaching the unreached in the 10/40 window: The missionary geoscience of race, difference, and distance. In Dittmer, J., & Sturm, T. (eds), *Mapping the End Times: American Evangelical Geopolitics and Apocalyptic Visions*, Farnham: Ashgate, 183–208.

Haynes, J. (1994). *Religion in Third World Politics*. Boulder, CO: Lynne Rienner.

Herb, G., & Kaplan, D. (eds) (1999). *Nested Identities: Nationalism, Territory, and Scale*. Lanham, MD: Rowman and Littlefield.

Huntington, S. (1996). *The Clash of Civilizations and the Remaking of World Order*. New York: Simon and Schuster.

Hurd, E. (2008). *The Politics of Secularizm in International Relations*. Princeton, NJ: Princeton University Press.

Hyndman, J. (2004). Mind the gap: Bridging feminist and political geography through geopolitics. *Political Geography* 23: 307–322.

Jameson, F. (1998). *Brecht and Method*. New York: Verso.

Johnston, D. (ed.) (2003). *Faith-Based Diplomacy: Trumping Realpolitik*. New York: Oxford University Press.

Katzenstein, P.J. (2010). *Civilizations in World Politics: Plural and Pluralist Perspectives*. New York: Routledge.

Lieven, A. (2004). *America Right or Wrong: An Anatomy of American Nationalism*. Oxford: Oxford University Press.

Little, D. (ed.) (2007). *Peacemakers in Action: Profiles of Religion in Conflict Resolution*. Cambridge: Cambridge University Press.

Marty, M.E. (2005). *When Faiths Collide*. Malden, MA: Blackwell.

Massumi, B. (2002). *Parables for the Virtual: Movement, Affect, Sensation*. Durham, NC: Duke University Press.

Mearsheimer, J.J., & Walt, S.M. (2007). *The Israel Lobby and U.S. Foreign Policy*. New York: Farrar, Straus, and Giroux.

McAlister, M. (2008). What is your heart for? Affect and internationalism in the evangelical public sphere. *American Historical Literature* 20(4): 870–895.

McConnell, F. (2013). The geopolitics of Buddhist reincarnation: Contested futures of Tibetan leadership. *Area* 45(2): 162–169.

Megoran, N. (2006). God on our side? The Church of England and the geopolitics of mourning 9/11. *Geopolitics* 11(4): 561–579.

Megoran, N. (2010). Towards a geography of peace: Pacific geopolitics and evangelical Christian Crusade apologies. *Transactions of the Institute of British Geographers* 35(3): 382–398.

Megoran, N. (2013). Radical politics and the Apocalypse: Activist readings of Revelation. *Area* 45(2): 141–147.

Mullin, C. (2010). Islamist challenges to the "liberal peace" discourse: The case of Hamas and the Israel–Palestine "peace process." *Millennium: Journal of International Studies* 39(2): 525–546.

Newman, D. (2012). Israel–Palestinian conflict, peace studies, borders, academia politics. Interview by L. van Efferink. *Exploring Geopolitics*. http://www.exploringgeopolitics.org/interview_newman_david_israel_palestinian_conflict_resolution_peace_studies_borders_academia_politics_territorial_boundaries_postmodernity/, accessed March 26, 2015.

Nyroos, L. (2001). Religeopolitics: Dissident geopolitics and the "fundamentalism" of Hamas and Kach. *Geopolitics* 6(3): 135–157.

Ó Tuathail, G. (1996a). *Critical Geopolitics*. Minneapolis, MN: University of Minnesota Press.

Ó Tuathail, G. (1996b). An anti-geopolitical eye: Maggie O'Kane in Bosnia 1992–1993. *Gender, Place, and Culture* 3(2): 171–185.

Ó Tuathail, G. (2000). Spiritual geopolitics: Fr. Edmund Walsh and Jesuit anti-Communism. In Dodds, K., & Atkinson, D. (eds), *Geopolitical Traditions*, London: Routledge, 187–210.

Ó Tuathail, G., & Agnew, J. (1992). Geopolitics and discourse: Practical geopolitical reasoning and American foreign policy. *Political Geography* 11(2): 190–204.

Philpott, D. (2009). Has the study of global politics found religion? *Annual Review of Political Science* 12: 183–202.

Renan, E. (1996 [1882]). "What is a nation?" In Eley, G., and Suny, R.G. (eds), *Becoming National: A Reader*, pp. 41–55. Oxford: Oxford University Press.

Sand, S. (2010). *The Invention of the Jewish People*. London: Verso.

Secor, A. (2001). Islamist politics: Antisystemic or post-modern movements? *Geopolitics* 6: 117–134.

Sharp, J. (2000). *Condensing the Cold War: Reader's Digest and American Identity*. Minneapolis, MN: University of Minnesota Press.

Sharp, J. (2009). Geography and gender: What belongs to feminist geography? Emotion, power and change. *Progress in Human Geography* 33(1): 74–80.

Smith, A.D. (2003). *Chosen Peoples*. Oxford: Oxford University Press.

Smith, C. (ed.) (1996). *Disruptive Religion: The Force of Faith in Social Movement Activism*. New York: Routledge.

Snyder, J.L. (ed.) (2011). *Religion and International Relations Theory*. New York: Columbia University Press.

Sturm, T. (2006). Prophetic eyes: The theatricality of Mark Hitchcock's premillennial geopolitics. *Geopolitics* 11(2): 231–255.

Sturm, T. (2010). Imagining apocalyptic geopolitics: American evangelical citationality of evil Others. In Dittmer, J., & Sturm, T. (eds), *Mapping the End Times: American Evangelical Geopolitics and Apocalyptic Visions*, Farnham: Ashgate, 133–154.

Sturm, T. (2013). The future of religious geopolitics: Toward a theory and research agenda. *Area* 45(2): 134–140.

Thomas, S.M. (2005). *The Global Resurgence of Religion and the Transformation of International Relations: The Struggle for the Soul of the Twenty-First Century*. New York: Palgrave Macmillan.

Chapter 27

Sexual Politics

Catherine J. Nash[1] *and Kath Browne*[2]

[1]*Brock University, St. Catharines, Ontario, Canada*
[2]*University of Brighton, England, UK*

Discussions of sexualities are frequently regarded as marginal to a consideration of the nature and constitution of political geographies. Nevertheless, a substantial body of geographical scholarship has examined the intertwined relationships between sexuality, space, and place, paying particular attention to the processes through which sexualities are spatialized as well as how space and place are sexualized: from the more intimate places of the home and the domestic sphere to the public spaces of the workplace, the neighborhood, the street, and the national terrain (e.g., Bell & Valentine 1995; Browne, Lim, & Brown 2007; Johnston & Longhurst 2010).

This chapter will argue for a consideration of the centrality of questions of sexuality to the constitution and understanding of political geographies. It will explore how the contestations over normative and non-normative sexual lives and practices are both enabled and constrained in and through various geographies and, in turn, are implicated in the constitution of various spaces and places. It does so through the utilization of three fundamentally critical terms lying at the heart of geographies of sexualities: heternormativity and homonormativity/homonationalism. These terms refer to more than homosexuality or heterosexuality, as they also attend to the reformation of normativities that frames the politics of sexualities. Our exploration necessarily not only entails a consideration of the places of the everyday, but also focuses attention on multiple spatialities and spatial scales, including the national, the intranational, and the transnational. Note that the predominantly global North, indeed mainly Anglo-American, focus reflects the positionalities and research areas of the authors, and needs explication in other areas, including considerations of, for example, the contestations of the French "*le mariage pour tous*" (Cervulle & Pailler 2013) and recent debates in Australia.

The chapter begins with a consideration of the nature and impact of various heteronormativities, since these were central to how questions about sexualities were organized in the late twentieth century and are particularly pertinent to engagements with political

The Wiley Blackwell Companion to Political Geography, First Edition.
Edited by John Agnew, Virginie Mamadouh, Anna J. Secor, and Joanne Sharp.
© 2015 John Wiley & Sons Ltd. Published 2017 by John Wiley & Sons Ltd.

geographies. Heteronormativity refers to the normalization of particular forms of both gender and sexuality within a heterosexual frame; that is, individuals are understood as either (and unambiguously) biologically male or female, exhibiting appropriately gendered behaviors as masculine or feminine, and experiencing sexual intimacy only with opposite-sex partners. Heternormative understandings continue to dominate political geographies today, even in those places where lesbian and gay rights are supposedly achieved, such as Canada or Great Britain. This understanding circulates in and constitutes spaces and places in ways that might be exclusionary to non-heternormative individuals, including other gay, bisexual, lesbian, and trans* people. Note that the term trans* is intended to signify in the broadest terms possible the multiple and various ways in which individuals experience a gendered sense of self or a gendered identity, and focuses mainly on the myriad of trans-related experiences (Stryker 2009). While some people now consider the * redundant, it is still predominantly used at the time of writing.

Two related concepts have also emerged since the beginning of the twenty-first century: the notion of "homonormativity" and that of "homonationalism." Homonormativity refers to the ways in which some lesbians and gay men have come to be recognized as reflecting the "proper" (or normative) ways to be homosexual, such as by participating in same-sex marriage, engaging in monogamous relationships, and exhibiting gender-appropriate behaviors in ways that are seen as reiterating the dominant social order, including neoliberal consumerism and white, middle-class aspirations and aesthetics. Homonationalism refers to those nations that, through social, political, and legislative initiatives, seem to support "tolerance" of sexual diversities and argue that they are therefore "more progressive," enlightened, and civilized than other nations. This positioning can then be used for numerous ends, including justifying wars, humiliating prisoners, and vilifying certain countries or peoples (Puar 2007, 2011; see also Duggan 2002; Brown 2009).

Heteronormativity

Early geographical scholarship began with the premise that all places were constituted as normatively heterosexual, meaning that the spaces of the home, work, and street presumed the presence of heterosexual subjects and practices only (Valentine 1993; Bell et al. 1994; McDowell 1995). Normative expectations in place worked to discipline those present to be understood as heterosexual (in terms of both sexual desire and gendered behaviors) and to make non-normative sexual (and gendered) subjects and identities invisible. Heterosexual privilege operates to normalize particular forms of heterosexuality, rendering others deviant and out of place. More than this, space itself was seen to be reformed as heterosexual. Work critical of this perspective argues that, in fact, there is nothing natural to the heterosexualization of space. Rather, performativity theorizations of the 1990s claim that space and place are created through the continuous performance of normatively gendered and sexual practices. As such, heterosexuality is an embodied practice that, through repetition, constitutes places as heterosexual. In the context of sexualities, space is constituted through normative forms of sexualities that are reiterated to produce the illusion of fixity and naturalness (Bell et al. 1994; Bell & Valentine 1995). Contemporary work has questioned the ways in which this conceptualization sees space as heterosexual until it is disrupted, challenged, or, in some terms, queered. All space is sexualized through various performativities, in ways that render it constantly in flux, but with the illusion of stability (Browne & Bakshi 2011).

The geographies of sexual citizens

Sexual citizenship has been key to understanding heteronormativities and how these are not only formed through the micro spaces of everyday life, but also pervade notions of statehood and national identity. "Citizenship" is understood as an exclusionary concept where questions about who meets the requirements for inclusion, particularly in terms of subject positions and identities, must be adjudicated within broader normative frameworks (Isin 2011; Hubbard 2013). In particular, states must set criteria to determine who is entitled to be considered a citizen as well as the rights and responsibilities that adhere to such a status. In modern democratic societies, the ideal citizen is normatively gendered and sexualized; a subject that exhibits the proper gendered (masculine or feminine) behaviors required of a biologically unambiguous body. In the late twentieth century, this ideal citizen was not only and unfailingly heterosexual, but experienced desire only within monogamous, opposite-sex, state-sanctioned marriage (Richardson 2000). These idealized expectations about the heteronormative citizen arguably work to exclude LGBT (lesbian, gay, bisexual, trans*) and queer individuals from full participation in important state institutions such as marriage, military service, and family benefits.

As a practical matter, this demonstrates that "all citizenship is sexual citizenship" and, as such, has distinctly spatial implications (Bell & Binnie 2000: 10). Sexual citizenship is not only about sexual practices in the private spaces of the home and domestic life, but is also about "the right to express and publicize particular sexual identities, and to have them acknowledged as legitimate in the public sphere" through inclusion in such state institutions as marriage (Hubbard 2013: 225). The sexual citizen given the most attention in the geographical literature is the homosexual citizen, who largely disrupts the presumed heterosexualities of the state and citizens through urban queer activism across a variety of scales.

Lesbian, gay, bisexual, and trans* (LGBT) geographies

Geographical scholarship has focused on how the broader processes of deindustrialization and decentralization in many downtown areas opened up the possibility for gays and lesbians to socialize in what became less desirable inner-city neighborhoods in the post–Second World War period (Castells 1983; Forest 1995; Armstrong 2002; Nash 2006). By the mid-1980s, distinct gay-identified residential and commercial districts had emerged, largely in the downtown core in North American, UK, European, and Australian cities (Cooper 1994; Whittle 1994; Forest 1995; Collins 2004). These locations provided much-needed and safe locations for gay men to meet and served as an organizing base for the growing gay and lesbian rights movement. Research on lesbian spatial networks and social organization not only found different spatialities, including less territorial control, but also suggested that lesbians were less inclined to want to control visible territories because of safety and financial limitations (Rothenberg 1995; Valentine 1995; Nash 2001; Podmore 2001, 2006).

The political possibilities for neighborhood areas to develop gay and lesbian concentrations are, not surprisingly, spatially specific. As Knopp (1998) noted, the electoral politics of places such as San Francisco, which in 1977 saw the election of its first out gay officer, Harvey Milk, differed from those in the United Kingdom, where gay territorial areas did not try to gain mainstream political leverage through the fielding of gay candidates in elections. Such differences reflect the historical political geographies of places where urban development, planning and electoral policies, and regulatory environments influenced the particular ways in which gay urban districts developed.

By the late 1970s, gay villages had become a fixture in the downtown core of many medium to large cities in the global North, although in many cases they continued to be viewed as marginal and suspect neighborhoods within more mainstream discourses (Knopp 1995, 1998; Nash 2005, 2006; Browne et al. 2007). By the mid-1990s, these well-established neighbourhoods served as the political, social, and economic hub for gay and lesbian communities and were caught up in the globalizing neoliberal processes transforming how cities were managed and administered. Neoliberal policies reworked the notion of the city from that of a mechanistic provider of infrastructure and welfare services to a more entrepreneurial and competitive entity needing to market and promote itself as a desirable location in which to live and work (Harvey 1989; Leitner, Peck, & Sheppard 2007; Boudreau, Keil, & Young 2009; Stenning et al. 2010). Cities in the global North, struggling to cope with the effects of deindustrialization and decentralization, began a rebranding effort to attract young, cosmopolitan professionals seeking vibrant and diverse cultural landscapes. Gay villages, proffered as evidence of a city's tolerant and edgy appeal, found themselves increasingly incorporated into newly constituted commodified landscapes (Florida 2002; see also Bell & Binnie 2000; Rushbrook 2002; Binnie & Skeggs 2004). Sexual others, including LGBT and queer people (and their progressively commodified locations, including gay villages, Pride, Mardi Gras, the Gay Games), were increasingly consumed while becoming consumers themselves of particular urban lifestyles within larger global circuits of commodification and tourism (Probyn 2000; Markwell 2002; Rushbrook 2002; Waitt 2003). Thus, gay areas of cities were no longer hidden, but instead actively commoditized and promoted for tourism and business investment and were no longer seen as challenging heteronormativities.

The commodification and consumption of gay spaces by an ever-widening variety of urbanites have led to a critical exploration of the potential "de-gaying" (and relatedly depoliticizing) of traditional gay commercial spaces as they are increasingly frequented by heterosexuals, who constitute a growing component of a district's residential population (Collins 2004; Ruting 2008). Moreover, the growth of the internet and social media is arguably making the need for actual physical meeting places obsolete (Usher & Morrison 2010; Mowlabocus 2012). The growing social, legal, and political acceptance of LGBT and queers offers the opportunity for greater visibility in places outside of the traditional village, and suggests that the gay village may be less central or even necessary to many LGBT and queer lives (Nash 2013; Nash & Gorman-Murray 2014; Gorman-Murray & Nash 2015).

Ruralities and migration

With the amount of research undertaken on urban neighborhoods as well as their visibility and political clout, the city came to be understood as the sole location where one could assume a homosexual identity within its mainly anonymous spaces. Early research tended to reify rural to urban migrations as central to the ability to experience a fully "gay" or "lesbian" life, and urban spaces were often portrayed as the only places where a gay or lesbian identity was even possible. For example, Bech (1997: 217) argued, "homosexual existence is a phenomenon of the city and not just something occurring in the city ... the city is the social world proper of the homosexual, his life space." Rural spaces, by contrast, were often depicted as exclusively heterosexual in ways that not only excluded and oppressed non-normative sexualities, but also suggested that homosexuals did not belong or even exist in rural locales (Little 2002, 2007).

Critiques of the urban-focused literatures took a more nuanced approach to this seemingly unidirectional and spatially deterministic view of the politics of LGBT and queer lives, and highlighted the varied possibilities for movement and relocation across local and regional landscapes. Explorations moved beyond the urban/rural binary to explore intranational migration in Canada, the United States, Britain, and Australia, finding that a focus on the singular move to urban spaces neglects the complexities of travel that can characterize life courses (Smith & Holt 2005; Gorman-Murray 2009; Lewis 2012). With social and political gains, some LGBT and queer people are moving more freely across a variety of spaces, prompting work in new mobilities literature around the transformations in urban and rural spaces arising from new LGBT and queer mobilities (Nash & Gorman-Murray 2014; Gorman-Murray & Nash 2015). Research also considers the legal, social, and economic limitations and constraints experienced through migration and resettlement in host cities (Luibhéid 2008). As we will argue later, in discussions of homonormativity, escaping heteronormative contexts does not necessarily facilitate liberation for all. These uneven experiences of migration and (im)mobility have been addressed through discussions of "queer diasporas" emerging from explorations of gay tourism, the experiences of LGBT people within ethnic diasporas, and globalizing LGBTQ political activism, to which we now turn (Gorman-Murray 2009; Lewis 2012).

Global queer rights claims

From the late 1990s on, there has been an increasing internationalization of LGBT human rights activism that has supported the emergence of global queer rights movements such as Human Rights Watch and Amnesty International (Kollman & Waites 2009; Swiebel 2009). Various international efforts supporting LGBT and queer rights claims have resulted in the signing of such documents as the Declaration of Montreal (2006) and the Yagyakarta Principles (2006). As national contexts such as the United Kingdom are understood as places where LGBT human rights are seen to be "won," attention in international human rights activism has focused on influencing internal domestic policies and citizenship debates within countries where LGBT rights and protections are not in place (Petchesky 2000; Stychin & Herman 2000; Buss & Herman 2003; Graupner & Tahmindjis 2005; Corrêa, Petchesky, & Parker 2008).

Despite the apparent benefits of international LGBT rights initiatives, some scholarship has been critical of this global proliferation and circulation of specifically Western-centric conceptualizations of gender, sexuality, and embodiment. Sexual and gendered meanings associated with the global North are often problematic in states where these LGBT identities do not translate easily to local understandings of sexual and gendered practices, relationships, and desires (Altman 1997, 2001; Drucker 2000; Grewal & Kaplan 2001). The globalization of LGBT rights claims is often linked to the processes and critiques of globalization. For example, Altman (1997) links capitalism's global reach with the constitution of global consumer societies, where a particularly global North conceptualization of homosexual identity ("the global gay") was both produced and consumed as a political and commercial identity. Nevertheless, others argue for a more complex and nuanced consideration of the notion of "queer" globalization, critiquing suggestions that there is an unproblematic, unidirectional, and uniform appropriation of Western gay and lesbian identities around the globe (e.g., Cruz-Malavé & Manalansan 2002). Povinelli and Chauncey (1999: 439) argue that we need to be attentive to "the effect of increasing transnational mobility of people, media, commodities, discourses and capital on local, regional and national modes of sexual desire, embodiment and subjectivity," not only in places such as Asia and Africa, but in Central and Eastern Europe as well (Kulpa & Mizielińska 2011). In other words, globalizing processes are multidirectional and uneven, ensuring that ideas, identities, and politics travel in complex and unpredictable ways.

Reaffirming heteronormativities

Despite the increasing legislative protections and recognitions of LGBT communities in some contexts and the globalization agendas in relation to human rights, there continues to be significant resistance to LGBT equalities. The heteronormativities of nation-states are implicit in the globalized calls for LGBT human rights. Homophobic nation-states are often regarded as those located in the global South, with "gay-friendly" regions found in the "developed," "civilized," and "progressive" global North. In this conceptualization, entire countries, regions, and even continents (particularly Africa) are labeled homophobic. Other regions are then associated with "freedom," such as the United States, despite the existence of repressive legislation as well as the lived experiences of LGBT people, and the presence of groups, state legislators, and others who are aggressively opposed to LGBT and queer equalities. In the United Kingdom, for example, 161 members of parliament voted against gay marriage in 2013; and in Canada, opposition groups continue actively to oppose anti-gay bullying legislation (Nash & Browne 2015; Browne & Nash forthcoming).

Thus, making clear-cut distinctions between states that are supportive of LGBT and queer people and those that are not is not only problematic because of the heterogeneity of states, but because it also ignores the transnational operation of contestations of LGBT rights. The successful passage of recent anti-gay legislation in Uganda has been closely tied to support from and intervention by Christian and conservative groups from the United States. In a related example, Valentine et al. (2012) illustrate the transnational flows within the Anglican commune, such that those who are opposed to LGBT equalities in places such as the United States and the United Kingdom have worked in South Africa in order to push for, and support, repressive legislation and church doctrine in these areas. Conversely, those who attend supposedly "homophobic" churches in New York have been shown to hold a variety of views and opinions on homosexuality and religion (Valentine et al. 2012).

Research by Weiss and Bosia (2013) has focused on the homogeneity and modular nature of homophobia globally, arguing that while much scholarship has paid attention to the "diffusion" of LGBT human rights, less attention has been paid to the diffusion of homophobia through forms of globalization. While agreeing that more work is needed on homophobia, Nash and Browne (2015) have argued against the homogeneity and globalized assertions of straightforward diffusions of homophobia. Exploring new manifestations of heteronormativities in countries where LGBT equalities have been gained, they have shown that the term "homophobia" is problematized by groups in Canada and the United Kingdom. So, while groups argue against, for example, the introduction of same-sex marriage, these groups contend that they are not bigoted or homophobic, but instead are seeking to preserve the privileged status of heterosexualities (Browne & Nash forthcoming). Given this, such groups assert that they are not "against gays" so much as claiming a privileged position for "one man, one woman marriage" as "best for society."

Homonormativity

As gays and lesbians have increasingly acquired inclusion in state institutions in some countries, a growing critique suggests that such assimilationist tendencies reflect a "sexual politics of neoliberalism," privileging those gays and lesbians who reproduce middle-class, monogamous, and normatively gendered values. As Lisa Duggan (2002) argues, this "new homonormativity" works to inscribe suitable norms of sexual and gendered expression within an acceptable (and arguably increasingly limited) "gay public sphere" (in Bell & Binnie 2004:

1811). In other words, it is argued that we are seeing the normalization of certain forms of "acceptable" homosexuality deemed suitable and appropriate within wider social and political contexts.

This contrasts with those who refute these norms, and the neoliberal sanitizations of gay identities, practices, and spaces. Those not fitting into these normalized gendered and sexual positionings or those not able to afford to frequent these commodified and normative gay and lesbian practices are excluded. For example, Taylor (2007) has discussed the ways in which working-class lesbians do not "fit" into gay and lesbian spaces, and Doan (2007, 2010) has pointed to the marginalizations and exclusions of trans* people from LGBT spaces (see also Nash 2010).

Non-normatively gendered and sexual individuals might also be informally excluded in order to make commodified gay and lesbian commercial locations compatible with new cosmopolitan consumers. As Binnie (2004) argues, those "queer unwanted" who do not conform with homonormative mainstream expectations can find themselves excluded from prepackaged and widely marketed, trendy gay venues, highlighting the "difference, unevenness and geographical specificity" of various urban spaces (Brown 2009: 1498). Bell and Binnie (2004) contend that the "interweaving of urban governance and sexual citizenship agendas produces particular kinds of sexual spaces," raising thorny questions about authenticity, commodification, and exclusion (2004: 1807), and creating the figure of the "good gay" versus the "bad gay."

Discussions of homonormativities disrupt the focus of political geographies on heteronormativities, highlighting the multiple ways in which sexualities operate in intersectional ways to create new normativities. Recent political movements that call for equalities through marriage and access to the military have pointed to the limits of homonormative politics, which refuse to critique neoliberal, class, and race orders, instead reproducing repressive and violent state institutions and practices (Warner 1999; Halberstam 2005; Spade 2013). Spade (2013) argues that trans* people should not utilize the same strategies for gaining equalities that lesbian and gay political movements have deployed, contending that these not only leave some queer people out in the cold (Sears 2005), they also shore up the power of the state by using state violence and incarceration.

Homonationalism

Recently, queer analytical approaches have decentered and denaturalized the assumed linkages between certain hetero- and homonormatively understood bodies, practices, and identities and their linkages to understandings of citizenship and global geopolitics (Isin 2002; Puar 2007; Payne & Davies 2012). The work by Jasbir Puar (2007, 2013) is ground-breaking in pushing forward political geographies of sexualities through her exploration of what she terms a "national homosexuality" arising in certain countries where LGBT and queer people enjoy full citizenship and state protections. In exploring this further, Puar develops the concept of "homonationalism" in order to work through the "complexities of how 'acceptance' and 'tolerance' for gay and lesbian subjects have become a barometer by which the right to and capacity for national sovereignty is evaluated" (2013: 366). In other words, claims for "tolerance" of sexual diversity and the provision of human rights protections for LGBT people is often positioned as an important value, demonstrating the modern civility and liberal democratic ideals largely associated with the democracies of the global North (Kirchick 2009; Hockberg 2010; Ahmed 2011; Puar 2011). Arguably, this reflects a form of homonationalism, where "gay-friendly" states are understood as "modern, cosmopolitan, developed, first-world ... and democratic," while those states regarded as "anti-gay" are represented as uncivilized and backward (Puar 2011: 138; see also Hockberg 2010). Using homonationalism as an analytical category is useful for understanding how and why a nation's status as "gay friendly" has become a positive strategic positioning within global geopolitics (Hyndman et al. 2010). For example, in her discussion of OutRage!, an LGBT organization in the United Kingdom, Puar notes that it draws on forms of homonationalism to condemn the actions of certain Muslim groups. As she notes, OutRage! argues

> that Muslims are an especial threat to homosexuals, that Muslim fundamentalists have deliberately and specifically targeted homosexuals and that the parameters of this opposition correlate with those of the war on terror: civilization versus barbarianism. (2007: 20)

One specific form of homonationalism is said to be visible in the actions of the state of Israel, in its drive to rebrand itself as "an oasis of liberal tolerance in a reactionary religious backwater" for LGBT people in the Middle East (Kirchick 2009). Termed "pinkwashing," these activities are a form of homonationalism in which a state such as Israel asserts its gay-positive attributes in order to deflect attention from what many regard as its violent degradation and containment of Palestinians. While some argue that LGBT and queer people need to defend Israel given Palestinian hostility to gays and lesbians (e.g., Kirchick 2009; StandWithUs 2010), others contest the use of gay and lesbian rights discourses to redirect attention away from oppressive policies and regimes (Ahmed 2011; Puar 2011).

Puar (2013) argues that pinkwashing should be understood as a practice arising from various forms of homonationalism, which highlights how scholars need to be attentive to the "global conditions of homonationalism that make a practice such as Israeli pinkwashing possible and legible in the first place" (Puar 2013: 337–338). Arguably, then, pinkwashing is one particular manifestation of homonationalism; there is the potential to consider its applicability to other contexts. What is apparent is that pinkwashing is not solely related to Israel, but can be applied more broadly to consider how nations use sexual and gendered equalities to argue for more curbs and controls on others. For example, in Britain the Conservative–Liberal Democrat coalition passed the Marriage (Same Sex) Act in 2013, while only months later the Home Secretary was arguing for repeal of the "unnecessary" Human Rights Act, in order to allow the UK government to more easily deport migrants and to neglect the rights of prisoners. The coalition's support of same-sex marriage, then, could be described as pinkwashing, where it supports same-sex marriage while undercutting legislation that offers protection to a broad range of individuals.

In another example, Deborah Cowen considers the implications of the Canadian military marching in Pride Toronto for the first time in 2008. While crowds cheered those queers in uniform, she argues that their presence raises difficult questions about how the most disciplinary and violent institution of the state has "been incorporated into a community that was historically constituted by contesting norms and making bold political claims" (quoted in Hyndman et al. 2010: 400; see also Lamble 2013). Cowen argues that we need to think carefully about the contingency and spatiality of homonationalism as it is practiced, given that "both the British and Canadian armed forces, but not the American ones, are marching in queer pride parades" (quoted in Hyndman et al. 2010: 400). Both the Canadian and British militaries are seeking to recruit LGBT and queer people through their presence at Pride at a time when volunteer armies are having trouble recruiting in states with largely urban, professional populations in advanced capitalist societies. In other words, understanding where, how, and for what purpose LGBT and queer people are incorporated into the nationalist project matters in determining the use to which LGBT and queer people are put in the service of nationalist interests.

The incorporation of visible lesbians and gay men also operates potentially to enhance the state's reputation on the global stage. Yet, more than this, "acceptance" of gay and lesbian citizens and troops can operate as a reason for invading other countries purportedly to "liberate" their homosexual populations. Discussions of Abu Ghraib (an American prison in Iraq where guards posed prisoners in degrading and sexualized positions) have pointed to how US guards served to conduct a "war of images on two fronts – against the Islamic terrorists abroad and emancipatory social movements at home" (Pease 2007: 82; see also Puar 2005, 2006, 2007). By placing Iraqi men in compromising positions suggestive of homosexual intimacy, the guards not only humiliated Muslim men, for whom this would be a grave sin, they also insulted LGBT and queer people by mocking their sexual relations. This illustrates how links between nationalism, homonationalism, and sexual citizens are complex, often incoherent sets of relations.

Conclusions

This chapter illustrates both the importance of political geographies to considerations of sexualities, and, conversely, that sexualities are important for understanding political geographies. Focusing on two axes of normativities (hetero- and homonormativities), we illustrate how these normativities are both formed through and also (re)form key objects of analysis for political geographers, including social movements, nation-states, transnational interactions, and globalizations. Significant transformations in how sexualities (and gender identities) have been taken up by geographers have promoted examinations of the centrality of sexual citizenship and considerations of the emergent question of the nature of homonormativities and homonationalisms. The concept of homonormativity engenders a consideration of how it is that some gays and lesbians are incorporated into the fabric of the nation, as citizens, while other queers remain excluded. An exploration of the various forms of homonationalism allows for a complex understanding of how sexualities are normalized or rendered abject in diverse ways. As individuals and groups continue to oppose LGBT and queer rights (including same-sex marriage) in places where LGBT and queer rights appear to have been "won," it is important to pay attention to how heteronormativities are reiterated and reproduced.

This chapter, of course, has not been able to do justice to the full range of possibilities of political geographies of sexualities, and indeed the areas that are addressed have much more depth and complexity than has been presented here. We have deliberately targeted this chapter at those unfamiliar with geographies of sexualities, to illustrate the ways in which political geographies are sexualized, with the hope that this whets the reader's appetite for more.

Acknowledgments

We would like to thank the reviewers for their insightful comments. This research is funded in part by a grant from the Canadian Social Sciences and Humanities Research Council (SSHRC grant #403-2012-0032).

References

Ahmed, S. (2011). Problematic proximities: Or why critiques of gay imperialism matter. *Feminist Legal Studies* 19: 119–132. doi:10.1007/s10691-011-9180-7

Altman, D. (1997). Global gaze/global gays. *GLQ: A Journal of Gay and Lesbian Studies*, 3(4): 417–436. doi:10.1215/10642684-3-4-417

Altman, D. (2001). *Global Sex*. Chicago, IL: University of Chicago Press.

Armstrong, E.A. (2002). *Forging Gay Identities: Organizing Sexuality in San Francisco, 1950–1994*. Chicago, IL: University of Chicago Press.

Bech, H. (1997). *When Men Meet: Homosexuality and Modernity*, trans. T. Mesquit & T. Davies. Cambridge: Polity Press.

Bell, D., & Binnie, J. (2000). *The Sexual Citizen*. Cambridge: Blackwell.

Bell, D., & Binnie, J. (2004). Authenticating queer space: Citizenship, urbanism and governance. *Urban Studies* 41(9): 1807–1820. doi:10.1080/0042098042000243165

Bell, D., & Valentine, G. (1995). *Mapping Desire: Geographies of Sexualities*. London: Routledge.

Bell, D., Binnie, J., Cream, J., & Valentine, G. (1994). All hyped up and no place to go. *Gender, Place and Culture* 1(1): 31–47. doi:10.1080/09663699408721199

Binnie, J. (2004). *The Globalization of Sexuality*. London: Sage.

Binnie, J., & Skeggs, B. (2004). Cosmopolitan knowledge and the production and consumption of sexualized space: Manchester's gay village. *Sociological Review* 52: 39–62. doi:10.1111/j.1467-954X.2004.00441.x

Boudreau, J.A., Keil, R., & Young, D. (2009). *Changing Toronto: Governing Urban Neoliberalism*. Toronto: University of Toronto Press.

Brown, G. (2009). Thinking beyond homonormativity: Performative explorations of diverse gay economies. *Environment and Planning A: Society and Space* 41 (6): 1496–1510. doi:10.1068/a4162

Brown, W. (2009). *Regulating Aversion: Tolerance in the Age of Identity and Empire*. Princeton, NJ: Princeton University Press.

Browne, K., & Bakshi, L. (2011). We are here to party? Lesbian, gay, bisexual and trans leisurescapes beyond commercial gay scenes. *Leisure Studies* 30: 179–196. doi:10.1080/02614367.2010.506651

Browne, K.A., & Nash, C.J. (forthcoming). Resisting LGBT rights where "we have won": Canada and England. *Journal of Human Rights*, special Issue, eds A. Langlois & C. Wilkinson. 13(3).

Browne, K., Lim, J., & Brown, G. (eds) 2007. *Geographies of Sexualities*. London: Ashgate.

Buss, D., & Herman, D. (2003). *Globalizing Family Values: The Christian Right in International Politics*. Minneapolis, MN: University of Minnesota Press.

Castells, M. (1983). *The City and the Grassroots*. Berkeley, CA: University of California Press.

Cervulle, M., & Pailler, F. (2013). Twitter pour tous: Debating same-sex unions in the French digital public sphere. Available from the authors.

Collins, A. (2004). Sexual dissidence, enterprise and assimilation: Bedfellows in urban regeneration. *Urban Studies* 41: 1789–1806. doi:10.1080/0042098042000243156

Cooper, D. (1994). *Sexing the City: Lesbian and Gay Politics within an Activist State*. London: Rivers Oram.

Corrêa, S., Petchesky, S., & Parker, R. (2008). *Sexuality, Health and Human Rights*. New York: Routledge.

Cruz-Malavé, A., & Manalansan, M.F. (2002). *Queer Globalizations: Citizenship and the Afterlife of Colonialism*. New York: New York University Press.

Doan, P.L. (2007). Queers in the American city: Transgendered perceptions of urban space. *Gender, Place and Culture* 14(1): 57–74. doi:10.1080/09663690601122309

Doan, P.L. (2010). The tyranny of gendered spaces – reflections from beyond the gender dichotomy. *Gender, Place and Culture* 17(5): 635–654. doi:10.1080/0966369X.2010.503121

Drucker, P. (ed.) (2000). Introduction: Remapping sexualities. In Drucker, P.R. (ed.), *Different Rainbows*, London: Gay Men's Press, 9–42.

Duggan, L. (2002). The new homonormativity: The sexual politics of neoliberalism. In Castronoo, R., & Nelson, D. (eds), *Materializing Democracy: Towards a Revitalized Cultural Politics*. New York: New York University Press, 175–194.

Florida, R. (2002). *The Rise of the Creative Class and How It's Transforming Work, Leisure, Community and Everyday Life*. New York: Basic Books.

Forest, B. (1995). West Hollywood as symbol: The significance of place in the construction of a gay identity. *Environment and Planning D: Society and Space* 13(2): 133–157. doi:10.1068/d130133

Gorman-Murray, A. (2009). Intimate mobilities: Emotional embodiment and queer migration. *Social and Cultural Geography* 10(4): 441–460. doi:10.1080/14649360902853262

Gorman-Murray, A., & Nash, C.J. (2015). Mobile places, relational spaces: Conceptualizing an historical geography of Sydney's LGBTQ neighbourhoods. *Environment and Planning D: Society and Space* 32(4): 622–641.

Graupner, H., & Tahmindjis, P. (2005). *Sexuality and Human Rights: A Global Overview*. New York: Harrington Park Press.

Grewal, I., & Kaplan, C. (2001). Global identities: Theorizing transnational studies of sexuality. *GLQ: A Journal of Gay and Lesbian Studies*, 7(4): 663–679. doi:10.1215/10642684-7-4-663

Halberstam, J. (2005). *In a Queer Time and Place: Transgender Bodies, Subcultural Lives*. New York: New York University Press.

Harvey, D. (1989). *The Condition of Postmodernity*. Oxford: Blackwell.

Hockberg, G.Z. (2010). Israelis, Palestinians, queers: Points of departure. *GLQ: A Journal of Lesbian and Gay Studies* 16(4): 493–516. doi:10.1215/10642684-2010-001

Hubbard, P. (2013). Kissing is not a universal right: Sexuality, law and the scales of citizenship. *Geoforum* 49: 224–232. doi:10.1016/j.geoforum.2012.08.002

Hyndman, J., Cowen, D., Oswin, N., Oza, R., & Kobayashi, A. (2010). Jasbir K. Puar, Terrorist Assemblages: Homonationalism in queer times. *Social and Cultural Geography* 11(4): 399–409. doi:10.1080/14649361003787714

Isin, E.F. (2002). *Being Political: Genealogies of Citizenship*. Minneapolis, MN: University of Minnesota Press.

Isin, E.F. (2011). Citizens without nations. *Environment and Planning D: Society and Space* 30(3): 450–467. doi:10.1068/d19210

Johnston, L., & Longhurst, R. (2010). *Space, Place, and Sex: Geographies of Sexualities*. Lanham, MD: Rowman and Littlefield.

Kirchick, J. (2009). Queers for Palestine? *Advocate* January 28. http://www.advocate.com/politics/2009/01/28/queers-palestine?page=full, accessed March 26, 2015.

Knopp, L. (1995). Sexuality and urban space: A framework for analysis. In Bell, D.J., & Valentine, G. (eds), *Mapping Desire: Geographies of Sexualities*, New York: Routledge, 149–160.

Knopp, L. (1998). Sexuality and urban space: Gay male identity politics in the United States, the United Kingdom, and Australia. In Fincher, R., & Jacobs, J.M. (eds), *Cities of Difference*, New York: Guilford Press, 149–176.

Kollman, K., & Waites, M. (2009). The global politics of lesbian, gay, bisexual and transgender human rights: An introduction. *Contemporary Politics* 15(1): 1–17. doi:10.1080/13569770802674188

Kulpa, R., & Mizieliñska, J. (2011). De-centring western sexualities: Central and eastern European perspectives. Farnham: Ashgate.

Lamble, S. (2013). Queer necropolitics and the expanding carceral state: Interrogating sexual investments in punishment. *Law Critique* 24(3): 229–253. doi:10.1007/s10978-013-9125-1

Leitner, H., Peck, J., & Sheppard, E. (eds) (2007). *Contesting Neoliberalism: Urban Frontiers*. Baltimore, MD: Johns Hopkins University Press.

Lewis, N. (2012). Remapping disclosure: Gay men's segmented journeys of moving out and coming out. *Social and Cultural Geography* 13(3): 211–231. doi:10.1080/14649365.2012.677469

Little, J. (2002). Rural geography: Rural gender identity and the performance of masculinity and femininity in the countryside. *Progress in Human Geography* 26: 665–670. doi:10.1191/0309132502ph394pr

Little, J. (2007). Constructing nature in the performance of rural heterosexualities. *Environment and Planning D: Society and Space* 25(5): 861–866. doi:10.1068/d2605

Luibhéid, E. (2008). Queer/migration: An unruly body of scholarship. *GLQ: A Journal of Lesbian and Gay Studies* 14(2–3): 169–190. doi:10.1215/10642684-2007-029

Markwell, K. (2002). Mardi Gras tourism and the construction of Sydney as an international gay and lesbian city. *GLQ: A Journal of Lesbian and Gay Studies* 8(1–2): 81–99.

McDowell, L. (1995). Body work: Heterosexual gender performances in city workplaces. In Bell, D., & Valentine, G. (eds), *Mapping Desire: Geographies of Sexualities*, London: Routledge, 75–98.

Mowlabocus, S. (2012). *Gaydar Culture: Gay Men, Technology and Embodiment in the Digital Age.* Burlington, VT: Ashgate.

Nash, C.J. (2001). Siting lesbians: Sexuality, space and social organization. In Goldie, T. (ed.), *Queer Country: Gay and Lesbian Studies in the Canadian Context*, Vancouver: Arsenal Press, 235–256.

Nash, C.J. (2005). Contesting identity: Politics of gays and lesbians in Toronto in the 1970s. *Gender, Place and Culture* 12(1): 13–135. doi:10.1080/09663690500083115

Nash, C.J. (2006). Toronto's gay village (1969 to 1982): Plotting the politics of gay identity. *Canadian Geographer* 50(1): 1–16. doi:10.1111/j.0008-3658.2006.00123.x

Nash, C.J. (2010). Trans geographies, embodiment and experience. *Gender, Place and Culture* 17(5): 579–595. doi:10.1080/0966369X.2010.503112

Nash, C.J. (2013). Queering neighbourhoods: Politics and practice in Toronto. *ACME: International E-Journal for Critical Geographies* 12(2): 193–213.

Nash, C.J., & Browne, K. (2015). Best for society? Transnational opposition to sexual and gender equalities in Canada and Great Britain. *Gender, Place and Culture* 22(4): 561–577. doi:10.1080/0966369X.2014.885893

Nash, C.J., & Gorman-Murray, A. (2014). LGBT neighbourhoods and "new mobilities": Towards understanding transformations in sexual and gendered urban landscapes. *International Journal of Urban and Regional Research* 38(3): 756–772. doi:10.1111/1468-2427.12104

Payne, R., & Davies, C. (2012). Introduction to the special section: Citizenship and queer critique. *Sexualities* 15(3/4): 251–256. doi:10.1177/1363460712436461

Pease, D.E. (2007). Between the homeland and Abu Ghraib: Dwelling in Bush's biopolitical settlement. In Dawson, A., & Schueller, M.J. (eds), *Exceptional State: Contemporary US Culture and the New Imperialism*. Durham, NC: Duke University Press.

Petchesky, R. (2000). Sexual rights: Inventing a concept. Mapping international practice. In Parker, R., Barbosa, R.M., & Aggleton, P. (eds), *Framing the Sexual Subject: The Politics of Gender, Sexuality and Power*. Berkeley, CA: University of California Press, 81–103.

Podmore, J. (2001). Lesbians in the crowd: Gender, sexuality and visibility along Montréal's Boul. St-Laurent. *Gender, Place and Culture* 8(4): 333–355. doi:10.1080/09663690120111591

Podmore, J. (2006). Gone "underground"? Lesbian visibility and the consolidation of queer space in Montréal. *Social and Cultural Geography* 7(4): 595–625. doi:10.1080/14649360600825737

Povinelli, E., & Chauncey, G. (1999). Thinking sexuality transnationally. *GLQ: A Journal of Lesbian and Gay Studies* 5(4): 439–450.

Probyn, E. (2000). Sporting bodies: Dynamics of shame and pride. *Body and Society* 6(1): 13–28. doi:10.1177/1357034X00006001002

Puar, J. (2005). On torture: Abu Ghraib. *Radical History Review* 93(1): 13–38. doi:10.1215/01636545-2005-93-13

Puar, J. (2006). Mapping U.S. homonormativities. *Gender, Place, Culture* 13(1): 67–88. doi:10.1080/09663690500531014

Puar, J. (2007). *Terrorist Assemblages: Homonationalism in Queer Times.* Durham, NC: Duke University Press.

Puar, J. (2011). Citation and censorship: The politics of talking about the sexual politics of Israel. *Feminist Legal Studies* 19(2): 133–142. doi:10.1007/s10691-011-9176-3

Puar, J. (2013). Rethinking homonationalism. *International Journal of Middle East Studies* 45(2): 336–339. doi:10.1017/S002074381300007X

Richardson, D. (2000). Claiming citizenship? *Sexualities* 3(2): 271–288.

Rothenberg, T. (1995). And she told two friends: Lesbians creating urban social space. In Bell, D., & Valentine, G. (eds), *Mapping Desire: Geographies of Sexualities*. London: Routledge, 165–181.

Rushbrook, D. (2002). Cities, queer space, and the cosmopolitan tourist. *GLQ: A Journal of Lesbian and Gay Studies* 8(1–2): 183–206.

Ruting, B. (2008). Economic transformations of gay urban spaces: Revising Collins' evolutionary gay district model. *Australian Geographer* 39(3): 259–269. doi:10.1080/00049180802270465

Sears, A. (2005). Queer anti-capitalism: What's left of lesbian and gay liberation? *Science and Society* 69(1): 95–112. doi:10.1521/siso.69.1.92.56800

Smith, D., & Holt, L. (2005). "Lesbian migrants in the gentrified valley" and "other" geographies of rural gentrification. *Journal of Rural Studies* 21(3): 313–322. doi:10.1016/j.jrurstud.2005.04.002

Spade, D. (2013). Under the cover of gay rights. *NYU Review of Law and Social Change* 79: 1–23.

StandWithUs (2010). LGBT rights under the Palestinian authority: Know the facts. 8 July. www.standwithus.com/pdfs/flyers/LGBT_booklet.pdf, accessed February 20, 2014.

Stenning, A., Smith A., Rochovska, A., & Swiatek, D. (2010). *Domesticating Neoliberalism: Spaces of Economic Practice and Social Reproduction in Post-socialist Cities*. Oxford: Wiley-Blackwell.

Stychin, C., & Herman, D. (2000). *Sexuality in the Legal Arena*. London: Athlone.

Stryker, S. (2009). *Transgender History*. London: Seal Press.

Swiebel, J. (2009). Lesbian, gay, bisexual and transgender human rights: The search for an international strategy. *Contemporary Politics* 15(1): 19–35. doi:10.1080/13569770802674196

Taylor, Y. (2007). *Working Class Lesbian Life: Classed Outsiders*. New York: Palgrave Macmillan.

Usher, N., & Morrison, E. (2010). The demise of the gay enclave, communication infrastructure theory, and the transformation of gay public space. In Pullen, C., & Cooper, M. (eds), *LGBT Identity and Online New Media*, New York: Routledge, 271–287.

Valentine, G. (1993). Heterosexing space: Lesbian perceptions and experiences of everyday spaces. *Environment and Planning D: Society and Space* 11(4): 395–413. doi:10.1068/d110395

Valentine, G. (1995). Out and about: Geographies of lesbian landscapes. *International Journal of Urban and Regional Research* 19: 96–112.

Valentine, G., Vanderbeck, R.M., Sadgrove, J., Andersson, J., & Ward, K. (2012). Transnational religious networks: Sexuality and the changing power geometries of the Anglican Communion. *Transactions of the Institute of British Geographers* 38(1): 50–56. doi:10.1111/j.1475-5661.2012.00507.x

Waitt, G. (2003). Gay games: Performing "community" out from the closet of the locker room. *Social and Cultural Geography* 4(2): 167–183. doi:10.1080/14649360309067

Warner, M. (1999). Normal and normaller: Beyond gay marriage. *GLQ: A Journal of Lesbian and Gay Studies* 5(2): 119–171.

Weiss, M.L., & Bosia, M.J. (2013). *Global Homophobia: States, Movements and the Politics of Pppression*. Champaign, IL: University of Illinois Press.

Whittle, S. (1994). *The Margins of the City: Gay Men's Urban Lives*. Manchester: Arena.

Chapter 28

The Rise of the BRICS

Marcus Power

University of Durham, England, UK

The BRICs acronym was first coined in 2001 by British economist Jim O'Neill, then working as the Head of Global Economics Research at the multinational global investment firm Goldman Sachs. Although his paper entitled "Building better global economic BRICs" (O'Neill 2001) did not cause an immediate splash and its main points were not seen as especially profound at the time (O'Neill 2011), he argued that Brazil, Russia, India, and China, which at the time controlled 8% of world gross domestic product (GDP), would see their share of the global economy grow significantly over the next decade, and that world policymaking forums (such as the G7 or Group of 7) would consequently need to be reorganized. The conceptual genealogy of the term BRICs can thus be traced to the investment discourses, financial industry narratives, statistical tools, and common classificatory regimes used by investment banks like Goldman Sachs (Wansleben 2013) and across the boundaries demarcated in these calculative devices and classificatory regimes between low-risk developed economies and high-risk developing economies, and through the assumptions regarding return expectations and the risks assigned to states by these distinctions. The BRICs acronym has since come into widespread use as a symbol of the apparently epochal shift in global economic power away from the developed G7 economies toward the "developing world", and the wider realignment of world economic and ultimately political power that would be engendered by the collective influence of these four countries. Foreign ministers from the BRICs countries met in New York in September 2006, beginning a series of high-level meetings before the first full-scale diplomatic meeting was held in Yekaterinberg in Russia in May 2008.

According to O'Neill, one of the major consequences of the BRICs story that he has sought to narrate has been the desire of other "emerging economies" with large populations to get into the BRICs "club" (O' Neill 2011). In 2005, in recognition of this, Goldman Sachs developed the idea of the "Next Eleven" (N-11) group of emerging economies: Bangladesh, Egypt, Indonesia, Iran, South Korea, Mexico, Nigeria, Pakistan, the Philippines, Turkey, and Vietnam (Goldman Sachs 2005). Four of these in particular (Mexico, Indonesia, South Korea, and

The Wiley Blackwell Companion to Political Geography, First Edition.
Edited by John Agnew, Virginie Mamadouh, Anna J. Secor, and Joanne Sharp.
© 2015 John Wiley & Sons Ltd. Published 2017 by John Wiley & Sons Ltd.

Turkey) were given their own acronym of "MIST," as they were regarded as sufficiently large and important to be considered "growth markets" along with the BRICs. South Africa began efforts to join the BRICs grouping in 2010 and the process for its formal admission began in August of that year, with South African President Jacob Zuma attending the 2011 BRICS summit as a full member. In 2013, O'Neill also endorsed the acronym MINT, which replaced South Korea (now regarded as already "developed") with the growing economy of Nigeria. Analysts from rival banks and investment firms have of course also sought to move beyond the BRICS concept as investment funds in search of new opportunities and pastures have proliferated. They have thus introduced their own groupings of "emerging markets", including the CIVETs from HSBC (Colombia, Indonesia, Vietnam, Egypt, Turkey, and South Africa), the EAGLES (emerging and growth-leading economies) from the Spanish bank BBVA, and the "7% Club" (including those countries that have averaged economic growth of at least 7% a year) from Standard Chartered Bank.

Differences between members of the original BRICs "club" remain substantial, however. Brazil and India do not want the Chinese renminbi as a global reserve currency as long as the government in Beijing still has a massive influence on the rate of exchange. Furthermore, not all the BRICs are equally committed to each other; nor is this commitment consistent across areas at issue. The relationship between India and China is undoubtedly the most difficult in the group, marked as it is by a history of one major war and persistent border disputes. Rather than inviting a comparison with other members of the original BRICs, in many ways it makes far more sense to compare Russia to the countries with which it shares the myriad after-effects of state socialism. Many commentators have also suggested that China is in a league of its own compared to the other BRICs countries. As Rothkopf (2009) has argued:

> Without China, the BRICs are just the BRI, a bland, soft cheese that is primarily known for the whine [sic] that goes with it. China is the muscle of the group and the Chinese know it. They have effective veto power over any BRIC initiatives because without them, who cares really? They are the one with the big reserves. They are the biggest potential market.

China's economy is larger than that of the three other original BRICs economies (Brazil, Russia, and India) combined. Moreover, China's exports and its official foreign exchange reserves are more than twice as large as those of the other BRICs combined. China alone accounts for more than 70% of the combined GDP growth generated by the BRICs countries from 1999 to 2010 and so, "if there is a 'BRICs miracle' it is perhaps first and foremost a Chinese one" (Firzli 2011: 1). With the inclusion of South Africa, the five BRICS countries now represent almost 3 billion people with a combined nominal GDP of US$16.039 trillion and an estimated US$4 trillion in combined foreign reserves (Marquand 2011; IMF 2013). Over the past decade, foreign direct investment (FDI) inflows to the BRICS countries have more than tripled, reaching an estimated US$263 billion in 2012 (UNCTAD 2013a). As a result, their share in world FDI flows kept rising even during the recent global economic crisis, reaching 20% in 2012, up from 6% in 2000. The BRICS countries have also become important investors – their outward FDI has risen from US$7 billion in 2000 to $126 billion in 2012, and now accounts for around 9% of world flows. Ten years before, that share was only 1%. Almost half (46%) of FDI flows to BRICS go to China, followed by Brazil (25%), Russia (17%), and India (10%; UNCTAD 2013a).

Despite the popularity and widespread use of the term, there remains a relatively limited amount of academic literature on the role of the BRICS in terms of global governance institutions, foreign policy, or development aid. In recent years, both the architecture of international governance and the established modes of development cooperation have been increasingly transformed by the (re)emergence of the BRICS as development donors, with important implications for the global geopolitical ecology of investment, production, and trade (Power & Mohan 2010). This chapter seeks to explore the changing political geographies resulting from the rise of the BRICS and is divided into three main sections. The first considers the historical rise of the BRICS countries (with particular reference to China, India, and Brazil) and their recent attempts to "go global" or to increase their influence across the global South. The second section goes on to consider what their rise means for the pre-existing hegemonic powers of the world system and the possibility of multipolarity, examining in particular the extent to which the BRICS have increasingly come to play an active role in the reform of global economic and political governance. The third and concluding section then discusses what the rise of the BRICS might mean for the contemporary study of international relations.

The rise of the BRICS in historical context

China

Since the late 1970s, China has experienced extraordinary economic growth (exceeding 9% per year over more than 20 years), brought on by the privatization and denationalization of economic enterprises, the opening of export markets, and the new partnerships of state cadres with transnational investors, and with local township, provincial, and national domestic entrepreneurs (Nonini 2008). What we are witnessing is "perhaps the most explosive and long-sustained period of economic advance that the world has ever seen" (Nolan 2004: 1). China's reform and "opening up" have seen exports grow from US$18.1 billion in 1978 to over US$2.21 trillion in 2013 (Nolan 2004; China Daily 2014). Together with high levels of public investment, FDI in China has been crucial to this growth and to the wider restructuring of the Chinese economy. From 1985 to 2013, FDI inflows rose from US$2 billion to US$117.59 billion (Xinhua 2014). There has also been a significant outflow of investments from China into the global economy over the past five years, a period that corresponded with a massive upsurge in foreign reserve accumulation and a significant expansion of China's domestic economy.

Following the death of Mao Zedong in 1976, China was on the verge of ruin and still coming to terms with the collective trauma that resulted from the failures of the Great Leap Forward and the Cultural Revolution, combined with the additional threat of impending aggression from Communist Vietnam. Chinese leaders initially identified markets as part of a necessary capitalist stage of freeing the forces of production in order to "re-socialize" them within a prosperous future socialist economy (Meisner 1999). Consequently, the Chinese state and its governing logics since 1978 have represented "a recombinant or hybrid assortment of oligarchic institutions, practices and disciplines of power" that have juxtaposed older elements of Maoist governance (e.g., central planning and an ideology of socialist paternalism toward "peasants" and "workers") with elements of market liberalization in a kind of "slow-tempo improvisation" (Nonini 2008: 156). This has been aimed simultaneously at developing China's forces of production, preserving the position and legitimacy of the Chinese Communist Party (CCP), and, particularly since the 1980s, consolidating the base of economic accumulation of China's "cadre-capitalist" class (So 2005). The "market socialism" that Deng Xiaoping ushered in has thus been combined with pre-existing Maoist developmentalist, nationalist, and socialist discourses and practices in an uneasy synthesis held together by the authoritarian rule of the CCP (Lin 2006).

A key part of what drove Deng's economic reforms was the crisis under state socialism (Wu 2008), whereby effective accumulation supported by state-led extensive industrialization had reached its limit. China's economic development has thus been co-produced by capital accumulation strategies both on the "inside" (the domestic restructuring of state-owned enterprises or SOEs and macroeconomic strategies) as well as on the "outside" (the global search for new markets and corresponding foreign demand for financial capital; Lim 2010). In 2001, the government released its Tenth Five-Year Plan (2001–05), which officially included a "going out" strategy aimed at raising the rate of outflow FDI. The central government would provide support to SOEs with a view to enhancing their economic competitiveness, ostensibly so that they could become "national champions", analogous to those in other East Asian economies. In many ways the "going out" strategy has been a success for China – in 1995 there were only three Chinese companies in the Fortune 500 list, but by 2015 that number had risen to 95.

China regularly claims to be sharing this recent experience with its partners in the name of South–South cooperation and it is often held up as a "model" of economic development and (as a consequence) poverty alleviation from which other countries from the South can learn. Other commentators have seen this "model" as an alternative to the "American model" or as "defying the conditionalities of the Bretton Woods institutions" (Manji & Marks 2007: 136). Joseph Stiglitz (2002) even touts China as a "model" for how developing nations should rise and escape the prescriptions of Anglo-American neoliberalism. Yet, there is no single, coherent "Chinese model" (Friedman 2009). Neither is there a unitary, singular, and coherent Chinese economic actor. Implying that there is considerably downplays the global interconnectedness and transnational dynamics that shape China's economic development strategies, both domestically and overseas. There is a danger here of overlooking the co-production of China's economic development (in which Africa is playing an increasingly important role) and missing the many ways in which the Chinese economy is intertwined with the fate of many other economies and corporations. We might also question the supposed "Chinese" nature of this "model", given the extent to which China has looked to and drawn from East Asian examples of state practice in pursuing "development" (Power & Mohan 2010), following in the tracks of Japanese and Korean mercantilism, the so-called developmental state model.

Thus, the singularity and coherence of a Chinese "model" have often been considerably overstated, despite the growing difficulties that the Chinese state is facing in managing the complex range of corporate agents now active overseas. Few of these Chinese private-sector firms share a sense of being part of an overall project or strategy, and most have followed commercial rather than policy objectives in deciding to establish themselves overseas (Gu 2009). As Gonzalez-Vicente (2011: 403) argues: "the political economy of China's development is not determined by centralized geo-graphing alone". Furthermore, China's primary oversight agencies do not enjoy direct lines of authority over Chinese corporations overseas (Gill & Reilly 2007), while the increasing autonomy of local states further complicates the management of "national" economic space. There has also been a considerable blurring of the lines that separate inside and outside, domestic and foreign, state and non-state, and public and private in contemporary China. The "China Inc." euphemism is therefore very problematic, particularly given the significant tensions between different actors within the Chinese state (Liou 2009). "China" can be and often is a mix of contradictory and confusing realities; not only is the country changing quickly, but its sheer geographical and population size defies simple descriptions (Ferchen 2013). We thus need to move beyond the simplistic notion of China as a monolithic strategic actor and disaggregate it and the increasingly diverse range of foreign policy actors involved in its "soft power" charm offensive (Power, Mohan, & Tan-Mullins 2012).

Brazil

Brazil experienced a serious financial crisis in the early 1980s and again at the end of that decade because of its increasingly unmanageable foreign debt exposure (Christensen 2012). It reacted with a major shift in its state-led development strategy toward neoliberalism (Lima & Hirst 2006). Rather than roll back its interventions in leading sectors of the economy, the state consolidated its presence not only as regulator, but also as owner and investor. Brazil has institutionalized a relatively open market economy, initially under President Cardoso (1995–2003) but more recently under President Lula (2003–11), in which the preference for an open economy was reinforced through the aggressive promotion of bilateral and multilateral trade agreements in the Americas and beyond (Baer 2008) and through Lula's efforts to establish a "new trade geography of the South". Brazil has since advanced economically in the past decade to become the sixth largest economy in the world and in 2010, as the developed capitalist core was struggling with anemic growth rates, double-digit unemployment, and failing companies, Brazil's economy grew by 7.5% (Ban 2013). Its leading corporations have been on a global acquisitions spree and, far from needing bailouts, its banks were building transnational empires and funding the internationalization of Brazilian firms.

Brazil's current socio-economic policy regime is neither a local replica of the Washington Consensus, nor a revolutionary departure from it. Instead, the evidence suggests that it is a hybrid made out of economically *liberal* policy goals and instruments associated with the Consensus, and policy goals and instruments that can be traced to the *developmentalist* tradition. The resulting neodevelopmentalism entails a new form of state activism that has a national capitalist development program at its core that is meant to guide the transition of developing countries away from the Washington Consensus (Bresser-Perreira 2007). In addition to industrial policy, the Brazilian government has directly intervened in the market by maintaining state control over sectors that it deems strategic for development, such as banks, oil, electricity, and aerospace. The majority of Brazil's biggest companies – those that possess the necessary financial resources to invest internationally – are specialized in the areas of civil construction and resources. Their activities are not a recent trend but date back to the 1970s, when the Brazilian military regime supported the internationalization of domestic companies as a means of securing resources and fostering the country's development. According to some observers, under President Rousseff (2011–present) Brazil stands to consolidate further the direct as well as indirect targeting of state-owned firms in its industrial strategy, with Chinese "best practices" emerging as explicit sources of inspiration (Ban 2013).

Since 2003, the Brazilian government has been actively engaged in a foreign policy strategy of "global power diffusion" (Christensen 2013), motivated by a desire to extend the foreign policy decision-making matrix across the global South and weight it toward Brazil and away from the United States and other northern countries (Burges 2013). Building on the strong *dependentista* understandings that pervade Brazilian public policy and adding an energy security dynamic, trade policymakers in the late 1980s embarked on a conscious policy of diversifying Brazil's trade linkages, using the South American trade bloc Mercosur as a platform for reducing dependence on North American and European markets. Brazil has regarded South America as a platform for its competitive insertion in the global economy and its political ambitions in the regional and global spheres (Christensen 2012). It has also sought to ensure its presence in all the relevant multilateral arenas of global economic and political governance, with the major aim of its inclusion as a permanent member of the United Nations (UN) Security Council. Relations with developing countries are in particular seen as having the potential to facilitate Brazil's rise and the realization of its foreign policy aims. The country's long-standing ambition to be viewed and accepted as a major world power has thus led to the construction of "a benign, conciliatory, consensus-creating persona for Brazilian diplomacy" (Burges 2013: 577), the goal of which is to manage the evolution of existing institutions and norms for Brazil's benefit. Brazil's foreign policy priorities after 2003 have thus taken on a "Third Worldist" orientation, similar to its independent foreign policy tradition before the introduction of an authoritarian government in 1964 (Almeida 2007). Brazil has become increasingly interested in other regions such as Africa and the Middle East, and consequently Brazilian business has since seen a significant period of internationalization focused largely on energy, engineering, and other natural resource–based activities, many supported by the Brazilian National Development Bank (BNDES). South America and the BRICS, however, have been the top priorities, thanks to their particular relevance to Brazil's economy and its ambition to rise in the global political hierarchy.

For the last 20 years, the Ministry of External Relations (Itamaraty) has thus been working to position Brazil as the leader of first South America and then the global South, with a foreign policy strategy of making it the conduit for regional coordination and wider pan-Southern interaction. The tactic that Brazil has most frequently used has been to locate itself as a North–South balancer by trumpeting its "southernness" and claiming to be uniquely positioned to bridge chasms and build consensus in international affairs, either as a representative of the global South (a position that is far from universally accepted) or by working to organize coalitions in the South around particular policy positions. This "diplomacy of development" has been an underlying theme in Brazilian foreign policy for many years, providing "a soft power glue for coalition building" (Dauvergne & Farias 2012: 908).

Brazil regularly states that its development cooperation is guided by principles of joint diplomacy, based on solidarity, non-interference in domestic issues of partner countries, demand-driven action, acknowledgment of local experience, no imposition of conditions, and no association with commercial interests (ABC 2011: 3). These principles are claimed to distinguish Brazilian cooperation from traditional forms of cooperation, particularly by reflecting a horizontal relationship between Southern countries. Brazil rejects being labeled a "donor", a term that it associates with the perceived vertical nature of North–South cooperation (Cabral et al. 2013). Instead, it prefers to portray its cooperation as a mutually beneficial relationship between partners (similar to the "win–win" rhetoric of Chinese foreign cooperation discourses). Brazil claims the advantage of having expertise and technologies that fit the needs of developing countries, due to greater proximity (vis-à-vis Northern donors) in terms of economic and institutional development, culture, and language (in the case of some African countries) and agro-climatic conditions. Brazil also claims that its approach to cooperation is guided by the principle of "solidarity diplomacy", which brings together elements of altruism (supporting those in need) and reciprocity (forging mutually beneficial partnerships) in a horizontal relationship between Southern peers.

Brazil has been attempting to engage with Africa in a big way in recent years. The current pro-Africa stance can be traced back to Brazilian diplomatic attempts in the 1960s and 1970s to establish economic, political, and cultural relations with the continent (Freitas-Barbosa,

Narciso, & Biancalana 2009). Following the progress of decolonization in Africa, Brazil's cooperation with the continent increased considerably, marked by a clear effort to depict relations as "special". This rhetoric is heavily based on the African heritage within Brazilian culture, on Brazil's historical debt to Africa for perpetuating slavery for centuries, as well as on the fact that African countries are also developing ex-colonies that have much to gain from cooperation with Brazil in multilateral fora. Brazilian companies are by no means on a level playing field with their Chinese competitors, however, as the latter's expansion in Africa relies heavily on a multiplicity of governmental mechanisms that Brazil is simply not in a position to offer. As a result, the presence of Brazilian companies in Africa cannot solely be credited to an efficient and pragmatic foreign policy regime (Freitas-Barbosa et al. 2009).

India

The largest democracy in the world has also become one of the fastest growing in recent decades, particularly since the commencement of India's market-oriented economic reforms in 1991. Average annual GDP growth rates reached 5.7% in the 1980s and the 1990s (Shrivastava 2009: 126) and 6.75% and above from 2001–12 (World Bank 2014), while FDI inflows increased from US$129 million in 1991–92 to almost US$36 billion in 2011 (UNCTAD 2013b). FDI out-flows from India stood at some US$8.6 billion in 2012 (UNCTAD 2013b). Since 2000, the overall thrust of Indian foreign policy has been to seek geopolitical partnerships in multiple directions to serve the national interest, and India has started to play a more involved role in a wide spectrum of multilateral and regional institutions. While Indian policymakers remain uncertain about what is and is not successful in terms of domestic development strategies, there is little likelihood of a state-led strategy to export "development" on a large scale to other countries. Instead, the Indian government has acted as an enabler for its private sector, with an important role being played by some state-owned firms, including energy and infrastructure companies. The Export-Import Bank of India (set up in 1982 for the purpose of financing, facilitating, and promoting foreign trade) extended lines of credit between 2007 and 2012 of US$8.16 billion, with the vast majority of this going to sub-Saharan Africa (US$3.74 billion) and South Asia (US$2.6 billion) (Exim, 2014). The Indian government set aside nearly US$1.3 billion for foreign assistance in 2013–14, a fourfold jump from 2003–04. Over the past four years, Indian foreign aid spending has grown annually by an average of 32%.

Traditionally, India has offered assistance to other countries through training and capacity-building, mainly through the Indian Technical and Economic Cooperation scheme (ITEC) founded in 1964 (Price 2011). The country has so far decided, like China, not to establish its own "aid department", so responsibility for overseas development aid programs is spread over several ministries and its assistance to the least developed countries (LDCs) is still predominantly bilateral. As with China and Brazil, India also remains reluctant to use the same language as Western donors in describing its assistance, preferring to talk instead of "South–South cooperation". As Narlikar (2013: 605) has argued:

> Even though India does not explicitly rely on the language of Third Worldism to justify its engagement with LDCs, a strong moralistic framing of global distributive justice permeates the demands for "policy space" and "development."

However, India is not necessarily perceived as a "champion" of the global South. It is also still politically difficult, domestically, to be seen to provide assistance to other countries when

India itself faces major development challenges around poverty and inequality. In spite of taking a leading role in South–South initiatives such as the G-20 in the World Trade Organization (WTO) and the India–Brazil–South Africa (IBSA) Dialogue Forum (discussed later), the Indian approach has arguably not been radical enough concerning the transformation of existing international economic or political structures, nor does it articulate a concrete agenda to promote Southern interests, or the interests of non-members of these initiatives (Alden & Vieira 2005). Furthermore, the Nehruvian notion of non-interference, which retains a considerable influence on India's policymaking, has much in common with the Chinese foreign cooperation principle of non-interference in the affairs of foreign states. Unlike China, however, Indian investment is notably more diverse in terms of its regional as well as sectoral spread (Naidu 2008). Indian NGOs and civil society are also becoming increasingly active at an international level (Price 2011) and India is currently the third largest contributor to UN peacekeeping missions (after Pakistan and Bangladesh). Over the last five years, nevertheless, India has also been the world's largest importer of weapons (mostly sourced from Russia).

As with China, India's alliances with Africa were forged during the 1950s and 1960s in the context of debates about non-alignment and decolonization (Dubey 1990; Beri 2003). India launched an integrated "Focus Africa" program in 2002 to enhance Indian trade with the African continent and there are currently 24 African countries included. It has also granted preferential market access to LDCs (33 of which are in Africa) through removing import duties on 85% of all tariff lines. There is considerable anxiety in India about China's growing influence on the continent and a growing desire to mimic the close economic and commercial ties that Beijing has forged in recent years. Thus, India has increasingly begun to offer trade financing, opening many embassies and holding frequent talks with African leaders while staging elaborate regional summits and underlining the virtues of "South–South cooperation". Indian officials are at pains to emphasize the long-standing continuities in India's African engagement, and its trade with Africa has soared from just under US$1 billion in 1991 to around US$70 billion in 2012 (The Economist 2013).

Energy and extractive industries, particularly oil, coal, and uranium, have been at the forefront of India's Africa strategy, with the continent seen as a possible source of raw materials and energy to fuel India's industrial growth and ensure its energy security. With rapid economic growth and industrialization, India is expected to double its energy consumption by 2030, overtaking Japan and Russia to become the world's third largest consumer, after the United States and China (Shrivastava 2009). An increasing proportion of India's oil and coal imports now comes from Africa and thus the significance of the Indian Ocean to India's economic development and security is regarded as particularly immense (Vines & Campos 2010).

The BRICS and reform of global economic and political governance

The global economic crisis has opened up additional space for the emerging economies of the global South to play an increasingly active role in the reform of global economic and political governance, to the extent that a "regime change" in global governance is now at least a distinct possibility (Gray & Murphy 2013). This has been characterized by some scholars as the beginning of a transition from a unipolar US hegemony to one of "emancipatory multipolarity" (Pieterse 2011), in which the countries that represent the majority of the world's peoples now have a position at the head table, or even as a broader underlying "global centre shift" or "hegemonic transition" (Gills 2011).

Reforming the governance of the International Monetary Fund (IMF) and the World Bank has been a central component in the strategy of "global power diffusion" that the BRICS have pursued. These countries have argued that the West is overrepresented in the IMF at the expense of developing countries, and have called for a greater share of votes and a change in what they see as the organization's obsolete governance. As they are becoming growing net contributors to the IMF, the BRICS are thus pressing for a greater voice within the institution, even threatening to hold back the additional financing that the IMF requested to fight the European debt crisis unless they gained greater voting power. There were also discussions at the BRICS 2012 summit in India about the creation of a parallel mechanism to the World Bank. This included proposals for a "BRICS development bank" that would lend to infrastructure projects and the facilitation of sustainable development in the countries of the grouping themselves, as well as other developing countries. More generally, the BRICS have also focused on the need for national policy autonomy and have been critical of the global economic governance frameworks that introduce rules and norms corresponding to dominant-country interests. The BRICS have also been instrumental in establishing regional development banks that have eroded the primacy of the IMF and World Bank as lenders in Asia and Latin America, and have agreed to use their own currencies when trading among themselves, effectively reducing their dependence on the US dollar as the main currency of trade. The BRICS summits have, however, struggled to move beyond "the platitudinous and empty promises that all the established and rising powers seem to make in various forums" (Narlikar 2013: 601).

There have also been a number of trilateral agreements and frameworks established among the BRICS, including the Shanghai Cooperation Organisation (member states include Russia and China, observers include India) and the IBSA Trilateral Forum. IBSA is a state-led developmental initiative between India, Brazil, and South Africa to promote South–South cooperation and exchange (Taylor 2009) and aims to win a permanent seat on the UN Security Council for its members. However, the forum seems not to go much beyond informal consultations, small-scale technical partnership agreements, or discussions about how to harmonize national statistics so as to make them comparable. To date, discussions about technical cooperation on specific projects and the alignment of positions on issues across the gamut of foreign policy questions have formed the substance of the IBSA grouping, and the depth of its dialogue around "development" remains rather shallow.

Conclusions: A BRICS challenge to the established global order?

Western modernity is no longer uncritically viewed as the future of developing countries. (Humphrey 2007: 16)

The rise and circulation of the BRICS acronym rest on and rework previous "emerging markets" discourses going back to the 1980s and represent a further shattering of prior metageographical demarcations (categories such as "developed" and "Third World"; (Sidaway 2012)). Such labels draw on the investment discourses, financial industry narratives, statistical tools, and common classificatory regimes used by investment banks and corporate analysts to narrate the risk and return expectations assigned to particular regions and groups of states. An exclusive focus on the BRICS, however, ignores important changes in the international system and the increasing influence of countries like South Korea, Turkey, and Mexico, who are among the upward movers in the global economic order, and also overlooks the emergence of a number of new Southern development donors like Malaysia or some of the Gulf States (Mawdsley 2012). These other economies are often presented as part of the BRICS subset, but the scale and population of the BRICS are seen to confer on these countries a special role as the leading non-Western economies and as future motors of global accumulation. Africa in particular has become a key focal point for the intensifying competition between the BRICS and other emerging powers. As part of the non-Western "rest", African countries are able to provide the "followership" that these rising powers seek and legitimize claims to greater power by securing for them the backing of large numbers (Narlikar 2013). The relationship between Africa and the rising powers therefore appears to be increasingly symbiotic.

Twenty years ago, it would have been difficult to imagine Brazil as the main regional leader in Latin America, India as a major player in the WTO, or China as the second largest economy in the world (Vom Hau, Scott, & Hulme 2012). What is becoming clear is that the rapid and steady intrusion and recognition of a set of major emerging economies are challenging the established global order, "wrenching global relations into flux" (Shaw, Cooper, & Chin 2009: 27). High and sustained growth rates among the BRICS and beyond have led to a partial decoupling of the global South from the Organisation for Economic Co-operation and Development (Vom Hau et al. 2012), although the markets of the United States and the European Union continue to be of immense importance. As these countries have sought a reorientation of power toward multipolarity, there have been important implications for the contemporary study of international relations.

The BRICS' preoccupation with resources has become very significant in their relations with Africa, Latin America, and the Asia Pacific region, to the point where it may even be framed as a security issue. Most of Russia's outward investment, for example, is dominated by large resource-based corporations that seek to gain greater access to the African markets of fuel, energy, and metallurgy; among the BRICS, resource diplomacy is gathering increasing momentum. The nature of the assistance that these countries provide to foreign "partners" is also quite different. Among the BRICS, China is probably the most active in using hybrid forms of financing tools, blurring the lines of distinction between aid, trade and FDI, although India and, to some extent, Brazil have made use of similar approaches. In many cases they have also sought to avoid the traditional donor–recipient relations and policy conditionalities favored by established Western donors (although many projects proposed by emerging partners, especially China, are often tied economically to the use of labor and input procurements). In this sense, the BRICS have posed a significant challenge to established modes of development cooperation (Six 2009), introducing competitive pressures into the existing world of international development assistance and weakening the bargaining position of Western donors by offering alternatives to aid-receiving countries (Woods 2008).

It could, however, be argued that the rise of these new powers is firmly located within the Western global hegemony, and that they have been reluctant to challenge the Euro–North American domination of world affairs (Palat 2008). As Schweller and Pu (2011: 53) put it, available evidence suggests that "China seeks a gradual modification of Pax Americana, not a direct challenge to it". More generally, each of the BRICS has adopted elements of a "neo-liberal" strategy for development focused largely on the facilitation of capital accumulation and market socialization. China's involvement in Africa may well permit the "revival of triangulation" (Large 2008), which means that African states can pursue relations with more than one external state, but it is important to differentiate what this new-found leverage vis-à-vis Northern development partners might mean for the variety of actors involved (not merely for states) or for key issues of social justice and poverty eradication. Palat (2009) argues that "neither the Chinese corporations nor the Chinese government has done anything to benefit trade unions or social justice movements in Africa". There can be little doubt, nevertheless, that China's international interests and influence in particular are not only growing, but have already had a significant impact on the global order (Breslin 2013). Far from encouraging a "rolling back" of the state, the BRICS' reassertion of the role of the state in development is "one of the most consequential events of the global economy" (Vezirgiannidou 2013: 250), as they have been able to protect and nurture their own industries and make them competitive within a global economy. More generally, it is impor-tant to recognize the increasing capacity of the BRICS states directly and indirectly to alter geopolitical and developmental spaces outside their borders, despite the many proclama-tions around "non-interference".

Articulated through an economics-led diplomacy, the BRICS have demanded a new set of international norms, a new trade agenda, and equitable representation in the multilateral arena. The traditional Western powers now increasingly need to negotiate matters of security and economic governance with these emerging powers. The strong discourse of "South–South cooperation" deployed by several members of the BRICS group might perhaps more accurately, then, be seen as rhetoric with which to negotiate a stronger bargaining position within the US-centered world order. Debates about the rising influence of the BRICS with the traditional Western powers have, however, often been characterized by attempts to portray them (partic-ularly China) as the "villainous other", to be socialized into the liberal way of doing development. IBSA (and similar groupings) will provide the platform for countries of the South to work around the familiar institutions of international relations in the years ahead; the creation of such dialogue forums demonstrates a frustration with institutions such as the IMF, World Bank, and WTO. Brazil, China, India, and South Africa not only espouse the cause of "developing countries", but are also vociferous in their assertions that they themselves belong to this group in ways that are reminiscent of older Third Worldist coalitions that some of these countries led at different points in the past (Narlikar 2013). Paradoxically, these new donors represent models of economic success, yet they have been, or are still, recipients of international aid. Although India, for example, is a donor, it was the world's eighth-largest recipient of official development assistance as recently as 2008 (to the value of US$2.1 billion) and fourth overall from 1995 to 2009. It is also worth remembering that considerable levels of poverty and inequality remain within each of the BRICS, despite the many claims that have been made about their "miraculous growth".

References

Agência Brasileira de Cooperaçao (ABC) (2011). *Brazilian Technical Cooperation*. Brasília: Agência Brasileira de Cooperação.

Alden, C., & Vieira, M.A. (2005). The new diplomacy of the South: South Africa, Brazil, India and trilat-eralism. *Third World Quarterly* 26(7): 1077–1095.

Almeida, P.R. (2007). O Brasil como ator regional e como emergente global: Estratégias de política externa e impacto na nova ordem internacional. *Cena Internacional* 9(1): 7–36.

Baer, W. (2008). *The Brazilian Economy: Growth and Development*. Boulder, CO: Lynne Rienner.

Ban, C. (2013). Brazil's liberal neo-developmentalism: New paradigm or edited orthodoxy? *Review of International Political Economy* 20(2): 298–331.

Beri, R. (2003). India's Africa policy in the post-Cold War era: An assessment. *Strategic Analysis* 27(2). http://www.idsa.in/system/files/strategicanalysis_rberi_0603.pdf, accessed February 4, 2014.

Breslin, S. (2013). China and the global order: Signalling threat or friendship? *International Affairs* 89(3): 615–634.

Bresser-Pereira, L.C. (2007). The new developmentalism and conventional orthodoxy. *Iberoamericana: Nordic Journal of Latin American and Caribbean Studies* XXXVII(1): 31–58.

Burges, S.W. (2013). Brazil as a bridge between old and new powers? *International Affairs* 89(3): 577–594.

Cabral, L., Shankland, A., Favareto, A., & Vaz, A.C. (2013). Brazil-Africa agricultural cooperation encounters: Drivers, narratives and imaginaries of Africa and development. *IDS Bulletin* 44(4): 53–68.

China Daily (2014). China's 2013 exports rise 7.9%, imports up 7.3%. *China Daily* January 10. http://www.chinadaily.com.cn/business/2014-01/10/content_17229362.htm, accessed March 26, 2015.

Christensen, S.F. (2012). South–South relations in Brazil's responses to the challenges of globalization. In Nilsson, M., & Gustafsson, J. (eds), *Latin American Responses to Globalization in the 21st Century*, Basingstoke: Palgrave Macmillan, 231–252.

Christensen, S.F. (2013). Brazil's foreign policy priorities. *Third World Quarterly* 34(2): 271–286.

Dauvergne, P., & Farias, D.B.L. (2012). The rise of Brazil as a global development power. *Third World Quarterly* 33(5): 903–917.

Dubey, A.K. (1990). *Indo-African Relations in the Post-Nehru Era (1965–1985)*. Delhi: Kalinga Publications.

Export-Import Bank of India (2014). Lines of credit. http://www.eximbankindia.in/, accessed July 23, 2014.

Ferchen, M. (2013). Whose China model is it anyway? The contentious search for consensus. *Review of International Political Economy* 20(2): 390–420.

Firzli, N.J. (2011). Forecasting the future: The BRICs and the China model. *Journal of Turkish Weekly* March 8. http://www.turkishweekly.net/op-ed/2799/forecasting-the-future-the-brics-and-the-china-model.html, accessed March 26, 2015.

Freitas-Barbosa, A., Narciso, F., & Biancalana, M. (2009). Brazil in Africa: Another emerging power in the continent? *Politikon* 36(1): 59–86.

Friedman, E. (2009). How economic superpower China could transform Africa. *Journal of Chinese Political Science* 14: 1–20.

Gill, B., & Reilly, J. (2007). The tenuous hold of China Inc. in Africa. *Washington Quarterly* Summer: 37–52.

Gills, B.K. (2011). Going South: Capitalist crisis, systemic crisis, civilisational crisis. *Third World Quarterly* 31(2): 169–184.

Goldman Sachs (2005). How solid are the BRICs? Global Economics paper 134. http://www.sdnbd.org/sdi/issues/economy/BRICs_3_12-1-05.pdf, accessed January 27, 2014.

Gonzalez-Vicente, R. (2011). The internationalization of the Chinese state. *Political Geography* 30(7): 402–411.

Gray, K., & Murphy, C.N. (2013). Introduction: Rising powers and the future of global governance. *Third World Quarterly* 34(2): 183–193.

Gu, J. (2009). China's private enterprises in Africa and the implications for African development. *European Journal of Development Research* 21: 570–587.

Humphrey, J. (2007). Forty years of development research: Transformations and reformations. *IDS Bulletin* 38: 14–19.

International Monetary Fund (IMF) (2013). *World Economic Outlook*. Washington, DC: International Monetary Fund.

Large, D. (2008). China and the contradictions of "non-interference" in Sudan. *Review of African Political Economy* 115: 93–106.

Lim, K.F. (2010). On China's growing geo-economic influence and the evolution of variegated capitalism. *Geoforum* 41: 677–688.

Lima, M.R.S., & Hirst, H. (2006). Brazil as an intermediate state and regional power: Action, choice and responsibilities. *International Affairs* 82(1): 21–23.

Lin, C. (2006). *The Transformation of Chinese Socialism*. Durham, NC: Duke University Press.

Liou, C.S. (2009). Bureaucratic politics and overseas investment by Chinese state-owned oil companies. *Asian Survey* 49(4): 670–690.

Manji, M., & Marks, S. (eds) (2007). *African Perspectives on China in Africa*. Cape Town: Fahamu.

Marquand, R. (2011). Amid BRICS' rise and "Arab Spring", a new global order forms. *Christian Science Monitor*, October 18. http://www.csmonitor.com/World/Global-Issues/2011/1018/Amid-BRICS-rise-and-Arab-Spring-a-new-global-order-forms, accessed March 26, 2015.

Mawdsley, E. (2012). *From Recipients to Donors: Emerging Powers and the Changing Development Landscape*. London: Zed Books.

Meisner, M. (1999). *Mao's China and After: A History of the People's Republic*, 3rd edn. New York: Free Press.

Naidu, S. (2008). India's growing African strategy. *Review of African Political Economy* 35(1): 116–128.

Narlikar, A. (2013). India rising: Responsible to whom? *International Affairs* 89(3): 595–614.

Nolan, P. (2004). *China at the Crossroads*. London: Polity Press.

Nonini, D.M. (2008). Is China becoming neoliberal? *Critique of Anthropology* 28(2): 145–176.

O'Neill, J. (2001). Building better global economic BRICs. Goldman Sachs Global Economics paper No. 66. http://www.goldmansachs.com/our-thinking/archive/archive-pdfs/build-better-brics.pdf, accessed March 26, 2015.

O'Neill, J. (2011). *The Growth Map: Economic Opportunity in the BRICs and Beyond*. Harmondsworth: Penguin.

Palat, R.A. (2008). A new Bandung? Economic growth vs. distributive justice among emerging powers. *Futures* 40: 721–734.

Palat, R.A. (2009). Rise of the global South and the emerging contours of a new world order. In Pieterse, J.N., & Rehbein, B. (eds), *Globalization and Emerging Societies: Development and Inequality*, Basingstoke: Palgrave, 39–60.

Pieterse, J.N. (2011). Global rebalancing: Crisis and the East–South turn. *Development and Change* 42: 22–48.

Power, M., & Mohan, G. (2010). Towards a critical geo-politics of China's engagement with African development. *Geopolitics* 15(3): 462–495.

Power, M., Mohan, G., & Tan-Mullins, M. (2012). *China's Resource Diplomacy in Africa: Powering Development?* Basingstoke: Palgrave.

Price, G. (2011). *For the Global Good: India's Developing International Role*. London: Chatham House.

Rothkopf, D. (2009). The BRICS and what the BRICS would be without China. *Foreign Policy* June 15.

Schweller, R., & Pu, X. (2011). After unipolarity: China's visions of international order in an era of U.S. decline. *International Security* 36(1): 41–72.

Shaw, T.M., Cooper, A.F., & Chin, G.T. (2009). Emerging powers and Africa: Implications for/from global governance. *Politikon* 36(1): 27–44.

Shrivastava, M. (2009). India and Africa: From political alliance to economic partnership. *Politikon* 36(1): 117–143.

Sidaway, J.D. (2012). Geographies of development: New maps, new visions? *Professional Geographer* 64(1): 49–62.

Six, C. (2009). The rise of postcolonial states as donors: A challenge to the development paradigm? *Third World Quarterly* 30(6): 1103–1121.

So, A. (2005). Beyond the logic of capital and the polarization model: The state, market reforms, and the plurality of class conflict in China. *Critical Asian Studies* 37(3): 481–494.

Stiglitz, J. (2002). *Globalization and Its Discontents*. New York: Norton.

Taylor, I. (2009). The South will rise again? New alliances and global governance: The India–Brazil–South Africa dialogue forum. *Politikon* 36(1): 45–58.

The Economist (2013). India and Africa: Elephants and tigers. *The Economist*, October 26. http://www. economist.com/news/middle-east-and-africa/21588378-chinese-businessmen-africa-get-attention-indians-are-not-far, accessed March 26, 2015.

UNCTAD (2013a). The rise of BRICS FDI and Africa. *Global Investment Trends Monitor* March 25. http://www.safpi.org/sites/default/files/publications/webdiaeia2013d6_en.pdf, accessed February 1, 2014.

UNCTAD (2013b). *World Investment Report 2013: Global Value Chains – Investment and Trade for Development*. New York: United Nations.

Vezirgiannidou, S.E. (2013). The United States and rising powers in a post-hegemonic world order. *International Affairs* 83(3): 635–651.

Vines, A., & Campos, I. (2010). China and India in Angola. In Cheru, F., & Obi, C. (eds), *The Rise of China and India in Africa: Challenges, Opportunities and Critical Interventions*, London: Zed Books, 193–207.

Vom Hau, M., Scott, J., & Hulme, D. (2012). Beyond the BRICs: Alternative strategies of influence in the global politics of development. *European Journal of Development Research* 24: 187–204.

Wansleben, L. (2013). Dreaming with BRICs. *Journal of Cultural Economy* 6(4): 453–471. doi:10.1080/17530350.2012.756826

Woods, N. (2008). Whose aid? Whose influence? China, emerging donors and the silent revolution in development assistance. *International Affairs* 84(6): 1205–1221.

World Bank (2014). GDP growth (annual %). http://data.worldbank.org/indicator/NY.GDP.MKTP.KD.ZG, accessed January 16, 2014.

Wu, F. (2008). China's great transformation: Neoliberalization as establishing a market society. *Geoforum* 39: 1093–1096.

Xinhua (2014). China's FDI inflows grows 5.25pct in 2013. *Xinhua* January 16. http://news.xinhuanet.com/english/china/2014-01/16/c_133050941.htm, accessed January 16, 2014.

Chapter 29

Social Media

Paul C. Adams

University of Texas at Austin, Texas, USA

In the introduction to *Political Matter*, Braun and Whatmore argue: "From cell phones to stem cells, stuff of all kinds increasingly makes us what we are" (2010: x). This is not to say that our relationship with the material world is deterministic, as if politics were controlled by matter itself, any more than humans are able to drive the entire people–matter interaction. The relationship is reciprocal: As we act on the world with things in our hands, over our eyes, and in our ears, we continually rediscover ourselves as agents. What it means to be a person changes in tandem with our material culture (including communication technologies) and there is no one in charge of this process except for an always already hybrid agency.

Viewed historically, the progression of artifacts from sharpened stones to touchscreens is, at the same time, a progression of actor networks (Couldry 2008). Understood in this context, media are not merely matter, however complicated – they are structured opportunities for things to happen. For political scientist Jane Bennett, more-than-human agency includes the efficacy of things as diverse as a cast-off bottle cap, an abandoned glove, and a dead rat caught on a storm drain, which together have an "energetic vitality" (2010: 39). Like Bennett, I find that political agency can coalesce around assemblages of matter and dead bodies, but my analysis sticks closer to conventional notions of political power, exploring how mediated agency outlasts the death of a human martyr's body. In particular, I show that social media have played a role in political processes by reconfiguring human agency in space and time, taking the Iranian protests and the Arab Spring as demonstrations of the political appropriation of new media to extend human agency across geographical borders, linguistic divides, and even the border between life and death.

The discussion starts with media in general. Next, it turns to social media, which offer some special, though hardly unique, affordances for political action. It concludes with a brief discussion of where political geography might be headed as agency becomes increasingly bound up in digitally mediated networks.

The Wiley Blackwell Companion to Political Geography, First Edition.
Edited by John Agnew, Virginie Mamadouh, Anna J. Secor, and Joanne Sharp.
© 2015 John Wiley & Sons Ltd. Published 2017 by John Wiley & Sons Ltd.

Media affordances

An affordance can be thought of as a way of getting things done, insofar as "tools do not determine action, they afford it" (Nielsen 2013: 174). In the words of Biekart and Fowler, technological advances "are not a cause as such, but they have certainly opened up innovative avenues for people to challenge existing configurations of power" (2013: 529). Media affordances facilitate new ways of organizing actions in space and time, and therefore enable the reworking of political action. Agency that involves humans acting in conjunction with technology has also been called *technicity* (Simondon 1992; Mackenzie 2002) and its subtle complexity pushes us to think beyond the binary logic of access/lack of access to technology. Technicity envisions a more complex and multidimensional set of relations between technologies and people, where appropriations of media are "an exteriority that is necessarily also an interiority" (Braun & Whatmore 2010: xix). What this means is that just as we circulate through and inhabit the nested containers that make up an architectural environment – including a building's outer and inner surfaces, entryways, halls, and rooms – we also circulate through and inhabit communication technologies, occupying each medium in a range of volitional, sometimes resistant, but always multivalent ways.

If we set aside the habitual division of the world into humans and non-humans and view the "person" as a heterogeneous network, the inhabitation of media can be seen as enabling the formation of dynamic, spatially extensible actors (Adams 2005; Latour 2005). Extending geographical agents through space and time – whether through the use of parchment, paper, Instagram, or Facebook – mediated technicity shapes what can (and cannot) become political, whether standard and "normal" for our time and place, or creative and resistant. Perhaps the "rise of 'digital activism' has shaken the existing relationships between state, markets, civil society and citizen action by developing new and networked ways of thinking" (Shah 2013: 666), but we should interpret these "new ways of thinking" in a modest fashion by attending to the precise changes occurring in how people achieve political goals, whether in terms of the articulation of grievances, discovery of opportunities, leveraging of resources, or formation of identities.

Political actors are always embodied, but precisely because of the political importance of the body we must avoid a false opposition between mediated experience and embodied experience, or, worse yet, a dichotomy wherein media are reduced to vectors of oppression while embodiment is cast as resistant. The embodiment of mediated communication and the communication of embodiment are flip sides of a hybridity that actively *attaches* scales to political projects (Sharp 1996, 2000; Whatmore 2002; Dittmer 2007: 405; Dittmer et al. 2011). It may help to think of the moral and ethical characteristics of new media in terms of "a human 'gesture' upon the space suggested by newer media" (Papacharissi 2010: 9), and in this media-configured space individual and collective political action "is not only about jumping pre-existing scales, but also about producing them" (Mamadouh 2004: 483). The most obvious recent example of this is in the Middle East and North Africa, where "a growing middle class with access to decentralized technologies, shaped through a global flow of ideology and capital, questions the sovereignty of the territorial state and looks at new forms of governance and organization" (Shah 2013: 666).

Political geographers have most often addressed the politics of media in connection with popular geopolitics (Ó Tuathail 1996). This work has directed worthwhile attention to the political roles of radio (Pinkerton & Dodds 2009), television (Luke & Ó Tuathail 1997), film (Dodds 2003; Carter & McCormack 2006), newspapers (Adams 2004), magazines

(Sharp 2000), comics (Dittmer 2007, 2013), and the internet (Dodds 2006), among other things. Hybrid politics are in no way restricted to new media, however. The fact that we have lived for a long time as media hybrids is generally written out of our accounts of history and modernity (Latour 1993: 13–48). I will therefore start to explore these hybrid, mediated politics by showing a few historical links between media innovation and political transformation. After that, I will engage with the political dynamics observed in the Middle East and North Africa (MENA) in the first decade of the twenty-first century, with particular attention to social media.

Old new media

As a social order based around one set of media gives way to a social order based around a different set of media, actors seek to increase their extensibility – the span of sensation and agency – through the creative use of new technologies. Inevitably, these replacements cause "distributional changes" (Deibert 1997: 31–37), new geographical manifestations of political action, by reinforcing certain forms of power at both organizational and individual levels while weakening other forms of power. We can think of these distributional changes as the political opportunities and limitations inherent in the spatio-temporal affordances of each new medium. James Carey (1988) adopts this perspective when he argues that to study the historical impact of the telegraph we should

> attempt to demonstrate how this instrument altered the spatial and temporal boundaries of human interaction, brought into existence new forms of language as well as new conceptual systems, and brought about new structures of social relations, particularly by fostering a national commercial middle class. (1988: 204)

Reaching further back, the changes we label "the rise of civilization" are often attributed to a social surplus that fed non-farmer specialists whose existence gave rise to a complex social hierarchy (Childe 1950). More subtly, civilization arose with the increased ability to extend actions through space and time (Innis 1951). Writing and its predecessors *fixed* language, making speech acts durable and portable (Adams 2009: 18–28), and the resulting actor networks incorporated not only scribes, emissaries, priests, and rulers, but also materials such as clay, stone, bone, and papyrus used to record meaning in symbolic form (Adams 2010). In Mesopotamia, Egypt, China, and Mesoamerica, the development of the fixed word (including what we might call predecessors to writing, such as Mesopotamian tokens and Inca *quipu*) led to distributional changes in power relations involving new spatio-temporal patterns of organization (Schmandt-Besserat 1996; Crowley & Heyer 2010).

A more familiar story explores the distributional changes associated with the printing press, which "favored the strategic interests of the Protestant Reformation and scientific humanism to the detriment of the papal-monastic network" (Deibert 1997: 203). The power of printing had much to do, of course, with what was written, but aside from questions of content (patriotism, dogma, treason, heresy), its political potential arose from a radical increase in the availability of books, which fostered an increase in literacy, in turn stimulating critical thinking, creativity, public intellectuals, and new models of leadership (Eisenstein 1979: 136–159). Benedict Anderson's influential study of nationalism and nation-building employs the term "print capitalism" to capture the way in which an emerging print-based economic order materialized particular meanings, values, and, above all, collective identities (1991). Distributional changes arose again with the diffusion of electric and electronic communications such as the telegraph and telephone (Marvin 1988: 107), and television subsequently helped change both official politics and the politics of everyday life (Meyrowitz 1985: 67).

This glimpse of the history of media suggests that if digital media have prompted a profound sense of novelty, then ironically this sense of newness is nothing new. Media diffusions repeatedly bring into focus a "perceptual and semiotic economy that they then help to transform" (Pingree & Gitelman 2003: xii). History recapitulates a paradox wherein the political agency of actants is extended, but also surpassed and transformed by the new media assemblages in which they are enrolled. Let us narrow this discussion to the so-called social media, keeping in mind that their disruptive character echoes the distributional changes introduced by older "new media."

A panoply of social media

Even if we start with the modest assumption that new digital media, and social media in particular, are "not so much creating new forms of protest [as] amplifying traditional forms of protest, such as street demonstrations" (Valenzuela 2013: 936), they still hold great political importance. Simply reposting or retweeting an image or snippet of information can be a political act (Lotan et al. 2011; Conover et al. 2013). Public engagement from private space is one of the hallmarks of social media (Papacharissi 2010). In addition, social media combine three distinct functions that have been identified by numerous writers: access to information, means of expressing opinions, and tools facilitating activism (Valenzuela 2013: 920). Theocharis (2013b: 1477) similarly refers to the informational, interactive, and mobilizing features of the websites used by students involved in the 2010 university occupations in the United Kingdom, while Mamadouh identifies three related functions in websites by and for Dutch Moroccans: "group news media targeting a specific group ... a platform for exchanges between group members" and a means to "increase the general public's awareness of the existence and the position of the minority group" (2003: 201). We can summarize the affordances of social media as sharing information and other meaningful communications, building and modifying social networks, and engaging in discussion and debate as part of multiple publics.

The first of these affordances is most evident in the online sharing of status updates, photographs, videos, jokes, and quotes. Examples of social media that emphasize sharing of content include Flickr, Instagram, YouTube, and Vimeo. This aspect of social media overlaps with the functionalities labeled "Web 2.0," indicating user-generated content posted online in a collaborative web environment.

The second is indicated through protocols such as "friending" and "following" (Wellman 2001). This affordance is most obvious on Facebook and Twitter, where content sharing is eclipsed by the function of maintaining a connection to others. Much attention has been devoted to the question of whether the weak social ties that dominate such networks are socially relevant, and indeed whether such "friends" are not in fact mere acquaintances. Yet, this debate misses the point. Insofar as digitally mediated social networks include both weak ties and strong ties, they permit the formation of "small world" networks, allowing information to percolate through social and physical space with remarkable ease (Granovetter 1973; Watts & Strogatz 1998).

Third, social media are used for the purposes of discussion and debate, as people reflect on current events, images, and ideas. Political discourse encompasses more than just commenting and forwarding, so social media political forums often leave something to be desired. Instead of critical thinking, one finds *ad hominem* critiques of other participants and diatribes "flaming" those who disagree. Nonetheless, the internet does offer forums for protracted discussion and, while most online discussion forums are exclusive (requiring membership), this does not make them entirely unlike the cafés that formed the archetypal public sphere (Habermas 1989 [1962]). Both physical and social contexts can support politically efficacious counter-publics (Fraser 1992; Warner 2003).

Particular social media emphasize one or another of the three affordances – social networking, information exchange, or debate/dialogue – but all three are generally present. Blogs, for example, primarily support political discussion, but they also permit the sharing of user-generated and borrowed materials, the development and maintenance of social networks, and linkage to other media. Facebook's primary affordances are in the areas of social networking and the sharing of content, but the site supports a limited sort of political debate.

Building on these arguments, what makes internet-based assemblages political is that they structure communication, knowledge, and action in what Langlois et al. (2009: 416) call "the coming into being of a public." A simple example is the creation of a "tweet" linked to an amateur video captured with a cell phone, showing protestors dancing in the street. In this case the act of physically occupying urban space, the act of making a video, and the act of tweeting an embedded link to this video are interdependent; each inflects the meaning of the others. This digitization of the public sphere involves a blurring of the "lines between production and consumption, between making media and using media, and between active or passive spectatorship of mediated culture" (Deuze 2007: 74). There is, of course, literal work involved in using any medium and people sometimes feel obligated to maintain their online social capital (Dijck 2013: 51), but social media facilitate pleasure and creativity as much as obligation and necessity (Kitchin & Dodge 2011: 111–134).

Likewise, if the essence of computing is technical control and "the technical is always political" (Galloway 2004: 214, quoted in Zook & Graham 2007: 471), then social media entail a new apparatus of state and corporate control. However, the "means of protest are simultaneously more controlled and anarchic" in the hybrid space that Zook and Graham call DigiPlace than in physical places (2007: 473). The political process in this hybrid space or place "is not simply one of human actors mobilizing communication technologies, but also of communication technologies enabling new patterns of political organization" (Langlois et al. 2009: 420). Digital media therefore alternate between the status quo–preserving function and a more progressive function (Hassid 2012). Novel political threats are mingled with novel opportunities (Castells 2001).

Consider, for example, the interactions of a "Wikipedian" who was involved in updating the Wikipedia entry for Ann Dunham, Barack Obama's mother, during the 2008 presidential campaign (Rainie & Wellman 2012). This person regularly "encountered vulgar insults about [Dunham's] sexual exploits and choice in men ... Although he had to deal with these allegations on a recurring basis, they were easy to fix with a one-key-click deletion" (Rainie & Wellman 2012: 203). Political contestation was supported by a digital context that brought together the bodies of Ann Dunham and Barack Obama, the agency of "Wikipedians," software codes, and coded infrastructure. While the preceding example demonstrates a political struggle *in* social media, increasingly there are struggles *over* social media. In 2007 Facebook was updated to include Beacon, a feature that monitored purchases made by Facebook users, then inundated their friends with ads for the supplying companies. A struggle ensued, led by MoveOn.org and joined by 50,000 Facebook users who signed a petition and joined in a class action lawsuit (Dijck 2013: 48–60). What was at stake was not only privacy but also a "double logic of sharing," since the site offers a way of sharing experiences with one's peers but also provides a means of harvesting valuable information about users, which is then shared (at a cost) with companies that want to target consumers via their social networks (Dijck 2013: 65).

Drawing on Jensen, Jorba, and Anduiza (2012: 8–9), I would argue that three sets of variables bring geographical specificity to these digitally mediated politics. First, there are spatial variations, as indicated by the familiar term "digital divide," involving not merely quantitative gaps in the availability, accessibility, and reliability of digital communication technologies, but also qualitative differences in the skills, uses, and motivations of individuals and groups in different places as they appropriate digital media for political purposes. Second, places vary with regard to government regulation. Regimes of control and permissiveness aimed at traditional print and broadcast media have been extended to digital media; such governmental control techniques include filtering, monitoring, denial of service, rerouting search queries, planting viruses on "enemy" computers, and attacks on outspoken bloggers (Warf 2013: 47). Third, social media vary with regard to contextual factors such as political party structure and legal frameworks. While movements in relatively undemocratic societies seek to achieve recognition or find a "voice" despite censorship and silencing, movements in democratic societies struggle to overcome the political apathy that comes with wealth and consumerism. Where "freedom of speech" is the rule, speech may nonetheless be limited by the market and the power of money to inundate people with propaganda and entertainment (Castells 2009). We find, therefore, a context-based continuum in tactics, with evasion of censorship and persecution at one extreme and the attempt to shout above the noise of consumerism at the other.

The question is no longer whether social media are involved in political processes, but rather "*how and under what conditions* these new digital platforms relate to citizen activism and protest politics" (Valenzuela 2013: 921, italics in original). The main theoretical challenges

> lie in understanding the uniqueness of social networking sites as assemblages where software processes, patterns of information circulation, communicative practices, social practices, and political contexts are articulated with and redefined by each other in complex ways. (Langlois et al. 2009: 416)

This social and geographical contingency suggests the need for detailed case studies. Here, I focus on the area that has yielded the most studies thus far: the Middle East and North Africa (MENA).

Microcelebrities and martyrs

Social media in MENA has quite strikingly illustrated what Theocharis refers to as "a myriad of offline and online tactics" (2013a: 38), small but incremental political activities that constitute publics as participants rather than merely audiences. These tactics include posting, reposting, converting, and "liking." Starting in 2009, people throughout the region captured quotes, photos, or bits of news in one digital format and forwarded them in the same or a different format, working around the fact that quantitatively speaking they were in a "lagging" region with regard to new media.

The scale-jumping dynamic set up during the Arab Spring could be characterized as "the emergence of highly networked, dense, but also hierarchical information ecology in which newly emergent microcelebrity activists gain access to, and become means of, flow of attention and visibility" (Tufekci 2013: 862). Intertwined in this "ecology" were various actors, multiple languages, heterogeneous communication media (including social media as well as broadcast media), and international media venues. The movement was less about democracy than government accountability, and its roots were more anti- than pro-American (Lynch 2006). Yet, it was celebrated in mainstream US and UK media, in part because of a widespread assumption that the internet is inherently democratic (Schulte 2013: 158).

Already, in the late 1990s, an Arab public sphere had begun to coalesce around Al Jazeera, with its international reach and sponsorship of impassioned argumentation (Lynch 2006). Interpersonal communication technologies brought new kinds of actors into this public sphere (Howard & Hussain 2013). Among them were rebel-celebrities, who maintained a consistent online presence as spokespersons for political movements and also as the embodiment of a more general vulnerability to arbitrary violence. For example, Zainab Al-Khawaja achieved a kind of online fame in Bahrain, serving a set of interlocking objectives: exposure of government-backed oppression, self-protection, domestic political mobilization, and international news dissemination. By early 2014 she had over 30,000 online followers, whose attention assisted her in spreading her political perspective while protecting her from the violent intimidation and long prison terms that the monarchy used to suppress dissent. In Ethiopia, the Facebook page "We are all Eskinder Nega" showed solidarity with a dissident journalist and blogger who had been persecuted and imprisoned. Social media helped such microcelebrities (to borrow Tufekci's term) circulate within various counterpublics (Warner 2003), break into "the public" via broadcast media, and cross geographical borders to gain sympathy and solidarity from foreign audiences. At some point, such political actors may transition from having local followings to becoming internationally recognized figures like Pakistani education activist Malala Yousafzai or Egyptian comedian Bassem Youssef.

Closely related to such celebrities are those whose damaged bodies are converted into digital form – grisly photographs taken just before or just after death. These images become testaments to arbitrary and unjust authority. The resulting actor networks can circulate a martyr whose identity continues to act in political ways although the body is no longer alive. Martyrs of this sort include Mohamed Bouazizi in Tunisia, Khaled Said in Egypt, Hamza Ali al-Khateeb in Syria, and Neda Agha-Soltan in Iran (Amin 2009). In regard to the latter:

> Bystanders captured her last moments on a cell phone, and within hours the grainy, low-resolution footage was uploaded to the Internet and soon spread virally across the globe. With links to the video posted on YouTube, Facebook, and Twitter, the amateur clip eventually harnessed the attention of the mainstream media, grabbing headlines on CNN and in the New York Times. (Amin 2009: 64)

A simple story of innocence and victimhood, an unjust end, a gut-wrenching image of a young woman at the moment of her death – such were the elements of Neda Agha-Soltan that circulated as a resistant actor network opposing the reelection of Mahmoud Ahmadinejad in 2009.

The technologies used to capture, reframe, and propel the stories and images of martyrs are heterogeneous, including the increasingly ubiquitous cell phone, the internet (as inaccessible and unreliable as it is in some parts of the world), and broadcast media such as Al Jazeera (Howard & Hussain 2013). Media convergence has political implications, as "multiplatform connectivity was clearly important to the success of the revolution and to the revolution's reception around the world" (Schulte 2013: 158). Collective life supplies the rest: the communication vehicles and discursive packages for martyrs who circulate in a kind of political afterlife, and the experiences of suffering that cease to be merely personal and become collective injuries that call out to be avenged.

Common social media practices associated with this circulation include capturing, selecting, forwarding, interpreting, and reinterpreting the image of the martyr, all of which accumulate fertile soil for subsequent political involvement. Forwarding via phone and internet are keys to maintaining the connection between actions on the street and coverage by formal news media such as Al Jazeera (Lotan et al. 2011). Gana captures the dynamic: "Neither social media and Al Jazeera nor WikiLeaks can fully account for the Tunisian people's grassroots revolt," but nonetheless:

> When the news of [Mohamed] Bouazizi's self-immolation and ensuing protests in Sidi Bouzid was downloaded on Facebook and on YouTube by field bloggers and Facebookers, and then by Al Jazeera, then the wave of protests quickly reached the major cities in Tunisia and eventually hit the capital, Tunis. (Gana 2013: 8–9)

Subsequently, the Syrian democratization movement created a YouTube channel called SyrBouazizi. This kind of communication activity can be encapsulated as a *martyrology*, sustained by "a feedback loop between actions in the street and actions online" (Ziter 2013: 117). Such appropriations of social media depend on continued action on the ground, as well as cross-border flows of inspiration, exemplification, and affect.

Facebook dominated this process, although other social media were appropriated for similar purposes. In Egypt, the Facebook page "We are all Khaled Said" turned the young Egyptian's brutal beating death into a cause célèbre in 2010 and had over 300,000 likes by early 2014 (although the term "like" in this case is a bit misleading, as it indicates collective disapprobation). The grisly photo of Khaled Said's mutilated head went viral online and on the streets (on handheld signs) in a diffusion that has been identified as a cause of the Egyptian uprisings of 2011 (Preston et al. 2011). Likewise, as Syrian martyrs in the reigning Baath Party were celebrated in official media to legitimate state power, insurgents elevated a different sort of martyr. Hamza Ali al-Khateeb was a 13-year-old boy captured by government forces during a rally in Daraa in April 2011. His mutilated body was returned to his family on the condition of silence, but activists created a Facebook page entitled "We are all the child martyr Hamza Ali al-Khateeb," which circulated horrific images among 60,000 friends and many more visitors, and was again printed on handheld signs.

In short, martyrs circulated as political catalysts throughout the region and beyond. Networked dead bodies participated in political actor networks. The transborder diffusion of resistance was standardized to a degree by the creation of Facebook pages in various places that followed the naming convention "We are all...," which at once signified local collective identities and broader solidarities. The naming convention was extended beyond the individual person as martyr. At least four different "We are all Palestine" pages were created as well as "We are all Al-'Arakeeb," a page dedicated to a Bedouin village that has been demolished repeatedly by the Israel Land Authority (Schejter & Tirosh 2012: 307). A Bedouin from this village reflected on these politics:

> I used to lecture and to call our youth that does nothing "Facebook kids" as a derogatory term ... then I used the same words as a compliment ... If the Facebook kids of Tunis were able to over-throw their dictator ... and then in Egypt, then we can with Facebook wake up our youth and wake up the country. (Schejter & Tirosh 2012: 309)

It cannot be stressed enough that these places are poor and technologically lagging, and yet the *relative* lack of access to technology does not constitute an absolute barrier to political mobilization enabled by the affordances of such technology. A low spot on the trend surface of digital media penetration does not necessarily imply that online activism is futile.

These observations in no way imply that the internet is replacing the face-to-face context of political mobilization. Place holds a singular importance in the relationships between political action and social media. For example, the group called Freedom Days posted mock political campaign posters around Damascus and made a short video showing hands past-ing up these posters. The video ended "with a tracking shot from a car window showing the oppositional posters on storefronts and walls, pasted over official Party posters" (Ziter 2013: 127). The reworking of place with posters and the circulation of images through an online video came together to constitute a single political action. The video flowed through a space of coded infrastructure, but its power derived from its place-based images and political tactics.

The fusion between place-based action and digital flow takes a particularly striking form in *I Will Cross Tomorrow*, the final footage shot by film-maker Basel Shehadeh in the Syrian city of Homs: "the camera makes us present in his frantic sprint across a roadway, we hear the explosion that presumably killed Shehadeh, and we share his dying vision of a cloudless sky" (Ziter 2013: 131). Testimonial to both a literal and a metaphorical crossing, this video garnered over 11,000 views on YouTube.

In summary, political action in the MENA region created a web of relationships between social media such as Facebook, Twitter, and YouTube and places such as city streets and squares. It integrated digital, social, and spatial flows of microcelebrities and martyrs. These events are emblematic of the ways in which "geographically referenced content ... intersects and helps shape the relationships that undergird our lived geographies" (Graham, Zook, & Boulton 2013: 465). Other forms of "augmented reality" are more subtly political (Graham et al. 2013; Zook & Graham 2007), but spectacular political acts such as dying online for a political cause and mundane political acts such as "liking" a restaurant are similar, in that both participate in an actor network that exceeds the agency of the individual person. Both demonstrate hybrid agency that mingles human sensations, thoughts, and interests with digi-tally mediated networks. And both manifest distributional changes reinforcing certain forms of power and weakening others.

Conclusion

Famous journalist Thomas Friedman argued: "What brought Hosni Mubarak down was not Facebook and it was not Twitter. It was a million people in the streets, ready to die for what they believed in" (2011). There is something geographically compelling about claims like this, but Friedman presents a false choice between the internet and the street, the virtual and the real. Human environments have hybridized over time. The slogan "Reclaim the streets, reclaim the code" from Indymedia more usefully puts "the reclaiming of the digital code by media activists on a par with the reclaiming of the streets by urban protesters" (Mamadouh 2004: 485) and reflects an understanding similar to what Zook and Graham mean by DigiPlace (2007). Manuel Castells corrects the view that the internet is "just a handy tool to be used because it is there" and argues that "it fits with the basic features of the kind of social movements emerging in the Information Age" (2001: 139). This is in part because these movements "need the legitimacy and support provided by their reliance on local groups, yet they cannot remain local or they lose their capacity to act upon the real sources of power in our world" (Castells 2001: 143).

Political geographers are, of course, uniquely sensitive to the fact that "one of the most significant threats to digital activism occurs when cyber protests replace protesting in the public square" (Amin 2009: 65). As Schulte warns, if we credit Facebook or other social media with the ability to act on people in distant places, guiding them "toward self-liberation and self-determination," then we fall prey to American exceptionalism and technological determinism, whitewash the corporations producing and marketing internet applications, and obscure the "bodies and bodily risk required for revolution" (2013: 163). However, the appropriation of digital technology can be highly effective in conjunction with embodied forms of protest. Political geography has shown that scale is constructed in multiple ways, and the study of media reveals how this construction occurs through both face-to-face and mediated communications, affording greater opportunities for activists to maneuver. Rather than choose between cyberspatial tactics or territorial tactics, the most effective political techniques are both mediated and territorial, occupying cyberplaces and urban places; they use a mixture of both to reconstitute the scale of political struggle.

Digital code embodies "a range of political, economic and cultural imperatives" (Zook & Graham 2007: 480) and people engage in active negotiation with those imperatives. This sensibility with regard to political action may be framed as "Internet-assisted" (Nielsen 2013), "digitally enabled" (Earl & Kimport 2011), or "digitally networked" (Bennett & Segerberg 2012). In any case, what Castells calls "mass self-communication" (2009) is an important force in the early twenty-first century.

Recognition of this situation is important, because the false dichotomy between real and mediated politics leads to the formulation of rather confused arguments. Jon Alterman's article "The revolution will not be tweeted" (Alterman 2011) exposes "five fallacies" in the argument that social media had an important role in the Arab Spring. However, the title is misleading because he eventually admits:

> They filmed events with their cell phones, they created photo montages, they swapped songs, and they combined them and recombined them in countless ways. In many ways, this was the true transformative effect of social media. (2011: 111)

What Alterman wants to reject is the idea that social media acted autonomously, but his argument shows that the acts people take in conjunction with communication technologies – capturing, combining, and forwarding various types of digital content, within networks of bodies, physical places, and media – are indeed revolutionary. These actions are so powerful precisely because they reconfigure scale. The political actor looks less and less like the introspective Cartesian self and increasingly like an extroverted, extensible, networked self.

References

Adams, P.C. (2004). The September 11 attacks as viewed from Quebec: The small-nation myth in geopolitical discourse. *Political Geography* 23(6): 765–795. doi:10.1016/j.polgeo.2004.04.006

Adams, P.C. (2005). *The Boundless Self: Communication in Physical and Virtual Spaces.* Syracuse, NY: Syracuse University Press.

Adams, P.C. (2009). *Geographies of Media and Communication.* Malden, MA: Wiley-Blackwell.

Adams, P.C. (2010). Networks of early writing. *Historical Geography* 38: 70–89.

Alterman, J.B. (2011). The revolution will not be tweeted. *Washington Quarterly* 34(4): 103–116. doi:10.1080/0163660X.2011.610714

Amin, R. (2009). The empire strikes back: Social media uprisings and the future of cyber activism. *Kennedy School Review* 10: 64–66.

Anderson, B. (1991). *Imagined Communities: Reflections on the Origin and Spread of Nationalism,* 2nd edn. London: Verso.

Bennett, J. (2010). Thing power. In Braun, B., & Whatmore, S.J. (eds), *Political Matter: Technoscience, Democracy, and Public Life,* Minneapolis, MN: University of Minnesota Press, 35–62.

Bennett, W.L., & Segerberg, A. (2012). The logic of connective action. *Information, Communication and Society* 15: 739–768. doi:10.1080/1369118X.2012.670661

Biekart, K., & Fowler, A. (2013). Transforming activisms 2010+: Exploring ways and waves. *Development and Change* 44(3): 527–546. doi:10.1111/dech.12032

Braun, B., & Whatmore, S.J. (2010). The stuff of politics: An introduction. In Braun, B., & Whatmore, S.J. (eds), *Political Matter: Technoscience, Democracy, and Public Life,* Minneapolis, MN: University of Minnesota Press, ix–xl.

Carey, J. (1988). *Communication as Culture: Essays on Media and Society.* Boston, MA: Unwin Hyman.

Carter, S., & McCormack, D.P. (2006). Film, geopolitics and the affective logics of intervention. *Political Geography* 25(2): 228–245. doi:10.1016/j.polgeo.2005.11.004

Castells, M. (2001). *The Internet Galaxy: Reflections on the Internet, Business, and Society.* Oxford: Oxford University Press.

Castells, M. (2009). *Communication Power.* Oxford: Oxford University Press.

Childe, V.G. (1950). The urban revolution. *Town Planning Review* 21(April): 3–17.

Conover, M.D., Davis, C., Ferrara, E., McKelvey, K., Menczer, F., & Flammini, A. (2013). The geospatial characteristics of a social movement communication network. *PLoS ONE* 8(3): e55957. doi:10.1371/journal.pone.0055957

Couldry, N. (2008). Actor network theory and media: Do they connect and on what terms? In Hepp, A., Krotz, F., Moores, S., & Winter, C. (eds), *Connectivity, Networks and Flows: Conceptualizing Contemporary Communications,* Cresskill, NJ: Hampton Press, 93–110.

Crowley, D., & Heyer, P. (eds) (2010). *Communication in History,* 6th edn. Boston, MA: Pearson.

Deibert, R.J. (1997). *Parchment, Printing, and Hypermedia: Communication in World Order Transformation.* New York: Columbia University Press.

Deuze, M. (2007). *Media Work.* Cambridge: Polity Press.

Dijck, J.V. (2013). *The Culture of Connectivity: A Critical History of Social Media.* Oxford: Oxford University Press.

Dittmer, J. (2007). "America is safe while its boys and girls believe in its creeds!": Captain America and American identity prior to World War 2. *Environment and Planning D: Society and Space* 25(3): 401. doi:10.1068/d1905

Dittmer, J. (2013). Captain America and the Nationalist Superhero: Metaphors, Narratives, and Geopolitics. Philadelphia, PA: Temple University Press.

Dittmer, J., Moisio, S., Ingram, A., & Dodds, K. (2011). Have you heard the one about the disappearing ice? Recasting Arctic geopolitics. *Political Geography* 30(4): 202–214. doi:10.1016/j.polgeo.2011.04.002

Dodds, K. (2003). Licensed to stereotype: Geopolitics, James Bond and the spectre of Balkanism. *Geopolitics* 8(2): 125–156. doi:10.1080/714001037

Dodds, K. (2006). Popular geopolitics and audience dispositions: James Bond and the Internet Movie Database (IMDb). *Transactions of the Institute of British Geographers* 31(2): 116–130. doi:10.1111/j.1475-5661.2006.00199.x

Earl, J., & Kimport, K. (2011). *Digitally Enabled Social Change: Activism in the Internet Age*. Cambridge, MA: MIT Press.

Eisenstein, E.L. (1979). *The Printing Press as an Agent of Change: Communications and Cultural Transformations in Early-Modern Europe*, Vols I and II. Cambridge: Cambridge University Press.

Fraser, N. (1992). Rethinking the public sphere. In Calhoun, C. (ed.), *Habermas and the Public Sphere*, Cambridge, MA: MIT Press, 109–142.

Friedman, T. (2011). Commencement remarks. Tulane University, May 12. http://tulane.edu/grads/speakers-thomas-friedman.cfm, accessed May 23, 2014.

Galloway, A.R. (2004). *Protocol: How Control Exists after Decentralization*. Cambridge, MA: MIT Press.

Gana, N. (2013). Tunisia. In Amar, P., & Prashad, V. (eds), *Dispatches from the Arab Spring: Understanding the New Middle East*, Minneapolis, MN: University of Minnesota Press, 1–23.

Graham, M., Zook, M., & Boulton, A. (2013). Augmented reality in urban places: Contested content and the duplicity of code. *Transactions of the Institute of British Geographers* 38(3): 464–479. doi:10.1111/j.1475-5661.2012.00539.x

Granovetter, M.S. (1973). The strength of weak ties. *American Journal of Sociology* 78(6): 1360–1380. http://www.jstor.org/stable/2776392, accessed May 23, 2014.

Habermas, J. (1989 [1962]). *The Structural Transformation of the Public Sphere*, trans. T. Burger. Cambridge, MA: MIT Press.

Hassid, J. (2012). Safety valve or pressure cooker? Blogs in Chinese political life. *Journal of Communication* 62: 212–230. doi:10.1111/j.1460-2466.2012.01634.x

Howard, P.N., & Hussain, M.M. (2013). *Democracy's Fourth Wave? Digital Media and the Arab Spring*. Oxford: Oxford University Press.

Innis, H.A. (1951). *The Bias of Communication*. Toronto: University of Toronto Press.

Jensen, M.J., Jorba, L., & Anduiza, E. (2012). Introduction. In Anduiza, A., Jensen, M.J., & Jorba, L. (eds), *Digital Media and Political Engagement Worldwide: A Comparative Study*, Cambridge: Cambridge University Press, 1–15.

Kitchin, R., & Dodge, M. (2011). *Code/Space: Software and Everyday Life*. Cambridge, MA: MIT Press.

Langlois, G., Elmer, G., McKelvey, F., & Devereaux, Z. (2009). Networked publics: The double articulation of code and politics on Facebook. *Canadian Journal of Communication* 34: 415–434.

Latour, B. (1993). *We Have Never Been Modern*. Cambridge, MA: Harvard University Press.

Latour, B. (2005). *Reassembling the Social: An Introduction to Actor-Network-Theory*. Oxford: Oxford University Press.

Lotan, G., Graeff, E., Ananny, M., Gaffney, D., Pearce, I., & boyd, d. (2011). The revolutions were tweeted: Information flows during the 2011 Tunisian and Egyptian revolutions. *International Journal of Communication* 5: 1375–1405. http://ijoc.org/index.php/ijoc/article/view/1246, accessed May 23, 2014.

Luke, T.W., & Ó Tuathail, G. (1997). On videocameralistics: The geopolitics of failed states, the CNN International and (UN) governmentality. *Review of International Political Economy* 4(4): 709–733. doi:10.1080/09672299708565789

Lynch, M. (2006). *Voices of the New Arab Public: Iraq, Al-Jazeera, and Middle East Politics Today*. New York: Columbia University Press.

Mackenzie, A. (2002). *Transductions: Bodies and Machines at Speed*. London: Continuum.

Mamadouh, V. (2003). 11 September and popular geopolitics: A study of websites run for and by Dutch Moroccans. *Geopolitics* 8(3): 191–216. doi:10.1080/14650040412331307762

Mamadouh, V. (2004). Internet, scale and the global grassroots: Geographies of the Indymedia network of independent media centres. *Tijdschrift voor Economische en Sociale Geografie* 95(5): 482–497. doi:10.1111/j.0040-747X.2004.00334.x

Marvin, C. (1988). *When Old Technologies Were New: Thinking about Electric Communication in the Late Nineteenth Century.* New York: Oxford University Press.

Meyrowitz, J. (1985). *No Sense of Place: The Impact of Electronic Media on Social Behavior.* Oxford: Oxford University Press.

Nielsen, R.K. (2013). Mundane internet tools, the risk of exclusion, and reflexive movements – Occupy Wall Street and political uses of digital networked technologies. *Sociological Quarterly* 54(2): 173–177. doi:10.1177/1461444810380863

Ó Tuathail, G. (1996). *Critical Geopolitics.* Minneapolis, MN: University of Minnesota Press.

Papacharissi, Z.A. (2010). *A Private Sphere: Democracy in a Digital Age.* Malden, MA: Polity Press.

Pingree, G.B., & Gitelman, L. (2003). Introduction: What's new about new media? In Gitelman, L., & Pingree, G.B. (eds), *New Media, 1740–1915,* Cambridge, MA: MIT Press, xi–xxii.

Pinkerton, A., & Dodds, K. (2009). Radio geopolitics: Broadcasting, listening and the struggle for acoustic spaces. *Progress in Human Geography* 33(1): 10–27.

Preston, J., Kirkpatrick, D., Fahim, K., & Shadid, A. (2011). Movement began with outrage and a Facebook page that gave it an outlet. *New York Times,* February 5: 6, 10.

Rainie, L., & Wellman, B. (2012). *Networked: The New Social Operating System.* Cambridge, MA: MIT Press.

Schejter, A.M., & Tirosh, N. (2012). Social media new and old in the Al-'Arakeeb conflict: A case study. *The Information Society* 28(5): 304–315. doi:10.1080/01972243.2012.708711

Schmandt-Besserat, D. (1996). *How Writing Came About,* abridged edition of *Before Writing, Vol. I: From Counting to Cuneiform.* Austin, TX: University of Texas Press.

Schulte, S.R. (2013). *Cached: Decoding the Internet in Global Popular Culture.* New York: New York University Press.

Shah, N. (2013). Citizen action in the time of the network. *Development and Change* 44(3): 665–681. doi:10.1111/dech.12036

Sharp, J.P. (1996). Gendering nationhood: A feminist engagement with national identity. In Duncan, N. (ed.), *Bodyspace: Destabilising Geographies of Gender and Sexuality,* New York: Routledge, 97–107.

Sharp, J.P. (2000). *Condensing the Cold War: Reader's Digest and American Identity.* Minneapolis, MN: University of Minnesota Press.

Simondon, G. (1992). The genesis of the individual. In Crary, J., & Kwinter, S. (eds), *Zone 6: Incorporations,* Cambridge, MA: Zone Books, 296–319.

Theocharis, Y. (2013a). The wealth of (occupation) networks? Communication patterns and information distribution in a Twitter protest network. *Journal of Information Technology and Politics* 10(1): 35–56. doi:10.1080/19331681.2012.701106

Theocharis, Y. (2013b). The contribution of websites and blogs to the students' protest communication tactics during the 2010 UK university occupations. *Information, Communication and Society* 16(9): 1477–1513. doi:10.1080/1369118X.2012.706315

Tufekci, Z. (2013). "Not this one": Social movements, the attention economy, and microcelebrity networked activism. *American Behavioral Scientist* 57(7): 848–870.

Valenzuela, S. (2013). Unpacking the use of social media for protest behavior: The roles of information, opinion expression, and activism. *American Behavioral Scientist* 57(7): 920–942.

Warf, B. (2013). *Global Geographies of the Internet.* Springer Briefs in Geography. Dordrecht: Springer.

Warner, M. (2003). *Publics and Counterpublics.* New York: Zone Books.

Watts, D.J., & Strogatz, S.H. (1998). Collective dynamics of "small-world" networks. *Nature* 393(6684): 440–442. doi:10.1038/30918

Wellman, B. (2001). Physical place and cyberplace: The rise of personalized networking. *International Journal of Urban and Regional Research* 25(2): 227–252. doi:10.1111/1468-2427.00309

Whatmore, S. (2002). *Hybrid Geographies: Natures Cultures Spaces*. Thousand Oaks, CA: Sage.

Ziter, E. (2013). The image of the martyr in Syrian performance and web activism. *Drama Review* 57(1): 116–136. doi:10.1162/DRAM_a_00238

Zook, M.A., & Graham, M. (2007). Mapping DigiPlace: Geocoded internet data and the representation of place. *Environment and Planning B: Planning and Design* 34(3): 466–482. doi:10.1068/b3311

Material Political Geographies

Chapter 30

More-Than-Representational Political Geographies

Martin Müller

University of Zurich, Switzerland

It all began inconspicuously enough. The eight pages of Nigel Thrift's "It's the little things" (2000) were tacked onto a weighty tome that explored the representations of geopolitics ranging across 100 years and several continents. Thrift framed his propositions as a modest afterthought, yet they were anything but that. What was to form the closing thoughts of a book about geopolitical traditions came with the ambition to stake out a new agenda. Thrift (2000: 381) lamented what he called "representationalism," "the mesmerized attention to texts and images" in political geography, and especially in the subfield of critical geopolitics. This fascination, he claimed, occluded the fact that geopolitics was made of a myriad of non-textual practices:

> some of the most potent geopolitical forces are, I suspect, lurking in the "little" "details" of people's lives … [for example] the deep, often unconscious aggressions … behind so much geopolitical "reasoning". (Thrift 2000: 384)

With this short intervention, Thrift threw down the gauntlet at political geographers' feet: He challenged the discipline to become more than representational.

What does it mean to be "more-than-representational"?

But what are we to understand by "more-than-representational"? At its most basic, the term refers to what people (and indeed animals or things) *do* and thus squarely engages with practices. More specifically, more-than-representational styles of research are united by an interest in

> how life takes shape and gains expression in shared experiences, everyday routines, fleeting encounters, embodied movements, precognitive triggers, practical skills, affective intensities, enduring urges, unexceptional interactions and sensuous dispositions. (Lorimer 2005: 84)

The Wiley Blackwell Companion to Political Geography, First Edition.
Edited by John Agnew, Virginie Mamadouh, Anna J. Secor, and Joanne Sharp.
© 2015 John Wiley & Sons Ltd. Published 2017 by John Wiley & Sons Ltd.

With this focus, more-than-representational research seeks to displace an exclusive concern with humans as the prime movers and shakers of the world and to disturb the preoccupation with cognitive processes of the production of meaning (Lorimer 2005, 2008; Thrift 2008; Anderson & Harrison 2010). As such, it positions itself against a purely semiotic understanding of the world, predicated on the analysis of human-made meaning. This semiotic understanding had been the founding rationale of the influential critical geopolitics, analyzing the writing of global space (Müller 2013). Critical geopolitics had emerged from the turn toward examining symbolic meaning production, which human geography underwent from the late 1980s and which spawned the new cultural geography (Cosgrove & Daniels 1988; Barnes & Duncan 1992). Yet, *more*-than-representational research does not abandon the representational entirely, but maintains that the more-than-representational – practices, affects, things – is intertwined with the production of meaning.

The theoretical inspirations feeding into more-than-representational research are diverse and eclectic. Among the most prominent are contributions on hybridity and vital materialism (Haraway 1991; Ingold 2000; Latour 2005; Bennett 2010), affect and assemblage thinking (Spinoza 1677; Deleuze & Guattari 1987; Deleuze 1988; Massumi 2002; Barad 2007), phenomenology (Merleau-Ponty 1962), feminist theory (Sedgwick 2003; Ahmed 2004; Grosz 2008), and practice theory (De Certeau 1984; Schatzki 2002; Wetherell 2012).

If we were to boil down more-than-representational research to a set of propositions, these could be:

- *The world is made through performative practice.* Practices are the basic unit of analysis. Their performative quality lies in constantly bringing into being new socio-material associations.
- *The world is always in the making.* There are no fixed stable states. The aleatory and the unexpected reign supreme.
- *The world is affective.* The lived immediacy of experience becomes present through our bodies. Affect underscores the importance of registers that exceed the cognitive and the conscious.
- *The world is more than human.* Things and animals partake in it as more than mere passive objects. They become enmeshed, tied up in hybrid, socio-material assemblages of humans and non-humans, and they co-define human experience.
- *More-than-representational research is experimental.* Attentiveness to multiple registers of sensation requires novel modes of presenting and presencing research to tackle the more-than-representational.

The term "world" here refers not to a pre-existing entity but to the continuous enactment of relations that produces action. For more-than-representational research, action emerges through the interplay of cognitive deliberation and meaning-making as well as embodied affordances and habits. The relation between humans and their surroundings is thus a two-way street: Humans act as much as they are acted on. They are always already "caught up in the fabric of the world" (Merleau-Ponty 1962: 256; cf. Anderson & Harrison 2010) and not detached from it or above it, as social constructivism would claim. It is this emergence of relations between very different elements to which more-than-representational research devotes its attention. It is important to emphasize that a more-than-representational approach does not understand practices as a product of discourse, but as constitutive of the world in their own right.

To be sure, the key tenets of more-than-representational research were not completely new or unheard of in human geography before Thrift's intervention. In its attention to the mundane, lived experience and the body, it shares many common concerns that feminist political geography already raised in the 1990s (e.g., Kofman & Peake 1990). In German-speaking geography, controversial engagements with phenomenology also foreshadowed many of the key claims of more-than-representational research, for example in the aesthetics of nature (e.g., Hasse 1993). However, language barriers being what they are, this minor knowledge went unacknowledged. Last but not least, the tradition of humanistic geography also placed emphasis on human everyday experience (e.g., Entrikin 1976; Tuan 1979). Yet, there are also important differences, as we will see in the following sections.

Although more-than-representational research entered human geography via cultural geography and its intellectual roots remain there, after some initial hesitation reception in political geography has been gathering speed. In recent years, the field has seen a growing number of contributions that seek to push political geography beyond the representational in different ways. Along three key axes, this contribution surveys how political geographers have taken up the more-than-representational challenge. First are questions of affect and emotion, which have commanded much attention, not least because they have always been central to feminist geography. Here, a key dispute revolves around the issue to what degree affect and emotions are shaped through differential social positionings and thus not entirely prediscursive or precognitive. Second are socio-material assemblages and the process of relating human and non-human elements into larger wholes. This focus has helped to develop a more processual perspective on entities such as the state or organizations and to bring non-human forces back into the analytical fold. Third, finally, are experiments with new ways of presenting or *presencing* – that is, making present – research that are called for if we take the first two axes seriously. The separation into three axes follows the main focus of the studies reviewed and helps to make the material less unwieldy. It goes without saying that there are, multiple interrelations between the three axes. The chapter concludes with an outlook toward a future research agenda for this nascent field of inquiry. In so doing, it suggests four lines of further engagement: the politics of the more than representational, the link between the representational and the more-than-representational, the move from micro to macro, and the vitality of matter.

Affect and emotion

Early modern philosopher Baruch Spinoza (1677) posited the existence of what he called the three primary affects: desire (*cupiditas*), joy (*laetitia*), and sadness (*tristitia*). From these three affects, he claims, all other affects can be derived. Spinoza lists 45 in addition to the three primary ones, among them wonder, contempt, love, hatred, devotion, fear, despair, envy, shame, and anger. A short definition could describe affect as intensities of feeling that influence behavior. While there is no shortage of debates on the definition and delimitation of affect in human geography (e.g., McCormack 2003; Thrift 2004; Thien 2005; Anderson 2006; Tolia-Kelly 2006; Lorimer 2008), "intensities of feeling that influence behavior" conveys two important notions. First, affect is something that *works in and through the body* and bodily experience. Second, affect *drives action* and produces visible conduct.

This takes us some way, but not all the way, toward an understanding of affect. A further two features are important. One is that affect is distributed and *exceeds the subject*. For it to come into existence, the human body must be entangled in a whole set of relations:

> Affect is distributed between, and can happen outside, bodies which are not exclusively human, and might incorporate technologies, things, non-human living matter, discourses or even, say, a swathe of noise or swarm of creatures. (Lorimer 2008: 552)

Affect thus means the capacity to affect manifold others and, in turn, to be affected by manifold others. Last and most contentious, affect is said to be *precognitive, preconscious, and irrational*. It is untainted by measured reflection and discursive symbolization. It has been characterized as "that indescribable moment before the registration of the audible, visual, and tactile transformations produced in reaction to a certain situation, event, or thing" (Colman 2010: 11). Affects are thus continuously emergent and fleeting.

It is primarily this last point that has provoked critical reactions. Feminist geographers in particular have pointed out that conceptualizing affect as prediscursive risks reinstating the body of the subject as a blank slate, unmarked by unequal differentiations such as race, gender, sexual orientation, or class (e.g., Thien 2005; Tolia-Kelly 2006; Colls 2012; see also Sharp 2009; Pile 2010). Yet, these discursive significations matter a great deal in how bodies experience affect: "a body that is signified as a source of fear through its markedness cannot be free to affect and be affected similarly to one that is not" (Tolia-Kelly 2006: 215). Not taking into account the multiple discursive subject positions would amount to a blindness to power and present a disembodied, ahistoricist stance on affects. Feminist geographers have productively employed the notion of emotion, closely related to affect, to describe precisely the interplay between the symbolic positioning of subjects and the intensities of feeling that it affords or denies (cf. Anderson & Smith 2001; Davidson, Bondi, & Smith 2007; see also Fluri this volume). In other words, the concept of emotion moves affect from a non-representational to a more-than-representational register and thus acts as an important corrective. In more concrete terms, a political speech can affect a person, but whether it incites joy, hate, or pride has much to do with whether this person is male or female, white or black, belongs to the minority or the majority ethnicity, and so on.

The political relevance of affect and emotion is as striking as it is manifold. An explicit engagement with the politics of affect has taken place both outside of human geography (e.g., Connolly 2002; Massumi 2002; Ahmed 2004; Protevi 2009) and within it (e.g., Thrift 2004; Barnett 2008; Pain 2009; Ruddick 2010; Anderson 2012, 2014). The consideration of the link between affect and politics has focused on a variety of arenas, where affect intervenes in decision-making or is even consciously engineered to serve political or economic goals. These different realms of affective politics, Thrift (2008: 248) claims, work against democratic expression, since they contribute to a style of democracy that is consumed but not practiced and bypass channels of rational, communicative deliberation. If affect is irrational and unable to be signified, it is also difficult to intervene in its workings.

Urban and public spaces are more and more designed to evoke affective responses through the use of design, lighting, events, music, or performance, providing "a new minute landscape of manipulation" (Thrift 2004: 66). Political campaigns make use of affects to create support for one or the other candidate (Thrift 2008; Schurr 2013). The visceral effects of rhythm and sound can help buttress political activism and mobilize people into action (Waitt, Ryan, & Farbotko 2014). The cultivation of fear to garner support for military interventions or new security measures has become particularly strong after the attacks of 9/11 and is an integral part of a geopolitics that divides the world into us/them (Hyndman 2007; Sparke 2007; Pain & Smith 2008; Pain 2009, 2010; Anderson 2010). Affect is involved in forms of bio-power, where affective lives become targets to achieve control and coordination of individuals and populations (Anderson 2012). The ubiquitous mediatization of most societies worldwide also calls attention to the imbrication of the media with affect and the implications for politics (Carter & McCormack 2006). Whether in films, newsreels, or images, the media are saturated with affect that is linked to symbolic content. Of particular interest has been the affective power of images to touch the body (MacDonald 2006; Carter & McCormack 2010).

One of the earliest explicit engagements with affect in political geography is still one of the most original. Gearóid Ó Tuathail's (2003) "Just out looking for a fight: American affect and the invasion of Iraq" puts forward the hypothesis that the United States' military intervention in Iraq was driven to a large extent by gut feelings of rage and pride after the terrorist attacks on September 11, 2001 that brought about an "affective economy of revenge" (Ó Tuathail 2003: 868).

> The affective tsunami unleashed by the terrorist attacks of 2001 is a broad and deep one that has set down a powerful somatic marker for most Americans. "9/11" is its shorthand, a phrase that has instant meaning for millions of Americans but more estranged resonance overseas. The calendar digits memorialize a moment in time that has become an affect-imbued memory bank for the media and political class in the United States and, consequently for the media-incited nation. (Ó Tuathail 2003: 859)

US television networks transformed 9/11 into a collective trauma of loss, multiplying the feeling of pain and stoking public outrage. As a result, a large part of the public in the United States, Ó Tuathail argues, experienced the events of 9/11 in a visceral way, as though they had been present on the spot. This affective economy allowed affect to overpower deliberative discourse and rational discussion. The war against terrorism in Iraq rested on dubious claims, lacked rational justification, and had thin support in the United Nations, yet it managed to enthuse a large part of the public and of the leadership of the United States – "a triumph of affect over intellect," as Ó Tuathail (2003: 863) puts it. What Ó Tuathail's piece manages particularly well is the upscaling of the microgeographies of affect from the level of the body to that of the globe. In so doing, he makes affectual scholarship relevant for an analysis of global geopolitical questions.

While Ó Tuathail examines affect from a critical distance and considers its significance for war and peace, Pain et al. (2010) move close to the emotional experiences of young people in the United Kingdom and New Zealand with the help of participatory research. This piece stands in the emerging tradition of an emotional geopolitics (Pain 2009) – probably the most vigorous reception of more-than-representational ideas in political geography (e.g., Smith 2012; Cuomo 2013; Faria 2014). Pain and her co-authors consider how geopolitical events and discourses are recast and trigger feelings in the everyday lives of their research subjects, focusing particularly on emotions of fear and hope. Their study drives home the point that emotions are refracted through the prism of dominant power relationships that characterize people's everyday lives. Emotions do not just trickle down from big geopolitical events – wars, terrorist attacks, financial crises – but are renegotiated: "closeness to or distance from sites of geopolitical risk has some relevance, but is generally less important than young people's own structural and social positions" (Pain et al. 2010: 980). Pain et al. thus destabilize dominant hierarchies of local/global and near/distant and show how symbolic meanings interact with and shape bodily experience. This kind of more-than-representational research seeks to occupy a promising middle ground between the extremes of a wholly precognitive, transhuman affect and the meaning-saturated discursive subject position.

Assemblages

Affects are an important force in the emergence of assemblages. Assemblage is the English translation of Deleuze and Guattari's (1987; see also DeLanda 2006) French term *agencement*. Deleuze defines assemblage as "a multiplicity constituted by heterogeneous terms and which establishes liaisons, relations between them" (Deleuze 2007: 52). It refers to the process of arranging and organizing heterogeneous materials to hold together for some time and create new actions. When Deleuze and Guattari write of "heterogeneous materials," they exhibit a broad conception of the social, encompassing both non-human and human elements. These could be mundane objects, animals, bacteria, plants, a sticky note, and so on – Thrift's "little things" taken literally. Assemblages result in a coalescence of forces and are productive of new realities. According to Deleuze and Guattari, what holds assemblages together is the force of desire, producing connections between initially separate elements. In other words, affects glue assemblages together. Similar to the notion of the actor-network, the concept of assemblage places an emphasis on emergence and multiplicity, acknowledging that the "world has to be built from utterly heterogeneous parts that will never make a whole, but at best a fragile, revisable, and diverse composite material" (Latour 2010: 474).

The notion of assemblage has become very much en vogue with human geographers over the past few years; see the edited collection by Anderson and McFarlane (2011) and the handbook article by Robbins and Marks (2010) for just two examples. This proliferation, however, has also resulted in an increasingly diffuse usage of the term. Within political geography, the ideas associated with assemblages have been put to several uses. One is that assemblages draw together distant locales and fold scales into each other in creating a topological space of relations. This argument links into the debate on the production of scales and proximity/distance as effects of networked relations: "scale is what actors achieve by *scaling*, *spacing*, and *contextualizing* each other through the transportation in some specific vehicles of some specific traces" (Latour 2005: 183–184). Thus, authors have examined the distributed organization of social movements across sites (McFarlane 2009; Davies 2012), the contested production of scales by international organizations and governments (Legg 2009), and the topological reach of state power, mediated through multiple connections that draw together dispersed elements (Allen & Cochrane 2010).

A second interpretation of assemblage thinking has resulted in greater attention to the role of the material world, the more than human, in shaping the spaces of politics (Latour 1994). No longer considered inanimate, passive matter, materiality is accorded a more active role in assemblages than social constructivist research would allow (Braun & Whatmore 2010; Meehan, Shaw, & Marston 2013; Dittmer 2014). "Matter matters" is the clarion call. Materials participate in the construction of information (Barry 2013), nature co-constitutes agency (Bakker & Bridge 2006), objects extend state power (Meehan 2014), citizenship is assembled in relation to other species and living organisms (Barker 2010), close encounters between bodies and objects can reinforce or disrupt geopolitical narratives (Sundberg 2008). What all these studies have in common is that materials experience an emancipation from their role as passive objects of political deliberation and prostheses of human agency. They co-articulate agency and shape political practices, the art of governing, the constitution of sovereignty in crucial, often unexpected ways.

A final avenue has emphasized the processual and composite aspect of assemblages. Although the concept of assemblage is not used in each instance, there is a clear alignment with its principal ideas. Such research has foregrounded the ways in which different elements are brought together so as to generate the capacity to act. An important strand has sought to combat reifications of the state as a unified entity and refocus attention on the mundane practices, bodily performances, and multiple objects through which states come to acquire the agency that defines them (Painter 2006). Such work is particularly illuminating in contexts of emergent state formation, such as in the post-Soviet republics (Schueth 2012). In a similar vein, other research has questioned the treatment of organizations as black boxes in political geography, and proposed to unravel how the actorness of organizations is constituted through continuous processes of ordering (Müller 2012). Such ordering is often temporarily stabilized through affective binds, such as when humor ties together the participants in a Model United Nations Session, aligning them with each other and providing a common orientation (Dittmer 2013).

Andrew Barry's (2013) monograph *Material Politics* is an excellent example of both the strengths of an assemblage-oriented approach and the ways in which it connects representational with more-than-representational modes of research. It revolves around the controversies connected to the construction of the 248 km stretch of the Baku–Tbilisi–Ceyhan oil pipeline passing through Georgia on the way from Azerbaijan to Turkey. Barry asks how and why particular materials and sites along the pipeline gained transnational significance, while others did not. He contends that this had much to do with the modalities of the assembling of information about materials and sites, which performatively enacted those materials and sites. The wide-ranging transparency initiative accompanying the construction process sought to delimit what belonged and what did not belong to the political space of the pipeline, thus making certain issues visible and others invisible. Whether an issue escalated or not depended on the stability of the assemblage. Controversies arose when materials or people behaved in unforeseen ways and thus necessitated the redefinition of the boundaries of the assemblage. In the course of this, the formerly delimited spaces of knowledge production became destabilized and had to be renegotiated. The pipeline as socio-material assemblage was thus mutable and assumed different forms and spatial extents, transforming, as a consequence, what issues could be bound up with it, what could be known about it, and what could be contested.

Presenting and presencing the more-than-representational

Affect, emotion, and socio-material assemblages are ways of conceptualizing more-than-representational political geographies. Yet, how are we to present the more-than-representational? Since academic work is commonly predicated on representational modes of communicating and evidencing research – texts, graphs, pictures – more-than-representational geographers are faced with a methodological dilemma here:

> the leanness of descriptive language comes up short of the manifold affective events and textures it seeks to speak up for. Or, dare I say it, meaningfully represent. Rather than glorying in the ringing refrain, often I am left keening for more varied words to express and explain geography being done otherwise. (Lorimer 2008: 557)

The "crisis of representation," diagnosed by anthropologists in the 1980s (Clifford & Marcus 1986), might well be called the starting point for the debate about the limits of descriptive language. It held that no account could be given of an objective reality to be discovered independently of our narratives and interpretations (Geertz 1973: 20, 23). This observation started to open the social sciences to more creative forms of constructing research narratives and to experimentation.

For more-than-representational geographies, there are two principal avenues of dealing with the limits of representation. One is to remain in a representational mode and attempt to *describe* the more-than-representational; the other is to *evoke* it. The first variant is similar to the classic interpretive-hermeneutic approach, only that it takes as its object the more than representational. It reports on emotions observed, socio-material networks assembled, and the consequences for the research question at hand. These paths were followed by the studies reported in depth earlier – Ó Tuathail's (2003) examination of affect in the invasion of Iraq, Pain et al.'s (2010) study of young people's fears and hopes, and Barry's (2013) tracing of the pipeline assemblage. Another approach in this vein employs minute descriptions of fleeting encounters and ordinary sitations that call attention to embodied practices and language use beyond explicit meaning production. Inspired by ethno-methodology, the key point here is that many aspects of language in routine situations are anything but representational, but fulfill a variety of other functions, whether to create empathy, start conversations, or express belonging (Laurier & Philo 2006).

Several scholars, however, have sought to push beyond the representation of the more-than-representational in a second avenue, which can be described as seeking to *evoke* the more-than-representational. This avenue is less about presenting than about presencing – about bringing something into being (cf. Law 2004: 11–12; Dirksmeier & Helbrecht 2008). It starts from the assumption that "representations of affect can only ever fail to represent affect itself" (Pile 2010: 8). In other words, the more-than-representational is irreducible to the meaning that we seek to give to it, when we decide to describe and interpret it (Harrison 2007).

Hence, efforts have attempted to push the boundaries of traditional formats and experiment with other forms. Some have opted to stay with writing as a genre, but engage with different forms of creative writing – something that geographers have been exhorted to do for quite some time (cf. Meinig 1983). Some geographers have mobilized the expressive force of poetry (e.g., Cresswell 2014) and creative prose (e.g., McCormack 2003; Cook 2004; Wylie 2005), often with an autoethnographic component, to capture and convey the passions and lived intensities of the more-than-representational. Arundhati Roy's novel *The God of Small Things* is an enthralling illustration of how writing can interweave big *P* politics with the emotional lives and desires of ordinary people – connecting Thrift's little things to the big ones. She captures power relations in the fleeting moment of their instantiation with an exceptional sensuous sensibility that does not fail to capture the reader. Consider this evocation of love and fear of its loss:

> "D'you know what happens when you hurt people?" Ammu said. "When you hurt people, they begin to love you less. That's what careless words do. They make people love you a little less."

> A cold moth with unusually dense dorsal tufts landed lightly on Rahel's heart. Where its icy legs touched her, she got goose bumps. Six goose bumps on her careless heart. (Roy 1997: 112)

Even without knowing the context of these lines, the powerful image of the moth with the icy legs cannot fail but grip most readers. It may send a shiver down one's spine, or perhaps even create those goose bumps on some skins. The affective, corporeal reaction takes place after a remark by 7-year-old Rahel on love and marriage, considered inappropriate in the religious and sociocultural context of postcolonial Kerala in which the novel is set. As such, the small things of embodied experience are embedded in broader societal concerns and prescribed roles tethered to gender, religion, and caste.

Instead of working with a density of words to get at the more-than-representational, another option is to examine the opposite: the absence or loss of words. The more-than-representational begins where research subjects fumble for words and established codes of communication break down. "How shall I say it…?" as a response in an interview, for example, then does not so much hint at the inability to put things into words, but at the different, more-than-representational registers at work that disrupt the smooth sheen of meaning production, but cannot be entirely separated from it at the same time (Hyams 2004; Harrison 2007; Proudfoot 2010).

A final path for presencing the more-than-representational is inspired more by the performing arts (Thrift 2008: 12; Lorimer 2010). It relies on moving images or performances to engage audiences, often in a more visceral way than would be possible with texts or talk. Film is one good way of offering an immediate window on embodied and non-textual performances and addressing different sensory registers, including the visual and the aural. Political geographers have used video clips embedded in text, for example, to examine how emotions are linked to hegemonic discourses in politicians' campaign performances (Schurr 2012). Performances – dance, music, theater – could create a more immediate sense of presence and being affected, but in most cases remain the objects of research that are then again cast into a familiar textual mold (cf. Pratt 2000).

Looking at the existing research, one cannot but diagnose that political geographers have been rather timid in abandoning the seductive security and comforting conventions of the cherished academic paper for livelier forms of presencing research that more-than-representational political geographies would warrant. Yet, many of the themes of political geography are open to this presencing. Rage, anger, fear, or pride are part of the mix in many kinds of political action, whether in the annexation of Crimea, the local protests against nuclear reprocessing plants, or the fervent articulation of nationalism. Video snippets or audio recordings can go a long way here and could indeed spice up the reciting of presentation manuscripts at conferences. Evocative field notes, such as when out doing research on protective accompaniment (Koopman 2012) or on the imbrication of bodies and territory (Smith 2012), can serve a similar purpose, as can the writing or indeed performing of plays (cf. Johnston & Pratt 2010). Material things, extracted from assemblages, can be brought home from fieldwork into the conference room or sketched for paper publication: the contentious pipeline coating (Barry 2013), the test tube and petri dish for the assisted reproduction of the nation (Schurr 2014), the invasive species (Barker 2010), the belongings left behind by migrants in the Sonora desert (Sundberg 2008), the standardized test as a tool of state power (Meehan et al. 2013), the mobile Zimbabwe bush pump (De Laet & Mol 2000), or indeed the manifold odors that create attachment and aversion to places (Hoover 2009). After all, these are the things that our research subjects encounter on a daily basis and so do we when we are on fieldwork. Reducing them to mere words on paper unneccesarily truncates their extensive materiality.

Future paths

If the recent swirl of interest is any indication, more-than-representational modes of research are going to be a major influence in political geography in the coming decade. They have the potential of opening up new lines of inquiry and shining new lights on established ones – both desirable attributes for a new paradigm. In the state of flux in which the more-than-representational engagement of political geography exists at the moment, four issues appear to be of particular relevance.

What is the politics of the more-than-representational?

The engagement with more-than-representational forms of doing research originated with cultural geography and was initially concerned with the everydayness of life. As a political agenda is gradually etched out, it should be the prime task of political geographers to weigh in on the debate and demonstrate that more-than-representational research is not *l'art pour l'art*. Who or what shapes assemblages? Why and how do assemblages nevertheless hold together, despite all the insistence on multiplicity and flux? How are injustices the result of assembling elements in a particular way? What political aims and purposes do affects and emotions buttress? How can the instrumental management of affect be critiqued and resisted? These are the sorts of questions that political geographers would be well placed to ask in order to move the agenda forward.

How are the representational and the more-than-representational tied together?

In the fascination with more-than-representational modes of research, care needs to be taken not to throw out the insights of representational research. It is perhaps an inevitable process that in establishing a new agenda, stark, sometimes caricaturing dividing lines need be drawn to distinguish it from the dominant paradigm. In so doing, the more-than-representational has sometimes been positioned in too strong an antagonistic relationship with the representational. Many social theories, however, have a place for both: Foucault with his concept of the dispositif, actor-network theory with the concept of material semiotics, Haraway's material-semiotic actor, the discourse theory of Laclau and Mouffe with the concept of dislocation, or psychoanalysis with the concepts of the Symbolic and the Real. Indeed, for giving the more-than-representational a political edge, exploring how the representational is tied up with the more-than-representational is critical (e.g., Barad 2007; Wetherell 2013).

How do we move from the micro to the macro?

More-than-representational research started from work on the microgeographies of the body and the object. Its attention to practices links it to the concrete and close at hand. However, how can it be harnessed to analyze the state, geopolitics, the media, nationalism – topics that are at the heart of political geography? To be sure, the opposition of micro and macro is to some degree artificial. After all, what appears to be macro is just a composite of many smaller assemblages (Callon & Latour 1981). Upscaling from the micro to the macro would then involve following the relations that tie elements together, considering how, why, and when they make more extensive associations durable.

How can we do justice to the vitality of matter?

With its emphasis on the material world, more-than-representational research comes with the great promise of recovering a more vital sense of what seems to have increasingly fallen by the wayside of political geography: nature and materials. The challenge of becoming more-than-representational thus offers the potential of reinvigorating the link with political ecology (see Meehan & Molden this volume). It calls on political geographers to acknowledge that nature and things can push humans in unexpected directions and affect them deeply in corporeal ways. Humans are bound up with the material world in a reciprocal relationship, which asks us to rethink the terms in which political geography has traditionally thought of animals, plants, resources, or objects (Braun & Whatmore 2010; Fall 2010).

The big task ahead for more-than-representational research, then, is perhaps not so much to push forward, for much has already been achieved there. Rather, it would be well advised to look around and look back. Where can connections be made? What of political geography has it left behind in the initial euphoria that could be useful along the more-than-representational journey? If that happens, the next edition of this *Companion* will not have to feature a separate chapter on more-than-representational political geographies – instead, all of political geography will have become a little more than representational.

References

Ahmed, S. (2004). *The Cultural Politics of Emotion*. London: Routledge.

Allen, J., & Cochrane, A. (2010). Assemblages of state power: Topological shifts in the organization of government and politics. *Antipode* 42(5): 1071–1089. doi:10.1111/j.1467-8330.2010.00794.x

Anderson, B. (2006). Becoming and being hopeful: Towards a theory of affect. *Environment and Planning D: Society and Space* 24(5): 733–752. doi:10.1068/d393t

Anderson, B. (2010). Morale and the affective geographies of the "war on terror." *Cultural Geographies* 17(2): 219–236. doi:10.1177/1474474010363849

Anderson, B. (2012). Affect and biopower: Towards a politics of life. *Transactions of the Institute of British Geographers* 37(1): 28–43. doi:10.1111/j.1475-5661.2011.00441.x

Anderson, B. (2014). *Encountering Affect: Capacities, Apparatuses, Conditions*. Aldershot: Ashgate.

Anderson, B., & Harrison, P. (2010). The promise of non-representational theories. In Anderson, B., & Harrison, P. (eds), *Taking-Place: Non-Representational Theories and Geography*, Farnham: Ashgate, 1–36.

Anderson, B., & McFarlane, C. (eds) (2011). Assemblage and geography. *Area* 43(2): 124–164. doi:10.1111/j.1475-4762.2011.01004.x

Anderson, K., & Smith, S.J. (2001). Editorial: Emotional geographies. *Transactions of the Institute of British Geographers* 26(1): 7–10. doi:10.1111/1475-5661.00002

Bakker, K., & Bridge, G. (2006). Material worlds? Resource geographies and the "matter of nature." *Progress in Human Geography* 30(1): 5–27. doi:10.1191/0309132506ph588oa

Barad, K. (2007). *Meeting the Universe Halfway: Quantum Physics and the Entanglement of Matter and Meaning*. Durham, NC: Duke University Press.

Barker, K. (2010). Biosecure citizenship: Politicising symbiotic associations and the construction of biological threat. *Transactions of the Institute of British Geographers* 35(3): 350–363. doi:10.1111/j.1475-5661.2010.00386.x

Barnes, T.J., & Duncan, J.S. (1992). Introduction: Writing worlds. In Barnes, T.J., & Duncan, J.S. (eds), *Writing Worlds: Discourse, Text and Metaphor in the Representation of Landscape*, London: Routledge, 1–17.

Barnett, C. (2008). Political affects in public space: Normative blind-spots in non-representational ontologies. *Transactions of the Institute of British Geographers* 33(2): 186–200. doi:10.1111/j.1475-5661.2008.00298.x

Barry, A. (2013). *Material Politics: Disputes along the Pipeline*. Oxford: Blackwell.

Bennett, J. (2010). *Vibrant Matter: A Political Ecology of Things*. Durham, NC: Duke University Press.

Braun, B., & Whatmore, S. (eds) (2010). *Political Matter: Technoscience, Democracy, and Public Life*. Minneapolis, MN: University of Minnesota Press.

Callon, M., & Latour, B. (1981). Unscrewing the big Leviathan: How actors macrostructure reality and how sociologists help them to do so. In Knorr, K., & Cicourel, A.V. (eds), *Advances in Social Theory and Methodology: Toward an Integration of Micro- and Macro-Sociologies*, Boston, MA: Routledge and Kegan Paul, 277–303.

Carter, S., & McCormack, D.P. (2006). Film, geopolitics and the affective logics of intervention. *Political Geography* 25(2): 228–245. doi:10.1016/j.polgeo.2005.11.004

Carter, S., & McCormack, D.P. (2010). Affectivity and geopolitical images. In MacDonald, F., Hughes, R., & Dodds, K. (eds), *Observant States: Geopolitics and Visual Culture*, London: IB Tauris, 103–122.

Clifford, J., & Marcus, G.E. (1986). *Writing Culture: The Poetics and Politics of Ethnography*. Berkeley, CA: University of California Press.

Colls, R. (2012). Feminism, bodily difference and non-representational geographies. *Transactions of the Institute of British Geographers* 37(3): 430–445. doi:10.1111/j.1475-5661.2011.00477.x

Colman, F.J. (2010). Affect. In Parr, A. (ed.), *The Deleuze Dictionary*, Edinburgh: Edinburgh University Press, 11–14.

Connolly, W.E. (2002). *Neuropolitics: Thinking, Culture, Speed*. Minneapolis, MN: University of Minnesota Press.

Cook, I. (2004). Follow the thing: Papaya. *Antipode* 36(4): 642–664. doi:10.1111/j.1467-8330.2004.00441.x

Cosgrove, D., & Daniels, S. (1988). *The Iconography of Landscape: Essays on the Symbolic Representation, Design and Use of Past Environments*. Cambridge: Cambridge University Press.

Cresswell, T. (2014). Geographies of poetry/poetries of geography. *Cultural Geographies* 21(1): 141–146. doi:10.1177/1474474012466117

Cuomo, D. (2013). Security and fear: The geopolitics of intimate partner violence policing. *Geopolitics* 18(4): 856–874. doi:10.1080/14650045.2013.811642

Davidson, J., Bondi, L., & Smith, M. (eds) (2007). *Emotional Geographies*. Aldershot: Ashgate.

Davies, A.D. (2012). Assemblage and social movements: Tibet support groups and the spatialities of political organisation. *Transactions of the Institute of British Geographers* 37(2): 273–286. doi:10.1111/j.1475-5661.2011.00462.x

De Certeau, M. (1984). *The Practice of Everyday Life*. Berkeley, CA: University of California Press.

De Laet, M., & Mol, A. (2000). The Zimbabwe bush pump: Mechanics of a fluid technology. *Social Studies of Science* 30: 225–263.

DeLanda, M. (2006). *A New Philosophy of Society: Assemblage Theory and Social Complexity*. London: Continuum.

Deleuze, G. (1988). *Spinoza: A Practical Philosophy*. San Francisco, CA: City Lights.

Deleuze, G. (2007). *Dialogues II*. New York: Columbia University Press.

Deleuze, G., & Guattari, F. (1987). *A Thousand Plateaus*. Minneapolis, MN: University of Minnesota Press.

Dirksmeier, P., & Helbrecht, I. (2008). Time, non-representational theory and the "performative turn" – towards a new methodology in qualitative social research. *Forum: Qualitative Sozialforschung/Forum: Qualitative Social Research* 9(2). http://www.qualitative-research.net/index.php/fqs/article/view/385, accessed March 26, 2015.

Dittmer, J. (2013). Humour at the Model United Nations: The role of laughter in constituting geopolitical assemblages. *Geopolitics* 18(3): 493–513. doi:10.1080/14650045.2012.742066

Dittmer, J. (2014). Geopolitical assemblages and complexity. *Progress in Human Geography* 38(3): 385–401. doi:10.1177/0309132513501405

Entrikin, J.N. (1976). Contemporary humanism in geography. *Annals of the Association of American Geographers* 66(4): 615–632. doi:10.1111/j.1467-8306.1976.tb01113.x

Fall, J.J. (2010). Grounding politics on nearness. *Geographische Zeitschrift* 98(4): 213–225.

Faria, C. (2014). Styling the nation: Fear and desire in the South Sudanese beauty trade. *Transactions of the Institute of British Geographers* 39(2): 318–330. doi:10.1111/tran.12027

Geertz, C. (1973). *The Interpretation of Cultures*. New York: Basic Books.

Grosz, E.A. (2008). *Chaos, Territory, Art: Deleuze and the Framing of the Earth*. New York: Columbia University Press.

Haraway, D. (1991). *Simians, Cyborgs and Women: The Reinvention of Nature*. London: Routledge.

Harrison, P. (2007). "How shall I say it...?" Relating the nonrelational. *Environment and Planning A* 39(3): 590–608. doi:10.1068/a3825

Hasse, J. (1993). *Ästhetische Rationalität und Geographie: sozialräumliche Prozesse jenseits kognitivistischer Menschenbilder*. Oldenburg: BIS Verlag.

Hoover, K.C. (2009). The geography of smell. *Cartographica: The International Journal for Geographic Information and Geovisualization* 44(4): 237–239. doi:10.3138/carto.44.4.237

Hyams, M. (2004). Hearing girls' silences: Thoughts on the politics and practices of a feminist method of group discussion. *Gender, Place and Culture* 11(1): 105–119. doi:10.1080/0966369042000188576

Hyndman, J. (2007). The securitization of fear in post-tsunami Sri Lanka. *Annals of the Association of American Geographers* 97(2): 361–372. doi:10.1111/j.1467-8306.2007.00542.x

Ingold, T. (2000). *The Perception of the Environment: Essays in Livelihood, Dwelling and Skill*. London: Routledge.

Johnston, C., & Pratt, G. (2010). Nanay (Mother): A testimonial play. *Cultural Geographies* 17(1): 123–133. doi:10.1177/1474474009350003

Kofman, E., & Peake, L. (1990). Into the 1990s: A gendered agenda for political geography. *Political Geography* 9(4): 313–336.

Koopman, S. (2012). Making space for peace: International protective accompaniment in Colombia (2007–2009). PhD thesis. Vancouver: University of British Columbia. http://circle.ubc.ca/handle/2429/42470, accessed March 26, 2015.

Latour, B. (1994). Esquisse d'un parlement des choses. *Écologie et Politique* 10: 97–115.

Latour, B. (2005). *Reassembling the Social*. Oxford: Oxford University Press.

Latour, B. (2010). An attempt at a "compositionist manifesto." *New Literary History* 41(3): 471–490.

Laurier, E., & Philo, C. (2006). Possible geographies: A passing encounter in a café. *Area* 38(4): 353–363. doi:10.1111/j.1475-4762.2006.00712.x

Law, J. (2004). *After Method: Mess in Social Science Research*. London: Routledge.

Legg, S. (2009). Of scales, networks and assemblages: The League of Nations apparatus and the scalar sovereignty of the Government of India. *Transactions of the Institute of British Geographers* 34(2): 234–253. doi:10.1111/j.1475-5661.2009.00338.x

Lorimer, H. (2005). Cultural geography: The busyness of being "more-than-representational." *Progress in Human Geography* 29(1): 83–94. doi:10.1191/0309132505ph531pr

Lorimer, H. (2008). Cultural geography: Non-representational conditions and concerns. *Progress in Human Geography* 32(4): 551–559. doi:10.1177/0309132507086882

Lorimer, J. (2010). Moving image methodologies for more-than-human geographies. *Cultural Geographies* 17(2): 237–258. doi:10.1177/1474474010363853

MacDonald, F. (2006). Geopolitics and "the vision thing": Regarding Britain and America's first nuclear missile. *Transactions of the Institute of British Geographers* 31(1): 53–71. doi:10.1111/j.1475-5661.2006.00196.x

Massumi, B. (2002). *Parables for the Virtual: Movement, Affect, Sensation*. Durham, NC: Duke University Press.

McCormack, D.P. (2003). An event of geographical ethics in spaces of affect. *Transactions of the Institute of British Geographers* 28(4): 488–507. doi:10.1111/j.0020-2754.2003.00106.x

McFarlane, C. (2009). Translocal assemblages: Space, power and social movements. *Geoforum* 40(4): 561–567. doi:10.1016/j.geoforum.2009.05.003

Meehan, K. (2014). Tool-power: Water infrastructure as wellsprings of state power. *Geoforum* 57: 215–224. doi:10.1016/j.geoforum.2013.08.005

Meehan, K., Shaw, I.G.R., & Marston, S.A. (2013). Political geographies of the object. *Political Geography* 33(March): 1–10. doi:10.1016/j.polgeo.2012.11.002

Meinig, D.W. (1983). Geography as an art. *Transactions of the Institute of British Geographers* 8(3): 314–328. doi:10.2307/622047

Merleau-Ponty, M. (1962). *The Phenomenology of Perception*. London: Routledge.

Müller, M. (2012). Opening the black box of the organization: Socio-material practices of geopolitical ordering. *Political Geography* 31(6): 379–388. doi:10.1016/j.polgeo.2012.06.001

Müller, M. (2013). Text, discourse, affect and things. In Dodds, K., Sharp, J.P., & Kuus, M. (eds), *The Ashgate Research Companion to Critical Geopolitics*, Farnham: Ashgate, 49–68.

Ó Tuathail, G. (2003). "Just out looking for a fight": American affect and the invasion of Iraq. *Antipode* 35(5): 856–870. doi:10.1111/j.1467-8330.2003.00361.x

Pain, R. (2009). Globalized fear? Towards an emotional geopolitics. *Progress in Human Geography* 33(4): 466–486. doi:10.1177/0309132508104994

Pain, R. (2010). The new geopolitics of fear. *Geography Compass* 4(3): 226–240. doi:10.1111/j.1749-8198.2009.00295.x

Pain, R., Panelli, R., Kindon, S., & Little, J. (2010). Moments in everyday/distant geopolitics: Young people's fears and hopes. *Geoforum* 41(6): 972–982. doi:10.1016/j.geoforum.2010.08.001

Pain, R., & Smith, S. (2008). *Fear: Critical Geopolitics and Everyday Life*. London: Ashgate.

Painter, J. (2006). Prosaic geographies of stateness. *Political Geography* 25(7): 752–774. doi:10.1016/j.polgeo.2006.07.004

Pile, S. (2010). Emotions and affect in recent human geography. *Transactions of the Institute of British Geographers* 35(1): 5–20. doi:10.1111/j.1475-5661.2009.00368.x

Pratt, G. (2000). Research performances. *Environment and Planning D: Society and Space* 18(5): 639–651. doi:10.1068/d218t

Protevi, J. (2009). *Political Affect: Connecting the Social and the Somatic*. Minneapolis, MN: University of Minnesota Press.

Proudfoot, J. (2010). Interviewing enjoyment, or the limits of discourse. *Professional Geographer* 62(4): 507–518. doi:10.1080/00330124.2010.501271

Robbins, P., & Marks, B. (2010). Assemblage geographies. In Smith, S.J., Pain, R., Marston, S., & Jones, J.P., III (eds), *The Sage Handbook of Social Geographies*, London: Sage, 176–195.

Roy, A. (1997). *The God of Small Things*. London: Harper.

Ruddick, S. (2010). The politics of affect: Spinoza in the work of Negri and Deleuze. *Theory, Culture and Society* 27(4): 21–45. doi:10.1177/0263276410372235

Schatzki, T. (2002). *The Site of the Social*. University Park, PA: Penn State Press.

Schueth, S. (2012). Apparatus of capture: Fiscal state formation in the Republic of Georgia. *Political Geography* 31(3): 133–143. doi:10.1016/j.polgeo.2011.11.004

Schurr, C. (2012). Visual ethnography for performative geographies: How women politicians perform identities on Ecuadorian political stages. *Geographica Helvetica* 67(4): 195–202. doi:10.5194/gh-67-195-2012

Schurr, C. (2013). Towards an emotional electoral geography: The performativity of emotions in electoral campaigning in Ecuador. *Geoforum* 49: 114–26.

Schurr, C. (2014). Serving the transnational surrogacy market as a development strategy? In Coles, A., Grey, L., & Momsen, J. (eds), *Handbook of Gender and Development*, London: Routledge, 236–244.

Sedgwick, E.K. (2003). *Touching Feeling: Affect, Pedagogy, Performativity*. Durham, NC: Duke University Press.

Sharp, J. (2009). Geography and gender: What belongs to feminist geography? Emotion, power and change. *Progress in Human Geography* 33(1): 74–80. doi:10.1177/0309132508090440

Smith, S. (2012). Intimate geopolitics: Religion, marriage, and reproductive bodies in Leh, Ladakh. *Annals of the Association of American Geographers* 102(6): 1511–1528. doi:10.1080/00045608.2012.660391

Sparke, M. (2007). Geopolitical fears, geoeconomic hopes, and the responsibilities of geography. *Annals of the Association of American Geographers* 97(2): 338–349. doi:10.1111/j.1467-8306.2007.00540.x

Spinoza, B. (1677). *Ethica, Ordine Geometrica Demonstrata*. Mimeo.

Sundberg, J. (2008). "Trash-talk" and the production of geopolitical boundaries in the USA–Mexico borderlands. *Social and Cultural Geography* 9(8): 871–890. doi:10.1080/14649360802441424

Thien, D. (2005). After or beyond feeling? A consideration of affect and emotion in geography. *Area* 37(4): 450–454. doi:10.1111/j.1475-4762.2005.00643a.x

Thrift, N. (2000). It's the little things. In Dodds, K., & Atkinson, D. (eds), *Geopolitical Traditions: A Century of Geopolitical Thought*, London: Routledge, 380–387.

Thrift, N. (2004). Intensities of feeling: Towards a spatial politics of affect. *Geografiska Annaler B* 86(1): 56–78. doi:10.1111/j.0435-3684.2004.00154.x

Thrift, N. (2008). *Non-Representational Theory: Space, Politics, Affect*. London: Routledge.

Tolia-Kelly, D.P. (2006). Affect – an ethnocentric encounter? Exploring the "universalist" imperative of emotional/affectual geographies. *Area* 38(2): 213–217. doi:10.1111/j.1475-4762.2006.00682.x

Tuan, Y.-F. (1979). *Landscapes of Fear*. New York: Pantheon.

Waitt, G., Ryan, E., & Farbotko, C. (2014). A visceral politics of sound. *Antipode* 46(1): 283–300. doi:10.1111/anti.12032

Wetherell, M. (2012). *Affect and Emotion: A New Social Science Understanding*. London: Sage.

Wetherell, M. (2013). Affect and discourse – what's the problem? From affect as excess to affective/discursive practice. *Subjectivity* 6(4): 349–368. doi:10.1057/sub.2013.13

Wylie, J. (2005). A single day's walking: Narrating self and landscape on the South West coast path. *Transactions of the Institute of British Geographers* 30(2): 234–247. doi:10.1111/j.1475-5661.2005.00163.x

Chapter 31

Resources

Kathryn Furlong[1] and Emma S. Norman[2]

[1] *Université de Montréal, Quebec, Canada*
[2] *Northwest Indian College, Bellingham, Washington State, USA*

The *Oxford English Dictionary* defines resources as "the collective means possessed by a country or region for its own support, enrichment or defense," including the "stocks or reserves of money, materials, people, or some other asset, which can be drawn on when necessary." The term "natural resources" is attributed to Adam Smith in *The Wealth of Nations* and is said to comprise "those materials or substances of a place which can be used to sustain life or for economic exploitation." Resources are to be both safeguarded and exploited. They are associated with territory and nation and are assumed to express a territory's relative wealth or potential, and thus its politics.

Many of these themes are debated and disassembled in the resource geographies literature. Among geographers, it has long been widely accepted that "resources" are as much a product of culture, politics, and economics as they are of the physical world. In the oft-cited words of Zimmerman, "resources are not: they become" (Zimmerman 1933). Building on this insight, geographers have drawn on a range of theories to demonstrate that *how, if*, and *which* resources "become" is far from a consensual process leading to stable results. The stabilization of resource categories – however temporary or spatially limited – thus involves multiple forms of exclusion and enclosure. These rely on the work of economic and political actors, states, and transnational bodies, as well as the work of governmentality, affect, and science. Resources and territory are not only about swimming fish and compressing fossils; they relate to populations, their control, their exclusion, their division, and their mobilization. Populations are not simply resources; they can also be obstacles to and a tool in resources extraction, as well as targets of resource consumption and commodification.

Drawing on these definitions, we examine how emerging debates about territory, state, economy, science, and nature are leading researchers to interpret resources in new ways. Having evolved over several decades, the breadth of the resource geographies literature cannot be encompassed in this short chapter.[1] We thus limit ourselves to emerging topics, while making an effort to avoid overlap with other recent reviews.[2] Specifically, we examine the

The Wiley Blackwell Companion to Political Geography, First Edition.
Edited by John Agnew, Virginie Mamadouh, Anna J. Secor, and Joanne Sharp.
© 2015 John Wiley & Sons Ltd. Published 2017 by John Wiley & Sons Ltd.

emerging scholarship on financialization, in terms of resource economies; on resource grabbing, in terms of conflict and violence; and on science, in terms of knowledge, power, and representation. Other key innovations in the literature that have yet to be systematically reviewed include the naturalization of scale in resource management (Molle 2009; Cohen 2012; Norman, Bakker, & Cook 2012), the rise of informal mining (Tschakert 2009; Nel, Binns, & Gibb 2014), the end-life of resources (Gregson & Crang 2010; Davies 2012; Moore 2012), and debates over resource "security" (Detraz & Betsill 2009; Cook & Bakker 2012).

Bringing resources into being

What comes to be defined as a resource and how that definition is mobilized is not *natural*, emanating from an agreed order of things. It is contested and within this contestation, particular visions, life-worlds, and systems of resource management are excluded and concealed. Bridge defines "state resources" as resources that are "naturalized" not because of shared ideologies and stable categories, but because the contestation that brought them into being is suppressed (Bridge 2014: 119). We see the state forest, but we do not see the violence, science, and economics that rendered it so (Hölzl 2010; Peluso & Vandergeest 2011).

On the one hand, a specific "thing" in the biophysical world may be conceived in fundamentally different ways, as it is "embedded in multiple knowledge systems" (Bridge 2009b: 266; Reo 2011). On the other hand, our relationships to what we perceive as resources also serve to define us culturally (Nadasdy 2003). As such, the very parameters of resource debate may rest on fundamentally different meanings, with important implications for those involved. In Gombay's work on the moral and economic geographies of food in Inuit territories in Quebec, the process of redefining resources includes not only adaptation to settler culture, but also the imposition of regulation, the redefinition of territory, and the criminalization of past practice (Gombay 2009, 2010).

Similarly, water – while apparently a universal substance essential to life – means different things to different people. For many indigenous communities, water is seen as something to be revered that plays centrally in creation stories (Reo 2011; Norman 2015). In dominant discourses, by contrast, water is defined as a resource for exploitation in its "abundance" and as one for commodification in its "scarcity" (Schmidt 2012). According to Barnes and Alatout, this is not only about multiple meanings; it is also about multiple materialities (2012: 484) that are often reduced to a single substance: H_2O (Swyngedouw 2005). The frequent attempts to universalize such framings have exacerbated problems of access and environmental degradation around the world (Perreault 2008; Boelens & Post Uiterweer 2013; Julien 2013).

Authors have divided divergent resource perspectives along different lines. Bridge locates tensions between "those who have a subsistence relationship to a resource, and those whose relationship is based on the logic of extraction or harvesting for exchange" (2009a: 1221). Others make the distinction between "land based" resources and "underground" resources to express how the desires of government and capital can conflict with local livelihoods (Vélez-Torres 2014). Different groups may be interested in the same extractive resources, but their methods have different impacts on surface resources and livelihoods.

While environmental racism or "ecocolonization" is often used to justify such tactics (Watson 2008), resource contestation and exclusion can occur within apparently homogenous communities. In such cases, the exploitation of distinct resources may come into conflict. This can emerge around interdependent resources, such as fresh-water fisheries and hydroelectric development (White 1996; Evenden 2004). Another example is fracking, whereby resource extraction that is said to generate wealth in the wider economy risks the water supplies of resident populations and contributes to global climate change through "flaring" (Howarth & Ingraffea 2011; Finewood & Stroup 2012; Fry et al. 2012).

Differences in the perception of resources and their value can also shift as resources move through space and time. Historical commodity chain analysis demonstrates that resources can be valued very differently between the spaces of extraction and consumption and from one temporal period to another. In *From Silver to Cocaine*, Topik, Marichal, and Frank (2006) show how a range of resources, including tobacco, coffee, sugar, hemp, rubber, and so on, go in and out of being "resources." Their enrollment in global trade is independent of local perceptions of their value, and trade itself transforms the meaning ascribed to resources across space (Topik et al. 2006). The history of the fur trade in North America raises similar issues (Dolin 2010). More recently, diamonds, teak, and gold are examples of the creation of elite goods through commercial practices that attempt to divorce the products from the precariousness, violence, and environmental damage that often characterize their extraction (Le Billon 2008; Swenson et al. 2011; Bryant 2013).

Many visions of a resource can coexist simultaneously across space and shift over time. Yet, sustaining the necessary simplifications to enable exploitation and the "triumph of one imaginary over others" requires continual work (Bridge 2009a: 1221). To explain this triumph, how it is maintained and enacted, researchers have mobilized the tools of governmentality (Agrawal 2005; Elden 2007). Lately, authors have also appealed for analyses that look beyond governmentality. Singh calls for an engagement with "affect, emotions and embodied practices in shaping human subjectivities" around resource management (2013: 189). Bouzarovski and Bassin (2011) examine the building of national identity around energy extraction in Russia, and Bryant (2013) calls for a focus on branding as "a *flexible* ordering device – promoting continuity in a context of perpetual change."

Exploiting and managing resources

In his most recent review of the resource literature, Bridge identifies "three core 'logics' of resource making ... economy, territory and subject formation" (2011: 820). Taking emerging aspects of each of these that have yet to have a central place in other reviews, we focus on issues of financialization, resource grabbing, and the role of science.

New economic concerns: Financialization

Much of the work on markets and resources has examined what market metaphors and management hide about resources, such as environmental externalities, social injustices, and diverse forms of meaning. Over the last 20 years, neoliberalization and its effects on resource management and allocation have been of significant concern. Studies have tackled a range of resources, from wetlands (Robertson 2004) to forests (McCarthy 2005) to fisheries (Mansfield 2004).[3] This literature has been charged with missing the agency of nature and its unruliness vis-à-vis economic policy (Bakker 2010: 3). In its sacrifice of political ecology for political economy, Bakker argues that the literature has neglected the diverse effects and counter-effects that arise in the relationship between particular neoliberal reforms and particular types of resources (Bakker 2010: 8).

The burgeoning literature on resource financialization could likewise be subject to this critique. Although many authors engage with how nature is reformed through financialization, few focus on the reciprocal effects of diverse natures. A notable exception is Çaliskan's work on cotton, whereby the substance itself and the environments in which it is produced influence market abstractions (Çaliskan 2010). For Pellizzoni, while the literature may not deal with nature's "unruliness," processes of financialization thrive on "uncertainty, contingency, instability" in the human–nature relationship (Pellizzoni 2011: 795). He argues that nature is no longer seen as a barrier to be overcome, but as something to be manipulated. With financialization, the fluidity of nature is no longer epistemic but ontological, entailing "an increase in its manipulability and controllability" (Pellizzoni 2011: 799).

Financialization can be defined as "the gravitational shift from productive to financial forms of capital accumulation" (Gunnoe & Gellert 2011: 266). This occurs at every scale, from the accumulation and trading of vast tracts of resource territory (Gunnoe & Gellert 2011), to the systematic division of resources into saleable pieces (Robertson 2012), to the reordering of nature at the molecular level (Pellizzoni 2011). At the territorial scale, for example, forestry companies have gone from owning timberlands for exploitation to selling them for investor portfolio diversification (Gunnoe & Gellert 2011). Timberland is traded, but forest resources are not extracted. As institutional investors accumulate more forest stock, they gain both political power and control over traditional forestry companies and risk producing a speculative bubble, with potentially significant repercussions for the "social and ecological systems that rely on US timberlands" (Gunnoe & Gellert 2011: 273).

With respect to resource segmentation, ecosystem services are a much-discussed example. The processes are dependent on techniques of financial abstraction that enable the quantification of value, including classification, categorization, unbundling, and stacking of so-called ecosystem services (Robertson 2012). These are seen as enabling the commodification of "nearly every aspect of its material existence, and at every scale from the atmospheric to the biochemical" (Robertson 2012: 387). Yet, they are diverse, operate through a range of logics, and are enacted by a range of parties (Dempsey & Robertson 2012; Wynne-Jones 2012).

In terms of the physical transformation of resources through financialization, Arsel and Büscher (2012) draw attention to "an intensifying friction" between environmental change and ecological limits in the capital–nature relation. They see capitalism as "thriving" on overcoming ecological limits in order to convert nature into "natural capital." Similarly, Smith (2007) sees the creation of ecosystem service commodities as involving not simply "a new calculus of *value*" and an appropriation of nature, but its recreation for the purposes of accumulation (Smith 2007: 33). Lansing (2013), however, cautions against placing undue emphasis on economic processes. He argues that the "long standing patterns of agricultural production and land use" into which processes of financialization are inserted are equally important in directing the formation of new natures (Lansing 2013: 103).

Alternatively, financialization can be used to create the perception of physical change through a change in market dynamics. This is the case with peak oil. For Bridge and Wood (2010), discourses of peak oil and declining reserves are related to issues that are "primarily above ground." These have to do with reserve ownership, "the politics of resources access and the changing structure of the international oil industry," rather than physical scarcity (Bridge & Wood 2010: 71). Similarly, Labban (2010) argues that physical scarcity or availability "has no direct bearing on the price of oil." Instead, "the relations between price, investment and production have been transformed by the financialization of trade and investment in oil" (Labban 2010: 542). This is because more "virtual" oil is exchanged through financial markets than produced oil is traded on spot markets, as the dominant companies increasingly turn to financial markets for short-term profits (Labban 2010: 542).

Labban goes further, arguing that finance renders "material production somewhat irrelevant to the accumulation of capital" (2010: 541). It has decoupled the circulation of oil in the market from its circulation in physical space. He draws on Karatani's idea that value is not inherent but is made through the process of exchange when it comes into relation with other products; it involves a *salto morale* (leap of faith) at the moment of sale (Karatani 2003). As the importance of finance increases, sales of oil futures exceed physical production; "[a]ccumulation of debt ... appears as the accumulation of capital" (Labban 2010: 550). As such, "resource crises have become financial crises" (Labban 2010: 550). This new situation means that metabolic considerations have become insufficient in accounting for resource flows.

Zalik (2010) gives further empirical depth to these arguments, asserting that notions of peak oil are not only about financialization, but also about *how* financialization is done. She argues that futures markets are shaped through the work of business agents and the interests of the industry "in the ongoing development of unconventional fossil fuel sources" (Zalik 2010: 553). In this way, multinationals like Shell produce scenarios that serve to construct the conditions of "speculative markets" (Zalik 2010: 554). These arguments reflect the detailed empirical findings of Çaliskan, whereby the financial and physical markets for cotton operate in parallel but not in sync. Several instruments – or "prostheses" like price tables and trade magazines – are mobilized to lessen this leap of faith, which is inevitably constructed through the action of powerful and experienced market players at a variety of scales (Çaliskan 2010).

Other key actors in financialization include governments, which provide the necessary "enabling conditions" through laws, standards, and policies (Arsel & Büscher 2012), and conservationists (Wynne-Jones 2012). For Gunnoe and Gellert (2011), detailed exploration of the role of institutional investors and shareholders in financialization is essential to dispelling the notion that finance is separate from the "real" (i.e., productive) economy. Similarly, in his wide-ranging study of cotton, Çaliskan argues that:

> the relations of exchange and production are simultaneous instances of the same relation of economization. The market does not just happen after the cotton is produced. Nor does the production terminate once the cotton is sold ... one cannot analytically identify the relations of exchange and production as separate fields of encounter. (Çaliskan 2010: 201)

Processes of financialization are seen as having negative distributive effects. Yet, this is an issue over which the literature remains largely preliminary. For McAfee (2012), the rhetoric of those who promote financialization assumes a win–win for society and nature that is unlikely to materialize. As such, she calls for more attention to the redistributive effects of trading nature. With respect to wetlands trading, several authors have pointed to the geographical divides between mitigation and benefit, which are likely to "deepen uneven development and intensify poverty" (Smith 2007: 23).

From resource curse to resource grabs

Debates about the theory of the "resource curse" have been ongoing since the 1990s. The violence that makes a resource a "curse" is said to occur in cases where there is an abundance of high-value minerals and the possibility of resource "capture" by armed groups (e.g., Humphreys, Sachs, & Stiglitz 2007). Such propositions have been widely challenged by resource geographers, who call for a focus on the politics of resources beyond armed actors

(Bridge 2009b: 265), as well as an engagement with the variety of possible forms of violence beyond civil war and rebellion (Watts 2004).

Recent discourses have turned from notions of abundance to notions of scarcity, such as "peak oil," to explain conflict in areas of oil extraction. Underlining scarcity as a driver of war turns out to be as simplistic as it is the opposite: the so-called curse of abundance. For Le Billon and Cervantes (2009), scarcity is mobilized for the production of prices. War can play a role in producing scarcity and increasing prices by inhibiting access to markets and reserves. Huber (2011) goes further. Taking the study of resource conflict out of the global South, he examines oil production in Texas in the 1930s, where the federal government introduced martial law to curb "overproduction" and halt the decline of oil prices. Here, violence was instrumental in creating "socially produced" scarcity that was "necessary for the market to function" (Huber 2011: 816). For Huber, violence ought to be seen as a generator rather than a consequence of scarcity.

Recent work on water resources has tackled this problem of abundance in relation to scarcity as coupled historical-political discourses. Alatout's historical papers on the mobilization of water over the course of the Israeli–Palestinian conflict show the alternative mobilization of abundance and scarcity to occupy and control Palestinian lands (Alatout 2008, 2009: 366). He demonstrates that, as circumstances shift, seemingly opposing resource narratives can be mobilized to serve analogous political ends.

Resource scarcity and abundance are linked not only discursively, but experientially. For most resource geographers, resource availability is relational; income inequality influences the experience of scarcity. Access to resources is generally produced through various forms of exclusion and enclosure, including illegal and extralegal measures. As Emel and Huber (2008) underscore, those who live proximate to sites of extraction often see little economic or social benefit from these activities. This is especially true in cases of resource grabbing, to which we turn next.

Concern with various forms of land, resource, and green grabbing have expanded engagement with resource violence beyond the typical focus on oil, minerals, and timber. Land grabbing involves the transfer of use rights, ownership, title, and so on of land from the poor to the powerful (Fairhead, Leach, & Scoones 2012). This need not involve the official transfer of land, but may be organized through the "restructuring of rules and authority over the access, use and management of resources ... that may have profoundly alienating effects" (Fairhead et al. 2012: 239). The increasing investment in biofuels on the part of China (as well as India and Brazil), through landholdings in Africa, shows how international "investments" and the liberalization of trade have impacts on local economies and socio-environmental landscapes (Robertson & Pinstrup-Andersen 2010; Vermeulen & Cotula 2010; Hall 2011).

Following Vélez-Torres (2014), it is not only about the rules and the discourses that uphold them; it is also about their discriminatory application. Between 1985 and 2001, Colombia was one of 90 countries to liberalize their legislative frameworks to attract foreign mining investment. In 1991, it also enshrined the protection of indigenous rights within its new constitution. Still, mining companies are advantaged through the uneven attention given to these two legalities, as well as the differential capabilities of each group to defend their interests through the law. Even so, violence has increased in mineral-rich areas, as lands that cannot be taken by law are taken by force (Vélez-Torres 2014).

These issues are tied to processes of speculative financialization. First, legislative change to enable grabbing is often tied to "fiscal policies designed to favour investment" over livelihood security (Fairhead et al. 2012: 245). In a striking similarity between mining land grabs in

Colombia and the financialization of timberlands in the United States, the acquisition of territory often proceeds without any extraction taking place (Gunnoe & Gellert 2011; Vélez-Torres 2014). This ties land grabbing to the more specific "green grabbing," in which environmental issues are the central focus of the grab. Green grabbing, for example, may involve reserving land for conservation, ecosystem services, biofuels, eco-tourism, and so on (Fairhead et al. 2012). What distinguishes this from "green imperialism" is its clear economic intent as a further step in market environmentalism (Fairhead et al. 2012).

Such processes have also been studied with respect to other resources, like water, as well as urban environments. In a special issue of *Water Alternatives*, the authors extend the land-grabbing discussion to engage with how water has been both a target and a driver of grabs, and how its very fluidity reveals the complex, shifting, and uneven nature of the phenomenon (Mehta, Veldwisch, & Franco 2012). In the case of urban environments, authors have explored how resource deprivation acts as a way of recolonizing slum areas in cities. In a study of Mumbai, Graham, Desai, and McFarlane (2013) argue that the "systematic dehydration" of vulnerable urban populations is employed to "discourage the formation, or force abandonment, of slums" (Graham et al. 2013: 129).

These processes of exclusion capitalize on pre-existing inequalities within communities and engage a range of actors and alliances across scales. Beyond governments and industry, authors emphasize the work of environmental actors and financial entrepreneurs (Fairhead et al. 2012). In Panama, Finley-Brook and Thomas (2011: 864) tie efforts to "extinguish indigenous land claims" to the work of international markets in carbon offsets, the Clean Development Mechanism (CDM), hydroelectric development, governments, and Latin American energy multinationals. Finding the same range of connections, Sonnenfield and Mol (2011) argue that resource grabs are increasingly expressions of interlinked global crises, whereby land grabbing in sub-Saharan Africa is "directly related to biofuel production as an answer to peak oil, leading to growing phosphate use in remaining, more intensively cultivated, agricultural lands, in turn contributing to further climate change" (Sonnenfeld & Mol 2011: 772). As such, "resource sovereignty cannot be territorially circumscribed"; it must be understood within the relation of "the global geography of non-state actors" that conditions resource access (Emel, Huber, & Makene 2011: 70).

As with resource grabs, geographers have long been concerned with how politics in the "metropole" are linked to resource exploitation in the "periphery." In two recent studies, Mitchell and Hecht have gone further by tying countries from the North and South together in the same processes of resource development and exploitation. For Mitchell, carbon extraction is not merely a factor in the political organization of extractive countries, it defines consuming countries as well. Dependence on carbon fuels as a concentrated source of energy cannot be divorced from the evolution of democracy, which is in fact carbon democracy (Mitchell 2011). In an opposite way, Hecht (2012) shows how studies of nuclear energy have focused on the sites of energy generation, neglecting the extractive countries that are equally part of a global network of "nuclearity."

Science, calculation, and representation

Considerations of science and technology have become increasingly important among resource geographers. Similarly, science and technology studies (STS) has taken a growing interest in resources. Much of this work engages with traditional resource issues like geopolitics. In Sneddon's work on damming in Southeast Asia, he follows the efforts of the American government to enlist engineering science in the achievement of geopolitical goals. Following Schatzki (2003), he argues that water resource development creates nature–society hybrids that become increasingly interdependent over time (Sneddon 2012). Specifically, while geopolitical forces seek to enlist technical ones in the remaking of nature, the technical forces have to be convinced. Once convinced, however, they may prove difficult to halt where the political goals lose traction (Sneddon 2012).

Similarly, Pritchard (2012) finds that the boundaries between "hydrologic 'science' and hydraulic 'technology' are fluid." In French hydroelectric projects in Africa, studies "often accompanied or even followed proposals for dams" (Pritchard 2012: 595). Reminiscent of Mitchell's work in Egypt and Chakrabarty's *Provincializing Europe* (Chakrabarty 2000; Mitchell 2002), Pritchard finds that dam building was often used to bring knowledge back to the metropole, as opposed to the application of "French knowledge" in the South. These practices served to "constitute, reproduce and ultimately deepen profoundly uneven forms of political and economic power in the modern era" (Pritchard 2012: 598).

Hecht (2012) is also concerned with the geopolitics of French resource politics. What she calls "nuclear things," such as bombs and reactors, were "meant to replace empire by generating geopolitical 'radiance' as the new means and emblem for projecting power" (Hecht 2012: 118). This meant putting a development face on the nuclear relationship with uranium-extracting countries like Gabon. Maintaining networks of nuclearity, however, was first and foremost the work of technoscience, dependent on tools of measurement and evaluation to create the semblance of markets.

The focus on tools of measurement and calculation in the production of and control over resources is a prominent topic in resource geography. Processes of calculation serve to define particular types of value, in what Callon and Law (2005) refer to as qualculation. This has been central in defining what types of exploitation should be prioritized and for whom (Crampton & Elden 2006); decisions that are fraught with colonial and neocolonial implications (Elden 2007). These processes simultaneously draw on and seek to erase other forms of knowledge. In their prize-winning historical work on rivers in western Canada, Loo and Stanley (2011) show how First Nations knowledge was instrumental in the production of the "Western" scientific knowledge that ultimately enabled the damming of rivers. Conversely, Budds has shown how "scientific" knowledge about water systems is used to exclude indigenous knowledge as well as indigenous rights claims (Budds 2009).

This work of calculation in suppressing certain forms of knowledge also serves to silence the very compromises on which it is based. In a study in France, Fernandez shows how indicators work to render invisible the "specific combinations of water, power and financial flows" that brought them into being. While indicators are produced through particular compromises in particular socio-scientific contexts, their effect over time is to "erase history and to naturalize rationales" (Fernandez 2014: 258). For Vetter (2010), such erasure permits the globalization of scientific "facts" by putting "observations and experiences of the natural world" into a form that can "circulate in an increasingly global, cosmopolitan science" (Vetter 2010: 483). The simplification of the natural word and the knowledge of it, however, should not be conflated with the simplification of its management. Hölzl (2010) finds that the simplification of the forest through species homogenization in Germany went hand in hand with making forest management more "complex, detailed and scientific" (Hölzl 2010: 447).

A similar process of nature's simplification enabling management complexity is found in the workings of financialization. Many authors have been concerned with the role that science plays in producing the necessary knowledge for resource commodification. They focus on how nature is simplified in the service of markets (e.g., Newell & Patterson 2010; Robertson 2012), which renders the actual management of resources more opaque. Certain authors have sought to nuance the state–science–market relationship. Nobert (2013) argues that, in Quebec, the government used scientific forestry to augment its power vis-à-vis industry. Erickson (2010) reverses the question about the relationship among scientific knowledge, resources, and markets, exploring what markets have revealed about resources. He argues that markets should be seen as part of "integrated social and scientific knowledge production systems" that serve to define environmental problems and to shape responses (2010: 546, 548). Drawing on Foucault, he shows how the science of wildlife populations was made possible through expanding trade in wildlife products, whereby "commodity market data, such as prices paid and quantities delivered, would prove to be the key to understanding population fluctuations and their causes" (2010: 534). In the 1980s, this data revealed shifts in elephant populations, leading to conservation measures as well as to the division of the "African elephant" into a variety of subspecies that could be individually monitored and protected (Erickson 2010: 545).

In the case of the Newfoundland cod fisheries, however, population science – in part derived from market data – precipitated the "annihilation" of the species that it purported to manage (Bavington 2010b). This was not a "tragedy of the commons"; the cod were literally managed out of existence (Bavington 2010b). Recalling the work of science in circumscribing knowledge, Bavington notes how "[t]he demographic paradigm empowered fisheries scientists but marginalized fishermen." Moreover, in its focus on simplification through aggregated statistical data, it also "marginalized empirical knowledge about fish biology, life history traits, and behaviour" (Bavington 2010a: 509). Despite these failures, population-based fisheries management still has legitimacy (Bavington 2010b).

However, such legitimacy may be in the simplification as opposed to the science. In Lave's (2012) work on stream restoration in the United States, science – at least as produced by scientists in universities – is losing some of its authority over resource management. Lave (2012) shows how authority over restoration practices has shifted from university-based scientists to the methods of David L. Rosgen, a consultant with little formal scientific training. The postmodernism of contemporary science, focused on complexity and contingency, offers few readymade solutions to those who wish to engage in stream restoration. Rosgen's available toolkit, "universally" applicable classifications, and short courses, by contrast, go some way toward democratizing the "scientific management" of rivers by enabling broader participation (Lave 2012).

Conclusions

The notion of a "resource" is tied to wider politics of territoriality, state, economy, science, and nature. Those that hold resources, hold power; and those that hold power, hold resources. Defining what constitutes a resource is fundamentally about world view. Similarly, identifying who benefits from those resources is fundamentally about power. The asymmetries and uneven development that arise from resource accumulation (Smith 1984, 2007) continue to remind us that a resource is not a *Ding an sich*, a thing in itself. Rather, resources are entangled with power dynamics and enacted through the tools of markets, law, politics, and science. Thus, continued and increased understanding of what a "resource" means in today's political, economic, and social climate depends on constant critical engagement with what appears as *natural*, what is suppressed, and what is taken as given.

Notes

1 See the special issue of *Development and Change*, edited by Anthony Bebbington, which gathers articles on extractive industry over 27 years (Bebbington, Bornschlegl, & Johnson 2013).
2 Gavin Bridge has provided several excellent reviews (Bridge 2004, 2009b, 2011, 2014). As these concentrate mostly on extractive resources, we have also made the effort to engage with a wider range of nature to resource conversions. For a review of the literature on Latin America, see Renfrew (2011).
3 Several review articles and debate pieces explore this topic (e.g., McCarthy & Prudham 2004; Bridge 2008; Castree 2008a, 2008b; Bakker 2009); as such, it is not addressed in detail here.

References

Agrawal, A. (2005). *Environmentality: Technologies of Government and the Making of Subjects*. Durham, NC: Duke University Press.

Alatout, S. (2008). "States" of scarcity: Water, space, and identity politics in Israel, 1948–59. *Environment and Planning D: Society and Space* 26(6): 959–982. doi:10.1068/d1106

Alatout, S. (2009). Bringing abundance into environmental politics: Constructing a Zionist network of water abundance, immigration, and colonization. *Social Studies of Science* 39(3): 363–394.

Arsel, M., & Büscher, B. (2012). Nature™ Inc.: Changes and continuities in neoliberal conservation and market-based environmental policy. *Development and Change* 43(1): 53–78. doi:10.1111/j.1467-7660.2012.01752.x

Bakker, K. (2009). Commentary: Neoliberal nature, ecological fixes, and the pitfalls of comparative research. *Environment and Planning A* 41(8): 1781–1787. doi:10.1068/a4277

Bakker, K. (2010). The limits of "neoliberal natures": Debating green neoliberalism. *Progress in Human Geography* 34(6): 715–735. doi:10.1177/0309132510376849

Barnes, J., & Alatout, S. (2012). Water worlds: Introduction to the special issue of *Social Studies of Science*. *Social Studies of Science* 42(4): 483–488. doi:10.1177/0306312712448524

Bavington, D. (2010a). From hunting fish to managing populations: Fisheries science and the destruction of Newfoundland cod fisheries. *Science as Culture* 19(4): 509–528.

Bavington, D. (2010b). *Managed Annihilation: An Unnatural History of the Newfoundland Cod Collapse*. Vancouver: UBC Press.

Bebbington, A., Bornschlegl, T., & Johnson, A. (2013). Political economies of extractive industry: From documenting complexity to informing current debates. *Development and Change* online before print. doi:10.1111/dech.12057

Boelens, R., & Post Uiterweer, N.C. (2013). Hydraulic heroes: The ironies of utopian hydraulism and its politics of autonomy in the Guadalhorce Valley, Spain. *Journal of Historical Geography* 41: 44–58. doi:10.1016/j.jhg.2012.12.005

Bouzarovski, S., & Bassin, M. (2011). Energy and identity: Imagining Russia as a hydrocarbon superpower. *Annals of the Association of American Geographers* 101(4): 783–794.

Bridge, G. (2004). Contested terrain: Mining and the environment. *Annual Review of Environment and Resources* 29: 205–259.

Bridge, G. (2008). Environmental economic geography: A sympathetic critique. *Geoforum* 39(1): 76–81.

Bridge, G. (2009a). Material worlds: Natural resources, resource geography and the material economy. *Geography Compass* 3(3): 1217–1244.

Bridge, G. (2009b). Natural resources. In Kitchin, R., & Thrift, N. (eds), *International Encyclopedia of Human Geography*, Oxford: Elsevier, 261–268.

Bridge, G. (2011). Resource geographies 1: Making carbon economies, old and new. *Progress in Human Geography* 35(6): 820–834. doi:10.1177/0309132510385524

Bridge, G. (2014). Resource geographies II: The resource–state nexus. *Progress in Human Geography* 38(1): 118–130. doi:10.1177/0309132513493379

Bridge, G., & Wood, A. (2010). Less is more: Spectres of scarcity and the politics of resource access in the upstream oil sector. *Geoforum* 41(4): 565–576.

Bryant, R.L. (2013). Branding natural resources: Science, violence and marketing in the making of teak. *Transactions of the Institute of British Geographers* 38(4): 517–530. doi:10.1111/tran.12006

Budds, J. (2009). Contested H$_2$O: Science, policy and politics in water resources management in Chile. *Geoforum* 40(3): 418–430.

Çaliskan, K. (2010). *Market Threads: How Cotton Farmers and Traders Create a Global Economy*. Princeton, NJ: Princeton University Press.

Callon, M., & Law, J. (2005). On qualculation, agency, and otherness. *Environment and Planning D: Society and Space* 23(5): 717–733.

Castree, N. (2008a). Neoliberalising nature: Processes, effects, and evaluations. *Environment and Planning A* 40(1): 153–173.

Castree, N. (2008b). Neoliberalising nature: The logics of deregulation and reregulation. *Environment and Planning A* 40(1): 131–152.

Chakrabarty, D. (2000). *Provincializing Europe: Postcolonial Thought and Historical Difference*. Princeton, NJ: Princeton University Press.

Cohen, A. (2012). Rescaling environmental governance: Watersheds as boundary objects at the intersection of science, neoliberalism, and participation. *Environment and Planning A* 44(9): 2207–2224.

Cook, C., & Bakker, K. (2012). Water security: Debating an emerging paradigm. *Global Environmental Change* 22(1): 94–102.

Crampton, J.W., & Elden, S. (2006). Space, politics, calculation: An introduction. *Social and Cultural Geography* 7(5): 681–685.

Davies, A.R. (2012). Geography and the matter of waste mobilities. *Transactions of the Institute of British Geographers* 37(2): 191–196. doi:10.1111/j.1475-5661.2011.00472.x

Dempsey, J., & Robertson, M.M. (2012). Ecosystem services: Tensions, impurities, and points of engagement within neoliberalism. *Progress in Human Geography* 36(6): 758–779. doi:10.1177/0309132512437076

Detraz, N., & Betsill, M.M. (2009). Climate change and environmental security: For whom the discourse shifts. *International Studies Perspectives* 10(3): 303–320.

Dolin, E.J. (2010). *Fur, Fortune and Empire: The Epic History of the Fur Trade in America*. New York: W.W. Norton.

Elden, S. (2007). Governmentality, calculation, territory. *Environment and Planning D: Society and Space* 25(3): 562–580. doi:10.1068/d428t

Emel, J., & Huber, M.T. (2008). A risky business: Mining, rent and the neoliberalization of "risk." *Geoforum* 39(3): 1393–1407.

Emel, J., Huber, M.T., & Makene, M.H. (2011). Extracting sovereignty: Capital, territory, and gold mining in Tanzania. *Political Geography* 30(2): 70–79.

Erickson, P. (2010). Knowing nature through markets: Trade, populations, and the history of ecology. *Science as Culture* 19(4): 529–551.

Evenden, M. (2004). *Fish versus Power: An Environmental History of the Fraser River*. Cambridge: Cambridge University Press.

Fairhead, J., Leach, M., & Scoones, I. (2012). Green grabbing: A new appropriation of nature? *Journal of Peasant Studies* 39(2): 237–261. doi:10.1080/03066150.2012.671770

Fernandez, S. (2014). Much ado about minimum flows...Unpacking indicators to reveal water politics. *Geoforum* 57: 258–271. doi:10.1016/j.geoforum.2013.04.017

Finewood, M.H., & Stroup, L.J. (2012). Fracking and the neoliberalization of the hydro-social cycle in Pennsylvania's Marcellus Shale. *Journal of Contemporary Water Research and Education* 147(1): 72–79. doi:10.1111/j.1936-704X.2012.03104.x

Finley-Brook, M., & Thomas, C. (2011). Renewable energy and human rights violations: Illustrative cases from indigenous territories in Panama. *Annals of the Association of American Geographers* 101(4): 863–872.

Fry, M., Hoeinghaus, D.J., Ponette-González, A.G., Thompson, R., & La Point, T.W. (2012). Fracking vs faucets: Balancing energy needs and water sustainability at urban frontiers. *Environmental Science and Technology* 46(14): 7444–7445.

Gombay, N. (2009). Sharing or commoditising? A discussion of some of the socio-economic implications of Nunavik's hunter support program. *Polar Record* 45(2): 119–132.

Gombay, N. (2010). Community, obligation, and food: Lessons from the moral geography of Inuit. *Geografiska Annaler: Series B, Human Geography* 92(3): 237–250.

Graham, S., Desai, R., & McFarlane, C. (2013). Water wars in Mumbai. *Public Culture* 25(1): 115–143. doi:10.1215/08992363-1890486

Gregson, N., & Crang, M. (2010). Guest editorial: Materiality and waste: Inorganic vitality in a networked world. *Environment and Planning A* 42(5): 1026–1032. doi:10.1068/a43176

Gunnoe, A., & Gellert, P.K. (2011). Financialization, shareholder value, and the transformation of timberland ownership in the US. *Critical Sociology* 37(3): 265–284.

Hall, R. (2011). Land grabbing in Southern Africa: The many faces of the investor rush. *Review of African Political Economy* 38(128): 193–214.

Hecht, G. (2012). *Being Nuclear: Africans and the Global Uranium Trade*. Cambridge, MA: MIT Press.

Hölzl, R. (2010). Historicizing sustainability: German scientific forestry in the eighteenth and nineteenth centuries. *Science as Culture* 19(4): 431–460.

Howarth, R.W., & Ingraffea, A. (2011). Should fracking stop? Yes, it is too high risk. *Nature* 477: 271–273.

Huber, M.T. (2011). Enforcing scarcity: Oil, violence, and the making of the market. *Annals of the Association of American Geographers* 101(4): 816–826.

Humphreys, M., Sachs, J.D., & Stiglitz, J.E. (2007). *Escaping the Resource Curse*. New York: Colombia University Press.

Julien, F. (2013). *La gestion intégrée des ressources en eau en Afrique Subsaharienne: Paradigm occidental, pratiques africaines*. Québec: Presses de l'Université du Québec.

Karatani, K. (2003). *Transcritique: On Kant and Marx*, trans. S. Kohso. Cambridge, MA: MIT Press.

Labban, M. (2010). Oil in parallax: Scarcity, markets, and the financialization of accumulation. *Geoforum* 41(4): 541–552.

Lansing, D.M. (2013). Understanding linkages between ecosystem service payments, forest plantations, and export agriculture. *Geoforum* 47: 103–112.

Lave, R. (2012). Bridging political ecology and STS: A field analysis of the Rosgen wars. *Annals of the Association of American Geographers* 102(2): 366–382.

Le Billon, P. (2008). Diamond wars? Conflict diamonds and geographies of resource wars. *Annals of the Association of American Geographers* 98(2): 345–372.

Le Billon, P., & Cervantes, A. (2009). Oil prices, scarcity, and geographies of war. *Annals of the Association of American Geographers* 99(5): 836–844.

Loo, T., & Stanley, M. (2011). An environmental history of progress: Damming the Peace and Columbia rivers. *Canadian Historical Review* 92(3): 399–427.

Mansfield, B. (2004). Rules of privatization: Contradictions in neoliberal regulation of north Pacific fisheries. *Annals of the Association of American Geographers* 94(3): 565–584.

McAfee, K. (2012). The contradictory logic of global ecosystem services markets. *Development and Change* 43(1): 105–131. doi:10.1111/j.1467-7660.2011.01745.x

McCarthy, J. (2005). Devolution in the woods: Community forestry as hybrid neoliberalism. *Environment and Planning A* 37(6): 995–1014.

McCarthy, J., & Prudham, S. (2004). Neoliberal nature and the nature of neoliberalism. *Geoforum* 35(3): 275–283.

Mehta, L., Veldwisch, J., & Franco, J. (2012). Introduction to the special issue: Water grabbing? Focus on the (re)appropriation of finite water resources. *Water Alternatives* 5(2): 193–207.

Mitchell, T. (2002). *Rule of Experts: Egypt, Techno-Politics and Modernity*. Berkeley, CA: University of California Press.

Mitchell, T. (2011). *Carbon Democracy: Political Power in the Age of Oil*. London: Verso.

Molle, F. (2009). River-basin planning and management: The social life of a concept. *Geoforum* 40(3): 484–494.

Moore, S.A. (2012). Garbage matters: Concepts in new geographies of waste. *Progress in Human Geography* 36(6): 780–799. doi:10.1177/0309132512437077

Nadasdy, P. (2003). *Hunters and Bureaucrats: Power, Knowledge, and Aboriginal-State Relations in the Southwest Yukon*. Vancouver: UBC Press.

Nel, E., Binns, T., & Gibb, M. (2014). Community development at the coal face: Networks and sustainability among artisanal mining communities in Indwe, Eastern Cape Province, South Africa. *Geographical Journal* 180(2): 175–184. doi:10.1111/geoj.12022

Newell, P., & Patterson, M. (2010). *Climate Capitalism: Global Warming and the Transformation of the Global Economy*. Cambridge: Cambridge University Press.

Nobert, S. (2013). Nationalism, (dis)simulation, and the politics of science in Québec's forest crisis. *Annals of the Association of American Geographers* 103(6): 1332–1347.

Norman, E.S. (2015). *Governing Transboundary Waters: Canada, the United States, and Indigenous Communities*. London: Routledge.

Norman, E.S., Bakker, K., & Cook, C. (2012). Introduction to the themed section: Water governance and the politics of scale. *Water Alternatives* 5(1): 52–61.

Pellizzoni, L. (2011). Governing through disorder: Neoliberal environmental governance and social theory. *Global Environmental Change* 21(3): 795–803.

Peluso, N.L., & Vandergeest, P. (2011). Political ecologies of war and forests: Counterinsurgencies and the making of national natures. *Annals of the Association of American Geographers* 101(3): 587–608.

Perreault, T. (2008). Custom and contradiction: Rural water governance and the politics of Usos y Costumbres in Bolivia's Irrigators' Movement. *Annals of the Association of American Geographers* 98(4): 834–854.

Pritchard, S.B. (2012). From hydroimperialism to hydrocapitalism: "French" hydraulics in France, North Africa, and beyond. *Social Studies of Science* 42(4): 591–615. doi:10.1177/0306312712443018

Renfrew, D. (2011). The curse of wealth: Political ecologies of Latin American neoliberalism. *Geography Compass* 5(8): 581–594.

Reo, N.J. (2011). The importance of belief systems in traditional ecological knowledge initiatives. *International Indigenous Policy Journal* 24(4). http://ir.lib.uwo.ca/iipj/vol2/iss4/8/, accessed March 26, 2015.

Robertson, B., & Pinstrup-Andersen, P. (2010). Global land acquisition: Neo-colonialism or development ppportunity? *Food Security* 2(3): 271–283.

Robertson, M.M. (2004). The neoliberalization of ecosystem services: Wetland mitigation banking and problems in environmental governance. *Geoforum* 35(3): 361–373.

Robertson, M. (2012). Measurement and alienation: Making a world of ecosystem services. *Transactions of the Institute of British Geographers* 37(3): 386–401. doi:10.1111/j.1475-5661.2011.00476.x

Schatzki, T.R. (2003). Nature and technology in history. *History and Theory* 42(4): 82–93.

Schmidt, J.J. (2012). Scarce or insecure? The right to water and the ethics of global water governance. In Sultana, F., & Loftus, A. (eds), *The Right to Water: Politics, Governance and Social Struggles*, Abingdon: Earthscan, 94–109.

Singh, N.M. (2013). The affective labor of growing forests and the becoming of environmental subjects: Rethinking environmentality in Odisha, India. *Geoforum* 47: 189–198.

Smith, N. (1984). *Uneven Development: Nature, Capital and the Production of Space*. Oxford: Blackwell.

Smith, N. (2007). Nature as accumulation strategy. *Socialist Register* 43: 19–41.

Sneddon, C. (2012). The "sinew of development": Cold War geopolitics, technical expertise, and water resource development in Southeast Asia, 1954–1975. *Social Studies of Science* 42(4): 564–590. doi:10.1177/0306312712445835

Sonnenfeld, D.A., & Mol, A.P.J. (2011). Social theory and the environment in the new world (dis)order. *Global Environmental Change* 21(3): 771–775.

Swenson, J.J., Carter, C.E., Domec, J.-C., & Delgado, C.I. (2011). Gold mining in the Peruvian Amazon: Global prices, deforestation, and mercury imports. *PLoS ONE* 6(4): e18875. doi:10.1371/journal.pone.0018875

Swyngedouw, E. (2005). Dispossessing H_2O: The contested terrain of water privatization. *Capitalism Nature Socialism* 16(1): 81–98.

Topik, S., Marichal, C., & Frank, Z. (2006). *From Silver to Cocaine: Latin American Commodity Chains and the Building of the World Economy, 1500–2000*. Durham, NC: Duke University Press.

Tschakert, P. (2009). Digging deep for justice: A radical re-imagination of the artisanal gold mining sector in Ghana. *Antipode* 41(4): 706–740.

Vélez-Torres, I. (2014). Governmental extractivism in Colombia: Legislation, securitization and the local settings of mining control. *Political Geography* 38: 68–78.

Vermeulen, S., & Cotula, L. (2010). Over the heads of local people: Consultation, consent, and recompense in large-scale land deals for biofuels projects in Africa. *Journal of Peasant Studies* 37(4): 899–916.

Vetter, J. (2010). Capitalizing on grass: The science of agrostology and the sustainability of ranching in the American West. *Science as Culture* 19(4): 483–507.

Watson, T.L. (2008). Ho`I Hou Iā Papahānaumoku: A history of ecocolonization in the Pu`Uhonua of Wai`Anae. PhD thesis. University of Hawaii, Honolulu.

Watts, M. (2004). Resource curse? Governmentality, oil and power in the Niger Delta, Nigeria. *Geopolitics* 9(1): 50–80.

White, R. (1996). *The Organic Machine: The Remaking of the Columbia River*. New York: Hill and Wang.

Wynne-Jones, S. (2012). Negotiating neoliberalism: Conservationists' role in the development of payments for ecosystem services. *Geoforum* 43(6): 1035–1044.

Zalik, A. (2010). Oil "futures": Shell's scenarios and the social constitution of the global oil market. *Geoforum* 41(4): 553–564.

Zimmerman, E. (1933). *World Resources and Industries*. New York: Harper and Brothers.

Chapter 32

Political Ecologies of the State

Katie Meehan and Olivia C. Molden
University of Oregon, Eugene, Oregon, USA

Like many stories in political ecology, this one starts with a bulldozer. In 2012, after several decades of armed conflict with ethnic Tamil minorities in the countryside, President Mahinda Rajapaksa, the leader of Sri Lanka, took up a new fight – this time, against nature (Figure 32.1).[1] Supported by a $213 million loan from the World Bank, the Sri Lankan government launched a massive infrastructure modernization project in its capital city, Colombo. Rather than retire military personnel, the Rajapaksa regime rechristened the Ministry of Defense and Urban Development (MDUD) and put soldiers to work taming the "frenetic" nature of Colombo: excavating wetlands, straightening rivers, and filling lakes and water tanks (reservoirs) across the metropolitan area. Ostensibly, the Metro Colombo Urban Development Project (MCUDP) aims to "rehabilitate, improve and maintain local infrastructure and services through selected drainage management," with a focus on flood control and urban beautification (World Bank n.d.). The Rajapaksa regime was more direct in its hydraulic ambitions, peppering the city with posters that declared: "Let us now build the nation" (Figure 32.2).

Bulldozers were accompanied by strategic discourses of climate-proofing the capital. On November 11, 2010, Colombo suffered widespread inundation after a record rainfall of 440 millimeters within 24 hours (Jones 2010). Losses reached US$50 million and nearly half a million people were affected (Perera 2013). While the low-lying city and surrounding coastal areas have been historically vulnerable to tropical storms, monsoonal flooding, and sea level rise (Policy Network of Sri Lanka 2013), the 2010 floods were used to justify a new political ecology of state power: the role of the *military* in producing urban nature. As Gotabaya Rajapaksa, head of the MDUD and brother of the president, explained in a press conference:

> As a historic multicultural, tropical city bordered by the sea, Colombo has many natural advantages that should be capitalised on. Its historic architecture, greenery and public open spaces must be heightened and its people-friendly nature fostered. It is with such improvements in mind that the Government has launched many initiatives through which it hopes to further develop Colombo. (Jayathilaka 2012)

The Wiley Blackwell Companion to Political Geography, First Edition.
Edited by John Agnew, Virginie Mamadouh, Anna J. Secor, and Joanne Sharp.
© 2015 John Wiley & Sons Ltd. Published 2017 by John Wiley & Sons Ltd.

Figure 32.1 Posters depicting President Rajapaksa and the "new" activities of the Sri Lankan armed forces. *Source*: Olivia Molden, May 2009.

Figure 32.2 The Rajapaksa regime announces postwar rebuilding efforts. *Source*: Olivia Molden, May 2009.

Despite World Bank projections that the MCUDP will benefit 1.6 million people, the project comes at significant environmental and social costs. To date, nearly 500,000 slum dwellers have been forcibly evicted to make room for construction projects, mainly in areas with high concentrations of Tamil Muslims and Hindus (Center for Policy Alternatives 2014). Alternative housing and compensation were promised, but most families have received minimal or no compensation (Center for Policy Alternatives 2014). Because of the military's central role in the project – following a history of intense state surveillance and known cases of disappearances – few people have stepped forward to protest the MCUDP.

However, nature is no easy ally of development. On Colombo's periphery, the Thalangama wetland – an ancient complex of ponds, canals, and paddy fields – has engineered new problems for state dreams. Soldiers cut riverbanks and removed trees and vegetation, which local farmers had maintained to protect irrigation supply. Given the region's endemic history of flooding and hydrologic activity, the Ministry may have found itself in a much larger battle than initially expected. As a retired engineer noted:

> It was embarrassing for me to disagree with proposed solutions, which looked something like the methods used by the police to control – using hand signals to direct the flow of traffic. Sadly no one can control floods the way we control traffic because flood waters cannot be controlled by intermittently blocking canals against the natural direction of flow of water especially during major floods. (Nanayakkara 2014)

In securing the waterscapes of Colombo, this story spotlights the *prosaic* nature of the state, what Joe Painter (2006: 753) calls the "intense involvement of the state in so many of the most ordinary aspects of social life" such as "giving birth, child rearing, schooling, working, housing, shopping, travelling, marrying, being ill, [and] dying." In contrast to reified understandings of the state – the idea that states are real, a priori, coherent entities, often with divine origins – the prosaic approach situates the state among the earthly exercises of power: dredging a wetland, compensating a flood victim, bulldozing a slum. Central to this approach is the notion of *practice*. Painter argues (2006: 755):

> By emphasizing the role of practices, the concept of prosaic stateness reveals the geographies of state power with greater complexity and subtlety ... not only extend[ing] the apparent spatial reach of state power, but also reveal[ing] its geographical unevenness.

Beyond political geography, the notion of prosaic stateness has especially taken hold in political ecology, a multidisciplinary field concerned with issues of socio-environmental change and power (Peet & Watts 2004; Biersack 2006; Robbins 2008, 2012). Because of its roots in field-based empiricism – a tradition dating back to Alexander von Humboldt and Peter Kropotkin (Robbins 2012; Wainwright 2013) – political ecology is particularly well equipped to drill down through the actual practices that constitute stateness: making maps, counting wildlife, eradicating insects, creating parks, building dams, measuring flows, or restoring nature, to name a few (e.g., Mitchell 2002; Neumann 2004; Harris 2012; Robbins 2012). Forged through practice, states produce new spatial arrangements that unevenly "celebrate the visions of elite networks, reveal the scars suffered by the disempowered, and nurture the possibilities and dreams for alternative visions" (Swyngedouw 2007: 24). In this approach, the state rarely appears monolithic, coherent, and temporally or spatially stable. Indeed, the state as an environmental actor is "arguably inherently more internally fractured than it is as a provider of welfare services, paver of roads, or even maker of wars" (Robbins 2008: 205).

In examining the prosaic state, however, political ecologists are also keenly attuned to the *non-human* dimensions of state power (Robbins 2000, 2008; Meehan, Shaw, & Marston

2013; Meehan 2014). Take water, for example: As it flows through the landscape, water transgresses geopolitical boundaries, defies jurisdictions, pits upstream against downstream users, and creates competition between economic sectors, both for its use and for its disposal (Bakker 2012). Here, practice *and* materiality matter to formations of power. From miniature mosquitoes to megawatt dams, political ecologists examine how non-human actors, actants, and technologies are woven in a dynamic web of materiality and social representations that constitute our understandings of "nature" (Braun & Castree 2001). Such natures are understood as hybrid – biopolitical, discursive, and material – and necessarily multiple (Castree & Braun 1998; Whatmore 2002).

Just as we recognize that the "social" cannot be thought of as separate from "things, nature, or machines" (Castree & Braun 1998: 31), in this chapter we advance conceptualizations of prosaic stateness in political geography by drawing on key tenets in political ecology. We synthesize characterizations of the state and state power in political ecology, and identify three major theses that anchor the field:

- The state is not a thing, but a *resonance chamber* of objects and ecologies.
- It enables the *calculus* and *legibility* of nature for capital.
- And in so doing, material practices and objects produce state *subjects* – who, in turn, often subvert their subjection.

In what follows, we suggest that these insights advance prosaic state theory beyond anthropocentric notions of power, place, and practice. Our review is necessarily partial, but deliberately catholic. In contrast to classic definitions of political ecology – in which critical scholars construct a "chain of explanation" that threads together different spatial scales and structural inequalities at the heart of human–environment problems (Wolf 1972; Blaikie & Brookfield 1987) – here we adopt a far more cosmic approach: political ecology as a community of practice, "something people do" rather than a coherent theory or body of knowledge (Robbins 2012: 5–6; see also Peet & Watts 2004). Certain commonalities exist: political ecology is fruitful in considering the spatio-temporalities of resources, inequality and differentiated access, the question of scale, and the materiality of natures (Harris 2012: 26). For all the necessity of political ecology's anthropocentrism, it also remains exceptionally ecocentric, enrolling the non-human world in its conceptual framework and analyses (Biersack 2006). This point, as we demonstrate next, emerges most prominently in our first thesis: The state is not a thing, but a resonance chamber.

The resonance chamber

Colombo is not the only city to grapple with infrastructure crises. By the end of the eighteenth century in Europe, rapid industrialization had led to dismal environmental conditions in many cities. Urban rivers became arteries of waste, sources of disease and death. The Thames was dubbed the "Great Stink" of London, while Athens was characterized as a "city in ruins" (Kaika 2005). In Manchester, Frederick Engels described the infernal conditions of working-class neighborhoods:

> the streets are generally unpaved, rough, dirty, filled with vegetable and animal refuse, without sewers and gutters, but supplied with foul, stagnant pools instead. Moreover, ventilation is impeded by the bad, confused method of building the whole quarter. (quoted in Merrifield 2002: 70–71)

Spurred by housing and ecological crises, city planners rolled out ambitious designs to modernize the city, constructing a universal water supply and sewerage, mass transport, and large-scale communication networks to "civilize" and control urban spaces and subjects (Kaika 2005). Municipal and federal governments played central roles in reworking urban nature, boosted by infusions of rapidly globalizing capital. Newly built highway networks and water conduits – examples of what David Harvey (1985) calls the "secondary circuit of capital" – had knock-on effects of facilitating further growth and accumulation, while simultaneously entrenching the state as service provider and guarantor of public order.

What these approaches share is a deeply *material* conceptualization of stateness and power, one literally built into the guts and bowels of the city. This is not to say the state is a thing. As Timothy Mitchell (2002) explains, the appearance of the nation-state as an "object" – an idea that emerged in tandem with capitalist modernization and development – is in fact the *effect* of power and knowledge. Mitchell (2002: 2) points out that the constructivist impulse to explain everything as social and cultural "left aside the existence of other spheres, the remainder or excess that the world of social construction works upon – the real, the natural, the nonhuman." Given the impossibility of treating the state as a "thing," he instead proposes examining the state as an effect.

> We tend to think of [language, imagery, space, and movement] as processes that merely mark out and represent the nation-state, as though the nation-state itself has some prior reality. In fact the nation-state is an effect of all these everyday forms of regulation and representation, which set it up in the appearance of an empirical object. (Mitchell 2002: 231)

Drawing on this framework, Harris (2012) explores state formation through major development and hydraulic schemes in southeast Turkey. While small-scale Turkish farmers celebrate new flows of irrigation water ("the state thinks about us now"), Kurds have been forced to relocate their villages for dam construction and wait in long queues ("the Arabs got water first"; Harris 2012: 31). These uneven experiences of state power are further differentiated by the biophysical realities of seasonal crops, electricity production, large-scale water diversions, and migrant labor.

Yet, if the state is not coherent or real, then what is it? Political ecologists have drawn on Deleuze to probe the complex geometries that compose stateness. Jeff Banister, for instance, explores the rhizomatic nature of state power in Mexico's Río Mayo valley:

> For Deleuze and Guattari "the State" is among other things a set of assemblages (official organizations, etc.) that attempt to capture, encode, and stratify (de-territorialize and re-territorialize) flows of people, ideas, and matter. Redirecting these flows toward a pivot point (e.g. a large irrigation bureaucracy in a regional or national capital), however, requires a mobility and velocity usually beyond the capacity of the top-heavy, centralized organs of the state. An irrigation bureaucracy must therefore avail itself of the potentialities of decentered assemblages that both exist within and outside of its ambit. (Banister 2014: 207)

Even in northern Mexico, where the government has spent decades consolidating state power through the manipulation of hydraulic rules, flows, and infrastructures, it remains impossible to pin down the "state" in a single dam, pipe, or bureaucrat (Banister 2011, 2014). As Deleuze and Guattari (1987: 433) suggest:

> the State is a phenomenon of *intraconsistency*. It makes points resonate together, points that are not necessarily already town poles but very diverse points of order, geographic, ethnic, linguistic, moral, economic, technological particularities.

In this way, the state is a process or condition, "a plurality of states, a multiplicity of localized orders of appearance" (Meehan, Shaw, & Marston 2014: 61). Meehan et al. (2014) explain:

> Building on Deleuze and Guattari, we call this phenomenon a resonance chamber. "The State is not a point taking all the others upon itself, but a resonance chamber for them all" (1987: 224). If we replace the word "point" for "object" in the preceding quote, then the state becomes "not an object taking all others upon itself, but a resonance chamber for them all." And while we previously wrote that the state is "located in the interaction or passage between objects," we find this idea of a resonance chamber a useful extension.

Central to the resonance chamber are *objects*, what Shaw and Meehan (2013: 218) call "an autonomous unit of reality that exists independent of its relations." At their core, objects defy anthropocentric conceptions of existence, and therefore power. Shaw and Meehan suggest (2013: 218):

> Reality at any one moment is like a flickering TV set split between what appears and what does not. Once an object is unleashed in the world, it produces and reproduces a localised order of appearance that can interrupt, and even master, the very "intentions" humans may have once engineered.

Consider, for example, the meter, a fairly mundane technology used to measure residential or industrial water consumption. In the South African city of Durban, so-called check meters were installed in households during the 1950s to assess "the extent of what the municipal administration assumed to be hedonistic water use" (Loftus 2006: 1030). As Durban began to industrialize and expand, high rates of water abstraction (primarily by the rich) started to threaten the existing water supply, even with the construction of South Africa's largest reservoir. Meters became ideal mechanisms for regulating water use, but their entry into private homes, Loftus points out, enabled entirely new modes of state surveillance and control:

> While from the 1970s onwards the water meter came to express and embody the emerging bureaucratic concerns of the local administration, it was with the formation of the semi-commercialised water board, and its desire to seek profits from the provision of water, that formal rationality came to be fused with commodity fetishism in a much more powerful combination. Here, the targets were those most endangering the realisation of these profits. Water meters became much more powerful agents in the households of poor residents in the city. (Loftus 2006: 1031–1032)

For Loftus (2006: 1042), technologies like the water meter embody state logics of rationality and accumulation, becoming flashpoints in the struggle between state institutions and subjects.

At the same time, the water meter is curiously *force-full*: "brimming with affect, productive of difference, and generative of power" (Shaw & Meehan 2013: 220). Meters regulate pressure, restrict consumption, and facilitate new flows of water and capital. As examples of "mediating technologies" – or simple additive devices meant to transform systems (Furlong 2011) – meters may bring an unforeseen malleability to networked infrastructure and alter socio-technical relationships in significant ways. Power is thus mobilized and transformed *through* artifacts (Monstadt 2009), including the spatial arrangements of humans and non-humans that facilitate a space of "stateness" (Meehan 2014). Technologies and practices not only crystallize state–society relations, they also contribute toward the policing of the world

(Shaw & Meehan 2013). The contours of this world, as we explore in the next section, are inevitably policed in the name of capital.

Capital and the calculus of nature

Deep in the tropical rainforests of Costa Rica, the jungle is increasingly inventoried for capital. In exchange for carbon payments – purchased on the internet by consumers and managed by a foreign company – local state authorities in Costa Rica agree to manage retired pasture in a way that encourages tree growth, thus fixing carbon dioxide (CO_2) and offsetting global emissions from elsewhere (Lansing 2012). In theory, this market-based system internalizes externalities (CO_2), produces local benefits (payment-for-ecological services, or PES), and introduces efficiency and parity into environmental management. However, in practice, making forests work for capital requires a considerable amount of calculation and sweat. Once a year, a group of state bureaucrats, non-governmental organization (NGO) members, company employees, and indigenous political leaders get together to examine and tour the reserve:

> Guided by GPS devices, they inspect the space's physical boundaries; they repair signs; recut trails; and take photographs of the site. GPS devices, maps, signs, trails, and cameras: these artifacts of calculation, measure, and inscription join the human actors of scientists, bureaucrats, politicians, farmers, businessmen, and consumers in the ongoing and iterative process of maintaining this precarious object of exchange – the carbon offset commodity. (Lansing 2012: 205)

Since the mid-1990s, the government has implemented several reforestation policies meant to transform parts of the nation's territory into sites of commodified carbon storage (Lansing 2010, 2011, 2012). Such calculative practices – hiking, observing, recording, reporting – help constitute the forest as an object of global economic exchange (Lansing 2012). With the growth of market-based tools and mechanisms used to manage the environment (Bakker 2005, 2010; Castree 2007), resources like wetlands, minerals, rivers, wildlife, and even genes are increasingly becoming more "organized by value-form" (Robertson & Wainwright 2013; see Bakker 2010 for a review).

Central to the market process is the role of the state. As Robertson and Wainwright (2013: 899) explain:

> The state in question is not a generic state but a capitalist one. That means that it must work to ensure accumulation … the state's talk is to thread a fine needle: to ensure that the capitalist value form expands but prevent it from becoming the sole means of determining the identification of objects and the measurements of their qualities. The state refuses to define value. This ambiguity and silence allows the state to continue to call on noncapitalist knowledge and resources to express nature in capitalist form.

Such accumulation strategies demonstrate some of the contradictory tendencies of capitalism (Jessop 2000). As class fractions attempt to resolve the crises of contraction, this in turn produces the capitalist state, meaning that the state is not merely the instrument of capitalism (Jessop 2000; Roberston & Wainwright 2013). Jessop's approach leads us to appreciate the capitalist state not as a thing but as a social relation; not as a coherent agent unified around a simple project of expanding capital, but as a stage on which struggles over capital accumulation play out (Jessop 2000; Robertston & Wainwright 2013).

Political ecologists examine the diverse ways in which states mobilize, contest, and negotiate "nature" as a central facet of state power and control, including efforts to make non-humans legible and controllable (Harris 2012). Ecosystem services, for example, arise through the same process of metabolic exchange involving living labor, controlling nature, and social abstraction that produces commodities (Robertson 2006). PES and ecosystem services require a calculus that supersedes and overcomes the specificities of price and ecology (Roberston & Wainwright 2013).

Yet, even in its attempt to inventory nature for capital, the state is not an external or monolithic force. In Peru, for example, the threat of melting glaciers prompted the government to construct drainage tunnels and floodgates during the 1980s. Technologies that were designed to protect local residents from hazardous flooding had the counter-intuitive effect of facilitating successful adaptation to changing climatic and hydrologic conditions, as well as creating new opportunities for capital exploitation and exportation of hydroelectric energy production (Carey, French, & O'Brien 2012). Moreover, the "localness" of people and knowledge systems does not create immunity from capital agendas. Indeed, "the state does not produce knowledge to the exclusion of local accounts, but instead seizes and reproduces locally powerful knowledges and enforces management through alliances with locally powerful groups" (Robbins 2000: 127). Transformations in socio-ecological systems are driven by internal changes and contradictions in the system, leading to struggles that create new ways of organizing nature and labor (Robbins 2000). Part of this control, as we explore in the next section, involves the production and control of state subjects.

State subjects and the environment

The complex nature of environmental subjects, defined by Agrawal (2005: xiv) as "people who have come to think and act in new ways in relation to the environment," is central to explanations of state power. Geographers have long argued that modern forms of state regulation and power function through networks of self-regulating agents – called subjects – who act as willing *accomplices*, rather than coerced footservants, in state logics and agendas (Peet & Watts 2004; Secor 2007). In this way, subject formation is "a productive dimension of the practices that produce power/knowledge," and not simply a negative effect of power (Gabriel 2014: 42). Building on the notion that governance is achieved through the spatial arrangement of people and objects, what Michel Foucault calls the "right disposition of things" (quoted in Gabriel 2014), geographers have also recently advanced theories of subjectivity as more-than-human (Gibson-Graham 2006; Whatmore 2006), as a "site of supple autonomy" (Smith 2012: 1513) that specifically emerges from new states of objects. In short, objects not only *discipline* their subjects, as Robbins (2007) so convincingly portrays in *Lawn People*, they also unlock other worlds and ways of being political.

While geographers widely understand human subjectivity as "*the* motive force in organizing politics" (Woodward, Jones, & Marston 2012: 204), political ecology has specifically questioned the role of non-humans in subject formation, state–society relations, and the spatial governance of modern life (Robbins 2007; Braun & Whatmore 2010; Gabriel 2014). At stake in these debates is the broader question of how subjects come to "be" in concert with their objects. For example, Robbins (2007) explores the ways in which "turfgrass subjects" arise from a historical synchronization of laws, aesthetics, neighborhood politics, petrochemicals, and state rule. Key here is the lawn itself: a crucial disciplinarian of suburban homeowners as much as a product of their chemical care and mowing.

Colombo and its new water infrastructure provide another example. Couched in the language of beautification, modern efficiency, and improvement, the MCUDP presents an intractable puzzle for citizen-subjects: How can one *not* desire development (Wainwright 2008)? In this mix, the World Bank inserts itself as an expert that can inform citizens about how to live and how to improve (Li 2005). Non-humans play a big role in schemes of "improvement": The collection and circulation of infrastructural objects like maps, blueprints, or demographics help legitimize projects like the MCUDP, which in turn craft new urban subjects while "relocating" others (Li 2005: 383; Scott 1998). Such strategies range from the forced eviction (at gunpoint) of targeted ethnic and class groups, to the highly visible construction of new parks and Navy-run boat navigation in the newly "cleaned" canals – the government's attempt to make Colombo "another Venice" (World Bank News 2012). Important to this process are the intended and also accidental imaginaries, objects, and moments in everyday life (Li 2005; Stoddart 2005; Robbins 2008). As Painter highlights:

> Governmentality draws attention to the construction of the objects of government, and to the logics, rationalities and technologies of rule, whereas prosaics highlights the unsystematic, the indeterminate and the unintended. On the other hand, the two perspectives do converge in their focus on mundane practices and the productive nature of discourse. (Painter 2006: 763)

The Colombo improvement scheme mirrors (post)colonial projects of subject-making more broadly. Across South Asia, colonial rulers saw local populations as a generally "deficient" group; however, in order to control populations and extract goods from the territories, colonizers distinguished certain groups within the population, using local leaders to assign particular values and designations to different groups (Li 2005: 388). Although exploiting tensions between different groups was initially strategic for colonial control, such arrangements aggravated conflict during the postindependence period, such as between the Tamils and Sinhalese in Sri Lanka, or the Indians and Tanzanians in Zanzibar (Larson & Aminzade 2008; Spencer 2008). The physical division and recording of people – alternately enabled and complicated by objects and technologies – helped produce goods for European economies and also cemented notions of development, aesthetics, and morality (Kaika 2005; Li 2005). Governing infrastructure and non-human natures is thus a powerful way to discipline and regulate state subjects (McFarlane & Rutherford 2008).

At the same time, power and subject formation unfold unevenly and at times unpredictably (Scott 1998; Li 2005; Stoddart 2005; Painter 2006). Colonial agents aimed to enroll certain groups as "civilized" based on their hygienic practices and reliance on specifically "modern" infrastructures (Gandy 2006; McFarlane & Rutherford 2008). Yet, this ideal often fell short in particular places and contexts: subjects often subverted state control by relying on multiple water and sanitation technologies (Kooy & Bakker 2008). People developed water technologies that empowered groups to oppose state mandates (Birkenholtz 2009), or cultivated alternatives that modified centralized infrastructure networks without abandoning them (Kooy 2014; Meehan 2014). As state and citizen agents interact, the abstraction of the state and complexity of bureaucracy lead to a variegated co-production of different official and civilian roles (Painter 2007: 606). In short, despite the permeation of Western notions of modernity and moral politics, the role of the subject in modern statecraft is not a done deal.

Conclusion

In recent years, prosaic and "everyday" approaches to explaining state power have emerged as powerful conceptual frameworks in geography. Yet, while the prosaic approach marks a significant advance, in this chapter we have tried to highlight and address several weak spots through a synthetic review of political ecology. In the first place, the prosaic approach is unquestionably anthropocentric: humans and their actions are the protagonists of power, the sources of statecraft, the "agents around whose actions and intentions the story is written" (Mitchell 2002: 29). Geography, like all social science, has been "founded upon a categorical distinction between the ideality of human intentions and purposes and the object world upon which these work, and which in turn may affect them" (Mitchell 2002: 29). By extension, scripting the state as a centrally human practice further extends this division between human labor and non-human objects (Shaw & Meehan 2013). Drawing on political ecology, we have instead argued that the state is best conceptualized as a *resonance chamber*: one that enables the legibility of nature for capital and produces state subjects who, in turn, often resist and subvert their subjection.

A second issue is methodological: We suggest that the focus on commonplace practices risks reducing the "everyday" to the innocuous. While actions such as birth, death, and paying taxes are normal and ubiquitous – replicated across time and space in ways that may cultivate or constrain stateness – when coalesced, their effects are potentially extraordinary, dazzling, or toxic. This scalar tension is perhaps most visible in political ecology research, where everyday practices of state power are often irreversible (emitting CO_2), broad-scale (Three Gorges Dam), morphological (river channelization), monstrous (GM crops), pervasive (payment for ecological services), with unanticipated or undesirable effects (suburban lawn chemicals, agrarian counter-reforms). State environmental practices may rewire webs of life, trigger multiplier effects, and hold dire implications for global communities of humans and non-humans. Political ecology, as a field-based mode of inquiry that often places non-human materialities at the core of its explanation, "perhaps offers a needed corrective to the heavy focus in recent work on the banal, everyday, and capillary experiences of states or power in ways that downplay processes of large-scale change or grand schemes of social engineering, centralization, or territorialization" (Harris 2012: 39).

In an era of increased securitization and geopolitical tension – both at home (intensively) and abroad (extensively) – we conclude by suggesting that political ecology adds great analytical traction in understanding the increasingly militarized and more-than-human nature of statecraft and power. Like bulldozers, accounts in political ecology often highlight historical and contemporary injustices, unanticipated results, and calculative efforts to remake reality. To whit, Colombo's redevelopment has come at great social and environmental costs: Undesired human and non-human habitats have been displaced and destroyed to fit a certain vision of a modern, world-class city. Redevelopment has remetabolized urban ecologies, erased and produced new forms of environmental knowledge, enrolled desired groups as stakeholders, and accumulated new forms of value. In an era of rapidly emerging technologies and techniques of statecraft – from communication to war to river channelization – new political ecologies of state power are emerging daily, and their urgency demands critical inquiry now more than ever.

Note

1 The United Nations and several human rights groups (including Human Rights Watch, Amnesty International, and Genocide Watch) have criticized President Rajapaksa for supporting unethical policy and actions during the war, including the deaths of nearly 40,000 people in its final weeks (BBC 2012).

References

Agrawal, A. (2005). *Environmentality: Technologies of Government and the Making of Subjects*. Durham, NC: Duke University Press.

BBC (2012). Amnesty accuses Sri Lanka of "post war abuses." 13 March. http://www.bbc.com/news/world-asia-17353223, accessed March 26, 2015.

Bakker, K. (2005). Neoliberalizing nature? Market environmentalism in water supply in England and Wales. *Annals of the Association of American Geographers* 95(3): 542–565.

Bakker, K. (2010). Debating green neoliberalism: The limits of "neoliberal natures." *Progress in Human Geography* 34(6): 715–735.

Bakker, K. (2012). Water: Political, biopolitical, material. *Social Studies of Science* 42(4): 616–623. doi:10.1177/0306312712441396

Banister J. (2011). Deluges of grandeur: Water, territory, and power on Northwest Mexico's Río Mayo, 1880–1910. *Water Alternatives* 4(1): 35–53.

Banister, J.M. (2014). Are you Wittfogel or against him? Geophilosophy, hydro-solidarity, and the state. *Geoforum* 57(1): 205–214.

Biersack, A. (2006). Reimagining political ecology. In Biersack, A., & Greenberg, J. (eds), *Reimagining Political Ecology*, Durham, NC: Duke University Press, 3–40.

Birkenholtz, T. (2009). Groundwater governmentality: Hegemony and technologies of resistance in Rajasthan's (India) groundwater governance. *Geographical Journal* 175(3): 208–220.

Blaikie, P., & Brookfield, H. (eds) (1987). *Land Degradation and Society*. London: Methuen.

Braun, B., & Castree, N. (eds) (2001). *Social Nature: Theory, Practice, and Politics*. Malden, MA: Wiley-Blackwell.

Braun, B., & Whatmore, S. (2010). *Political Matter: Technoscience, Democracy and Public Life*. Minneapolis, MN: University of Minnesota Press.

Carey, M., French, A., & O'Brien, E. (2012). Unintended effects of technology on climate change adaptation: An historical analysis of water conflicts below Andean glaciers. *Journal of Historical Geography* 38(2): 181–191.

Castree, N. (2007). Neo-liberalising nature: Processes, outcomes and effects. *Environment and Planning A* 40(1): 153–173.

Castree, N., & Braun, B. (1998). The construction of nature and the nature of construction: Analytical and political tools for building survivable futures. In Braun, B., & Castree, N. (eds), *Remaking Reality: Nature at the Millennium*, New York: Routledge, 3–42. Center for Policy Alternatives (2014). *Forced Evictions in Colombo: The Ugly Price of Beautification*. Colombo: Sri Lanka.

Deleuze, G., & Guattari, F. (1987). *A Thousand Plateaus: Capitalism and Schizophrenia*. Minneapolis, MN: University of Minnesota Press.

Furlong, K. (2011). Small technologies, big change; Rethinking infrastructure through STS and geography. *Progress in Human Geography* 35(4): 460–482.

Gabriel, N. (2014). Urban political ecology: Environmental imaginary, governance, and the non-human. *Geography Compass* 8(1): 38–48.

Gandy, M. (2006). Planning, anti-planning and the infrastructure crisis facing metropolitan Lagos. *Urban Studies* 43(2): 371–396.

Gibson-Graham, J.K. (2006). *A Postcapitalist Politics*. Minneapolis, MN: University of Minnesota Press.

Harris, L.M. (2012). State as socionatural effect: Variable and emergent geographies of the state in southeastern Turkey. *Comparative Studies of South Asia, Africa and the Middle East* 32(1): 25–39. doi:10.1215/1089201X-1545345

Harvey, D. (1985). *The Urbanization of Capital*. Baltimore, MD: Johns Hopkins University Press.

Jayathilaka, K. (2012) Colombo building a world class city. *Business Today* July. http://businesstoday.lk/article/php?article=7297, accessed March 26, 2015.

Jessop, B. (2000). The crisis of the national spatio-temporal fix and the tendential ecological dominance of globalizing capitalism. *International Journal of Urban and Regional Research* 24(2): 323–360.

Jones, B. (2010). Extreme weather around the world. *The Guardian* Nov. 16. http://www.theguardian.com/news/2010/nov/16/weatherwatch-sri-lanka-canada-china-russia, accessed Mrch 26, 2015.

Kaika, M. (2005). *City of Flows: Modernity, Nature, and the City*. New York: Routledge.

Kooy, M. (2014). Developing informality: The production of Jakarta's urban waterscape. *Water Alternatives* 7(1): 35–53.

Kooy, M., & Bakker, K. (2008). Technologies of government: Constituting subjectivities, spaces, and infrastructures in colonial and contemporary Jakarta. *International Journal of Urban and Regional Research* 32(2): 375–391.

Lansing, D.M. (2010). Carbon's calculatory spaces: The emergence of carbon offsets in Costa Rica. *Environment and Planning D: Society and Space* 28(3): 710–725.

Lansing, D.M. (2011). Realizing carbon's value: Discourse and calculation in the production of carbon forestry offsets in Costa Rica. *Antipode* 43(3): 731–753.

Lansing, D.M. (2012). Performing carbon's materiality: The production of carbon offsets and the framing of exchange. *Environment and Planning A* 44(1): 204–220. doi:10.1068/a44112

Larson, E., & Aminzade, R. (2008). Nation-building in post-colonial nation-states: The cases of Tanzania and Fiji. *International Social Science Journal* 59(192): 169–182.

Li, T.M. (2005). Beyond "the state" and failed schemes. *American Anthropologist* 107(3): 383–394.

Loftus, A. (2006). Reification and the dictatorship of the water meter. *Antipode* 38(5): 1023–1045.

McFarlane, C., & Rutherford, J. (2008). Political infrastructures: Governing and experiencing the fabric of the city. *International Journal of Urban and Regional Research* 32(2): 363–374.

Meehan, K. (2014). Tool-power: Water infrastructure as wellsprings of state power. *Geoforum* 57(1): 215–224.

Meehan, K.M., Shaw, I.G.R., & Marston, S.A. (2013). Political geographies of the object. *Political Geography* 33(1): 1–10.

Meehan, K.M., Shaw, I.G.R., & Marston, S.A. (2014). The state of objects. *Political Geography* 39(1): 60–62.

Merrifield, A. (2002). *Metromarxism: A Marxist Tale of the City*. New York: Routledge.

Mitchell, T. (2002). *Rule of Experts: Egypt, Techno-Politics, Modernity*. Berkeley, CA: University of California Press.

Monstadt, J. (2009). Conceptualizing the political ecology of urban infrastructures: Insights from technology and urban studies. *Environment and Planning A* 41(8): 1924–1942.

Nanayakkara, K.D.A.H. (2014) Flooding Kotte to save Colombo? *The Island* February 21. http://www.island.lk/index.php?page_cat=article-details&page=article-details&code_title=98359, accessed March 25, 2015.

Neumann, R.P. (2004). Nature-state-territory: Toward a critical theorization of conservation enclosures. In Peet, R., & Watts, M. (eds), *Liberation Ecologies: Environment, Development, Social Movements*, London: Routledge, 195–217.

Painter, J. (2006). Prosaic geographies of stateness. *Political Geography* 25(7): 752–774.

Painter, J. (2007). Stateness in action. *Geoforum* 38(4): 605–607.

Peet, R., & Watts, M. (2004). *Liberation Ecologies: Environment, Development, Social Movements*, 2nd edn. London: Routledge.

Perera, A. (2013). Sri Lanka rolls out plan to reduce flood losses in Colombo. *Thomson Reuters Foundation* February 18. http://www.trust.org/item/?map=sri-lanka-rolls-out-plan-to-reduce-flood-losses-in-colombo, accessed March 25, 2015.

Policy Network of Sri Lanka (2013). Climate Net: Climate Change Policy Network of Sri Lanka – Resources. February 14. http://www.ips.lk/climatenet/resources/index.html, accessed March 26, 2015.

Robbins, P. (2000). The practical politics of knowing: State environmental knowledge and local political economy. *Economic Geography* 76(2): 126–144.

Robbins, P. (2007). *Lawn People: How Grasses, Weeds, and Chemicals Make Us Who We Are.* Philadelphia, PA: Temple University Press.

Robbins, P. (2008). The state in political ecology: A postcard to political geography from the field. In Cox, K.R., Low, M., & Robinson, J. (eds), *The SAGE Handbook of Political Geography*, Thousand Oaks, CA: Sage, 205–218.

Robbins, P. (2012). *Political Ecology: A Critical Introduction*, 2nd edn. Malden, MA: Wiley-Blackwell.

Robertson, M.M. (2006). The nature that capital can see: Science, state, and market in the commodification of ecosystem services. *Environment and Planning D: Society and Space* 24(3): 367–387.

Robertson, M.M., & Wainwright, J.D. (2013). The value of nature to the state. *Annals of the Association of American Geographers* 103(4): 890–905.

Scott, J.C. (1998). *Seeing Like a State: How Certain Schemes to Improve the Human Condition Have Failed*. New Haven, CT: Yale University Press.

Secor, A.J. (2007). Between longing and despair: State, space, and subjectivity in Turkey. *Environment and Planning D* 25(1): 33–52.

Shaw, I.G.R., & Meehan, K. (2013). Force-full: Power, politics and object-oriented philosophy. *Area* 45(2): 216–222.

Smith, S. (2012). Intimate geopolitics: Religion, marriage, and reproductive bodies in Leh, Ladakh. *Annals of the Association of American Geographers* 102(6): 1511–1528.

Spencer, J. (2008). A nationalism without politics? The illiberal consequences of liberal institutions in Sri Lanka. *Third World Quarterly* 29(3): 611–629.

Stoddart, M. (2005). The Gramsci-Foucault nexus and environmental sociology. *Alternate Routes* 21: 40–63.

Swyngedouw, E. (2007). Technonatural revolutions: The scalar politics of Franco's hydro-social dream for Spain, 1939–1975. *Transactions of the Institute of British Geographers* 32(1): 9–28.

Wainwright, J.D. (2008). *Decolonizing Development: Colonial Power and the Maya*. Malden, MA: Blackwell.

Wainwright, J. (2013). *Geopiracy: Oaxaca, Militant Empiricism, and Geographical Thought*. New York: Palgrave Macmillan.

Whatmore, S. (2002). *Hybrid Geographies: Natures Cultures Spaces*. London: Sage.

Whatmore, S. (2006). Materialist returns: Practising cultural geography in and for a more-than-human world. *Cultural Geographies* 13(4): 600–609.

Wolf, E.R. (1972). Ownership and political ecology. *Anthropological Quarterly* 45: 201–205.

Woodward, K., Jones, J.P., III, & Marston, S.A. (2012). The politics of autonomous space. *Progress in Human Geography* 36(2): 204–224.

World Bank (n.d.). Metro Columbo Urban Development Project. *Projects and Operations*. http://www.worldbank.org/projects/P122735/metro-colombo-urban-development-project?lang=en, accessed May 9, 2014.

World Bank News (2012). Improving infrastructure and services in Colombo. http://www.worldbank.org/en/news/video/2012/11/07/improving-infrastructure-services-colombo-sri-lanka, accessed May 10, 2014.

Chapter 33

Environment: From Determinism to the Anthropocene

Simon Dalby

Wilfrid Laurier University, Waterloo, Ontario, Canada

A century and a half ago, George Perkins Marsh began his classic book on *Man and Nature* (1965 [1864]) with a description of the Roman Empire and what he argues was a "happy combination of physical advantages" consisting of a reasonable climate, abundant agricultural possibilities, and mineral resources that allowed the empire to thrive for many centuries. The physical circumstances when Marsh wrote a millennia and a half after the demise of the western empire were dramatically different, the fecundity of its ecosystems much less than those that supported Rome at its zenith. Human degradation of the natural ecology, notably by the removal of forests, had transformed the landscapes in ways that suggested destruction of many valuable attributes. While ignorance of the ways of nature and geological causes of change were part of these processes, Marsh (1965 [1864]: 11) emphasizes that

> the primitive source, the *causa causarum*, of the acts and neglects which have blasted with sterility and physical decrepitude the noblest half of the empire of the Caesars, is, first, the brutal and exhausting despotism which Rome herself exercised over her conquered kingdoms, and even over her Italian territory; then, the host of temporal and spiritual tyrannies which she left as her dying curse to all her wide dominium, and which in some form of violence or of fraud, still brood over almost every soil subdued by the Roman legions.

Marsh goes on to suggest that nineteenth-century geographers (he singles out Humboldt, Ritter, and Guyot in his introduction) had opened up interesting lines of inquiry into "how far external physical conditions, and especially the configuration of the earth's surface, and the distribution, outline, and relative position of land and water, have influenced the social life and social progress of man" (Marsh 1965 [1864]: 13). However, having noted that point, he then goes on to emphasize the other side of the matter: "[M]an has reacted upon organized and inorganic nature, and thereby modified, if not determined, the material structure of his earthly home" (1965 [1864]: 13).

The Wiley Blackwell Companion to Political Geography, First Edition.
Edited by John Agnew, Virginie Mamadouh, Anna J. Secor, and Joanne Sharp.
© 2015 John Wiley & Sons Ltd. Published 2017 by John Wiley & Sons Ltd.

Thus, in the 1860s, Marsh focuses his reader's attention squarely on the human imprint on the physiognomy of the earth, and links it to the violence of empire and the extraction of resources to feed imperial ambition. That humanity's political arrangements and modes of rule are a key part of these processes is not a very original insight judged from the early years of the twenty-first century, but it is worth noting that this is part of a long tradition of human thought. Marsh himself was influential in the nineteenth century, not least in early conservation efforts in North America that were concerned to manage forests in particular in the face of the landscape transformations caused by the rapid expansion of European colonization in the "New World." Even if the gender-specific formulations and nineteenth-century rhetorical flourishes seem slightly strange to twenty-first-century readers, the argument is very clear indeed.

What has only become clear much more recently is the fact that climate, too, is part of that "material structure of his earthly home." It is time, in Daniel Deudney's (1999) terms, to bring nature back into geopolitical discussions. Nevertheless, how to do this is not so simple given the intellectual history of environmental determinism, its outright rejection by modern scholarship, and its persistence in some political spheres where "geography as destiny" is a very useful geopolitical formulation. Still, as the current discussion of the Anthropocene makes clear, climate change in particular is very much a matter of contemporary geopolitics.

Environment and milieu

A century after Marsh's influential text, Clarence Glacken (1967) structured his overview of Western thinking about the earth and human habitation thereon in terms of three very persistent questions. These questions were, first, whether the earth was designed by some divine power to facilitate human life and the building of civilizations; second, what influences climate, topography, and the geographical configurations of the continents have had on human societies; and third, what the impacts of humanity as a geographical agent have been on the world. While Marsh was little concerned with either of the first two questions, being primarily involved in the third, it is interesting to note that he does not use the term environment; in Glacken a century later, the second question is generalized in terms of "environmental" influences.

The emergence of a category of environment is mostly a twentieth-century phenomenon, but there is no simple narrative that can adequately deal with the emergence of a term that has become ubiquitous by the early twenty-first century. In part the rise of ecological sciences, with their basic distinction between particular entities and that which environs them, has shaped this vocabulary. Indeed, such thinking has been key to the emergence of social science teaching programs in many universities called "environmental studies." The ambiguities about what is natural and artificial when related to humanity continue to cause confusion, with terms such as "natural environment" proliferating from the 1960s onward.

One attempt to clarify all this, contemporaneous with Glacken's writing, is Harold and Margaret Sprout's (1965) landmark effort to reformulate matters in terms of "milieu" while trying to rethink the ecological premises implicit in so many geopolitical texts. In their classification, the political environment is both natural and artificial. Crucially, they emphasize that environments present possibilities for human actions; specific contexts or, in their terms, "milieus" present possibilities for polities. This is especially so when matters of international relations are considered, because particular states are in given locations that inevitably constrain and shape the policy options available to political leaders. Humanity is not unconstrained, and some actions may be probable in given circumstances and others unlikely; there is an element of human political agency that is unavoidable and hence key to geopolitical analysis. Thus, the Sprouts (1965) formulate matters in *The Ecological Perspective on Human Affairs* in terms of a philosophical position called "possibilism."

What is clear is that the third of Glacken's questions, that concerning the consequences of humanity as a geographical agent, has now become the key issue, one made all the more urgent by the extraordinary expansion of human powers in the second half of the twentieth century. The world is being remade, not only degraded, by the forces that concerned Marsh. As the discussion of climate engineering in particular and the debate about the new geological epoch of the Anthropocene in general make clear (Dalby 2013), planetary change is happening in ways that dwarf the deleterious actions of the Roman empire that had grabbed Marsh's attention just as industrial powers fueled by fossil fuels were beginning to transform the human condition radically. Yet, as this chapter concludes, environmental arguments that attribute powerful causal factors to "nature" are still in circulation, despite the last two centuries wherein it has become clear that Marsh's original concerns with the disruptions caused by humanity are the most important matter for geopolitical consideration.

Colonizing nature

A century ago, the literature that gave rise to political geography, as it came to be known, was much more concerned with the second of Glacken's questions, the influences of climate and environment on human affairs. Indeed, political geography has often simply been understood as being concerned with this question (Agnew & Muscara 2012). As Marsh emphasized, the questions of "influence" and what subsequently became discussed (and frequently derided) as determinism were in vogue in the nineteenth century, not least because the expansion of European power and the colonization of much of the planet were accompanied by a growth of exploration, mapping and cataloguing the new lands conquered and annexed by European empires. The study of geography was concerned with appropriating territory, extracting resources to feed the industries of metropolitan states, and documenting its peoples and places to facilitate their administration. It was also about grappling with very different flora and fauna, and human inhabitants that did not obviously follow modes of conduct that made sense in the cultural categories of European observers (Gregory 2001). This also involved the transformation of those landscapes as modern ways of agriculture, transport, and economy were imposed on the new colonies, frequently with disastrous consequences for local inhabitants, human and otherwise.

Furthermore, these processes moved European settlers, plantation administrators, farmers, soldiers, and colonial functionaries into climates with which they were unfamiliar and which they frequently misinterpreted. Related to all these phenomena are matters of political economy and the understandings of land not cultivated and owned in European terms as effectively in a legal state of *terra nullius*, and hence available for appropriation so that it could be used, "improved," and made productive in terms of the expanding modern economy. Understood as a resource, nature was there for the taking by those with the will and abilities to do so. Simultaneously, in the nineteenth century transportation innovations allowed for food and other commodities to be shipped around the world in ways that incorporated peripheral locations into global markets for cotton, grain, and crops grown in many places (Hornborg, McNeill, & Martinez-Alier 2007).

Ideologies of class were interconnected with those of race, and assumptions about the virtue of labor on the part of the poor produced discourses that facilitated the abandonment of impoverished populations should climate and market conditions conspire to deprive people of the means of subsistence. As Mike Davis (2001) shows so vividly, such imperial ideologies coupled with then nascent sciences of meteorology allowed imperial administrators to evade their responsibilities to people in their charge during a series of late Victorian famines that radically disrupted many colonial societies and gave rise to assumptions of their being in need of development. Partly this was because the inhabitants were supposedly fated to suffer from a fickle nature, given their propensity to breed too quickly in conditions that apparently could not be relied on to provide subsistence. As Marsh had already pointed out, such thinking ignores the prior human disruptions of environments and economies, and allows for the framing of matters in terms of the influence of climate to the exclusion of prior human actions.

Simultaneously, these issues of imperial administration gave rise to many discourses of untouched pristine nature, of natural "Edens" that needed to be protected from human encroachment and mismanagement (Grove 1995). Notions of wilderness, and the need to protect it from human encroachment, gave rise to the national parks movement in North America as well as wildlife preservation efforts in many parts of Africa, where parks destined for animals frequently required the forceful removal of the human inhabitants. With hunting designated as poaching and traditional peoples reframed as a threat to nature when they tried to maintain their traditional modes of life, this dispossession was integral to the imperial management practices that used geographical designations to separate indigenous peoples from their traditional territories.

The romantic reaction to the brutality of industrialism in Europe and the rapidly urbanizing parts of North America was linked loosely to these formulations. While in part this was about aesthetics, it was also about the need for urban reform, and the provision of municipal sewage systems as well as parks and recreational spaces for the working classes, warehoused in cramped, crowded, and unsanitary housing in industrial cities. From here, town planning movements and park designs for cities were linked to the larger reimaginations of what was "natural" in contrast to human productions. Ebenezer Howard's (1899) "*Garden Cities of Tomorrow*" formulated town planning in terms of arranging things spatially so that the virtues of both city life and "nature" were incorporated directly into urban design. Hence, artificial environments could, so the arguments went, change the moral character of inhabitants, supposedly for everyone's benefit. It is not much of an exaggeration to argue that town planning was effectively artificial environmental causation on a small scale.

Environmental determinism

In the latter part of the nineteenth century and the first decades of the twentieth, doctrines of climate influence on human affairs gained traction in the academy in what in retrospect is frequently termed environmental determinism. Viewed through these imperial optics, nature was both something that had to be managed, albeit often by "managing" local populations, and also a source of both threats and possibilities that shaped the human condition. Ellsworth Huntington's *Mainsprings of Civilization* (1959 [1945]) suggests that much of the geographical diversity of humanity, and the rise of European power in particular, is related to climate factors. The supposedly beneficent circumstances of Europe were posed as key to the rise of civilization, in contrast to the apparent difficulties of severe tropical climates.

Friedrich Ratzel (1897), widely understood as the "founder" of political geography, was enamored with notions of the state and its growth as key to the human project. His concern with the expansion of states was heavily influenced by the imperial ethos of the late nineteenth century; while his analogies of states with organisms drew from crude interpretations of biology, he did look to environmental factors in developing his arguments about states and the limitations that resources might provide to their progress. His student Ellen Semple, who brought many of Ratzel's themes to North American audiences, was convinced of the importance of geographical context in understanding the rise of particular societies. One of her noteworthy texts (Semple 1911) was explicitly titled *Influences of Geographic Environment, on the Basis of Ratzel's System of Anthropo-Geography*.

Halford Mackinder, one of the most cited authors in classical geopolitics, also argued in various ways that geographical matters determined politics. Most famous for his pithy formulation that "Who rules East Europe rules the Heartland; Who rules the Heartland commands the World Island; Who rules the Heartland commands the World," Mackinder (1996 [1919]: 106) looked to environmental matters as key to the rise of empires. While his concerns with sea power and land power and the rivalries over Asia were key to his discussions of geopolitics, the environmental causation at the heart of his writing is a significant point that works to justify imperial modes of rule. Providing "natural" explanations for the rise of European empires has powerful political effects (Kearns 2009). If climate is destiny, then human responsibilities for the violence of colonization can be discounted. Malthusian fears of overpopulation can be invoked when farming systems fail to provide for the poor; it is nature's fault, not the failings of imperial administration.

In the aftermath of the Second World War, resources were a major concern in the superpowers' strategic planning. The huge military production systems that had brought victory to the Soviet Union and the United States had rapidly expanded the industrial capacities of both; in the case of the United States, its rapidly expanded international capacities were related to a global trading system and the possibilities of importing commodities from numerous places. The huge transformations of environments begun by the expansion of imperial power were accelerated by the world wars and then by the growth of what became accurately called the consumer society, first in the United States, and then, as automobile culture and suburbia spread, in many other parts of the world (McNeill 2000).

Determinist arguments, especially those connected to geopolitics and tainted by Nazi notions of *Lebensraum* (ideas sometimes associated with Ratzel), were largely abandoned. In the aftermath of the Second World War, the breakup of European empires and the self-determination of former colonies made arguments that supposedly explained the superiority of European civilization redundant too. New states looked to develop their societies rapidly by economic modernization and infrastructure-building; the colonial legacy had to be overcome quickly if modern states and the benefits of economic growth were to lift the former colonies from poverty. "Development" became the new priority. In the United States in particular, geography did not fare well in the early years of the Cold War. While geopolitical assumptions were embedded in the new discourses of national security and strategic studies, the earlier environmental causation arguments had little to say about the technological world set in motion by the rapid expansion of industrial forces. Above all, the Cold War geopolitical culture emphasized the possibilities of new technologies and the behavioral sciences to remake societies and the world (Farish 2010).

Environmentalism

The consequences of "development" generated environmental concerns that picked up some of Marsh's themes. Harrison Brown (1954) posed matters of food, energy, and resources in terms of *The Challenge of Man's Future*. The landmark conference in the mid-1950s on these themes was aptly called "Man's Role in Changing the Face of the Earth" (Thomas 1956). The accelerated transformation of the earth gave rise to a variety of political campaigns concerned to limit the destruction, which are now collectively known as environmentalism. Nevertheless, this was not about environmental influences on humanity, it was all about Glacken's third question, and Marsh's main concern, the deleterious effects of human actions on what was increasingly called "the environment," in terms of both the exhaustion of natural resources and the problems of pollution.

While there have long been legal attempts to protect people from what in today's terms is understood in terms of pollution, modern attempts to deal comprehensively with pollution problems are usually understood to have begun in British endeavors to deal with coal fire–generated smogs in London in particular in the 1950s (O'Riordan 1976). While such measures had been part of the political life of the capital city for centuries, the clean air measures finally worked to end the smog hazard from coal fires. Extended elsewhere, such technological innovations as clean-burning combustion have produced an approach to environmental matters that has come to be called "ecological modernization," under which ever more sophisticated technological innovations can supposedly eradicate pollution by cleanly producing products (Mol 2003).

The 1950s was the beginning of what earth system scientists now call the period of the great acceleration, when the global economy, increasingly powered by petroleum and subsequently natural gas, expanded rapidly (Steffen et al. 2011). This was also the start of the nuclear age, and while nuclear power plants never produced electricity that was too cheap to meter, as their early boosters suggested they would, nuclear technology provided both weapons of awesome destructive capabilities and a legacy of radioactive isotopes in the biosphere that threatened health around the world. Campaigns to curtail weapons testing in the atmosphere gave rise to the Partial Test Ban Treaty in 1963, which in part can be understood as the first global environmental treaty (Soroos 1997). The attention to technological hazards on a planetary scale emphasized the importance of pollution prevention; Rachel Carson's (1962) manifesto *Silent Spring* against the unrestricted deployment of pesticides, which galvanized people into action, dates from the same period.

In the 1960s, ecological models of animal populations and the possibilities of population crashes following the overexploitation of food supplies were used to raise the alarm about human population growth. Paul Ehrlich's *The Population Bomb* (1968) drew on field studies in animal populations in which overgrazing led to dramatic reductions in animal numbers to suggest that humanity faced disaster because of looming food shortages. Coupled with famine in India and international efforts to ship grain to the subcontinent at the time, fears of Malthusian futures galvanized political activists, leading in part in the United States to the first "Earth Day" in 1970, and two years later to the United Nations Conference on the Human Environment in Stockholm (Robertson 2012). The unofficial conference report warned that there was *Only One Earth* (Ward & Dubos 1972). Simultaneously, NASA photographs of the earth taken during the Apollo missions to the moon made the point graphically. These "earth rise" and "whole earth" images became iconic symbols of a vulnerable, isolated planet in need of protection (Cosgrove 1994).

While only a few national government heads made it to the conference in Stockholm, Indian Prime Minister Indira Gandhi grabbed the headlines with a speech suggesting that the most important pollutant needing attention was poverty, and that environmental efforts to constrain economic growth had to be resisted if they foreclosed development among the poorest nations of the world. Discussions of the "limits to growth" (Meadows et al. 1974) driven by environmental concerns were subsequently widely resisted, on the grounds that this was a ploy to maintain the privileges of the wealthy few at the expense of the majority of humanity. Finding ways to make development compatible with environment subsequently developed into the discussion of "sustainable development," codified by the World Commission on Environment and Development in 1987. Its report, entitled *Our Common Future*, echoed themes that ran through much of the earlier literature concerning the dangers of conflict caused by environmental shortages.

The links between warfare and global atmosphere were accentuated in 1983, when the first clear indications that nuclear warfare might have unforeseen climate effects generated discussions of "nuclear winter" (Turco et al. 1983). In turn, this fed rising concern about global warming and human alterations to the atmosphere, which were to give rise to the United Nations Convention on Climate Change at the Rio de Janeiro "Earth Summit" in 1992. Here, sustainable development became the watchword for the future, even if ambitious plans to tackle climate change and forest destruction have not lived up to expectations.

Determinism redux

Notwithstanding the rising concerns with climate change and the growing recognition that abrupt change in coming decades is likely, themes of environmental determinism have remained and reappeared, especially in popular literature dealing with world politics. Despite the contemporary scholarship dealing with political ecology in specific places and global environmental change generally, ideas that geography is destiny and that politics is shaped by a given natural environment persist, especially in American thinking about geopolitics. However, as with the blanket denunciations of earlier formulations of geography and environmental causation, caution needs to be exercised in dismissing sometimes complicated arguments. Climate matters greatly in human affairs, and has, by periodically causing disruptions and food shortages, shaped the modern world in many ways. Yet, as contemporary environmental history shows, this has not happened in the geographical patterns that the determinist arguments supporting empire used to suggest (Parker 2013).

In the last two decades in particular, Jared Diamond's long books on environment and world history have stirred controversy, and claims that he has reinvented determinism have been widespread (Correia 2013). Nonetheless, it is easy to read Diamond's books as a more complicated argument reviving possibilism rather than determinism. His discussions of the domestication of animals in Eurasia (Diamond 1997), and the emergence of diseases such as small pox from the interactions of people and cattle, suggest that the shape of civilization was influenced by the available fauna and flora, but that what humans did in the process of domestication was to remake nature in many ways. When Europeans subsequently encountered American populations, their diseases destroyed native societies by killing them off en masse. Europeans had much less success in Africa and Asia, where diseases worked against easy conquest.

His subsequent analysis (Diamond 2005) of the environmental factors in the collapse of various societies frequently suggests that they were remarkably persistent in the face of great difficulties, but also that in some cases political considerations prevented adaptations that would have been necessary for survival. Diamond argues that such historical investigations provide lessons that are applicable to the global situation that humanity now faces; that collapse is not inevitable; and that politics matters crucially in remaking the increasingly artificial environments that humanity now inhabits. However, he warns that there are constraints on human possibilities that have to be taken seriously; not all options are available in the biosphere.

More directly linked to matters of geopolitics are the ruminations of Robert Kaplan (2012). His popular books return repeatedly to the geopolitical writers of earlier generations to invoke geographical verities as the context for political action. Rather than looking to the future and the transformations that an urbanized global economy has made to the geographical context, he prefers to look back to Mackinder in particular, as well as to other writers who argued that European and American dominance were provided by the superior environmental contexts that shaped their societies. While Kaplan sometimes recognizes the importance of climate change and the growth of Asian powers (Kaplan 2010), nonetheless the larger transformation of human circumstances is downplayed in favor of discussions of the rivalries of great powers and the invocation of geographical factors as destiny.

While this might all seem rather quaint and nostalgic in the twenty-first century, it remains a powerful series of rhetorical devices in the hands of national security policymakers in Washington. Many strategic thinkers seem more comfortable with the military verities of earlier times than the uncertain situations presented by the transformations of globalization. Insisting that geography is a permanent state of affairs, rather than a matter that is increasingly being transformed by the global economy, reprises tropes of national rivalry and inevitable conflict among political elites (Dalby 2014). Thus, geographical language is substituted for political analysis, and in the process obscures the consequences of prior human dislocations by evading historical investigation. The increasingly artificial circumstances of the present make such reasoning both politically dangerous as well as intellectually dubious; it completely ignores contemporary analyses of both geography and environmental politics (Death 2014).

The Anthropocene

Earth system science has made it clear that the sheer scale of human activity is transforming many of the geological parameters of the biosphere (Steffen et al. 2004). Humanity is now the largest geomorphic agent shaping the configuration of the physiognomy of the planet. What in Marsh's analysis the Roman empire did to the Mediterranean area of the earth, industrial humanity is now doing to the whole world, and doing so very quickly.

This is a matter not only of climate change, which gets most of the attention in the early twenty-first century, but of many other simultaneous transformations too, such that the world's biosphere is now entering a new phase (Barnosky et al. 2012). The list of transformations that, combined, are causing this phase shift is long, as documented by a series of reports on the *Global Environmental Outlook* (United Nations 2012). The dramatic changes include the construction of vast urban spaces; concrete and asphalt are effectively new geological strata. The nitrogen cycle has been changed by the widespread artificial production of nitrogen-based fertilizers, which have been key to the rapid land use changes caused by the expansion of agriculture, which has in turn fed the swift growth of urban populations (Ellis 2011). The massive pollution of the oceans and the simultaneous overfishing of marine systems that threaten their ecological functioning are now being exacerbated by the acidification of the ocean due to rising carbon dioxide levels in the atmosphere. The extirpation of numerous species on land, mainly due to habitat change as farming and plantation forestry expand, although partly through hunting too, is causing a global extinction event on the scale of earlier ones in geological history (Kolbert 2014). Terrestrial river systems are being completely reengineered by dams, irrigation systems, canals, and urban water use arrangements. Many river waters no longer consistently reach their estuaries.

Put together, all these human changes are of such a scale that geologists are now discussing the present in terms of a new geological epoch called the Anthropocene (Steffen et al. 2011). Some of the changes, especially climate change, may bring very dangerous consequences for at least some parts of humanity, although precisely how these things will play out is uncertain, despite the extraordinary efforts of climate modelers to address such questions (Edwards 2010). Extreme events are unfolding in the increasingly artificial circumstances of urban humanity; floods, tropical storms, and droughts are apparently becoming more frequent, with human casualties and economic disruptions the consequence (Scheffran et al. 2012).

While these events have got the attention of the financial and insurance industries in particular (Potsdam Institute 2012), given both the loss of life and the disruption to economic activities, it is also clear that, at least so far, political leaders have been reluctant to move to dramatically reinvent energy supplies and decarbonize industrial economies and urban systems. Despite 20 years of discussions under the auspices of the United Nations Framework Convention on Climate Change, formulated at the Rio "Earth Summit" conference in 1992 and subsequently ratified by most states in the UN system, the accumulation of carbon dioxide in the atmosphere continues to rise relentlessly. This is now leading to discussions of possible technical fixes to "engineer" the climate, given the failure of negotiations to establish an international agreement to curb emissions, never mind reduce greenhouse gas levels (Hamilton 2013). Climate change is more and more shaping politics as the realization of the increasingly artificial vulnerabilities that face humanity spreads in academic and policy discussion (Dalby 2009).

Once again, climate is important in geopolitical reasoning, although now, unlike the discussions a century ago, it is very clear that this is not about natural influences on particular societies: It is about the artificial consequences of modernity at the global scale. Environment is now all to do with Glacken's third question; in the Sprouts' terms, humanity is remaking its milieu. Political rhetoric that is used to justify contemporary geopolitics may suggest once again that the climate "requires" certain forms of violent human conduct in the face of contemporary disruptions that threaten metropolitan consumption (Welzer 2012). Yet, unlike earlier historical invocations of climate as a determinant of human behaviour, now climate change and its consequences are clearly understood as something that humans have caused. George Perkins Marsh's concerns with the human impacts on nature a century and a half ago have turned out to be prescient, albeit, as climate change now indicates, on a scale very much larger than even his historical discussion suggested.

References

Agnew, J., & Muscara, L. (2012). *Making Political Geography*. Plymouth: Rowman and Littlefield.
Barnosky, A.D., Hadly, E.A., Bascompte, J., et al. (2012). Approaching a state shift in Earth's biosphere. *Nature* 486(7 June): 52–58. doi:10.1038/nature11018
Brown, H. (1954). *The Challenge of Man's Future*. New York: Viking.

Carson, R. (1962). *Silent Spring*. Boston, MA: Houghton Mifflin.

Correia, D. (2013). F**k Jared Diamond. *Capitalism Nature Socialism* 24(4): 1–6. doi:10.1080/1045575 2.2013.846490

Cosgrove, D. (1994). Contested global visions: One-world, whole-earth, and the Apollo space photographs. *Annals of the Association of American Geographers* 84(2): 270–294.

Dalby, S. (2009). *Security and Environmental Change*. Cambridge: Polity Press.

Dalby, S. (2013). The geopolitics of climate change. *Political Geography* 37: 38–47.

Dalby, S. (2014). Rethinking geopolitics: Climate security in the Anthropocene. *Global Policy* 5(1): 1–9. doi:10.1111/1758-5899.12074

Davis, M. (2001). *Late Victorian Holocausts: El Nino Famines and the Making of the Third World*. London: Verso.

Death, C. (ed.) (2014). *Critical Environmental Politics*. New York: Routledge.

Deudney, D. (1999). Bringing nature back in: Geopolitical theory from the Greeks to the global era. In Deudney, D., & Matthew, R. (eds), *Contested Grounds: Security and Conflict in the New Environmental Politics*, Albany, NY: State University of New York Press, 25–57.

Diamond, J. (1997). *Guns, Germs and Steel*. New York: Norton.

Diamond, J. (2005). *Collapse: How Societies Choose to Fail or Succeed*. New York: Viking.

Edwards, P.N. (2010). *A Vast Machine: Computer Models, Climate Data, and the Politics of Global Warming*. Cambridge, MA: MIT Press.

Ehrlich, P. (1968). *The Population Bomb*. New York: Ballantine.

Ellis, E.C. (2011). Anthropogenic transformation of the terrestrial biosphere. *Philosophical Transactions of the Royal Society A* 369: 1010–1035.

Farish, M. (2010). *The Contours of America's Cold War*. Minneapolis, MN: University of Minnesota Press.

Glacken, C. (1967). *Traces on the Rhodian Shore*. Berkeley, CA: University of California Press.

Gregory, D. (2001). (Post)colonialism and the production of nature. In Castree, N., & Braun, B. (eds), *Social Nature: Theory, Practice, and Politics*, Oxford: Blackwell, 84–111.

Grove, R.H. (1995). *Green Imperialism: Colonial Expansion, Tropical Island Edens, and the Origins of Environmentalism, 1600–1800*. Cambridge: Cambridge University Press.

Hamilton, C. (2013). *Earthmasters: The Dawn of the Age of Climate Engineering*. New Haven, CT: Yale University Press.

Hornborg, A., McNeill, J.R., & Martinez-Alier, J. (eds) (2007). *Rethinking Environmental History: World System and Global Environmental Change*. Lanham, MD: Altamira.

Howard, E. (1899). *Garden Cities of Tomorrow*. London: Faber.

Huntington, E. (1959 [1945]). *Mainsprings of Civilization*. New York: Mentor.

Kaplan, R.D. (2010). *Monsoon: The Indian Ocean and the Future of American Power*. New York: Random House.

Kaplan, R.D. (2012). *The Revenge of Geography: What the Map Tells Us about the Coming Conflicts and the Battle against Fate*. New York: Random House.

Kearns, G. (2009). *Geopolitics and Empire: The Legacy of Halford Mackinder*. Oxford: Oxford University Press.

Kolbert, E. (2014). *The Sixth Extinction: An Unnatural History*. New York: Henry Holt.

Mackinder, H.J. (1996 [1919]). *Democratic Ideals and Reality: A Study in the Politics of Reconstruction*. Washington, DC: National Defense University. Original London: Constable.

Marsh, G.P. (1965 [1864]). *Man and Nature: Or, Physical Geography as Modified by Human Action*. Cambridge, MA: Belknap Press.

McNeill, J.R. (2000). *Something New under the Sun: An Environmental History of the Twentieth Century World*. New York: Norton.

Meadows, D.H., Meadows, D.L., Randers, J., & Behrens, W.W., III (1974). *The Limits to Growth*. London: Pan.

Mol, A. (2003). *Globalization and Environmental Reform: The Ecological Modernization of the Global Economy*. Cambridge, MA: MIT Press.

O'Riordan, T. (1976). *Environmentalism*. London: Pion.

Parker, G. (2013). *Global Crisis: War, Climate Change and Catastrophe in the Seventeenth Century*. New Haven, CT: Yale University Press.

Potsdam Institute for Climate Impact Research and Climate Analytics (2012). *Turn Down the Heat: Why a 4°C Warmer World Must Be Avoided*. Washington, DC: World Bank.

Ratzel, F. (1897). *Politische Geographie*. Munich: R. Oldenbourg.

Robertson, T. (2012). *The Malthusian Moment: Global Population Growth and the Birth of American Environmentalism*. New Brunswick, NJ: Rutgers University Press.

Scheffran, J., Brzoska, M., Brauch, H.G., Link, P.M., & Schilling, J. (eds) (2012). *Climate Change, Human Security and Violent Conflict: Challenges for Societal Stability*. Berlin: Springer-Verlag.

Semple, E.C. (1911). *Influences of Geographic Environment, on the Basis of Ratzel's System of Anthropo-Geography*. New York: Henry Holt.

Soroos, M.S. (1997). *The Endangered Atmosphere: Preserving a Global Commons*. Columbia, SC: University of South Carolina Press.

Sprout, H., & Sprout, M. (1965). *The Ecological Perspective on Human Affairs*. Princeton, NJ: Princeton University Press.

Steffen, W., Sanderson, A., Tyson, P.D., et al. (2004). *Global Change and the Earth System: A Planet under Pressure*. Berlin: Springer.

Steffen, W., Grinewald, J., Crutzen, P., & McNeill, J. (2011). The Anthropocene: From global change to planetary stewardship. *Ambio* 40: 739–761.

Thomas, W. (ed.) (1956). *Man's Role in Changing the Face of the Earth*. Chicago, IL: University of Chicago Press.

Turco, R., Toon, O.B., Ackerman, T.P., Pollack, J.B., & Sagan, C. (1983). Nuclear winter: Global consequences of multiple nuclear explosions. *Science* 222: 1283–1292.

United Nations Environment Programme (2012). *Global Environmental Outlook GEO5: Environment for Development*. Nairobi: UNEP.

Ward, B., & Dubos, R. (1972). *Only One Earth*. Harmondsworth: Penguin.

Welzer, H. (2012). *Climate Wars*. Cambridge: Polity Press.

World Commission on Environment and Development (1987). *Our Common Future*. Oxford: Oxford University Press.

Chapter 34

Financial Crises

Brett Christophers

Uppsala University, Sweden

This chapter takes as its point of departure two observations on the eurozone's high-profile travails offered, in 2012, at what was the very height of the region's own particular manifestation of the wider global financial crisis that began in 2008. "The eurozone financial crisis," Stuart McMillan claimed in April 2012, "has turned, remarkably quickly, into a major political crisis." Two months later, Faisal Islam put an alternative twist on this nexus of politics, finance, and crisis: "This is a political crisis, not a financial crisis," he averred (cited in Wedeman 2012).

What are we to make of these observations? Had a financial crisis indeed turned into a political crisis? Or was it in fact simply a political one all along? The argument advanced in this chapter is that neither statement gets things quite right, either for the eurozone crisis specifically or for financial crises – our topic of concern – more generally. Financial crises cannot "turn into" political crises because, the chapter maintains, such crises are inherently political, by their very nature, from the outset. To present things in either/or terms, à la Islam, is thus also misleading, for such crises are financial *and* political. Not only that, but the geography of such crises is intrinsically material to their political-financial constitution and dynamics. In sum, the chapter's argument is that *financial crises are always and everywhere crises of political geography*. It demonstrates this both empirically and conceptually, and by moving iteratively between these two levels.

In doing so, it recognizes and aims to underline the veracity and importance of Agnew's (2012: 568) argument that "a fuller focus on politics is very much in the interest of making [economic geography] more realistic in accounting for the very economic phenomena that remain its focus of description and analysis." Equally, however, this chapter would lay a comparable explanatory burden at the door of *political* geography, insofar as it insists that a fuller focus on economy in general – and finance in particular – is indispensable to making *that* subdiscipline more realistic and thus more relevant. The chapter's strategic approach to substantiating both of these propositions is to view the question of financial crises and their political geographies through the lens of (geographical) political economy (Sheppard 2011;

The Wiley Blackwell Companion to Political Geography, First Edition.
Edited by John Agnew, Virginie Mamadouh, Anna J. Secor, and Joanne Sharp.
© 2015 John Wiley & Sons Ltd. Published 2017 by John Wiley & Sons Ltd.

Mann 2012). The nature of this lens will become clearer as the chapter progresses, but two linked dimensions warrant foregrounding. First, it conceives of political economy much more along "classical" lines – where the concern is with the structural and systemic dynamics of capitalism as a mode of socio-economic production and reproduction – than those typically trodden by contemporary "international" or "comparative" political economists. And it does so because, second, it fully concurs with Charles Gillespie's (1988: 351) injunction:

> Political economy should be more than economics with a sprinkling of references to policies, and more than political sociology with a bias towards explanations rooted in classes and economic sectors.

If political economy is its preferred optic, what, then, is the chapter's subject matter? What *is* (a) "financial crisis"? The label is deceptively simple. As Mary Poovey asks:

> Does such a crisis emanate from an insufficiency in the money *supply*? Is it the result of a decrease in the *value* of money? Does it reflect the loss of value of *banks* and other financial *institutions*? … Is a financial crisis simply a function of *confidence* – or the lack thereof? Does it come from too little monetary *liquidity*? Is "a stock market crash" an *example* of a crisis or the cause – or possibly the result – of one? (2012: 140, emphasis in original)

At the risk of being accused of evasiveness, this chapter's provisional answer is: A financial crisis can be all, or any, of the above. Secondarily, however, it would also posit, again after Poovey (2012: 140), that a financial crisis invariably entails significant diversion from a more or less established historical pattern of financial dynamics and structures, not least inasmuch as "no one – not even the most seasoned Wall Street trader, not even a Nobel laureate – can identify financial or commercial activities as a 'crisis' without some context that establishes a norm, which the crisis is then said to violently disrupt." And thirdly, it should be evident to all readers, though it often seems to be forgotten by influential commentators and politicians alike, that the "reality" of financial crisis is always subjective as well as socially and spatially varied (Christophers 2015).

Finally, a word is in order about the empirical scope of the chapter. While it ranges relatively broadly across historical and geographical examples of financial crises, its primary focus is on the most recent global financial crisis, and especially on its crystallization in the eurozone. Given space constraints, selectivity is essential, and the most recent crisis justifies prioritization on numerous grounds. Partly these could arguably be described as superficial – the scale of the crisis makes it stand out, its recentness (and, for many, tangible impact) makes it familiar to readers – and indeed, while there is a temptation for each new crisis to be seen as unique, in reality financial crises distributed widely across time and space often share much in common (Reinhart & Rogoff 2009; Christophers 2014). Yet, as Poovey (2012: 142), among others, observes, there *is* something distinctive about the recent financial crisis that marks it out for particular scrutiny, for "while there have been many banking crises in the past, as well as many liquidity freezes and curtailments of credit," a financial crisis of the kind witnessed since 2008 has only been *possible* since the late 1970s and especially since "banks began to market financial instruments whose primary purpose was to facilitate speculation about future changes in the prices of financial instruments." Why? Because such instruments "amplify the swings in supply and demand, greed and fear, that characterize the operations of any commercial market," and because of "the speed and volume with which they can be traded by investors" (Poovey 2012: 142).

The chapter starts, empirically, with the issue of the causes of financial crises, before turning to political responses to crisis and the implications of crisis for political structures and their territorial expressions. It then considers the salient fact that financial crises very often are – or become – crises in the finances of political institutions themselves. The chapter then takes a more conceptual turn, asking why it is that finance and politics are so intimately intertwined, not only in relation to crisis conditions but much more generically. In answering this question it encourages us to think closely and critically about the nature of the socio-economic institution that is the lifeblood of finance: money. Last, in attempting to tie together its various threads, the chapter returns to where it began, and the emblematic case that is the geographical political-financial crisis of the contemporary eurozone. There follows a brief conclusion, in which another of the chapter's starting points – the question of what actually constitutes a financial crisis in the first place – is critically revisited in the light of its key findings.

Geographies of crisis in the political-financial nexus

"Almost every financial crisis," argues Raghuram Rajan (2010: 7), "has political roots, which no doubt differ in each case but are political nevertheless." This observation in itself appears to offer compelling grounds for emphasizing the political dimensions of financial crises. And while it is open to question whether the *roots* of (most) financial crises are indeed of a political nature, it is certainly the case, this chapter suggests, that it would be hard to explain any financial crisis without reference to contributory political-geographical factors. The most recent global financial crisis is a case in point.

Consider first of all the historical regulatory context within which the core ingredients of the US subprime fiasco – the crucible of the wider financial crisis – were brewing in the years and decades leading up to 2007. It is now widely recognized that increasingly lax lending standards played a pivotal role in encouraging the emergence of the financial instabilities that subsequently precipitated this housing-focused meltdown. However, how should we understand this relaxation of lending standards? Wallison (2009) makes the case that explicitly political initiatives ultimately underpinned this development. And if the historical weakening of regulation is explained by politics, so too, argues Thompson (2009), is the failure, in latter years, of eleventh-hour efforts to strengthen such regulation. In short, the US case appears to offer a striking example of the significance of what Nolan McCarty and co-authors, in the same geographical context, have termed "political bubbles" (McCarty, Poole, & Rosenthal 2013): the blowing of financial bubbles through market behaviors predisposed to generate instability and specifically fostered by political projects and ideologies.

More generally, Calomiris and Haber (2014) argue that the incidence and nature of financial – or in their treatment, specifically banking – crises cannot help but be politically shaped. Banks, they point out (2014: x, 4), represent "an institutional embodiment – a mirror of sorts – of the political system," their respective "strengths and shortcomings" being the product of "political bargains" that are themselves structured by the "fundamental political institutions of a society" and by the distribution of power within it. The terminology "political bargains" is carefully chosen here, for the authors claim, convincingly, that in formatting and operating monetary and financial systems the state faces endemic conflicts of interest that it can never entirely overcome. They see three main types thereof: first, states regulate banks, but also often rely on them for funding; second, states enforce debt contracts, while relying on debtors (among others) for political support; and third, states also rely for such support on banks' creditors (including, notably, depositors), but meanwhile are frequently required to allocate losses among them in the event of bank failures. Calomiris and Haber's study shows that through history different countries have suffered banking crises of widely varying intensity and with widely varying frequency; the United States, for example, has been "highly crisis prone" (2014: 5), featuring 14 major crises over the past 180 years, in stark contrast to, say, Canada's 2 minor ones. Whether this variance can be laid largely at the door of the aforementioned political bargains and their degree of success in overcoming conflicts of interest is moot; what are not in doubt, however, are the existence and materiality of such conflicts and bargains.

None of this emphasis on the political sphere, needless to say, is to absolve private-sector actors of responsibility for financial crises. Indeed, the argument could be made that even where responsibility seemingly lies in the political and regulatory sphere, private-sector interests are nonetheless heavily implicated, to the extent that they can be shown to have effectively "captured" politics and the regulatory process. The close links between the commanding heights of politics and finance in numerous countries are legion, and are manifested in the empirical facts of, inter alia, campaign financing and the revolving doors between the finance industry on the one hand and political parties and regulators on the other. Whether in David Harvey's scathing denunciation of "the Party of Wall Street" (2010: 11) or Johnson and Kwak's (2010) 13 Bankers, the United States, in respect of the most recent financial crisis, once again represents a case in point. The broader relevance of "regulatory capture" to our understanding of the crisis is now a well-established theme (Etzioni 2009; Baker 2010). Igan, Mishra, and Tressel (2011) have even shown that insofar as those US financial institutions that were the heaviest lobbyists took – or were allowed to take? – the greatest risks in the period 2000–07, such lobbying contributed directly to the crisis. Not only are financial crises political phenomena, the implication seems to be, but the explanation of such crises lies – at least in part – in the merging *of* politics and finance.

There were and are, moreover, vital geographies to this merging. Eric Helleiner (2011: 76) offers in this regard a very helpful summary of the work of prominent scholars of international political economy – Mark Blyth, Jonathan Kirshner, and Robert Wade among others – in demonstrating that the power of private interests to shape regulation did indeed have a distinctive political geography:

> Because of the international importance of their financial markets, the United States and Britain had unique power to determine international regulatory outcomes … They also had strong representation and influence in many of the key international forums in which regulatory issues were discussed. U.S. and British support for precrisis regulatory developments was attributed partly to the strong influence in these countries of private financial interests … [S]cholars also argued that policy makers in these states believed that these trends would benefit their states disproportionately.

Meanwhile, even where politics is not actively implicated in the origination of financial crisis, governments can scarcely not concern themselves with crisis fallout, and politics can hardly not shape the response *to* crisis. Why might this be the case? There are all manner of reasons why financial crisis management is always overtly political. Partly, such politicization is a function of what we have already seen; namely, the deep interpenetration of politics and finance at a functional level. With the recent growth in the phenomenon of sovereign wealth funds (SWFs), such interpenetration is arguably more pronounced than at any stage in recent history. As Clark, Dixon, and Monk (2013: 46) note, much of the lively debate over such funds has focused on the "geopolitics of investment: the extent to which SWFs are the instruments of their sponsoring nation-states with certain geopolitical goals and objectives."

More fundamentally, the modern state has ordinarily recognized – although admittedly, it has occasionally overlooked the fact – that finance is far too important to be left to financiers. Viewed alongside other sectors of the economy, finance is regarded as unique and vital "in the sense of both mediating wider rhythms of capital accumulation and serving as a principal site of regulatory leverage within the economy at large" (Christophers 2013: 9). As a consequence of this presumed vitality, the finance sector, as Cohen (1988: 12) has observed, "is typically subject to more regulation than other sectors of the economy." The state, that is to say, has usually – at least under capitalism – been actively involved in configuring finance-sector structures and dynamics, often to the point of taking full ownership and control. Much of its concern is based on a perceived need for financial stability (to allow smooth capital circulation and economic and social reproduction more broadly), and this concern inevitably intensifies at moments of crisis and thus of financial stress. Closely concerned to ensure stability even at the best of times, it would be inconceivable for the state not to intervene to such an end in the event of crisis irruption.

Perhaps most importantly of all, however, in the sense of being specific to the question of crisis, the management of crisis is perforce (geo)political because its effects are always (geographically) uneven (Kitson, Martin, & Tyler 2011). Different social factions, located in different places, are inevitably affected by crisis in different ways. Add to that the fact that different social factions, again often in different places, will typically be seen to bear different degrees of responsibility for causing the crisis in the first place, and one has a ready-made cocktail of inherently political contestation to shape crisis response. Such response simply can never be entirely consensual. The case of the contemporary eurozone, to which we return later, is an obvious exemplar. So too, however, is the contemporary United Kingdom, where most commentators have discerned a marked geopolitical divide between a (City of) London identified as responsible for the recent crisis and a United Kingdom excluding London bearing the brunt of its aftermath (see especially Ertürk et al. 2011).

The ways in which politics and political geography shape and couch the response to crisis are, of course, myriad, and it is impossible to generalize. The literature is full of examples, ranging from the national to the international scale. Woodruff (2005), for instance, provides a fascinating study of how a political desire to avoid harming powerful business interests influenced – almost certainly for the worse – management of currency-focused financial crises in Russia and Argentina in 1998 and 2001, respectively. In the Argentinian case, Woodruff's thesis extends those developed previously by scholars such as Javier Corrales, who (2002: 39) had argued shortly after the crisis began that even if the "onset of Argentina's recession had both economic and political causes," its "longevity and depth have had mostly political causes," of which one was "infighting between the executive and the ruling party." This infighting constituted one of two "political shocks" – we shall encounter the second shortly – that "killed convertibility" (of the peso into the US dollar at a fixed exchange rate, a policy in place since 1991), thus turning "a mild recession, in a country whose economic fundamentals were no worse than those of many other Latin American nations, into a massive depression" (Corrales 2002: 39).

Crisis management, then, is intimately (geo)political. Related to this, and indeed partly because of it, it is clear that financial crises can also directly cause – or, at least, create the space and impetus for – not only national political-economic models to be revisited and rewritten, but regional or even global configurations of political-economic power to be transformed. These two outcomes/possibilities are, furthermore, inevitably connected to one another.

If we start with the question of domestic political-economic models, the first critical insight is that financial crises can and do precipitate all manner of questioning of the status quo. How and why did "we" end up in this mess? Was and is it not the responsibility of the state to ensure financial stability? Why did our government make the decision to open up our borders to international capital flows when history shows just how volatile such flows can be, and just how threatening to domestic financial robustness? All such questions have been raised once again, and with a renewed urgency, in the wake of the most recent global financial crisis, but they have a long lineage. They were, for instance, and returning to the Argentinian case, asked widely and forcefully in the wake of financial crisis not only in the early 2000s (Epstein & Pion-Berlin 2006) – when such a crisis elicited a widespread fall in public confidence in domestic political institutions, the period between mid-2000 and late 2002 seeing enormous popular protest and two presidential resignations – but also in the early 1980s (Hartlyn & Morley 1986), which notably was likewise a crisis, as the work of indigenous writers such as Guillermo O'Donnell and Marcelo Cavarozzi has shown, with deeply political roots.

Scholarly studies have demonstrated, however, that if and when political-economic models are revised in the light of such questioning, it is imperative that we understand such revision in relation to wider geopolitical forces. As Öniş (2009) emphasizes, crises do indeed present opportunity spaces for domestic political change, but always in the context of more broadly based political geographies. His example, now resonant more than ever, is Turkey, where political-economic restructuring in the 2000s occurred against the backdrop not only of financial crises (in 2000 and 2001), but also of a long-running bid to accede to the European Union (EU). Along similar lines, Clark and Jones (2012) critically examine prospects for – and internal elite discourses of – political reconfiguration in present-day Iceland, where the global financial crisis caused a catastrophic loss of public trust and political legitimacy, and where the issue of EU accession is no less politically acute. However, once again Argentina specifically, and the global South more generally, are especially illuminating. Wider geopolitical forces – namely, "the toughen-as-you-sink policy experiment undertaken by the IMF and the U.S. Treasury in 2001" – represented the second of Corrales's (2002: 39) political shocks deepening the 2001–02 Argentinian crisis. Meanwhile, Corbridge (1984), drawing on the ideas of Riccardo Parboni, had earlier argued that US geopolitical interests substantially explained the slow progress toward "resolution" of the international debt crisis in the global South in the early 1980s.

The fact that wider political-geographical considerations and constituencies are often implicated in the political change associated with financial crisis is, needless to say, vital, for it speaks to the most fundamental of political issues: issues such as sovereignty, representation, and democracy. The ongoing crisis in the eurozone has raised these issues anew. Hare (2012: 461; cf. Derviş 2013; Streeck 2014) points to the "surrendering of national sovereignty over fiscal policy in exchange for financial assistance" in questioning the democratic legitimacy of euro-area sovereign bailouts and the role of a technocratic economic directorate (the "Troika"), with, says Watkins (2013), "no legitimation beyond the emergency itself"; while Pianta (2013), in the same context, writes more definitively of "democracy lost." However, there are also precedents from elsewhere from which to learn. Aihwa Ong (2000) observes that the Asian financial crisis of 1997–98 triggered changes to the model of "graduated sovereignty" favored by certain Southeast Asian states, with Indonesia and Thailand heading in one political direction and Malaysia in another – the former both submitting "to the economic prescriptions of the IMF," the latter resisting and instead "reimposing its territorial state sovereignty" (2000: 69). Lee (2003: 892–893) assesses South Korea's experience in similar terms to Ong's account of Indonesia/Thailand, identifying the conditionality attached to the IMF's "rescue" loan as an erosion of sovereignty insofar as it covered "policies and decisions that were normally within the purview of purely domestic governance" (2003: 877).

Whether connected to changes in domestic political models or not, transformations in regional or ultimately even global configurations of political-economic power are often also routinely mooted as actual or potential consequences of financial crises. Again, the basic reasons should be clear: Crises can and do weaken some states more than others (among whom there may even be direct beneficiaries), while also highlighting discrepancies between perceptions and realities of states' political-economic strength. This is why it has conventionally been assumed that financial crises serve to strengthen regionalism as states seek out alliances for their "defensive potential against the vagaries of global capitalism" (Phillips 2000: 385). Phillips does, however, dispute this particular reading, arguing that although one may see stronger regionalism in the longer term, in the short term "possibilities for collective action are weakened and regional projects are as likely to fragment as to coalesce" (2000: 384–385; cf. Grimes 2012). Once more, it is hard to avoid thinking of the contemporary eurozone here.

Meanwhile, at a much wider geographical scale, there is no shortage of commentators prognosticating on the possible ramifications for global power balances, and global governance regimes, of the most recent financial crisis, even though more cautious voices warn that "it is undoubtedly premature to attempt to reach a verdict on the overall impact and significance of the financial crash of 2007–09, in terms of a new or future political economy" (Baker 2013: 112). Both Woods (2010) and Xinbo (2010), for example, foresee the emergence out of the crisis, in coming decades, of multilateral/multipolar governance and power structures, with US geopolitical preeminence in decline; and arguably the G20's displacement of the G7/G8 since the crisis (Cooper 2010) represents a harbinger of things to come.

Before turning, in the next section, to a more conceptual perspective on financial crisis and political geography, one final important observation of an empirical nature requires elaborating. For, in a chapter on crisis, finance, and politics, it is evidently significant that many financial crises are crises explicitly of political finance – of, that is to say, *political* institutions and their financial viability. This is true at a range of geographical and temporal scales. One of the most high-profile financial crisis stories of 2013 – a year not short on such tales – has been that of the US city of Detroit, which, in declaring bankruptcy in July of that year, became the largest US city ever to do so (see, for example, The Economist 2013). With many US city centers having seen their tax base eviscerated by deindustrialization and suburbanization in the postwar era, municipal fiscal crises have been a recurrent feature of the US urban political economy for more than four decades. This history pales, however, beside the much longer history of fiscal crises at the scale of states – whether city-states or, more commonly, nation-states. Ever since the initial emergence of public debt through the issue of bonds by the city-states of northern Italy in the fourteenth century, states have been serial debtors and defaulters, with the sovereign debt crises of today's eurozone thus representing only the latest in a very long line of materializations of this crisis phenomenon.

A great deal could be said about the political geographies of financial crisis in this particular guise, not least concerning the indissolubility of politics and finance that it entails. However, perhaps most important for our purposes here is the fact that financial crises frequently *end up* being crises of public (and thus political) finances, even if they do not start out that way. Consider the following, startling statistic from Reinhart and Rogoff's exhaustive study of the 800-year history of financial crises around the world: "On average, government debt rises by 86 percent during the three years following a banking crisis" (2009: xxxii). To be sure, correlation does not imply causality, and rises in government debt do not always assume crisis proportions; but it would be hard to find more suggestive empirical evidence of the links between crises in the finances of nominally "financial" institutions on the one hand and nominally "political" institutions on the other.

The past six years have, of course, shone an especially harsh light on those links. What began, with subprime and the credit crunch, as a financial crisis for banks (but also, of course, for subprime households being foreclosed on) soon became a financial crisis for numerous nation-states, Ireland and Spain among them. Two mechanisms of "transmission" appear to have been material. First, there have been cases of direct transmission, where the state's bailing-out of insolvent banks, through stake-taking or otherwise, literally socializes accrued losses and transfers liabilities to the public sector. Second, and less transparent, are mechanisms of indirect transmission, where crisis in the financial sector generates recessionary tendencies, with concomitant implications for tax receipts and state finances. In each case, the political dimensions of finance and financial crisis are illuminated in a crucial, incontrovertible fashion.

Structural political-economic foundations

How, therefore, might we strive to understand all of this conceptually? We have essentially seen that wherever and however one may care to look – whether it be at the causes of financial crisis, the responses to it, its implications for configurations of power, or the identity of the entity experiencing crisis – the political geographies of financial crises tend to be writ large. Nevertheless, are there generic, deeper structures of the political-geographical economy to which we can appeal in order to grasp more clearly this perennial emergence of the political in the financial? This section argues, relatively summarily, that there are.

We can go some way to appreciating this by recognizing first of all that our preceding emphasis on crisis is, in a sense, a distraction, to the extent that it is merely contingent. Financial crisis may "have" characteristic political geographies, but in reality it merely intensifies – and/or brings into sharper focus – a much more deep-seated and fundamental intertwining of finance and politics, money and state. Indeed, this is true not only of crisis but of those other phenomena to which commentators often point in order to highlight such intertwining. Sovereign wealth funds are a good example here. In reference to these, Clark et al. (2013: 2) write that "the prospect of political localities being motivated by the same market incentives as the firms located within these jurisdictions augurs a degradation of the boundaries that previously separated states from markets." Still, the word "previously" is a misnomer. Markets and states, finance and politics, have *never* existed separately from one another. For finance per se, and not just financial crisis (or financial entities like SWFs), is always and everywhere politically geographically constituted.

What, first of all, *is* finance – and therefore what is "financial crisis" a crisis in and of? Consult only the most rudimentary of dictionaries, and one is confronted with a welter of definitions: inter alia, finance can comprise monetary resources or funds; it can be the supplying of such funds; it can be the management of these funds; and it can be the "science" of such management. However, at the core of all such definitions is the "stuff" of finance in its various incarnations: money. Finance is thus a heterogeneous constellation of assets and practices that cohere around money and that realize and extend its materiality in the world. Financial crisis, in turn, represents crisis in the ongoing social viability and credibility of such assets and practices.

So if finance coheres around money, what is money? This is not an easy question to answer. Indeed, few questions elicit such a wide range of responses. This chapter, however, makes the case for a broadly sociological theory of money along the lines of that advanced by Geoffrey Ingham (2004), in contra-distinction to the "commodity" theories of money more closely associated with economic theories of various types (cf. Christophers 2011). It does so because such a sociological understanding can better help us grasp the essential political-geographical nature of finance.

The crux of the sociological theory lies in the fact that money always has to be *issued* – it has to originate somewhere, in the moment of issue by a social entity of some type. This is important because of what money's issue creates: specifically (Ingham 2004: 12), "a claim or credit against the issuer." (All subsequent quotations in this paragraph are also taken from Ingham [2004: 12], any emphasis in the original.) When money is issued, the issuer incurs a debt that can be settled with that money, and the value of which the money denotes. Money's essential promise is that the holder can always go back to the issuer and redeem this value. As such, money is "a provisional 'promise' to pay": "Money cannot be said to exist without the simultaneous existence of a debt that it can discharge." Herein lies much of its inherent power – "in the promise between the issuer and the user of money." (Power also derives from money's transferability; it is "*transferable debt*.") And that promise, critically, is always *socially* constituted: There is an issuer, and a user, and money establishes their relation. It is for this reason that Ingham defines money as "a 'claim' or 'credit' that is constituted by social relations" – as, that is, "a social relation of credit and debt."

The fact that we can envision money as a social relation is, this chapter argues, vital to understanding why finance, and financial crisis, are necessarily political-geographical phenomena. For what is a social relation if not always a relation of power, and thus a relation of political significance? One does not have to be a Foucauldian to appreciate that, as Tony Lawson (2012: 368) writes, all "rights and obligations" (financial or otherwise) are "forms of power," and thus "all social relations are power relations." And, equally, what is a social relation if not also a relation of geography? One of the key insights of geographers' revolutionary engagements with critical social theory in the 1970s and 1980s, after all, was not only that – in the words of Ramesh Dikshit (1999: 10) – "spatial relations are indeed social relations," but that "all social relations are spatially constituted." Social relations are relations of power *and* space, and hence the social relation that is money must be seen as an inherently political-geographical construct.

Money and finance's geopolitical essence is of course epitomized under, if not unique to, the political economy of modern, nation-state capitalism. The rise to prominence of the nation-state as the dominant scale of political-territorial organization cemented and made much more explicit the political nature of money and finance. Recall, again after Ingham (2004: 12), that money is "a claim or credit against the issuer." Historically, this issuer has assumed many different forms – "monarch, state, bank and so on." However, when the nation-state specifically issues money, as it increasingly did in Europe from the seventeenth century onward, it does so "in the form of a promise to accept it in payment of taxes," which serve in turn to fund state expenditures *and* to service the interest on the sovereign debts also being created from this period. Hence, modern monetary societies with state-backed legal tender "are held together by networks of credit/debt relations that are underpinned and constituted by sovereignty." Indeed, "[m]oney *is* a form of sovereignty, and as such it cannot be understood without reference to an authority."

This, however, does not fully explain the singular significance of *national* money for much of the history of modern capitalism. In the early, mercantilist period of capitalism, private monies issued by banks circulated relatively freely alongside such public monies. How and why was it that, certainly by the early twentieth century and often considerably earlier, most nations had a government monopoly on legal tender (circulating currency), albeit currency into which private, bank-issued debt can be readily converted? The answer is that private bank credit and state currency were successfully fused together. Ingham (2008: 32–34, 70–74), following Max Weber, identifies this fusion as the core component of the "memorable alliance" forged between the state and the capitalist bourgeoisie in capitalism's formative stages. This fusion was made possible by each side's "acceptance of mutual dependence" (2008: 73). "The state," writes Mann (2013: 71), "recognized the importance of a banking system that can create an elastic supply of credit-money," while the bourgeoisie recognized "that money could only enjoy the requisite confidence across space and time if it was secured by a strong territorial state." Ever since, capitalist finance has been locked together with the state "in a relationship of mutual advantage in which each has an interest in the survival and prosperity of the other" (Ingham 2008: 73).

This is not to say that the relationship in question has always been harmonious or the two forces always in balance. Plenty of historical research attests to the fluctuating influences of the public and private sectors in the shaping of the monetary and financial nexus, not least Ingham's own work. In his examination (1994) of the varying ways in which sterling, and then the US dollar, were made into effective world monies during different eras of modern history, he appeals explicitly to "different combinations of state and market power resources" (1994: 32). Nor does recognition of the preeminence of territorialized state money mean that the dominant alignment of money with the nation-state, in the aftermath of the memorable alliance, has always been smooth and uncontested – among other studies, the essays in Gilbert and Helleiner (1999) provide ample testimony to the complex and often conflictual relations between monies and nation-states.

Our brief analysis *is* intended to insist, however, that under capitalism money and finance are irredeemably political, and that this embedding of finance in political processes and structures assumes crucial scalar configurations. Seen in this light, our earlier empirical findings – that financial crises are at once crises of political geography, and perhaps especially that such crises very often become crises of *political*, state finance, even where they do not begin as such – are much more susceptible to theoretical understanding and generalization. What other outcome would one expect in the particular Western financial system bequeathed to us by capitalist history? In times of financial crisis, when trust dissipates and the networked claims on which the monetary system is based begin to unravel, the greater the stress inevitably becomes on the sovereign debt that constitutes the pivot of the network and to which the whole financial system is ultimately, however distantly, connected.

It all comes together (or falls apart) in the eurozone

If the conceptualization in the previous section appears altogether too abstract, or perhaps too confined to history, then consider in more detail the case of the particular financial crisis with which the chapter began, that of the contemporary eurozone. It would be hard to find a superior exemplar of the intertwining of finance and geopolitics than the eurozone, and thus it is entirely appropriate that this region became the epicenter of contemporary crisis – albeit a crisis that "started," of course, elsewhere.

To begin to try to unpack some of the key dimensions of this crisis and its political geographies insofar as they pertain to Europe, some basic background is vital (see Van Middelaar

2013). In particular, how should we understand the historical collective decision to replace the legacy national currencies of the eurozone with the new single currency from 2002? Scholars have persuasively argued that this decision evinced a highly distinctive melding of money and politics, or of money in the service *of* politics. The main goals, Shore (2012: 6) argues, were "to advance political integration and social cohesion": "Political union was the goal; money was the instrument for its attainment." Mann (2013: 207) concurs, identifying the currency's creation as "perhaps the most audacious attempt yet to make monetary policy and money markets the preeminent instrument of political power and capitalist governance."

For such a project to have been possible in the first place, more than just political will was clearly required. A transnational capitalist class also had to be persuaded; Weber's famous "alliance" may be less memorable today, but it still exists and requires ongoing nourishment, especially when the terms of its joining are fundamentally reconfigured. Murphy (2013: 5) thus emphasizes that monetary union required the support of economic elites, who seemingly believed that the euro "would further European integration and, in the process, enhance the region's economic clout and political standing."

Yet, the geopolitical implications were, and are, daunting, not least in regard to sovereignty and its monetary manifestation: The euro, recognizes Mann (2013: 207), "created a situation in which a healthy European circulatory system necessitated the end of 'traditional' state sovereignty in eleven nation-states." The preceding conceptual discussion gives us useful tools to understand the significance of this shift, particularly in relation to questions of geographical-political-economic stability. How could it *not* be relevant, for example, that the euro signaled a decisive break in the correspondence that had historically existed between the geopolitical scale at which money was issued and the scale at which the "promise" (national debt and tax collection) ultimately existed to back that issue? And this, it transpires, is just one of the multiple spatial/scalar contradictions inherent to the project of European integration and exposed as such by the recent crisis (Hadjimichalis 2011: 269–270).

The preceding background, of course, gives the lie to both of the quotes that opened this chapter, since in the era of the euro, European finance and European politics have been rendered more interdependent than ever before. Nevertheless, the global financial crisis did become more *obviously* (geo)political with the crystallization of sovereign debt problems in various countries – most notably Portugal, Ireland, Italy, Greece, and Spain – beginning in late 2009.

Perhaps the single most important analytical observation we can make about these sovereign debt problems is that they had very different sources and mechanisms of formation from place to place. It is difficult and hazardous to generalize. The Irish state, for instance, was brought to its knees by a very different combination of underlying maladies – high levels of bank and household debt, in particular, both tied directly to runaway property speculation – than, say, the Greek state. In some places crisis was largely sovereign from the start; in others, it rapidly became that way as states stepped in to bail out insolvent private-sector actors. And everywhere, "financial" and "political" roles have freely interchanged: Political institutions have been compelled to prioritize their financial being, while financial institutions have been minded to observe explicitly political imperatives. The only generalization we can safely make regarding euro-area crisis formation is that geopolitics – and especially the geopolitics of straitjacketing *existing* patterns of uneven geographical development within the euro monetary framework (Hadjimichalis 2011) – played a vital, but often underappreciated, role.

Geopolitics has also, demonstrably, shaped the political response to the crisis at every conceivable level, from the national to the supranational. The effects of the crisis have been highly uneven at all scales, and hence every major decision about financial (and political) remedies has been characterized by high-level disagreement, debate, and, at least from some quarters, compromise. What have we learned, in the process, about the state of eurozone geopolitics and its rooting in financial realities? Reijer Hendrikse (2012: 129) reflects the view of many when he puts it like this:

> although *de jure* eurozone governance is ambivalent, as the crisis continues the *de facto* balance of power becomes increasingly clear: unless Germany pulls the plug the European Central Bank calls the shots. As a result, Brussels' political bickering increasingly looks like a sideshow to legitimize decisions taken in Frankfurt.

Watkins (2013) comes to a similar conclusion, arguing that Germany has in effect come to "stand in" for "absent sovereign power."

Meanwhile, and equally significantly, the crisis has also strongly reaffirmed one of our other main findings, concerning the role of crisis in opening up spaces of opportunity for – or threats of – political-geographical reconfiguration at the regional scale. Bickerton (2011) has made this case succinctly: The eurozone crisis has thrown the very project of European political and economic integration into flux, with new possibilities emerging just as fast as others are closed down. There is, as Murphy (2013: 12) observes, a strong sense of déjà vu in this respect, as, *once again*, "the fundamental question for the European Union is cast in modernist territorial terms: will the EU be one large superstate or a collection of states that cede certain powers to a supranational body?"

The signs are rather ominous for supporters of the European integrationist project – just as Phillips (2000), more than a decade ago, warned that they often will be for regionalist initiatives in the post–financial crisis short term. There is, for example, "little question that the crisis has heightened Euroscepticism" (Murphy 2013: 11). "In Spain," reports Watkins (2013), "the number expressing 'mistrust' of the EU has risen from 23 to 72 per cent over the past five years; in Germany, France, Austria and the Netherlands the figure is around 60 per cent." Moreover, secessionary noises have been emanating from EU countries – most notably the United Kingdom – *outside* the eurozone, and thus sheltered from much of its direct financial stress, as well as within it.

However, perhaps the most telling signs of all, appropriately enough, are found in the shifting geographies of eurozone state finance per se. Ever since 2011, pointedly, there has been a marked and progressive domestication of sovereign debt; in the first half of 2013, Spanish banks absorbed 60% of their government's net debt issuance, with the comparative figure even higher, at 75%, in Italy (Unmack 2013). This trend is significant for what, more broadly within contemporary Europe, it portends: a retreat to the nation-state – one that is occurring culturally, politically, and financially, if not (yet) monetarily. Or, as Hadjimichalis (2011: 269), following Nancy Fraser, puts it, the crisis seems to be re-establishing "the 'Westphalian' political-geographical imagination of bounded national spaces" that "those who established the euro were trying to overcome."

Conclusion

This chapter has encouraged us to try to think about the political geographies of finance and financial crisis from a theoretical as well as purely empirical standpoint. That financial crises are deeply geopolitical is, from an empirical perspective, obvious. The conceptual reasons for this geopolitical dimension are much less obvious, however; and this chapter has argued for the merits of a political-economic approach in attempting to tease them out. Money and finance, it showed, represent and are embedded in particular sets of socio-spatial relations. In that sense, we might argue that periods of financial calm (or of "non-crisis") are characterized above all by a lack of questioning of the representational work that finance performs; such work is not questioned, more specifically, because the existing social structures of finance are reproduced so unproblematically in daily economic practice that they are typically not even recognized for the sociality they bear. In such periods, finance appears purely "technical."

If this is an approximately correct reading, we are now in a better position, among other things, to answer productively the question with which we began: What *is* financial crisis? It is, we can posit, an event or period during which, at the very least, the financial system's existing mechanisms for producing and reproducing socio-spatial relations are substantially disturbed and – it now being recognized that finance *is* social – come to be questioned as such. The more severe a financial crisis, the more likely it is that this disturbance and questioning of finance's social relations and their territorial expressions extend to the actual restructuring thereof. This is not a snappy definition, by any means, but it is hopefully a meaningful one – and, in any event, enough of one to indicate that the contemporary eurozone crisis, for all its undoubted specificity, is very much an exemplary case of crisis in modern money and the financial systems assembled around it.

In this regard, it is both striking and sobering that the bailout "solution" to the eurozone crisis, as formulated and effected by the Troika of the European Commission, the European Central Bank, and the International Monetary Fund, was positioned as a strictly technocratic one – thus serving to mask the complicated and conflictual socio-spatial relations underlying the euro and euro-denominated financial instruments. This was clearly an attempt to depoliticize what we know, on the contrary, to be fundamentally (geo)political issues; to represent messy and controversial socio-political interventions into European sovereign spaces as clean and merely technical ones. With finance's sociality having been uncomfortably exposed to view by the crisis, it was now to be summarily hidden once more; not for nothing, after all, had Marx (1981: 515–516) asserted that it is in money that the social relations of capital achieve their "most superficial and fetishized form," and that credit's "mystifications" are capitalism's "most flagrant." Such mystification and fetishism are essential *to* capitalism, and especially to finance capitalism; identifying and understanding the political-geographical dynamics that they veil must be central to any critical geography, and has been the aim of this chapter.

References

Agnew, J. (2012). Putting politics into economic geography. In Barnes, T., Peck, J., & Sheppard, E. (eds), *The Wiley-Blackwell Companion to Economic Geography*, Oxford: Wiley-Blackwell, 567–580.

Baker, A. (2010). Restraining regulatory capture? Anglo-America, crisis politics and trajectories of change in global financial governance. *International Affairs* 86(3): 647–663.

Baker, A. (2013). The new political economy of the macroprudential ideational shift. *New Political Economy* 18(1): 112–139.

Bickerton, C. (2011). Crisis in the Eurozone: Transnational governance and national power in European integration. *Political Geography* 30(8): 415–416.

Calomiris, C., & Haber, S. (2014). *Fragile by Design: The Political Origins of Banking Crises and Scarce Credit*. Princeton, NJ: Princeton University Press.

Christophers, B. (2011). Follow the thing: Money. *Environment and Planning D: Society and Space* 29(6): 1068–1084. doi:10.1068/d8410

Christophers, B. (2013). *Banking across Boundaries: Placing Finance in Capitalism*. Oxford: Wiley-Blackwell.

Christophers, B. (2014). Geographies of finance I: Historical geographies of the crisis-ridden present. *Progress in Human Geography* 38: 285–293. doi:10.1177/0309132513479091

Christophers, B. (2015). Geographies of finance II: Crisis, space and political-economic transformation. *Progress in Human Geography* 39(2): 205–213. doi:10.1177/0309132513514343

Clark, G., Dixon, A., & Monk, A. (2013). *Sovereign Wealth Funds: Legitimacy, Governance, and Global Power*. Princeton, NJ: Princeton University Press.

Clark, J., & Jones, A. (2012). After "the collapse": Strategic selectivity, Icelandic state elites and the management of European Union accession. *Political Geography* 31(2): 64–72.

Cohen, B. (1988). *In Whose Interest? International Banking and American Foreign Policy*. New Haven, CT: Yale University Press.

Cooper, A. (2010). The G20 as an improvised crisis committee and/or a contested "steering committee" for the world. *International Affairs* 86(3): 741–757. doi:10.1111/j.1468-2346.2010.00909.x

Corbridge, S. (1984). Political geography of contemporary events V: Crisis, what crisis? Monetarism, Brandt II and the geopolitics of debt. *Political Geography Quarterly* 3(4): 331–345.

Corrales, J. (2002). The politics of Argentina's meltdown. *World Policy Journal* 19(3), 29–42.

Derviş, K. (2013). Balancing the technocrats. *Project Syndicate* 10 May. http://www.project-syndicate.org/commentary/the-false-promise-of-technocratic-governance-by-kemal-dervi, accessed March 26, 2015.

Dikshit, R. (1999). *Political Geography: The Spatiality Of Politics*, 3rd edn. Noida: Tata McGraw-Hill Education.

Epstein, E., & Pion-Berlin, D. (eds) (2006). *Broken Promises? The Argentine Crisis and Argentine Democracy*. Lanham, MD: Lexington Books.

Ertürk, I., Froud, J., Johal, S., Leaver, A., Moran, M., & Williams, K. (2011). City state against national settlement: UK economic policy and politics after the financial crisis. CRESC Working Paper No. 101. h http://www.cresc.ac.uk/medialibrary/workingpapers/wp101.pdf, accessed March 26, 2015.

Etzioni, A. (2009). The capture theory of regulations – revisited. *Society* 46(4): 319–323.

Gilbert, E., & Helleiner, E. (eds) (1999). *Nation-States and Money: The Past, Present and Future of National Currencies*. New York: Routledge.

Gillespie, C. (1988). Latin American Political Economy: Financial Crisis and Political Change. Edited by Jonathan Hartlyn and Samuel A. Morley. *American Political Science Review* 82(1): 351–353.

Grimes, W. (2012). Financial regionalism after the global financial crisis: Regionalist impulses and national strategies. In Grant, W., & Wilson, G. (eds), *The Consequences of the Global Financial Crisis: The Rhetoric of Reform and Regulation*, Oxford: Oxford University Press, 88–108.

Hadjimichalis, C. (2011). Uneven geographical development and socio-spatial justice and solidarity: European regions after the 2009 financial crisis. *European Urban and Regional Studies* 18(3): 254–274. doi:10.1177/0969776411404873

Hare, S. (2012). The democratic legitimacy of Euro-area bailouts. *Mediterranean Politics* 17(3): 459–465.

Hartlyn, J., & Morley, S. (1986). *Latin American Political Economy: Financial Crisis and Political Change*. Boulder, CO: Westview Press.

Harvey, D. (2010). *The Enigma of Capital and the Crises of Capitalism*. London: Profile.

Helleiner, E. (2011). Understanding the 2007–2008 global financial crisis: Lessons for scholars of international political economy. *Annual Review of Political Science* 14: 67–87. doi:10.1146/annurev-polisci-050409-112539

Hendrikse, R. (2012). On crisis, currencies and the rise of the Phoenix. *Political Geography* 31(2): 127–130.

Igan, D., Mishra, P., & Tressel, T. (2011). A fistful of dollars: Lobbying and the financial crisis. NBER Working Paper No. 17076. http://www.nber.org/papers/w17076, accessed March 26, 2015.

Ingham, G. (1994). States and markets in the production of world money: Sterling and the dollar. In Corbridge, S., Thrift, N., & Martin, R. (eds), *Money, Power and Space*, Oxford: Blackwell, 29–48.

Ingham, G. (2004). *The Nature of Money*. Cambridge: Polity Press.

Ingham, G. (2008). *Capitalism*. Cambridge: Polity Press.

Johnson, S., & Kwak, J. (2010). *13 Bankers: The Wall Street Takeover and the Next Financial Meltdown*. New York: Pantheon.

Kitson, M., Martin, R., & Tyler, P. (2011). The geographies of austerity. *Cambridge Journal of Regions, Economy and Society* 4(3): 289–302. doi:10.1093/cjres/rsr030

Lawson, T. (2012). Ontology and the study of social reality: Emergence, organisation, community, power, social relations, corporations, artefacts and money. *Cambridge Journal of Economics* 36(2): 345–385. doi:10.1093/cje/ber050

Lee, C. (2003). To thine ownself be true: IMF conditionality and erosion of economic sovereignty in the Asian financial crisis. *University of Pennsylvania Journal of International Economic Law* 24(4): 875–904.

Mann, G. (2012). Release the hounds! The marvelous case of political economy. In Barnes, T., Peck, J., & Sheppard, E. (eds), *The Wiley-Blackwell Companion to Economic Geography*, Oxford: Wiley-Blackwell, 61–73.

Mann, G. (2013). *Disassembly Required: A Field Guide to Actually Existing Capitalism*. Edinburgh: AK Press.

Marx, K. (1981). *Capital*. Vol. 3. London: Pelican.

McCarty, N., Poole, K., & Rosenthal, H. (2013). *Political Bubbles: Financial Crises and the Failure of American Democracy*. Princeton, NJ: Princeton University Press.

McMillan, S. (2012). Eurozone crisis turns into political crisis. *National Business Review* 28 April. http://www.nbr.co.nz/article/eurozone-crisis-turns-political-crisis-117591, accessed March 26, 2015.

Murphy, A.B. (2013). Trapped in the logic of the modern state system? European integration in the wake of the financial crisis. *Geopolitics* 18(3): 705–723.

Ong, A. (2000). Graduated sovereignty in south-east Asia. *Theory, Culture and Society* 17(4): 55–75. doi:10.1177/02632760022051310

Öniş, Z. (2009). Beyond the 2001 financial crisis: The political economy of the new phase of neo-liberal restructuring in Turkey. *Review of International Political Economy* 16(3): 409–432.

Phillips, N. (2000). Governance after financial crisis: South American perspectives on the reformulation of regionalism. *New Political Economy* 5(3): 383–398.

Pianta, M. (2013). Democracy lost: The financial crisis in Europe and the role of civil society. *Journal of Civil Society* 9(2): 148–161.

Poovey, M. (2012). Introduction: Financial panics and crises. *Journal of Cultural Economy* 5(2): 139–146.

Rajan, R. (2010). *Fault Lines: How Hidden Fractures Still Threaten the World Economy*. Princeton, NJ: Princeton University Press.

Reinhart, C., & Rogoff, K. (2009). *This Time Is Different: Eight Centuries of Financial Folly*. Princeton, NJ: Princeton University Press.

Sheppard, E. (2011). Geographical political economy. *Journal of Economic Geography* 11(2): 319–331. doi:10.1093/jeg/lbq049

Shore, C. (2012). The euro crisis and European citizenship: The euro 2001–2012 – celebration or commemoration? *Anthropology Today* 28(2): 5–9. doi:10.1111/j.1467-8322.2012.00859.x

Streeck, W. (2014). The politics of public debt: Neoliberalism, capitalist development and the restructuring of the state. *German Economic Review* 15(1): 143–165. doi:10.1111/geer.12032

The Economist (2013) Motown's blues continue. *The Economist* July 19. http://www.economist.com/blogs/democracyinamerica/2013/07/detroit-files-bankruptcy?fsrc=scn/tw_ec/motowns_blues_continue, accessed March 26, 2015.

Thompson, H. (2009). The political origins of the financial crisis: The domestic and international politics of Fannie Mae and Freddie Mac. *Political Quarterly* 80(1): 17–24. doi:10.1111/j.1467-923X.2009.01967.x

Unmack, N. (2013). "Home bias" for bonds put euro zone at risk. *International Herald Tribune* 19 July.

Van Middelaar, L. (2013). *The Passage to Europe: How a Continent Became a Union*. New Haven, CT: Yale University Press.

Wallison, P. (2009). Cause and effect: Government policies and the financial crisis. *Critical Review* 21(2–3): 365–376.

Watkins, S. (2013). Vanity and venality. *London Review of Books* 35(16): 17–21. http://www.lrb.co.uk/v35/n16/susan-watkins/vanity-and-venality, accessed March 26, 2015.

Wedeman, N. (2012). A ray of sunshine in the black cloud surrounding Europe. *Cronkite 2012: New Media, Old Empires* 15 June. http://cronkiteuro12.wordpress.com/2012/06/15/a-ray-of-sunshine-in-the-black-cloud/, accessed March 26, 2015.

Woodruff, D. (2005). Boom, gloom, doom: Balance sheets, monetary fragmentation, and the politics of financial crisis in Argentina and Russia. *Politics and Society* 33(1): 3–45.

Woods, N. (2010). Global governance after the financial crisis: A new multilateralism or the last gasp of the great powers? *Global Policy* 1(1): 51–63. doi:10.1111/j.1758-5899.2009.0013.x

Xinbo, W. (2010). Understanding the geopolitical implications of the global financial crisis. *Washington Quarterly* 33(4): 155–163. doi:10.1080/0163660X.2010.516648

Chapter 35

Migration

Michael Samers

University of Kentucky, Lexington, Kentucky, USA

This chapter concerns what I will call simply the "political geography of migration." I will return to this idea in a moment, but in the interim, it is necessary to clarify how "migration" is being circumscribed. The focus here will be on *international* migration and not *internal* migration (migration *within* countries) or other forms of "mobility." After all, international migration is itself a sufficiently vast and diverse subject to preoccupy us in this chapter. Furthermore, the analysis will be limited largely to the control and facilitation of migration (entry) rather than policies of *immigration* (that is, citizenship and settlement, which are covered to some extent in Chapter 13). Last, the emphasis will be on the relationship between migration, "space," and state practices (or those non-state or quasi-state governance practices that mirror or substitute for state policies).

While political geographers may sometime tackle questions of space with more explicit intent, they do not have a monopoly on using spatial metaphors to explain migration-related issues. This presents an immediate challenge for illuminating a "political geography of migration," since *inter alia* anthropologists, political scientists, and sociologists have all produced seemingly endless research on migration, especially from a nation-state perspective, as well as from other "scales" or "territories." True, other social scientists might not necessarily interrogate spatial metaphors with the same fervor, but insofar as national territories or municipalities, for example, are forms of spatial metaphors used by, let us say, political scientists, the work of a broad range of social scientists needs to be taken into account. Thus, this chapter will offer a brief survey of such research, principally over the last two decades. It is therefore a thoroughly selective cut through some of the developments in the vast migration/immigration-oriented literature.

In undertaking this overview, I organize my discussion via the territorial lens through which research takes place. That is, some studies begin with "place" or "municipality," "locality" or "the local," others with the "nation-state," and still others with interstate relations or "global governance." Some might label these the "scales" of analysis, but, perhaps at odds with other political geographers, I avoid the use of the term "scale" or "scalar," since "scale" seems a rather confused concept, encompassing a range of meanings (the "spatial extent of a process" or "a territory," for example; Samers 2010b). Rather, I focus here on territory *as jurisdiction*, consciously avoiding other spatial metaphors such as "landscape" or "place." Nonetheless, this chapter assumes three aspects of spatial metaphors: that privileging a particular metaphor over another is not necessarily fruitful (see Leitner, Sheppard, & Sziarto 2008); that "territory," for example, is constructed through *discourses* and *imaginations* as well as practices; and that the invocation of spatial metaphors is intimately related to the political strategies of research (Dikeç 2012; cf. Massey 2005). In any case, such spatialities (e.g., those of territoriality) do not simply regulate or even constitute migration and migrants; rather, migrants shape these spatialities. And as Feldman (2012: 8) writes, "like any other subject-position, the 'migrant' does not exist ontologically but rather as someone who is enticed into a social role and negotiates it in a historically constituted field of political, economic, and social relations."

Migration, territory, and state processes

Guiraudon and Lahav (2000) refer to changes in the governance of migration during the 1990s as the combined processes of "upscaling," "downscaling," and "out-scaling." Political geographers might refer synonymously to these first two processes as "scale-jumping" (for example, from "the national" to "internationalization," or from "the national" to "localization"), and the latter might simply be understood as "privatization" or the increasing prevalence of non-state actors. Now, whatever the difficulties of their language of scale, the argument that governance involves tendencies such as localization, for example, is not only plausible, but one that punctuates the literature on migration, especially that by political geographers. We can therefore first discuss – if conventionally – the relationship between municipal or metropolitan states and migration, then address the relationship with national states, and finally consider the relationship between migration and interstate relations or "global governance."

Migration and local states

If we equate "the local" or "local states" with municipal policies and practices, then let us turn to how social scientists have addressed such local political geographies of migration. One might begin with the relationship between national and municipal policies and politics. In political science this is a vast literature, which explores, for example, the "political opportunities" that are available at the intersection of national policy and metropolitan policymaking (e.g., Cinalli & El Hariri 2012 in a European context). In the United States, there is now a substantial literature by political geographers that explores the discord between the federal government and the policies of states and municipalities, and the consequences of these tensions for immigrants and undocumented immigrants in particular. In this respect, much of the focus is on the implications of Section 287g of the 1996 IIRIRA (Illegal Immigration Reform and Responsibility Act), which allows the Department of Homeland Security to authorize state and local officers to carry out the responsibilities of federal agents. Coleman (2009, 2012) and Varsanyi (2008a, 2011) draw out the broader point that "place" or municipalities (and their respective non-state local actors) matter for the way in which federal policies are translated into local (securitarian) practices, which are decidedly detrimental to undocumented immigrants. Varsanyi et al. (2012) refer to this enmeshing of jurisdictional practices as a "multi-layered jurisdictional patchwork," which seems a better term than "re-scaling" (cf. Varsanyi 2008b; Samers 2010b).

A second and very much related body of work by political geographers with regard to localizing processes focuses on "re-bordering" in the United States, or what might be called "border internalization"; that is, practices of internal enforcement that move beyond the external boundaries of the United States and connect national with local immigration policies (e.g., Wright & Ellis 2000; Cravey 2003; Mountz 2004, 2010; Coleman 2007, 2012; Winders 2007; Varsanyi 2008a, b, 2011; Varsanyi et al. 2012). As Coleman (2007) observes, "border enforcement is everywhere," and for Coleman (2012), national immigration policies (such as those concerning deportation) are not simply "downloaded" from the national to the local, national policies are translated into what he calls "site-specific" local practices ("site" seems to be a synonym for both "place" and "territory"). Put differently, he is especially concerned with bridging the gap between legislation and actual enforcement practices. Following Cravey (2003), Coleman argues that site-specific immigration practices are intervening in the spaces of reproduction (e.g., traveling to work, shopping, etc.) by non-federal immigration officers asking for proof of citizenship. This means that immigrants are as exposed in the spaces of *reproduction* as they are in the spaces of *production* (e.g., working informally; see also earlier work by Mitchell 1995).

A third literature concerns what might be called the exclusionary politics of localities, and is often combined with the insights of the border internalization literature. The former literature is vast and centers on the racialization/ethnicization and criminalization of (undocumented) immigrants and asylum-seekers across the world. Such anti-immigrant politics and policies have been the subject of considerable attention, whether this examines the relationship with neoliberalism (or neoliberal governmentalities) and "illegality" (e.g., Varsanyi 2008b, 2011; Hiemstra 2010); or the ways in which the public resist, for example, asylum "accommodation centers" on the dubious grounds of supporting the welfare of asylum-seekers themselves; or, more believably, the perception among local citizens that such centers pose a threat to their "rural" way of life that hides racist feelings toward "not quite white" subjects (Hubbard 2005).

Last, we can cite work on the protests of migrants themselves against deportations and other poor housing and working conditions. In "Western" countries there is a long history of such protests. Perhaps over the last 20 years, we could mention, among many examples, the protests of Malian undocumented migrants in Paris in 1996, Zimbabweans in detention in the United Kingdom (McGregor 2011), or migrants at the Oakington Reception Centre near Cambridge and Campsfield House near Oxford (Schuster 2003; Hayter 2004; Gill 2009). Other social movements in the United States, such as the immigrant workers' freedom rides in 2005 (Leitner et al. 2008), or immigrant political activism in Washington, DC (Leitner & Strunk 2014), demonstrate the significance of either putatively localized or nationally networked migrant protests against their political and social conditions.

Migration and national states

In understanding the relationship between national states and migration, one might distinguish between the research objectives of especially political scientists and political sociologists on the one hand, and those of political geographers on the other. The former seem to be more concerned with *explaining the politics of* migration and immigration policies ("outputs") or their consequences for migration ("outcomes"; Hollifield 2008) by assuming the continual salience of the nation-state (e.g., Hollifield, Martin, & Orrenius 2014), while the latter are preoccupied (though not exclusively) with *critiquing national state effects,* or even the ontological status of the nation-state and its supposed territorial effects on migration (e.g., Gill 2010). With respect to the former, we might begin with Hollifield's (2004) notion of an "emerging migration state." While this idea might be historically myopic (at least if we take into account the twentieth century in, say, Australia, New Zealand, or the United States), the idea that states must balance openness in terms of trade, investment, and people with the closure of "security" seems reasonable enough. After all, national governments do *seem* to practice these apparently contradictory processes. Nonetheless, as Cornelius (1994) points out with his "gap hypothesis," states do not necessarily achieve their desirable migration and immigration outcomes, although some have questioned whether states have really "lost control" of immigration (e.g., Zolberg 1999). In contrast, Brown (2010) sees a certain decoupling of states from sovereignty, arguing counter-intuitively that border walls and fences are symbolic of the waning of sovereignty, rather than its strength. Thus, if there is an "emerging migration state," it is one that involves a diminished sovereignty. With these matters in mind, let us review the way in which political geographers, political scientists, and (political) sociologists have theorized migration through the lens or metaphor of "the national."

Marxist political economic analysis of migration policies were especially popular in the 1970s in European countries, in which scholars argued that immigrants (understood in both governmental and popular imagination as migrant *workers without families*) served as an "industrial reserve army" to increase capital accumulation by lowering the costs of "social reproduction." That is, many migrant workers were single adults and Western European countries did not have to educate them to adulthood. Furthermore, they were often housed in low-cost accommodation, which in turn lowered inflation (see Samers 2010a for a review). There was a "structural functionalism" inherent in these arguments, insofar as the state and migration policies served the interests of the (national) capitalist classes and capitalist accumulation. Variations on this argument sought to emphasize the racialization of migrant workers and the relationship between accumulation and legitimation with a more neo-Marxist conception of the state (Samers 1999). Whatever the merit of Marxist-inspired arguments, however, these theoretical currents seemed to disappear, and the 1980s saw little work on the political *economy* of immigration (but see Cohen 1987). It was not until the mid-1990s that a certain revival of the political economy of immigration gathered steam, but from a very different epistemological starting point.

That is, a range of political scientists were clearly more taken by "liberal," pluralist, and institutionalist forms of political economy and political sociology to understand migration politics and policies. Freeman (1995) stimulated a substantial debate with his idea of "client politics," in which he argued that "clients" are "the small and well organized groups intensely interested in policy … [who] develop close working relationships with those responsible for it" (1995: 886). The "strongest" clients will shape migration policy through lobbying. Since minority and ethnic groups have become increasingly vocal, and employers (such as agribusiness firms) are interested in securing an adequate number of migrant workers, migration policy is likely to be "expansionary"; that is, more "liberal." This is reinforced, he argues, because the benefits are concentrated among ethnic minority groups and employers, while the "costs" are diffuse; that is, they are spread throughout the electorate and therefore are not represented by clearly defined lobbies. This suggests that political leaders are likely to favor concentrated interests. Furthermore, Freeman maintains, "advanced liberal democracies" entail an "embedded liberalism" where an "anti-populist" norm prevails, and therefore a wide array of political parties will seek a consensus on immigration, while at the same time preventing the incorporation of immigration into party politics.

Freeman's thesis quickly drew a number of criticisms, especially from scholars observing the politics of immigration in *European* countries. Guiraudon (1998) argues that in at least France, Germany, and the Netherlands in the late twentieth century, the costs were not as diffuse as Freeman contends, as political parties attempted to appease right-leaning or right-drifting political parties. This would support the argument that Freeman's thesis seems to be more relevant for labor migration, rather than other types of migration that might attract more resistance from citizens (Hamsphire 2013). Joppke (1998) also argues that Freeman's analysis makes little sense for European countries, especially in relation to what he called "unwanted family reunification" from the 1970s onward. European countries do not have the same practices of lobbying as the United States, and the reluctance of governments to accept family reunification in the 1970s and 1980s has been contradicted by the liberal jurisprudence of judges based on the moral and legal rights of labor migrants, and not the lobbying of "clients." Thus, while Freeman focuses his efforts on the *political* process, Joppke emphasizes the *legal* process and the role of courts in accepting family migrants and blocking deportations. Thus, it is not "globalization" that is eroding the sovereignty of national states, but rather national states themselves that are eroding their own sovereignty. Joppke refers to this as "self-limited sovereignty." Indeed, the majority of undocumented immigrants are *not* deported in liberal democratic states for reasons of cost, or bureaucratic and logistical challenges, or because of opposition to the violation of civil rights and norms by a range of immigration actors and activists (e.g., Hampshire 2013; Chauvin 2014).

Yet, Joppke's argument itself needs to be interrogated. Thus, what Gibney (2008) calls a "deportation turn" reflects the willingness of European governments to rely on detentions and deportations (or "returns") for *at least* symbolic reasons; that is, to appease publics and to dissuade further migration. Beyond European countries, the practices of detention and deportation in Canada (Mountz 2010) and the United States (Peutz 2006; Hagan, Eschbach, & Rodriguez 2008; Peutz & DeGenova 2010; Anderson, Gibney, & Paoletti 2011; Martin 2012a, b), as well as the actions of, let us say, the Malaysian or United Arab Emirates governments toward Indonesian and South Asian domestic migrant workers, respectively (Silvey 2004, 2007; Silvey & Olson 2006; Hedman 2008; Buckley 2013), or those of the South African government's "non-racial xenophobia" toward Zimbabweans during the late 1990s (Klotz 2000), all call into question the generalizability of Joppke's argument across time and space. In fact, there are grounds for arguing that political (or political economic) theory with respect to migration rests far too heavily on the analysis of "liberal democracies," although Silvey (2009) is wary of the tendency to separate out the "global North" from the "global South" in terms of migration.

Much of the debate inspired by Freeman's work within political science on the politics (or the political economy) of migration is surprisingly devoid of criticism from a gender perspective. In fact, most of the work that involves a gender-centric critique of governance is conducted by social scientists not working in political science departments. However, there is now a large literature by political geographers and others on the gendering of migration (or refugee) policies, and how this produces vulnerabilities for women in particular (e.g., Hyndman 2000; Kofman 2004; Silvey 2004, 2006; Walton-Roberts 2004; Dannecker 2005; Kofman & Raghuram 2006; Piper 2006; Gabriel 2008; Fitzgerald 2010); the range of themes is vast. For example, Dannecker (2005) ties men's patriarchal vision of Bangladeshi Muslim women and the fears of their sexual promiscuity and consumerism abroad to migration policies. More specifically, Dannecker examines the discourses and policies of migration to Malaysia, a significant destination for Bangladeshi women. The Bangladeshi government sought to ban most Bangledeshi women from leaving the country in 1981, except for "professional" Bangladeshi women. The government overturned this in 1988, which dramatically increased migration, but closed the exit doors again in 1997, this time banning *all* emigration by women without an accompanying man.

The debate on the politics of immigration moved increasingly during the 2000s to matters of security. Well before the events of September 11, 2001, a literature emerged in the European Union that, with a few exceptions (see, e.g., Tesfahuny 1998), much of this work ignored the racialized, gendered, classed, criminalizing, and securitizing nature of immigration policies in Europe. After 9/11, this expanded into a far more extensive literature within and beyond Europe, from Foucauldian-inspired or other critical/heterodox perspectives (e.g., Huysmans 2000, 2008; Walters 2004, 2008; Sparke 2006; Hyndman & Mountz 2008; Bigo et al. 2010; Feldman 2012; El Qadim 2014) to less epistemologically daring analyses, which offered none-theless pointed insights (e.g., Boswell 2008; Givens, Freeman, & Leal 2009). To begin with, Walters (2004) underlines the significance of a "domopolitics," in which state discourses speak of a threat to the "homeland," with its attendant implications for surveillance. And yet, perhaps counter-intuitively, both Boswell (2008) and Huysmans (2008) find little evidence for links between migration and security/terrorism in political discourse in several European countries and the United Kingdom, respectively (compare with D'Aoust 2013 in the context of North America). Thus, Boswell (2008) argues that we need to draw a distinction between highly restrictive immigration policies and the securitization of *discourse*. As with Huysmans, she arrives at what is perhaps a surprising conclusion: In France, Germany, and the United Kingdom after 9/11, there has been little evidence of the establishment of links between ter-rorism and immigration in *political discourse*, even if one can speak of the *securitization of practice* (read highly restrictive immigration policies reinforced by surveillance technologies).

During the 2000s, a notable set of papers stressed a political sociological approach (e.g., Sciortino 2000; Guiraudon 2003; Boswell 2008) aimed explicitly or implicitly at what they saw as the limitations of a political economic analysis of immigration. For example, writing with respect to immigration politics in Italy, and more broadly the European Union, Sciortino argues that immigrants themselves rarely shape migration and immigration policy, and this is even more the case for undocumented migrants. Thus, migrants do not constitute particularly powerful "clients." Moreover, Sciortino argues that "information" from clients does not simply enter the political system *qua* "information" and is not subsequently processed "objec-tively" by political actors. Rather, scholars need to understand the interaction between political actors and any potential "clients" within the spaces of, let us say, Brussels' wine bars and restaurants.

In contrast to the debates on the political process by political scientists and political soci-ologists (for a comprehensive review of these debates, see Hampshire 2013), political geogra-phers arrived quite late to the study of the relationship between immigrants and states, let alone private or quasi-state actors. This might be divided into two literatures: one concerned with questioning the discourses and practices of "migration management" and the intensifica-tion of bordering or "gating" in the shadow of 9/11, and the other with the consequences of (neoliberal) forms of governmentality.

In the 2000s, the idea of "migration management" became popular among policymakers and received critical attention from a number of scholars (e.g., Morris 2002; Bauder 2008;

Kofman 2008; Geiger & Pecoud 2010; Ashutosh and Mountz 2011). As is well known, this refers to the rise of an amalgam of policies that attempt to manage different forms of migration (asylum/refugee, family, student, or work oriented, etc.) *together* (Kofman 2008) through often complex policy schemes. Migration management may be seen as the migration armature of "neoliberalization" (depending on one's definition of neoliberalization); as the distillation of neoliberal and neonational tendencies (Feldman 2012); as the intersection of neoliberalism, an international human rights regime, and the nation-state (Ashutosh & Mountz 2011); or even (I would argue) as "neo-Schumpeterian" (Jessop 1993), where migrants are viewed and selected in terms of their contribution to technological innovation in a global labor market of *apparently* scarce "talent." It may also be seen as reflecting what Hampshire (2013) calls a "politics of closure" and "a politics of openness," or what Hollifield (2004) calls a "liberal paradox" – the need to balance the demands of markets with human rights and security issues (see also Coleman 2005; Sparke 2006; Van Houtum & Pijpers 2007; Nevins 2010).

The (effects of) governmentality have also drawn enormous attention from political geographers and other social scientists not strictly concerned with the relationship with security. Foucauldian in perspective, some of this literature focuses on the place of non- or quasi-state institutions in the governance of migration (such as the work cited earlier on detention), and others emphasize the biopolitical production of the "migrant subject." To take but one example of the former, Hedman (2008) recounts the heinous consequences of paramilitary vigilante groups and the associated criminalization of especially Indonesian migrants in Malaysia. In terms of the latter, Fitzgerald (2010) examines the biopolitical production of sexually trafficked women into the United Kingdom, and how this casting of the sexually trafficked migrant as "vulnerable" shapes the racialized and sexualized contours of migration policy more generally. Other work has suggested a move from research on "illegal migrants" to the (biopolitical) construction of "illegality" (for a review of this literature, see Chauvin 2014).

Migration, interstate relations, and global governance

We might begin this discussion by asking, as Hollifield (1992) does, why there is no "global liberal order" for migration as there is for trade or finance. An immediate and seemingly "common-sense" response to this might be the demands for sovereignty in the face of neonationalist fears of "cultural dilution," unemployment, the protection of welfare states, and security concerns. Yet, such explanations remain too facile and raise more questions than they answer, not least because of the instances of (neoliberalized?) cooperation among states. In this respect, political scientists and sociologists, at least since the 1980s, have addressed Hollifield's question indirectly through a migration-related literature from the perspective of Marxism, realist (or neo-Realist) international relations (IR), and international (or global) political economy (IPE/GPE; for reviews, see Mitchell 1989; Heisler 1992; Hollifield, 1992, 2008; Leitner 2003; Betts 2012). In contrast, political geographers have remained relatively silent on interstate relations and global governance in favor of focusing on local political geographies or explaining international migration without a comprehensive treatment of interstate relations. Nonetheless, over the previous decade or so, a small literature by geographers has embraced what is now called "critical geopolitics," and in the following I review each of these broad currents.

In an earlier section, I discussed the basic tenets of Marxist analysis within the context of national states. If most Marxist analyses of immigration *on a "world scale"* concerning immigration rested on "*dependencia*," unequal exchange, and world systems analyses during the 1970s, then such approaches left considerable room for understanding complex state actions that might connect Marxism's often abstract critique of global capitalism with the messiness of immigration, migration, and emigration policies, as well as their consequences. At the same time, realist and neorealist IR theorists had by and large neglected migration until the 1980s (Mitchell 1989; Heisler 1992), and these research traditions could sometimes be accused of "methodological nationalism," "the territorial trap," "embedded nation-statism," "state rationality," or "state essentialism" (for a similar discussions, see Gill 2010; Samers 2010a). Nonetheless, a few authors attempt to connect a political sociological approach with neo-Realist analysis (where, at a minimum, states are assumed to be situated within a complex interstate system that involves asymmetries of power; e.g., Delano 2009 on NAFTA and Mexican–US relations). Yet, if neither realism nor neo-Realism is suited to the task of understanding interstate relations with respect to international migration (Heisler 1992), what is striking is then how much unexamined discussions of "neoliberalism" and "migration management" in the twenty-first century do in fact entail a certain "Realism," "state essentialism," and "methodological nationalism." In any case, by the 1990s, IR seemed to be partially eclipsed by other cognate perspectives, which I detail next.

The first actually involves a "liberal" political-institutional or political sociological (rather than political economy) literature in political science concerned with which institutions were responsible for decision-making on migration and immigration (and especially at what "level" of governance); a second, and equally liberal, body of work on "global governance" draws on the questions that arise from asymmetric power relations; and third is a more critical literature associated with political geographers and a sprinkling of political scientists that embrace "critical geopolitics," including anything from diasporic or transnational political movements, to (transnational) governmentality approaches informed or not by poststructuralism (e.g., more recently, Lafleur 2013; Pearlman 2014), to a specific concern for the classist, racist, and gendered discourses of governance.

With regard to the first, it is in the context of the European Union that an institutionalist and political sociological approach to the politics of migration seemed to garner the most attention. To take but two related examples, Guiraudon (2000) and Guiraudon and Lahav (2000) captured changes in policymaking with respect to migration in European countries. In a process that she called "venue-shopping," Guiraudon (2000) argued that domestic policymakers in European member states increasingly removed migration from public debate in the 1990s and shifted policymaking into intergovernmental cooperative fora at the European level, which were dominated by the security agenda of national interior ministers. Similarly, while Guiraudon and Lahav (2000) argued more generally for changes in the "scales" of governance (see the previous discussion), at the same time Geddes (2008) asked whether and to what extent migration, immigration, and citizenship policies were becoming truly supranational in the European Union – that is, decided on by EU-level institutions – or rather "communautarized" – that is, shifted from national *fora* to intergovernmental decision-making within EU institutions such as the Council of Ministers. Geddes concluded that migration policy in the European Union demonstrated more a process of communautarization than supranationalization.

Similarly, but in the context of the North American Free Trade Agreement (NAFTA), Pellerin (1999, 2008) and Gabriel (2008) question how or to what extent NAFTA has led to a supranational migration regime between Canada, Mexico, and the United States with respect to either "low-skilled" workers or the migration of nurses. If it is supranational at all, then it is limited to temporary visas for four categories of "business persons," including "professionals" such as nurses. This has two consequences: NAFTA does not include provisions for the protection of "low-skilled" migrant workers; and the system of visas seems to privilege Canadian rather than Mexican nurses, because of the complex overlapping territorial demands for professional certification.

Questions of "asymmetric power relations," competition, and cooperation between states with respect to migration have given rise to discussions of the "global governance" of migration (or "global mobility regimes") from a liberal perspective (see especially Koslowski 2011; Betts 2012). These two edited volumes seek to explore how and to what extent global institutions matter for migration. Betts defines global governance as the "norms, rules, principles and decision-making procedures that regulate the behavior of states (and other transnational actors)" (2012: 4). Yet, what is important for Betts is not whether the governance of migration occurs at the "level" of bilateralism, multilateralism, or "supranationalism," but whether or not "it is constraining of or constitutive of the behaviour of states (and transnational actors)" (2012: 4). He acknowledges immediately that there is no "multilateral global governance framework," but that this does not mean that "there is no global governance" (2012: 6–7). The global governance of migration is in fact a mixture of formal and informal institutions that operate through different actors and levels of governance, and different institutions regulate the various forms of migration (asylum, labor, family, and so forth) through not only direct but indirect policies such as those of the International Labour Organization, which are aimed at protecting migrant workers. Betts calls this "thin multilateralism." Some levels of governance are important for certain migration issues, but not for others. For example, he argues that while refugee protection is regulated multilaterally, highly skilled migration is governed multilaterally. Furthermore, he distinguishes between "thin multilateralism" ("regulating states' behavior in relation to refugees, international travel, and labour migration," 2012: 12) and "facilitative multilateralism," which does not entail "formal multilateral cooperation but is designed to encourage states to become involved in dialogue and the sharing of information" (2012: 12).

In contrast to the global governance literature, human geographers and more critical political scientists have embraced "critical geopolitics," once again heavily influenced by Foucault and "governmentality," "transnational studies," critical security studies, and feminist critique. First, over the last decade, a greater emphasis has been visible on non- or quasi-state actors in the interstate governance of migration, including airport officials, airlines, ferries, and other carriers, labor intermediaries, international human rights organizations, and so forth (e.g., Lahav & Guiraudon 2006; Silvey 2007; Ashutosh & Mountz 2011; Menz & Caviedes 2011). This literature is complemented by the innovative discussion of Walters (2010), who offers a critique of what he calls "migration cartographies"; that is, the unexamined visual representations of the geography of migration trajectories that underscore migration policies. Second, initially based on what Zolberg (2002) called "remote control," attention has turned over the last decade to the way in which border control and the "securitization" of migration have been shifted well beyond the boundaries of nation-states, variously described as the "re-scaling of migration control" in relation to development assistance (Samers 2004), "excisement" in the context of Australia (Hyndman & Mountz 2008), the "external dimension" or "externalization" (Boswell 2003; Lavanex 2006), "an archipelago of enforcement controls" (Mountz 2010), or "border externalization" (e.g., Lavenex & Schimmelfennig 2009; Casas-Cortes, Cobarrubias, & Pickles 2010, 2013; Mountz 2010; Van Houtum 2010;

Bialasiewicz 2012). In short, the state is argued to be a "transnational actor" whose tentacles can extend well beyond its sovereign territorial borders (Mountz 2010).

Another set of authors has focused on the gendering of immigration (especially citizenship) policies, or their relationship to other related processes, such as the worldwide crisis of social reproduction (Silvey 2009). However, unlike much of the political science literature, this work not only emphasizes the gendered characteristics of immigration policies, but focuses on the *consequences* of these policies in terms of family migration, gender, and sexual relations among migrants in the European Union (e.g., Kofman 2004; Simmons 2008; D'Aoust 2013). While Kofman (2004) and certainly others underline how EU and member-state immigration policies tend to see women as the ephiphemenon of labor migration policies based on the figure of the male worker, D'Aoust (2013) wishes to focus on the "emotional technologies" (the conceptualization of love, for example) that connect governmentality with migration.

Finally, in what might be seen as an implicit critique of Wallersteinian "world systems theory" or other Marxist-inspired arguments that might situate migration within an unequal capitalist system broadly demarcated along global South/global North lines, El Qadim (2014) wishes to "decolonize" the study of migration policies by "looking for the agency of state actors in so-called countries of origin" (2014: 242).

Conclusions

The political geography of migration is in fact quite a "young" endeavor, probably only dating to the 1990s. No doubt there are some earlier exceptions, but most geographers interested in migration and immigration were concerned with the structural (economic) or "agentic" explanations of the migration process, and especially socio-spatial patterns of immigrant settlement. A more complex analysis by political geographers of the (il)logics of governance and their effects remained largely untreated until the twenty-first century. This is no longer the case, and political geographers (among a whole range of other social scientists) have embraced critiques of neoliberal, securitarian, and biopolitical processes with vigor, not least perhaps because of more intensive and extensive securitarian practices since especially 9/11, or because of persistent and pervasive, xenophobic responses to immigration in countries that extend from Austria to Malaysia. At the same time, the de-essentialization of the state popular with political geographers seems to echo a long-standing tradition found in political sociology, even if the latter itself arrived late to the study of migration.

In spatial terms, much of this research by political geographers is split between a focus on local practices of governance on the one hand and "border externalization" processes on the other. This seems to reflect an uncomfortableness with "methodological nationalism," but also a certain distance from IPE as practiced by political scientists. Ironically, however, given geographers' apparent predilection for "spatial correctness," political geographers have not interrogated spatial metaphors with (I would argue) the necessary energy, whether this concerns the problematic references to "scale" or to "transnationalism"; unreflexively referring to national state power and rationality when it suits an author's particular political standpoint; or specifying exactly what "local" means (see Samers 2010b). Such energy is legion because, as Dikeç (2012) insists in a discussion critical of spatial metaphors as a "mirror of nature," "space" is instead a "mode of political reasoning." Whether or not one accepts this point, more attention needs to be placed on how space, migration, and governance are intertwined. That is the task ahead.

References

Anderson, B., Gibney, M.J., & Paoletti, E. (2011). Citizenship, deportation and the boundaries of belonging. *Citizenship Studies* 15(5): 547–563.

Ashutosh, I., & Mountz, A. (2011). Migration management for the benefit of whom? Interrogating the work of the International Organization for Migration. *Citizenship Studies* 15(1): 21–38.

Bauder, H. (2008). Neoliberalism and the economic utility of immigration: Media perspectives of Germany's immigration law. *Antipode* 40(1): 55–78.

Betts, A. (ed.) (2012). *Global Migration Governance*. Oxford: Oxford University Press.

Bialasiewicz, L. (2012). Off-shoring and out-sourcing the borders of Europe: Libya and EU border work in the Mediterranean. *Geopolitics* 17(4): 843–866.

Bigo, D., Carrera, S., Guild, E., & Walker, R. (eds) (2010). *Europe's 21st Century Challenge: Delivering Liberty and Security*. London: Ashgate.

Boswell, C. (2003). The "external dimension" of EU immigration and asylum policy. *International Affairs* 79(3): 619–638.

Boswell, C. (2008). Evasion, reinterpretation and decoupling: European Commission responses to the "external dimension" of immigration and asylum. *West European Politics* 31(3): 491–512.

Brown, W. (2010). *Walled States, Waning Sovereignty*. New York: Zone Books.

Buckley, M. (2013). Locating neoliberalism in Dubai: Migrant workers and class struggle in the autocratic city. *Antipode* 45(2): 256–274. doi:10.1111/j.1467-8330.2012.01002.x

Casas, M., Cobarrubias, S., & Pickles, J. (2010). Stretching borders beyond sovereign territories? Mapping EU and Spain's border externalization policies. *Geopolítica(s)* 2(1): 71–90. doi: 10.5209/rev_GEOP.2011.v2.n1.37898

Casas-Cortes, M., Cobarrubias, S., & Pickles, J. (2013). Re-bordering the neighbourhood: Europe's emerging geographies of non-accession integration. *European Urban and Regional Studies* 20(1): 37–58. doi:10.1177/0969776411434848

Chauvin, S. (2014). Becoming less illegal: Deservingness frames and undocumented migrant incorporation. *Sociology Compass* 8(4): 422–432.

Cinalli, M., & El Hariri, A. (2012). Contentious opportunities: Comparing metropolitan policy-making for immigrants in France and Italy. In Carmel, E., Cerami, A., & Papadopoulos, T. (eds), *Migration and Welfare in the New Europe*, Bristol: Policy Press, 213–226.

Cohen, R. (1987). *The New Helots: Migrants in the International Division of Labour*. Aldershot: Gower.

Coleman, M. (2005). U.S. statecraft and the U.S.–Mexico border as security/economy nexus. *Political Geography* 24(2): 185–209.

Coleman, M. (2007). Immigration geopolitics beyond the Mexico–US border. *Antipode* 39(1): 54–57.

Coleman, M. (2009). What counts as the politics and practice of security, and where? Devolution and immigrant insecurity after 9/11. *Annals of the Association of American Geographers* 99(5): 904–913.

Coleman, M. (2012). The "local" migration state: The site-specific devolution of immigration enforcement in the U.S. South. *Law and Policy* 34(2): 159–190. doi:10.1111/j.1467-9930.2011.00358.x

Cornelius, W. (1994). Spain: Uneasy transition from labor exporter to labor importer. In Cornelius, W., Martin, P., & Hollifield, J. (eds), *Controlling Migration: A Global Perspective*, Stanford, CA: Stanford University Press, 331–370.

Cravey, A. (2003). Toqua una ranchera, por favor. *Antipode* 35(3): 603–621.

Dannecker, P. (2005). Transnational migration and the transformation of gender relations: The case of Bangladeshi labour migrants. *Current Sociology* 53(4): 655–674.

D'Aoust, A.-M. (2013). In the name of love: Marriage migration, governmentality, and technologies of love. *International Political Sociology* 7(3): 258–274.

Delano, A. (2009). From "shared responsibility" to a migration agreement? The limits for cooperation in the Mexico–United States case (2000–2008). *International Migration* 50: e42–e59.

Dikeç, M. (2012). Space as a mode of political reasoning. *Geoforum* 43(4): 669–676.

El Qadim, N. (2014). Postcolonial challenges to migration control: French–Moroccan cooperation practices on forced returns. *Security Dialogue* 45(3): 242–261.

Feldman, G. (2012). *The Migration Apparatus*. Stanford, CA: Stanford University Press.

Fitzgerald, S. (2010). Biopolitics and the regulation of vulnerability: The case of the female trafficked migrant. *International Journal of Law in Context* 6(3): 277–294.

Freeman, G. (1995). Modes of immigration politics in liberal democracies. *International Migration Review* XXIX(4): 881–902.

Gabriel, C. (2008). A "healthy" trade? NAFTA, labour mobility and Canadian nurses. In Gabriel, C., & Pellerin, H. (eds), *Governing International Labour Migration*, London: Routledge, 112–127.

Geddes, A. (2008). *Immigration and European Integration: Towards Fortress Europe?* Manchester: Manchester University Press.

Geiger, M., & Pecoud, A. (eds) (2010). *The Politics of International Migration Management*. Basingstoke: Palgrave Macmillan.

Gibney, M.J. (2008). Asylum and the expansion of deportation in the United Kingdom. *Government and Opposition* 43(2): 146–167.

Gill, N. (2009). Presentational state power: Temporal and spatial influences over asylum sector decision-makers. *Transactions of the Institute of British Geographers* 34(2): 215–233.

Gill, N. (2010). New state theoretic approaches to asylum and refugee geographies. *Progress in Human Geography* 34(5): 626–645.

Givens, T.E., Freeman, G.P., & Leal, D.L. (eds) (2009). *Immigration Policy and Security: US, European and Commonwealth Perspectives*. Oxford: Routledge.

Guiraudon, V. (1998). Citizenship rights for non-citizens: France, Germany, and the Netherlands. In Joppke, C. (ed.), *Challenge to the Nation State*, Oxford: Oxford University Press, 272–318.

Guiraudon, V. (2000). *Les politiques d'immigration en Europe: Allemagne, France, Pays-Bas*. Paris: L'Harmattan.

Guiraudon, V. (2003). The constitution of a European immigration policy domain: A political sociology approach. *Journal of European Public Policy* 10(2): 263–282.

Guiraudon, V., & Lahav, G. (2000). The state sovereignty debate revisited: The case of migration control. *Comparative Political Studies* 33(2): 163–195.

Hagan, J., Eschbach, K., & Rodriguez, N. (2008). US deportation policy, family separation, and circular migration. *International Migration Review* 42(1): 64–88.

Hampshire, J. (2013). *The Politics of Immigration: Contradictions of the Liberal State*. Cambridge: Polity Press.

Hayter, T. (2004). *Open Borders: The Case against Immigration Controls*. London: Pluto Press.

Hedman, E.-L. (2008) Refuge, governmentality and citizenship: Capturing "illegal migrants" in Malaysia and Thailand. *Government and Opposition* 43(2): 358–383.

Heisler, M. (1992). Migration, international relations and the new Europe: Theoretical perspectives from institutional political sociology. *International Migration Review* 26(2): 596–622.

Hiemstra, N. (2010). Immigrant "illegality" as neoliberal governmentality in Leadville, Colorado. *Antipode* 42(1): 74–102. doi:10.1111/j.1467-8330.2009.00732.x

Hollifield, J. (1992). *Immigrants, Markets, and States: The Political Economy of Postwar Europe*. Cambridge, MA: Harvard University Press.

Hollifield, J. (2004). The emerging migration state. *International Migration Review* 38(3): 885–912.

Hollifield, J.F. (2008). The politics of international migration. In Brettell, C.B., & Hollifield, J.F. (eds), *Migration Theory: Talking across Disciplines*, New York: Routledge, 183–237.

Hollifield, J., Martin, P.L., & Orrenius, P.M. (eds) (2014). *Controlling Immigration: A Global Perspective*. Stanford, CA: Stanford University Press.

Hubbard, P. (2005). Accommodating otherness: Anti-asylum centre protest and the maintenance of white privilege. *Transactions of the Institute of British Geographers* 30(1): 52–65.

Huysmans, J. (2000). The European Union and the securitization of migration. *Journal of Common Market Studies* 38(5): 751–777.

Huysmans, J. (2008). Politics of exception and unease: Immigration, asylum and terrorism in parliamentary debates in the UK. *Political Studies* 56(4): 766–788.

Hyndman, J. (2000). *Managing Displacement*. Minneapolis, MN: University of Minnesota Press.

Hyndman, J., & Mountz, A. (2008). Another brick in the wall? Neo-*refoulement* and the externalization of asylum by Australia and Europe. *Government and Opposition* 43(2): 249–269.

Jessop, B. (1993). Towards a Schumpeterian workfare state? Preliminary remarks on post-Fordist political economy. *New Political Economy* 40(Spring): 7–39.

Joppke, C. (1998). Immigration challenges the nation-state. In Joppke, C. (ed.), *Challenge to the Nation-State*. Oxford: Oxford University Press, 5–46.

Klotz, A. (2000). Migration after apartheid: Deracializing South African foreign policy. *Third World Quarterly* 21(5): 831–847.

Kofman, E. (2004). Family-related migration: A critical review of European studies. *Journal of Ethnic and Migration Studies* 30(2): 243–262.

Kofman, E. (2008). Managing migration and citizenship in Europe: Towards an overarching framework. In Gabriel, C., & Pellerin, H. (eds), *Governing International Labour Migration*, London: Routledge, 13–26.

Kofman, E., & Raghuram, P. (2006). Gender and global labour migrations: Incorporating skilled workers. *Antipode* 38(2): 282–303.

Koslowski, R. (ed.) (2011). *Global Mobility Regimes*. New York: Palgrave Macmillan.

Lafleur, J.-M. (2013). *Transnational Politics and the State: The External Voting Rights of Diasporas*. New York: Routledge.

Lahav, G., & Guiraudon, V. (2006). Actors and venues in immigration control: Closing the gap between political demands and policy outcomes. *West European Politics* 29(2): 201–223.

Lavanex, S. (2006). Shifting up and out: The foreign policy of European immigration control. *West European Politics* 29(2): 329–350.

Lavanex, S., & Schimmelfennig, F. (2009). EU rules beyond EU borders: Theorizing external governance in European politics. *Journal of European Public Policy* 16(6): 791–812.

Leitner, H. (2003). The political economy of international labor migration. In Sheppard, E., & Barnes, T. (eds), *A Companion to Economic Geography*, Oxford: Basil Blackwell, 450–467.

Leitner, H., Sheppard, E., & Sziarto, K.M. (2008). The spatialities of contentious politics. *Transactions of the Institute of British Geographers* 33(2): 157–172.

Leitner, H., & Strunk, C. (2014). Spaces of immigrant advocacy and liberal democratic citizenship. *Annals of the Association of American Geographers* 104(2): 348–356.

Martin, L.L. (2012a). Governing through the family: Struggles over US noncitizen family detention policy. *Environment and Planning A* 44(4): 866–888.

Martin, L.L. (2012b). "Catch and remove": Detention, deterrence, and discipline in US noncitizen family detention practice. *Geopolitics* 17(2): 312–334.

Massey, D. (2005). *For Space*. London: Sage.

McGregor, J. (2011). Contestations and consequences of deportability: Hunger strikes and the political agency of non-citizens. *Citizenship Studies* 15(5): 597–611. doi:10.1080/13621025.2011.583791

Menz, G., & Caviedes, A. (eds) (2011). *The Changing Face of Labour Migration in Europe*. Basingstoke: Palgrave Macmillan.

Mitchell, C. (1989). International migration, international relations, and foreign policy. *International Migration Review* 23(3): 681–708.

Mitchell, D. (1995). *The Lie of the Land: Migrant Workers and the California Landscape*. Minneapolis, MN: University of Minnesota Press.

Morris, L. (2002). *Managing Migration: Civic Stratification and Migrant's Rights*. London: Routledge.

Mountz, A. (2004). Embodying the nation-state: Canada's response to human smuggling. *Political Geography* 23(3): 323–345.

Mountz, A. (2010). *Seeking Asylum: Human Smuggling and Bureaucracy at the Border*. Minneapolis, MN: University of Minnesota Press.

Nevins, J. (2010). *Operation Gatekeeper and Beyond: The War on "Illegals" and the Remaking of the U.S.–Mexico Boundary*. London: Routledge.

Pearlman, W. (2014). Competing for Lebanon's diaspora: Transnationalism and domestic struggles in a weak state. *International Migration Review* 48(1): 34–75. doi:10.1111/imre.12070

Pellerin, H. (1999). The cart before the horse: The coordination of migration policies in the Americas and the neoliberal project of economic integration. *Review of International Political Economy* 6(4): 468–493.

Pellerin, H. (2008). Governing labour migration in the era of GATS: The growing influence of *lex mercatoria*. In Gabriel, C., & Pellerin, H. (eds), *Governing International Labour Migration*, London: Routledge, 27–42.

Peutz, N. (2006). Embarking on an anthropology of removal. *Current Anthropology* 47(2): 217–241.

Peutz, N., & DeGenova, N. (2010). Introduction. In Peutz, N., & DeGenova, N. (eds), *The Deportation Regime: Sovereignty, Space, and the Freedom of Movement*, Durham, NC: Duke University Press, 1–31.

Piper, N. (2006). Gendering the politics of migration. *International Migration Review* 40(1): 133–164.

Samers, M. (1999). "Globalization", migration and the geopolitical economy of the "spatial vent." *Review of International Political Economy* 6(2): 166–199.

Samers, M. (2004). An emerging geopolitics of "illegal" immigration in the European Union. *European Journal of Migration and Law* 6: 27–45.

Samers, M. (2010a). *Migration*. London: Routledge.

Samers, M. (2010b). The socio-territoriality of cities: A framework for understanding the incorporation of migrants in urban labor markets. In Glick-Schiller, N., & Caglar, A. (eds), *Locating Migration: Rescaling Cities and Migrants*, Ithaca, NY: Cornell University Press, 42–59.

Schuster, L. (2003). Sangatte: Smoke and mirrors. *Global Dialogue* 4(4): 57–68.

Sciortino, G. (2000). Toward a political sociology of entry policies: Conceptual problems and theoretical proposals. *Journal of Ethnic and Migration Studies* 26(2): 213–238.

Silvey, R. (2004). Transnational domestication: State power and Indonesian migrant women in Saudi Arabia. *Political Geography* 23(3): 245–264.

Silvey, R. (2006). Geographies of gender and migration: Spatializing social difference. *International Migration Review* 40(1): 64–81.

Silvey, R. (2007). Unequal borders: Indonesian transnational migrants at immigration control. *Geopolitics* 12(2): 265–279.

Silvey, R. (2009). Development and geography: Anxious times, anemic geographies, and migration. *Progress in Human Geography* 33(4): 507–515.

Silvey, R., & Olson, E. (2006). Guest editorial: Transnational geographies: Rescaling development, migration and religion. *Environment and Planning A* 38(5): 805–808.

Simmons, T. (2008). Sexuality and immigration: UK family reunion policy and the regulation of sexual citizens in the European Union. *Political Geography* 27(2): 213–230.

Sparke, M. (2006). A neo-liberal nexus: Economy, security and the biopolitics of citizenship on the border. *Political Geography* 25(2): 151–180.

Tesfahuny, M. (1998). Mobility, racism, and geopolitics. *Political Geography* 17(5): 499–515.

Van Houtum, H. (2010). Human blacklisting: The global apartheid of the EU's external border regime. *Environment and Planning D* 28(6): 957–976. doi:10.1068/d1909

Van Houtum, H., & Pijpers, R. (2007). The European Union as a gated community: The two faced border and immigration regime of the EU. *Antipode* 39(2): 291–309.

Varsanyi, M. (2008a). Immigration policing through the backdoor: City ordinances, the "right to the city", and the exclusion of undocumented day laborers. *Urban Geography* 29(1): 29–52.

Varsanyi, M. (2008b). Rescaling the "alien", rescaling personhood: Neoliberalism, immigration and the state. *Annals of the Association of American Geographers* 98(4): 877–896.

Varsanyi, M. (2011). Neoliberalism and nativism: Local anti-immigrant policy activism and an emerging politics of scale. *International Journal of Urban and Regional Research* 35(2): 295–311.

Varsanyi, M., Lewis, P., Provine, M., & Decker, S. (2012). A multilayered jurisdictional patchwork: Immigration federalism in the United States. *Law and Policy* 34(2): 138–158.

Walters, W. (2004). Secure borders, safe haven, domopolitics. *Citizenship Studies* 8(3): 237–260.

Walters, W. (2008). Anti-illegal immigration policy: The case of the European Union. In Gabriel, C., & Pellerin, H. (eds), *Governing International Labour Migration*, New York: Routledge, 43–59.

Walters, W. (2010). Anti-political economy: Cartographies of "illegal immigration" and the displacement of the economy, In Best, J., & Paterson, M. (eds), *Cultural Political Economy*, London: Routledge, 113–138.

Walton-Roberts, M. (2004). Rescaling citizenship: Gendering Canadian immigration policy. *Political Geography* 23(3): 265–281.

Winders, J. (2007). Bringing back the (b)order: Post-9/11 politics of immigration, borders, and belonging in the contemporary U.S. South. *Antipode* 39(5): 920–942. doi:10.1111/j.1467-8330.2007.00563.x

Wright, R.A., & Ellis, M. (2000). Race, region and the territorial politics of immigration in the U.S. *International Journal of Population Geography* 6(3): 197–221.

Zolberg, A. (1999). Matters of state: Theorizing immigration policy. In Hirschman, C., Kasinitz, P., & DeWind, J. (eds), *The Handbook of International Migration: The American Experience*, New York: Russell Sage, 71–93.

Zolberg, A. (2002) Guarding the gates. In Calhoun, C., Price, P., & Timmers, A. (eds), *Understanding September 11*, New York: New Press, 285–299.

Chapter 36

Everyday Political Geographies

Sara Fregonese

University of Birmingham, England, UK

The war has turned into a siege … I live on the eighth floor of a building that might tempt any sniper, to say nothing of a fleet now transforming the sea into one of the fountainheads of hell. The north face of the building, made of glass, used to give tenants a pleasing view over the wrinkled roof of the sea. But now it offers no shield against stark slaughter. Why did I choose to live here? What a stupid question! I've lived here for the past ten years without complaining about the scandal of glass. (Darwīsh 2013: 6)

Beirut 2008: The little things

At about 4 pm on May 7, 2008, a journalist for a Lebanese newspaper looks at the clock in his office near Beirut's seaside and decides to leave early. After driving in his car through the city, he gets home and phones his mother and friends before turning on the news. The sound of automatic fire resonates from nearby neighborhoods. He is trapped in his flat. Meanwhile, a correspondent for a foreign press agency jumps in a taxi from Beirut's southern periphery to reenter the city center. Bullets are being shot in the air and raining back down. He hears the shots intensifying instead of decreasing as the taxi approaches the city center. He takes refuge in his office and prepares a news release before escaping home on foot, away from the source of the fire. In a different part of town, an NGO worker frantically tries to contact friends while on her way home, having heard what sounded definitely not like the usual celebratory shots that people in Beirut fire at weddings, football matches, or at political rallies, but like confrontation fire. She reaches home after passing a checkpoint manned by individuals in unrecognizable uniforms. She stays indoors for three days.[1]

In the previous hours, Sayyed Hassan Nasrallah, the leader of Lebanon's party-and-militia Hezbollah, gave an inflammatory press conference from his base in south Beirut. His speech responded to the Lebanese government's decision to outlaw a private telecommunications network owned by Hezbollah and extending through several areas of the country. The network,

The Wiley Blackwell Companion to Political Geography, First Edition.
Edited by John Agnew, Virginie Mamadouh, Anna J. Secor, and Joanne Sharp.
© 2015 John Wiley & Sons Ltd. Published 2017 by John Wiley & Sons Ltd.

aimed at coordinating paramilitary operations without being intercepted by the Israeli army and intelligence, was now declared a violation of sovereignty by the government. In his speech, Nasrallah stated that the government had crossed a "red line" and compared its decision to a declaration of war against Hezbollah (Al Jazeera English, see Vineyardsaker 2008 for an English translation). This controversy started a week of armed clashes in Beirut between armed militants affiliated to two party coalitions representing the government majority and the Hezbollah-led opposition. By late May, after spreading outside Beirut, the clashes had killed 69 people, including several civilians, and injured more than 180.[2] Following Nasrallah's speech, the clashes profoundly changed the inhabitants' everyday experience of the city. During that week, as in its long aftermath, the everyday life of residents in the Lebanese capital was remapped according to the violent geographies of the clashes. Beirutis drastically changed their daily routines; some moved house, avoided exposure to fire, managed vulnerability in public space, and negotiated the new territorialities established by irregular armed militants affiliated to official political parties.

This chapter is concerned with the everyday practices that have reshaped the urban political geography of Beirut leading up to, during, and in the aftermath of the urban clashes of May 2008. I use this case to argue that there are two underexplored realms – the *more-than-human* and the *more-than-state* – where political geographers can engage more comprehensively with the everyday elements of the relations between space and politics. There has been, in the past 15 years, a growing interest in political geography and critical geopolitics for the everyday experience of the state's mechanisms of power, and the ways in which geopolitics matter in people's everyday lives beyond official statecraft and diplomacy. Despite Nigel Thrift's (2000) encouragement to study "the little things" in geopolitics, the "actual practices … and their attendant territorializations within which geopower ferments and sometimes boils over" (2000: 385), it is only more recently that tangible agendas for researching the everyday are being proposed by political geographers.

First, I will consider some political geography literature that has engaged with everyday practices of power and situate urban conflict as the empirical realm in which to observe everyday dynamics of space and power evolving. Secondly, consideration is given to the realm of the more-than-human and the ways in which it has characterized academic debates in political geography and geopolitics so far. The case of the management of Beirut's urban security to face and improve socio-political polarization offers an empirical platform for a non-anthropocentric look at the everyday political geographies of the city. Here, urban conflict is managed not only through political rhetoric, but also through the deployment of material artifacts that regulate everyday spatial relations in the city.

Thirdly, the attention shifts to another understudied realm of engagement with the everyday: the more-than-state. Emerging literature on the role of non-state actors in political geography is briefly analyzed, before taking the case of Beirut's Rafiq Hariri International Airport (RHIA) as an example of how irregular non-state actors contest and redefine state sovereignty through their claims and specific uses of ordinary urban infrastructure.

Toward everyday political geographies: The case of urban conflict

A number of agendas in political geography and especially in critical geopolitics are currently disturbing the predominance of representation and discursive deconstruction of the rhetoric of statecraft in favor of more-than-discursive and everyday manifestations of (geo)politics. The argument that these agendas share is that, far from being relegated to the elite sphere of official statehood and international politics, geopolitics is instead connected in complex ways with people's daily lives, actively shaping their perceptions, experience, and management of power, risk, fear, violence, and so on. Political geographers have focused on the non-elite domains and practices that comprise the ways in which we can experience geopolitics, such as place (Nicley 2009), the body (Hyndman 2007), displacement and exile (McConnell 2009; Ramadan 2009), the built environment (Fregonese 2009), peacemaking (Koopman 2011), and popular culture (Dodds 1996).

Cities and their built environment are at the center of a growing corpus of scholarly work investigating the connections between specific urban sites and wider processes of geopower, for example the militarization and pervading surveillance of urban space. This interest has been shared by a multidisciplinary range of scholars who, using different theoretical frameworks and methodologies, have attempted to unpack those connections under the term "urban geopolitics" (Demarest 1995; Vincent & Warf 2002; Douzet 2003; Graham 2004; Amar 2009; Fregonese 2009; Ramadan 2009; Yacobi 2009). Urban geopolitics' main tenet is that the city and its architectures are not mere backdrops to conflict and violence, but also its deliberate targets, as well as part of the mechanism of geopower.

One risk with the perspective of some urban geopolitics scholarship is its limitation to the most dramatic actions against urban architectures – for example, state-driven operations against refugee camps (Graham 2002; Ramadan 2009) and slums (Amar 2009). Without diminishing the importance of these dramatic events, I argue elsewhere (Fregonese 2012b) for the need for an *everyday urban geopolitics* that focuses not only on big-scale events of urban militarism and targeting, but also on the more banal practices and embodiments of geopolitics that constantly populate the lives of ordinary dwellers.

In geography and architecture, an increasing number of contributions is accounting for less dramatic and more ordinary urban spatialities and their involvement in the making of social conflict, war, security, and peace. These spatialities include, for example, the micro-resilience of Palestinian street vendors in the most tense areas of Jerusalem (Pullan 2006), the targeting and destruction of ordinary physical spaces (such as private homes and post offices) where a diverse community interacted in former Yugoslavia (Coward 2009), and the more individual everyday experiences of Palestinian refugees in Lebanon, researched by Ramadan (2009). All of these accounts emphasize the importance of the socio-spatial, architectural, and generally more-than-human aspects that shape the ways in which geopower and territoriality are experienced in daily life. They also point to spatialities and everyday practices that do not emanate directly from the state, but from constellations of non-state actors such as refugees, private capital, insurgencies, and local communities.

The everyday as more-than-human

Human geographers have regained interest in the relations between humans and objects in a response to a dematerializing "cultural turn," which emphasizes text and representation over human encounters with objects, nature, and technologies, and over the spatiality of the lived and emotional body (Anderson & Tolia-Kelly 2004). Far from merely doing away with discourse, the "materialist return" (Whatmore 2006) has instead contributed important debates about the complex interplays of discourse, materiality, and affect (Anderson & Wylie 2009), which also resonate in critical geopolitics (Müller 2008).

While political geographers appreciate the everyday manifestations of the distribution of state power throughout human lives (Painter 2006) and the role of mundane practices in shaping national communities (Billig 1995), these analyses concentrate on social relations, leaving out "the nonhuman traffickings of power" (Meehan, Shaw, & Marston 2013: 2) involved in the functioning of states. It is only more recently that a substantial debate in political geography and geopolitics has developed around the role of objects and materiality in the functioning of state power and global politics, which considers objects not only as containers of power, but as mediators of it, as they "enable, disable and transform state power, beyond just reflecting it" (Meehan et al. 2013: 2).

Developing the "focus on the everyday intersection of the human body with places, environments, objects, and discourses linked to geopolitics" advocated in earlier writings on methodologies for critical geopolitics (Dittmer & Gray 2010), Dittmer (2014) also calls for a "materialist return" in geopolitics to obviate the focus of textual discourse that has predominated since the 1990s through assemblage and complexity theories. Dittmer follows Escobar's (2007) argument to replace the anthropocentrism of geopolitics with a more inclusive notion of the political, so that more complex connections between micro and macro processes stand out.

These arguments resonate indirectly in the studies of conflict and violence by scholars who advocate stepping beyond anthropocentrism and overcoming assumptions that, in war, the damage to material objects and buildings is subordinate to the damage to human subjects, and instead needs to be considered as a distinctive form of violence (urbicide; Coward 2006), or at least as one intimately connected to the human experience of genocide (Shaw 2004).

A more-than-human political geography is illustrated here through the example of conflict and its regulation in Beirut, in a period of intense polarization between the assassination of former prime minister Rafiq al Hariri in 2005 and the aftermath of the 2008 clashes described at the beginning of this chapter. The discussion follows the approach of science and technology studies (STS), which – although still underemployed in the study of politics and conflict (Fregonese & Brand 2009) – can speak to both to the materialist (re)turn in political geography and the appreciation of architectural and non-anthropocentric elements in studies of conflict. The STS approach breaks down the cognitive split between human subjects and non-human objects, and, in this case, can be used to consider holistically the social and material processes that shape conflict. The case of Beirut shows that conflict does not happen in a void, but is mediated instead through the built environment of cities and, most importantly for this chapter, it is often non-dramatic and subtle everyday elements that signal the worsening (or improving) of urban tension, polarization, and conflict. Rather than simply reflecting an open state of war, the built environment and urban artifacts are powerful embodiments – and mediators – of daily, slow-burning tension, or gradual return to everyday coexistence (Brand & Fregonese 2013).

(De)radicalizing Beirut

After a phase of relatively peaceful postwar redevelopment (1990–2005), the explosion that, on February 14, 2005, assassinated former prime minister Rafiq al-Hariri on central Beirut's waterfront was possibly the biggest shock to the political landscape of post–civil war Lebanon. Mass demonstrations occupied the reconstructed city center (Rafiq al Hariri's own financial and real estate project), demanding and obtaining the withdrawal of the prime suspects for the killing – the Syrian army and intelligence, present in Lebanon since 1976. What started as a unified civic protest movement soon polarized along new political alliances. In a country exhausted by the Israeli invasion of 2006, the presidential vacuum of 2007, the occupation of the perimeter of the government palace by cabinet opposition protesters, and numerous clashes between opposed armed groups and activists at sensible urban frontlines, socio-political tension culminated in deadly clashes in early May 2008. This phase marked the worst intrastate violence since the civil war of 1975–90. While a Qatari-brokered peace agreement ended the presidential stalemate and brought some degree of stability, the uprising in Syria and its degeneration into civil war since 2012 have exacerbated sectarian and political conflict in Lebanon. These wider geopolitical processes are intimately connected with specific urban divisions that have emerged in several Lebanese cities.

However, as Fawaz and Deboulet (2011) notice, open conflict is only one type of manifestation of identity politics in the city, and instead:

> it is possible to investigate how these politics penetrate the context of everyday urban life ... through silent ruse and subterfuge, within quotidian practices of everyday life. (2011: 116)

Even before the bloody 2008 clashes everyday life in the Lebanese capital was shaped by objects and spaces that not only reflected slow-burning social tension, but were also invested with the power of defusing it. These include defensive urbanism, the use of memorials and of of political paraphernalia, and spaces of leisure and retail designed to promote social mix.

Everyday urban life in contemporary Beirut is characterized by an extensive apparatus of security and surveillance artifacts (Fawaz, Harb, & Gharbieh 2011). Far from being exclusive to conflict zones, these artifacts are part of a military urbanism that pervades many cities globally (Graham 2010). However, in Beirut this translates locally into multiple networks of private security infrastructures parallel to state security, functioning for the protection of political leaders and their areas of residence, and owned and managed by a range of non-state armed groups affiliated with political parties. These artifacts are not only embedded in the daily practices and mobilities of urban residents (think for example of road closures and "private" checkpoints in residential neighborhoods), but also reflect and catalyze socio-political inequality and division (Fawaz, Harb, & Gharbieh 2012).

The recently reconstructed Beirut Central District (BCD), the supposedly "neutral" postwar site of urban encounter, has not escaped the process of securitization following the post-2005 political turmoil. A hard, concrete-rich defensive urbanism, reminiscent of the "rings of steel" around Belfast and London during the Troubles, came into being in the already highly privatized and securitized space of the BCD. Concrete blocks were erected around the government palace, closing down one lane of the adjacent road to keep traffic at a distance; guarded metal barriers were installed at the pedestrian entrances to the city center; chicanes for vehicles were organized outside public buildings and security checkpoints installed at their entrances; undercar inspection mirrors regularly used at the entrances of buildings and car parks; even trash bins closed for fear, as BCD managers explained to us in 2009, of bombs being hidden inside by terrorists; and all of this happening under the gaze of dozens of CCTV cameras. The ring of steel was an indicator of the country's socio-political instability, but also had an impact on the BCD's financial stability[3] and – most importantly – changed the everyday direct economy of BCD workers, shop owners, and employees, who, between 2005 and 2009, found their businesses surrounded by security checks. Consequently, the clientele decreased, intimidated by a heavily securitized BCD, especially at nighttime and with the need to walk across a large sit-in of protesters encamped in tents. As a result, at least 40 businesses closed permanently in the BCD (Khazen.org n.d.).

Another remarkable change in the landscape of central Beirut that reflected the socio-political fluctuations was the installation – in a space originally designed to be neutral of all political expressions (Khalaf & Khoury 1993) – of numerous memorials and commemo-rating posters of the victims of a chain of political assassinations that started in 2004 with the attempted murder of former economy minister Marwan Hamade, included the killing of former PM Rafiq al Hariri in 2005, and continued until the most recent assassination: that of former finance minister Mohammed Shatah in late 2013. A large tent on Martyrs' Square hosts the shrine of Rafiq al Hariri and his bodyguards. At a few meters' distance, a giant poster of *Al-Nahar* newspaper editor and MP Gebran Tueni, assassinated in December 2005, hangs from the newspaper headquarters building. Next to this stands a memorial to slain historian and journalist Samir Kassir. On the opposite side of Martyrs' Square hangs another giant poster of MP Pierre Amine Gemayel, also killed in a bomb attack in November 2006. These more or less makeshift memorials, funded with private money, do not only *represent* the breakdown of political relations in Lebanon, but also *reproduce* and *reinforce* this breakdown: Their prominent occupation of space creates new practices, encounters, and affects in the supposedly non-partisan territory of the BCD. Besides occupying a prominent site in the BCD, Hariri's shrine attracts groups of visitors who come to pay their respects and habitual practices of pilgrimage, while a digital clock displays the days passed since Hariri's assassination, for which no one has yet been convicted. The clock and the other giant posters are not state-recognized memorials, as revealed by their makeshift nature (plastic tents, hanging posters), but are instead the material means through which specific political and sectarian communities in Lebanon negotiate memory and history. They are also everyday sights for residents, through which deep social wounds and divisions are kept open and prevented from healing.

If some objects mediate the continuation of tension and prevent the wounds of division from closing, there are other non-human elements of Beirut's built environment that instead have the power to diffuse tension. Since the civil war (Chakhtoura 2005; Maasri 2008), political posters, flags, graffiti, and portraits of dead and living politicians have been a constant presence in the city streets and neighborhoods, signaling the real or intended dominance of a particular political party or celebrating localized versions of the political past in the absence of a state-recognized collective memory of the civil war. As part of reconciliation and de-escalation efforts in the aftermath of the 2008 clashes, the government majority ordered the removal of all political posters and signs from municipal Beirut. It was common, in the follow-ing months, to walk around the city and spot big patches of white paint on walls, signs of a very mundane attempt to de-escalate social tension.

The state is not the sole actor to devise ways to de-escalate everyday tensions: Private capital has also aimed at keeping communities together through mundane interventions in Beirut. A particular retail space, Beirut Mall, went to great lengths to neutralize community-specific references as well as convey a sense of class equality. According to its manager, careful interior design choices were made to follow a modernist and minimal design without specific ornamentations or arabesques. Particular attention is dedicated to avoid displaying openly religious decoration during festivities like Ramadan and Christmas, giving preference to generally festive signs like snowflakes and stars (interview, January 2009).

Beirut is an example of how more-than-human elements mediate the relations between space politics and power on a day-to-day basis. Political escalation or de-escalation is at once reflected and mediated by a plethora of artifacts and physical infrastructures comprising everyday urban spaces.

The everyday as more-than-state

Despite considering the prosaic and non-human mechanisms underpinning power, analyses like those by Painter (2006) and Meehan et al. (2013) do not venture beyond the workings of the official and legitimate nation-state. Although increasingly inclusive of the everyday and the more-than-human, political geography and critical geopolitics remain predominantly state-centered and mostly confined to the discursive agendas of sovereign states at war with each other (Kearns 2008).

The view of the world as composed by a range of fully sovereign and immobile territories has long been challenged (Agnew 1994) for its exclusion of several other entities other than the state, which, far from developing in total opposition to or independently from it, instead weave complex relationships with the state. Contemporary changes in the way in which war is waged globally, for example, imply "the unraveling of the national model of citizenship and service at the center of geopolitical forms of state and population security" (Cowen & Smith 2009: 36), where several actors other than the state – militias, drug cartels, terrorist cells, and others – are involved in conflict with states and produce everyday spaces and practices of power whose grounded mechanisms remain understudied.

However, bar a number of quantitative datasets overviewing the presence of non-state actors in wars (Sundberg, Eck, & Kreutz 2012; Cunningham, Gleditsch, & Salehyan 2013), ethnographic research into the actual mechanisms through which non-state actors conduct war is less developed, and "the literature is still in need of a better understanding of the micro-geographies of small wars" (Korf 2011: 733). A political geography of the everyday, therefore, ought to keep its attention on the more-than-state.

Unofficial and irregular actors, however, do not thrive in isolation from the state: They are *more-than-state* because their power derives from a different relation *with* the state. In this sense, more-than-state geopolitics go beyond Gerry Kearns' (2008) notion of "progressive geopolitics" as simply an exploration of the ways in which other-than-state actors make peace. More-than-state political geographies conceptualize non-state actors within a relational rather than oppositional link to the state, emphasizing the mutual shaping between everyday practices, performances, rhetoric, and landscapes of non-state actors and the power mechanisms of the state.

A number of political geographers are emphasizing the relation of more-than-state actors to the state, such as in deploying mundane practices of sovereignty: those of governments in exile (McConnell 2009) that "mimic" the diplomatic bureaucracies and performances of recognized states, thus collapsing traditional notions of state/non-state, proper or legitimate/illegitimate. Recent work has also drawn attention to the multiplicity of intra- and transnational actors that negotiate – often through unequal interactions – with the state sovereignty over resources (Emel, Huber, & Makene 2011), revealing "the everyday processes at work beyond the state-centered bureaucratic, elitist, disembodied realms" (Lunstrum 2013: 9). However, political geographers still know very little, for example, about how non-state actors – both those outsourced by the state, like private companies, and those who oppose it, like revolutionary militias – interact with everyday military landscapes (Woodward 2014). Besides their role in political violence, non-state actors ought to be considered also for their everyday engagement with state policies, as in the case of the strategy of international protective accompaniment (Koopman 2011), where more-than-state knowledges and practices are developed by volunteers who often interact with state actors (through physical presence, persuasion, mediation, and enhancement of the visibility of threatened communities) in ways that "ground geopolitics to everyday life" (2011: 280).

Beirut's more-than state urban geopolitics

Including the more-than-state in political geographical and geopolitical analysis is essential when considering Lebanon, a country where the boundary between state and non-state, official and "irregular" is blurred. The notion of "hybrid sovereignties" (Fregonese 2012a) applies the hybridity approach in geography (Whatmore 2002) to the subject of sovereignty, in order to trace the cross-contaminations and spaces of contact between the state-sanctioned and the irregular in situations of political violence, as well as in everyday urban governance. This view considers the practice of sovereignty as a negotiation between constellations of actors who, rather than subsisting independently from the state, operate instead *across* classic state/non-state dichotomies and entail complex relations with the state. These, in Lebanon, have included resistance and violent confrontation, but also less dramatic interactions like daily negotiation and even coordination with the state (Fregonese 2012a).

In a city like Beirut, where conflict is engrained in the urban everyday, as already seen, sovereignty is reproduced through negotiations over use and control of urban everyday infrastructure and materialities that involve complex relations between both state and non-state actors. As Mona Fawaz and Agnes Deboulet (2011) note, city-dwellers and other non-state actors "contribute ... to redefining legitimacies and sometimes sovereignties" (2011: 119) over and through urban space, by contesting how urban infrastructures should be run and who can make claim to them.

Beirut's Rafiq al Hariri International Airport (RHIA) is an example of how crucial national infrastructure is a terrain for the renegotiation of contested sovereignty by both state and non-state actors. RHIA is situated in the southern suburb of Bir Hassan. Inaugurated on June 6, 1939, its construction was made possible by the flattening of the sand dunes in the area, the sand from which was then used to refurbish the docks in the city center port (Yacoub 2010). An additional runway was added on sea embankments in 1998, and in 2005 the airport was renamed after slain PM Rafiq al Hariri. Although connected to the city center by a new motorway separated from the surroundings by high walls and tunnels, Beirut airport's status as crucial national infrastructure is made problematic by its location in the city's southern suburbs (al-Dahya). This is a Hezbollah-dominated area, hosting its political headquarters and characterized by a mixture of informal settlements and a Hezbollah-controlled urbanism, developed after the extensive destruction of the area by the Israeli army in 2006 (Fawaz 2007). The current controversy over RHIA was at the basis of the 2008 violence, when government officials accused Hezbollah of using their private surveillance cameras to prepare attacks against private jets using one of the runways. Following the accusations, the government removed RHIA's security chief. Allegedly close to Hezbollah, the security chief would have tolerated the party's installation of these cameras.

The infrastructure of the airport then became a terrain for redefining sovereignty, as an infrastructure owned by the state, but at the same time including non-state "observant practices" (MacDonald, Hughes, & Dodds 2010). The RHIA controversy also shows not only how local sovereignties are negotiated through the control of specific infrastructure, but also how geopolitics are referred to and grounded in the everyday practices concerning its management. In a crucial press conference in response to the government's accusations against Hezbollah and the removal of RHIA's security chief, Hezbollah leader Sayyed Hassan Nasrallah declared, for example, that the Lebanese government wanted "to transform the airport into a base for the F.B.I. and C.I.A. and Israeli Mossad" (Al Jazeera English, see Vineyardsaker 2008 for English translation).

The 2008 spying row and following violence is a dramatic yet not isolated episode in the contested sovereignty over Beirut's airport. The hybrid sovereignties that to this day govern RHIA were never as evident as on September 18, 2010, when a convoy of Hezbollah suburban utility vehicles (SUVs) without license plates drove onto the runway and welcomed former head of

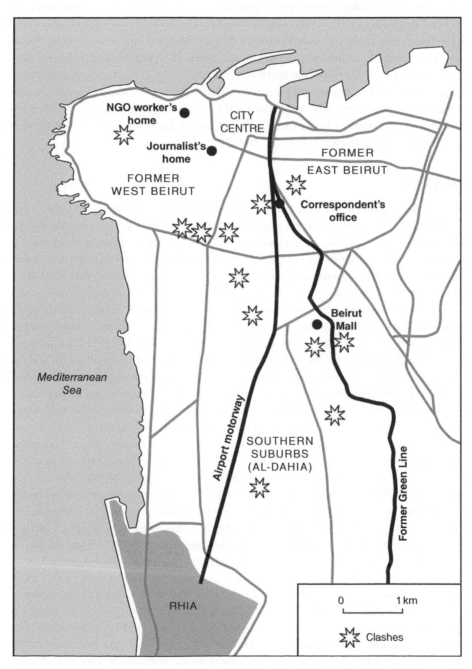

Figure 36.1 Map of greater Beirut showing the locations of the May 2008 clashes, the places mentioned in the interviewees' oral histories, and the artifacts mentioned in the text.

Lebanese security services General Jamil al-Sayyed, just returned from Paris and still suspected of being involved in the Hariri assassination in 2005. The convoy escorted al-Sayyed from his plane to the airport's VIP lounge, from where he gave a press conference in a space usually reserved for state representatives. He was finally escorted to his home by the same convoy.

Again, in August 2013, the control of the state over the airport and its surroundings was put in doubt as two Turkish airline pilots were kidnapped not far from an army checkpoint on the road that leads to RHIA. This event reignited the controversy over the degree of control that Hezbollah effectively has over Lebanon's only commercial airport and about the supposed lack of tangible action by government and army. The possibility of using the Renee Moawad airport near Tripoli in the north of the country has been discussed (Luca 2012), although this is seen as a drastic and potentially divisive political as well as logistical gesture.

Infrastructure and materiality are instrumental to producing Beirut's political urban geographies. Sovereignty appears here as a collective of knowledges and practices that thrive not in total isolation or opposition to the state, but alongside it.

Conclusions

This chapter has focused on two realms where political geographers can engage with the everyday workings of space and (geo)power: the more-than-human elements that comprise the everyday spatial practice and experience of (geo)politics; and the complex relations between state and non-state actors that comprise the everyday practice of territoriality, sovereignty, and violence. These two realms of engagement are by no means completely separate. I chose the case of Beirut's airport to highlight one possible way in which political geographers can explore and interpret both how non-state actors operate alongside the state and how specific materialities shape their relationship. Cases like this can tell us much about the everyday workings of sovereignty and territoriality in contemporary states. There is still a vast debate to develop in political geography and critical geopolitics around the role of objects, technologies, and materialities for state and non-state actors, not only as containers of power, but as mediators for its deployment. Empirically, the case of Beirut illustrates how political escalation as well as de-escalation are at once reflected and mediated by several artifacts and physical infrastructures that shape everyday urban life. In Beirut as in other contested cities, urban physical and infrastructural elements of conflict are often the ones affecting residents' everyday lives. And yet, although urban territory and its management are becoming more known in conflict studies, too little attention is paid by the practice of diplomacy and conflict resolution to how urban environments are geopolitically relevant.

As seen here, Beirut's urban infrastructure is also the terrain on which state sovereignty is redefined and legitimacy claimed by non-state actors, who develop their own territorialities and securities not in isolation from, but alongside, the state. Political geography's often still state-centered approach also needs to pay more attention to more-than-state knowledges and practices, and to the ways in which specific materialities shape these knowledges and practices – not only in Beirut, but also elsewhere. Possible avenues to achieve this include engaging with non-anthropocentric theories, as reviewed in this chapter; increasing the presence and importance of ethnographic fieldwork and methods in political geography and geopolitics to allow for the grasping of socio-material dynamics that are often subtle and/or context specific; and, finally, conducting analysis that focuses on the relevance of (geo) political practices to everyday life. In sum, theoretical, methodological, and analytical engagements with *matter and what matters* can produce political geographies of the everyday with potentially broader impact.

Acknowledgment

The section "(De)radicalizing Beirut" is based on Chapter 5 ("The Beirut case") in Brand & Fregonese (2013).

Notes

1 This narration is based on a selection of interviews conducted by the author during her fieldwork in Beirut between October and December 2010. The anonymity of the interviewees has been maintained on the basis of the release forms they signed.
2 Cited in United Nations Security Council (2008).
3 The BCD (owned by the company Solidere) has become known as one of the fastest-growing business and real estate enterprises in the Mediterranean and Middle East. However, according to the 2006 Solidere annual report, company share prices decreased from US$26 to $16 that year (mainly due to Israel's military campaign in Lebanon) and no land sales were concluded that year.

References

Agnew, J. (1994). The territorial trap: The geographical assumptions of international relations theory. *Review of International Political Economy* 1(1): 53–80.

Amar, P. (2009). Operation Princess in Rio de Janeiro: Policing "sex trafficking", strengthening worker citizenship, and the urban geopolitics of security in Brazil. *Security Dialogue* 40(4–5): 513–541. doi:10.1177/0967010609343300

Anderson, B., & Tolia-Kelly, D. (2004). Matter(s) in social and cultural geography. *Geoforum* 35(6): 669–674. doi:10.1016/j.geoforum.2004.04.001

Anderson, B., & Wylie, J. (2009). On geography and materiality. *Environment and Planning A* 41(2): 318–335. doi:10.1068/a3940

Billig, M. (1995). *Banal Nationalism*. London: Sage.

Brand, R., & Fregonese, S. (2013). *The Radicals' City: Urban Environment, Polarisation, Cohesion*. Design and the Built Environment. Farnham: Ashgate.

Chakhtoura, M. (2005). *La Guerre Des Graffiti. Liban 1975–1977*. Beirut: Dar an-Nahar.

Coward, M. (2006). Against Anthropocentrism: The destruction of the built environment as a distinct form of political violence. *Review of International Studies* 32(3): 419–437. doi:10.1017/S0260210506007091

Coward, M. (2009). *Urbicide: The Politics of Urban Destruction*. Oxford: Routledge.

Cowen, D., & Smith, N. (2009). After geopolitics? From the geopolitical social to geoeconomics. *Antipode* 41(1): 22–48. doi:10.1111/j.1467-8330.2008.00654.x

Cunningham, D.E., Skrede Gleditsch, K., & Salehyan, I. (2013). Non-state actors in civil wars: A new dataset. *Conflict Management and Peace Science* 30(5): 516–531. doi:10.1177/0738894213499673

Darwīsh, M. (2013). *Memory for Forgetfulness: August, Beirut, 1982*, trans. I. Muhawi. Berkeley, CA: University of California Press.

Demarest, G. (1995). Geopolitics and urban armed conflict in Latin America. *Small Wars and Insurgencies* 6(1): 44–67. doi:10.1080/09592319508423098

Dittmer, J. (2014). Geopolitical assemblages and complexity. *Progress in Human Geography* 38(3): 385–401. doi:10.1177/0309132513501405

Dittmer, J., & Gray, N. (2010). Popular geopolitics 2.0: Towards new methodologies of the everyday: Popular geopolitics 2.0. *Geography Compass* 4(11): 1664–1677. doi:10.1111/j.1749-8198.2010.00399.x

Dodds, K. (1996). The 1982 Falklands War and a critical geopolitical eye: Steve Bell and the If... cartoons. *Political Geography* 15(6–7): 571–592. doi:10.1016/0962-6298(96)00002-9

Dodds, K. (2007). Steve Bell's eye: Cartoons, geopolitics and the visualization of the "War on Terror." *Security Dialogue* 38(2): 157–177. doi:10.1177/0967010607078536

Douzet, F. (2003). Immigration et géopolitique urbaine: La question du logement à Oakland, Californie. *Espace, Populations, Sociétés* 21(1): 103–115. doi:10.3406/espos.2003.2067

Emel, J., Huber, M.T., & Makene, M.H. (2011). Extracting sovereignty: Capital, territory, and gold mining in Tanzania. *Political Geography* 30(2): 70–79. doi:10.1016/j.polgeo.2010.12.007

Escobar, A. (2007). The "ontological turn" in social theory: A commentary on "Human Geography without Scale", by Sallie Marston, John Paul Jones II and Keith Woodward. *Transactions of the Institute of British Geographers* 32(1): 106–111. doi:10.1111/j.1475-5661.2007.00243.x

Fawaz, M. (2007). Beirut: The city as a body politic. *ISIM Review* 20(1): 22–23.

Fawaz, M., & Deboulet, A. (2011). Contesting the legitimacy of urban restructuring and highways in Beirut's irregular settlements. In Davis, D., & Libertun de Duren, N. (eds), *Cities and Sovereignty: Identity Politics in Urban Spaces*, Bloomington, IN: Indiana University Press, 117–151.

Fawaz, M., Harb, M., & Gharbieh, A. (2011). Beirut and its dwellers: The security map. Published as addendum to *Al-Akhbar Daily*, May [in Arabic].

Fawaz, M., Harb, M., & Gharbieh, A. (2012). Living Beirut's security zones: An investigation of the modalities and practice of urban security: Living Beirut's security zones. *City and Society* 24(2): 173–195. doi:10.1111/j.1548-744X.2012.01074.x

Fregonese, S. (2009). The urbicide of Beirut? Geopolitics and the built environment in the Lebanese Civil War (1975–1976). *Political Geography* 28(5): 309–318. doi:10.1016/j.polgeo.2009.07.005

Fregonese, S. (2012a). Beyond the "weak state": Hybrid sovereignties in Beirut. *Environment and Planning D: Society and Space* 30(4): 655–674. doi:10.1068/d11410

Fregonese, S. (2012b). Urban geopolitics 8 years on: Hybrid sovereignties, the everyday, and geographies of peace. *Geography Compass* 6(5): 290–303. doi:10.1111/j.1749-8198.2012.00485.x

Fregonese, S., & Brand, R. (2009). Polarisation as sociotechnical phenomenon: A bibliographical review. *Journal of Urban Technology* 16(2–3): 9–34. doi:10.1080/10630730903278546

Graham, S. (2002). Bulldozers and bombs: The latest Palestinian–Israeli conflict as asymmetric urbicide. *Antipode* 34(4): 642–649. doi:10.1111/1467-8330.00259

Graham, S. (2004). *Cities, War, and Terrorism: Towards an Urban Geopolitics*. Malden, MA: Blackwell.

Graham, S. (2010). *Cities under Siege: The New Military Urbanism*. London: Verso.

Hyndman, J. (2007). Feminist geopolitics revisited: Body counts in Iraq. *Professional Geographer* 59(1): 35–46. doi:10.1111/j.1467-9272.2007.00589.x

Kearns, G. (2008). Progressive geopolitics. *Geography Compass* 2(5): 1599–1620. doi:10.1111/j.1749-8198.2008.00125.x

Khazen.org (n.d.) Beirut restaurants face new challenges as sit-in ends. http://www.khazen.org/index.php?option=com_content&view=article&id=1771&Itemid=234, accessed March 26, 2015.

Khalaf, S., & Khoury, P.S. (ed.) (1993). *Recovering Beirut: Urban Design and Post-War Reconstruction*. Leiden: Brill.

Koopman, S. (2011). Alter-geopolitics: Other securities are happening. *Geoforum* 42(3): 274–284. doi:10.1016/j.geoforum.2011.01.007

Korf, B. (2011). Resources, violence and the telluric geographies of small wars. *Progress in Human Geography* 35(6): 733–756. doi:10.1177/0309132510394120

Luca, A.M. (2012). A new airport for Lebanon. *NOW*, December 20. https://now.mmedia.me/lb/en/reportsfeatures/a_new_airport_for_lebanon, accessed April 3, 2015.

Lunstrum, E. (2013). Articulated sovereignty: Extending Mozambican state power through the Great Limpopo Transfrontier Park. *Political Geography* 36: 1–11. doi:10.1016/j.polgeo.2013.04.003

Maasri, Z. (2008). *Off the Wall: Political Posters of the Lebanese Civil War*. London: IB Tauris.

MacDonald, F., Hughes, R., & Dodds, K. (2010). *Observant States: Geopolitics and Visual Culture*. International Library of Human Geography 16. London: I. B. Tauris.

McConnell, F. (2009). De facto, displaced, tacit: The sovereign articulations of the Tibetan government-in-exile. *Political Geography* 28(3): 343–352.

Meehan, K., Shaw, I.G.R., & Marston, S.A. (2013). Political geographies of the object. *Political Geography* 33: 1–10. doi:10.1016/j.polgeo.2012.11.002

Müller, M. (2008). Reconsidering the concept of discourse for the field of critical geopolitics: Towards discourse as language and practice. *Political Geography* 27(3): 322–338. doi:10.1016/j.polgeo.2007.12.003

Nicley, E.P. (2009). Placing blame or blaming place? Embodiment, place and materiality in critical geopolitics. *Political Geography* 28: 19–22. doi:10.1016/j.polgeo.2009.04.001

Painter, J. (2006). Prosaic geographies of stateness. *Political Geography* 25(7): 752–774. doi:10.1016/j.polgeo.2006.07.004

Pullan, W. (2006). Locating the civic in the frontier: Damascus Gate. In Miessen, M., & Basar, S. (eds), *Did Someone Say Participate? An Atlas of Spatial Practice*, Cambridge, MA: MIT Press, 109–122.

Ramadan, A. (2009). Destroying Nahr El-Bared: Sovereignty and urbicide in the space of exception. *Political Geography* 28(3): 153–163. doi:10.1016/j.polgeo.2009.02.004

Shaw, M. (2004). New wars of the city: Relationships of "urbicide" and "genocide." In Graham, S. (ed.), *Cities, War, and Terrorism*, Malden, MA: Blackwell, 141–153.

Sundberg, R., Eck, K., & Kreutz, J. (2012). Introducing the UCDP Non-State Conflict Dataset. *Journal of Peace Research* 49(2): 351–362. doi:10.1177/0022343311431598

Thrift, N. (2000). It's the little things. In Atkinson, D., & Dodds, K. (eds), *Geopolitical Traditions: A Century of Geopolitical Thought*, London: Routledge, 380–387.

United Nations Security Council (2008). *Report of the Secretary-General on the implementation of Security Council resolution 1701 (2006)*. S/2008/425. New York: United Nations.

Vincent, P., & Warf, B. (2002). Eruvim: Talmudic places in a postmodern world. *Transactions of the Institute of British Geographers* 27(1): 30–51. doi:10.1111/1475-5661.00040

Vineyardsaker (2008). Translation of Sayyid Hassan Nasrallah Speech on Thursday. *The Vineyard of the Saker*. http://vineyardsaker.blogspot.co.uk/2008/05/translation-of-sayyid-hassan-nasrallah.html, accessed March 26, 2015.

Whatmore, S. (2002). *Hybrid Geographies: Natures, Cultures, Spaces*. London: Sage.

Whatmore, S. (2006). Materialist returns: Practising cultural geography in and for a more-than-human world. *Cultural Geographies* 13(4): 600–609. doi:10.1191/1474474006cgj377oa

Woodward, R. (2014). Military landscapes: Agendas and approaches for future research. *Progress in Human Geography* 38(1): 40–61. doi:10.1177/0309132513493219

Yacobi, H. (2009). Towards urban geopolitics. *Geopolitics* 14(3): 576–581. doi:10.1080/14650040802694091

Yacoub, G. (2010). *A history of architecture in Lebanon, 1875–2010*. Beirut: Alphamedia. In English, French, and Arabic.

Doing Political Geography

Chapter 37

Academic Capitalism and the Geopolitics of Knowledge

Anssi Paasi

University of Oulu, Finland

The "pure" universe of even the "purest" science is a social field like any other, with its distribution of power and its monopolies, its struggles and strategies, interests and profits ... Every scientific "choice" the choice of the area of research, the choice of methods, the choice of the place of publication ... is in one respect – the least avowed, and naturally the least avowable – a political investment strategy, directed, objectively at least, towards maximization of strictly scientific profit, i.e. of potential recognition by the agent's competitor-peers. (Bourdieu 1975: 19, 22–23)

Refereed journals are at the core of the contemporary academic enterprise, probably more so now than even before: most new knowledge, however defined, is carried in their pages. At the same time, there is intense pressure for individual academics to publish in "high-quality" journals, to meet both individual (career progression) and collective (departmental evaluation) goals ... Along with the charisma/status of publishing in a prestigious outlet, however, there is an issue of audience. Academics want their papers to be read and then cited, to be used as exemplars in later works. (Johnston 2005: 2, 4)

Some time ago I received an email from China, sent by an Anglophone scholar who worked there as a visiting university professor. He stated that one of his tasks was to help the local faculty prepare publications for major SSCI (Social Science Citation Index) journals. These are periodicals that a US-based firm, Thomson Reuters, has accepted in its Web of Science (WoS) databases. The WoS covers only a thin sample of journals printed in the world. It is overwhelmingly dominated by journals produced by major Anglophone publishers and no less than 95% of them are published in English (Meeus, Schuermans, & de Maesschalck 2011). Even though the WoS is both spatially and linguistically very uneven, its indexed journals have become synonyms of "quality" around the world.

This email reflects much of the politics of knowledge and current forms of hegemony in contemporary academia. It refers implicitly to the uneven power relations characterizing the globalizing publishing business, but also to the evolving neoliberal, competition-oriented rationalities that shape the conduct of researchers around the world toward standardized publishing

The Wiley Blackwell Companion to Political Geography, First Edition.
Edited by John Agnew, Virginie Mamadouh, Anna J. Secor, and Joanne Sharp.
© 2015 John Wiley & Sons Ltd. Published 2017 by John Wiley & Sons Ltd.

patterns, in which indexed journals and their "impact factors" are seen as new norms for "quality" (Agnew 2009; Sheppard 2013). Journals are also gradually beginning to resemble each other, since publishers and editors vigorously modify them to achieve a visible position in the market (Tienari 2012). Simultaneously, the market for publishing monographs has collapsed in many contexts around the world (Ward et al. 2009; Billig 2013). While monographs are still read and cited, the institutional valuation is increasingly attached to journal articles.

The email also demonstrates the growing significance of international consultation in *policy transfer* (Dolowitz & Marsh 1996), which leads to the diffusion of "best practices" in global space. Global policy networks – boundary-crossing networks of influence – increasingly shape political and policy decisions (Prince 2012). They bring together advisers, consultants, policymakers, and the "guiding hand" of supra-national actors and foreign governments, as well as legitimating transnational communities of experts (2012: 189). National governments are increasingly guided in their policy choices by foreign and transnational agencies. This also affects the shaping of "international" standards for academic work in the purported "knowledge societies."

Taken superficially, these tendencies are consistent with the ideal of "international science," where communication across borders is a virtue and science is to a great extent a network-based activity, occurring at and across various scales (Frickel & Gross 2005). Montgomery (2003: 1) declares: "There are no boundaries, no walls, between the doing of science and the communication of it; communicating is the doing of science." Knowledge is made as it circulates; it is never made completely in one place and then simply consumed elsewhere (Agnew 2007: 146).

Despite the noble ideal of a "borderless academy" and freely traveling ideas, academic communication, publishing, and exchange of ideas always reflect power relations and contextuality. Knowledge is incessantly under negotiation and tends to transform when it circulates (Agnew & Livingstone 2011). A number of human geographers, international relations, management, and postcolonial scholars, for example, have commented on power–knowledge relations since the turn of the millennium, and have argued that not only spatial contexts but also *languages* condition the making of scientific knowledge. Indeed, there is an uneven geopolitics of knowledge embedded in communication. Academic fields largely exist through publication forums that are structured asymmetrically in global space. Scholars operating in "peripheral" writing spaces often face substantial hindrances to publishing (cf. Canagarajah 2002; Tietze & Dick 2009; Waever 1998).

This has several consequences. First, knowledge is not a "thing" or a "product" that can be evaluated independently of its production, circulation, and use (Sayer 1992). Second, power relations are uneven and structured in and between the spaces in which knowledge is produced, distributed, and consumed. Third, circulation has a direction: The mobility of hegemonic knowledge tends to occur from "cores" to "peripheries," which is reinforced by the hierarchies of writing spaces and by the growing importance of English as a lingua franca. The core itself, and peripheries as well, are heterogeneous in terms of power. The production of knowledge is thus in many ways uneven. Fourth, power relations are not fixed; they can be resisted and transformed. Fifth, ethics is a highly important issue in international publishing.

The aims and structure of the chapter

This chapter will problematize the globalization of academia and the current politics of knowledge. The focus is on the debate that began in human and political geography 10–15 years ago on so-called Anglo-American or Anglophone hegemony and on the dominance of the English language in "international" publishing spaces. I will use this debate as research material to reflect both the roles of wider material and ideological processes, as well as the contextual factors that characterize globalizing academia. The hegemony debate showed the uneven constitution of the "international" in geography and raised questions about the impacts of such unevenness on research practices in the increasingly competitive academic market that emphasizes journal publishing.

Contemporary neoliberal rationalities stress the market-like logic of knowledge production and new "entrepreneurial" academic subjectivities, phenomena that can be labeled *academic capitalism* (Slaughter & Leslie 1997). Therefore, an investigation of the hegemony debate must be contextualized against the political economies of globalizing capitalism and the emerging "knowledge economy," and against the struggle over "symbolic capital" (Bourdieu 1975). Researchers face these aspects of the current "attention economy" in calls to be more international, more productive, and more visible. Cronin (2005: 5, 130) argues that the growing significance of symbolic capital in academic life has led to the "political economy of citation" and to the "economics of fame" (cf. Sheppard 2013). As the competition for attention intensifies, Cronin argues, the tools for "tracking eyeballs" and measuring impact are becoming progressively more sophisticated (2005: 185).

The previous tendencies – and the key themes of the hegemony debate to be discussed in what follows – can be approached as a "wheel of power" in which the globalization of academia, political economies, the state and major organizations (like the Organisation for Economic Co-operation and Development [OECD] and the European Union), policy transfer, neoliberal rationalities, academic capitalism, evaluation/indexing/ranking cultures, the clamor to be "international" (publications and conferences as its key manifestations), and the rise of English as a lingua franca all come together (Figure 37.1). The Anglophone hegemony debate is therefore related to the processes, events, and institutions identified in the wheel, but also to the struggle over the geopolitics/geoeconomy of knowledge, uneven writing spaces, and the operation of the publishing industrial complex.

Figure 37.1 helps illustrate the complexity of neoliberal rationality's role in formulating the subjectivities of scholars toward "entrepreneurial selves" (cf. Sheppard 2006). Researchers are not merely passive victims or compliant subjects under such rationality; they may develop various strategies to operate under the new market-like conditions or to resist them. Academics are to an increasing degree forced to evaluate their activities and to participate in "academic competition" over symbolic capital – both personal and institutional. Yet, they are not mere homogenous, individualistic, rational calculators (Bourdieu 2004). Indeed, for some scholars the features shaped in the wheel may provide an impetus for critical assessment; for others they may arouse a desire to contribute to major conferences or publishing in "best journals"; for some "peripheral others" they may represent a strategy to network with scholars from the "center" to lower publishing barriers, for example. The features thus position scientific practice in a wider relational space and condition the conduct of researchers, their self-evaluation and strategic choices. The wheel therefore makes visible the complicated power relations and struggles embedded in the current, neoliberal forms of academic subjectification. The wheel should not be seen as a "causal chain" or as a set of successive stages. Rather, it distinguishes for analytical purposes some key dimensions that are related to the operation of contemporary academia.

This chapter proceeds as follows. The next section analyzes the structural and individual circumstances behind the upsurge of the hegemony debate and the rise of English into the position of a lingua franca. I will not use the hegemony debate merely as a "case study" to test

Figure 37.1 The wheel of power: A conceptual framework for understanding the Anglophone hegemony debates and the contemporary academic rat race.

the wheel of power empirically. Rather, the debate has been crucial in outlining this wheel via the dialogue within the debate itself. The key themes of the debate are then excavated to problematize such key questions as where "relevant knowledge" is produced and who has a right to speak (and to be heard) about their own and other contexts. Some illustrations of the endurance of Anglophone dominance in political geography will then be provided and conclusions will follow.

Globalization of science, neoliberalism, and the rise of the hegemony debate

A rapid globalization of academia is occurring, resulting from fundamental transformations in social life, such as changes in capitalist production, innovations in information and communication technologies, the spread of rationalism as a major knowledge framework, and the rise of forms of governance that render possible new regulatory bases in societies (Scholte 2000). The dynamics of science are a critical feature in these processes: Knowledge is crucial for the forces of production in the ostensible "knowledge societies" and is a substantial element in the internationalization, competition, and regulation of globalization around the world (Slaughter

& Leslie 1997). Higher education, academic research, and "society" have gradually become interrelated, as witnessed by the rise of a new vocabulary: Expressions such as academic capitalism, entrepreneurial university, McUniversity, or Mode 1/Mode 2 are frequently used to characterize the new links (Paasi 2005).

What seems to be common in the previous tendencies is "neoliberalism." Neoliberalism is not a monolithic political project that is associated only with the preference for a "minimalist state." All conceptualizations of neoliberalism need to recognize historical contingency, geographical specificity, and political complexity – context makes a difference (Larner & Le Heron 2005; Jessop 2013). In academia, neoliberalism means that knowledge is increasingly treated as a commodity, whose value is to be determined in the market; faculty at universities are increasingly assessed for their entrepreneurial acumen: "How much grant money you have raised? How many patents have you produced? How many citations have you garnered?" (Sheppard 2013). Recent transformations in the political, institutional, and funding settings have augmented the power of management and weakened the autonomy of professional academics.

Scientific communication across borders and distances is still a lauded ideal, but the ethos of ranking scholars, departments, universities, and even "nations" on the basis of research performance and citations has become crucial in academia (Agnew 2009). A variety of university-, national-, and international-level ranking exercises – often drawing on the WoS data – are the order of the day (Jöns & Hoyler 2013). Such rankings have become pervasive since the beginning of the twenty-first century and are unreservedly monitored by political elites, ministries, and university managers to figure out where one's "own state" or "university" is positioned in relation to other national or international "competitors" (Paasi 2013). Ranking lists powerfully modify the social reality and practices of academia. As a result, the global higher education map has become more structured: Lists have become policy instruments for the global governance of higher education (Kauppi & Erkkilä 2011). They have also created some paradoxes in monitoring knowledge production, such as "academic nationalism." The claims for the need to *internationalize* national science are often made in the name of *national* competiveness (Paasi 2013). Thus, a state-centric territorial order stubbornly conditions the internationalization of academia in "knowledge societies."

An additional international element of neoliberal rationality and policy transfer is the invention of new keywords in science policies that shape research practices, publication strategies, and ultimately the conduct of scholars. In addition to the general demands encouraging researchers to be more international, a new language of performativity has been adopted in science policies, drawing on such expressions as "world class," "world leading," or "top quality." The new calculative message implied by these buzzwords is aimed at having an impact on both the practices of academics and their publishing spaces. The new practices govern and disrupt conventional patterns of institutional organization and individual behavior (Larner & Le Heron 2005).

Rankings and citations are not a new invention – Bourdieu (1990: 76) referred to citations as the "most objectified of the indices of symbolic capital." The citation index was developed by Eugene Garfield in the North American Cold War reality of the 1950s. The geopolitical rivalry between the United States and the Soviet Union provided a background for these developments as the competition for control over the flows of knowledge was becoming fiercer. This effect was accelerated by the space race and the launch of the Soviet *Sputnik* satellite in 1957 (Risteli 2003; Paasi 2005). At first, the recording of publications was presumed to be logical in managing the skyrocketing amount of knowledge and in assisting academics to track new publications and information. Gradually, the citation index has become an end itself, however. Garfield wrote: "I think it is important not only to be literature minded, but to develop citation consciousness. I'm not sure how you teach this. It requires indoctrination by informed mentors" (cited in Hargittai 1999: 28). Today, such "mentoring" is done by ministries and university leaders who control the efficiency of universities, "the bright satanic mills of the emerging global knowledge economy" (Harding et al. 2007).

X-raying the key themes of the debate

Such indoctrination, audit culture, compliance, need for recognition, and feeling of being marginalized, for example, form the complex structural and subjective backgrounds of the hegemony debate. The Gramscian idea of hegemony means consent and approval of dominant practices and values by the subjugated, rather than a condition where force or violence is used to make them dominated (Williams 1988). Hegemony is thus related to *subjectification*; that is, to various techniques that are mobilized to influence people's thinking, behavior, and action so that they will follow certain patterns.

What could the "Anglophone hegemony" in geography mean from these angles? At times it has been regarded as a causal force: "the Anglo-American hegemony forces us [i.e., non-Anglophone scholars] to write papers in UK and US journals" (Aalbers 2004: 321). This statement implies that scholars are passive marionettes, ignores the role of symbolic capital as both an individual and a structural feature in academia, and disregards academic cultures and power structures. "Anglophone hegemony" is not claiming that non-Anglophone scholars should publish their works in English, in the "best" Anglophone journals, or participate in major conferences. Hegemony is embedded in much more complicated ways in academic and societal power relations (see Figure 37.1). There is also the *will* of the actors to participate in the "game" and the inherent struggle over symbolic capital. Hegemony must thus be understood partly through the national clamor about internationalization, mediated by policy transfer, the new audit culture, and the prestige of WoS journals; partly it is also based on the motives, conditioned by noted institutions, of subjects to participate in the new culture – a factor that has so far been absent in the analysis of the Anglophone hegemony debate. As Gramsci stated, through establishing a new cultural hegemony, "you can occupy people's heads, and their hearts and hands will follow" (Davies & Bansel 2010: 5).

Marginalization

The hegemony debate was initiated by scholars working in the former colonial peripheries of the Anglophone world (Berg & Kearns 1998; Olds 2001) and in continental Europe (Minca 2000). It soon attracted participants from other European states and Singapore and became more nuanced. Timár (2004), a Hungarian scholar, argued that it is better to speak about a "Western hegemony" than merely about an "Anglo-American" one. The key arguments were often related to the experience of marginalization and even discrimination. These manifested themselves first in "downgrading" places:

> the fact that its [social science knowledge] production is geographically dominated by the Anglo-American academic world means that knowledge of certain regions and places is likely to be less "relevant" and less cited, irrespective of the author of this knowledge and their institutional affiliations. (Yeung 2002: 2099)

Yeung's comment also points to the second issue, the emerging power of citations and evaluations, which shifted the debate on to the hegemony of Anglophone journals. It is thus obvious that the practices related to journal publishing rather than to books motivated the critique. Sometimes critics expressed their frustration about the behavior of referees in Anglophone "quality journals." The reviewers often labeled submissions from the non-Anglophone world peripheral in terms of both "geography" and "topics" and irrelevant for "major geographical discourses" (Berg & Kearns 1998). Likewise, many non-English-speaking commentators were disenchanted with how Anglophone editors and even colleagues did not recognize how difficult it was for non-Anglophone geographers to enter English-language markets and such journals. Sometimes the critique was based on personal experiences of uneven treatment in the editorial process on the basis of "poor use of English." The critics thus felt that they were being marginalized as linguistic Others.

Some commentators noted that skewed power relations were not simply based on the fact that non-native speakers have a language disadvantage and native speakers a language advantage in academic markets dominated by the English language, but also stemmed from the fact that the most-respected WoS journals were primarily run by Anglophone scholars (Garcia-Ramon 2003). Similarly, the monopolization of the "international" by Anglophone journals was seen as a major ethical problem and an expression of hegemony. Many journals marketed themselves as "international flagships," but empirical evidence showed that the editors and members of international advisory boards often represented only Anglophone departments (Gutiérrez & López-Nieva 2001).

The third form of experience of marginalization was related to the supposedly inferior quality of knowledge produced in non-Anglophone social spaces, downgrading such knowledge and, implicitly, non-Anglophone intellectual subjectivity itself. Many critics also argued more particularly that "international geography" was dominated by US/UK-based "theories," and that work done in other countries was considered irrelevant or incorrect, or, at best, was regarded as exotic, local case studies incapable of producing innovative theory or of framing the epistemological boundaries of disciplinary practices (Gregson, Kirsten, & Vaiou 2003). Simonsen (2002; cf. Olds 2001) contended that there has been at work a kind of "center–periphery" assumption in which the "center" becomes the referent for philosophical and theoretical reflection and the Anglo-American academic the major progenitor of theory formation. Non-Anglophone geographers thus questioned the division of the world into an academic core – that is, the Anglophone world – where relevant geographical theory is produced, and peripheries – the non-Anglophone world – where the receiver-consumers of such theories dabble.

Some non-Anglophone debaters themselves represented a theoretical research orientation, which doubtless partly explains why they were worried about non-Anglophone theoretical work being ignored in international geography. The debate often implied a rather unspecified idea of theory; something that is simply moved from one place to another. A more active view would probably accentuate *conceptualization* and the creation of abstractions in concrete contexts (Sayer 1992). Such "conceptual invariances," rather than fixed, ordering framework–type theories, are much more likely to travel also from "peripheries" to "centers."

The hegemony debate did not receive much feedback from the Anglophone community, but some scholars have been self-critical. Berg (2012) and Kitchin (2005) in particular have contributed much to uncovering the forms of hegemony. Outside of the hegemony debate, O'Loughlin, Raento, and Sidaway (2008) argue that most political geographers in the North American and European core are not very familiar with the evolution of the field in "relative peripheries." Sharp (2013) criticizes critical geopolitics for remaining a particularly Western way of knowing that has been less attentive to other traditions of thinking. She also notes the earlier critiques of (political) geography. Perry (1987), for example, claimed that Anglo-American political geography follows a restricted and impoverished version of the discipline, largely ignoring the political concerns of four-fifths of humankind. Slater (1998) proposed that it is crucial to move beyond the image of the Third World as a "conceptually empty space" to be filled with Western knowledge. Robinson (2003), for her part, suggests that political geography is perpetually subject to parochial forms of theorizing, while Bonnett (2003: 59) argues that various "turns" in English-speaking geography that have occurred since the 1960s have been turns away from the international and global toward "the West" or, more accurately, the English-speaking West. We can add here Berg's (2012) – indeed, globally valid – critique of Anglophone geography as a largely masculine and white enterprise.

English as a lingua franca

One major issue in the hegemony debate relates to the power of English as a lingua franca and the abduction of the idea of "international" into this linguistic realm. This hegemony is based on the assumption that English, as the dominant language, carries higher prestige and validity than other languages (Tietze & Dick 2009). This has strengthened the power of Anglophone publishing houses and native writers through creating asymmetric writing spaces. English-language publishers and academics have become major gatekeepers in determining who and what gets published in English and in regulating the "international," theoretical, or conceptual "cutting edge," which has made the circulation of knowledge in human geography unidirectional (Kitchin 2005). Tietze and Dick (2009: 120) argue that English acts as an invisible, hegemonic force that has a strong impact on the framing and ordering of the global world and organizations constituent of it.

An interesting question is how and when such hegemony became institutionalized. The present supremacy has long roots. It started with the voyages to the Americas, Asia, and the Antipodes, as well as with the growth of the British empire and its colonies in Africa and the South Pacific in the nineteenth century. The associations between English and the domains of trade, business, and commerce have been close for centuries. The spread of English and linguistic imperialism was consequently connected to the spread of Western capitalism, and was boosted by the intensification of colonialism (Canagarajah 2002; Tietze & Dick 2009).

In Europe, new post–Second World War geopolitical and cultural relations between the United States and Europe were crucial in the spread of English-speaking academia. The global geopolitical divide and the increasing rivalry between capitalist and socialist camps gave rise to novel forms of international connections in the Western world. The Marshall Plan, the European Recovery Program, was shaped in America after the war to help Western European states recuperate and to encourage the ideals of freedom. The Organization for European Economic Cooperation (OEEC), the predecessor of the OECD, was established partly to implement the Marshall Plan. The OECD, for its part, became the main economic adviser to the capitalist West in the Cold War against the Soviet Union (Kallo 2009), and is now one of the key international advisory institutions behind the demands for increasing competiveness in academia and policy transfer (cf. Kitchin 2005). This organization regularly carries out international assessments of publication activities and citations between "nations." It uses the WoS data as an instrument, hence sustaining an attitude of "competition" that is bolstered and coordinated by national higher education systems and individual universities (Paasi 2013).

One important institution that brought a number of European and US scholars into closer contact was the network of diverging grant systems. Pre-eminent among these was the American Fulbright scholarship system, established in 1946. This has been seen as a "cultural variant of the Marshall Plan" (Frijhoff & Spies 2004: 69). Likewise, several other US-based philanthropic institutions, such as the Ford, Carnegie, and Rockefeller Foundations, were important in the networking of the global academic elite and in the production of a "consensual hegemony"; that is, scientific dominance by the United States (Krige 2006; Parmar 2012). This hegemony not merely supported the interests of postwar recovery in Europe, but also abetted the maintenance of American leadership and Americanized scientific practices in many fields, from hard sciences to social sciences. "International science" became a vehicle to promote American values and interests in the postwar world, in fact more so since the US authorities tried to reconfigure the European scientific landscape (Krige 2006) and to create scientific networks around the globe (Parmar 2012).

Whatever has happened to American values and their spread to Europe and elsewhere, the triumph of English as *the* language of science has continued around the world. In spite of the emerging critiques, these developments are today strongly supported and even encouraged by many national ministries of science outside of the English-speaking world (Ammon 2001). The English language has become a naturalized part of neoliberal rationality in the management of scientific practices. National ministries and governments use OECD reports to steer their science policies and national evaluations, in many cases even commissioning such reports to justify their decisions (Kallo 2009).

Minca (2013) notes that the gradual move to using English as the only language of so-called international debate, and the parallel trend of some key sites in the production of geographical knowledge ("best departments in the UK and US") to show less and less interest in what happens in different contexts and in different languages, have taken on new meaning in light of the widespread adoption of neoliberal practices in Europe and beyond. The ongoing internationalization has not, he argues, meant academic interaction and exchange of knowledge, but the dominance of the leading Anglophone journals in which international debate occurs and gains recognition (2013: 8). As one of the original critics of Anglophone hegemony, Minca nonetheless seems to be quite optimistic about the intensifying (neoliberal) evaluation cultures and sees that they allow short-tracking of "the careers of the most talented and ambitious scholars" (the Research Assessment Exercise [RAE] in the United Kingdom) and reward, via recruitment, "the best scholars on the basis of merit and actual impact of the work" (in Italy; Minca 2013: 10, 14).

Scholars in the United States and the United Kingdom certainly look at the science produced in continental Europe, but, in contrast to the past, it is nowadays sociological and philosophical works rather than geographical ones that seem to motivate them – and this often takes place through English translations. References to the translations of the works of Foucault, Deleuze, Schmitt, or Agamben, for example, are now popular in Anglophone political geographical literature. Many of these authors were until very recently virtually absent, for instance in Francophone geography (Fall 2014). Indeed, the circulation of such theoretical ideas seems to be accelerating and new subtribes of geographers appear to be adopting new philosophers, while continually ignoring the non-Anglophone geographical tradition. Minca (2013) suggests that in order to reproduce itself successfully, the Anglophone monolingual machine seems to need to import "original ideas" from somewhere else, and that this new mobility of theoretical-philosophical ideas actually gives an opportunity for non-Anglophone geographers to act as natural suppliers of these philosophical and theoretical thoughts, and perhaps to establish "theory hubs" in continental European geography.

Such adoption of influences from "elsewhere" has deep roots, but has also been denounced in geography. Agnew and Duncan (1981) wrote decades ago that there is a need for geographers to display more critical acuity in borrowing ideas from outside the field. Such circulation of ideas may be understood as part of the neoliberal struggle for symbolic capital, in which scholars try to profile themselves in geographical and wider social science debates. While such borrowing provides an opportunity for the ever more rapid – mostly unidirectional – adoption and circulation of "new" ideas (which have often been published in their original languages decades earlier), it is still far removed from a genuine interchange of conceptual ideas in international geography (cf. Fall 2014).

Insiders and outsiders

One crucial question related to the geographies of knowledge and hegemony is what we can say about *other* linguistic-cultural contexts and who is allowed to study and speak about them. As Alcoff (1991) states, the problem of representation underlies all cases of speaking for another, and while representations may have material effects (as well as material origins), they are always mediated in complicated ways by location, discourse, and power. This issue has been fundamental in geography and anthropology since the Age of Exploration and the colonial period, in which the roots of the hegemonic position of the English language can be situated. This issue, implying/challenging a state-centric (territorial) viewpoint with an apparently sharp inside–outside divide, is crucial in problematizing the making and circulation of knowledge, which are largely relational processes.

The exchange between Moisio and Harle (2006) and Antonsich (2006) is a useful political geographical example of this issue and of the methodological difficulties that an insider–outsider divide may engender. Moisio and Harle criticized Antonsich for studying Finnish identity and identity politics discourses "externally," by using English-language documents, without knowing "deeply enough" the language and context. They drew on the metaphor of "geopolitical remote sensing" that I had used previously to criticize the tendency of critical geopolitics scholars to study contexts and related discourses at a distance, operating on the basis of written texts and other secondary materials. I suggested that an expansion of the "geo" was needed to appreciate also the social practices and geopolitics of everyday life, which require "local knowledge" (Paasi 2000).

The problem is that language is not merely a formal category (English, Finnish, etc.), but also a contextual form of reasoning (Moisio & Harle 2006). Achieving an understanding of the latter is a demanding process and often requires "first-hand local knowledge rather than half-baked generalizations culled from second-hand sources" (Agnew 2013: 162). In spite of this fact, such "remote sensing" has of course been extensive in the history of geopolitics and political geography. Often the maps and texts that geopoliticians produce have been guided by ideological motives, or have been distorted, or have divided the world in rather haphazard but perhaps politically well-motivated ways. Similarly, regional geographers continually divide the earth into meta-regions and much world regional geography teaching consistently draws on such divisions. Geographers have to be cautious with this issue, but perhaps they can never solve it, since it derives from the basic nature of geography: to make sense of the wider world and its places and regions. For critical geographers this is an anti-imperialist and non-nationalist effort.

The persistence of Anglophone hegemony

Following the critiques of Anglophone hegemony, many publishers and editors have taken steps to change the editorial boards of geographical journals to reflect a more international complexion or have named non-Anglophone editors to journals. At the same time, the WoS has expanded and the number of indexed journals has doubled during the last 10 years or so, and many non-Anglophone journals have also been indexed (Paasi 2013). Have these steps diversified authorship in international political geography? As an empirical example, I will look at a leading journal, *Political Geography*, which now has one non-Anglophone editor and almost ten non-Anglophone members on its international advisory board. Table 37.1, based on data from the WoS (accessed November 24, 2013), shows 12 leading countries from which published articles originate during two periods, 1992–2002 and 2003–2013. It is of course important to recognize the continual "brain circulation" between countries that partly challenges the "methodological nationalism" implied by the WoS data. Yet, this data can be used as a simple "proxy" to uncover possible steps toward internationalization.

Political Geography is open to submissions from all countries (O'Loughlin et al. 2014), but during both periods the published articles came predominantly from the United States, the United Kingdom, and other English-speaking countries, even though the absolute number of published articles (N) increased. Only a minority of authors from other countries seem to break this rule. The power of the English language is strengthened by the fact that many of the other countries also have English-language universities or at least partly English-language curricula (e.g., Singapore, South Africa, Israel). Yet, there is no clear tendency toward a wider internationalization of authorship during the two periods. The number of states from which the published articles originate expanded slightly during the second period. The dominance of English-language countries/authors in this forum is deeply rooted and has thus not changed

Table 37.1 Articles published in *Political Geography* by country of origin.

1992–2002		2003–2013	
Country	% of titles	Country	% of titles
USA	45.9	USA	38.2
England	23.9	England	31.1
Israel	6.2	Canada	7.1
Scotland	5.0	Norway	4.4
Canada	4.1	Australia	3.9
Wales	2.9	Scotland	2.7
North Ireland	2.3	North Ireland	2.4
Ireland	2.5	Netherlands	2.6
Singapore	2.4	Singapore	2.4
Netherlands	2.0	Israel	2.2
Ireland	1.5	Wales	2.2
Singapore	1.8	South Africa	1.7
Other countries (20)	5.9	Other countries (20)	6.4
N	338	N	408

Note: In the case of several co-authors for an article, Thomson Reuters counts each country only once. The WoS also presents the data for England, Scotland, Wales, and North Ireland separately.

much during the last 20 years, in spite of the hegemony debates. Even if we note the power of the "brain drain" and the mobility of academics between countries, the big picture remains. Similar observations have been made regarding several other major Anglophone geography journals: over 70% of articles were written by US- or UK-based authors (Banski & Mariola 2013; cf. Kitchin 2005).

Conclusions

This chapter has scrutinized the globalization of knowledge production and the uneven power relations embedded in this production. The debate on Anglophone hegemony in geography was analyzed in order to unearth the complexity of current politics of knowledge. This debate reveals the problems that non-Anglophone academics have confronted in "internationalizing academia," especially in publishing. The debate tells us about the uneven structural features in academic publishing, the regulatory role of "evaluation cultures," and the power of neoliberal rationalities. It also reveals the struggle over symbolic capital in a situation where the idea of "international" in geography was monopolized by English-language journals and where non-Anglophone scholars were given marginalized subject positions. In casting light on the hegemony and power of the English language, such debates have been a successful project: Many Anglophone journals have become more international, with non-Anglophone members being nominated to editorial boards and international advisory boards. Yet, recent evidence shows that the authors of major international journals are still largely Anglophone.

Some scholars have questioned the validity of the Anglophone vs. non-Anglophone dualism and even the hegemony of the English language. Johnston and Sidaway (2004) questioned the idea of a unified Anglo-American geography and argued that the two are at variance; national scales, traces, and traditions are hybrids that are intricately tangled up with other contexts. For Rodríguez-Pose (2004), the use of English is pivotal in ensuring the survival of diverse geographical traditions, and is the only way to guarantee that these traditions are exposed to and interact with each other. For him, English is the key to increasing interaction and to reaching out beyond national "endogamous and hierarchical cocoons" (2004: 2).

It is therefore a simplification to see the "cores" as located *only* in the West (or in the United Kingdom and United States) and the "peripheries" elsewhere. In order to understand contemporary power relations in science, we have to recognize that both cores and peripheries are also scaled in complex ways vertically within states, universities, and even within single departments, and that these all are located in various ways within the wider "wheel of power."

Livingstone (2003: 7) has noted that every social space has a range of possible, permissible, and intelligible utterances and actions; that is, "things that can be said, done, and understood." The hegemony debate illustrates that the social spaces of academic work and writing are not equal, that knowledge does not flow in a "borderless world," and that when it does flow, the directions of such flows are typically uneven. The debate also shows that in order to resist the ongoing homogenization of publishing cultures, we have to fight back against Kuhn's (1970 [1962]: 18–21) prediction. He suggested that fields that have passed the "pre-paradigmatic" stage habitually publish their results in journals, and no longer in reports/monographs. Kuhn predicted that a scientist who writes books "is more likely to find his professional reputation impaired than enhanced" (1970 [1962]: 20). The current culture, centered on journals (and impact factors and citations), is fast becoming hegemonic around the world. It has to be resisted in the name of diversity and academic freedom: Journals, monographs, and edited collections must have their place in geography and the flow of ideas must be multidirectional.

References

Aalbers, M. (2004). Creative destruction through the Anglo-American hegemony: A non-Anglo-American view on publications, referees and language. *Area* 36(3): 319–322. doi:10.1111/j.0004-0894.2004.00229.x

Agnew, J. (2007). Know-where: Geographies of knowledge of world politics. *International Political Sociology* 1(2): 138–148. doi:10.1111/j.1749-5687.2007.00009.x

Agnew, J. (2009). The impact factor. *AAG Newsletter* 44(6): 3.

Agnew, J. (2013). Why criticizing grand regional narratives matters. *Dialogues in Human Geography* 3(2): 160–162. doi:10.1177/2043820613490132

Agnew, J., & Duncan, J. (1981). The transfer of ideas into Anglo-American human geography. *Progress in Human Geography* 5(1): 42–57.

Agnew, J., & Livingstone, D. (2011). Introduction. In Agnew, J., & Livingstone, D. (eds), *The Sage Handbook of Geographical Knowledge*, London: Sage, 1–17.

Alcoff, L. (1991). The problem of speaking for others. *Cultural Critique* 20(Winter): 5–32.

Ammon, U. (2001). *The Dominance of English as a Language of Science*. New York: Mouton de Gruyter.

Antonsich, M. (2006). In defence of "geopolitical remote sensing": Reply to Moisio and Harle. *Eurasian Geography and Economics* 47(2): 211–215. doi:10.2747/1538-7216.47.2.211

Banski, J., & Mariola, F. (2013). "International" or "Anglo-American" journals in geography. *Geoforum* 45(March): 285–295. doi:10.1016/j.geoforum.2012.11.016

Berg, L. (2012). Geographies of identity I: Geography – (neo)liberalism – white supremacy. *Progress in Human Geography* 36(4): 508–517. doi:10.1177/0309132511428713

Berg, L., & Kearns, R. (1998). America unlimited. *Environment and Planning D: Society and Space* 16(2): 128–132. doi:10.1068/d160128

Billig, M. (2013). *Learn How to Write Badly: How to Succeed in Social Science*. Cambridge: Cambridge University Press.

Bonnett, A. (2003). Geography as the world discipline: Connecting popular and academic geographic imaginations. *Area* 35(1): 55–63. doi:10.1111/1475-4762.00110

Bourdieu, P. (1975). The specificity of the scientific field and the social conditions of the progress of reason. *Social Science Information* 14(6): 19–47. doi:10.1177/053901847501400602

Bourdieu, P. (1990). *Homo Academicus*. Cambridge: Polity Press.

Bourdieu, P. (2004). *Science of Science and Reflexivity*. Cambridge: Polity Press.

Canagarajah, A.S. (2002). *A Geopolitics of Academic Writing*. Pittsburgh, PA: University of Pittsburgh Press.

Cronin, B. (2005). *The Hand of Science: Academic Writing and Its Rewards*. Lanham, MD: Scarecrow Press.

Davies, B., & Bansel, P. (2010). Governmentality and academic work: Shaping the hearts and minds of academic workers. *Journal of Curriculum Theorizing* 26(3): 5–20.

Dolowitz, D., & Marsh, D. (1996). Who learns what from whom: A review of the policy transfer literature. *Political Studies* 44(2): 343–357. doi:10.1111/j.1467-9248.1996.tb00334.x

Fall, J.F. (2014). Writing (somewhere). In Lee, R., Castree, N., Kitchin, R., Lawson, V., Paasi, A., Radcliffe, S., & Withers, C. (eds), *The SAGE Handbook of Progress in Human Geography*, London: Sage, 296–315.

Frickel, S., & Gross, N. (2005). A general theory of scientific/intellectual movements. *American Sociological Review* 70(2): 204–232. doi:10.1177/000312240507000202

Frijhoff, W., & Spies, M. (2004). *Dutch Culture in a European Perspective*. Basingstoke: Palgrave Macmillan.

Garcia-Ramon, M.-D. (2003). Globalization and international geography: The questions of languages and scholarly traditions. *Progress in Human Geography* 27(1): 1–5. doi:10.1191/0309132503ph409xx

Gregson, N., Simonson, K., & Vaiou, D. (2003). Writing (across) Europe: On writing spaces and writing practices. *European Urban and Regional Studies* 10: 5–22. doi:10.1177/0969776403010001521

Gutiérrez, J., & López-Nieva, P. (2001). Are international journals of human geography really international? *Progress in Human Geography* 25(1): 53–69. doi:10.1191/030913201666823316

Harding, A., Scott, A., Laske, S., and Burtscher, C. (2007). *Bright Satanic Mills: Universities, Regional Development and the Knowledge Economy.* Aldershot: Ashgate.

Hargittai, I. (1999). Deeds and dreams of Eugene Garfield: Interview by Istvan Hargittai. *Chemical Intelligencer* October: 26–31.

Jessop, B. (2013). Putting neoliberalism in its time and place: A response to the debate. *Social Anthropology* 21(1): 65–74. doi:10.1111/1469-8676.12003

Johnston, R. (2005). On journals. *Environment and Planning A* 37(1): 2–8. doi:10.1068/a3811

Johnston, R., & Sidaway, J. (2004). The trans-Atlantic connection: "Anglo-American" geography reconsidered. *Geojournal* 59(1): 15–22. doi:10.1023/B:GEJO.0000015434.52331.72

Jöns, H., & Hoyler, M. (2013). Global geographies of higher education: The perspective of world university rankings. *Geoforum* 46: 45–59. doi:10.1016/j.geoforum.2012.12.014

Kallo, J. (2009). *OECD Education Policy.* Jyväskylä: Finnish Educational Research Association.

Kauppi, N., & Erkkilä, T. (2011). The struggle over global higher education: Actors, institutions, and practices. *International Political Sociology* 5(3): 314–326. doi:10.1111/j.1749-5687.2011.00136.x

Kitchin, R. (2005). Disrupting and destabilizing Anglo-American and English-language hegemony in geography. *Social and Cultural Geography* 6(1): 1–15. doi:10.1080/1464936052000335937

Krige, J. (2006). *American Hegemony and the Postwar Reconstruction of Science in Europe.* Cambridge, MA: MIT Press.

Kuhn, T. (1970 [1962]). *The Structure of Scientific Revolutions.* Chicago, IL: University of Chicago Press.

Larner, W., & Le Heron, R. (2005). Neoliberal spaces and subjectivities: Reinventing New Zealand Universities. *Organization* 12(6): 843–862. doi:10.1177/1350508405057473

Livingstone, D. (2003). *Putting Science in Its Place: Geographies of Scientific Knowledge.* Chicago, IL: University of Chicago Press.

Meeus, B., Schuermans, N., & de Maesschalck, F. (2011). Is there a world beyond academic geography: A reply to Ben Derudder. *Area* 43(1): 113–114. doi:10.1111/j.1475-4762.2010.00987.x

Minca, C. (2000). Venetian geographical praxis. *Environment and Planning D: Society and Space* 18(3): 285–289. doi:10.1068/d1803ed

Minca, C. (2013). (Im)mobile geographies. *Geographica Helvetica* 68(1): 7–16. doi:10.5194/gh-68-7-2013

Moisio, S., & Harle, V. (2006). The limits of geopolitical remote sensing. *Eurasian Geography and Economics* 47(2): 204–210. doi:10.2747/1538-7216.47.2.204

Montgomery, S. (2003). *The Chicago Guide to Communicating Science.* Chicago, IL: University of Chicago Press.

Olds, K. (2001). Practices for "process geographies": A view from within and outside the periphery. *Environment and Planning D: Society and Space* 19(2): 127–136. doi:10.1068/d1902ed

O'Loughlin, J., Raento, P., & Sidaway, J. (2008). Editorial: Where the tradition meets its challengers. *Political Geography* 27(1): 1–4. doi:10.1016/j.polgeo.2007.11.003

O'Loughlin, J., Raento, P., Sharp, J.P., Sidaway, J.D., & Steinberg, P.E. (2014). Editorial: Predators, sanctions and *Political Geography*. *Political Geography* 38(1): A1–A3. doi:10.1016/j.polgeo.2013.11.003

Paasi, A. (2000). Rethinking critical geopolitics. *Environment and Planning D: Society and Space* 18(2): 282–284. doi:10.1068/d1802rvw

Paasi, A. (2005). Globalization, academic capitalism and the uneven geographies of international journal publishing spaces. *Environment and Planning A* 37(5): 769–789. doi:10.1068/a3769

Paasi, A. (2013). *Fennia*: Positioning a "peripheral" but international journal under the conditions of academic capitalism. *Fennia* 191(1): 1–13. doi:10.11143/7787

Parmar, I. (2012). *Foundations of the American Century: The Ford, Carnegie, and the Rockefeller Foundations in the Rise of American Power*. New York: Columbia University Press.

Perry, P. (1987). Editorial comment. Political Geography Quarterly: A content (but discontented) review. *Political Geography Quarterly* 6(1): 5–6.

Prince, R. (2012). Policy transfer, consultants and the geographies of governance. *Progress in Human Geography* 36(2): 188–203. doi:10.1177/0309132511417659

Risteli, L. (2003). Sitaatioindeksin alkuvuosikymmenet ja oppihistorialliset yhteydet (The early decades of citation index and its historical connections, in Finnish). MA thesis. University of Oulu.

Robinson, J. (2003). Postcolonizing geography: Tactics and pitfalls. *Singapore Journal of Tropical Geography* 24(3): 273–289. doi:10.1111/1467-9493.00159

Rodríguez-Pose, A. (2004). On English as a vehicle to preserve geographical diversity. *Progress in Human Geography* 28(1): 1–4. doi:10.1191/0309132504ph467xx

Sayer, A. (1992). *Method in Social Science*. London: Routledge.

Scholte, J. (2000). *Globalization*. New York: Palgrave.

Sharp, J. (2013). Geopolitics at the margins? Reconsidering genealogies of critical geopolitics. *Political Geography* 37: 20–29. doi:10.1016/j.polgeo.2013.04.006

Sheppard, E. (2006). Geographies of research assessment: The neoliberalization of geography? *Progress in Human Geography* 30(6): 761–764. doi:10.1177/0309132506071518

Sheppard, E. (2013). Geography and neoliberalizing academy. *AAG Newsletter* May 31. http://www.aag.org/cs/news_detail?pressrelease.id=2330, accessed March 26, 2015.

Simonsen, K. (2002). Editorial: Global–local ambivalence. *Transactions of the Institute of British Geographers NS* 27(4): 391–394.

Slater, D. (1998). Post-colonial questions for global times. *Review of International Political Economy* 5(4): 647–678. doi:10.1080/096922998347417

Slaughter, S., & Leslie, L. (1997). *Academic Capitalism: Politics, Policies, and the Entrepreneurial University*. Baltimore, MD: Johns Hopkins University Press.

Tienari, J. (2012). Academia as financial markets? Metaphoric reflections and possible responses. *Scandinavian Journal of Management* 28(3): 250–256. doi:10.1016/j.scaman.2012.05.004

Tietze, S., & Dick, P. (2009). Hegemonic practices and knowledge production in the management academy: An English language perspective. *Scandinavian Journal of Management* 25(1): 119–123. doi:10.1016/j.scaman.2008.11.010

Timár, J. (2004). More than "Anglo-American", it is "Western": Hegemony in geography from a Hungarian perspective. *Geoforum* 35(5): 533–538. doi:10.1016/j.geoforum.2004.01.010

Waever, O. (1998). The sociology of a not so international discipline: American and European developments in international relations. *International Organization* 52(4): 687–727. doi:10.1162/002081898550725 DOI:10.1162/002081898550725#_blank

Ward, K., Johnston, R., Richards, K., et al. (2009). The future of research monographs: An international set of perspectives. *Progress in Human Geography* 33(1): 101–126. doi:10.1177/0309132508100966

Williams, R. (1988). *Marxismi, kulttuuri ja kirjallisuus*. Tampere: Vastapaino.

Yeung, H. (2002). Deciphering citations. *Environment and Planning A* 34(12): 2093–2106. doi:10.1068/a3412com

Index

The Wiley Blackwell Companion to Political Geography, First Edition.
Edited by John Agnew, Virginie Mamadouh, Anna J. Secor, and Joanne Sharp.
© 2015 John Wiley & Sons Ltd. Published 2017 by John Wiley & Sons Ltd.